Distributed Systems and Networks

Distributed Systems and Networks

William Buchanan
Napier University

THE McGRAW-HILL COMPANIES

London · Burr Ridge IL · New York · St Louis · San Francisco · Auckland
Bogotá · Caracas · Lisbon · Madrid · Mexico · Milan
Montreal · New Delhi · Panama · Paris · San Juan · São Paulo
Singapore · Sydney · Tokyo · Toronto

Published by
McGraw-Hill Publishing Company
SHOPPENHANGERS ROAD, MAINDENHEAD, BERKSHIRE, SL6 2QL, ENGLAND
Telephone: +44 (0) 1628 502500
Fax: +44 (0) 1628 770224
Web site: http://www.mcgraw-hill.co.uk

British Library Cataloguing in Publication Data
A catalogue record for this book is available from the British Library

ISBN 007 709583 9

Library of Congress Cataloguing-in-Publication Data
The LOC data for this book has been applied for and may be obtained from
The Library of Congress, Washington, D.C.

Further information on this and other McGraw-Hill titles is to be found at
http://www.mcgraw-hill.co.uk

Authors Website address: http://mcgraw-hill.co.uk/textbooks/buchanan

Publisher:	Alfred Waller, David Hatter
Sponsoring Editor:	David Hatter
Editorial Assistant:	Sarah Douglas
Marketing Manager:	Jackie Harbor
Production Manager:	Penny Grose
Cover by:	Hybert Design
Printed by:	Bell & Bain Ltd., Glasgow

1 2 3 4 5 BB 5 4 3 2 1 0

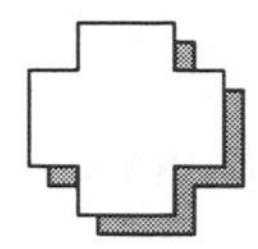

Contents

Preface

This is one of the most exciting times in technology, ever, and the Internet has the potential to change the way that people work and play, in a way that few technological areas have ever done before. It will soon become part of the fabric of our life, in the same way that the motor car, the telephone and the television have done in the past. The greatest problem is that Internet technology is moving so fast that it is difficult to keep up with it. Thus it is important to understand the key underlying principles of it, which allows everyone to learn new developments. I have seen so many people who struggle with new technological developments as they do not have a proper foundation in the subject area. This book and the associated WWW site will hopefully help provide this foundation.

The amount of transmitted information over networks increases by a large factor every year (over the Internet, traffic doubles every 100 days), and the demand for bandwidth seems unlimited. Unfortunately there are many different types of networks, from low-speed single computer connections, to high-speed multiple computer networks. There are also many different types of computer systems, there are different protocols, and so on. It is an exciting area, but also a difficult area to keep up-to-date with. Thus, one of the main aims of this book is to cover many of the important networking and distributed system areas, from networking technologies to data encryption. It splits into ten main areas, these are:

1. **Distributed Systems.** Distributed System Elements, Operating Systems, Processes and Scheduling, Distributed Processing and Distributed File Systems.
2. **Data Communications.** Data Communications and Compression.
3. **Networking Technologies.** Network Types and Cables, Ethernet (including Fast and Gigabit Ethernet), ATM and Routers.
4. **Networking Protocols.** TCP/IP, ICMP, DNS, ARP, Bootp, IP multicasting, UDP, WinSock, IP Version 6, SPX/IPX and HTTP.
5. **World Wide Web.** HTTP, Client/server architecture, Web browsers, Internet resources, URLs, URI, Web browser design, SSL, S-HTTP, Content advisor, Security zones.
6. **Network Security/Intranets.** Proxy servers, Firewalls, Filtering routers, Passwords, Hacking methods, Hacker problems and Hardware security.
7. **Data Encryption and Authentication Principles.** Cryptography, Legal issues, Cracking the code, Message hash, Private-Key Encryption, Public-key Encryption, RSA and PGP.
8. **Network Operating Systems.** Microsoft Windows, UNIX, Novell NetWare and NDS.
9. **Electronic Mail.** Architecture, Email addresses, SMTP, X.400 and MIME.
10. **Appendices.** Extensive Glossary and Abbreviations, ASCII tables, and a quick reference guide.

Many computer networks are now a hybrid of different types, typically a mixture of two or more different operating systems. These systems must successfully integrate for the complete system to operate properly, whether it is in terms of its compatibility, its security or its ease-of-use. Thus, the book includes an in-depth look at the configuration, architecture and networking of three of the most popular networking operating systems (NOS), which are Microsoft Windows, UNIX and Novell NetWare.

Further information on the book can be found at:

http://www.dcs.napier.ac.uk/~bill/dist.html
http://ceres.dcs.napier.ac.uk/staff/bill/dist.html

Sample teaching strategies are available at:

http://www.dcs.napier.ac.uk/~bill/cnds.html
http://www.dcs.napier.ac.uk/~bill/nos.html

These WWW sites also contain:

- **Sample examination questions, with solutions.** This includes sample examination questions from previous years, along with sample solutions.
- **Teachers' notes.** This includes an example teaching schedule, with highlighted areas.
- **Projects.** This includes examples of projects which could be undertaken along with the teaching, and fits with a module which includes a mixture of coursework and examination.
- **Worksheets.** If possible, the module should contain practical sessions. Example worksheets can be found on the WWW site.
- **On-line multiple-choice questions.** This includes on-line tests which are available at the end of each chapters, with many additional questions.
- **Animated graphics.** This includes animation of the key networking principles.
- **Presentation files.** This includes PowerPoint and HTML graphics from each of the chapters of the book.
- **RFC files.** This includes the key RFC files.
- **Additional chapters.** This includes chapters which appeared in an earlier draft of the book, and were not included in the final version.
- **Information on the Cisco Networking Academy.** This includes information on becoming Cisco Certified.

Help from myself can be sought using one of the following email addresses:

w.buchanan@napier.ac.uk   bill@dcs.napier.ac.uk
w_j_buchanan@hotmail.com

As much as possible I have tried to make the book as readable as possible, and still cover important concepts. For this I have included numerous text inserts, and have tried to illustrate key concepts either in figures, or through animated graphics on the WWW site.

Finally, I would personally like to thank Penny Grose (Production Editorial Manager) and Dave Hatter (Acquisitions Editor) at McGraw-Hill for their hard work and their continued support. Also, I would like to thank my family, Julie, Billy, Jamie and David for their love and understanding.

Dr William Buchanan [http://www.dcs.napier.ac.uk/~bill or http://ceres.dcs.napier.ac.uk/staff/bill]

Senior Lecturer, School of Computing, Napier University,
219 Colinton Road, Edinburgh, UK.

Trademarks

- Microsoft, XENIX, Visual Basic, Win32, Win32s, DOS, MS-DOS, Windows, Windows 95, Windows 98, Windows 2000, Windows NT, Internet Explorer, Microsoft Word, Microsoft Office, Microsoft PowerPoint and Microsoft Excel are trademarks of Microsoft Corporation.
- Adobe and PostScript are trademarks of Adobe Systems Incorporated.
- Intel, Intel Inside, i386, i486, Intel386, Intel486, IntelDX2, IntelDX4 and Pentium are registered trademarks of Intel Corporation.
- CompuServe is a registered trademark of CompuServe Incorporated.
- Novell, NDS, IPX/SPX, ODI and NetWare are trademarks of Novell Incorporated.
- Java, Sun, Sun Microsystems, Java, NFS, Network File System, Sun Workstation are trademarks of Sun Microsystems Incorporated.
- Lotus Notes is a registered trademark of Lotus Development Corporation.
- Xerox and Ethernet are trademarks of Xerox Corporation.
- ANSI is a trademark of American National Standards Institute.
- Netscape, Netscape Navigator and Netscape Communicator are trademarks of Netscape Communications Incorporated.
- IBM, PS/2, OS/2 and PC/XT are trademarks of International Business Machines.
- UNIX is a registered trademark, exclusively licensed through X/Open Company, Ltd.
- SPARC is a trademark of SPARC International, Inc. Products bearing SPARC trademarks are based upon an architecture developed by Sun Microsystems, Inc.
- Cisco IOS, Cisco, Cisco Systems, are registered trademarks of Cisco Systems, Inc.
- cc:Mail is a trademark of cc:Mail, a subsidiary of Lotus Development.
- WordPerfect is a registered trademark of WordPerfect Corporation.
- Compaq, Alpha and OpenVMS are trademarks of Compaq Computer Corporation.
- OSF, OSF/1, OSF/Motif, Motif, and Open Software Foundation are trademarks of the Open Software Foundation.
- Motorola 68000 is a registered trademark of Motorola, Inc
- COBOL is a trademark of MicroFocus, Ltd.
- Apple, AppleTalk and Mac are trademarks of the Apple Corporation Inc.
- Dell is a registered trademark of Dell Computer Corporation.
- HP-UX and HP are registered trademarks of Hewlett-Packard Company.
- Other product names mentioned herein may be trademarks and/or registered trademarks of their respective companies.

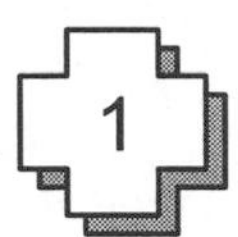

Introduction

> **THE FIRST OF MANY**
>
> The world's first large electronic computer (1946), containing 19 000 values was built at the University of Pennsylvania by John Eckert during World War II. It was called ENIAC (Electronic Numerical Integrator and Computer) and it ceased operation in 1957. By today's standards, it was a lumbering dinosaur and by the time it was dismantled it weighed over 30 tons and spread itself over 1500 square feet. Amazingly, it also consumed over 25 kW of electrical power (equivalent to the power of over 400, 60 W light bulbs), but could perform over 100 000 calculations per second (which is reasonable, even by today's standards). Unfortunately, it was unreliable, and would only work for a few hours, on average, before a valve needed to be replaced.

1.1 Introduction

The heavy technical material will follow this chapter, so before we get to that, please excuse this gentle introduction to the areas of computer systems, networks and the Internet. Computers are everywhere these days (many of them are invisible to the eye as they are embedded into electronic equipment). A crucial increase in their power, and their acceptance has been the development of computer networks, and the Internet.

Computers have been responsible for many great technological developments that would not have been possible without advanced computing technology and networking, such as in Air Traffic Control Systems and Electronic Banking. Their power has steadily increased, and the number of devices which connect to them increase by the day.

The Internet is basically a worldwide infrastructure which uses a standardized communications protocol, known as TCP/IP (Transmission Control Protocol/Internet Protocol). Many people often confuse the Internet with the World Wide Web (WWW), but the WWW is just one of the applications of the Internet. Some of its other uses are outlined in Figure 1.1 and Figure 1.2.

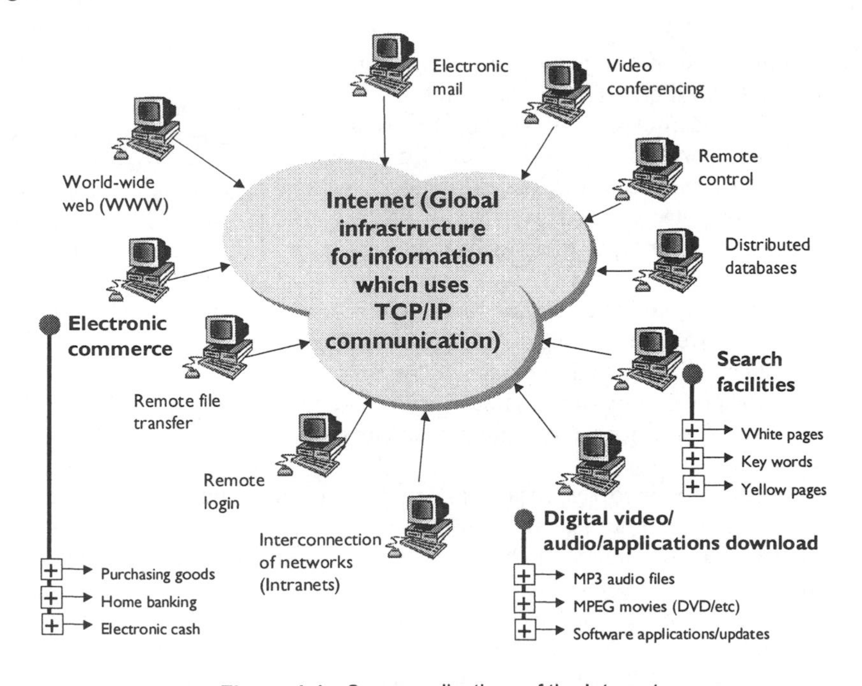

Figure 1.1 Some applications of the Internet

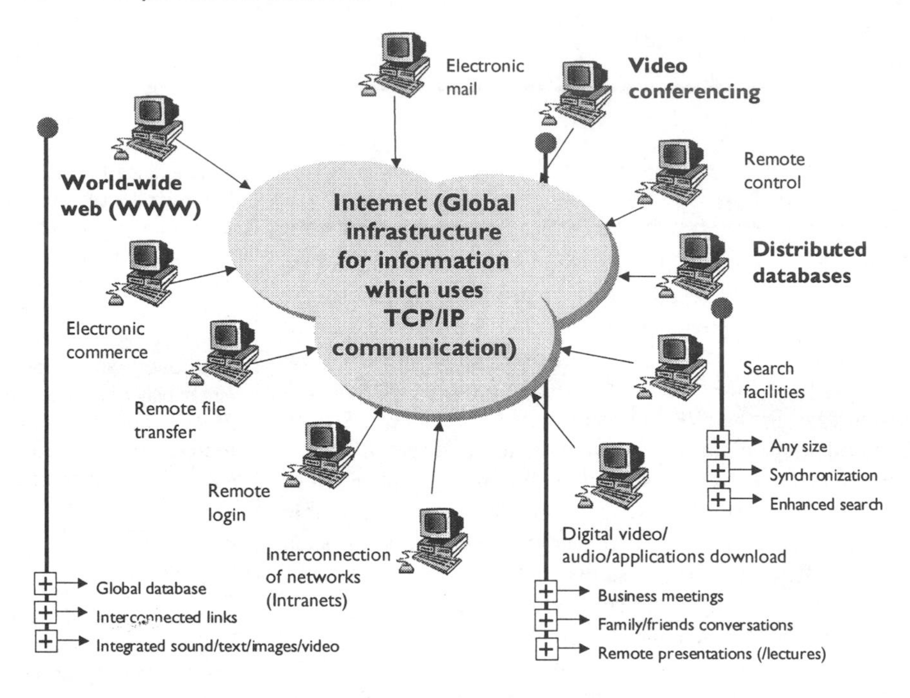

Figure 1.2 Some applications of the Internet

> **ALL TIME GREATS**
>
> One of the great revolutions of all time occurred in December 1948 when William Shockley, Walter Brattain, and John Bardeen at the Bell Labs produced a transistor that could act as a triode. It was made from a germanium crystal with a thin p-type section sandwiched between two n-type materials. Rather than release its details to the world, Bell Laboratories kept its invention secret for over seven months so that they could fully understand its operation. Unfortunately, as with many other great inventions, it received little public attention and even less press coverage (the *New York Times* gave it 4½ inches on page 46).

The Internet, networks and increased computing power will have great effects on all areas of life, whether they are in commerce, in industry or in home life. Figure 1.3 outlines how commerce, industry and the home may be influenced by the increasing usage of computers, the Internet and networks. The standardization of networking technology has allowed for the standardization of systems, especially in electronic mail, and remote working. The key of this success is the worldwide acceptance of the TCP/IP protocol, which allows different computer systems over the world to communicate, no matter their type, their architecture, or their operating system.

The Internet is likely to have a great effect on how companies do business. Over the coming years many companies will become reliant on electronic commerce for much of their business, whether it is by direct sales over the Internet or the integration of their financial operation in an electronic form.

Electronic commerce involves customers using electronic communications to purchase goods, typically using the Internet. This will change the way that many businesses do business, and the way that consumers purchase their goods. Figure 1.4 illustrates how society is moving from a cash based society to a cashless society. Most consumers now use ATMs (Automatic Telling Machines) for cash withdrawals, and debit and credit cards to purchase goods. The future is likely to see an increase in consumers using electronic methods to pay for their goods. An important key to the acceptance of Internet-based purchases is that they must be secure, and cannot be used by criminals to make false purchases, or criminals setting up companies which take payments for incorrect services.

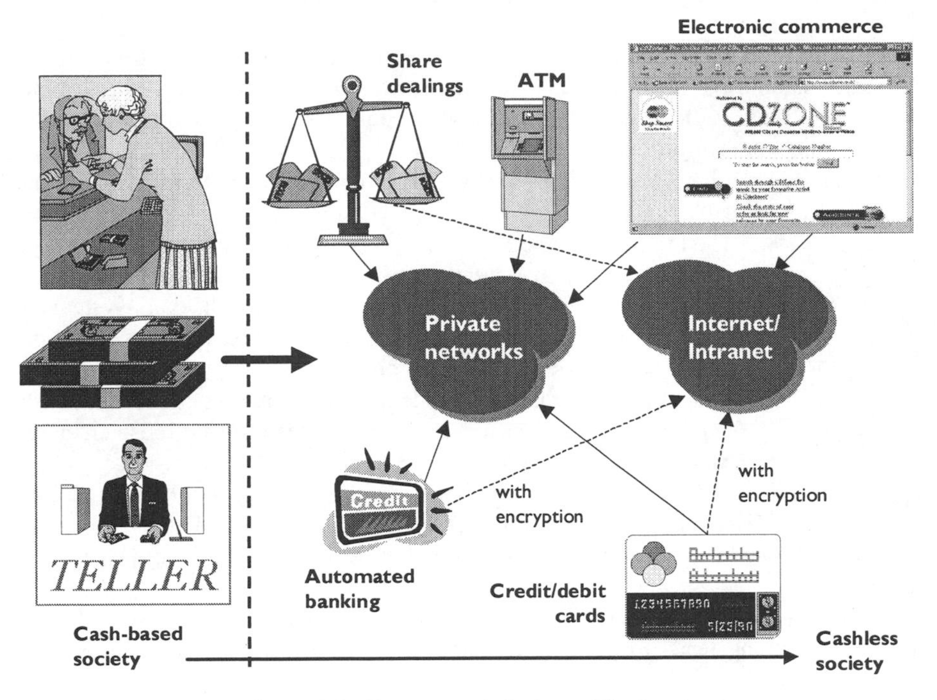

Figure 1.3 Commerce, industry and the home

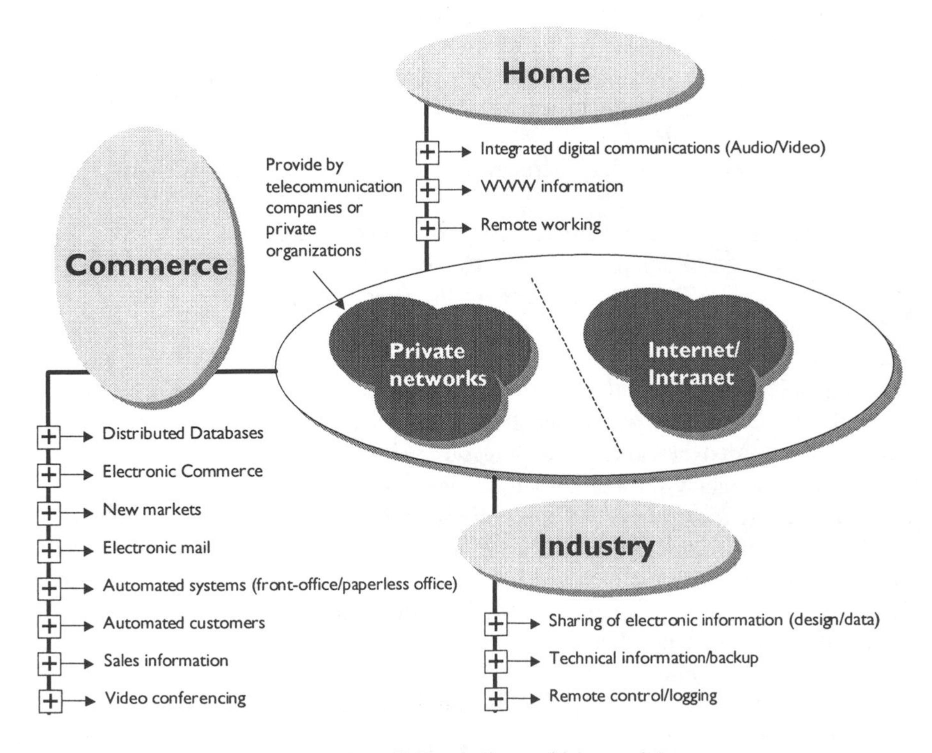

Figure 1.4 Towards the cashless society

Internet shopping has many advantages over traditional shopping, such as:

- If the goods are in an electronic format, such as digital audio or video, the customer can typically sample the material before buying it. A good example of this is listening to tracks on a music CD, or view chapters from a book. This allows a consumer to make a better choice on their products.
- Customers do not have transport problems on the Internet (apart from TCP/IP transport problems). 
- The Internet is open 24 hours a day, 365 days per year, and is not affected by holidays, weather, traffic and parking problems, and industrial strikes (not directly). It also allows for backup systems to be installed over geographically wide areas, so that companies do not have to rely on a single source of supply.
- Apart from a computer, a modem, and an Internet connection, all that is required is a credit or a debit card. No need to find loose change, or to wait in a queue.
- The Internet is not affected by the weather (well, not directly, of course), while traditional shopping involves bracing all kinds of weather (although, some weather is pleasurable, most of it is okay, but the rest is pretty bad).
- It can be based anywhere, in virtually any country, in any town or in any building. Traditional shopping tends to group shops around key shopping areas, such as cities, or shopping centers.
- Internet-based companies can quickly update their stock provision, at the press of a button, worldwide. Price changes can also be quickly reported.
- Internet-based companies can typically offer a much greater supply of goods, as they do not necessarily have to have the goods they display, actually in stock. They can simply take an order, and then quickly order the goods from the supplier.

With all these advantages, of course, there are also many disadvantages, such as:

- Internet shopping typically depends on postal service delivery, which can take days to deliver a single package.
- Internet shopping is highly dependent on the speed of delivery from the supplier. Problems with suppliers can lead to lengthy time delays.
- Post and packaging can considerably add to the cost of goods. Typically goods are cheaper on the Internet, but when post and packaging are added they end up being less of a bargain.
- Consumers are quickly put-off by bad service, and slow delivery times. Typically, a consumer will only allow delivery times of a few days; once it is greater than this they may become annoyed, and never purchase over the Internet again.
- Internet shopping is becoming swamped with too many suppliers, which makes it difficult to differentiate the good ones from the bad ones.
- For non-electronic goods, such as clothes and jewelry, the consumer cannot properly feel, look-at, or touch the goods, and will simply waste time in sampling them (similar to buying through a mail-order company).

AN INTEGRATED CIRCUIT?

In 1952, GW Dummer, a radar expert from Britain's Royal Radar Establishment had presented a paper proposing that a solid block of materials could be used to connect electronic components, without connecting wires. This would lay the foundation of the integrated circuit.

1.2 Networks

Computer networks are a crucial part of many organizations and many users now even have a network connection in their own home. Without networks there would be no electronic mail, no Internet access and no networked applications. It is one of the fastest growing technological areas and brings benefits to virtually every country in the world. With the interconnection of networks to the Internet, the world has truly become a Global Village. For many people, especially children, the first place to search for a given topic is the World Wide Web (WWW).

Who would believe the pace of technology over ten short years, such as:

- From networks of tens of computers operating at speeds of thousands of bits per second, to networks with thousands of computers operating at billions of bits per second.
- From organizations that passed paper documents back and forward, to the totally paperless organizations.
- From people who sent one letter each month, to people who send tens of electronic mails every day.
- From sending letters around the world which would take days or weeks, to arrive to the transmission of information around the world within a fraction of a second.
- From businesses that relied on central operations, to ones that can be distributed around the world, but can communicate as if they were next door.
- From the transmission of memos which people and organizations could view who were not meant to read the message, to the transmission of messages which can only be read by the intended destination (and maybe, by space aliens). Not even the CIA can decrypt these messages.
- From written signatures that can be easily forged, to digital signatures which are almost impossible to forge, and authenticate both the sender and the contents of a message.
- These days virtually every computer in a company is networked and networks are a key element to the effective working of an organization. Without them, few people could work effectively. They provide us with:

  - ✦ Electronic mail.
  - ✦ Remote connections.
  - ✦ Networked video conferencing.
  - ✦ Remote data acquisition.
  - Networked application software.
  - Shared printers.
  - Remote control of remote equipment.
  - Shared disk resources.

1.3 TCP/IP - The world's most important protocol

As the PC and other computers have become more powerful, there was a demand for increased remote access of information and interconnection. This required a common communication language (network protocol). For this many protocols were developed, but the most important was developed by the US Defense Advanced Research Projects Agency (DARPA). They named it TCP/IP and it has since allowed computers around the world to intercommunicate, no matter their operating system, their type or their network connection. DARPA's initial aim was simply to connect a number of universities and other research establishments to its own network. This network was designed by the military so that a military strike on one or more parts of the network would still support intercommunications, as illustrated in Figure 1.5. The resultant interconnected network (internet) is now known as the Internet, and has since outgrown its original application and many commercial organizations and home users now connect to it.

The Internet uses TCP/IP as a standard to transfer data and each node on the Internet is assigned a unique network address, called an IP address. Unfortunately IP addresses are difficult to remember, thus Domain Name Services (DNS) are used to allow users to use symbolic names rather than IP addresses. DNS computers on the Internet determine the IP address of the named destination resource or application program. This dynamic mapping has the advantage that users and application programs can move around the Internet and are not fixed to an IP address. An analogy of this relates to the public telephone service. A telephone directory contains a list of subscribers and their associated telephone number. If someone looks for a telephone number, first the user name is looked up and their associated telephone number found. The telephone directory listing thus maps a user name (symbolic name) to an actual telephone number (the actual address). On the Internet, when a user enters a domain name (such as `www.fred.co.uk`) into the WWW browser then the local DNS server must try and resolve the domain name to an IP address, which can then be used to send the data to it. If it cannot resolve the IP address then the DNS server interrogates other DNS servers to see if they know the required IP address, as illustrated in Figure 1.6. If it cannot be resolved then the WWW browser displays an error message.

The Internet naming structure uses labels separated by periods (full stops); an example is `dcs.napier.ac.uk`. It uses a hierarchical structure where organizations are grouped into primary domain names, such as `com` (for commercial organizations), `edu` (for educational organizations), `gov` (for government organizations), `mil` (for military organizations), `net` (Internet network support centers) and `org` (other organizations). The primary domain name may also define the country in which the host is located, such as `uk` (United Kingdom), `fr` (France), and so on. All hosts on the Internet must be registered to one of these primary domain names.

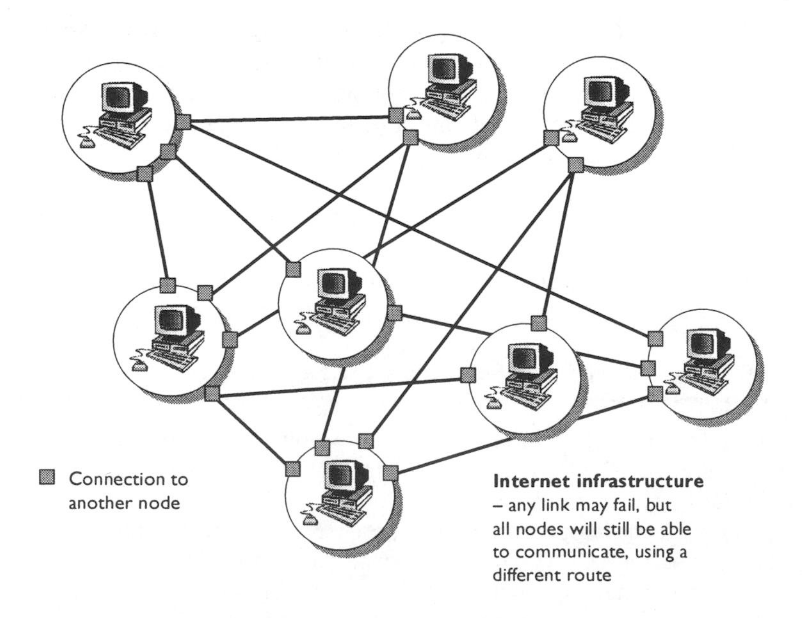

Connection to
another node
Internet infrastructure
– any link may fail, but
all nodes will still be able
to communicate, using a
different route

Figure 1.5 Internet infrastructure

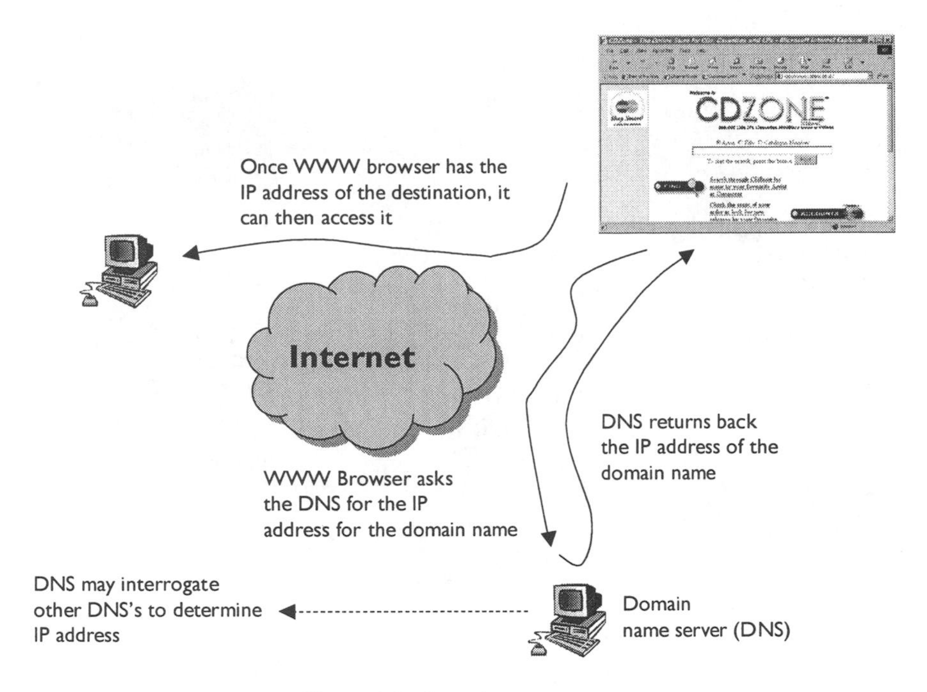
Once WWW browser has the
IP address of the destination, it
can then access it
Internet
DNS returns back
the IP address of the
domain name
WWW Browser asks
the DNS for the IP
address for the domain name
DNS may interrogate
other DNS's to determine
IP address
Domain
name server (DNS)
CDZONE

Figure 1.6 Domain name server

What's my IP
address?

Domain name labels after the primary field describe the subnetworks (subnets) within the network. For example, in the address `dcs.napier.ac.uk`, the `ac` label relates to an academic institution within the `uk`, `napier` to the name of the institution and `dcs` the subnet within that organization.

TCP/IP has been unbelievably successful and has outgrown its original application. Unfortunately, its addressing structure only allows for up to 4 billion addresses (many of which will be unused). There is thus a need for a new addressing structure which provides many more addresses, as the number of devices which connect to the Internet increases by the day (not only computers, but printers, fax machines, mobile phones, and so on, can be allocated IP addresses). For this, the IP Version 6 protocol is being developed, which will, in the future, replace the existing IP Version 4 protocol and provide for 128-bit addresses. This will give:

3,400,000,000,000,000,000,000,000,000,000,000,000,000 $[3.40 \times 10^{38}]$

different addresses (which should be enough for everyone in the world to have their own IP address). In fact, with 128-bit IP addresses it is possible to grant an IP address to every piece of electronic equipment in the world (or possibly, if there are aliens, the known Universe). This even includes toasters and tea makers. Just imagine being able to communicate directly with any electronic device that you want. All that is required is a connection onto the Internet. It really changes the notation of 'Getting onto the Internet', as illustrated in Figure 1.7.

The IP protocol, itself, allows for the routing and addressing of the transmitted data, whereas TCP supports the communication between application programs, using a stream of data. The main function of TCP is to provide a robust and reliable transport of the data.

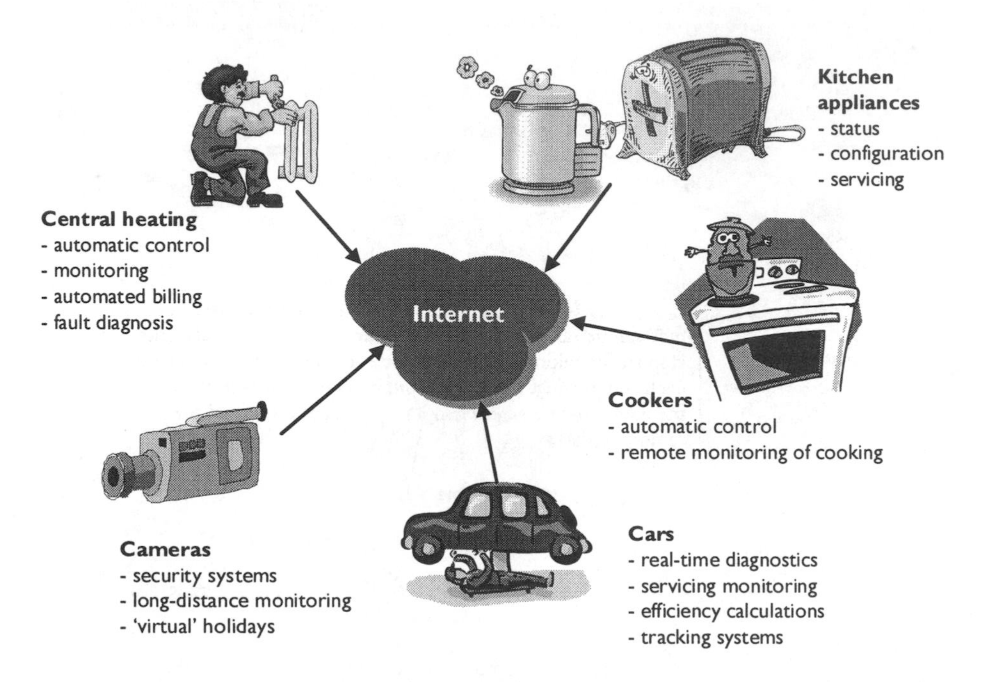

Figure 1.7 Getting on the Internet

1.4 Java, the WWW and Intranets

Along with the growth of the Internet came the World Wide Web (WWW). It is unlikely that the Internet would have taken off as fast as it did without the WWW, and that the WWW could have not existed without the Internet. So the two required each other. Along with electronic mail, the WWW was the killer application that was required to create the massive, worldwide, acceptance of the Internet.

The WWW was initially conceived in 1989 by Tim Berners-Lee at CERN, the European particle physics research laboratory in Geneva, Switzerland. Its main objective was to allow various different types of information, such as text, graphics and video, to be integrated together in an easy-to-use manner, typically distributed over a geographically wide area. The result was the world-wide acceptance of the protocols and specifications used (especially TCP/IP, HTTP and HTML). A major part of its success was the full support of the National Center for Supercomputing Applications (NCSA), which developed a family of user interface systems known collectively as Mosaic. Netscape and Microsoft have since developed excellent WWW browsers which have an easy-to-use interface to the WWW.

Typical modern enhancements are:

- **Search facilities.** Browsers now support many search engine connections (such as Yahoo and Alta-Vista).
- **History of recently visited WWW pages.** See Figure 1.8 for an example of a history folder using Internet Explorer.
- **Favorites list.** This allows users to add WWW pages to a favorites list or folder. See Figure 1.9 for an example of a favorites folder using Internet Explorer.

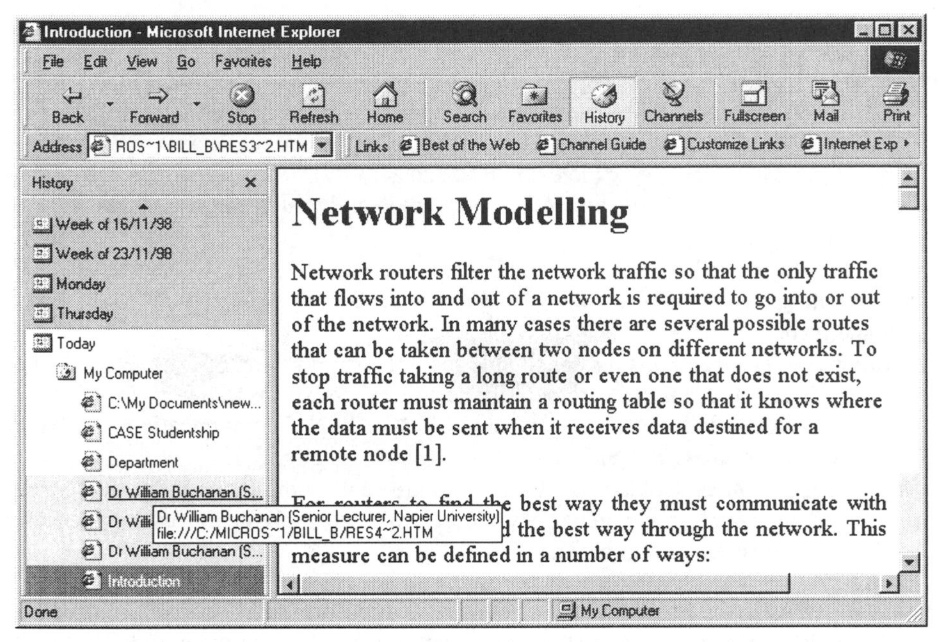

Figure 1.8 History folder

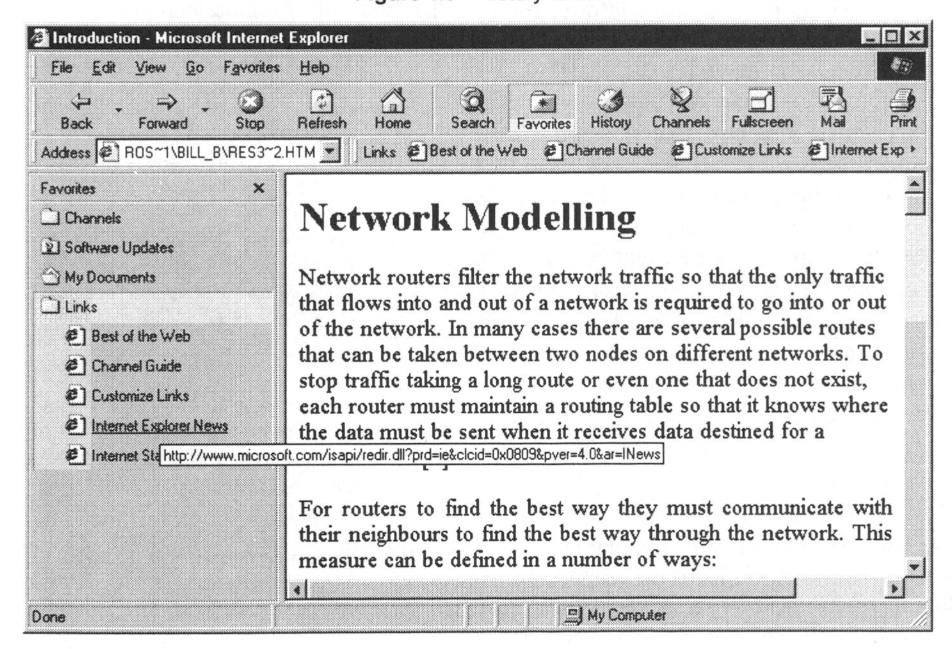

Figure 1.9 Favorites folder

- **Increased security.** This allows Internet sites to be zoned into security levels, such as High (most secure), Medium and Low (least secure). In the most secure level, the browser excludes any material (called content) that could damage the computer.
- **Content advisors.** Microsoft Internet Explorer uses a rating system which was developed by the Recreational Software Advisory Council (RSAC). This is based on the work of Dr Donald F. Roberts of Stanford University, who studied the effects of media on

humans for nearly 20 years. In this, the content of the material is graded into four main levels, which is rated for:

- o Language (from inoffensive slang to explicit or crude language).
- o Nudity (from no nudity and provocative frontal nudity).
- o Sex (from no sex to explicit sexual activity).
- o Violence (from no violence to wanton and gratuitous violence).

- **Usage of site certificates.** These allow sites to positively identify themselves. Certificate granting authorities include ATT Certificate Services, Microsoft Root Authority and Verisign.
- **Support for secure transmissions.** This includes support for SSL2 (Secured Sockets Layer, Level 2), the standard protocol for secure transmissions. Secure WWW sites support this protocol and some also support PCT (Personal Communications Technology) which is more secure than SSL2.
- **Support for cookies.** Cookies are files which are sent by the Internet site and are stored locally on the user's computer. They store information about the user's identity and preferences when revisiting that site.

The WWW, or Web, is basically an infrastructure of information, which is stored on the WWW by Web servers and uses the Internet to transmit data around the world. These servers run special programs that allow information to be transmitted to remote computers which are running a Web browser.

The standard language developed was HTML (HyperText Markup Language), which is a text-based language that contains certain formatting tags. These tags are identified within a less than (<) and a greater than (>) symbol. Most have an opening and closing version; for example, to highlight bold text, the bold opening tag is `<B>` and the closing tag is `</B>`. When a greater than or a less than symbol is required then a special character sequence is used. HTML is acceptable for low quality and medium print quality documents, but it is difficult to produce any high-quality printed material. It is likely that new versions of HTML will support enhanced presentation.

HTML is a rather limited language, and includes limited user interaction and a lack of much of the functionality of a programming language, such as not having expressions, loops, or decisions. A short-term fix is the JavaScript language which integrates some of the features of a high-level language into an HTML page. JavaScript also allows some functionality for developing client and server Internet applications. It can be used to respond to user events, such as mouse clicks, form input and page navigation. A major advantage of JavaScript over HTML is that it supports the use of generic pieces of code (functions) without any special declarative requirements.

Top 20 Computer People of All Time

1. DAN BRICKLIN (VisiCalc)
2. BILL GATES (Microsoft)
3. STEVE JOBS (Apple)
4. ROBERT NOYCE (Intel)
5. DENNIS RITCHIE (C Programming)
6. MARC ANDREESSEN (Netscape Communications)
7. BILL ATKINSON (Apple Mac GUI)
8. TIM BERNERS-LEE (CERN/WWW)
9. DOUG ENGELBART (Mouse/Windows/etc)
10. GRACE MURRAY HOPPER (COBOL)
11. PHILIPPE KAHN (Turbo Pascal)
12. MITCH KAPOR (Lotus 123)
13. DONALD KNUTH (TEX)
14. THOMAS KURTZ
15. DREW MAJOR (NetWare)
16. ROBERT METCALFE (Ethernet)
17. JARNE STROUSTRUP (C++)
18. JOHN WARNOCK (Adobe)
19. NIKLAUS WIRTH (Pascal)
20. STEVE WOZNIAK (Apple)

– Byte, Sept 1995

JavaScript was a short-term fix for the WWW, but as the power of computers increased, it became possible to run a computer program within a browser. For this, Sun Microsystems developed the Java programming language. It was first released in 1995 and was quickly adopted as it integrates well with Internet-based programming. Java 1.0 introduced the concept of an applet, which is a machine-independent program which runs within a WWW browser. It was quickly followed by Java 1.1 which gave faster interpretation of Java applets and included many new features (and has now been replaced with Java 1.2). Java is a general-purpose, concurrent, class-based, object-oriented language and has been designed to be relatively simple to built complex applications. Java evolved from C and C++, but some parts of C++ (mainly the most difficult parts, such as pointers and parameter passing) have been dropped and others added. It was also the first, truly object-oriented programming language.

Java has the great advantage over conventional software languages in that it produces code which is computer hardware independent. This is because the compiled code (called bytecode) is interpreted by the WWW browser. Unfortunately, this leads to slower execution, but, as much of the time in a graphical user interface program is spent updating the graphics display, then the overhead is, as far as the user is concerned, not a great one. Another advantage that Java has over conventional software languages is that it directly supports the Internet and networking, especially for the transmission of compressed image, audio and video formats. The small 'black-box' networked computer is one of the founding principles of Java, and it is hoped that, in the future, small Java-based computers could replace the complex PC/workstation for general-purpose applications, like accessing the Internet or playing network games.

The WWW uses TCP/IP for the transmission of the data, and HTTP (Hyper Text Transmission Protocol) to allow the WWW browser to communicate with the WWW server. HTTP is a stateless protocol where each transaction is independent of any previous transactions. The advantage of this is that it allows the rapid access of WWW pages over several widely distributed servers. HTTP uses the TCP protocol to establish the connection between the client and the server for each transaction then terminates the connection once the transaction completes. An important element of HTTP is that it supports many different formats of data, where a client issues a request to a server which may include a prioritized list of formats that it can handle. This allows new formats to be easily added and also prevents the transmission of unnecessary information.

A client's WWW browser (the user agent) initially establishes a direct connection with the destination server which contains the required WWW page. To make this connection the client initiates a TCP connection between the client and the server. After this, the client then issues an HTTP request, such as, the specific command (the method), the URL (Universal Resource Locator, which is the full name of the requested data), and possibly extra information such as request parameters or client information. When the server receives the request, it attempts to perform the requested action. It then returns an HTTP response, which includes status information, a success/error code, and extra information. After the client receives this, the TCP connection is closed.

Since the growth in the Internet, there has been great investment in local area networks (LANs), wide area networks (WANs) and metropolitan area networks (MANs). The demand for bandwidth and faster applications increases as the applications of networks, the Internet and the WWW increase by the day. Along with this growth has come the introduction of many different types of networks, such as ISDN, Modems, Ethernet, Token Ring, ATM, FDDI, Frame relay and X.25. The great achievement, though, has been the worldwide acceptance of TCP/IP which has allowed nodes on these networks to intercommunicate.

> **SUPERSONIC**
>
> Compaq Computers, in 1981, generated $110 million in their first year, a further two years on it was $503.9 million, and two years after that it was $1 billion. The following year it was $2 billion. From zero to $2 billion, in six years (a world record, at the time).

Many organizations, though, are wary of the Internet as it allows external users to 'hack' into their network and possibly steal secrets or damage their computer systems. Thus many companies have set-up a special internet, known as an intranet. These are in-house, tailor-made internets for use within the organization and provide limited access (if any) to outside services and also limit the external traffic into the intranet (if any). An intranet might have access to the Internet, but there will be no access from the Internet to the organization's intranet (unless through carefully controlled portals).

Intranets normally have firewalls, which filter incoming and outgoing network traffic. This can be set-up to limit access from external and internal users, and also to block access to certain external systems, or block certain types of network traffic (such as video conferencing or remote file transfer).

1.5 Basic computer systems

In the past, it has been relatively easy to define what a computer system was, as it was basically an electronic device which processed information in some way to give some form of output. The processing unit must have some form of stored program, which contains the information on how to process the input data, as illustrate in Figure 1.10. The stored program can either be permanently embedded in the processing unit (fixed function) or it can be changed using software (programmable function). Examples of these are:

- **Fixed function.** These have a fixed function which is either set up in the device when it is produced or is set up electronically by the user.
- **Programmable functions.** These run software which can be modified, if required. An example of this is a general-purpose microprocessor, such as the Intel Pentium.

Typically, fixed function devices are faster than general-purpose devices. High-speed applications typically use special processors, which are optimized for a specific application. For example, graphics processors are optimized for graphics applications, mathematical processors are optimized for mathematical transformation operations, and so on.

It would be very difficult for the user to directly interface to the computer hardware, thus the system requires an operating system. It is often also more convenient for the user to use some form of interface to communicate with the operating system, thus a user interface is required. Figure 1.11 shows a layered model of a computer system, the layers are:

- **Hardware.** This provides for the processing, storage and interfacing (Input/Output).
- **Operating system.** This provides the interface between the user, and/or the user interface, and the hardware. Typical operating systems are: DOS (Disk Operating System), Microsoft Windows, Unix and Linux (PC-based version of Unix). Its main functions are: creating a file system, copying/deleting/moving files, multitasking programs (running more than one program, at a time), starting-up the computer (booting), interfacing with the hardware (typically with software drivers), networking, program intercommunication, and so on.
- **User interface.** This provides the interface between the user and the operating system. Typically this is a GUI (Graphical User Interface) which uses a WIMPs (Windows, Icons, Menus and Pointers) approach. Typical GUIs are: Microsoft Windows and X-Windows.

- **Application software.** These are programs which perform specific tasks, such as word processing, spreadsheets, database analysis, business presentation, and so on.

A layered approach allows software and hardware designers to develop systems which are designed to communicate through various standard interfaces. For example, a hardware designer could design a new system which communicated with the operating system in a way that was standardized by the interface between the operating system and the hardware.

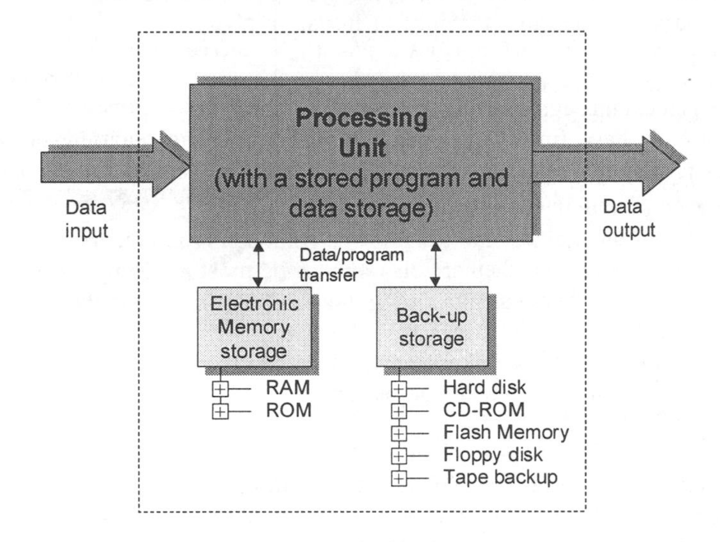

Figure 1.10 Processing unit

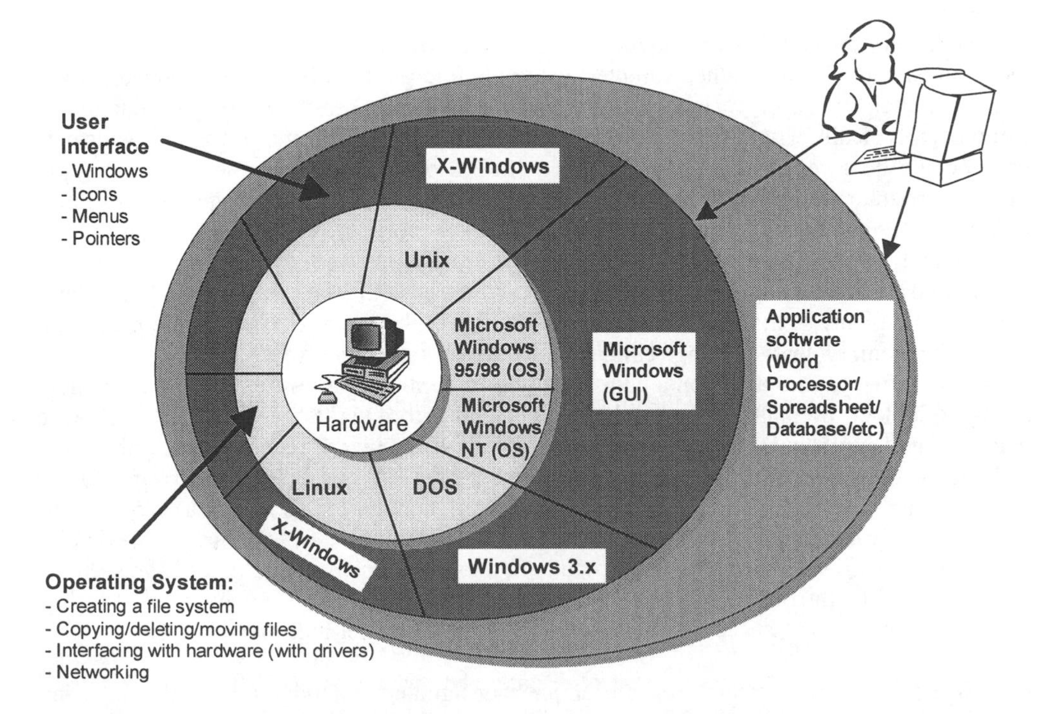

Figure 1.11 Model of layers of a computer system

SAME OLD STORY

In the late 1940s, IBM had a considerable share of the computer market, so much so that a complaint was filed against them alleging monopolistic practices in its computer business, in violation of the Sherman Act. By January 1954, the US District Court made a final judgment on the complaint against IBM. For this, a 'consent decree' was then signed by IBM, which placed limitations on how IBM conducts business with respect to 'electronic data processing machines'.

1.6 Top 15 Achievers/Top 15 under-achievers

Most of the chapters in this book are technically biased, but this chapter is slightly less so. It is intended to give a gentle introduction to computer systems, networks, the Internet and distributed systems. As a precursor to the more technical material in the following chapters, I've included my Top 15 achievers and under-achievers in the computer industry. Please excuse the usage of technical jargon, before it is introduced in later units. Most of the terms, such as TCP/IP and Ethernet, are well-known terms and are even used in the broadcasting industry.

The PC is an amazing device, and has allowed computers to move from technical specialists to, well, anyone. However, they are also one of the most annoying pieces of technology of all time, in terms of their software, their operating system, and their hardware. If we bought a car and it failed at least a few times every day, we would take it back and demand another one. When that failed, we would demand our money back. Or, imagine a toaster that failed half way through making a piece of toast, and we had to turn the power off, and restart it. We just wouldn't allow it.

So why does the PC lead such a privileged life. Well it's because it's so useful and multi-talented, although it doesn't really excel at much. Contrast a simple games computer against the PC and you'll find many lessons in how to make a computer easy-to-use, and easy to configure. One of the main reasons for many of its problems, though, is the compatibility with previous systems both in terms of hardware and software (and lots of dodgy software, of course).

The history of the PC is an unbelievable story, full of successes and failures. Many people, who have used some of the computer systems before the IBM PC was developed, wipe a tear from their eyes, for various reasons, when they remember their first introduction to computers, typically with the Sinclair Spectrum or the Apple II. In those days, all your programs could be saved to a single floppy disk; 128 KB of memory was more than enough to run any program, and the nearest you got to a GUI was at the adhesives shelf at your local DIY store. It must be said, though, that computers were more interesting in those days. Open one up, and it was filled with processor chips, memory chips, sound chips, and so on. You could almost see the thing working (a bit like how it was in the days of valves). These days, computers lack any soul; one computer system is much like the next. There's the processor, there's the memory, that's a bridge chip, and, oh, there's the busses, that's it.

As we move to computers on a chip, they will, in terms of hardware, become even more boring to look at. But, maybe I'm just biased. Oh, and before the IBM PC, it was people who made things happen in the computer industry; people like Steve Wozniak, Steve Jobs, Kenneth Olson, Sir Clive Sinclair, Bill Gates, and so on. These days it is large teams of software and hardware developers who move the industry. Well, enough of this negative stuff. The PC is an extremely exciting development, which has changed modern life. Without its flexibility, its compatibility, and, especially, its introduction into the home, we would not have seen the fast growth of the Internet.

Here are my Top 15 successes (in rank order) in the computer industry:

- **IBM PC** (for most), which was a triumph of design and creativity. One of the few computer systems to ever to be released on time, within budget, and within specification.

Hooray to Hollerith

One of the first occurrences of computer technology occurred in the USA in the 1880s. It was due to the American Constitution demanding that a survey is undertaken every 10 years. As the population in the USA increased, it took an increasing amount of time to produce the statistics. By the 1880s, it looked likely that the 1880 survey would not be complete until 1890. To overcome this, Herman Hollerith (who worked for the Government) devised a machine which accepted punch cards with information on them. These cards allowed a current to pass through a hole when there was a hole present.

Hollerith's electromechanical machine was extremely successful and used in the 1890 and 1900 Censuses. He even founded the company that would later become International Business Machines (IBM).

Bill Gates must take some credit in getting IBM to adopt the 8088 processor, rather than the relatively slow 8080. After its success, every man and his dog at IBM had a say in what went into it. The rise of the bland IBM PC is an excellent example of an open-system triumphing over a closed-system. Companies that have quasi-monopolies are keen on keeping their systems closed, while companies that openly compete against other competitors prefer open systems. The market, and thus, the user, prefer open-systems.

- **TCP/IP**, which is the standard protocol used by computers communicating over the Internet. It has been designed to be computer-independent and operating system independent, thus any type of computer can talk to any other type (as long as they both use TCP/IP communications). It has withstood the growth of the Internet with great success. Its only problem is that we are now running out of IP addresses to grant to all the computers that connect to the Internet. It is thus a victim of its own success. TCP/IP has proved the foundation for all the Internet applications, such as the World Wide Web, video conferencing, file transfer, remote login and electronic mail. It has also been followed by domain names (such as fred.com), which map symbolic names to IP addresses.
- **Electronic mail**, which has taken the paperless office one step nearer. Many mourned the death of letter writing, as TV and the telephone had suppressed its form. With e-mail it is back again, stronger than ever. It is not without its faults, though. Many people have sent e-mails in anger, or ignorance, and then regretted them later. It is just too quick, and does not allow for a cooling off period. My motto is: 'If you're annoyed about something, sleep on it, and send the e-mail in the morning'. Also, because e-mail is not a face-to-face communication, or a voice-to-voice communication, it is easy to take something out of context. So another motto is: 'Carefully read everything that you have written, and make sure there is nothing that is offensive or can be misinterpreted'. Only on the Internet could e-mail addressing (such as, fred@bloggs.com) be accepted, worldwide, in such a short time.
- **Microsoft**, which made sure that it could not lose in the growth of the PC, by teaming up with the main computer manufacturers, such as IBM (for DOS and OS/2), Apple (for Macintosh application software) and for its own operating system: Windows. Luckily, for Microsoft, it was its own operating system which became the industry standard. With the might of having the industry-standard operating system (DOS, and then Microsoft Windows), Microsoft captured a large market for industry-standard application programs, such as Word and Excel. For a company that never specialized in application software, it has done well to capture a larger market share than all of its competitors put together (many of whom specialize in application software).
- **Intel**, which was gifted an enormous market with the development of the IBM PC, but has since invested money in enhancing its processors, but still keeping compatibility

with its earlier ones. This compatibility caused a great deal of hassle for software developers, but had great advantages for users. With processors, the larger the market you have, the more money you can invest in new ones, which leads to a larger market, and so on. Unfortunately, the problem with this is that other processor companies can simply copy your designs, and change them a little so that they are still compatible. This is something that Intel have fought against, and, in most cases, have succeeded in regaining their market share, either with improved technology or with legal action. The Pentium processor was a great success, as it was technologically superior to many other processors on the market, even the enhanced RISC devices. It has since become faster and faster.

First Microprocessor

Around the late 1960s, the electronics industry was producing cheap pocket calculators, which led to the development of affordable computers, when the Japanese company Busicom commissioned Intel to produce a set of between eight and 12 ICs for a calculator. Then instead of designing a complete set of ICs, Ted Hoff, at Intel, designed an integrated circuit chip that could receive instructions, and perform simple integrated functions on data. The design became the 4004 microprocessor. Intel produced a set of ICs, which could be programmed to perform different tasks. These were the first ever microprocessors and soon Intel (short for *Int*egrated *El*ectronics) produced a general-purpose 4-bit microprocessor, named the 4004.

- **6502** and **Z80** (joint award), the classic 16-bit processors which became a standard part of most of the PCs available before the IBM PC. The 6502 competed against the mighty Motorola 6800, while the Z80 competed directly with the innovative Intel 8080.
- **Apple II**, which took computing out of the millionaires' club, and into the classroom, the laboratory, and, even, the home.
- **Ethernet**, which has become the standard networking technology. It is not without its faults, but has survived because of its upgradability, its ease-of-use, and its cheapness. Ethernet does not cope well with high capacity network traffic, because it is based on contention, where nodes must contend with each other to get access to a network segment. If two nodes try to get access at the same time, a collision results, and no data is transmitted. Thus the more traffic there is on a network, the more collisions there are. This reduces the overall network capacity. However, Ethernet had two more trump cards up its sleeve. When faced with network capacity problems, it increased its bit rate from the standard 10 Mbps (10BASE) to 100 Mbps (100BASE), which gave ten times the capacity and reduced contention problems. For networks backbones it also suffered because it could not transmit data fast enough. So, it played its next card: 1000BASE, which increased the data rate to 1 Gbps (1000 MBps). Against this type of card player, no other networking technology had a chance.
- **WWW**, which is often confused with the Internet, and is becoming the largest database ever created (okay, 99% of it is rubbish, but even if 1% is good then it is all worthwhile). The WWW is one of the uses of the Internet; others include file transfer, remote login and electronic mail.
- **Apple Macintosh**, which was one of few PC systems which properly competed with the IBM PC. It succeeded mainly because of its excellent operating system (MAC OS), which was approximately 10 years ahead of its time. Possibly, if Apple had spent as much of its time in developing application software rather than for their operating system it would have considerably helped the adoption of the Mac. Apple also refused, until it was too late, to license its technology to other manufacturers. For a long time it thus stayed a closed-system.
- **Compaq DeskPro 386**. Against all the odds, Compaq stole the IBM PC standard from

To the 4004, and beyond

The 4004 caused a revolution in the electronics industry as previous electronic systems had a fixed functionality. With this processor, the functionality could be programmed by software. Amazingly, by today's standards, it could only handle four bits of data at a time (a nibble), contained 2000 transistors, had 46 instructions and allowed 4 KB of program code and 1 KB of data. From this humble start, the PC has since evolved using Intel microprocessors. Intel had previously been an innovative company, and had produced the first memory device (static RAM, which uses six transistors for each bit stored in memory), the first DRAM (dynamic memory, which uses only one transistor for each bit stored in memory) and the first EPROM (which allows data to be downloaded to a device, which is then permanently stored).

the creators, who had tried to lead the rest of the industry up a dark alley, with MCA.

- **Sun SPARC**, which succeeded against the growth of the IBM PC, because of its excellent technology, its reliable Unix operating system, and its graphical user interface (X-Windows). Sun Microsystems did not make the mistakes that Apple had made, and allowed other companies to license its technology. They also supported open systems in terms of both the hardware and software. Sun is probably the main reason that Unix is still alive, and thriving.
- **Commodore**, which bravely fought on against the IBM PC. It released many great computers, including the Vic range and the Commodore Amiga, and was responsible for forcing down the price of computers.
- **Sinclair Research**, which, more than any other company, made computing affordable to the masses. Okay, most of its computers had terrible membrane keyboards, memory adaptors that wobbled, took three fingers to get the required command (Shift-2nd Function-Alt-*etc*), required a cassette recorder to upload a program, would typically crash after you had entered one thousand lines of code, and so on. However, all of this aside, in the Sinclair Spectrum they found the right computer, for the right time, at the right price. Sometimes success can breed complacency, and so it turned out with the Sinclair QL and the Sinclair C-5 (the electric slipper).
- **Compaq**, for startling growth, that is unlikely to ever be repeated. From zero to one billion dollars in five years, which it achieved, not by luck, but by shear superior technology, time-after-time, and by sharing its technology with others (which, at the time, was the only way to compete against the might of IBM).

Other contenders include:

- Unix, mainly for providing the communications protocol for the Internet: TCP/IP, and for being so reliable, and long-lasting in a short-term industry. Also for being one of the strongest rivals to Microsoft Windows. For the technically minded, Unix allows the user to view the complete system, which is often hidden in Microsoft Windows.
- X-Windows, for lots of things, including its openness, and ability to share with others (*good old human attributes*).
- Hewlett-Packard, for its range of printers and their brand strength.
- CISCO, for its networking products and providing the backbone of the Internet (with CISCO routers).
- Java, for ignoring computer architecture, the type of network connection, and, well, everything.
- Power PC, for trying to head off the PC, at the pass, but no quite succeeding.
- Dell, for, like Compaq, achieving unbelievable growth, and creating a new market niche in selling computers directly from the factory.

Oh, and the Intel 80386, the Intel 8088, the Intel Pentium, Microsoft Visual Basic (for bringing programming to the masses), Microsoft Office, Microsoft Windows 95, Microsoft Windows NT, and so on. Okay, Windows 95, Windows NT, the 80386 and the Pentium would normally be in the Top 15, but, as Microsoft and Intel are already there, I've left them out. Here's to the Wintel Corporation. We are in their hands. One false move and they will bring their world around themselves. Up to now, Wintel have made all the correct decisions.

When it comes to failures, there are no failures really, and it is easy to be wise after the event. Who really knows what would have happened if the industry had taken another route. So, instead of the Top 15 failures, I've listed the following as the Top 15 under-achievers (please forgive me for adding a few of my own, such as DOS and the Intel 8088):

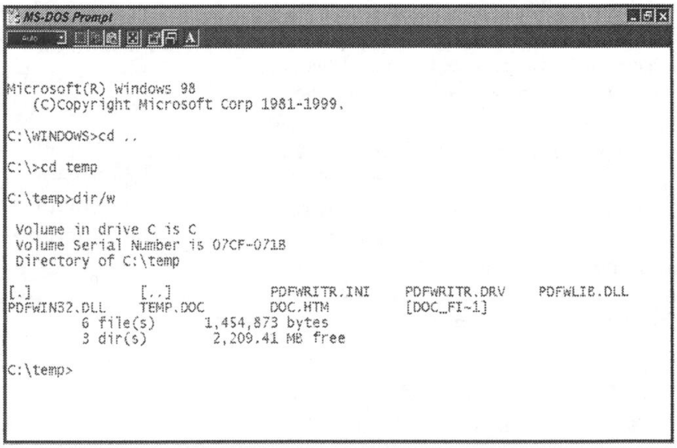

```
MS-DOS Prompt

Microsoft(R) Windows 98
   (C)Copyright Microsoft Corp 1981-1999.

C:\WINDOWS>cd ..

C:\>cd temp

C:\temp>dir/w

 Volume in drive C is C
 Volume Serial Number is 07CF-0718
 Directory of C:\temp

[.]             [..]            PDFWRITR.INI    PDFWRITR.DRV    PDFWLIB.DLL
PDFWIN32.DLL    TEMP.DOC        DOC.HTM         [DOC_FI~1]
        6 file(s)      1,454,873 bytes
        3 dir(s)        2,209.41 MB free

C:\temp>
```

- **DOS**, which became the best selling, standard operating systems for IBM PC systems. Unfortunately, it held the computer industry back for at least ten years. It was text-based, command-oriented, had no graphical user interface. It could also only access up to 640 KB of memory, at 16 bits at a time. Many users with a short memory will say that the PC is easy-to-use, and intuitive, but they are maybe forgetting how it used to be, before Microsoft Windows. With Windows 95 (and to a lesser extent with Windows 3.x), Microsoft made computers much easier to use. From then on, users could actually switch on their computer without having to register for a higher degree in Computing (*sic*). DOS would have appeared fine, as it was compatible with all its previous parents, but the problem was MAC OS, which showed everyone how a user interface should operate. Against this competition, it was no contest. So, what was it that made the PC a success? It was application software. The PC had application software coming out of its ears.
- **Intel 8088**, which became the standard PC processor, and thus the standard machine code for all PC applications. So why, after being such a success, is it in the failures list? Well, like DOS, it's because it was so difficult to use, and was a compromised system. While Amiga and Apple programmers were writing proper programs which used the processor to its maximum extent, PC programmers were still using their processors in 'sleepy-mode' (8088-compatible mode), and could only access a maximum of 1 MB of memory (because of the 20-bit address bus limit for 8088 code). The big problem with the 8088 was that it kept compatibility with its father: the 8080. For this Intel decided to use a segmented memory access, which is fine for small programs, but a nightmare for large programs (basically anything over 64 KB).
- **Alpha** processor, which was DEC's attack on the processor market. It had a blistering performance, which blew every other processor out of the water (and still does in many cases). Unfortunately, it has never been properly exploited, as there was a lack of application software and development tools for it. The Intel Pentium proved that it was a great all-comer and did many things well, and was also willing to improve the areas that it was not so good at.

Xerox and PARC

In the early 1970s, the Xerox Corporation gathered a team at the Palo Alto Research Center (PARC) and gave them the objective of creating 'the architecture of information.' It would lead to many of the great developments of computing, including personal distributed computing, graphical user interfaces, the first commercial mouse, bit-mapped displays, Ethernet, client/server architecture, object-oriented programming, laser printing and many of the basic protocols of the Internet. Few research centers have ever been as creative and forward thinking as PARC was over those years. So why didn't Xerox fully exploit their research. Well, maybe it was because they had seen themselves as being a 'paper-based' organization, and distributing information by electronic methods went against this core business.

- **Z8000** processor, which was a classic case of being technically superior, but was not compatible with its father, the mighty Z80, and its kissing cousin, the 8080. Few companies have given away such an advantage with a single product. Where are Zilog now? Head buried in the sand, probably.
- **DEC**, which was one of the most innovate companies in the computer industry. It developed a completely new market niche with its minicomputers, but it refused, until it was too late, to believe that the microcomputer would have a major impact on the computer market. DEC went from a company that made a profit of $1.31 billion in 1988, to a company which, in one quarter of 1992, lost $2 billion. Its founder, Ken Olsen, eventually left the company in 1992, and his successor brought sweeping changes. Eventually, though, in 1998 it was one of the new PC companies, Compaq, which bought DEC. For Compaq, DEC seemed a good match, as DEC had never really created much of a market for PCs, and had concentrated on high-end products, such as Alpha-based workstations, batch processing and network servers.
- **Fairchild Semiconductor**. Few companies have ever generated so many ideas and incubated so many innovative companies, and got so little in return.
- **Xerox**. Many of the ideas in modern computing, such as GUIs and networking, were initiated at Xerox's research facility. Unfortunately, Xerox lacked commitment in their great developments. Maybe this was because it reduced Xerox's main market, which was, and still is, very much based on paper.
- **PCjr**, which was another case of incompatibility. IBM lost a whole year in releasing the PCjr, and lost a lot of credibility with its suppliers (many of whom were left with unsold systems) and their competitors (which were given a whole year to catch-up with IBM).
- **OS/2**, which was IBM's attempt to regain the operating system market from Microsoft. In it conception it was a compromised operating system, and its development team lacked the freedom of the original IBM PC development. Too many people and too many committees were involved in its development. It thus lacked the freedom, flair and independence of the Boca Raton development team who developed the IBM PC. At the time, IBM's mainframe divisions were a powerful force in IBM, and could easily stall, or veto a product if it had an effect on their profitable market.
- **CP/M**, which many believed would become the standard operating system for microcomputers. Digital Research had an excellent opportunity to make it the standard operating system for the PC, but Microsoft overcame it by making its DOS system so much cheaper.
- **MCA**, which was the architecture that IBM tried to move the market with. It failed because Compaq, and several major PC manufacturers, went against it, and kept developing using the existing x86 architecture to support the 80386 processor.
- **Seattle Computer Products**, which sold the rights of its QDOS program to Microsoft, and thus lost out on one of the most lucrative markets of all time.

- **Sinclair Research**, which after the success of the ZX81 and the Spectrum, threw it all away by releasing a whole range of under-achievers, such as the QL, and the C-5.
- **MSX**, which was meant to be the technology that would standardize computer software on PCs. Unfortunately, it hadn't heard of the new 16-bit processors, and most of all, the IBM PC.
- **Lotus Development**, which totally misjudged the market, by not initially developing its Lotus 1-2-3 spreadsheet for Microsoft Windows. It instead developed it for OS/2, and eventually lost the market leadership to Microsoft Excel. Lotus also missed an excellent opportunity to purchase a large part of Microsoft when it was still a small company. The profits on that purchase would have been gigantic.

The Address Bus

The size of the address bus inside a computer defines the amount of physically addressable memory. It has an 8-bit address bus, the processor can access up to 256 different memory locations (2^8). Other addressing capabilities are:

Bus size	Addressable memory
1	2B
2	4B
8	256B
10	1024B (1 KB)
11	2K
14	16KB
16	64KB
20	1M (limit for original PC)
24	16MB
32	4GB (limit for Pentium)

1.7 History of Computer Systems

The highlights in the development of computer systems, the Internet and networks are:

1614 John Napier discovered logarithms, which allowed the simple calculation of complex multiplications, divisions, square roots and cube roots.
1642 Blaise Pascal built a mechanical adding machine.
1801 Joseph-Maire Jacuard developed an automatic loom controlled by punched cards.
1822 Charles Babbage designed his first mechanical computer, the first prototype for his difference engine. His model would be used in many future computer systems.
1880s Hollerith produced a punch-card reader for the US Census.
1896 IBM founded (as the Tabulating Machine Company).
1906 Lee De Forest produces the first electronic value.
1946 ENIAC built at the University of Pennsylvania.
1948 Manchester University produces the first computer to use a stored program (the Mark I). William Shockley (and others) invent the transistor.
1954 Texas Instruments produces a transistor using silicon (rather than germanium). IBM produces the IBM 650 which was, at the time, the workhorse of the computer industry. MIT produces the first transistorized computer: the TX-O.
1957 IBM develops the FORTRAN (FORmula TRANslation) programming language.
1958 Jack St Clair Kilby proposes the integrated circuit.
1959 Fairchild Semiconductor produces the first commercial transistor using the planar process. IBM produces the first transistorized computer: the IBM 7090.
1960 ALGOL introduced which was the first structured, procedural, language. LISP (LISt Processing) was introduced for the Artificial Intelligence applications.
1961 Fairchild Semiconductor produces the first commercial integrated circuit.
COBOL (COmmon Business-Orientated Language) developed by Grace Murray Hopper.
1963 DEC produce its first minicomputer.
1965 BASIC (Beginners All-purpose Symbolic Instruction Code) was developed at Dartmouth College. IBM produced the System/360, which used integrated circuits.
1968 Robert Noyce and Gordon Moore start-up the Intel Corporation.
1969 Intel began work on a device for Busicom, which would eventually become the first microprocessor.

1970 Xerox creates the Palo Alto Research Center (PARC), which would become one of the leading research centers of creative ideas in the computer industry. Intel releases the first RAM chip (the 1103), which had a memory capacity of 1 Kbits (1024 bits).

1971 Intel release the first microprocessor: the Intel 4004. Bill Gates and Paul Allen start work on a PDP-10 computer in their spare time. Ken Thompson, at Bell Laboratories, produces the first version of the UNIX operating system. Niklaus Wirth introduces the Pascal programming language.

1973 Xerox demonstrates a bit-mapped screen. IBM produces the first hard disk drive (an 8 inch diameter, and a storage of 70 MB).

1974 Intel produces the first 8-bit microprocessor: the Intel 8008. Bill Gates and Paul Allen start-up a company named Traf-O-Data. Xerox demonstrates Ethernet. MITS produces a kit computer, based on the Intel 8008. Xerox demonstrates WYSIWYG (What You See Is What You Get). Motorola develops the 6800 microprocessor. Brian Kerighan and Dennis Ritchie produced the C programming language.

1975 MOS Technologies produces the 6502 microprocessor. Microsoft develops BASIC for the MITS computer.

1976 Zilog releases the Z80 processor. Digital Research copyrighted the CP/M operating system. Steve Wozniak and Steve Jobs develop the Apple I computer, and create the Apple Corporation. Texas Instruments produces the first 16-bit microprocessor: the TMS9900. Cray-1 supercomputer released, the first commercial supercomputer (150 million floating point operations per second).

1977 FORTRAN 77 introduced.

1978 Commodore released the Commodore PET.

1979 Intel releases the 8086/8088 microprocessors. Zilog introduced the Z8000 microprocessor and Motorola releases the 6800 microprocessor. Apple introduced the Apple II computer, and Radio Shack releases the TRS-80 computer. VisiCalc and WordStar introduced.

1981 IBM releases the IBM PC, which is available with MS-DOS supplied by Microsoft and PC-DOS (IBM's version).

1982 Compaq Corporation founded. Commodore releases the Vic-20 computer and Commodore 64. Sinclair releases the ZX81 computer and the Sinclair Spectrum. TCP/IP communications protocol created. Intel releases the 80286, which is an improved 8088 processor.
WordPerfect 1.0 released.

1983 Compaq releases their first portable PC. Lotus 1-2-3 and WordPerfect released. Bjarn Stroustrup defines the C++ programming language. MS-DOS 2.0 and PC-DOS 2.0 released.

1984 Apple releases the Macintosh computer. MIT introduce the X-Windows user interface.

1985 Microsoft releases the first version of Microsoft Windows, and Intel releases the classic 80386 microprocessor. Adobe Systems define the PostScript standard which is used with the Apple LaserWriter. Philips and Sony introduce the CD-ROM.

1986 Microsoft releases MS-DOS 3.0. Compaq releases the Deskpro 386.

1987 Microsoft releases the second version of Microsoft Windows. IBM releases PS/2 range. Model 30 uses 8088 processor, Model 50 and Model 60 use 80286, and Model 80 uses 80386 processor. VGA standard also introduced. IBM and Microsoft release the first version of OS/2.

1988 MS-DOS 4.0 released.

1989 WWW (World Wide Web) created by Tim Bernes-Lee at CERN, European Particle Physics Laboratory in Switzerland. Intel develops the 80486 processor. Creative Laboratories release Sound Blaster card.

1990 Microsoft releases Microsoft Windows 3.0.

1991 MS-DOS 5.0 released. Collaboration between IBM and Microsoft on DOS finishes.

1993 Intel introduces the Pentium processor (60 MHz). Microsoft releases Windows NT, Office 4.0 (Word 6.0, Excel 5.0 and PowerPoint 4.0) and MS-DOS 6.0 (which includes DoubleSpace, a disk compression program).

1994 Netscape 1.0 released. Microsoft withdraws DoubleSpace in favor of DriveSpace (because of successful legal action by Stac which claimed that parts of it were copies of its program: Stacker). MS-DOS 6.22 would be the final version of DOS.

1995 Microsoft release Windows 95 and Office 95. Intel releases the Pentium Pro, which has speeds of 150, 166, 180 and 200 MHz **(400MIPs)**. JavaScript developed by Netscape.

1996 Netscape Navigator 2.0 released (the first to support Java Script). Microsoft releases Windows 95 OSR 2.0, which fixed the bugs in the first release and adds USB and FAT 32 support.

1997 Intel release Pentium MMX. Microsoft releases Office 97, which creates a virtual monopoly in office application software for Microsoft. Office 97 is fully integrated and has enhanced version of Microsoft Word (upgraded from Word 6.0), Microsoft Excel (upgraded from Excel 5.0), Microsoft Access, Microsoft PowerPoint and Microsoft Outlook. IBM's Deep Blue beats Gary Kasparov (the World Chess Champion) in a chess match. Intel releases the Pentium II processor (233, 266 and 300 MHz versions). Apple admits serious financial trouble. Microsoft purchases 100,000 non-voting shares for $150 million. One of the conditions is that Apple drops their long running court case with Microsoft for copying the Mac interface on Microsoft Windows (although Apple copied its interface from Xerox). Bill Gate's fortune reaches $40 billion. He has thus, since 1975 (the year that Microsoft were founded), earned $500,000 per hour (assuming that he worked a 14 hour day), or $150 per second.

1998 Microsoft releases Microsoft Windows 98. Legal problems arise for Microsoft, especially as its new operating system includes several free programs as standard. The biggest problem is with Microsoft Internet Explorer, which is free compared with Netscape, which must be purchased.

1999 Linux Kernel 2.2.0 released, and heralded as the only real contender in the PC operating market to Microsoft. Intel releases Pentium III (basically a faster version of the Pentium II). Microsoft Office 2000 released. Bill Gates' wealth reaches $100 billion (in fact, $108 billion in September 1999).

2000 Millennium bug bites with false teeth.

and on Microsoft release Windows NT Version 5/2000 in three versions: Workstation, Server and SMP Server (multiprocessor). It runs on DEC Alpha's, Intel *x*86, Intel IA32, Intel IA64 and AMD K7 (which is similar to an Alpha). Microsoft releases Office 2000, but lose court case.

1.8 Exercises

Using a WWW search engine (such as www.yahoo.com or www.altavista.com) or your own general knowledge, determine the answers to the following questions. Please select from a–d.

1.8.1 Which computer helped aid the British Government to crack codes in World War II:
(a) ENIAC (b) Harvard Mk I
(c) IBM System/360 (d) Colossus

1.8.2 What is ENIAC an acronym for:
(a) Electronic Numerical Integrator and Computer
(b) Electronic Number Interface Analysis Computer
(c) Electronic Number Interface and Computer
(d) Electronic Numerical Interchange Computer

1.8.3 Which computer was the first to use integrated circuits:
(a) Apple (b) IBM System/360
(c) IBM PC (d) DEC PDP-11

1.8.4 Which one of the following formed Intel:
(a) Bill Gates and Paul Allen
(b) Robert Noyce, Gordon Moore and Andy Grove
(c) Jerry Sanders (d) Steve Wozniak and Steve Jobs

1.8.5 Which one of the following formed Microsoft:
(a) Bill Gates and Paul Allen
(b) Robert Noyce, Gordon Moore and Andy Grove

(c) Jerry Sanders (d) Steve Wozniak and Steve Jobs

1.8.6 Which one of the following formed Apple Computers:
(a) Bill Gates and Paul Allen
(b) Robert Noyce, Gordon Moore and Andy Grove
(c) Jerry Sanders (d) Steve Wozniak and Steve Jobs

1.8.7 Which company did Kenneth Olsen help form:
(a) Compaq (b) DEC
(c) Microsoft (d) IBM

1.8.8 Which company developed the first microprocessor:
(a) Texas Instruments (b) Motorola
(c) Zilog (d) Intel

1.8.9 Which company was the first to demonstrate the usage of windows, mouse and keyboard:
(a) IBM (b) Xerox
(c) Microsoft (d) DEC

1.8.10 Which company was the first to demonstrate the WYSIWYG concept:
(a) IBM (b) Xerox
(c) Microsoft (d) DEC

1.8.11 What was the name of the Xerox famous research centre:
(a) PARC (b) XRES
(c) PERC (d) RESP

1.8.12 Which company did Bill Gates and Paul Allen initially create:
(a) Micro-Traffic (b) Traf-O-Data
(c) Traffic Software (d) Gates & Allen

1.8.13 Who developed the C programming language:
(a) Bill Gates and Paul Allen (b) Brian Kernighan and Dennis Ritchie
(c) Niklaus Wirth (d) Steve Wozniak and Steve Jobs

1.8.14 Who developed the Pascal programming language:
(a) Bill Gates and Paul Allen (b) Brian Kernighan and Dennis Ritchie
(c) Niklaus Wirth (d) Steve Wozniak and Steve Jobs

1.8.15 Which was the first ever commercial microprocessor:
(a) 4000 (b) 4004
(c) 8080 (d) 1000

1.8.16 Which processor did the Apple II use:
(a) Zilog Z80 (b) MOS Technology 6502
(c) Intel 8080 (d) NEC 780-1

1.8.17 Which processor did the Commodore PET use:
(a) Zilog Z80 (b) MOS Technology 6502
(c) Intel 8080 (d) NEC 780-1

1.8.18 Which processor did the TRS-80 use:
(a) Zilog Z80 (b) MOS Technology 6502
(c) Intel 8080 (d) NEC 780-1

1.8.19 Which processor did the ZX80 use:

(a)	Zilog Z80	(b)	MOS Technology 6502
(c)	Intel 8080	(d)	NEC 780-1

1.8.20 How did the Motorola 68000 gain its name:

(a)	No reason	(b)	It was sold for $680.00
(c)	It sounded like the 8008	(d)	It had 68,000 transistors

1.8.21 Which company produced the VAX range of computers:

(a)	IBM	(b)	DEC
(c)	Compaq	(d)	Apple

1.8.22 Which IBM product quickly failed because it was incompatible with its PC:

(a)	PC AT	(b)	PC XT
(c)	PCNext	(d)	PCjr

1.8.23 Which company released the first IBM PC-compatible portable:

(a)	Compaq	(b)	IBM
(c)	Radio Shack	(d)	Commodore

1.8.24 Which architecture did IBM try to develop an industry standard with:

(a)	RS	(b)	MCA
(c)	OS/2	(d)	PCI

1.8.25 Which IBM operating system failed to gain a large hold of the market:

(a)	PC-DOS	(b)	OS/2
(c)	Windows	(d)	Unix

1.8.26 Which company was the fastest growing of all time:

(a)	IBM	(b)	Compaq
(c)	DEC	(d)	Sun

1.8.27 Which processor did the first IBM PC use:

(a)	8086	(b)	8088
(c)	8085	(d)	8080

1.8.28 What was the clock speed of the first IBM PC:

(a)	1 MHz	(b)	4.77 MHz
(c)	8 MHz	(d)	16 MHz

1.8.29 Which company did Apple reach an agreement with about their name:

(a)	Apple Corps Limited	(b)	Apple Beatles Limited
(c)	Apple System Limited	(d)	Apple Records Limited

1.8.30 Which computer was used in the UK by the BBC to teach microcomputers:

(a)	Acorn, BBC micro	(b)	Sinclair, ZX81
(c)	Osborne, Osborne 1	(d)	Commodore, Vic-20

1.8.31 Which company is thought to be responsible for the first WYSIWYG application:

(a)	Apple	(b)	IBM
(c)	Microsoft	(d)	Xerox

1.8.32 Which was the first spreadsheet:

(a)	Excel	(b)	VisiCalc
(c)	Lotus 1-2-3	(d)	Top-Plan

1.8.33 What is Sun (as in Sun Microsystems) an acronym for:

(a) Sale Unicode Network (b) Safe Universal Network
(c) Stanford University Network (d) Salford University Network

1.8.34 Which company developed many of the standards for the 3.5-inch floppy disk:

(a) IBM (b) Sony
(c) Microsoft (d) Xerox

1.8.35 Which operating system did DEC initially use for its VAX range:

(a) Mac OS (b) AEGIS
(c) VMS (d) Unix

1.8.36 Which operating system did Apollo initially use for its workstations:

(a) Mac OS (b) AEGIS
(c) VMS (d) Unix

1.8.37 Which standard did several Japanese companies develop that was meant to be a standard for PC software:

(a) DOS (b) 1-2-3
(c) MSX (d) SCSI

1.8.38 Which university developed the X-Windows system:

(a) Stanford (b) MIT
(c) UMIST (d) New York

1.8.39 Which organization originally developed the Internet:

(a) DARPA (b) ISO
(c) CERN (d) IEEE

1.8.40 Which organization originally developed the initial specifications for the WWW:

(a) DARPA (b) ISO
(c) CERN (d) IEEE

1.8.41 Who invented the transistor:

(a) Bill Gates (b) Herman Hollerith
(c) William Shockley (d) Lee De Forest

1.8.42 Which company did William Shockley form:

(a) Shockley Semiconductor (b) Shockley Devices
(c) Shockley Electronics (d) Shockley Electrics

1.8.43 Which company first proposed the integrated circuit:

(a) IBM (b) Texas Instruments
(c) Motorola (d) Fairchild Semiconductors

1.8.44 Which company developed the first 8-bit microprocessor:

(a) NEC (780-1) (b) Motorola (6800)
(c) Zilog (Z80) (d) Intel (8008)

1.8.45 Which company developed the first 16-bit microprocessor:

(a) Texas Instruments (9900) (b) Motorola (68000)
(c) Zilog (Z8000) (d) Intel (8086)

1.8.46 Which processor did Zilog produce:

(a) Z80 (b) 6502
(c) 8080 (d) 6800

1.8.47 Which processor did MOS Technology produce:

(a) Z80 (b) 6502
(c) 8080 (d) 6800

1.8.48 Which processor did Motorola produce:

(a) Z80 (b) 6502
(c) 8080 (d) 6800

1.8.49 Which company is thought to be responsible for the first GUI:

(a) Apple (b) IBM
(c) Microsoft (d) Xerox

1.8.50 Which company developed the XENIX operating system:

(a) AT&T (b) IBM
(c) Microsoft (d) HP

1.8.51 Which company tried to standardize Unix with System V:

(a) AT&T (b) IBM
(c) Microsoft (d) HP

1.8.52 Who first produced an integrated circuit:

(a) John Cocke (b) Robert Noyce
(c) Gordon Moore (d) William Shockley

1.8.53 Create your own Top 15 computer industry achievers of all time, and your own personal under-achiever list, and complete Table 1.1. Possible ideas are computer games (such as Sonic the Hedgehog, Doom, and so on), or application packages (such as Microsoft Word, Lotus 1-2-3, and so on), Internet applications (such as Internet Explorer or Mirabillis ICQ), Computer Systems (such as Sony PlayStation or Sinclair Spectrum, and so on), and so on. Table 1. 2 gives an example.

Table 1.1 Top 15 over-achievers/under-achievers

Position	Over-achiever	Under-achiever
1		
2		
3		
4		
5		
6		
7		
8		
9		
10		
11		
12		
13		
14		
15		

Table 1.2 Top 15 over-achievers/under-achievers (example)

Po.	*Over-achiever*	*Under-achiever*
1	IBM PC (for creating a global market, and changing modern life)	DOS (for being such as horrible, nasty operating system, that failed to use the full potential of the PC)
2	TCP/IP (for connecting computers to the Internet)	Intel 8088 (for having such a difficult internal architecture, and being so difficult to program for)
3	Electronic Mail (for being the best application, ever)	DEC Alpha (for failing to reach its potential)
4	Microsoft (for making all the right choices, and winning in virtually every market that it competed in)	Zilog Z8000 processor (for failing to be compatible with the Z80 processor)
5	Intel (for keeping the industry-standard for PC processors)	DEC (for missing the PC)
6	6502/Z80 processors (for providing excellent processors)	Fairchild Semiconductor (for failing to cash-in on its ideas)
7	Apple II (for being an excellent computer)	Xerox (as Fairchild Semiconductor)
8	Ethernet (for its ease of use, its robustness, its **upgradability,** and so on)	PCjr (for completely failing to follow the success of the IBM PC)
9	WWW (for creating a global database)	OS/2 (for missing the point and trying to be an operating system which could be used on mainframes, minicomputers and PC)
10	Apple Macintosh (for a computer that was 10 years ahead of the PC)	CP/M (for missing the PC operating market)
11	Compaq DeskPro 386 (for its excellent specification, and stealing the market from IBM)	MCA (for failing to create a new standard, and losing IBM a great market share)
12	Sun SPARC (for its openness, its excellent specification, its Unix, and X-Windows)	Seattle Computer Products (for selling DOS to Microsoft)
13	Commodore Amiga (for being an excellent computer)	Sinclair Research (for the QL and C-5)
14	Sinclair Research (for the Sinclair Spectrum)	MSX (for failing to create a standard for PC software)
15	Compaq (for making all the right decisions, at the right time)	Lotus Development (for missing the market for Microsoft Windows)

1.8.54 Investigate the current value of Bill Gates' wealth. If possible, also determine some of the charitable organizations that Bill Gates has given money to.

Distributed System Elements

2.1 Information

Information is available in an analogue form or in a digital form, as illustrated in Figure 2.1. Computer-generated data can be easily stored in a digital format, but analogue signals, such as speech and video, must first be sampled at regular intervals and then converted into a digital form. This process is known as digitization and has the following advantages:

- Digital data is less affected by noise, as illustrated in Figure 2.2. Noise is any unwanted signal and has many causes, such as static pick-up, poor electrical connections, electronic noise in components, cross-talk, and so on. It makes the reception of a signal more difficult and can produce unwanted distortion on the received signal.
- Extra information can be added to digital signals so that errors can either be detected or corrected.
- Digital data tends not degrade over time.
- Processing of digital information is relatively easy, either in real-time or non real-time.
- A single type of media can be used to store many different types of information (such as video, speech, audio and computer data being stored on tape, hard-disk or CD-ROM).
- A digital system has a more dependable response, whereas an analogue system's accuracy depends on operation parameters such as component tolerance, temperature, power supply variations, and so on. Analogue systems thus produce a variable response and no two analogue systems are identical.
- Digital systems are more adaptable and can be reprogrammed with software. Analogue systems normally require a change of hardware for any functional changes (although programmable analogue devices are now available).

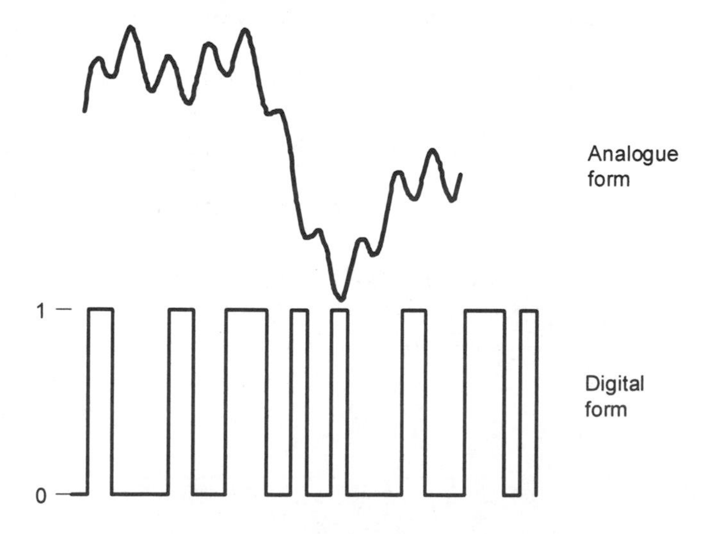

Figure 2.1 Analogue and digital format

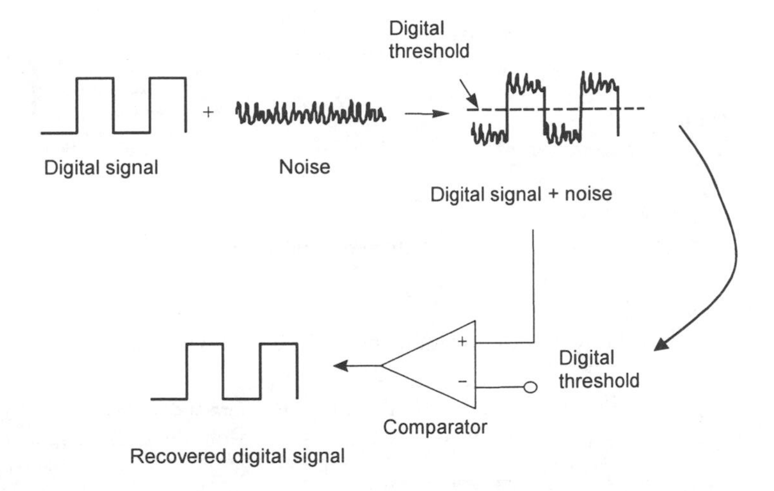

Figure 2.2 Recovery of a digital signal with noise added to it

As the analogue signal must be sampled at regular intervals, digital representations of analogue waveforms require large storage space. For example, 70 minutes of hi-fi quality music requires over 600 MB of data storage. The data once stored tends to be reliable and will not degrade over time. Typically, digital data is stored as either magnetic fields on a magnetic disk or as pits on an optical disk. A great advantage of digital technology is that once the analogue data has been converted to digital then it is relatively easy to store it with other purely digital data. Once stored in digital form it is relatively easy to process the data before it is converted back into analogue form. Analogue signals are relatively easy to store, such as video and audio signals are stored as magnetic fields on tape and a picture is stored on photographic paper. These media, though, tend to add noise (such as tape hiss) during storage and recovery.

> 'The use of COBOL cripples the mind; its teaching should, therefore, be regarded as a criminal offense.'
>
> Dijkstra

The accuracy of a digital system depends on the number of bits used for each sample, whereas an analogue system's accuracy depends on the specification of the components used in the system. Analogue systems also produce a differing response for different systems whereas a digital system has a dependable response. It is very difficult (if not impossible) to recover the original analogue signal after it is affected by noise (especially if it is affected by random noise). Most methods of reducing this noise involve some form of filtering or smoothing of the signal.

2.2 Conversion to digital

Figure 2.3 outlines the conversion process for digital data (the upper diagram) and for analogue data (the lower diagram). The lower diagram shows how an analogue signal (such as speech or video) is first sampled at regular intervals of time. These samples are then converted into a digital form with an ADC (analogue-to-digital converter). It can then be compressed and/or stored in a defined digital format (such as WAV, JPG, and so on). This digital form is then converted back into an analogue form with a DAC (digital-to-analogue converter). When data is already in a digital form (such as text or animation) it is converted into a given data format (such as BMP, GIF, JPG, and so on). It can be further compressed before it is stored, transmitted or processed.

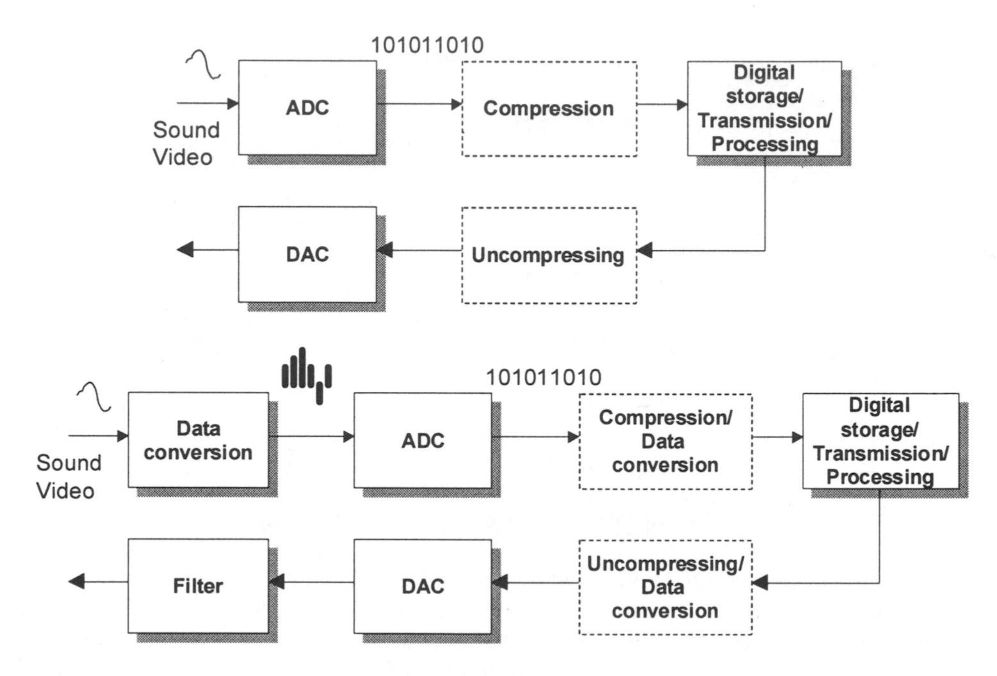

Figure 2.3 Information conversion into a digital form

2.3 Communications model

> **Winchester disks**
>
> Hard disks were at one time known as Winchester disks as they were initially available from IBM made from two spindles with 30MB on each spindle. The 30-30 became known as Winchester disks after the 30-30 rifle.

Figure 2.4 shows a communications model in its simplest form. An information source transmits signals to a destination through a transmission media. This transmission can either be with a direct connection (using a physical connection or wireless communication) or through an indirect connection (through a number of physical or wireless communications).

The information, itself, can either be directly sent using an electrical cable, or it can be carried on an electromagnetic wave. Typical carrier types are:

- **Radio waves.** The lower the frequency of a radio wave the more able it is to bend around objects. Defense applications use low frequency communications as they can be transmitted over large distances, and over and round solid objects. The trade-off is that the lower the frequency the less the information that can be carried. LW (MF) and AM (HF) signals can propagate large distances, but FM (VHF) requires repeaters because they cannot bend round and over solid objects such as trees and hills.
- **Microwaves.** Microwaves have the advantage over optical waves (light, infra-red and ultra-violet) in that they propagate well through water and thus can be transmitted through clouds, rain, and so on. If they are of a high enough frequency they can even propagate through the ionosphere and out into outer space. This is the property that is used in satellite communications where the transmitter bounces microwave energy off a satellite, which is then picked up at a receiving station. Their main disadvantage is that they will not bend round large objects, as their wavelength is too small.

- **Infra-red.** Infra-red is used in optical communications. When used as a carrier for information, the transmitted signal can have a very large bandwidth because the carrier frequency is high. Infra-red is extensively used for line-of-site communications, especially in remote control applications.
- **Light.** Light is the only part of the spectrum that humans can 'see'. It is a very small part of the spectrum and ranges from 300 to 900 nm (a nanometer is one billionth of a meter). Colors contained are Red, Orange, Yellow, Green, Blue, Indigo and Violet (ROY.G.BIV or **R**ichard **O**f **Y**ork **G**ave **B**attle **I**n **V**ain).
- **Ultra-violet.** As with infra-red, it is used in optical communications. In high enough exposures, it can cause skin cancer. Fortunately for humans, the ozone layer blocks out much of the ultra-violet radiation from the sun.

'It is practically impossible to teach good programming style to students that have had prior exposure to BASIC; as potential programmers they are mentally mutilated beyond hope of regeneration.' – Dijkstra

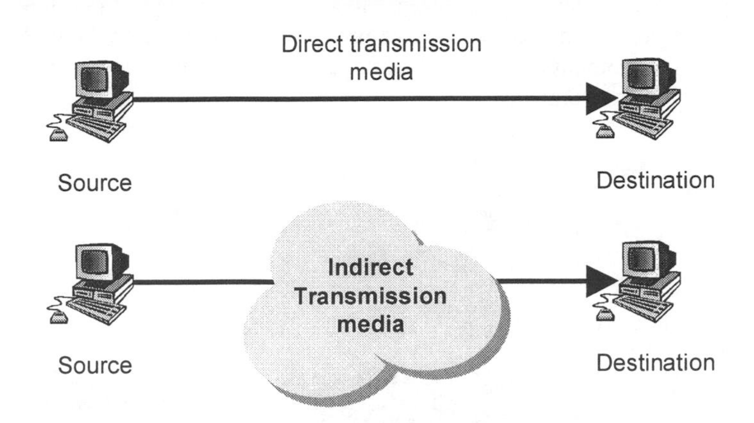

Figure 2.4 Simple communications model

2.4 Cables

The cable type used to transmit the data over a network depends on several parameters, including:

- The **reliability** of the cable.
- The maximum **length** between nodes.
- The possibility of electrical **hazards**.
- **Power** loss in the cables.
- Tolerance to **harsh** conditions.
- **Expense** and general **availability** of the cable.
- Ease of **connection** and **maintenance**.
- Ease of **running** cables, and so on.
- The signal **bandwidth**. The amount of information that can be sent directly relates to the bandwidth of the system, and the main limitation on the bandwidth is the channel between the transmitter and the receiver. With this, the lowest bandwidth of all the connected elements defines the overall bandwidth of the system (unless there are alternative paths for the data).

2nd Generation Computer Companies (Transistorized)

1. IBM
2. Univac
3. Burroughs
4. NCR
5. Honeywell
6. Control Data Corporation
7. Siemens
8. Fuji
9. Bendix
10. Librascope

The main types of cables used for networking are illustrated in Figure 2.5, they are:

- **Unshielded twisted-pair copper (UTP).** Twisted-pair and coaxial cables transmit electric signals, whereas fiber-optic cables transmit light pulses. Unshielded Twisted-pair cables are not shielded and thus interfere with nearby cables. Public telephone lines generally use twisted-pair cables. In LANs they are generally used up to bit rates of 100 Mbps and with maximum lengths of 100 m. UTP cables are typically used to connect a computer to a network. Cat-3 cables have a lower specification than Cat-5 cables. Cat-5 cables can transmit up to 100 Mbps (100,000,000 bits per second), while Cat-3 transmits at a maximum of 16 Mbps (16,000,000 bits per second).
- **Coaxial.** Coaxial cable has a grounded metal sheath around the signal conductor. This limits the amount of interference between cables and thus allows higher data rates. Typically, they are used at bit rates of 100 Mbps for maximum lengths of 1 km.
- **Fiber-optic.** The highest specification of the three cables is fiber-optic, and allows extremely high bit rates over long distances. Fiber-optic cables do not interfere with nearby cables and give greater security. They also provide more protection from electrical damage by external equipment and greater resistance to harsh environments, as well as being safer in hazardous environments.

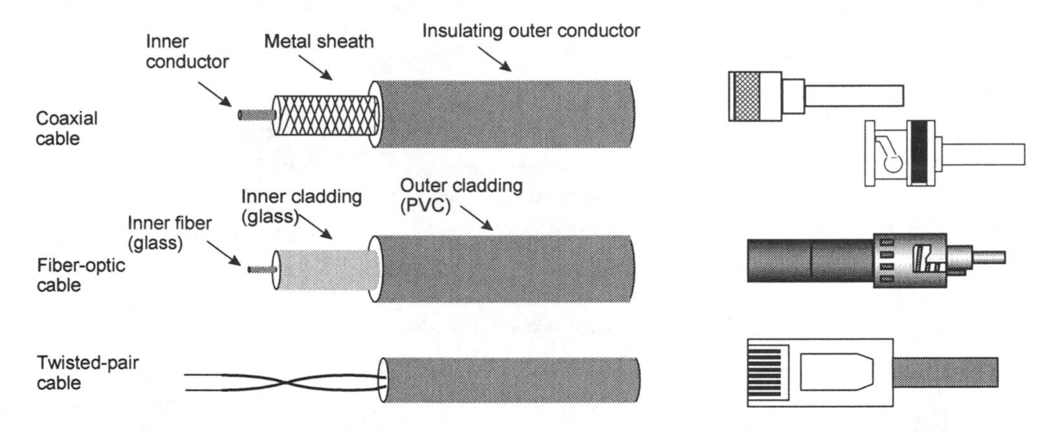

Figure 2.5 Types of network cable and their connector

2.4.1 Cable characteristics

The main characteristics of cables are:

- **Attenuation.** Attenuation defines the reduction in the signal strength at a given frequency for a defined distance. It is normally specified in decibels (dB) per 100 m. For example, an attenuation of 3 dB/100 m gives a signal voltage reduction of 0.5 for every 100 m.
- **Crosstalk.** Crosstalk is an important parameter as it defines the amount of signal that crosses from one signal path to another. This causes distortion on the transmitted signal. Shielded twisted-pair cables have less crosstalk than unshielded twisted-pair cables.
- **Characteristic impedance.** The characteristic impedance (as measured in Ω) of a cable and its connectors are important, as all parts of the transmission system need to be matched to the same impedance. This impedance is normally classified as the characteristic impedance of the cable. Any differences in the matching results in a reduction of signal power and can produce signal reflections (or ghosting). For example, twisted-pair

One of the first personal computers was the Altair which was named after a destination for the Star Ship Enterprise in an episode of Star Trek.

cables have a characteristic impedance of approximately 100 Ω, and coaxial cable using in networking has a characteristic impedance of 50 Ω (or 75 Ω for TV systems).

2.5 Peer-to-peer and client/server

An important concept is the differentiation between a peer-to-peer connection and a client-server connection. A peer-to-peer connection allows users on a local network to access a local computer. Typically, this might be access to:

Local printers. Printers, local to a computer, can be accessed by other users if the printer is shareable. This can be password protected, or not. Shareable printers on a Microsoft network have a small hand under the icon.

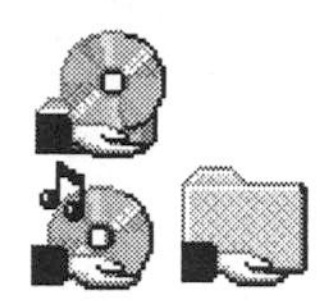

Local disk drives and folders. The disk drives, such as the hard disk or CD-ROM drives can be accessed if they are shareable. Normally the drives must be shareable. On a Microsoft network a drive can be made shareable by selecting the drive and selecting the right-hand mouse button, then selecting the Sharing option. User names and passwords can be set-up locally or can be accessed from a network server. Typically, only the local computer grants access to certain folders, while others are not shared.

These shared resources can also be mounted as objects on the remote computer. Thus, the user of the remote computer can simply access resources on the other computers as if they were mounted locally. This option is often the best when there is a small local network, as it requires the minimum amount of set-up and does not need any complicated server set-ups. Figure 2.6 shows an example of a peer-to-peer network where a computer allows access to its local resources. In this case, its local disk drive and printer are shareable.

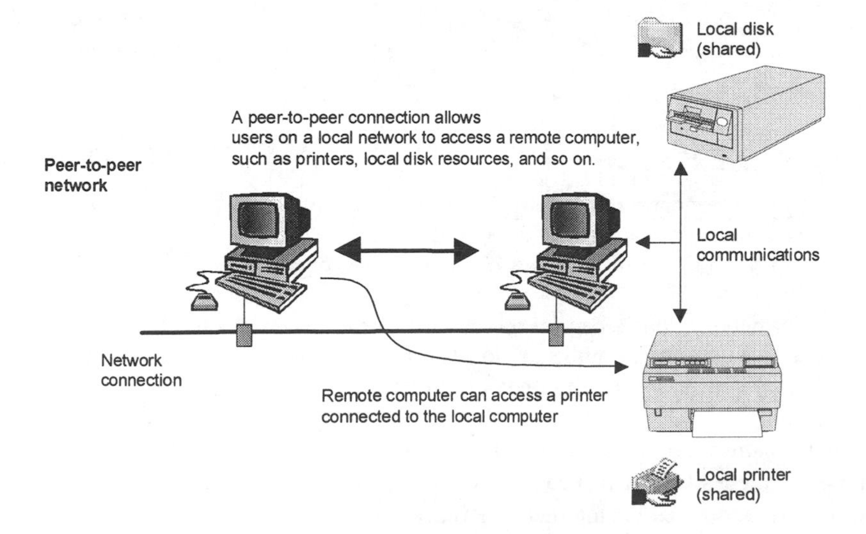

Figure 2.6 Peer-to-peer network

Normally a peer-to-peer network works best for a small office environment. Care must be taken, though, when setting up the attributes of the shared resources. Figure 2.7 shows an example of the sharing setting for a disk drive. It can be seen that the main attributes are:

Integration Generations

1st 1 bit/module
2nd 1 Register per module
3rd Register-on-a-Chip
4th Processor-on-a-Chip
5th System-on-a-Chip

- **Read-only.** This should be used when the remote user only requires to copy or execute files. The remote user cannot modify any of the files.
- **Full.** This option should only be used when the remote user has full access to the files and can copy, erase or modify the files.
- **Depends on Password.** In this mode, the remote user must provide a password to get either read-only access or full access.

If the peer-to-peer network has a local server, such as Novell NetWare or Windows NT/2000 then access can be provided for certain users and/or groups, if they provide the correct password.

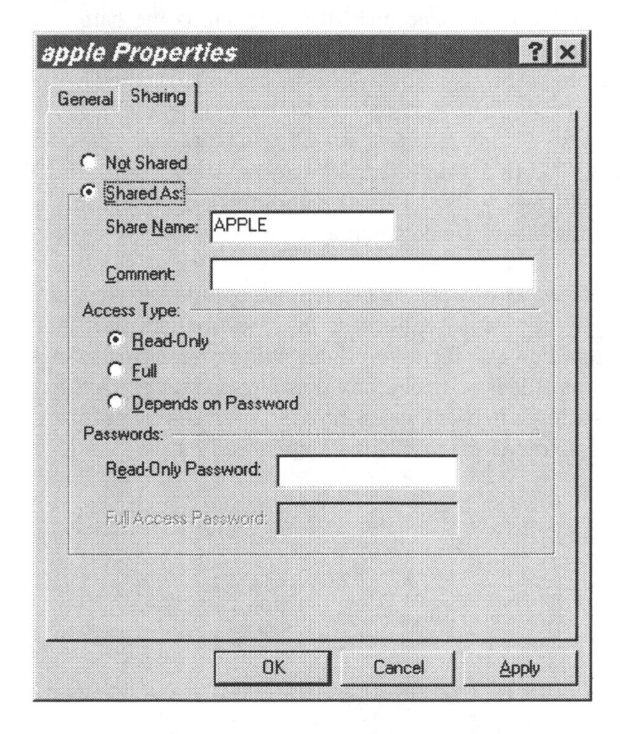

Computer Generations

1st Valves (ENIAC)
2nd Transistors (PDP-1)
3rd Integrated Circuits/ Time sharing (IBM System/360)
4th Large Scale Integration (ZX81)
5th System-on-a-Chip (Pentium)

Figure 2.7 File access rights

A client-server network has a central server which proves services to clients, as illustrated in Figure 2.8. These clients can either be local to the network segment, or from a remote network. The server typically provides one or more of the following services:

- Store usernames, group names and passwords.
- Run print queues for networked printers.
- Allocate IP addresses for Internet accesses.
- Provide system back-up facilities, such as CD-R disk drives and DAT tape drives.

- Provide centralized file services, such as networked hard disks or networked CD-ROM drives.
- Centralize computer settings and/or configuration.
- Provide access to other centralized peripherals, such as networked faxes and dial-in network connections.
- Provide WWW and TCP/IP services, such as remote login and file transfer.

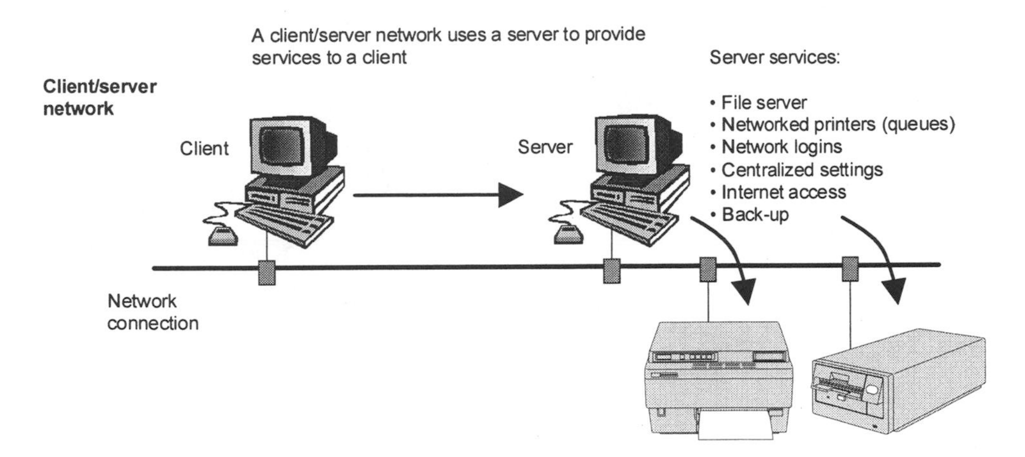

Figure 2.8 Client/server network

> '640 K ought to be enough for anybody.'
>
> Bill Gates, 1981

A network operating system server typically provides file and print services, as well as storing a list of user names and passwords. Typical network operating systems are Windows NT/2000, Novell NetWare and UNIX.

Internet and WWW services are typically run from an Internet server, which runs Internet services, such as:

- **HTTP** (Hyper Text Transfer Protocol), for WWW (World-Wide Web) services. On the WWW, WWW servers and WWW clients pass information between each other using HTTP. A simple HTTP command is GET, which a WWW client (the WWW browser) sends to server in order to get a file.
- **FTP** (File Transfer Protocol), which is a standard protocol and used to transfer files from one computer system to another. In order for the transfer to occur the server must run an FTP server program.
- **TELNET**, which is used for remote login services (see Figure 2.9).
- **SMTP** (Simple Mail Transport Protocol), which is used for electronic mail transfer.
- **TIME**, which is used for a time service.
- **SNMP** (Simple Network Management Protocol), which is used to analyze network components.
- **NNTP** (Network News Transfer Protocol), which is used for network news.

Figure 2.10 shows an example network which has two local network servers. One provides file and print services, while the other supports Internet services. The local computer accesses each of these for the required service. It can also access a remote Internet server through a router. This router automatically determines that the node is accessing a remote node and routes the traffic out of the local network.

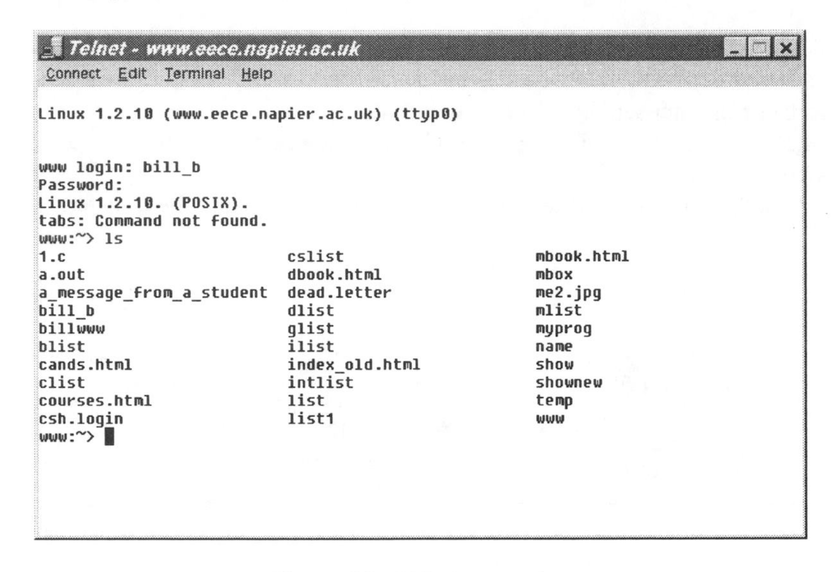

Figure 2.9 Telnet connection

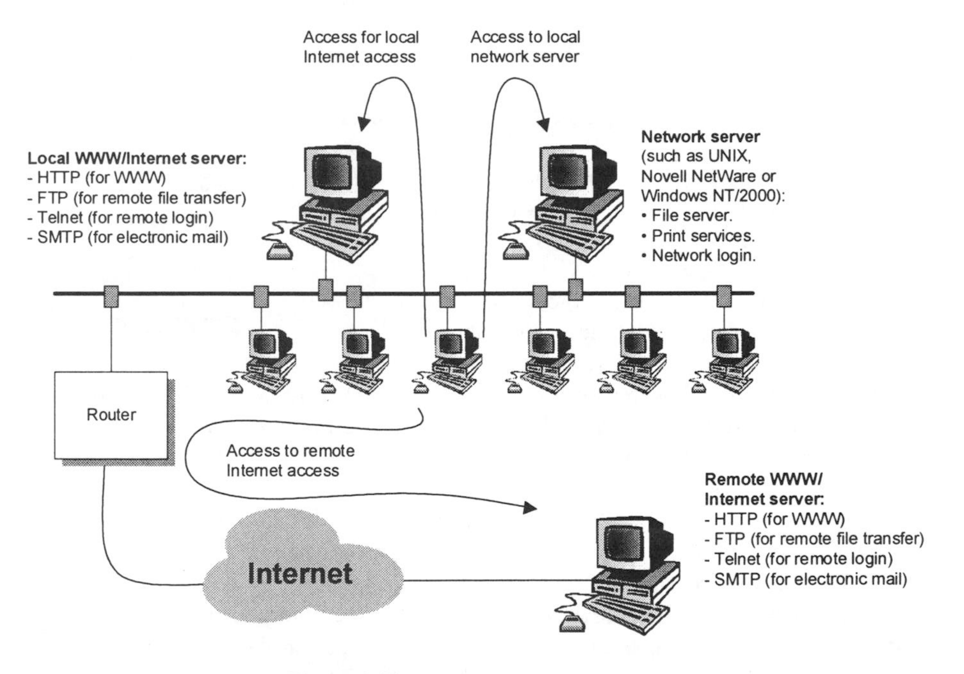

Figure 2.10 Local and remote servers

2.6 Exercises

The following questions are multiple choice. Please select from a–d.

2.6.1 Which type of communication transmission is typically used for remote transmission for short distances:

(a) Infra-red (b) Microwaves
(c) Light (d) Ultra-violet

2.6.2 Which type of communication transmission is typically used for transmission over long distances:

(a) Infra-red (b) Radio
(c) Light (d) Ultra-violet

2.6.3 Which type of communication transmission is typically used for transmission of satellite communications:

(a) Infra-red (b) Radio
(c) Light (d) Ultra-violet

2.6.4 Which type of communication transmission is typically used for transmission within fiber-optic cables:

(a) Infra-red (b) Radio
(c) Light (d) Ultra-violet

2.6.5 Which of the following best describes attenuation:

(a) The amount of signal that crosses from one signal path to another
(b) The reduction in signal strength
(c) The boosting of signal strength
(d) The reflection of signals at the receiver

2.6.6 Which of the following best describes crosstalk:

(a) The amount of signal that crosses from one signal path to another
(b) The reduction in signal strength
(c) The boosting of signal strength
(d) The reflection of signals at the receiver

2.6.7 What advantage does Cat-5 cable have over Cat-3 cable:

(a) It is lighter in weight
(b) It allows a high bit rate
(c) It is compatible with more systems
(d) It is less expensive

2.6.8 What advantage does shielded twisted-pair cables have over unshielded twisted-pair cables:

(a) They are more compatible with transmission systems
(b) They are lighter in weight
(c) They produce less attenuation in the signal
(d) They are less susceptible to crosstalk

2.6.9 Which of the following best describes a client/server network:

(a) Resources are centralized on a server
(b) Local resources, such as memory and processor, are shared between users
(c) Local resources, such as disk drives and printers, are shared between users
(d) Internet connections are allocated centrally

2.6.10 What type of cable connectors to the connector given below:

(a) Twisted-pair cable (b) Fiber-optic cable
(c) Telephone cable (d) Coaxial cable

2.6.11 What type of cable connectors to the connector given below:

(a) Twisted-pair cable
(b) Fiber-optic cable
(c) Telephone cable
(d) Coaxial cable

2.6.12 Which of the following icons represents a shared folder on a Microsoft Windows network:

(a) (b)

(c) (d)

2.6.13 Which of the following icons represents a shared CD-ROM drive on a Microsoft Windows network:

(a) (b)

(c) (d)

2.6.14 Investigate different forms of information and identify if they contain digital or analogue information. From this, complete Table 2.1 (one example has already been completed).

Table 2.1 Links

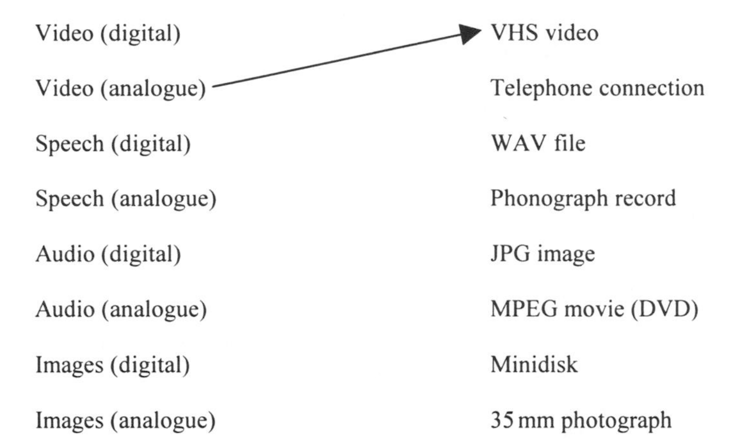

Video (digital)	VHS video
Video (analogue)	Telephone connection
Speech (digital)	WAV file
Speech (analogue)	Phonograph record
Audio (digital)	JPG image
Audio (analogue)	MPEG movie (DVD)
Images (digital)	Minidisk
Images (analogue)	35 mm photograph

2.6.15 Outline the main advantages of converting analogue information into a digital format.

2.6.16 Investigate a typical (or maximum) amount of information (in Bytes) that the following media contain:

Media	*Typical amount of information stored (B)*
Standard CD-ROM (ISO-9660 format)	
DVD drive	
MiniDisk audio	
Floppy disk (3.5 inch)	
ZIP drive	

2.6.17 Investigate different forms of information and identify if they contain digital or analogue information. From this, complete Table 2.2 (one example has already been completed).

Table 2.2 Statements

Statement	*Mostly true*	*Mostly false*	*Unproven*
Digital information does not degrade over time	✓		
Analogue information can be easily processed in non-real-time			
Analogue systems are more dependable in their response than a digital system			
Digital systems are always more accurate than analogue systems			
Errors in an analogue system can be easily corrected, if required			
Errors in a digital system can be easily corrected, if required			
Digital images are always better quality than analogue images			
Once an analogue signal is digitized, and converted back into an analogue, it will never be identical to the original			

2.6.18 Investigate the type of services of the protocols given in Table 2.3, and thus complete the table.

Table 2.3 Services

Media	*Typical amount of information stored (B)* .
HTTP	WWW server (Hyper Text Transfer Protocol)
FTP	
TELNET	
NFS	
POP	
SMTP	
SNMP	
FINGER	

2.6.19 If possible, connect to the Internet and find some Internet sites which respond to the following services in Table 2.4 (a few examples have already been completed).

Table 2.4 Services

Media	*Site name*
HTTP	www.microsoft.com
FTP	ftp.microsoft.com
TELNET	
NFS	
POP	pop.freeserve.co.uk
SMTP	smtp.freeserve.co.uk
FINGER	

2.6.20 If possible, locate a local area network (LAN), and complete Table 2.5.

Table 2.5 LAN

Type	*Options*	
Cables used	Coaxial	[ ]
	Fiber-optic	[ ]
	Twisted-pair cable	[ ]
	Other (please specify):	
Server type	Microsoft Windows NT/2000 Server	[ ]
	UNIX	[ ]
	Novell NetWare	[ ]
	Other (please specify):	
Services	File server	[ ]
	Printer server	[ ]
	WWW server	[ ]
	Main server	[ ]
	Telnet	[ ]
	FTP	[ ]
	Other (please specify):	

2.7 Note from the Author

This chapter has covered some of the fundamental issues of distributed systems. A key element is the understanding of the communication model, which has a source and a destination. In a perfect world we can send what we like and it will always be received perfectly (obviously in a perfect world we would also have world peace and no starvation). But we do not live in a perfect world, and all of our transmitted signals, whether they be analogue or digital, are affected in some way by the transmission media, and by noise. Noise when added to an analogue signal is normally difficult to get rid of, as it normally requires some form of filtering which will affect the contents of the original signal. Digital signals, on the other hand, are less likely to be affected as they only have two states (on or off, 1 or 0, true or false), and are less prone to noise and distortion as they must force the signal out of one of the binary states into the other state. But even if this happens we have many ways to overcome this. For example, an audio CD has a built-in error correction code, which will detect when bits are in error, and actually correct them. Most of the time though, we simply want to detect if there has been an error, and contact the sender so that it can retransmit the data that was in error. This is something that we will see often in this book, especially when we cover TCP communications (which, along with IP, is the backbone of the Internet).

The drive towards converting real-time data, such as speech, audio, video and image is one that will continue, and is the movement towards the creation of a totally integrated digital communications network. With this computer data and real-time data will be sent transparently into the network and be routed around the world, without care about the type of data that it is (although some special parameters may be defined, such as reliability, minimum bandwidth requirements, and so on).

Full circle

The IBM System/360 was one of the all-time classic computers. It was used by many large organizations in the 1960s and the 1970s. It gained its name because it was aimed at the full circle (360°) of customers, from business to science.

ANSI ASCII code

In 1963, ANSI defined the 7-bit ASCII standard code for characters. At the same time IBM had developed the 8-bit EBCDIC code which allowed for up to 256 characters, rather than 128 characters for ASCII. It is thought that the 7-bit code was used for the standard as it was reckoned that eight holes in punched paper tape would weaken the tape. Thus the world has had to use the 7-bit ASCII standard, which is still popular in the days of global communications, and large-scale disk storage.

Snow White and the Seven Dwarfs

In the 1960s and the 1970s, IBM was so powerful that it held a larger market than all of its competitors put together (at least 70%). There were seven main rivals; thus the term Snow White and the Seven Dwarfs. The Dwarfs were General Electric, RCA, Burroughs, UNIVAC, NCR, Control Data and Honeywell. In the face of the IBM System/370, RCA left the computer market, UNIVAC became a division of Sperry, GE sold its computer business to Honeywell, and AT&T was bought over by NCR (which changed the meaning of its name from National Cash Register to National Computing Resources). In 1986, Burroughs bought UNIVAC from Sperry and, at the time, became the second largest computer company, and, in the same year, AT&T spun off NCR into an independent company. Control Data was eventually taken over by a finance company called Commercial Credit, who never really exploited the dynamic nature of Control Data.

It is to IBM's credit that they are still one of the largest computing companies, while the other dwarfs have mainly left the mainstream computer business. In the era of the PC, the RCA's, UNIVAC's, and Burrough's, were replaced by the Commodore's, the Radio Shack's, the Osborne's, who have since been replaced by the Dell's, the Compaq's and the Packard Bell's.

Introduction to OS's

> 'We could conceivably put a company out of business with a bug in a spreadsheet, database, or word processor.'
>
> Chris Mason, Microsoft

3.1 Introduction

A computer system is typically made-up of hardware, an operating system and a user interface, as illustrated in Figure 3.1. The hardware includes the central processing unit (CPU), memory, and input/output devices. An operating system allows an easy interface between the user and the hardware. A good operating system should hide the complexity of the hardware from the user, and most modern operating systems automatically scan the connected hardware and configure the system when it is booted. The user interface provides an easy-to-use interface between the operating system and the user. Many older operating systems, such as DOS and UNIX, used text commands which the user had to enter in order for the system to run programs. Most systems, these days, provide a graphical user interface (GUI) in which the user uses Windows, Icons, Menus and Pointers (WIMPs) to run programs and organize the system.

An operating system provides an interface between the user and the hardware of the computer, and is designed to operate on a maximum number of bits at a time. This normally defines the classification of the operating systems. Typical classifications are: 16-bit, 32-bit and 64-bit (although some super computers use 128-bit operating systems). The limitation on the number of bits that the operating system can operate with, limits the software that can run on the operating system. For example 32-bit software (such as Microsoft Office 2000) will not run on a 16-bit operating system (such as DOS or Windows 3.x). Typically, the more bits that the operating systems uses, at a time, the faster it is.

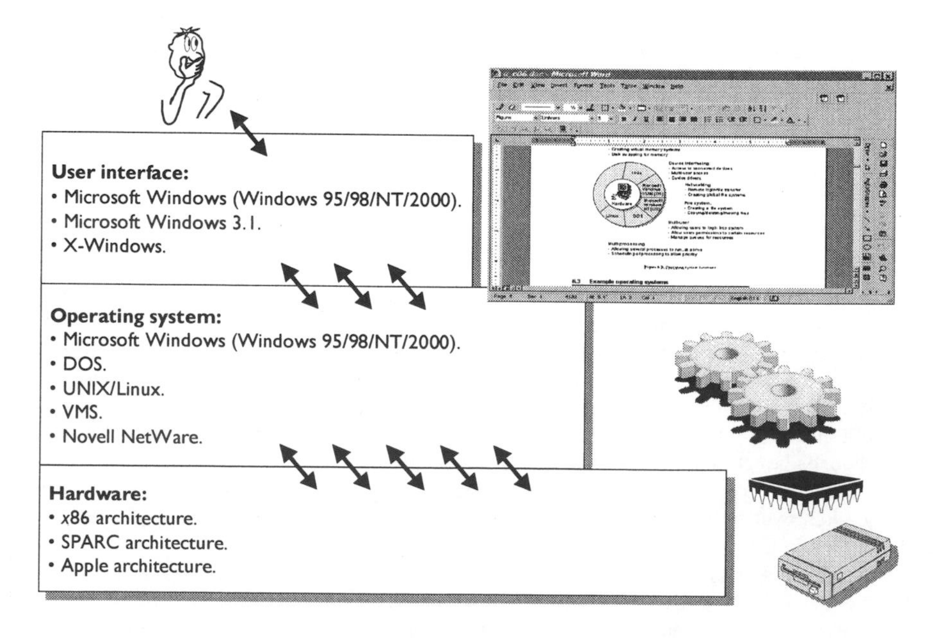

Figure 3.1 User interface, operating system and hardware

The main operating systems which are currently available are:

- **DOS.** The standard operating system for the PC for many years, but now becoming almost redundant as it is so limited. Networking is not built into its system and additional programs are required to run to create any network functions. It is also single tasking (that is, it can only run one program at a time). DOS uses 16-bit software, even although the processor might be capable of running 32-bit software.
- **Microsoft Windows 3.x.** This is a graphical user interface (GUI), and does not really work as an operating system, as it uses DOS for much of its operating system tasks. Windows 3.x uses 16-bit software.
- **Microsoft Windows 95/98/NT/ 2000.** These are complete operating systems and also a graphical user interface. These use 32-bit software.
- **UNIX.** A powerful and robust operating system, which is typically used in high-powered workstations. Normally a GUI, such as X-Windows, is required to interface to the main operating system. These use 32-bit or 64-bit software. UNIX has provided the world with its greatest gift: TCP/IP, which has grown-up around UNIX, and has now become the standard networking protocol for the Internet.
- **Novell NetWare.** A networked operating system that provides access to networked resources, such as print queues, file services, and so on.
- **VMS.** As with UNIX, a powerful and robust operating system. It is excellent at running batch processes, which are automatically started when the system is started. They then run quietly without little user input (typically described as batch processes).
- **Linux.** A version of UNIX for the PC.

The operating system basically splits into two parts: the kernel component and the main operating system component, as illustrated in Figure 3.2. The operating system and kernel components can make access to user account databases, which contain the names of users who are allowed to log into the system, and their password. They also allow access to file systems and other resources.

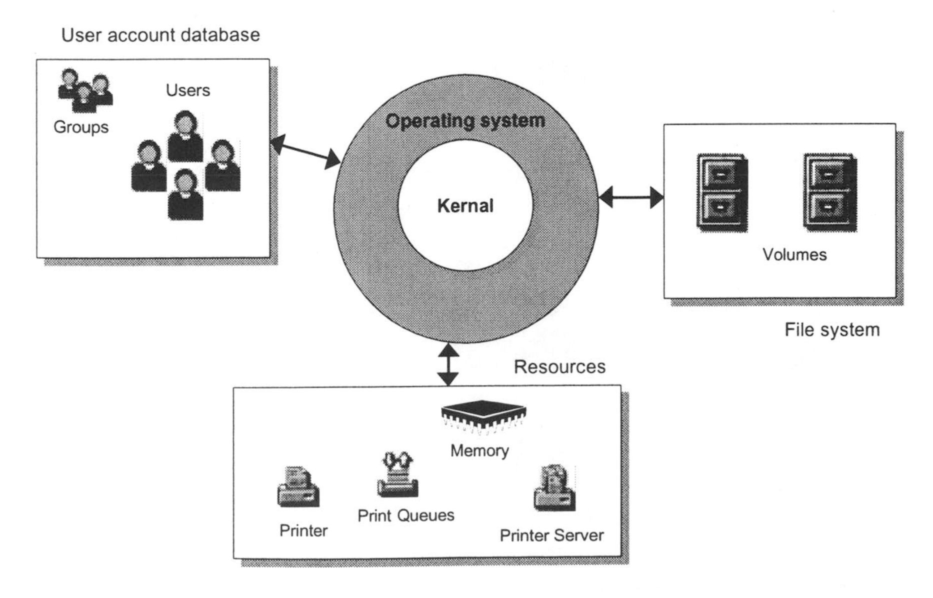

Figure 3.2 Operating system components

> 'I'm glad to be out of that bag', 'Hello, I am Macintosh. Never trust a computer you cannot lift.'
>
> Quotes from the Macintosh computer when it, introduced itself

Operating systems are typically differentiated, as illustrated in Figure 3.3, by:

- **Single user** v. **Multi-user.** Single user systems only allow a single user to login into the system at a time. They have no user account database, and have a low level of security, as users cannot protect their files from being viewed, copied or deleted. Typical single user systems are DOS and Windows 95/98. Multi-user files have a user account database, which defines the rights that users have on certain resources. Multi-user systems are normally more secure than single user systems, as the access to resources can be limited.
- **Networked** v. **Stand-alone.** A stand-alone computer does not connect to a network, and thus cannot access any networked resources. This mode is obviously more secure as remote users cannot log into the computer. A networked operating system uses a standard communications protocol, such as TCP/IP (for UNIX networks and over the Internet) or IPX/SPX (for Novell NetWare networks). Networked systems are obviously less secure than stand-alone systems, and should thus be protected in some way (typically by creating user accounts). Typical networked operating systems are: Windows NT/2000, Windows 95/98, Novell NetWare and UNIX.
- **Multi-tasking** v. **Single-tasking.** Multi-tasking operating systems allow one or more programs to be run, at a time. Typically, this is achieved by giving each process a prioritized amount of time on the processor. Single-tasking systems only allow one program to run, at a time. They are generally faster than multi-tasking systems, as some time is taken switching between programs, but multi-tasking systems are more efficient as they allow other tasks to run when a task is not performing any operations. Single-tasking systems are more robust, as multi-tasking programs often require to communicate with each other, which can cause synchronization problems (in some cases, deadlock).

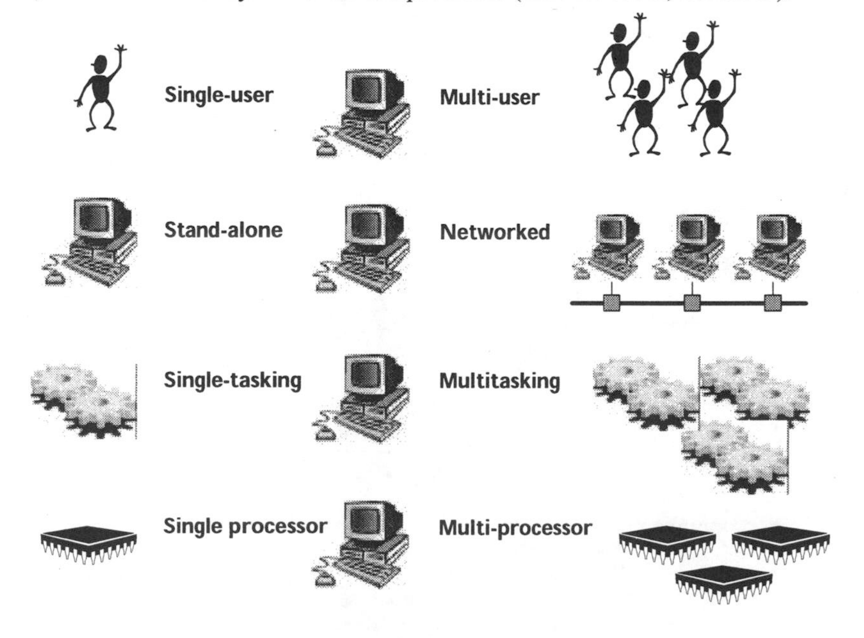

Figure 3.3 Operating system characteristics

- **Multiprocessor** v. **Single processor.** Some operating systems allow for more than one processor to be used on the system. This allows more than one task to be run, at a time, on different processors. Windows NT/2000 supports multiprocessors (up to four processors). Single processor, multi-tasking involves running each of the processes for a given time slice on a single processor, whereas multiprocessor systems allow processes to be run at the same time.

Figure 3.4 lists some of the basic functions that an operating system should implement. These functions are:

- **File system.** All operating systems create and maintain a file system, where users can create, copy, delete and move files around a structured file system. Many systems organize the files in directories (or folders). In multi-user systems these folders can have associated user ownership, and associated access rights.
- **Device interfacing.** Operating systems should try and hide the complexity of interfacing to devices from user programs and the user. Typically, an operating system should also try and configure devices at start-up (rather than getting the user to set them up). The operating system can also set up queues on devices, typically for printers, so that multiple accesses can occur when a device is busy.
- **Multi-user.** This allows one or more user to log into the system. For this, the operating system must contain a user account database, which contains user names, default home directory, user passwords, user rights, and so on.
- **Multi-tasking.** This allows one or more tasks to run on a system, at a time.
- **Multiprocessing.** This allows two or more processors to be used, at a time. The operating system must decide if it can run the different processes on individual processors. It must also manage the common memory between the processors.

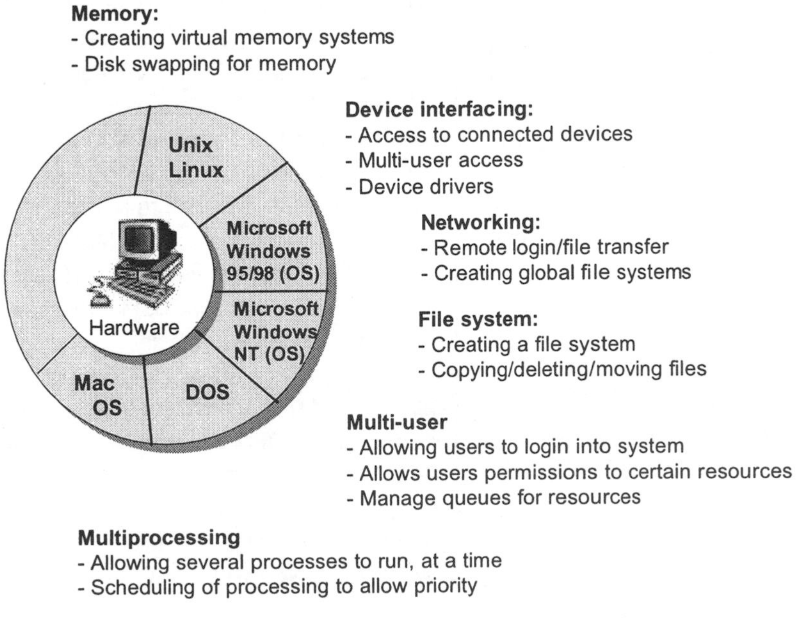

Memory:
- Creating virtual memory systems
- Disk swapping for memory

Unix
Linux
Microsoft
Windows
95/98 (OS)
Microsoft
Windows
NT (OS)
DOS
Mac
OS
Hardware

Device interfacing:
- Access to connected devices
- Multi-user access
- Device drivers

Networking:
- Remote login/file transfer
- Creating global file systems

File system:
- Creating a file system
- Copying/deleting/moving files

Multi-user
- Allowing users to login into system
- Allows users permissions to certain resources
- Manage queues for resources

Multiprocessing
- Allowing several processes to run, at a time
- Scheduling of processing to allow priority

Figure 3.4 Operating system functions

- **Multi-threaded applications.** Processes are often split into smaller task, named threads. These threads allow for smoother operation.
- **Managing memory.** This involves allocating memory to processes, and often to create a virtual memory for program. Two techniques are: paging (organizing programs so that the program data is loaded into pages of memory) and swapping (which involves swapping the contents of memory to disk storage).

3.2 Example operating systems

The section outlines several different types of operating systems for their historical development, and their basic system specification.

3.2.1 UNIX

The UNIX operating system was initially developed in 1971 by Ken Thompson, at AT&T's Bell Laboratories. UNIX is an extremely popular operating system and dominates in the high-powered, multitasking workstation market. It is relatively simple to use and to administer, and also has a high degree of security. UNIX computers use TCP/IP communications to mount disk resources from one machine onto another. UNIX's main characteristics are:

- **Multi-user.**
- **Pre-emptive multi-tasking.**
- **Multiprocessing.**
- **Multithreaded applications.**
- **Memory management with paging** (organizing programs so that the program is loading into pages of memory) **and swapping** (which involves swapping the contents of memory to disk storage).

IBM and Microsoft, a marriage made in heaven (for Microsoft)

Unlike DEC, IBM could see the growing power of the personal computer market. So, their Corporate Management Committee gave permission to William Lowe to start Project Chess, which would have an extremely small team of just 12 engineers. The time scales were short and IBM could not conceivably write their own operating system within the time limit. IBM then approached Bill Gates and Steve Ballmer, of Microsoft, and Gary Kildall at Digital Research. Bill Gates accepted the request from IBM to write the operating system for the IBM PC, but Gary Kildall rejected the offer for his CP/M-86 operating system.

After an initial prototype was produced, the Corporate Management Committee gave the go-ahead for full development. The code name was Acorn (which would indeed lead to greater things).

Bill Gates spent a great deal of time getting IBM to use the 16-bit 8086, rather than the 8-bit 8080 processor. IBM eventually agreed, and William Lowe assembled the engineers for Project Chess in Boca Raton, Florida.

Microsoft had previously had a good background in producing software computers such as FORTRAN, BASIC (Radio Shack and Apple) and COBOL, and had just released their version of the UNIX operating system for the PC, called XENIX. As Microsoft's main strengths were in software compilers, Paul Allen of Microsoft contacted Tim Patterson at Seattle Computer Products to get the rights to sell their DOS (86-DOS) to an unnamed client (IBM). It was the deal of the century, and Microsoft only paid $100 000 for the rights. After this, Bill Gates, Paul Allen, and Steve Ballmer met with IBM to propose that Microsoft be put in charge of the entire software development for IBM PC. Microsoft and IBM then signed a contract for Microsoft to develop certain software products for IBM's microcomputer. IBM also asked Microsoft to produce BASIC, FORTRAN, COBOL and Pascal compilers for their PC. In 1981, IBM released DOS 1.0, and the rest is history. The following year, Microsoft released MS-DOS versions of FORTRAN, BASIC (GW-BASIC) and COBOL.

The two main families of UNIX are UNIX System V and BSD (Berkeley Software Distribution) Version 4.4. System V is the operating system most often used and has descended from a system developed by the Bell Laboratories; it was recently sold to SCO (Santa Cruz Operation). System V was the first real attempt at unifying UNIX into a single standardized operating system, and succeeded in merging Microsoft XENIX, SunOS and UNIX 4.3 BSD. Unfortunately, after this there was still a drift by hardware manufacturers to move away from the standard (and define their standards). Although it has been difficult to standardize UNIX, it true strength is its communications protocols, especially TCP/IP, which are now worldwide standards for communicating over the Internet. The biggest challenge to UNIX has been from Windows NT, which has tried to create a hardware independent operating system. UNIX has, in the main, survived because of its simplicity and its reliability. The next threat will come from Windows 2000, which has moved towards UNIX in its network security methods.

Popular UNIX systems are:

- **AIX** (on IBM workstations and mainframes).
- **HP-UX** (on HP workstations).
- **Linux** (on PC-based systems).
- **OSF/1** (on DEC workstations).
- **Solaris** (on Sun workstations).

An initiative by several software vendors has resulted in a common standard for the user interface and the operation of UNIX. The user interface standard is defined by the common desktop environment (**CDE**), which allows software vendors to write calls to a standard CDE API (application program interface). The common UNIX standard has been defined as Spec 1170 APIs (Application Program Interfaces). Compliance with the CDE and Spec 1170 API are certified by X/Open, which is a UNIX standard organization.

'I don't think it's that significant.'

Tandy president John Roach, on the IBM PC.

3.2.2 Linux

Another important UNIX-like operating system is Linux, which was developed by Linus Torvalds at the University of Helsinki in Finland. It was first made public in 1991 and most of it is available free-of-charge. The most widely available version was developed by the Free Software Foundation's GNU project. It runs on most Intel-based, SPARC and Alpha-based computers, but its major problem is that it does not support as many hardware devices as Microsoft Windows does.

3.2.3 CP/M (Deceased)

'Welcome IBM. Seriously.' Headline, produced by Apple, for the full page advert in the *Wall Street Journal*

In 1973, before the widespread acceptance of PC-DOS, the future for personal computer operating systems looked to be CP/M (Control Program/Monitor), which was written by Gary Kildall of Digital Research. One of his first applications of CP/M was on the Intel 8008, and then on the Intel 8080. At the time, computers based on the 8008 started to appear, such as the Scelbi-8H, which cost $565 and had 1KB of memory.

On previous computers, IBM had written most of their programs for their systems. For the PC they had a strict time limit, so they first went to Digital Research (and Microsoft). Digital Research proposed that they use CP/M as it looked as if it would become the standardized operating system for microprocessors. Unfortunately, for Digital Research and Gary Kildall, they were unable to reach a final deal because they could not sign a strict confidentiality agreement.

3.2.4 DOS

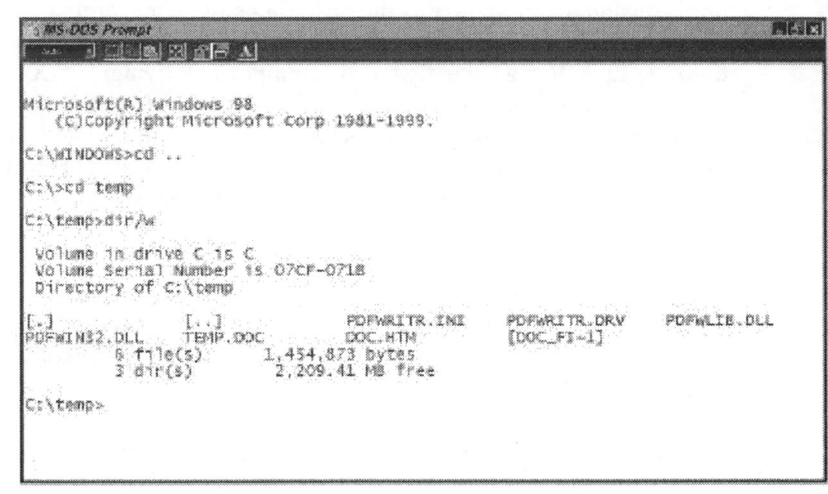

After IBM had communicated with Digital Research, they then went to a small computer company called Microsoft. For this Bill Gates bought a program called Q-DOS (often called the Quick and Dirty Operating System) from Seattle Computer Products. Q-DOS was similar to CP/M, but totally incompatible. Microsoft paid less than $100,000 for the rights to the software, and they allowed IBM to release it on the PC as PC-DOS. Soon Microsoft released their own version called MS-DOS, which has since become the best selling software in history.

To give users some choice in their operating system, the IBM PC was initially distributed with three operating systems: PC-DOS (provided by Microsoft), Digital Research's CP/M-86 and UCSD Pascal P-System. Microsoft understood that to make their operating system the standard, they must provide IBM with a good deal. Thus, Microsoft offered IBM the royalty-free rights to use Microsoft's operating system, forever, for $80000. This made PC-DOS much cheaper than the other two (such as $450 for P-System, $175 for CP/M and $60 for PC-DOS). Microsoft was smart in that they allowed IBM to use PC-DOS for free, but they held the control of the licensing of the software. This was one of the greatest pieces of business ever conducted. Eventually CP/M and P-System died off, while PC-DOS become the standard PC operating system.

The developed program was hardly earth shattering, but has since gone on to make billions of dollars. It was named the Disk Operating System (DOS) because of its original purpose of controlling the disk drives. Compared with some of the work that was going on at Apple and at Xerox, it was a very basic system. It had no graphical user interface and accepted commands from the keyboard and displayed them to the monitor. These commands were interpreted by the system to perform file management tasks, program execution and system configuration. Its function was to run programs, copy and remove files, create directories, move within a directory structure and to list files. To most people this was their first introduction to computing, but for many, DOS made using the computer to difficult, and it would not be until proper graphical user interfaces, such as Windows 95, that PCs would truly be accepted, and used by the majority.

3.2.5 Mac OS/System 8

Microsoft have relied on gradual improvement to their software, especially in their operating systems, while Apple got it right first time with their Mac OS system. Many reckon that MAC OS was 10 years ahead of all other systems. With the PC there were many difficulties, such as keeping compatibility with the original DOS, difficulties with the Intel 8086/88 memory model, support for a great number of peripherals, and the slowness of its internal busses (including the graphics adaptor). These all made software development of the PC difficult.

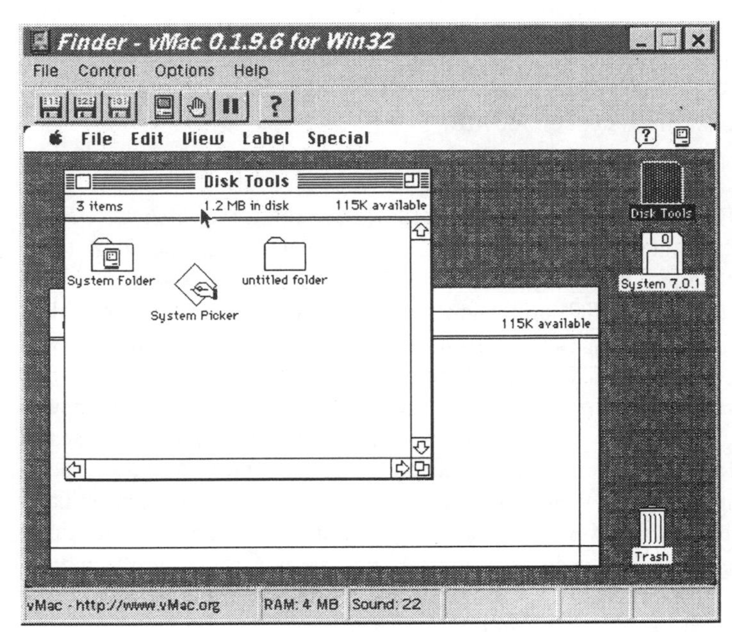

Apple did not have compatibility problems to deal with when they designed the Macintosh as they had a blank piece of paper to work with. They spent as much money on the operating system as they did on developing the hardware for the Macintosh. It was designed from a user's point-of-view, and borrowed many ideas from the Xerox system used at PARC. Mac OS made the computer, for the first time, friendly. Most Apple users actually enjoyed using their computer, which was not the same for the IBM PC-compatible market. At the time, PCs were still DOS based, with text commands, while Mac OS had a GUI which allowed users to operate the computer with graphical images and use a mouse to guide themselves around the system. Many Mac users wondered, why, in DOS, should someone have to locate a file with the CD command, and then use the DEL command to delete it, when, with the Mac, they could simply locate it by clicking the mouse pointer on folders, and then drag the file to the wastepaper basket (then undelete it if they had made a mistake). Many business analysts now acknowledge that Apple should have ported their operating system onto other systems, especially the IBM PC-compatible systems and to workstations, but Apple engineers thought that the Macintosh hardware could not be divorced from its beloved Mac OS. Mac OS was renamed System 7, in 1990, after a major upgrade.

3.2.6 Aegis (Deceased)

> 'We're just sitting here trying to put our PCjrs in a pile and burn them. And the damn things won't burn. That's the only thing IBM did right with it – they made it flameproof.'
>
> William Boman, Spinnaker Software chairman

Apollo burst onto the computer market in the 1980s with high-end workstations based on the Motorola 68000 processor (as in Apple Mac). They were aimed at serious users, and their main application area was in computer-aided design (CAD). One of the first to be introduced was the DN300, which was based around the excellent Motorola 68000 processor. It had a built-in mono monitor, an external 60MB hard disk drive, an 8-inch floppy drive, built-in ATR (Apollo Token Ring) network card, and 1.5MB RAM. It even had its own multi-user, networked operating system called Aegis. Unfortunately, for all its power and usability, Aegis never really took off, and when the market demanded standardized operating systems, Apollo switched to Domain/IX (which was a UNIX-clone). It is likely that Apollo would have captured an even larger market if they had changed to UNIX at an earlier time, as Sun (the other large workstation manufacturer) had done. The changeover of operating systems caused terrible problems for system administrators, and many users resisted the move away from their beloved Aegis. For a while most Apollo systems ran two different operating systems.

Aegis, as UNIX does, supported a networked file system, where a global file system could be built-up with local disk resources. Thus, a network of 10 workstations, each with 50 MB hard disks allowed for a global file system of 500 MB.

3.2.7 VMS

In the 1970s and 1980s, DEC carved out a large market for minicomputers (which were much cheaper than mainframe computers). Their first range of computer was the PDP (Programmed Data Processor) series, which become the foundation of many scientific and engineering groups. After the PDP, DEC took a great gamble and invested a great deal of research and development into the VAX (Virtual Address eXtension) computer range. It was an instant success, and the series provided a wide range of systems from basic terminals up to large mainframe computers. For the first time, DEC produced every part of the computer system: the operating system, the hardware and the software. One of the great successes of the VAX range was the VMS operating system (produced by David Culter). It allowed computer programmers to create programs which had more memory than the computer actually had (a virtual memory), and allowed several programs to run at the same time (multi-tasking). VMS is still popular in market niches, such as in the oil and gas industry (where reliability and batch processing are important). The VAX range also supported the Ethernet networking standard, which it had helped develop along with Intel and Xerox.

3.2.8 Microsoft Windows

DOS has long been the Achilles heel of the PC and has limited its development. It has also been its strength in that it provides a common platform for all packages. The first two versions of Microsoft Windows were pretty poor, as they had a poor interface, and still had all of the constraints of DOS (which are, of course, great). Microsoft Windows 3.0 changed all this with an enhanced usage of icons and windows, greater integration and increased memory usage. It was still 16-bit, but started to make full use of the processor, and gave a hint of the forthcoming multi-tasking (it could run two processes at a time, using fixed-rate time sharing). Other enhancements were: OLE; True Type fonts; and drag-and-drop commands. After Windows 3.0, Microsoft released Windows for Workgroups and was their first attempt at trying to network PCs together. Unfortunately it was too complicated to set-up, and not very powerful. Soon Windows 3.1 was released, along with a 32-bit version of Microsoft Windows.

The debate about whether Microsoft Windows 95 was just a copy of the Mac operating system will continue for years, but as Windows 3.x was a quantum leap from the previous versions, so Windows 95 was to Windows 3.x. The main problems with Windows 3.x was that it still used DOS as a basic operating system, and it also used 16-bit code (and most PCs used 32-bit processors, such as the 80386 and 80486 processors). Windows 95 was totally re-written using, mostly, 32-bit software, and thus used the full power of the processor, and also the full memory addressing capabilities. It could now address up to 4 GB of virtual memory, and supported a great deal more devices, such as CD-ROMs and back-up resources. It also had networking properly built into it.

At the time networking was becoming one of the key elements of a computer system, and Microsoft played a massive trump card, as it supported most of the widely available network protocols, such as TCP/IP (for Internet traffic), IPX/SPX (for Novell NetWare traffic), AppleTalk and IBM DCL. It could thus live with any type of network (and, Microsoft hoped, would eventually replace the existing network with one based on Microsoft networking, and not the existing network – which is known as gradual network migration). Windows 95/98

Workstation allow for a peer-to-peer network were computers can share resources, where as Windows NT/200 allows for the creation of a network server, which gives network-wide login, administration, global file systems, and so on. Windows NT 4.0, integrated the networking and robust process execution from Windows NT 3.0 and the classic user interface from Windows 95.

DOS and Windows 3.x operated in a 16-bit mode and had limited memory accessing. Windows 3.0 provided a great leap in PC systems as it provided an excellent graphical user interface to DOS. It suffered from the fact that it still used DOS as the core operating system. Windows 95/98 and Windows NT finally moved the PC away from DOS and operate as full 32-bit protected-mode operating systems. Their main features are:

- Runs both 16-bit and 32-bit application programs, with a 32-bit kernel for improved memory management, process scheduling and process management.
- Allow access to a large virtual memory (up to 4 GB).
- Support for pre-emptive multi-tasking and multi-threading of Windows-based and MS-DOS-based applications.
- Support for multiple file systems, including 32-bit installable file systems such as VFAT, CDFS (CD-ROM) and network redirectors. These allow better performance, use of long file names, and are an open architecture to support future growth.
- Support for 32-bit device drivers which give improved performance and intelligent memory usage.
- Enhanced robustness and clean-up when an application ends or crashes, and an enhanced dynamic environment configuration.

> 'The previous stars – Digital Research and Microsoft – may soon find themselves playing cameo roles as AT&T and IBM take center stage.'
>
> *ComputerWorld*, 1984

Microsoft Windows comes in many flavors; the main versions are outlined below and Table 3.1 lists some of their attributes.

- Microsoft Windows 3.x. 16-bit PC-based operating system with limited multi-tasking. It runs from MS-DOS and thus still uses MS-DOS functionality and file system structure.
- Microsoft Windows 95/98. *Robust* 32-bit multi-tasking operating system (although there are some 16-bit parts in it) which can run MS-DOS applications, Microsoft Windows 3.x applications and 32-bit applications.
- Microsoft Windows NT Version 4. Robust 32-bit multi-tasking operating system with integrated networking. Networks are built around NT servers and clients. As with Microsoft Windows 95/98 it can run MS-DOS, Microsoft Windows 3.x applications and 32-bit applications, although it is less tolerant of DOS programs.
- Windows NT Version 5/2000. This is available in three flavors: Workstation, Server and SMP Server (multiprocessor). It runs on Alphas, Intel *x*86, Intel IA32, Intel IA64 and AMD K7 (which is similar to an Alpha).

> All computers wait at the same speed.
>
> 2 + 2 = 5 for extremely large values of 2.

Windows NT/2000 and 95/98 provide excellent network support as they can communicate directly with many different types of networks, protocols and computer architectures. They can create networks to make peer-to-peer connections and also connection to servers for access to file systems and print servers.

Windows NT/2000 Server has more security in running programs than Windows 95/98 as programs and data are insulated from the operation of other programs. The operating system parts of Windows NT/2000 and Windows 95/98 run at the most trusted level of privilege of the Intel processor, which is ring zero. Application programs run at the least trusted level of privilege, which is ring three. These programs can use either a 32-bit flat mode or any of the memory models, such as large, medium, compact or small.

Table 3.1 Windows comparisons

	Windows 3.1	*Windows 95/98*	*Windows NT/2000*
Pre-emptive multi-tasking		✓	✓
32-bit operating system		✓	✓
Long file names		✓	✓
TCP/IP	✓	✓	✓
32-bit applications		✓	✓
Flat memory model		✓	✓
32-bit disk access	✓	✓	✓
32-bit file access	✓	✓	✓
Centralized configuration storage		✓	✓

3.2.9 Windows 95 and Windows 98

New features in Microsoft Windows operating systems include:

- **Advanced Plug and Play** and **Automatic hardware detection.** USB devices can be added to the computer without rebooting it.
- **Enhanced power management.** This allows the monitor and hard disks to be turned off, when they are not in use after a certain time period. The system board can also be powered-down after a certain time, and there is also enhanced support for notebooks, as power shutdown can save the battery drainage. The wakeup for the power-down can be mouse movement, or a keyboard press (or even the ring from a modem).
- **Increased WWW integration.** WWW page creation, integrated email, channels, and so on.
- **Windows updates.** This facility allows for a single source to update system drivers, system files and operating system programs, such as service packs.
- **System file checker.** The facility checks for system files and recovers old system files. It also checks the integrity of the operating system files and if necessary restores them or extracts them from the installation disks.
- **Maintenance wizard.** This facility allows tasks to be run at given time intervals.
- **Multiple monitors.** The facility allows the computer to display to multiple monitors. Different parts of the screen can be sent to the connected monitors, and thus expand the physical size of the desktop area.
- **Support for new devices.** Windows 98 supports many new hardware devices, such as: Universal Serial Bus (USB), IEEE 1394, Accelerated Graphics Port (AGP) and DVD. IEEE 1394 defines a class of hardware that makes it easy to add serial devices to a computer. The AGP interface is an enhanced video card interface which gives enhanced support for 3-D animation. DVD drives play software and music CDs.

'... the "irresistible tide" of AT&T's Unix now threatens to engulf the current microcomputer operating system standard, MS-DOS'
– *Datamation*, 1984

3.2.10 Windows NT/2000

Microsoft Windows NT is an excellent network operating system, and Windows 2000 has built on this. It has many benefits, such as:

- **Multiple platform support.** Windows NT Version 4.0/2000 runs on many types of systems, such as Intel 486/Pentium Pro/Pentium II-based PCs, DEC Alpha-based systems and PowerPC systems.
- **Multi-tasking and multi-threading.** Multi-tasking involves running several processes at a time. Multi-tasking programs split into a number of parts (threads) and each of these is run on the multi-tasking system (multi-threading). A program which is running more than one thread at a time is known as a multithreaded program.
- **Support for 16-bit, 32-bit and MS-DOS programs.** Windows NT has robust support for 16-bit and 32-bit applications, and gives some support for MS-DOS programs. Each MS-DOS program runs within its own virtual space, where it is allocated the resources it needs.
- **Multiple file system support.** Windows NT supports FAT (Version 4.0 only supports 16-bit FAT, whereas Windows 95 and Windows 98 support 32-bit VAT, or VFAT), HPFS (which is a UNIX-style file system) and NTFS (which is an enhanced file support that has built-in security).
- **Enhanced reliability.** Windows NT only allows access to the hardware through system drivers. These are reliable, well-tested files which stop application programs from incorrect hardware operation.
- **Built-in networking.** Windows NT supports many different types of networking technology. Client services include: Windows networks and Novell NetWare networks, network adapters (typically using standard interfaces such as ODI and NDIS) and network protocols (including TCP/IP and SPX/IPX).

'The 32-bit machine would be "overkill" for a personal computer'
– Sol Libes, ByteLines

Microsoft Windows is covered in Chapter 22.

3.2.11 Novell NetWare

Novell NetWare is one of the most popular network operating systems for PC LANs and provides file and print server facilities. Its default network protocol is normally SPX/IPX. This can also be used with Windows NT to communicate with other Windows NT nodes and with NetWare networks.

At the time that Microsoft was developing Windows 3.0, which had no networking, Novell released NetWare 3.0 which allowed DOS and Windows-based computers to communicate with a network server. It basically ran a program which read the commands from the operating system, and then decided if the command should be sent to the network server, or to the core operating system. For a long time, Novell had no competition in the PC networking market, as IBM and Microsoft failed to produce easy-to-use and robust networking operating systems. The biggest threat to NetWare came from Windows NT, which was a radical redesign and had networking at its core.

Novell NetWare 3.0 was a classic network operating system, and was the most popular PC-based networking operating system, before Windows NT started to carve a large market. Novell have since developed it to NetWare 4.0 and NetWare 5.0. With NetWare 4.0, Novell considerably enhanced the system using NDS (Novell Directory Services), which allows organizations to create large interconnected networks. NDS creates an organization-wide structure that allows users to get access to resources that are not physically connected to the local server. Many large organization use NetWare as it is robust and very reliable, and provides the main services that they require: print servers, file services and user accounts. Many others are in the phase of migrating their network operating systems away from NetWare, towards a totally Microsoft-based network.

3.3 Exercises

The following questions are multiple choice. Please select from a–d.

3.3.1 Which of the following operating systems does not directly support networking:
(a) Linux (b) UNIX
(c) DOS (d) Windows 95

3.3.2 Which of the following is not a function of the operating system:
(a) Multi-tasking (b) Device interfacing
(c) Creating a file system (d) Displaying windows

3.3.3 Which company initially developed UNIX:
(a) AT&T (b) Microsoft
(c) Sun Microsystems (d) Apple

3.3.4 Which operating system competed with DOS as the standard for microcomputer systems:
(a) Aegis (b) CP/M
(c) VMS (d) HP-UX

3.3.5 Which of the following is now classed as a deceased operating system:
(a) Aegis (b) Linux
(c) VMS (d) UNIX

3.3.6 Which of the following is a popular networking operating system:
(a) Novell NetWare (b) DOS
(c) CP/M (d) PRIME

3.3.7 What classification of operating system is Microsoft Windows 3.0:
(a) 8-bit (b) 16-bit
(c) 32-bit (d) 64-bit

3.3.8 Contrast the following operating systems and their main attributes:

Operating system	*Single-user/ multi-user*	*Networked/ Stand-alone*	*Multi-tasking/ Single-tasking*	*Multiprocessor/ Single processor*
DOS				
Windows 3.1				
Windows 95				
Windows NT				
UNIX				
Mac OS				

3.4 Note from the Author

This chapter is intended to be a gentle introduction to operating systems. I've covered the basic theory of OS's, but I've also looked at their history, as it's important to realize the choices that have been made in getting where we are today. An important point is that not always the best technology wins, and sometimes it's all about being in the right place at the right time. Look at DOS. How did it survive for so long, without ever really changing? We went through six major versions, and each time we paid up and basically got the same old software. But all of this changed with two major pieces of software; Microsoft Windows 3.0 and Microsoft Windows 95. Microsoft Windows 3.0 brought a useable graphical user interface which built itself on top of the horrible DOS. As long as DOS stayed there, Windows 3.0 was never going to be anything more than a graphical user interface. With Windows 95, Microsoft started from scratch and built a solid kernel which could run several processes, at a time, and also had a most amazing user interface. With Windows 95, computers could now be used by children, and non-computer specialists. The days of remembering the command for printing (it's PRINT, by the way) or for listing directories (it's DIR, in case you forgot, or are too young to remember DOS) were replaced with user menus, and graphical icons. So the star of the OS market must be Microsoft Windows, in its 32-bit form, that is, Microsoft Windows 95/98, Windows NT/2000. An honorable mention must go to UNIX, who has quietly fathered many of the major networking protocols, without ever demanding payment from the world. Without UNIX there would no TELNET and FTP, no TCP and IP, and no SMTP (e-mail), and NFS (distributed file systems), and NIS (global logins), and, well, I'd better stop there, because I could fill ten pages with its achievements. Oh, and it's got a proper kernel, and allows different graphical user interfaces to be bolted on top of it (honest, I'm not getting at anyone here, but, in Microsoft Windows, is it possible to split the kernel from the graphical user interface, and run the kernel on its own? Please send me an e-mail if you know.

3.5 Microsoft v. General Motors

At a recent COMDEX, Bill Gates compared the computer industry to the automobile industry and stated: 'If GM kept up with technology like the computer industry has, we would be driving $25 cars that got 1000 miles to the gallon.'

A slightly aggrieved General Motors immediately issued a press release stating: 'If GM had developed technology like Microsoft, we would all be driving cars with the following characteristics:

- For no reason whatsoever your car would crash twice a day.
- Every time they repainted the lines on the road, you would have to buy a new car.
- Occasionally your car would die on the freeway for no reason, and you would just accept this, restart and drive on.
- Occasionally, executing a maneuver such as a left turn would cause your car to shut down and refuse to start, in which case you would have to reinstall the engine.
- Only one person at a time could use the car, unless you bought "Car95" or "CarNT". But then you would have to buy more seats.
- Apple would make a car that was powered by the sun, reliable, five times as fast, and twice as easy to drive, but would only run on 5% of the roads.
- The oil, water temperature and alternator warning lights would be replaced by a single "general car default" warning light.
- New seats would force everyone to have the same size butt.
- The airbag system would say "Are you sure?" before deploying.
- Occasionally for no reason whatsoever, your car would lock you out and refuse to let you in until

you simultaneously lifted the door handle, turn the key, and grabbed hold of the radio antenna.

- GM would require all car buyers to also purchase a deluxe set of Rand McNally road maps (now a GM subsidiary), even though they neither need them nor want them. Attempting to delete this option would immediately cause the car's performance to diminish by 50% or more. Moreover, GM would become a target for investigation by the Justice Department.
- Every time GM introduced a new model, car buyers would have to learn how to drive all over again because none of the controls would operate in the same manner as the old car.
- You'd press the "Start" button to shut off the engine.

Operating Systems

4.1 Introduction

> By the middle of the 1990s, Intel and Microsoft were so profitable that they accounted for nearly half of the entire profits in the worldwide PC industry (which was worth over $100 billion, each year).

These days, networking and operating systems are almost intertwined, and most operating systems now directly support networking as part of their functionality. Without networks, operating systems could not provide the required user functionality, such as access to networked resources (printers, file systems, and so on), connection to the Internet or an Intranet, transferring of files, and so on. Typical network operating systems are Novell NetWare, Windows NT/2000 and UNIX. Thus to be able to properly understand distributed systems and networks, it is important to understand the component parts of an operating system.

The boundaries where processes run have now expanded from only running on local computers, to being distributed over networks. In the most extreme case it is possible to run processes over a large geographical area. This leads to the concept of distributed processing. Normally, each process must communicate with another, and there thus must be some mechanism for synchronization between them. Local computers use techniques such as hardware and software interrupts to generate events, and then signals or messages to pass the information between the processes.

Figure 4.1 shows an example of data passed between processes. Initially a process on a computer sends an interrupt to the computer to identify that it is ready to transmit data. When the interrupt is received on the local computer, or a remote computer, it informs the destination process that a process wants to communicate with it. A message or a signal is then sent between the two processes to identify the type of data to be passed (if any). The data can then be passed between the processes.

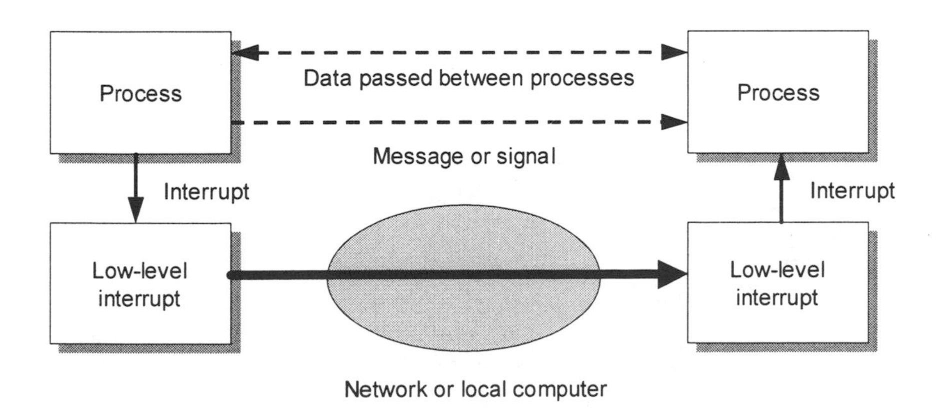

Figure 4.1 Information passed between processes

4.2 Multitasking and threading

Multitasking involves running several tasks at the same time. It normally involves running a process for a given amount of time, before releasing it and allowing another process a given amount of time. The two forms of multitasking are illustrated in Figure 4.3 and Figure 4.4; and are:

- **Pre-emptive multitasking.** This type of multitasking involves the operating system controlling how long a process stays on the processor. This allows for smooth multitasking and is used in 32-bit Microsoft Windows programs and the UNIX operating system.
- **Co-operative multitasking.** This type of multitasking relies on a process giving up the processor. It is used with Windows 3.x programs and suffers from processor hogging, where a process can stay on a processor and the operating system cannot kick it off.

The logical extension to multitasking programs is to split a program into a number of parts (threads) and run each of these on the multitasking system (multi-threading). A program that is running more than one thread at a time is known as a multi-threaded program. These have many advantages over non-multi-threaded programs, including:

- They make better use of the processor, where different threads can be run when one or more threads are waiting for data. For example, a thread could be waiting for keyboard input, while another thread could be reading data from the disk.
- They are easier to test, as each thread can be tested independently of other threads.
- They can use standard threads, which are optimized for given hardware.

They also have disadvantages, including:

- The program has to be planned properly so that threads know on which other threads they depend.
- A thread may wait indefinitely for another thread which has crashed or terminated.

The main difference between multiple processes and multiple threads is that each process has independent variables and data, while multiple threads share data from the main program, as illustrated in Figure 4.2.

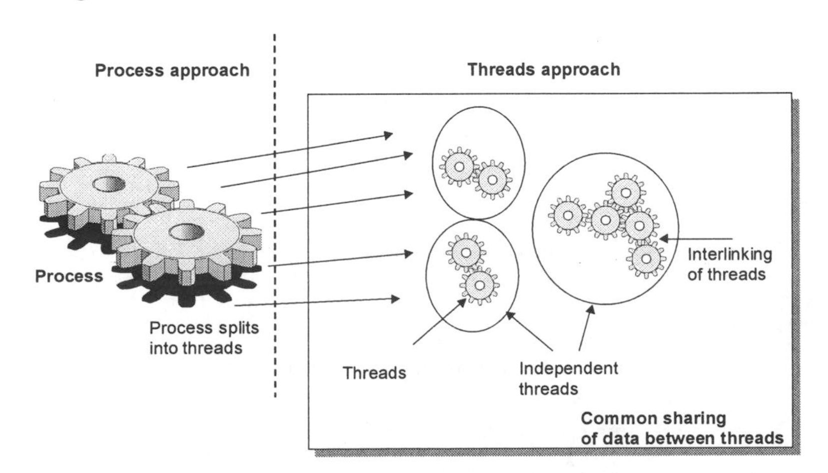

Figure 4.2 Process splitting into threads

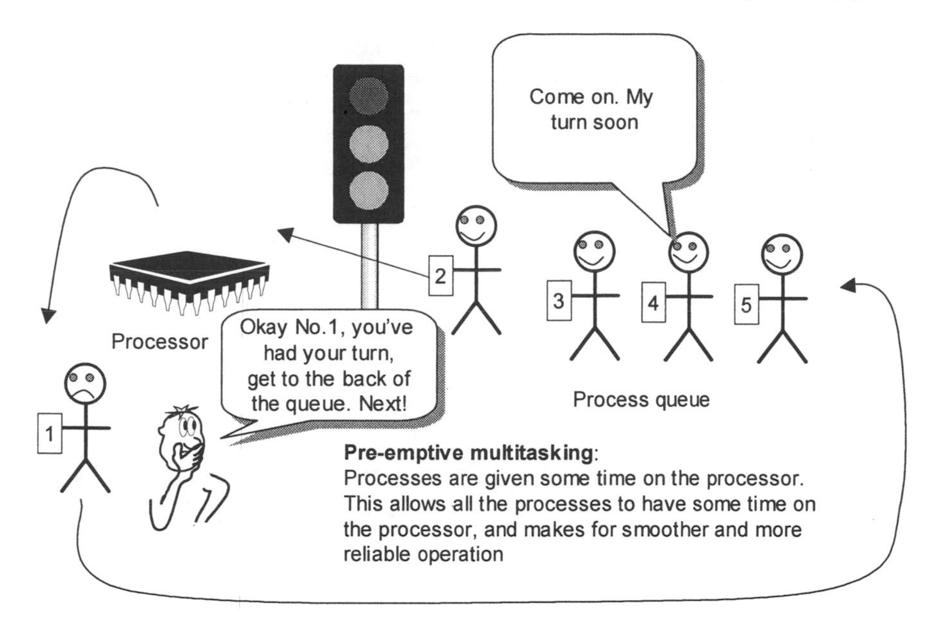

Figure 4.3 Pre-emptive multitasking

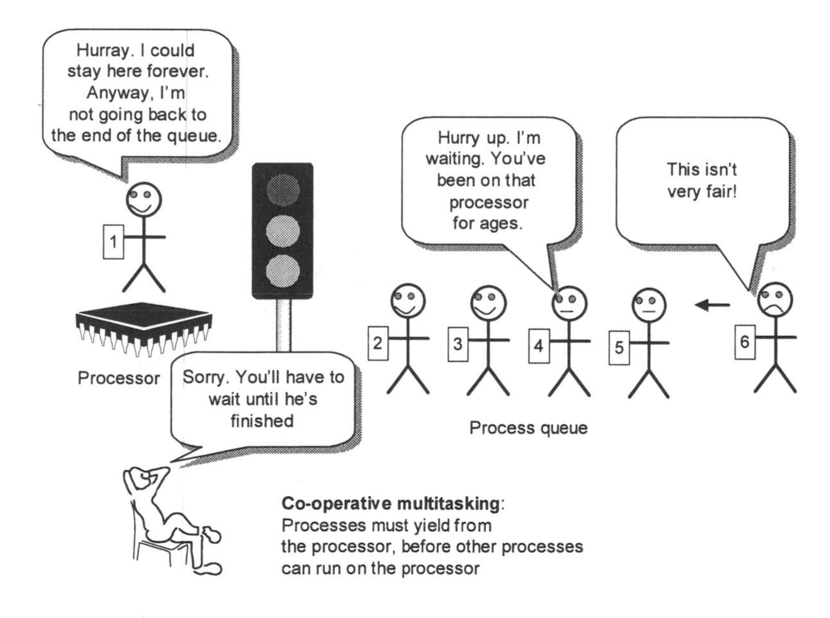

Figure 4.4 Co-operative multitasking

4.3 Interrupts (on PC systems)

> Computer message: (A)bort, (R)etry, (T)ake down entire network?, – Anon.

An interrupt allows a program or an external device to interrupt the execution of a program, and can occur by hardware (hardware interrupt) or software (software interrupt). When an interrupt occurs an interrupt service routine (ISR) is called. For a hardware interrupt the ISR then communicates with the device and processes any data. When it has finished the program execution returns to the original program. A software interrupt causes the program to interrupt its execution and goes to an interrupt service routine. Typical software interrupts include reading a key from the keyboard, outputting text to the screen and reading the current date and time. The operating system must respond to interrupts from external devices, as illustrated in Figure 4.5.

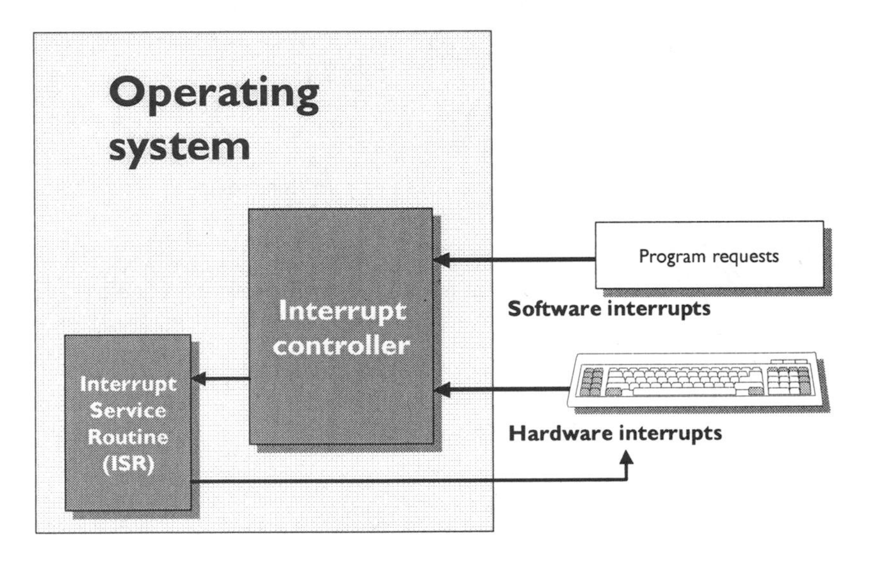

Figure 4.5 Interrupt service routine

4.3.1 Software Interrupts

The Basic Input/Output System (BIOS) communicates directly with the hardware of the computer. It consists of a set of programs which interface with devices such as keyboards, displays, printers, serial ports and disk drives. These programs allow the user to write application programs that contain calls to these functions, without having to worry about controlling them or about which type of equipment is being used. Without BIOS the computer system would simply consist of a bundle of wires and electronic devices.

There are two main parts to BIOS. The first is permanently stored in non-volatile memory (the ROM BIOS), and is this part that starts the computer (or bootstrap) and contains programs which communicate with resident devices. The second stage is loaded when the operating system is started.

An operating system allows the user to access the hardware in an easy-to-use manner. It accepts commands from the keyboard and displays them to the monitor. The Disk Operating System, or DOS, gained its name from its original purpose of providing a controller for the computer to access its disk drives. The language of DOS consists of a set of commands which are entered directly by the user and are interpreted to perform file management tasks, program execution and system configuration. It makes calls to BIOS to execute these. The main functions of DOS are to run programs, copy and remove files, create directories, move within a directory structure and to list files. Microsoft Windows calls BIOS programs directly.

BIOS was a key element in making the PC cloneable, as it contained all the hardware specific information in a single place. Compaq successfully rewrote its version of BIOS, rather than copying it from the original IBM PC, which stopped any legal action from IBM.

4.3.2 Hardware Interrupts

Computer systems either use polling or interrupt-driven software to service external equipment. With polling, the computer continually monitors a status line and waits for it to become active. An interrupt-driven device sends an interrupt request to the computer, which is then serviced by an interrupt service routine (ISR). Interrupt-driven devices are normally better in that the computer is thus free to do other things while polling slows the system down, as it must continually monitor the external device. Polling can also cause problems in that a device may be ready to send data but the computer is not watching the status line at that point. Figure 4.6 illustrates polling and interrupt-driven devices.

The generation of an interrupt can occur by hardware or software, as illustrated in Figure 4.7. If a device wishes to interrupt the processor it informs the programmable interrupt controller (PIC). The PIC then decides whether it should interrupt the processor. If there is a processor interrupt then the processor reads the PIC to determine which device caused the interrupt. Then, depending on the device that caused the interrupt, a call to an ISR is made, which then communicates with the device and processes any data. When it has finished, the program execution returns to the original program. Each PIC allows access to eight interrupt request lines. Most PCs use two PICs which gives access to 16 interrupt lines.

Hardware interrupts allow external devices to gain the attention of the processor. Depending on the type of interrupt the processor leaves the current program and goes to the ISR. This program communicates with the device and processes any data. After the ISR has completed its task then program execution returns to the program that was running before the interrupt occurred.

First Microprocessor

Around the late 1960s, the electronics industry was producing cheap pocket calculators, which led to the development of affordable computers, when the Japanese company Busicom commissioned Intel to produce a set of between eight and 12 ICs for a calculator. Then instead of designing a complete set of ICs, Ted Hoff, at Intel, designed an integrated circuit chip that could receive instructions, and perform simple integrated functions on data. The design became the 4004 microprocessor. Intel then produced a set of ICs, which could be programmed to perform different tasks. These were the first ever microprocessors and soon Intel (short for *Int*egrated *El*ectronics) produced a general-purpose 4-bit microprocessor, named the 4004.

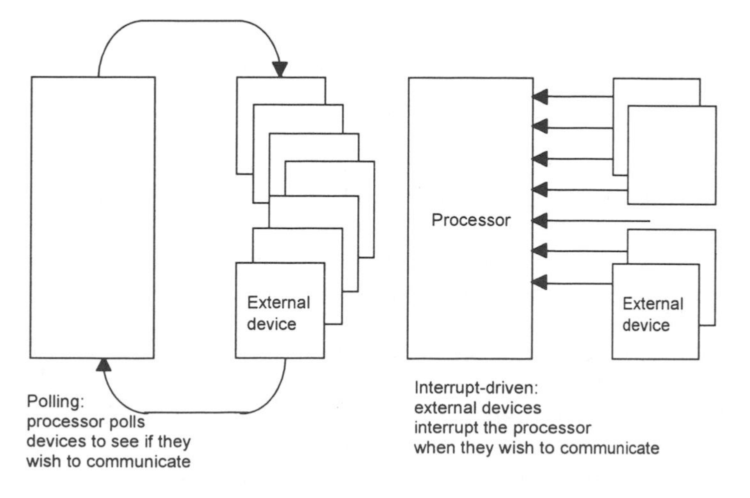

Figure 4.6 Polling or interrupt-driven communications

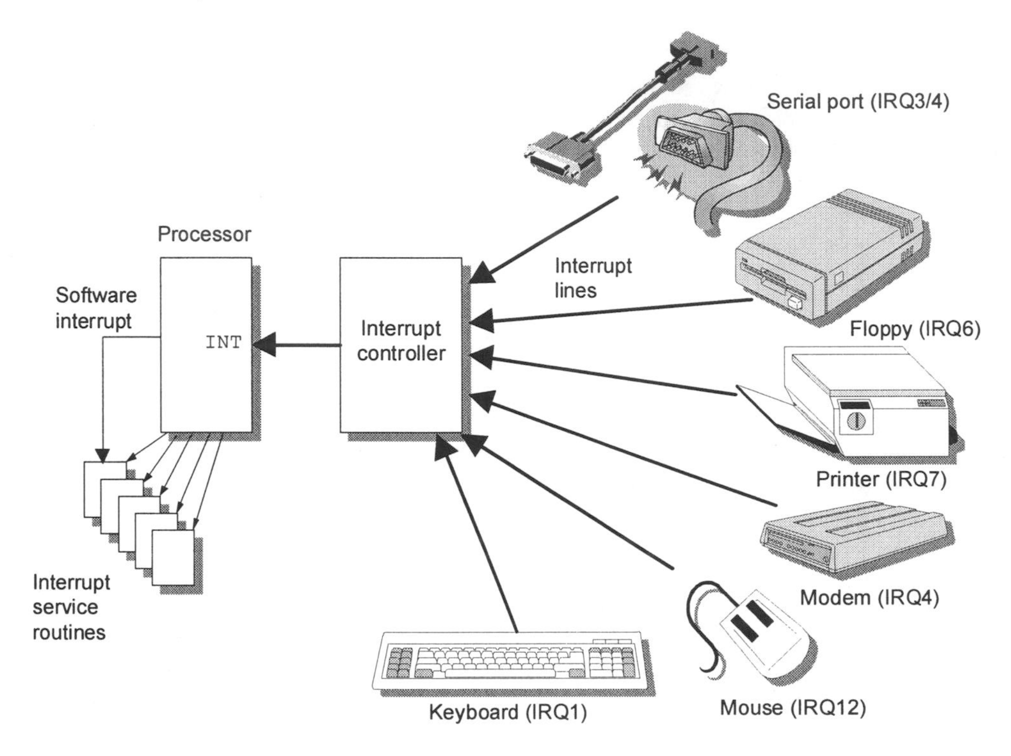

Figure 4.7 Interrupt handling

Interrupt vectors

On a PC, each device that requires to be 'interrupt-driven' is assigned an IRQ (interrupt request) line. Table 4.1 outlines the usage of each of these interrupts. `IRQ0` normally connects to the system timer, `IRQ1` to the keyboard, and so on. The system timer interrupts the processor 18.2 times per second, and is used to update the system time. When the keyboard has data it interrupts the processor with the `IRQ1` line.

Data received from serial ports interrupts the processor with `IRQ3` and `IRQ4` and the parallel ports use `IRQ5` and `IRQ7`. If one of the parallel, or serial, ports does not exist then the IRQ line normally assigned to it can be used by another device, such as for a sound card, which has a programmable IRQ line which is mapped to an IRQ line that is not being used.

Note that several devices can use the same interrupt line. A typical example is `COM1:` and `COM3:` sharing `IRQ4` and `COM2:` and `COM4:` sharing `IRQ3`. If they do share then the ISR must be able to poll the shared devices to determine

To the 4004, and beyond

The Intel 4004 processor caused a revolution in the electronics industry as previous electronic systems had a fixed functionality. With this processor, the functionality could be programmed by software. Amazingly, by today's standards, it could only handle four bits of data at a time (a nibble), contained 2000 transistors, had 46 instructions and allowed 4 KB of program code and 1 KB of data. From this humble start, the PC has since evolved using Intel microprocessors. Before the 4004, Intel had been an innovative company, and had produced the first memory device (static RAM, which uses six transistors for each bit stored in memory), the first DRAM (dynamic memory, which uses only one transistor for each bit stored in memory) and the first EPROM (which allows data to be downloaded to a device, which is then permanently stored).

which of them caused the interrupt. If two different types of device (such as a sound card and a serial port) use the same IRQ line then there may be a contention problem as the ISR may not be able to communicate with different types of interfaces.

Figure 4.8 shows a sample window displaying interrupt usage. In this case it can be seen that the system timer uses `IRQ0`, the keyboard uses `IRQ1`, the PIC uses `IRQ2`, and so on. Notice that a Sound Blaster is using `IRQ5`. This interrupt is normally reserved for the secondary printer port. If there is no printer connected then `IRQ5` can be used by another device. These interrupt lines are a legacy from the original PC, and are likely to be phased-out in the coming years, as hub based technology, such as USB, allows many devices to use the same interrupt line.

Table 4.1 Interrupt handling

Interrupt name	*Generated by*
System timer	IRQ0
Keyboard	IRQ1
Reserved	IRQ2
Serial communications (COM2:)	IRQ3
Serial communications (COM1:)	IRQ4
Reserved	IRQ5
Floppy disk controller	IRQ6
Parallel printer (LPT1:)	IRQ7
Real-time clock	IRQ8
Math co-processor	IRQ13
Hard disk controller (primary)	IRQ14
Hard disk controller (secondary)	IRQ15

Typical uses of interrupts are:

IRQ0: System timer — The system timer uses `IRQ0` to interrupt the processor 18.2 times per second and is used to keep the time-of-day clock updated.

IRQ1: Keyboard data ready — The keyboard uses `IRQ1` to signal to the processor that data is ready to be received from the keyboard. This data is normally a scan code.

IRQ2: Redirection of IRQ9 — The BIOS redirects the interrupt for `IRQ9` back here.

IRQ3: Secondary serial port (COM2:) — The secondary serial port (`COM2:`) uses `IRQ3` to interrupt the processor. Typically, `COM3:` to `COM8:` also use it, although `COM3:` may use `IRQ4`.

IRQ4: Primary serial port (COM1:) — The primary serial port (`COM1:`) uses `IRQ4` to interrupt the processor. Typically, `COM3:` also uses it.

IRQ5: Secondary parallel port (LPT2:) — Typically, it is used by a sound card on PCs which have no secondary parallel port connected.

IRQ6: Floppy disk controller — The floppy disk controller activates the `IRQ6` line on completion of a disk operation.

IRQ7: Primary parallel port (LPT1:) — Printers (or other parallel devices) activate the `IRQ7` line when they become active. As with `IRQ5` it may be used by another device, if there are no other devices connected to this line.

IRQ9: Redirected to `IRQ2` service routine.
IRQ12: PS/2-style mouse.
IRQ13: Maths co-processor.
IRQ14/IRQ15: Hard disk (IDE0/IDE1).

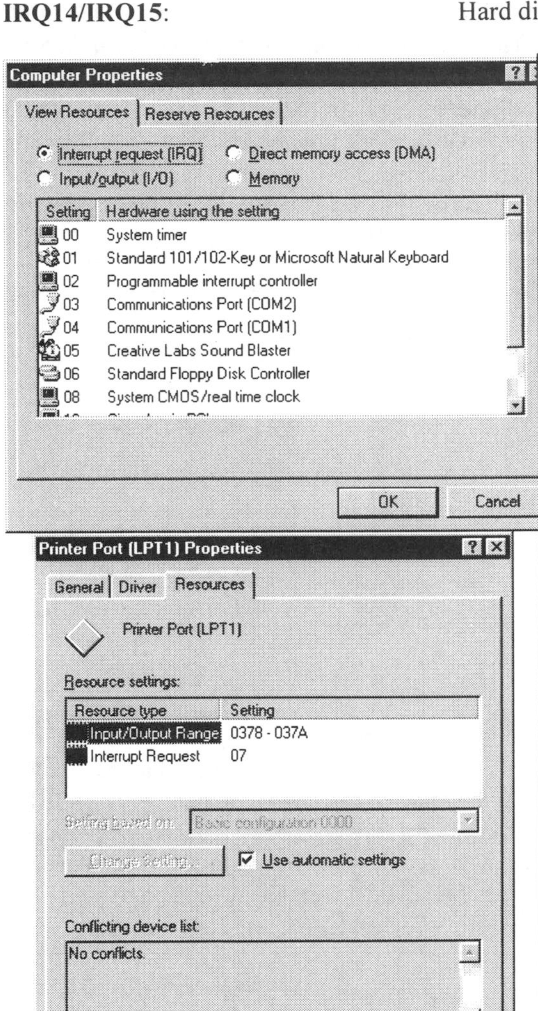

Figure 4.8 Example usage of IRQ lines

Open v. Closed Systems

In 1985, Apple was having difficult times. The sales of the Macintosh were not as great as expected, and the Apple II was facing a great deal of competition from other manufacturers. Many people at the time, including Bill Gates, were advising Apple to open-up its Macintosh computer by allowing others to build their own systems, under strict license arrangements. Bill Gates had advised that it become involved with companies such as HP and AT&T. However, Apple held onto both its Mac operating system, and its hardware. The two it believed were intertwined. Apple shot the GEM product of Digital Research out of the water. Digital Research developed GEM for the PC, and had borrowed the look-and-feel of the Mac operating system, but not the actual technology. Apple's lawyers, in 1985, visited Digital Research and threatened it with court action. At the time, IBM had been keen to license the GEM system for its products. IBM eventually backed away from the product over the fear of litigation, and that was the end of GEM.

Apple then turned to Microsoft to head off its attempt at producing a GUI. Bill Gates, though, had much greater strength against Apple, and he argued that the true originator of the GUI was Xerox. It was Xerox's ideas that were to be used for Windows. Microsoft had a trump card: If Apple was going to stop Microsoft from producing Windows then Microsoft would stop producing application software for the Macintosh. Apple knew that it needed Microsoft more than Microsoft needed Apple. They then both signed a contact which stated that Microsoft would:

*'have a non-exclusive, worldwide, royalty-free, perpetual, nontransferable license to use derivate works in **present and future software programs**, and to license them to and through third parties for use in their software programs'*

which basically gave Microsoft carte blanche for all future versions of its software, and was quite free to borrow which ever features it wanted. John Scully at Apple signed it, and gave away one of the most lucrative markets in history. Basically, Apple was buying peace with Microsoft, but it was piece with a long-term cost.

4.4 Win32

Win32 is a standard programming model which allows a Windows program to run as a full 32-bit program. It also gives access to a great deal of advanced Windows functions. Its main advantages are:

- A 32-bit programming model for Windows 3.x that shares binary compatibility with Windows NT and Windows 95/98.
- The ability to produce an application program that can be used with Windows NT, Windows 95/98, and Windows 3.x.
- Full OLE (object linking and embedding) support, including 16-bit/32-bit interoperability. OLE allows the application program to share data where an OLE server provides information for an OLE client.
- Improved performance with 32-bit operations.
- Access to a large number of Win32 APIs (application programming interfaces) for Windows NT/2000 and Windows 95/98 (such as Windows, Menus, Resources, Memory Management, Graphics, File Compression, and so on).
- Win32 semantics for the application programming interface (API).

4.5 Core system components

A software library contains useful routines which can either be compiled with a program (static library), or can be called up when the program is run (dynamic library). With a static library the routines are added to the program by the linker, which searches the static libraries for the functions that require to be linked into the program. This extra code is included in the executable program. The advantages that DLLs have over static libraries are:

- DLLs are only loaded when, and if, they are required.
- DLLs can be easily changed and upgraded.

Their main disadvantages are: that they can be easily tampered with and can be easily deleted or replaced with a previous version. One well-know virus actually replaces WINSOCK.DLL with its own DLL, and thus intercepts any Internet communications (with worrying consequences). This is illustrated in Figure 4.9. The WINSOCK.DLL dynamic library is responsible for any TCP/IP communications, thus an external hacker could make a copy of all of the Internet communication and send them to another computer, where they could be monitored. It also gives a port for an external hacker to gain access to the computer, and thus interrogate its contents, and monitor its activities.

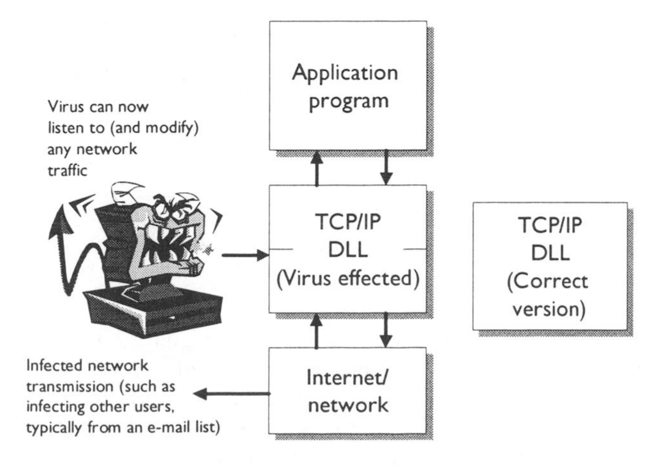

Figure 4.9 Virus attack on a DLL

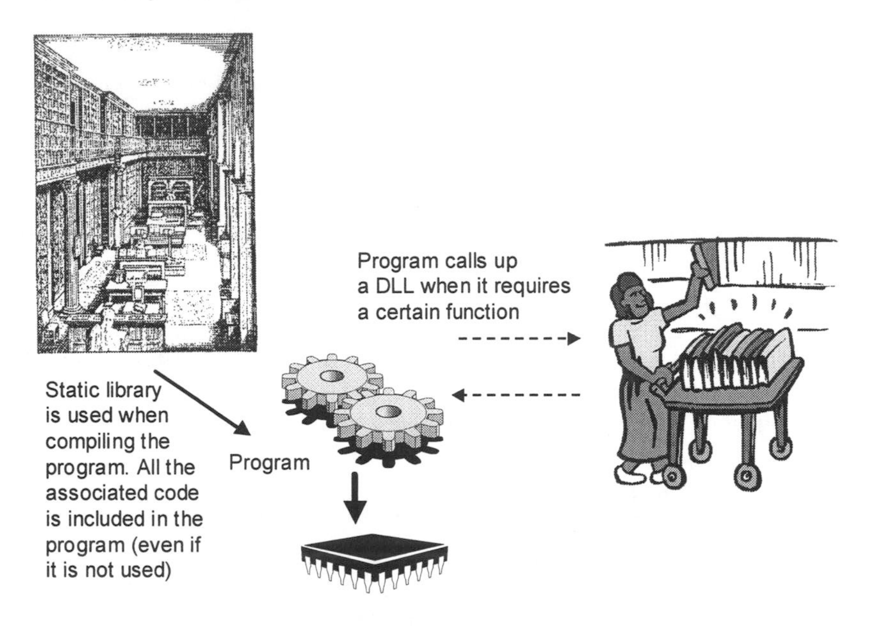
Program calls up
a DLL when it requires
a certain function
Static library
is used when
compiling the
program. All the
associated code
is included in the
program (even if
it is not used)
Program

Figure 4.10 Dynamic link libraries

The core of Windows 95/98 has three components: User, Kernel, and GDI (graphical device interface), each of which has a pair of DLLs (one for 32-bit accesses; the other for 16-bit accesses). The 16-bit DLLs allow for Win16 and MS-DOS computability. A DLL (Dynamic Link Library) contains a number of services, which can be called up when a program runs, as illustrated in Figure 4.10. Typical services include access to networking services, input/output functions and graphical functions.

4.5.1 User

The User component provides input and output to and from the user interface. Input is from the keyboard, mouse, and any other input device and the output is to the user interface. It also manages interaction with the sound driver, timer, and communications ports.

Win32 applications and Windows 95/98 use an asynchronous input model for system input. With this, devices have an associated interrupt handler (for example, the keyboard interrupts with IRQ1) that converts the interrupt into a message. This message is then sent to a raw input thread area, which then passes the message to the appropriate message queue. Each Win32 application can have its own message queue, whereas all Win16 applications share a common message queue.

4.5.2 Kernel

The Kernel provides for core operating system components including file I/O services, virtual memory management, task scheduling and exception handling, such as:

- File I/O services.
- Exceptions. These are events that occur as a program runs and call additional software which is outside the normal flow of control. For example, if an application generates an exception, the Kernel is able to communicate that exception to the application to perform the necessary functions to resolve the problem. An example exception is caused by a divide-by-zero error in a mathematical calculation, an exception routine can be designed so that it handles the error and does not crash the program.

- Virtual memory management. This resolves import references and supports demand paging for the application.
- Task scheduling. The Kernel schedules and runs threads of each process associated with an application.
- Provides services to both 16-bit and 32-bit applications by using a thunking process which is the translation process between 16-bit and 32-bit formats. It is typically used by a Win16 program to communicate with the 32-bit operating system core.

Virtual memory allows processes to allocate more memory than can be physically allocated. The operating system allocates each process a unique virtual address space, which is a set of addresses available for the process's threads. This virtual address space appears to be 4 GB in size, where 2 GB are reserved for program storage and 2 GB for system storage.

Figure 4.11 illustrates where the system components and applications reside in virtual memory. Its contents are:

- 3 GB–4 GB. All Ring 0 components.
- 2 GB–3 GB. Operating system core components and shared DLLs. These are available to all applications.
- 4 MB–2 GB. Win32-based applications, where each has its own address space. This memory is protected so that other programs cannot corrupt or otherwise hinder the application.
- 0–640 KB. Real-mode device drivers and TSRs (Terminate and Stay Resident).

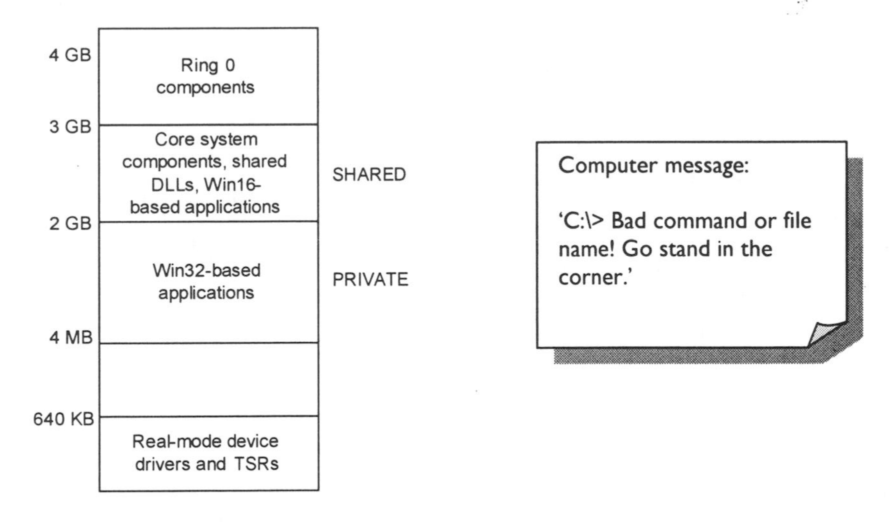

Figure 4.11 System memory usage

4.5.3 GDI

The Graphical Device Interface (GDI) is the graphical system that:

- Manages information that appears on the screen.
- Draws graphic primitives and manipulates bitmaps.
- Interacts with device-independent graphics drivers, such as display and printer drivers.

The graphics subsystem provides input and output graphics support. Windows uses a 32-bit graphics engine (known as DIB, Device-independent Bitmaps) which:

- Directly controls the graphics output on the screen.
- Provides a set of optimized generic drawing functions for monochrome, 16-color, 16-bit high color, 256-color, and 24-bit true color graphic devices. It also supports Bézier curves and paths.
- Support for Image Color Matching for better color matching between display and color output devices.

Ethernet (n): something used to catch the etherbunny

The Windows graphics subsystem is included as a universal driver with a 32-bit mini-driver. The mini-driver provides only for the hardware-specific instructions.

The 32-bit Windows 95/98 printing subsystem has several enhancements over Windows 3.x. These include:

- They use a background thread processing to allow for smooth background printing.
- Smooth printing where the operating system only passes data to the printer when it is ready to receive more information.
- They send enhanced metafile (EMF) format files, rather than raw printer data. This EMF information is interpreted in the background and the results are then sent to the printer.
- Support for deferred printing, where a print job can be sent to a printer and then stored until the printer becomes available.
- Support for bi-directional communication protocols for printers using the Extended Communication Port (ECP) printer communication standard. ECP mode allows printers to send messages to the user or to application programs. Typical messages are: 'Paper Jam', 'Out-of-paper', 'Out-of-Memory', 'Toner Low', and so on.
- Plug-and-play.

4.6 Virtual Machine Manager (VMM)

The perfect environment for a program is to run on a stand-alone, dedicated computer, which does not have any interference from any other programs and can have access to any device when it wants. This is the concept of the Virtual Machine. In Microsoft Windows the Virtual Machine Manager (VMM) provides each application with the system resources when it needs them. It creates and maintains the virtual machine environments in which applications and system processes run, as illustrated in Figure 4.12.

The VMM is responsible for three areas:

- **Process scheduling.** This is responsible for scheduling processes, and allows for multiple applications to run concurrently and also for providing system resources to the applications and other processes that run. Process scheduling allows multiple applications and other processes to run concurrently, using either co-operative multitasking or pre-emptive multitasking.
- **Memory paging.** Microsoft Windows uses a demand-paged virtual memory system, which is based on a flat, linear address space accessed using 32-bit addresses. The system allocates each process a unique virtual address space of 4 GB. The upper 2 GB is shared, while the lower 2 GB is private to the application. This virtual address space is divided into equal blocks (or pages).
- **MS-DOS Mode support.** Provides support for MS-DOS-based applications which must have exclusive access to tne hardware. When an MS-DOS-based application runs in this

mode then no other applications or processes are allowed to compete for system resources. The application thus has sole access to the resources.

Windows 95/98 has a single VMM (named System VMM) in which all system processes run. Win32-based and Win16-based applications run within this VMM. Each MS-DOS-based application runs in its own virtual machine.

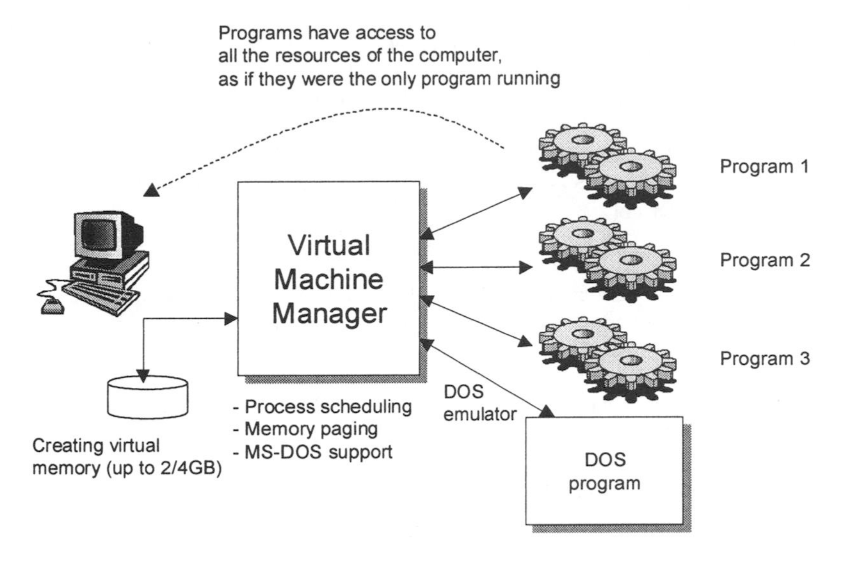

Figure 4.12 Virtual Machine Manager

4.6.1 Process scheduling and multitasking

> Three kinds of people: those who can count and those who can't.

This allows multiple applications and other processes to run concurrently, using either co-operative multitasking or pre-emptive multitasking (Windows 3.x, applications ran using co-operative multitasking). This method requires that applications check the message queue periodically and give up control of the system to other applications. Unfortunately, applications that do not check the message queue at frequent intervals can effectively 'hog' the processor and prevent other applications from running. As this does not provide effective multi-processing, Microsoft Windows uses pre-emptive multitasking for Win32-based applications (but also supports co-operative multitasking for computability reasons). Thus, the operating system takes direct control away from the application tasks.

Win16 programs need to yield to other tasks in order to multitask properly, whereas Win32-based programs do not need to yield to share resources. This is because Win32-based applications (called processes) use multi-threading, which provides for multi-processing. A thread in a program is a unit of code that can get a time slice from the operating system to run concurrently with other code units. Each process consists of one or more execution threads that identify the code path flow as it is run on the operating system. A Win32-based application can have multiple threads for a given process. This enhances the running of an application by improving throughput and responsiveness. It allows processes for smooth background processing.

4.7 Exercises

The following questions are multiple choice. Please select from a–d.

4.7.1 Which of the following operating systems is not multitasking:
(a) Linux (b) UNIX
(c) DOS (d) Windows 95

4.7.2 Which type of operating system allows processes to yield themselves from the processor:
(a) Pre-emptive multitasking (b) Co-operative multitasking
(c) Thread-based multitasking (d) Interrupt multitasking

4.7.3 Which of the following best describes a thread:
(a) A part of a process (b) A part of the hardware
(c) An interrupt (d) A signal

4.7.4 Which program responds to an interrupt:
(a) Interrupt Service Routing (b) Interrupt Control Routine
(c) Interrupt Service Routine (d) Interrupt Vector

4.7.5 How does a PC determine which device has caused a hardware interrupt:
(a) Hardware handshaking (b) Software handshaking
(c) Interval processing (d) IRQ lines

4.7.6 On a PC, which interrupt line is used by the keyboard:
(a) IRQ0 (b) IRQ1
(c) IRQ4 (d) IRQ7

4.7.7 On a PC, which interrupt line is used by the mouse:
(a) IRQ0 (b) IRQ4
(c) IRQ7 (d) IRQ12

4.7.8 On a PC, what file extension does a dynamic link library use:
(a) LIB (b) COM
(c) DYN (d) DLL

4.7.9 On a PC, what file extension does a dynamic link library use:
(a) LIB (b) COM
(c) STA (d) DLL

4.7.10 Differentiate between a low-level interrupt and a message/signal.

4.7.11 Outline the differences between pre-emptive multitasking and co-operative multitasking. Which is the most desirable?

STRANGE, BUT TRUE

A PC customer was asked to send a copy of its faulty diskettes. After a few days, a letter arrived from the customer along with a photocopy image of the floppy disk.

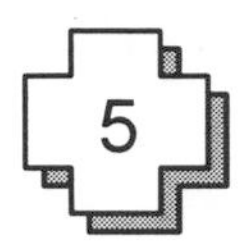

Processes and Scheduling

5.1 Scheduling

Multitasking operating systems can run many programs at the same time. Process schedulers must allow each process some time on the processor. A badly designed scheduler simply allows each of the processes the same period of time, whereas well-designed schedulers allow process priority, and can make decisions on which processes should be run, at a given time.

The scheduler operates on a queue of processes, each of which can either be:

> **STRANGE, BUT TRUE**
>
> A customer phoned a customer support line, and stated that they were not able to send a fax. After many minutes, the support line operator finally traced the fault. The user had been holding the document to the screen and then pressing the send button.

- **Running.** This is where the process is actually currently running on the processor.
- **Waiting.** This is where the process is waiting on another process to run and provide it with some data, or if a process is waiting to access a resource. A waiting process can sometimes turn into a zombie process, where a process terminates for some reason, but whose parent process has not yet waited for it to terminate. A zombie process is not a big problem, as it has no resources allocated to it.
- **Ready.** This is where the process is ready to be run on the processor, and is not waiting for any other processes or has terminated.
- **Terminated.** This is where a process has finished its run, and all resources that have been allocated to it must be taken away from it.

These concepts are illustrated in Figure 5.1. The scheduler thus makes a decision whenever a change occurs, such as:

- Running to waiting.
- Waiting to ready.
- Running to ready.
- Running to terminated.

A preemptive scheduler uses a timer to allow each process some time on the processor and coordinates access to shared data. Along with this, it requires a kernel designed to protect the integrity of its own data structures.

5.1.1 Scheduling queues

There are three main system queues: Job Queue – incoming jobs; Ready Queue and Device Queues (blocked processes). Normally the type of scheduler chosen depends on the type of system that is required, such as:

- **Long-term (Job) scheduler.** This type of scheduler is used in batch systems.
- **Short-term scheduler.** This type of scheduler typically uses a FIFO (First In, First Out) queue, or a priority queue.
- **Medium-term scheduler.** This type of scheduler swaps processes out to improve job mix. Normally it schedules on the following:

- o Time since swapped in or out. Swapping involves moving the running process to some temporary storage space (such as the local disk).
- o Processor time used.
- o Size.
- o Priority.

Processes often require different processing requirements, such as:

- Processor I/O burst cycle. Normally involves a large number of short bursts, along with a number of longer processor bursts.
- Processor bound. Involves long processor bursts.
- I/O bound. Involves short processor bursts.

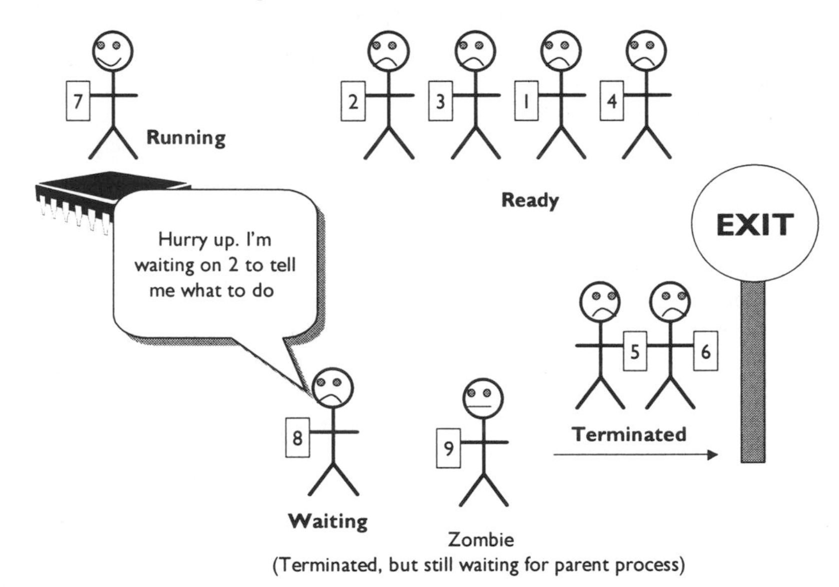

Figure 5.1 Running, ready, waiting and terminated

5.1.2 Scheduling Algorithms

Every scheduler must be fair in the way that it assigns tasks, much of which should be automatic, and should not require user input. The main objectives are:

Objectives	*Description*	*Scheduler must try to:*
Fairness	Each process has a fair share of the processor	Maximize
Efficiency	Efficient use of the processor	Maximize
Throughput	Number of processes completed	Maximize
Turnaround	Time taken for processes to complete	Maximize
Waiting time	Time taken in the queue	Minimize
Predictability	Allowing a dependable response	Maximize
Response time	Time to react to actions	Minimize

The two main classifications for scheduling are:

- Policy. Sets priority on individual processes.
- Mechanism. Implements a scheduling policy.

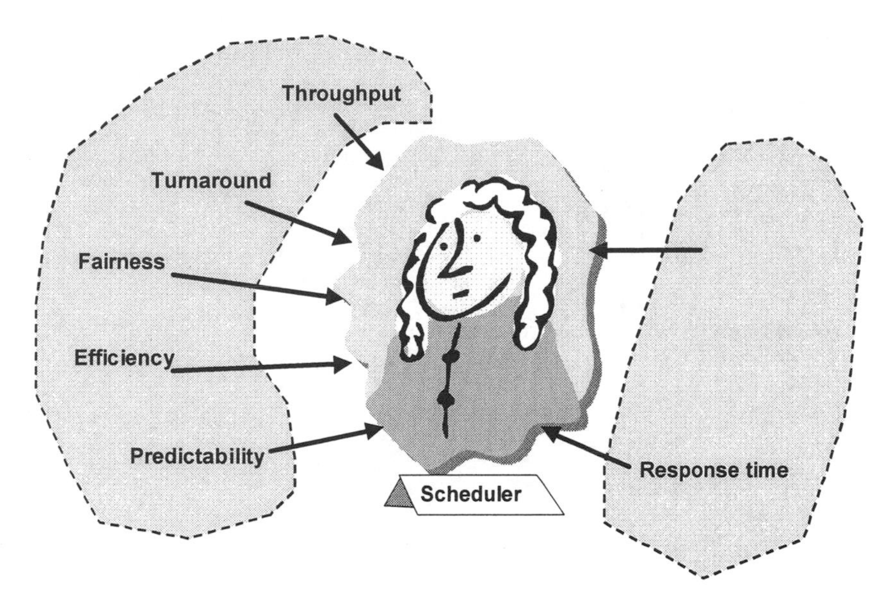

Figure 5.2 Decisions for a scheduler

There are various methods that the scheduler can use to implement a scheduling algorithm:

- **First-Come, First-Served** (FCFS). This type of scheduling is used with non-pre-emptive operating systems, where the time that a process waits in the queue is totally dependent on the processes which are in front of it, as illustrated in Figure 5.3. The response of the system is thus not dependable.
- **Shortest-Job-First** (SJF). This is one of the most efficient algorithms, and involves estimating the amount of time that a process will run for, and taking the process which take the shortest time to complete.
- **Priority Scheduling.** This is the typical used in general-purpose operating systems (Microsoft Windows and UNIX). It can be used with either pre-emptive or non-pre-emptive operating systems. The main problem is to assign a priority to each process, where priorities can either be internal or external. Internal priorities related to measurable system resources, such as time limits, memory requirements, file input/output, and so on. A problem is that some processes might never get the required priority and may never get time on the processor. To overcome this, low-priority waiting processes have their priority increased, over time (known as ageing).
- **Round-Robin** (RR). This is first-come, first-served with pre-emption, where each process is given a finite time slice on the processor. The queue is circular, thus when a process has been run, the queue pointer is moved to the next process, as illustrated in Figure 5.4. This is a relatively smooth schedule and gives all processes a share of the processor.

STRANGE, BUT TRUE

A customer phoned a customer support line, and complained that their computer would not work. All the wires were plugged in the right place, and it had been plugged-in. After many other questions the operator asked the user, 'What happens when you press the power switch', 'What's a power switch?', came the reply.

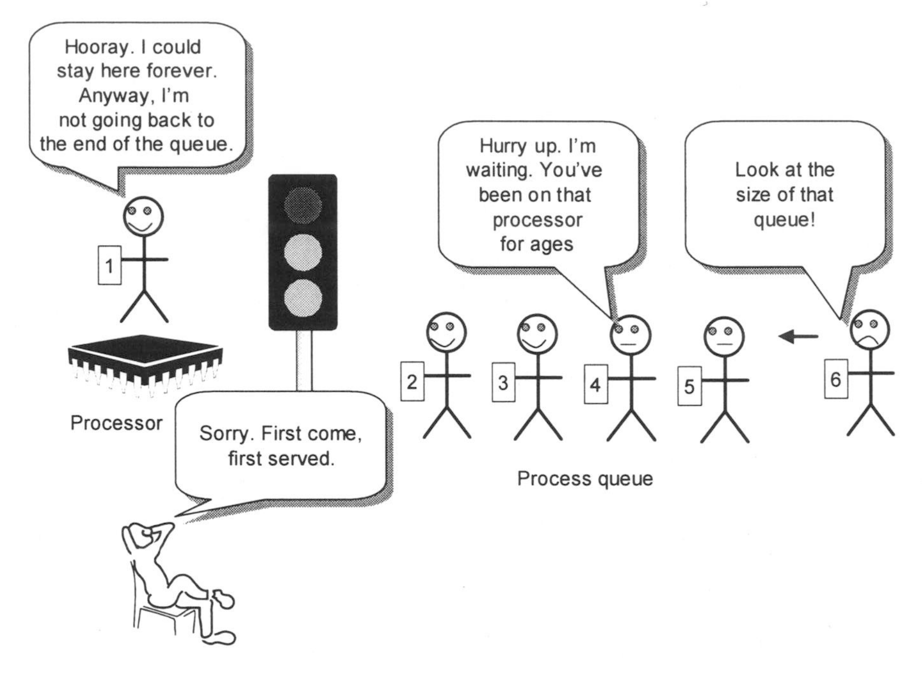

Figure 5.3 First come, first served

Figure 5.4 Round robin

- **Multilevel Queue Scheduling.** This scheme supports several different queues, and sets priorities for them. For example, a system could run two different queues: foreground (interactive) and background (batch), as illustrated in Figure 5.5. The foreground task could be given a much higher priority over the background task, such as 80%–20%. Each of the queues can be assigned different priorities. Windows NT runs a pre-emptive scheme where certain system processes are given a higher priority than other non-system processes. An example priority might be (in order of priority):

1. System processes. Top priority. This must have top priority as the system could act unreliably if they were not executed with a given time.
2. Interactive processes. These are processes which require some user input, such as from the keyboard or mouse. It is important that user must feel that these processes are running with a high priority, otherwise they may try to delete them, and try to rerun the process.
3. Interactive editing processes. These processes tend to run without user input for long periods, but occasionally require some guidance on how they run.
4. Batch processes. Lowest priority. These tend to be less important processes which do not require any user input.

- **Multilevel Feedback Queue Scheduling.** This scheme is the most complex, and allows processes to move between a number of different priority queues. Each queue has an associated scheduling algorithm. To support this there must be a way to promote and relegate processes for their current queues.

Figure 5.5 Multilevel queue scheduling

5.1.3 Multi-processor scheduling

The two main classifications of multiprocessor scheduling are:

- **Heterogeneous.** This is where there are a number of different types of processors, each with their own instruction set. This is typically the case in distributed computing, where a process is run over a number of computers on a network. For it to work there must be a high-level protocol that allows communication between the computers. Java is an excellent programming language for this, as it can produce machine-independent code.
- **Homogeneous.** This is where all the processors can run the same code. In the scheduler there is either a common queue for all processors (and they take processes from the

queue once they have completed a process) or there is a separate queue for each process (this involves a scheduler deciding on what processes should be given to each of the processors). The processing can either be asymmetric multiprocessor where a single processor looks after scheduling, I/O processing and other system activities, or symmetric multiprocessing, where each processor is self-scheduling or has a master-slave structure.

5.1.4 Real-time scheduling

Real-time systems normally have key processes which should be serviced before other processors. The two main classifications are:

- **Hard real-time.** This is where a critical process is completed within a critical time, and cannot be stored in second memory.
- **Soft real-time.** This is where critical processes receive a higher priority than less critical processes. Soft real-time can lead to process starvation, and also an unfair allocation of system resources. This type of real-time system is typically implemented in systems which require high-speed data transfer, especially in multimedia (such as MPEG movies) and high-speed, processor intensive graphics (such as 3D graphics).

5.2 Higher-level primitives

Processes often require to identify that they are waiting for another process to give them data or are busy, waiting for some I/O transfer. They must thus support higher-level primitives. On shared memory systems the following are used:

- **Semaphores.** This involves setting flags, which allow or bar other processes from access certain resources. An analogy of a semaphore is where two railway trains are using a single-track railway line. When one train enters the single-track line, it sets a semaphore which disallows the other train from entering the track. Once the train on the single track has left the single track, it resets the semaphore flag, which allows the other train to enter the single track.
- **Signals.** Signals are similar to interrupts, but are implemented in software, rather than hardware. This is a primitive interrupt handler and involves a signal handler which controls process signals.

On non-shared memory systems the following are used:

- Message passing. Messages are sent between processes, such as SEND and RECEIVE.
- Pipes. Pipes allow data to flow from one process to another, in the required way.
- Remote procedure calls (RPC).

5.3 Signals, pipes and task switching (UNIX-implementation)

UNIX does not implement a sharing system. In a sharing system, like Microsoft Windows and MAC OS, the operating system only changes to a different process when the current process identifies that it is ready to swap tasks. UNIX implements its scheduling using signals and pipes.

REAL STUDENT ANSWER

Gravity was invented by Isaac Walton. It is chiefly noticeable in the autumn when the apples are falling off the trees.

5.3.1 Signals

UNIX uses signals in a similar way to interrupts, but they are implemented at a higher-level. Events rather than hardware devices generate these interrupts. Typically, software interrupt service routines are called on certain signals. The signal handler sets the status of a process, but a process may also put itself in a sleep mode, waiting for a signal. One problem with this is that signals may get lost if they are sent before the process goes into a waiting mode. One solution is to set a flag in the process whenever it receives a signal. The process can then test this flag before it goes into the wait state, if it is set; the wait operation does not block the process.

REAL STUDENT ANSWER

The sun never set on the British Empire because the British Empire is in the East and the sun sets in the West.

5.3.2 Pipes

Pipes allow data to flow from one process to another, in the required way. Typically they are implemented with a fixed size storage area (a buffer) in which one process can write to it, while the other reads from it (when the data is available). UNIX implements pipes with a file-like approach, and uses the same system calls to write data to a pipe and read data from a pipe as those for reading and writing files. Each process which creates a pipe receives two identifiers: one for the reading and one for the writing. Typically, the creating process forks-off two child processes, one of which looks after one end of the pipe, and the other on the other end of the pipe. The two child processes can then communicate.

Figure 5.6 shows an example of a pipe. Pipes can also be implemented on two remote computers. This is normally defined as a connection, and the ends of the pipes are known as sockets.

5.3.3 Task switching

The dispatcher receives orders from the scheduler as to the processes that are to be run. It is part of the kernel and has privileges on process information. One of its main tasks is to extract information on the previous state of the process, such as processor registers, stack pointers, and so on. It then switches the processor from kernel mode to user mode (basically enabling the hardware memory protection), and the process begins to run.

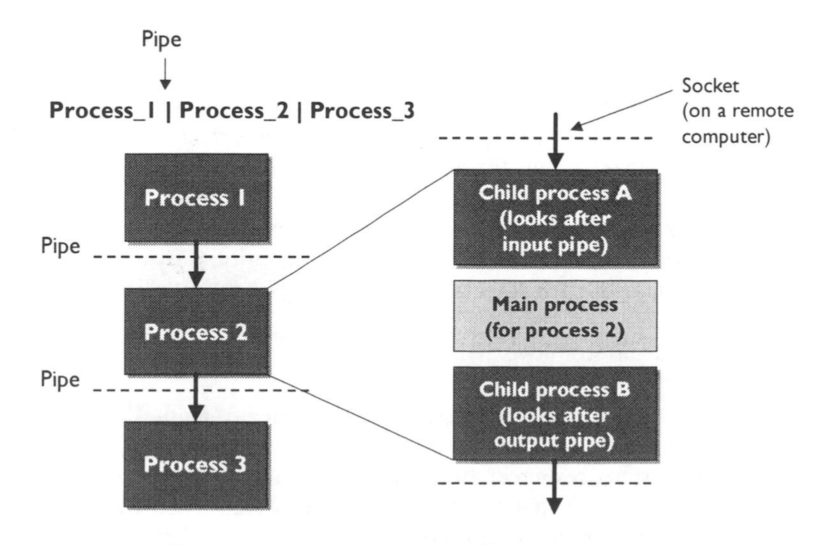

Figure 5.6 Pipes

5.4 Messages

The best method of interprocess communication is messages. These allow information on the actual process to be passed between processes. These messages can be of a fixed length, but are most generally of any length, and typically are unstructured. This is the method that Microsoft Windows uses to pass data between processes.

In a message system, each process communicates with a port (or message port), which is a data structure in which messages are sent to. Most systems have a single port, but others can have several message ports. In most cases, the system implements two system calls: SEND and RECEIVE.

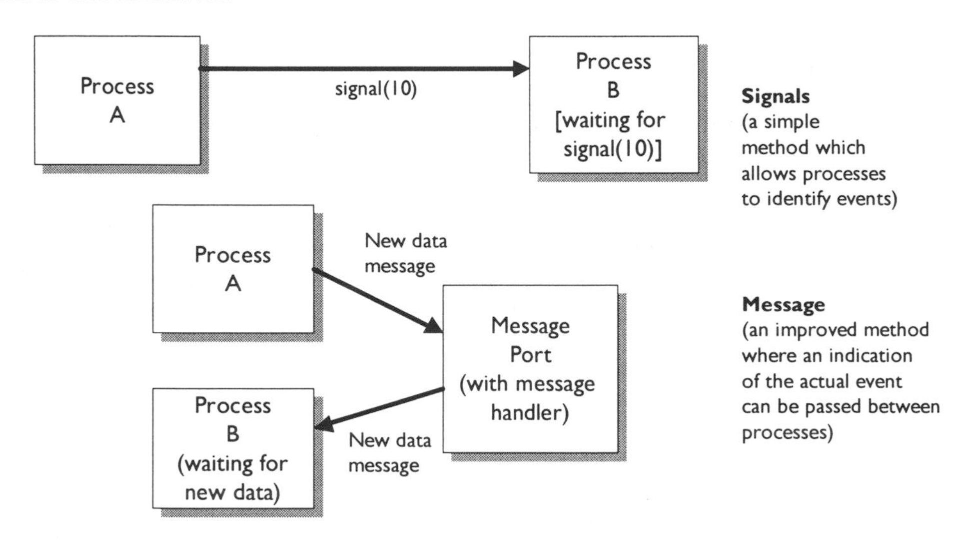

Figure 5.7 Message passing

5.5 Microsoft Windows scheduling

Scheduling involves determining which thread should be run on the processor at a given time. This element is named a time slice, and its actual value depends on the system configuration.

Each thread currently running has a base priority, which is set by the programmer who created the program. It defines how the thread is executed in relation to other system threads, and the thread with the highest priority gets use of the processor.

Microsoft Windows has 32 priority levels. The lowest priority is 0 and the highest is 31. A scheduler can change a threads base priority by increasing or decreasing it by two levels, thus changing the thread's priority.

The scheduler is made up from two main parts:

> **REAL STUDENT ANSWER**
>
> The nineteenth century was a time of a great many thoughts and inventions. People stopped reproducing by hand and started reproducing by machine.

- **Primary scheduler.** This scheduler determines the priority numbers of the threads which are currently running. It then compares their priority and assigns resources to them, depending on their priority. Threads with the highest priority are executed for the current time slice. When two or more threads have the same priority then the threads are put on a stack. One thread is run and then put to the bottom of the stack, then the next is run and it is put to the bottom, and so on. This continues until all threads with the same priority have been run for a given time slice.

- **Secondary scheduler.** The primary scheduler runs threads with the highest priority, whereas the secondary scheduler is responsible for increasing the priority of non-executing threads (which are all other threads apart from the currently executed thread). It is thus important for giving low priority threads a chance to run on the operating system. Threads which are given a higher or lower priority are:
  - A thread which is waiting for user input has its priority increased.
  - A thread that has completed a voluntary wait also has its priority increased.
  - Threads with a computation-bound thread get their priorities reduced. This prevents the blocking of I/O operations.

Apart from these, all threads get a periodic increase. This prevents lower-priority threads hogging shared resources that are required by higher-priority threads.

5.6 Windows NT Task Manager

Windows NT uses the Task Manager to show the currently running processes, and, if required, stopping them. It can be called by pressing Ctrl-Alt-Del and then selecting Task Manager. Figure 5.8 shows an example of some processes running. The window icon with gray indicates a program, and a window icon with white indicates a status window. The open file icon indicates an open folder.

The processes window (as shown in Figure 5.8) gives an indication of:

- **Image name.** Name of the process.
- **PID.** Process Identification.
- **CPU Time.** Total time that the process has used the processor.
- **Memory usage.** Total memory usage..

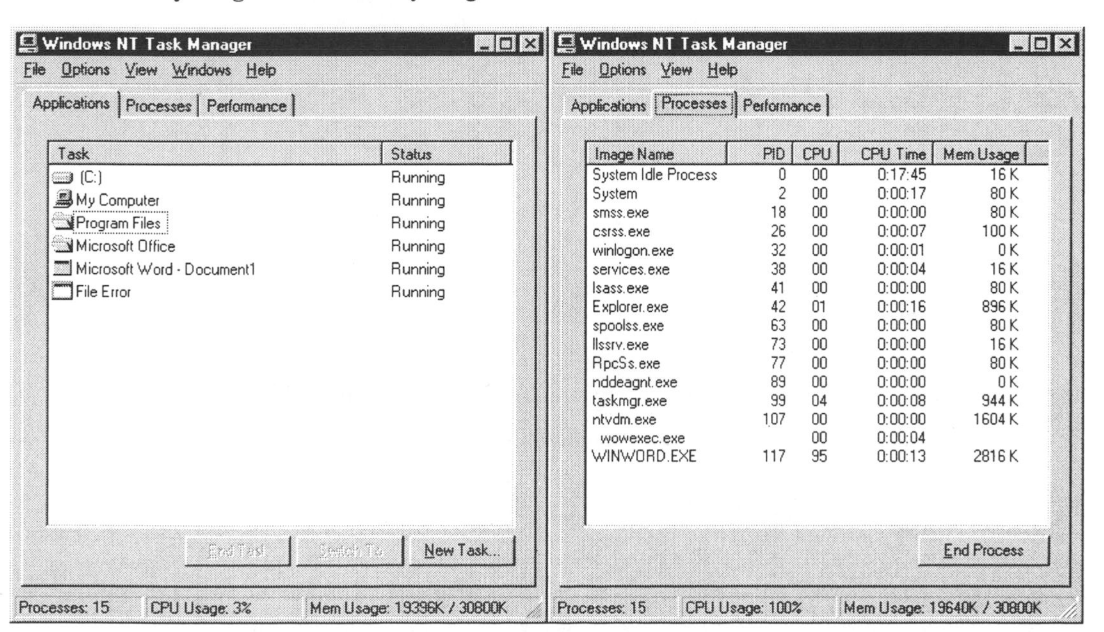

Figure 5.8 Applications and Processes

5.7 UNIX process control

REAL STUDENT ANSWER

Handel was half German half Italian and half English. He was very large.

UNIX is a multitasking, multi-user operating system, where many tasks can be running at any given time. Typically there are several processes which are started when the computer is rebooted; these are named daemon processes and they run even when there is no user logged into the system, as illustrated in Figure 5.9. Only the system administrator can kill these processes.

UNIX uses special characters (called metacharacters) to define how a process runs, these are:

- **Redirect output.** The '>' operator (greater-than sign) redirects the output from the standard output (normally the user's screen) to another output, such as to a file.
- **Redirect input.** The '<' operator (less-than sign) redirects the input from the standard input (normally the keyboard) to another input, such as from a file.
- **Background task.** The '&' operator (ampersand sign) sends a process into the background.
- **Pipe.** The '|' character (vertical bar) is used to pipe data from one process into another.

Some of these operators will be used in the following UNIX sessions.

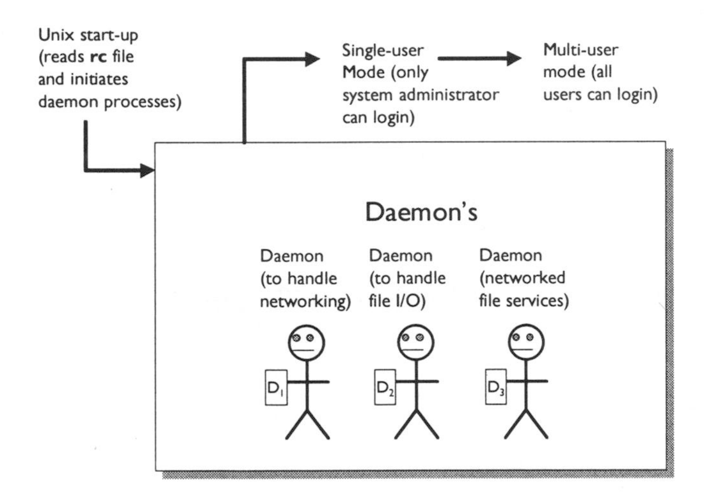

Figure 5.9 Daemon processes

5.7.1 ps (process status)

The `ps` command prints information about the current process status. The basic `ps` list gives a list of the current jobs of the user. An example is given in Sample session 5.1.

Sample session 5.1

```
% ps
  PID  TTY  TIME  CMD
   43   01  0:15  csh
   51   01  0:03  ls -R /
  100   01  0:01  ps
```

The information provided gives:

- **PID.** The unique process identification number.
- **TTY.** Every connected computer is identified with a unique name. The TTY name identifies the place at which the process was started.
- **TIME.** Identifies the amount of CPU time that has been used. Typically the format is HH:MM (for min:sec).
- **CMD.** Identifies the actual command line that was used to run the process.

REAL STUDENT ANSWER

Johann Bach wrote a great many musical compositions and had a large number of children. In between he practiced on an old spinster which he kept up in his attic.

A process can be stopped using the `kill` command.

A long listing is achieved using the `-l` option and for a complete listing of all processes on the system the `-a` option is used, as shown in Sample session 5.2.

Sample session 5.2

```
% ps -al
F S UID PID PPID CPU PRI NICE ADDR SZ WCHAN TTY TIME CMD
1 S 101 43    1   3  30   20   3211 12 33400 01  0:15 csh
1 S 104 44    2   2  27   20   4430 12 51400 04  0:08 sh
1 S 104 76   32   3  30   20   3223 12 33400 04  0:03 vi tmp
1 S 104 89    1   3  30   20  10324 02 44103 04  0:01 ls
1 R 101 99   55  43  52   20   4432 12 33423 01  0:01 ps
```

The main additional columns are:

- **UID.** This is used to identify the process owner (User ID), which is generated from the value for the user in the `/etc/passwd` file.
- **PRI.** This is used to define the priority of the process, where the higher the number the lower the priority it has.
- **F.** This identifies the flags that are associated with the process (0 – swapped, 1 – in core, 2 – system process, and so on).
- **STIME.** Start time for the process (the date is printed instead, if the process has been running for more than 24 hours).
- **ADDR.** Memory address of the process.
- **NI.** Nice value, which is used to determine process priority. See Section 5.7.3.
- **S.** This identifies the state of the process. An S identifies that the process is sleeping (the system is doing something else); W specifies that the system is waiting for another process to stop and R specifies that the process is currently running. In summary:
  - **R.** Process is running.
  - **T.** Process has stopped.
  - **D.** Process is in disk wait.
  - **S.** Process is sleeping (that is, less than 20 secs).
  - **I.** Process is idle (that is, longer than 20 secs).

5.7.2 kill (send a signal to a process, or terminate a process)

The `kill` command sends a terminate signal to a process. The general format is:

`kill` *–sig processid*

The *processid* is the number given to the process by the computer, which can be found by using the ps command. The *sig* value defines the amount of strength that is given to the kill process. A value of 9 is the strongest value, others are: 1 (hang up); 2 (interrupt); 3 (quit); 4 (alarm); 5 (terminate) and 6 (abort). The owner of a process can kill his own, but only the system administrator can kill any process. Sample session 5.3 gives an example session.

STRANGE, BUT TRUE

A support line operator asked a customer to put the floppy disk back in the drive and close the door. After a short delay and the padding of footsteps, the customer came back to the phone and told the operator that the floppy was back in the drive and the door was now closed.

Sample session 5.3

```
% ps
   PID  TTY  TIME  CMD
   112  01   1:15  csh
   145  01   0:05  lpr temp.c
   146  01   0:01  ps
% kill -9 145
% ps
   PID  TTY  TIME  CMD
   112  01   1:15  csh
   146  01   0:01  ps
% find / -name "*.c" -print > listing &
% ps
   PID  TTY  TIME  CMD
   112  01   1:15  csh
   177  01   0:03  find -nam
   179  01   0:01  ps
% kill -9 177
```

5.7.3 nice (run a command at a low priority)

The nice command runs a command at a low priority. The standard format is as follows:

nice *–number command* [*arguments*]

The lowest priority is –20 and the default is –10. Sample session 5.4 gives a sample session.

Sample session 5.4

```
% nice -15 ls -al
% nice -20 find / -name "*.c" -print > Clistings &
```

5.7.4 at (execute commands at later date)

The at command when used in conjunction with another command will execute a command at some later time. The standard format is:

at *time* [*date*] [*week*]

where *time* is given using from 1 to 4 digits, followed by either 'a', 'p', 'n' or 'm' for am, pm, noon or midnight, respectively. If no letters are given then a 24-hour clock is assumed. A 1- or 2-digit time is assumed to be given in hours, whereas a 3- or 4-digit time is assumed to be hours and minutes. A colon may also be included to separate the hours from the minutes.

The *Date* can be specified by the month followed by the day-of-the-month number, such as Mar 31. A *Week* can be given instead of the day and month. Session 5.4 shows a session where a program is compiled at quarter past eight at night.

Sample session 5.5

```
% at 20:15
   cc - test test.c
   ^D
   520776201.a at Tue May 26 20:15:00 1997
```

REAL STUDENT ANSWER

Louis Pasteur discovered a cure for rabbis.

and to send `fred` a message at 14:00:

Sample session 5.6

```
% at 14:00
   echo "Time for a tea-break" | mail fred
   ^D
   520777201.a at Mon Jun 4 14:00:00 1989
```

To remove all files with the .o extensions from the current directory on September 9th at 1 noon.

Sample session 5.7

```
% at 1n sep 9
  rm *.o
  ^D
  520778201.a at Sat Sep 9 13:00:00 1989
```

To list all jobs that are waiting to be executed at some later time use the `-l` option.

Sample session 5.8

```
% at -l
  520776201.a   Mon Jun 4 20:15:00 1989
  520778201.a   Sat Sep 9 13:00:00 1989
  520777201.a   Mon Jun 4 14:00:00 1989
```

To remove jobs from the schedule the `-r` option can be used, giving the job number.

Sample session 5.9

```
  % at -r 520777201.a
```

5.7.5 Example daemons

See Section 7.4.4.

5.8 Finite-state machines

Finite-state machines (FSM) are at the heart of most computer systems. They define the system as a finite number of states, each of which is linked by a series of events. Often complex systems can be easily modeled in this way. Figure 5.10 shows an example of a FSM for a traffic light controller. The system is started in State 1 (Red light ON, and Don't Walk ON, which is a safe starting state). After this, the system then goes from State 2 to State 3 and to State 4 (with a finite time delay between each state). When leaving State 4, the system goes back to either State 1 if the pedestrian button has not been pressed, or State 5, if it has been pressed. If the system goes into State 5, the traffic light goes to RED and the Don't Walk is still ON. Next, in State 6, the pedestrian light goes to Walk, and so on. Unfortunately, there is no state for a safe start-up (traffic lights OFF) and a shutdown state. Figure 5.11 overcomes this, with a safe shutdown from State 1. Only in State 1 can the system be shutdown (as it is unsafe to shut it down in any other state).

REAL STUDENT ANSWER

Milton wrote *Paradise Lost.* Then his wife died and he wrote *Paradise Regained.*

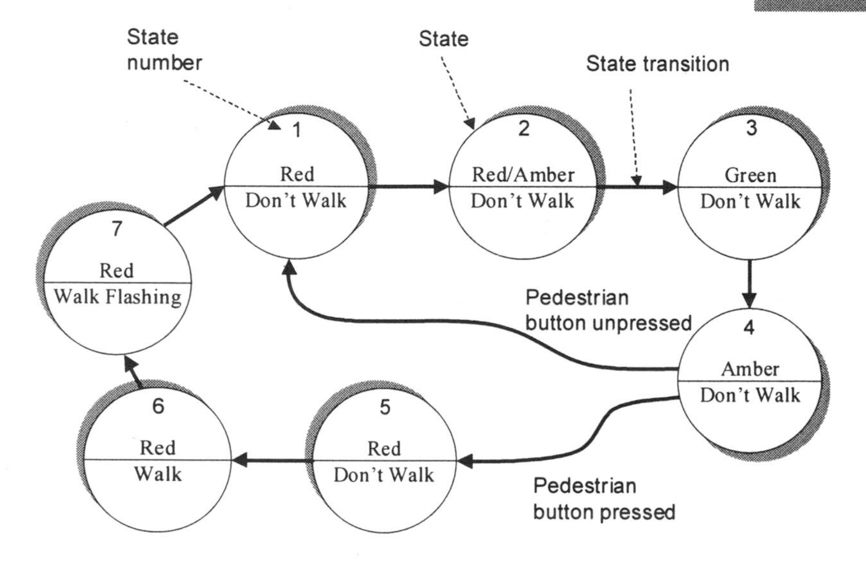

Figure 5.10 State transition for a traffic light controller

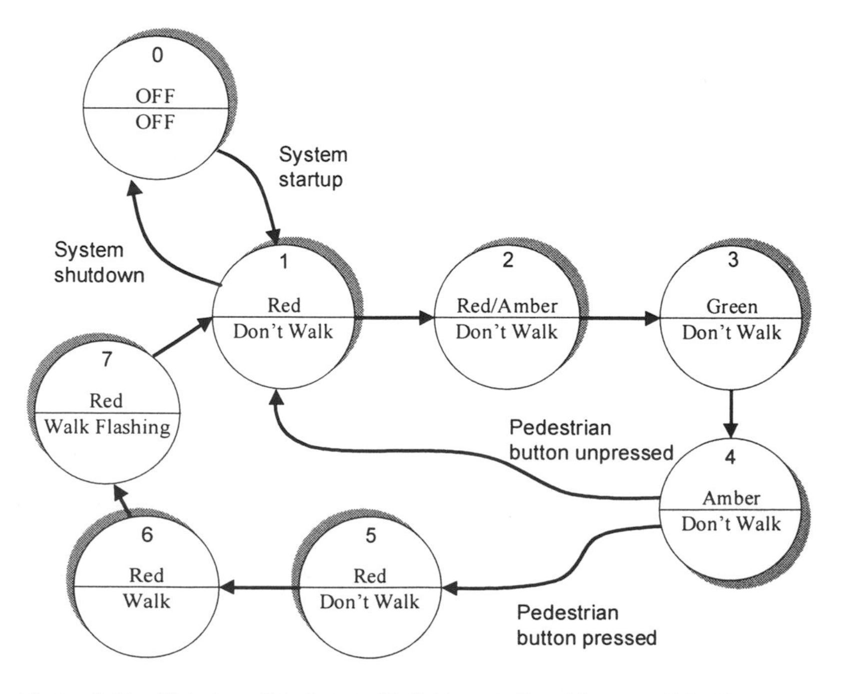

Figure 5.11 State transition for a traffic light controller with start-up/shutdown state

5.9 Exercises

STRANGE, BUT TRUE

Some computer companies are considering changing the message: 'Press any key to continue' as some users phone up their support lines to ask where the 'Any' key is.

The following questions are multiple choice. Please select from a–d.

5.9.1 Which of the following is not a process state:
(a) Running (b) Waiting
(c) Synchronizing (d) Ready

5.9.2 Which is the UNIX metacharacter for running a process in the background:
(a) | (b) &
(c) < (d) >

5.9.3 Which type of scheduler is used for batch systems:
(a) Long-term scheduler (b) Short-term scheduler
(c) Medium-term scheduler (d) Swapping scheduler

5.9.4 Which of the following best describes a zombie process:
(a) A terminated process, but still waiting for parent process
(b) A virus process
(c) A process that cannot be killed
(d) A process that is hogging the processor

5.9.5 Which of the following is not a major objective of a scheduler:
(a) Fairness (b) Throughput
(c) Predictability (d) Allowing user intervention

5.9.6 Which scheduling algorithm has a circular process queue:
(a) Shortest-job-first (b) Round-robin
(c) Multilevel queue (d) First-come, first-served

5.9.7 Which of the following is used in a non-pre-emptive operating system:
(a) Shortest-job-first (b) Round-robin
(c) Multilevel queue (d) First-come, first-served

5.9.8 Which is the UNIX metacharacter for redirecting the input to a program:
(a) | (b) &
(c) < (d) >

5.9.9 Which of the following is the fairest to all the running processes:
(a) Shortest-job-first (b) Round-robin
(c) Multilevel queue (d) First-come, first-served

5.9.10 Which of the following is the best for allowing key system processes to run smoothly:
(a) Shortest-job-first (b) Round-robin
(c) Multilevel queue (d) First-come, first-served

5.9.11 Which is the UNIX command which halts a process:
(a) kill (b) end
(c) delete (d) stop

5.9.12 Which is the UNIX command that allows a user to reduce the priority of a process:
(a) less (b) kind
(c) polite (d) nice

5.9.13 What is the term used in a multiprocessor system which has the same types of processors:
(a) Uniform (b) Homogeneous
(c) Collective (d) Heterogeneous

5.9.14 Which is the UNIX metacharacter for sending the output of data from one process to another:
(a) | (b) &
(c) < (d) >

5.9.15 How are UNIX processes uniquely identified:
(a) Process ID (b) Process name
(c) Process time (d) Process command

5.9.16 How is the user identified with a process:
(a) User ID (b) User name
(c) Workstation address (d) Terminal name

5.9.17 Which is the UNIX command which allows a user to run a process at a given time:
(a) by (b) at
(c) on (d) time

5.9.18 If UNIX had the user ID of a process, where would UNIX look to find the actual name of the user:
(a) /etc/group (b) /etc/password
(c) /etc/users (d) /etc/names

5.9.19 In UNIX, what are system processes called that run in the background and stay active, even when no users are logged in:
(a) Daemons (b) Batch programs
(c) Monitors (d) Vectors

5.9.20 What is the term used in a multiprocessor system which has different types of processors:
(a) Uniform (b) Homogeneous
(c) Collective (d) Heterogeneous

5.9.21 Outline the main scheduling algorithms, and discuss their advantages and drawbacks.

5.9.22 Explain the main differences between a signal and message. Which is the best, and why?

5.9.23 If possible, log into a UNIX-based computer and determine the processes that are running.

5.10 Note from the Author

Well. There you go. A brief introduction to processes and scheduling. It's amazing to think that UNIX has survived over the years, even after it was tipped to takeover the operating system market from DOS in the mid-1980s. So what happened? Why has it survived? Why don't more people use it? Well it's survived because it's totally robust and reliable. It's well trusted and relatively secure. It has always supported networking. And... So, what's the problem? Well its big problem has always been that it requires a skilled computer administrator to set it up and keep it running. There is no way that most home users, or small businesses could support this level of support. Another problem is that it is very difficult to recruit, and keep, good UNIX administrators. I have seen this first hand, as I used to be responsible for an Electronic Computer Aided Design (ECAD) network. It ran extremely well and was based on Apollo and HP workstations. Unfortunately we started to lose our system

administrators as they were offered much higher salaries than my university could afford. Soon we were left with no properly trained UNIX administrators. For a while the system ran reasonably well and was patched when something went wrong, but it was not properly maintained, and there were no backups. Soon the on-line CD manuals became unavailable, next printer queue kept failing, next some of the computers failed to log users in, and so on, until eventually there was a lightning strike which sent an electrical spike through the computers. Unfortunately it blew up one of the disk drives. The network never really recovered from this, as the damaged drive was the one that contained most of the ECAD software, which was so complicated to setup it was almost impossible to recover its original state. From that day forward my department vowed to move its ECAD software towards PCs, as they were so much easier to setup and administrate. The students could even install software on their computers at home (which in those days were lumbering 80386-computers, with 1 MB of memory, and a VGA monitor).

So ask anyone who has used UNIX and they will tell you that it tends to be much more reliable than a PC system using Microsoft Windows. So why is this? Well it's probably because UNIX machines tend not to be based on legacy type systems, and use peripherals which have robust interfaces, and cost is normally not a major factor.

So why do more users not adopt UNIX? Well, until recently, it was still very much text command based, where users must enter text commands at a user prompt (just as DOS did). Most technically trained users actually prefer this type of mechanism to run commands, but home users can never remember the required command, or the options to use with it. Thus UNIX has always been seen as a 'techie' operating system, as it allows users to carefully control and monitor the operation of the system. By the way, the word 'techie' is my least favorite word, and I think it is totally disrespectful to people who have a deep understanding of technology. I've seen messages such as:

'The e-mail system is very easy to use and you can press the mouse key to read your messages if you want. For you techies, it is based on a POP-3 server.'

'We have changed the cables in the campus (for the techies, they use Cat-5 cable).'

It's as if people with a technical knowledge are some alien force who would take over the planet if they had half a chance. Maybe we should, just for the fun of it. Where would the Company Directors be if the 'techies' brought down their entire IT infrastructure, or where would Stock Market Dealers be if the 'techies' crashed the stock market computer (or even failed to back up the data at regular intervals), or where would the Government be if they did not have computer systems which kept track of taxes, and made payments. So, need it go on? I don't think so. With the Internet, electronic mail, data communications, and so on, it is really the 'techies' who have the power. So the next time you see an e-mail with the word 'techie' in it, immediately put it in your recycle bin, and then trample on it a few times. In fact, just setup your electronic mail system to automatically delete any messages with the word in it.

Okay, I'm sorry. Back to UNIX. So, in the face of the all-powerful Microsoft Windows, what has saved UNIX from an early grave? Well apart from Sun Microsystems, it must be Linux, which guides the user through the steps of setting up the operating system. With a basic PC, you can end up with a WWW server, an FTP server, a TELNET server, an electronic mail server, a domain name server, and so on. But, its big problem is that it doesn't have the same support for peripheral devices as Microsoft Windows has. I've had to field lots of questions from students who could not setup their networking card or their video adaptor to properly install (me too!). My advice was always: 'use an older version of the device, as it's more likely to be supported'. But some people love all these problems. Microsoft Windows is really like buying a video with a big button that says PLAY, TUNE STATIONS, FORWARD, REWIND and another that says RECORD. For most people this is all the functionality they require. Others would like to be able to change the way that the video recorder operates, such as having the following buttons: PAUSE, FAST PLAY, MANUAL STATION TUNING, MONITOR SIGNAL STRENGTH, and so on.

So where would we be without UNIX? Well I don't think that the Internet would have even existed without it. UNIX carefully allowed TCP and IP to grow, adding all the other services that they required. And where did WWW browsers come from? And distributed processing and distributed file systems? Oh, and electronic mail? And TELNET? And FTP? Well, Microsoft Windows depends on the support of the dollars that it generates which go to fund software developers and vendors to add support for new device drivers. Linux, though, depends on individuals who, for the love of computing and the belief that there should be an alternative to Microsoft Windows, decide that they would like to develop a driver for a certain device. This type of system will always lag behind a commercial system, as a developer who sits in a research laboratory on a big salary and has eight hours every day to spend on it will, on average, produce better software than a developer who sets up a lab in a little room in their own home, who has coffee stains on the keyboard and modem cable that is slung around the doorway (or is it?).

6 Distributed Processing

> **CLASSIC SOFTWARE**
>
> **1. VisiCalc** [Original authors: Dan Bricklin, Bob Frankston]. VisiCalc is the classic case of a software package that helped sell a computer. It was released in 1979, and helped the Apple II to become quickly adopted. VisiCalc was the first ever spreadsheet, and preceded Lotus 1-2-3 by three years.

6.1 Introduction

Most modern operating systems now run multiprocesses. These processes can either run locally within a computer system, or can be run over a network, such as in a distributed system. When several processes run at the same time, there must be some mechanism for them to intercommunicate and pass information. Another requirement is when processes share the same resource. This tends to be reasonably easy when the resource can be shareable, but problems can occur when the resource must be dedicated to one process at a time. This type of situation can lead to deadlock where resources, which are dedicated to processes, do not yield to other processes which are waiting on them. One of the most common mechanisms for running remote processes over a network is RPC (Remote Procedure Call) on which a server waits for a request from a client. When a request is received it runs the process and returns back the results to the client, as illustrated in Figure 6.1.

Some processes can be distributed over a network, while others require to be run locally. The main criteria for determining if a process can be distributed are the communications overhead. If the communications channel is relatively slow compared with speed of processing the task, the distribution of the processing can be inefficient. One great advantage of distributing processes is when processing moves from a server to a client. This allows the server to perform high-level operations, while the client does most of the processing. An example of distributing a process is when a user runs a Word Processor from a server. The files that are executed reside on the server, but the actual running of the program occurs on the client.

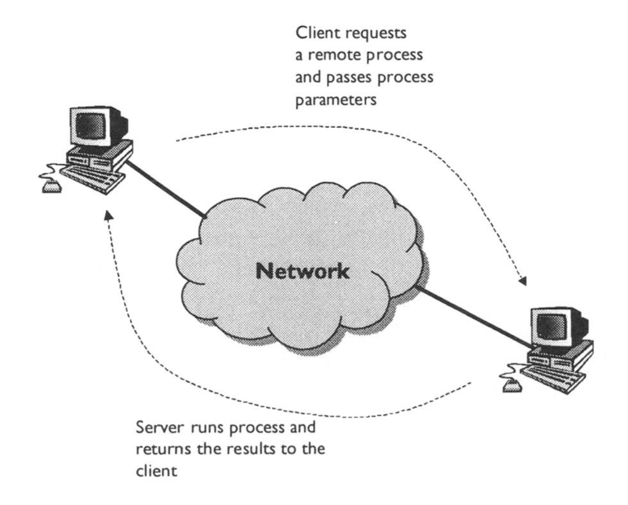
Client requests
a remote process
and passes process
parameters
Network
Server runs process and
returns the results to the
client

Figure 6.1 Distributed processing

6.2 Interprocess communication

Interprocess communication (IPC) is a set of interfaces that allow programmers to communicate between processes, and allow programs to run concurrently. Figure 6.2 illustrates some of the methods, these include:

- **Pipes.** Pipes allow data to flow from one process to another. Data only flows in the one direction, typically from the output of one process to the input of another. The data from the output is buffered until the input process receives it. In UNIX the single vertical bar character (|) is used to represent a pipe, and operates in a similar way to a pipe system call in a program. Two-way communication can be constructed with two pipes, one for each direction. Pipes must have a common process origin.
- **Named pipe.** A named pipe uses a pipe which has a specific name for the pipe. Unlike unnamed pipes, a named pipe can be used in processes that do not have a shared common process origin. Typically a named pipe is known as a FIFO (first in, first out), as the data written to a pipe is read first.
- **Message queuing.** Message queues allow processes to pass messages between themselves, using either a single message queue or several message queues. The system kernel manages each message queue, and puts messages on the queue which gives the identity of the message (message type). Messages can vary in length and be assigned different types or usages. A message queue can be created by one process and used by multiple processes that read and/or write messages to the queue. Application programs (or their processes) create message queues and send and receive messages using an application program interface (API).
- **Semaphores.** These will be discussed in this chapter, and are used to synchronize events between processes. They are integer values which are greater than or equal to zero. A zero value puts a process to sleep, and a non-zero value causes a sleeping process to awaken (when certain operations are performed). A signal is sent to a semaphore when an event occurs which then increments the semaphore.
- **Shared memory.** Shared memory allows processes to interchange data through a defined area of memory. For example, one process could write to an area of memory and another could read from it. To do this the writing process must check to see if the read process is reading from the memory, at the time, and vice-versa. If these are occurring the other process must wait for the other process to complete. This is implemented using semaphores, where only one process is allowed to access the memory at a time.
- **Sockets.** These are typically used to communicate over a network, between a client and a server (although peer-to-peer connections are also possible). Sockets are end points of a connection, and allow for a standard connection which is computer and operating system independent.

CLASSIC SOFTWARE

2. WordStar [Original author: Seymour Rubinstein]. For a time in the 1980s, WordStar was the standard word processing package. It was released in 1979, and quickly became the standard word processor. Often other word processors imitated WordStar's keystrokes, such as Cntrl-Q-F to find a piece of text.

CLASSIC SOFTWARE

3. dBase II [Original author: Wayne Ratliff]. Initially written for CP/M, dBase II was a massive success in its DOS version. It was bought by Aston-Tate, but has since lost its large market share, especially due to the bug-ridden dBase IV (1988) and their slowness to develop a Windows version (1994).

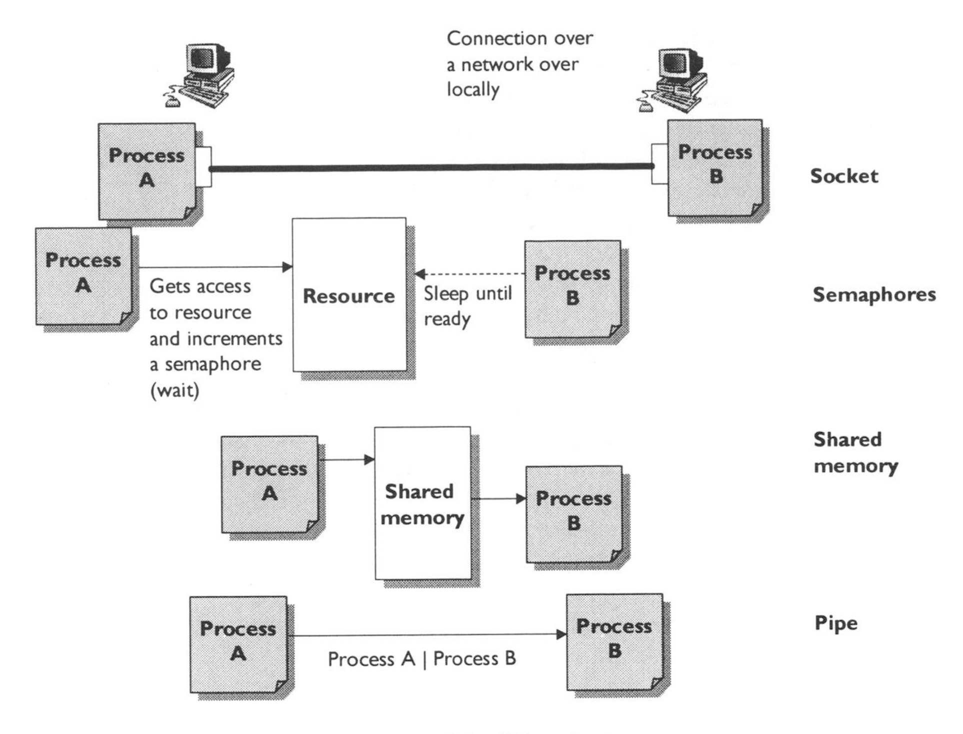

Figure 6.2 IPC methods

6.3 Flags and semaphores

Flags are simple variables which take on a binary state (0 or a 1), and are used to identify that an event has occurred, or to pass binary information. Semaphores are positive integer values which can take on any range of values, but can also be binary values (for mutual exclusion applications). An example of using a semaphore is when a process uses a resource, and sets a semaphore flag to indicate that it is currently accessing the resource. Any device, which then accesses the resource while the semaphore flag was set, will know that the resource is still being used, and must thus wait until the flag is unset. Semaphores were initially developed by Dijkstra and are implemented in IPC. Two common uses of semaphores are:

> **CLASSIC SOFTWARE**
>
> **4. AutoCAD** [Developer: Autodesk]. Initially written for CP/M, and ported to DOS and Windows. It allowed designers to access powerful mechanical design software, which was previously only available on mini and mainframe computers (1994).

- Memory. This allows processes to share a common area of memory.
- File access. This allows processes to share access to files.

A semaphore could simply be a memory location in which processes go to and test the value. If the semaphore is set to a given value, a process may have to sleep until the value is changed to a value which allows orderly access to the resource. Semaphores can also be used to define *critical code regions*, which are parts of a process which must wait for code in another process to implement. A practical example of this is when we imagine two trains (two processes) approaching a region of single-track rail, where only one train can gain access to the single-track line, at a time. The trains must then have some way of signaling to the other train that it is now on the track. A signal (the semaphore) could be setup so that it set either a red or a green light at the entrance of the track. When a train enters the track the signal will change to red, and stop any other trains from entering the track. When it has left the track the signal will be set to green, and the other train can enter the track. In software terms, the trains could be processes which require exclusive access to a resource. When one of the processes gets access to it, all other processes must wait for the resource to be released.

CLASSIC SOFTWARE

5. Lotus 1-2-3 [Developer: Lotus]. Lotus 1-2-3 became the best selling application of all time. It considerably enhanced the VisiCalc program, and was written specifically for the IBM PC.

Most operating systems run more than one process at a time (using time multiplexing). This causes many problems, especially in synchronizing activities and allowing multiple processes access to the resources of the computer. One method of overcoming this problem is semaphores, which are operating system variables that each process can check and change, if required. They are basically a counter value which can be zero or positive, but never negative. There are only two operations on the semaphore:

- UP (signal). Increments the semaphore value. It is named a signal operation, as it is used to wake up a process which is waiting on the semaphore.
- DOWN (wait). Decrements the semaphore value. If the counter is zero there is no decrement. Processes are blocked until the counter is greater than zero.

Figure 6.3 shows an example of two processes running with mutually exclusive code. Each piece of protected code is surrounded with a `wait()` at the start and a `signal()` at the end. The `wait()` operation decrements the semaphore value (which has an initial value of 1), and the `signal()` operation increments the semaphore value. Process A is the first process to execute the mutually exclusive code, and decrements the semaphore so that it is zero. When Process B tries the wait, it tests the semaphore, and since it is zero, the process will go into sleep mode. It will not waken until Process A has executed the `signal()` operation. When Process B is awoken it executes the `wait()` which sets the semaphore to 1, thus Process A cannot execute the mutually exclusive code. When finished, Process B will set the semaphore back to a 1 with the `signal()`.

6.3.1 Semaphore values

A signal is like a software interrupt, and can be viewed as a flag, as it does not give any indication on the number of events that have occurred, it is only possible to know that it has occurred. Whereas, a semaphore can be regarded as a generalized signal, which has an integer counter to record the number signals that have occurred. Processes can put themselves to sleep while waiting for a signal.

Semaphores are operated on by signal and wait operations. A wait operation decrements the value of a semaphore and a signal operation increments it. The initial value of a semaphore identifies the number of waits which may be performed on the semaphore. Thus:

CLASSIC SOFTWARE

6. The Norton Utilities [Original author: Peter Norton]. Early computers were difficult to use, and were often unreliable. DOS was always a very basic operating system which lacked a great deal of system tools. Norton Utilities allowed users to gain access to the hardware and firmware of the computer. A typical utility was one which recovered data from a corrupted hard disk (which often happened, as hard disks had to be parked before they were moved).

$$V = I - W + S$$

where I is the initial value of the semaphore.
W is the number of completed wait operations performed on the semaphore.
S is the number of signal operations performed on it.
V is the current value of the semaphore (which must be greater than or equal to zero).

As $V \geq 0$, then $I - W + S \geq 0$, which gives:

$$I + S \geq W$$

or

$$W \leq I + S$$

Thus, the number of wait operations must be less than or equal to the initial value of the semaphore, plus the number of signal operations. A binary semaphore will have an initial value of 1 ($I = 1$), thus:

$$W \leq S + 1$$

In mutual exclusion, waits always occur before signals, as waits happen at the start of a critical piece of code, with a signal at the end of it. The above equations state that no more than one wait may run to completion before a signal has been performed. Thus, no more than one process may enter the critical section at a time, as required.

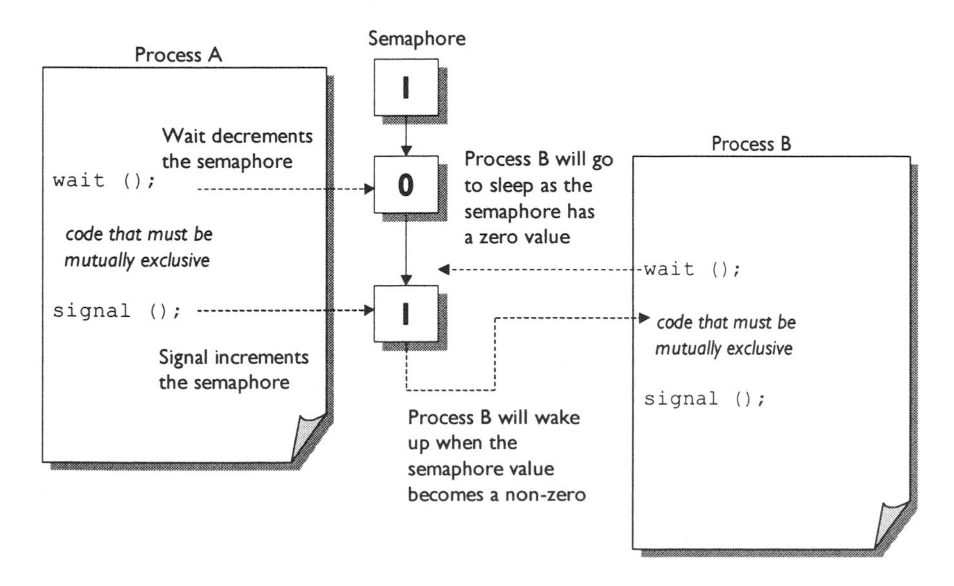

Figure 6.3 Example usage of semaphore in mutually exclusive code

6.3.2 P and V operations

A simple process can either be in RUNNING, SLEEPING or WAKEUP mode. A process is put to sleep if it is waiting for a resource which is currently being used by another process. When the resource is released, and no other processes require access to the resource, a sleeping process can be sent a WAKEUP, where it will reactivate itself. Semaphores with a value of zero can identify that there are no processes waiting on a resource, and the process can gain access to it. A positive value can then identify that there are a number of wakeups pending.

So why did Dijkstra use 'P' (for DOWN) and 'V' (for UP) for the operations? Well, one theory is that the lifts on the 13th floor of the Livvy tower is engraved with a 'P' instead of an arrow.

Two operations, P and V, are generalizations of SLEEP and WAKEUP and can be used to operate on the semaphore as follows:

- P operation (the DOWN operation). This checks the semaphore, and if the value of the semaphore is greater than zero, it decrements its value. A zero value puts the process to sleep without completing the P operation.
- V operation (the UP operation). This increments a referenced semaphore, and identifies that there are one or more processes that require some processing time (as they have been put to sleep with an earlier P operation). A sleeping process is chosen at random, and is allowed to complete a P operation.

There must be no interrupts when checking a semaphore, changing a semaphore and waking-up a process, and it must be done in a single indivisible operation (an atomic action). This overcomes timing hazards (see Section 6.3.3) as no other process can get access to the semaphore until the process has completed or is blocked.

6.3.3 Producer–consumer problem

The producer–consumer problem involves two processes sharing a common, fixed-size buffer. The producer puts information into the buffer, and the consumer process reads and removes information from the buffer. This is an exclusion problem as the consumer could be reading the buffer when the producer tries to write to the buffer. The solution to this is to put the producer to sleep when the consumer is reading and removing the data from the buffer, and then is awoken when complete. When the consumer wants to read and remove the data, and the producer is writing to the buffer, it must go to sleep, and then is awoken when the producer is finished.

Program 6.1 shows an example program. There can be a number of items in the buffer, which is identified with a variable called `buffer_count`, and the maximum number of items that can be stored in the buffer is `MAX_BUFF`. In the program, the producer keeps filling the buffer with data. When, if ever, the buffer is full it will go to sleep (`if (buffer_count== MAX_BUFF) sleep();`). It will then wait on the consumer to wake it up when it has read at least one item from the buffer (`if (buffer_count==MAX_BUFF-1) wakeup(producer_buffer);`). The consumer goes to sleep when there are no items in the buffer (`if (buffer_count==0) sleep();`), and will be woken-up when the producer has put at least one item in the buffer (`if (buffer_count==1) wakeup(consumer);`).

Unfortunately there is a timing problem in the code. This

CLASSIC SOFTWARE

7. MS-DOS 2.0 [Developer: Microsoft]. DOS was the solid foundation that the PC was built on. It was never a startling operating system, and just did enough to get by. Its true strength was its basic compatibility with all previous versions (this makes it extremely unusual in the computer industry). Programs were almost guaranteed to run on earlier versions of DOS.

can happen with an empty buffer and when the operating system scheduler has just run the consumer but stops it before it can check the empty buffer. The scheduler then runs the producer which then adds an item to the buffer, and thinks that the consumer should be sleeping (as the buffer was empty). This signal will be lost on the consumer, as it is not sleeping. When the scheduler runs the consumer again, it will have an incorrect count value of zero, as it has already checked the count, and will thus go to sleep. The producer will thus not wake the consumer up, as the producer will fill the buffer up, and then go to sleep. Both consumer and producer will be sleeping, awaiting the other to wake them up. One solution to this is to have a bit which defines that there is a wakeup waiting, which is set by a process which is still awake. If the process then tries to go to sleep, it cannot as the wakeup-waiting bit is set. The process would thus stay awake, and reset the wakeup-waiting bit.

CLASSIC SOFTWARE

8. Flight Simulator [Developer: Microsoft]. By today's standards, Microsoft's Flight Simulator was hardly startling, but in the 1980s its graphics were state-of-the-art. It led to a great market for supplying different terrains and cities to fly over. For many it was their first sight of New York from the sky.

Program 6.1

```
#define MAX_BUFF 100            /* maximum items in buffer              */
int buffer_count=0;             /* current number of items in buffer    */

int main(void)
{
      /* producer_buffer();    on the producer   */
      /* consumer_buffer();    on the consumer   */
}

void producer_buffer(void)
{
 while (TRUE){                            /* Infinite loop */
  put_item();                             /* Put item*/
  if (buffer_count==MAX_BUFF) sleep();    /* Sleep, if buffer full */
  enter_item();                           /* Add item to buffer*/
  buffer_count = buffer_count + 1;        /* Increment number of items in the
                                             buffer */
  if (buffer_count==1) wakeup(consumer);  /*was buffer empty?*/
 }
}

void consumer_buffer(void)
{
 while (TRUE) {                           /* Infinite loop */
  if (buffer_count==0) sleep();           /* Sleep, if buffer empty */
  get_item();                             /* Get item */
  buffer_count = buffer_count - 1;        /* Decrement number of items in the
                                             buffer*/
  if (buffer_count==MAX_BUFF-1) wakeup(producer_buffer);
                                          /* if buffer not full
                                             anymore, wake up producer*/
  consume_item();                         /*remove item*/
 }
}
```

6.3.4 Deadlock

Deadlock is a serious problem when running processes over a distributed system, or in a local operating system. It occurs when a process is waiting for an event that will never occur. This typically occurs when:

- **Resource locking.** This is where process is waiting for a resource which will never become available. Some resources are pre-emptive, where processes can release their access on them, and give other processes a chance to access them, but others are non-pre-emptive, where a process must be given full rights to the resource, and no other processes can get access to it until the currently assigned process is finished with it. An example of this is with the transmission and reception of data on a communication system. It would not be a good idea for a process to send some data that required data to be received, in return, to yield to another process which also wanted to send and receive data. The non-pre-emptive resources would thus be locked so that no other processes could access it. This can cause a problem when the resource which is accessing the resource never gets the event which will release the lock, or if the process crashes.
- **Starvation.** This is where other processes are run, and the deadlocked process is not given enough time to catch the required event. This can occur when processes have a low priority compared with other ones. The higher priority tasks tend to have better chances to hog the required resources.

CLASSIC SOFTWARE

9. Novell NetWare [Developer: Novell]. NetWare became the first company to properly bring together PC to make a local area network. For this it used its own, propriety protocol: IPX/SPX, and is the only networking operating system which can compete again the strength of Unix (which uses the TCP/IP protocol) and Windows 95/98/ NT/2000. It held an almost monopoly on networks in the commercial market, but has suffered recently due to the might of Microsoft Windows, as Windows can support many different protocols.

Resource deadlock occurs when Process 1 holds Resource A, and Process 2 holds Resource B, but Process 1 wants to gain access to Resource B, and vice-versa. Each process is waiting for the other to yield their exclusive access to their resource. This is a deadly embrace. A typical problem can occur when data buffers can become full. For example a print spooler can be setup so that it must receive the full contents of a print file, before it will actually send it to the printer. If print buffer is receiving print data from several sources, it can fill up the buffer before any of the print jobs have completed. The only way round this problem would be to increase the data buffer size, which can be difficult.

The four conditions that must occur for deadlock to occur are:

- **Mutual exclusion condition.** This is where processes get exclusive control of required resources, and will not yield the resource to any other process.
- **Wait for condition.** This is where processes keep exclusive control of acquired resources while waiting for additional resources.
- **No pre-emption condition.** This is where resources cannot be removed from the processes which have gained them, until they have completed their access on them.
- **Circular wait condition.** This is a circular chain of processes on which each process holds one or more resources that are requested by the next process in the chain.

CLASSIC SOFTWARE

10. Unix System V [Distributor: AT&T]. System V was the first real attempt at unifying Unix into a single standardized operating system, and succeeded in merging Microsoft XENIX, SunOS, Unix 4.3 BSD. Unfortunately, after this there was still a drift by hardware manufacturers to move away from the standard (and define their standards). Although it has been difficult to standardize Unix, its true strength is its communications protocols, such as TCP/IP, which are now world-wide standards for communicating over the Internet. The biggest challenge to Unix has been from Windows NT/2000, which has tried to create a hardware independent operating system. Unix has, in the main, survived because of its simplicity and its reliability.

6.3.5 Deadlock avoidance

If possible processes should run without the problem of deadlock, as systems normally require a reboot to clear the problem. One of the best-known avoidance algorithms is the Banker Algorithm, where resources refer to resources of the same type. It can be extended to resource pools with differing resource types.

In this algorithm, the operating system has a number of resources of a given type (N), which are allocated in a number of users (M). The operating system is told by each process the maximum number of resources that it requires (n), which must be less than N. The operating system gives access to one of the resources of a process, one at a time. Thus processes can be guaranteed access to one of the resources within a given time. The condition is seen as safe if one of the processes can complete with the amount of resources that are left unallocated. For example, if the operating system allocated memory to processes. If the operating system has a total of 100 MB (N=100), and there are four processes running (M=4). Each process tells the operating system about the maximum amount of memory that it will require (n). Processes must ask the operating system for an allocation of the resources. The algorithm checks to see if there is enough allocation left, after the new allocation has been granted, so that at least one of the processes with allocated resources can complete, even if it asks for its maximum allocation. The best way to illustrate this is with an example. I.e.:

Process A requires a maximum of 50 MB. Process B requires a maximum of 40 MB.
Process C requires a maximum of 60 MB. Process D requires a maximum of 40 MB.

The current state would be safe:

Process	*Current allocation*	*Maximum allocation required*
A	40	50
B	20	40
C	20	60
D	10	40
Resource unallocated	10	

This is safe as Process A can still complete, as there is still 10 MB to be allocated. This will be enough to complete this process, but no other processes would be given any more resources as all of the unallocated memory must be reserved for Process A. Process B possibly requires another 20 MB, Process C also possibly requires another 40 MB, and Process D possibly requires another 30 MB.

An unsafe condition would be:

Process	*Current allocation*	*Maximum allocation required*
A	15	50
B	30	40
C	45	60
D	0	40
Resource unallocated	5	

This is unsafe as there is only 5 MB of memory left, and this is not enough for any of the processes to complete. Thus we can have deadlock (unless a process is willing to give up its memory allocation). The operating system would reject any allocation which took it into the unsafe region. In summary the algorithm assumes:

- Each resource has exclusive access to resources that have been granted to it.
- Allocation is only granted if there is enough allocation left for at least one process to complete, and release its allocated resources.
- Processes which have a rejection on a requested resource must wait until some resources have been released, and that the allocated resource must stay in the safe region.

The main problems with the Banker Algorithm are:

CLASSIC SOFTWARE

11. Mac OS and System 7 [Developer: Apple]. Microsoft has relied on gradual improvement to its software, especially in its operating systems, while Apple got it right first time with its Mac OS system. Many reckon that it was 10 years ahead of all other systems. With the PC there were many difficulties, such as keeping compatibility with the original DOS, difficulties with the Intel 8086/88 memory model, support for a great number of peripherals, and the slowness of its internal buses (including the graphics adaptor). These all made software development of the PC difficult. Apple did not have compatibility to deal when it designed the Macintosh as it had a blank piece of paper to work with. It spent as much money on the operating system as it did on developing the hardware for the Macintosh.

- Requires processes to define their maximum resource requirement.
- Requires the system to define the maximum amount of a resource.
- Requires a maximum amount of processes.
- Requires that processes return their resources in a finite time.
- Processes must wait for allocations to become available. A slow process may stop many other processes from running as it hogs the allocation.

6.3.6 Deadlock detection and recovery

The main technique that is used to detect a deadlocked situation is the existence of a circular wait. This detection process has a time overhead on the operation system, but the operating system can try and release deadlocked resources, rather than the user rebooting the system. A typical technique is to use resource allocation graphs, which indicate resource allocations and requests. An arrow from a process to a resource maps the request currently under consideration, and an arrow from a resource to a process indicates the resource has been allocated to that process. Squares represent processes, large circles represent classes of identical devices, and small circles drawn inside large circles indicate the number of identical devices of each class. This graph can be used to determine the processes that can complete their execution and the processes that will remain deadlocked. The graph will reduce for a process when all the requests have been granted, and will release the resources. If a graph cannot be reduced for a set of processes, deadlock occurs.

To undeadlock a system, one of the four deadlock conditions must be broken. This normally involves determining the deadlocked processes (which is often a difficult task). Once identified it is often necessary to kill one or more of the deadlocked processes, and release the resources which are allocated to it.

6.4 RPC

The Remote Procedure Control is (RPC) defined in RFC1050, and was initially defined by Sun Microsystems Inc. It defines:

- Servers. This is software which implements the network services.
- Services. This is a collection of one or more remote programs.
- Programs. These implement one or more remote procedures.
- Procedures. These define the procedures, the parameters and the results of the RPC operation.
- Clients. This is the software that initiates remote procedure calls to services.
- Versions. This allows servers to implement different versions of the RPC software, in order to support previous versions.

CLASSIC SOFTWARE

12. Netscape (Navigator/Communicator)/ Internet Explorer [Developer: Netscape Communications Corporation (Navigator/Communicator)/Microsoft (Internet Explorer)]. Web browsers interpret special hypertext pages which consist of the hypertext markup language (HTML) and JavaScript. They then display it in the given format. Netscape Navigator is one of the most widely used WWW browsers and is available for PCs (running Microsoft Windows), Unix workstations and Macintosh computers. It has since become the standard WWW browser and has many add-ons and enhancements, which have been added through continual development by Netscape, and other software developers (the source code is now freely available to anyone who wants it).

Remote Procedure Call (RPC) provides the ability for clients to transparently execute procedures on remote systems of the network. RPC sits fits into the session layer of the OSI model, as illustrated in Figure 6.4. It can fit on top of any transport and network layer, such as TCP/IP, UDP/IP or SPX/IPX, but typically uses TCP/IP as the transport/network layer, as this allows for reliable communications. TCP/IP allows for a virtual connection between two hosts and the data is checked for errors, whereas UDP/IP does not setup a connection, and has no guarantee that the data has been received correctly by the session layer of the OSI model.

In a local procedure call model, a calling program inserts parameters into a predefined location, and then transfers control to the procedure, which reads the parameters from the predefined location. Eventually the calling procedure will regain control, and reads from a predefined location for the results of the called procedure. An RPC is similar to this, with a caller process on the client, and a server process on the server. The operation of the client and server is illustrated in Figure 6.5, and is as follows:

- The caller process sends a call message, with all the procedure's parameters, to the server process and waits for a reply message.
- The server runs a process which is dormant, and is woken-up with the arrival of a call message. When this occurs the server process extracts the procedure's parameters and runs the process.
- The server sends a reply message with the procedure's results. Once the reply message is received, the results of the procedure are extracted, and caller's execution is resumed.

The server process then waits, dormant, for the next call message.

6.4.1 Transports, semantics and authentication

RPC does not provide any form of reliability, as it assumes that the transport layer provides this, which is the reason that TCP/IP is typically used, as it provides for reliability. Each client request has a transaction ID, which is used by the client to keep track of requests (as a client can initiate more than one request, at any given time). Transaction IDs do not have to be unique and can be used with different requests (obviously the previous request would have to be completed, before the same ID is used again). The server has no choice on the ID, and must only use it to identify its response to the client.

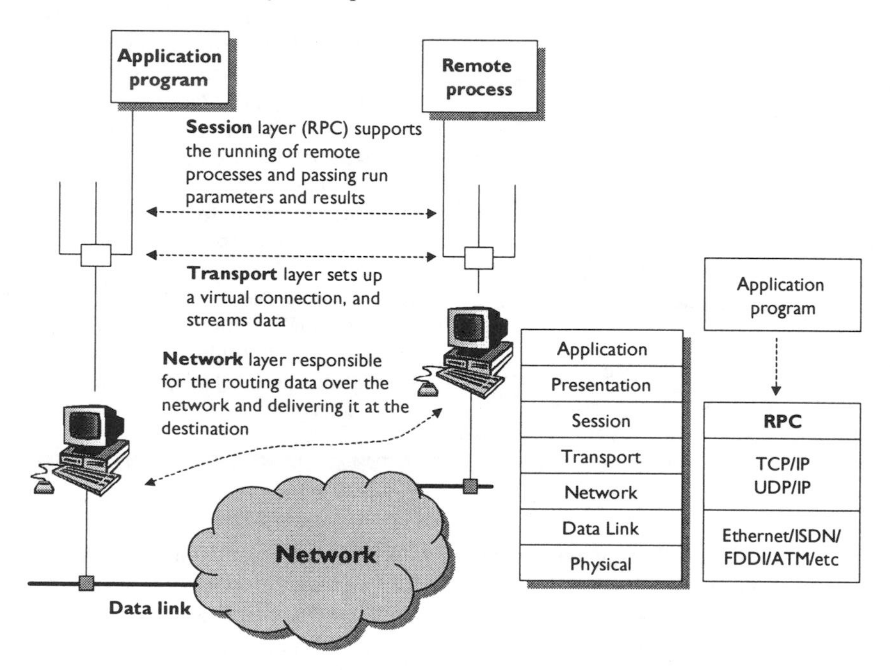

Figure 6.4 OSI model with RPC

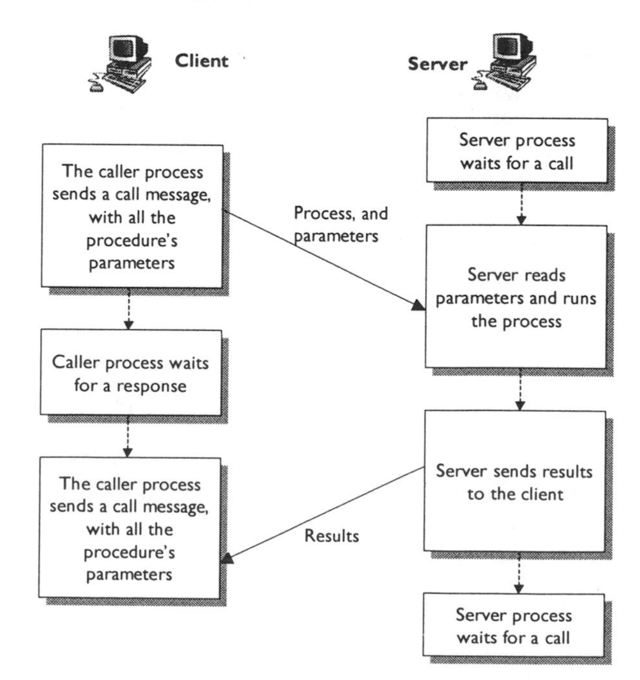

Figure 6.5 Operation of RPC

An important part of RPC is authentication, as a server would be open to abuse if any client was allowed to remotely run processes on it. This is a typical attack on a system, and, if too many processes are run on a computer, it will eventually grind to a halt, and typically requires to be rebooted. RPC supports various different types of authentication protocols.

CLASSIC SOFTWARE

13. SideKick 1.0 [Developer: Borland International]. Borland developed SideKick to provide a pop-up notepad, calendar, and calculator whenever the user wanted them. It used a TSR, and a hotkey, so that the user could quickly move from its current application to SideKick. In days of single-tasking DOS, around 1984, this was revolutionary, but in these days of multitasking Windows, it does not seem so.

6.4.2 RPC protocol

The RPC protocol provides:

- A unique specification of the called procedure.
- Mechanism for matching response parameters with request messages.
- Authentication of both callers and servers. The call message has two authentication fields (the credentials and verifier), and the reply message has one authentication field (the response verifier).
- Protocol errors/messages (such as incorrect versions, errors in procedure parameters, indication on why a process failed and reasons for incorrect authentication).

RPC has three unsigned fields which uniquely identify the called procedure:

- **Remote program number.** These are numbers which are defined by a central authority (like Sun Microsystems).
- **Remote program version number.** This defines the version number, and allows for migration of the protocol, where older versions are still supported. Different versions can possibly support different message calls. The server must be able to cope with this.
- **Remote procedure number.** This identifies the called procedure, and is defined in the specification of the specific program's protocol. For example, file service may define that an 8 defines a read operation and a 10 defines a write operation.

The reply message can give some indication of the cause of an error, including:

- Version number not supported. The returned message contains the upper and lower version of the version number that is supported.
- Remote program is not available on the remote system.
- Requested procedure number does not exist.
- Parameters passed are incorrect.

The RPC message has the following format:

- Message type. This is either CALL (0) or REPLY (1).
- Message status. There are two different message status fields, depending on whether it is a CALL or a REPLY. These are:
  - CALL. Followed by a field which gives the status of: SUCCESS (executed successful – 0), PROG_UNAVAIL (remote does not support the requested program – 1), PROG_MISMATCH (cannot support version – 2), PROC_UNAVAIL (procedure unavailable – 3).

- oREPLY. Followed by MSG_ACCEPTED (0) and MSG_DENIED (1). If the message was denied the field following defines the reason, such as: RPC_MISMATCH (Version mismatch – 0) or AUTH_ERROR (cannot authenticate the caller – 1).

Call messages then have:

- Rpcvers. RPC Version number (unsigned integer).
- Prog, vers and proc. Specifies the remote program, its version number and the procedure within the remote program (all unsigned integers).
- Cred. Authentication credentials.
- Verf. Authentication verifier.
- Procedure specific parameters.

CLASSIC SOFTWARE

14. Excel for the Macintosh [Developer: Microsoft]. Microsoft stole the thunder from VisiCalc and Lotus 1-2-3 when it produced a graphical version of a spreadsheet. Microsoft saw the potential of a graphical spreadsheet and quickly ported it to the PC. In the end, Lotus was too slow to convert Lotus 1-2-3 to Windows (going initially for an OS/2 version), and lost a considerable market share that would never be recovered.

6.4.3 RPC authentication

Authentication is important as it should authenticate both the caller and the server. This stops invalid callers from getting access to the server, and vice-versa. The call message has two authentication fields (the credentials and verifier), and the reply message has one authentication field (the response verifier). This can either be:

- No authentication (AUTH_NULL). No authentication is made when callers do not know who they are or when the server does not care who the caller is. This type of method would be used on a system that did not have external connections to networks, and assumes that all the callers are valid.
- Unix authentication (AUTH_UNIX). Unix authentication uses the Unix authentication system, which generates a data structure with a stamp (arbitrary ID which the caller machine may generate), machine name (such as 'Apollo'), UID (caller's effective user ID), GID (the caller's effective group ID) and GIDS (an array of groups which contain the caller as a member).
- Short authentication (AUTH_SHORT).
- DES authentication (AUTH_DES). Unix authentication suffers from two problems: the naming is too UNIX oriented and there is no verifier (so credentials can easily be faked). DES overcomes this by addressing the caller using its network name (such as 'unix.111@mycomputer.net') instead of by an operating system specific integer. These network names are unique on the Internet. For example unix.111@mycomputer.net identifies user ID number 111 on the mycomputer.net system.

Apart from providing a unique network name, DES authentication also provides authentication of the client, and vice-versa. It does this by the client generating a 128-bit DES key which is then passed to the server in the first RPC call. In any communications, the client then reads the current time, and encrypts it with the key. The server will then be able to decrypt the encrypted timestamp, as it knows the encryption key. If the decrypted timestamp is close to the current time, then the server knows that the client must be valid. Thus it is important that both the client and server keep the correct time (perhaps by consulting an Internet Time Server at regular intervals). After the initial timestamp has been validated, the server then authenticates following timestamps in that they have a later time than the previous timestamp and that the timestamp has not expired. This timestamp window is defined in the first RPC call, and thus defines the lifetime of the conversation.

The server authenticates itself to the client by sending back the encrypted timestamp it received from the client, minus one second. If the client gets anything different than this, it will reject it.

CLASSIC SOFTWARE

15. PageMaker [Developer: Aldus]. A classic package which has been used in millions of publications. Its approach was simple; it made the computer invisible, and made the user feel as if they were using traditional page design methods.

6.4.4 RPC programming

RPC divides into three layers:

- **Highest layer.** At this level the calls are totally transparent to the operating system, the computer type and the network. With this the programmer simply calls the required library routine, and does not have to worry about any of the underlying computer type, operating system or networking. For example, the `rnusers` routine returns the number of users on a remote computer (as given in Program 6.2).
- **Middle layer.** At this level the programmer does not have to worry about TCP sockets, the UNIX system, or other low-level implementation mechanisms, and just makes a remote procedure call to routines on other computers. It is the most common implementation as it gives increased amount of control over the RPC call. These calls are made with `registerrpc` (which obtains a unique system-wide procedure identification number), `callrpc` (which executes a remote procedure call), and `svc_run`. The middle layer, in some more complex applications, does not allow for timeout specifications, choice of transport, no UNIX process control, or error flexibility in case of errors. If these are required, the lower layer is used.
- **Lowest layer.** At this level there is full control over the RPC call, and can create robust and efficient connections.

Program 6.2

```
#include <stdio.h>
int main(int argc, char *argv[ ])
{
    int users;
    if (argc != 2) {
        fprintf(stderr, "Use: rnusers hostname\n");
        return(1);
    }
    if ((users = rnusers(argv[1])) < 0) {
        fprintf(stderr, "Error: rnusers\n");
        exit(-1);
    }
    printf("There are %d users on %s\n", users, argv[ 1] );
    return(0);
}
```

At the highest layer the programmer simply uses a call to the RCP library. In Unix this library is typically named `librpcsvc.a`, and the program is compiled with:

```
cc progname.c -l lrpcsvc
```

Example RPC server library routines are:

- `rnusers`. Returns number of users on remote machine.
- `rusers`. Returns information about users on remote machine.

- havedisk. Determines if remote machine has a disk.
- rstats. Gets performance data from remote kernel.
- rwall. Writes to specified remote machines.
- yppasswd. Updates user password in Yellow Pages.

CLASSIC SOFTWARE

16. Adobe Type [Developer: Adobe]. An important change in desktop publishing was the use of enhanced fonts, such as the Type 1 PostScript fonts available from Adobe. Adobe has also produced one of the standard methods to transfer files around computer systems, without losing any layout information. This form is named PDF (Portable Document Format), and is now well integrated with the WWW.

At the next level, the middle layer, the programmer has more control over the RPC call using callrpc and registerrpc. Program 6.3 determines the number of remote users, and uses the callrpc routine which has the following parameters:

argv(1)	Remote server name
RUSERSPROG	Program
RUSERSVERSION	Version
RUSERSPROCVAL	Procedure number. Together with the program and version numbers, this defines the procedure to be called.
xdr_void	Defines the data type for the next argument (which is the parameter to be sent to the remote procedure). As there are no arguments to be sent the data type is void. Other XDR types for basic data types are: xdr_bool, xdr_char, xdr_u_char, xdr_enum, xdr_int, xdr_u_int, xdr_long, xdr_u_long, xdr_short, xdr_u_short and xdr_wrapstring.
0	An argument to be encoded and passed to the remote procedure.
xdr_u_long	Defines the return type for next variable to be a long integer (users).
&users	Pointer to the users variable, in which the number of users is returned to.

If the routine is successful the returned value will be zero, otherwise it will contain a status value, which is defined in clnt.h.

Program 6.3

```
#include <stdio.h>
#include <rpc.h>
#define RUSERSPROG          10002 /* Program number    */
#define RUSERSVERSION       2     /* Version number    */
#define RUSERPROCVAL        1     /* Procedure number  */
int main(int argc, char *argv[ ] ) {
unsigned long  users;
int            rtn;
    if (argc != 2) {
        fprintf(stderr, "Use: nusers hostname\n"); exit(-1);
    }
    if (rtn = callrpc(argv[1], RUSERSPROG, RUSERSVERSION, RUSERSPROCVAL,
                    xdr_void, 0, xdr_u_long, &users) != 0) {
          clnt_perrno(stat); return(1);
    }
    printf("There are %d users on %s\n", users, argv[ 1] );
    return(0);
}
```

Typically a server registers all its RPC calls, and then goes into an infinite wait loop. Program 6.4 shows an example of a program on a server which registers the `nuser` RPC call. The program assumes that the `nuser` routine exists in the RPC library. The `svc_run` routine responds to remote calls and initiates the remote procedure.

Program 6.4

```
#include <stdio.h>
#include <rpc.h>
#define RUSERSPROG          10002 /* Program number   */
#define RUSERSVERSION        2    /* Version number   */
#define RUSERPROCVAL         1    /* Procedure number */

char  *nuser();
int   main(void)
{
    registerrpc(RUSERSPROG, RUSERSVERS, RUSERSPROC_NUM, nuser,
                                              xdr_void, xdr_u_long);
    svc_run();
    fprintf(stderr, "Error: server terminated\n");
    return(1);
}
```

In UNIX, the /etc/rpc file contains a listing of the RPC services (Notice that 100002 corresponds to rusers). The first column defines the RPC process name, the second the procedure number, and the third defines an alias for the process (the fourth column has been added to give extra information). An example is:

`portmapper`	`100000`	`portmap sunrpc`	*Port mapper*
`rstatd`	`100001`	`rstat rstat_svc rup perfmeter`	*Remote stats*
`rusersd`	**`100002`**	**`rusers`**	*Number of users*
`nfs`	`100003`	`nfsprog`	*Network File System (NFS)*
`ypserv`	`100004`	`ypprog`	*Network Information Service (NIS)*
`mountd`	`100005`	`mount showmount`	*Mount daemon*
`ypbind`	`100007`		*NLS binder*
`walld`	`100008`	`rwall shutdown`	*Shutdown message*
`yppasswdd`	`100009`	`yppasswd`	*yppasswd server*
`etherstatd`	`100010`	`etherstat`	*Ether stats*
`rquotad`	`100011`	`rquotaprog quota rquota`	*Disk quotas*
`sprayd`	`100012`	`spray`	*Spray packets*
`selection_svc`	`100015`	`selnsvc`	*Selection service*
`database_svc`	`100016`		*Remote database access*
`rexd`	`100017`	`rex`	*Remote execution*
`sched`	`100019`		*Scheduling service*
`llockmgr`	`100020`		*Local lock manager*
`nlockmgr`	`100021`		*Network lock manager*
`statmon`	`100023`		*Status monitor*
`bootparam`	`100026`		
`ypupdated`	`100028`	`ypupdate`	
`keyserv`	`100029`	`keyserver`	
`tfsd`	`100037`		

6.5 Multi-processor systems

Computer systems have generally evolved around a single centralized processor with an associated area of memory. This main processor performs most of the operations within the computer and also controls reads and writes to and from memory. This type of arrangement is useful in that there is little chance of a conflict when addressing any peripheral as only the single processor can access it. With the evolution of microelectronics it is now possible to build computers with many processors. It is typical on modern computers to have several processors, apart from the central processor. For example many computers now have dedicated processors to control the graphical display, processors to controls input/output functions of the computer, processors to control the hard-disk drive, and so on.

Computer systems, especially servers, are also now being designed with several processors that run application programs. Each of these processors can access its own localized memory and/or a shared memory. This type of multi-processor system, though, leads to several problems, including device conflicts and processor synchronization. Figure 6.6 illustrates the two types of systems.

A memory conflict occurs when a process tries to read from or write to an area of memory at the same time as another is trying to access it. Normally, multi-processor systems have mechanisms that lock areas of memory when a processor is accessing it.

Parallel systems require processor synchronization because one or more processors may require data from other processors. This synchronization can either be hard-wired into the system using data and addressing buses, or by a master controlling processor that handles the communication among slave processors (processor farms). They may also be controlled by the operating system software.

CLASSIC SOFTWARE

17. Windows 3.x [Developer: Microsoft]. The first two versions of Microsoft Windows were pretty poor. They had a poor interface, and still had all of the constraints of DOS (which are, of course, great). Microsoft Windows 3.0 changed all that with an enhanced usage of icons and windows, greater integration, increased memory usage. It was still 16-bit, but started to make full use of the processor, and gave a hint of the forthcoming multi-tasking (two processes at a time, using time sharing). Other enhancements were OLE, True Type fonts, and drag-and-drop commands. Windows for Workgroups was Microsoft's first attempt at trying to network PCs together, but it was too complicated to setup, and not very powerful.

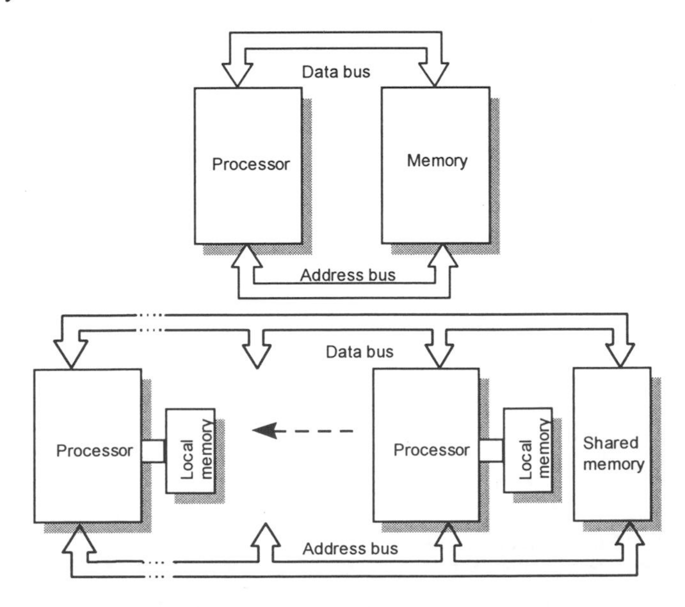

Figure 6.6 Single and multi-processor systems

6.5.1 Parallel Techniques

There are two main methods used when dividing computational tasks to individual processors. Computations are either divided into stages in a pipeline or they are divided into parallel streams, as illustrated in Figure 6.7. A mixed method uses a mixture of pipelines and parallel streams.

The pipeline method is preferable when there is a large number of computations on a small amount of data. Distributing data between streams can be awkward, since calculations often involve two or more consecutive items of data. Parallel streams are preferable for simple operations on large amounts of data, such as mathematical processing operations.

A major problem with pipelines is that it is difficult to ensure that all the processors have an equal loading. If one processor has a heavier workload than its neighbors then this processor holds-up the neighbors while they are waiting for data from the burdened processor.

It is always important to recognize the inherent parallelism in the problem and whether to allocate fast processors to critical parts and slower ones for the rest, or to equalize the workload, called load balancing.

CLASSIC SOFTWARE

18. Microsoft Office. The revenues from DOS allowed Microsoft to continue developing its application packages, such as Word, Excel, Access and PowerPoint, when many of its first attempts were inferior to its competitors, and typically were bug ridden. A classic example of this was Word 3.0, which had over 700 bugs, some of which caused the system to destroy data, or caused the computer to crash. Microsoft eventually had to give customers free upgrades, at a cost of over $1 million. Classic versions were Word for Windows 2.0, Word 6.0 and Excel 5.0.

6.5.2 Processor Farms

Processor farming is a technique for distributing work with automatic load balancing. It uses a master processor to distribute tasks to a network of slaves. The slave processors only get tasks when they are idle.

It is important in a parallel system that processor tasks are large enough because each task has its overheads. These include the handling overhead of the master controller and also the inter-processor communication. If the tasks are too small then these overheads take a significant amount of time and cause bottlenecks in the system.

6.6 Exercises

The following questions are multiple choice. Please select from a–d.

6.6.1 In which application is a binary semaphore used:
(a) Sockets (b) Mutual exclusion
(c) Message handling (d) Producer-client

6.6.2 In which application is a non-binary semaphore value typically used:
(a) Sockets (b) Mutual exclusion
(c) Message handling (d) Producer-client

6.6.3 Which of the following is not a communication device used in IPC:
(a) Sockets (b) Message handling
(c) Synchronization (d) Semaphores

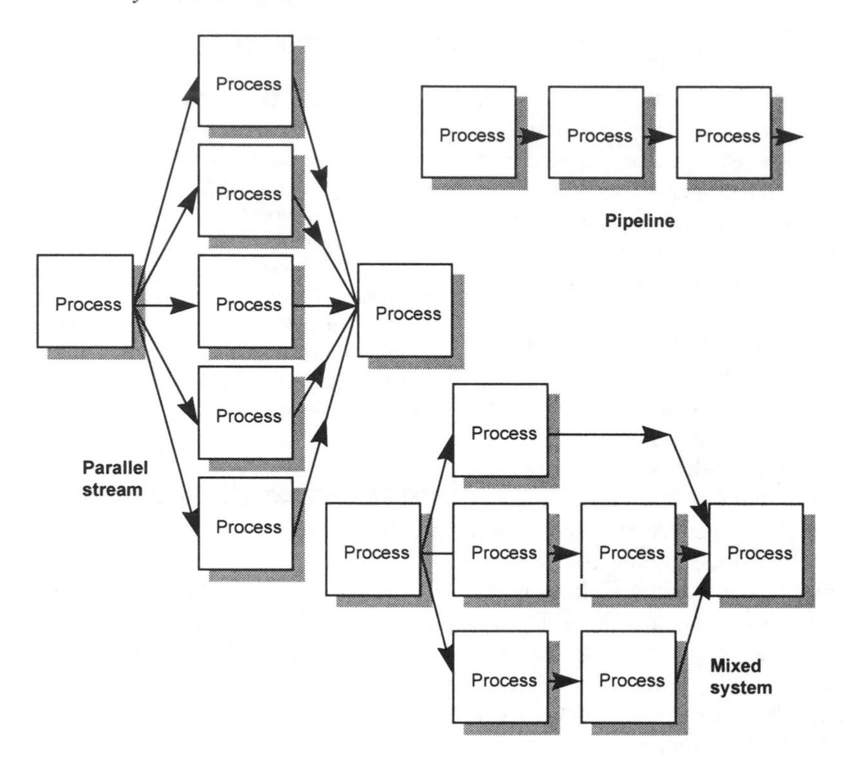

Figure 6.7 Pipeline, parallel stream and mixed systems

6.6.4 What does a signal operation do to a semaphore:

(a) Increment it (b) Decrement it
(c) Read it (d) Negate it

6.6.5 What does a wait operation do to a semaphore:

(a) Increment it (b) Decrement it
(c) Read it (d) Negate it

6.6.6 What value can a semaphore never take:

(a) Negative (b) Zero
(c) One (d) Two

6.6.7 What are the two types of RPC messages:

(a) REQUEST and REPLY (b) SEND and RECEIVE
(c) SEND and RECEIVE (d) CALL and REPLY

6.6.8 What three fields uniquely identify the called procedure in RPC:

(a) Procedure program number, program version and procedure number
(b) Procedure program name, program version and procedure number
(c) Procedure program number, program version and procedure name
(d) Procedure program name, program version and procedure name

6.6.9 In RPC using DES authentication, how does the caller authenticate itself to the server:

(a) A special bit sequence
(b) A special password which both caller and server agree on
(c) It sends its name and password
(d) It sends an encrypted value of the current time, which the server decrypts

6.6.10 In RPC using DES authentication, how does the server authenticate itself to the caller:

(a) It sends the same encrypted value back that it received from the caller
(b) A special password which both caller and server agree on
(c) It sends an encrypted value of the current time, which the server decrypts
(d) It sends an encrypted value of the time it received from the caller, minus one second

6.6.11 Which of the following is true:

(a) The caller and server set up a permanent connection, and remotely controls the remote procedure
(b) The caller requests a remote procedure from the server, which is then run on the server, and the results returned to the caller
(c) The server requests a remote procedure from the caller, which is then run on the server, and the results returned to the caller
(d) The server can call the caller, and vice-versa.

6.6.12 Which RPC server routine determines the number of users on a remote computer:

(a) `who` (b) `nusers`
(c) `rnusers` (d) `rusers`

6.6.13 Which RPC server routine writes to a remote computer:

(a) `send` (b) `rwall`
(c) `rwrite` (d) `write`

6.6.14 Which are the three RPC calls:

(a) `rpc`, `gorpc` and `run`
(b) `register`, `call` and `svc`
(c) `registerrpc`, `callrpc` and `svc_run`
(d) `rpcregister`, `rpccall` and `runsvc`

6.6.15 Explain the procedure that the caller and server go through with the RPC.

6.6.16 Outline the parameters used in Unix-style authentication.

6.6.17 Identify the two problems with Unix-type authentication, and how they are overcome with DES authentication.

6.6.18 How does DES authenticate the server to the caller, and vice-versa.

6.6.19 Outline how RPC fits into the OSI model.

CLASSIC SOFTWARE

19. Lotus Notes 3.0. Lotus had been losing a large market share for both its word processors (to Word), and for their spreadsheets (to Excel). It thus had to come-up with something that was innovative and also easy to use, but powerful, in a business environment. Its answer was truly inspirational: Lotus Notes. It successfully integrates electronic mail and other standard applications in an easy-to-use manner.

CLASSIC SOFTWARE

20. Microsoft Windows 95/98/NT/2000 The debate about whether Microsoft Windows 95 was just a copy of the Mac operating system will continue for years, but as Windows 3.x was a quantum leap from the previous versions, so Windows 95 was to Windows 3.x. The main problems with Windows 3.x is that it still used DOS as a basic operating system, and it also used 16-bit code (and most PCs used 32-bit processors, such as the 80386 and 80486 processors). Windows 95 was totally rewritten using, mostly, 32-bit software, and thus used the full power of the processor, and also the full memory addressing capabilities. It could now address up to 4 GB of virtual memory, and supported a great deal more devices, such as CD-ROMs and back-up resources. It also had networking properly built into it. The trump card for Microsoft, though, was that its networking supported many different network protocols, such as TCP/IP (for Internet traffic), IPX/SPX (for Novell NetWare traffic), and AppleTalk. It could thus live with any type of network (and, Microsoft hoped, would eventually replace the existing network with one based on Microsoft networking, and not the existing network (this is called gradual network migration). Windows 95/98 Workstation allow for a peer-to-peer network where computers can share resources, whereas Windows NT/2000 allows for the creation of a network server, which give network-wide login, administration, global file systems, and so on. Windows NT 4.0 integrated the networking from Windows NT 3.0 and the user interface found in Windows 95. Its other enhancements were support for pre-emptive multitasking and multithreading of Windows-based and MS-DOS-based applications, enhanced robustness and clean up when an application ends or crashes, and enhanced dynamic environment configuration. Windows 2000 builds on the success of NT.

POSIX (Portable Operating System Interface)

- Set of standard operating system interfaces based on the UNIX operating system.
- UNIX selected as a standard operating system, as it is not tied to any commercial companies (unlike Microsoft Windows).
- Manufacturer independence. Code can be moved between different POSIX-based operating systems.
- POSIX.1 – standard for application program interface in the C language.
- POSIX.2 – standard shell and utilities interface.
- POSIX.4 – thread management, have been developed or are being developed.
- The Open Group owns the UNIX trademark, and can brand UNIX-standard computers.

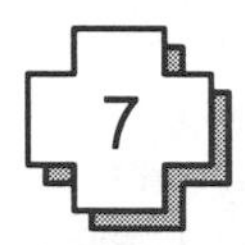

Distributed File Systems

7.1 Introduction

Files systems store information, program files and configuration data. In the past, most information was stored locally on a computer, or centrally on a mainframe computer, but there is now a trend to distribute information around networks. This has many advantages over traditional localized information:

- Many users can have access to a single source of information.
- Several copies of the same information can be stored, and when any one of them is updated they are synchronized to keep each of them up-to-date.
- Users can have access to a local copy of data, rather than accessing a remote copy of it.
- Large file systems can be built from a network of connected disk drives.
- Administrators can easily view the complete file system.
- Interlinking of databases to create large databases, which can be configured for a given application.

Most file systems have a hierarchical file structure, which has directories that contain subdirectories files and devices. The file system, itself, is a tree on a single server (normally a single disk or a physical partition) which has a specified root. Some systems use a mount system that mounts file systems onto a single tree, while others use a forest of file systems, where the file systems appear individually. If possible, the mounting of a drive system should be transparent to the user, and should be done automatically so that the user treats the mounted drive just as a local resource. A major problem, though, is the security of the remotely connected drive, thus each mounted drive must have strict rules on the access rights for the local user.

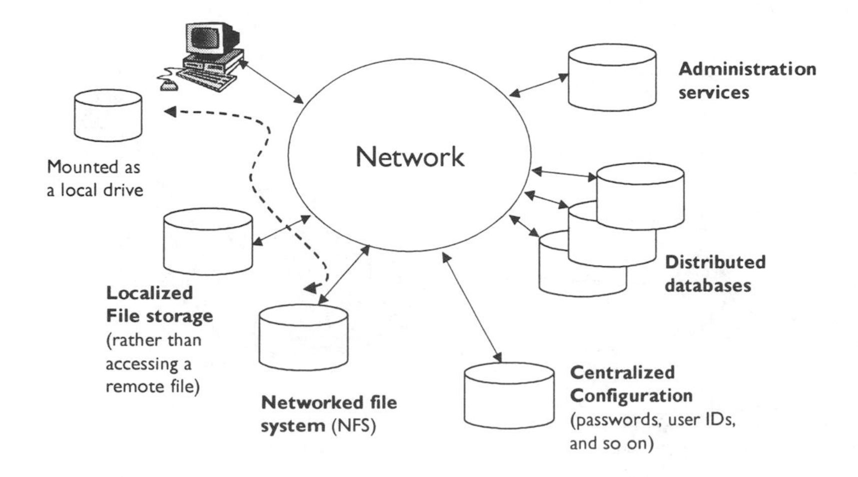

Figure 7.1 Distributed file system

Drives can either be mounted locally to a computer as a single tree (as Unix) or as a forest of drives (as used with Microsoft Windows). In Figure 7.2, one of the computers has created a single tree that uses its local drives to create the /etc/ and /user directories, and then mounts two networked drives to give /progs and /sys. The global file system will then be mounted onto the common tree, with four subdirectories below the top-level directory (/). The advantages of this type of system are:

- The structure of the file system, and the drives that are mounted are transparent to the user. As far as the user is concerned the complete file system is viewable.
- Every user can view the complete file system, if required.
- The file system is consistent around the network, and can be setup on a per computer basis.

With the forest of disks, a disk drive is mounted locally as if it is a local drive. In the example in Figure 7.2, the remote drives have been mounted as E: and F:. Its main advantage over the global file system is that:

- It is easier to determine if the remote drive is mounted, as it will appear as a mounted resource. With a single tree it is often difficult to determine if a drive is loaded onto the global file system as the basic structure still exists.
- Less complex than a global file, and easier to mount drives, but can become complex to setup if there are many remote drives to be mounted.

Its main disadvantage is it is more difficult to setup than the single tree system as the local mount drive must be specified, along with the path. In the global file system, files are mounted on the system in a consistent way, such as with E:\FREDS_DRIVE. If the local system does not mount the remote drive onto the required disk partition, there may be problems in the configuration of the system.

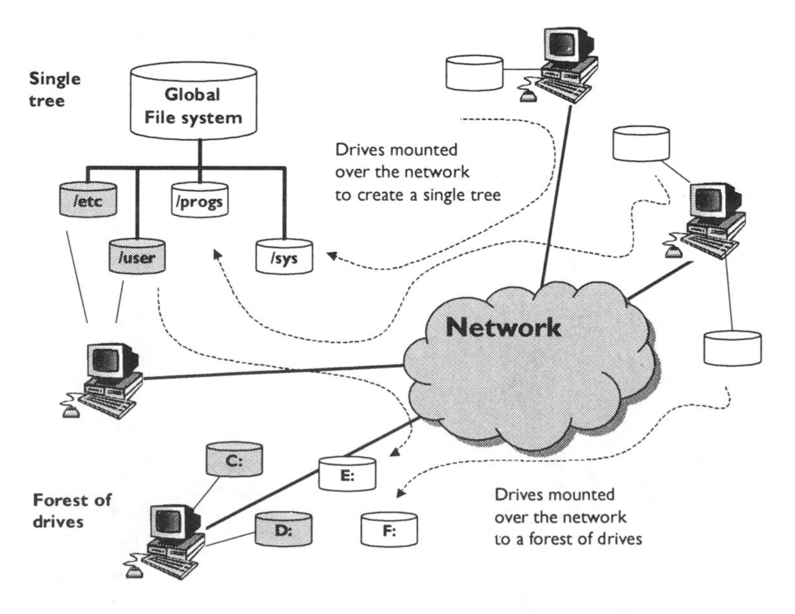

Figure 7.2 Distributed file system

7.2 Representation of data

The representation of data types is always a problem, as different computer systems use different ways to store and represent data. For example, the PC, which is based on Intel microprocessors, uses the little endian approach of representing a floating-point value, where the least significant byte is inserted into memory with the lowest-order byte first, and the highest-order byte last. The big endian approach always starts with the high-order byte and ends with the lowest-order byte. The Intel *x*86 family uses the little endian approach, whereas the Motorola 68000-series uses a big endian approach. For example with little endian, the value to store the 16-bit integer values of 4 (000 0000 0000 0100b), 5,241 (0001 0100 0111 1001b) and 10,342 (0110 0110 0010 1000b):

Memory location	*Contents (hex)*	*Contents (binary)*	*Value*
00	04	0000 0100	4
01	00	0000 0000	
02	79	0111 1001	5,241
03	14	0001 0100	
04	28	0010 1000	10,342
05	66	0110 0110	

Whereas, in big endian it would be stored as:

Memory location	*Contents (hex)*	*Contents (binary)*	*Value*
00	00	0000 0000	4
01	04	0000 0100	
02	14	0001 0100	5,241
03	79	0111 1001	
04	88	0110 0110	10,342
05	28	0010 1000	

Thus a program which has been written for a PC would incorrectly read data which has been written for a big endian program (typically for a UNIX workstation), and vice-versa. Another particular problem is that different computer systems represent data (such as numeric values) in different formats. For example an integer can be represented with either 16 bits, 32 bits, 64 bits, or even, 128 bits. The more bits that are used, the larger the integer value that can be represented.

All these problems highlight the need for a conversion technique that knows how to read the value from memory, and convert it into a standard form that is independent of the operating system or the hardware of the computer. This is the function of eXternal Data Representation (XDR), which represents data in a standard format. In XDR the basic data types are:

- **Unsigned integer and signed integer.** An unsigned and signed integer uses a 32-bit value. The unsigned value uses the range from 0 to $2^{32}-1$ (4,294,967,295), whereas the signed integer uses 2's complement which gives a range of –2,147,483,648 (1111 1111 1111 … 1111 1111) to +2,147,483,647 (0111 1111 1111 … 1111).
- **Single-precision floating point.** A single-precision floating-point value uses a 32-bit IEEE format of a floating-point value. An example is given next. The range is from $\pm 3.4\times 10^{-38}$ to $\pm 3.4\times 10^{38}$.

- **Double-precision float point.** A double-precision floating-point value uses a 64-bit IEEE format of a floating-point value. The range is from $\pm 1.7 \times 10^{-308}$ to $\pm 1.7 \times 10^{308}$.
- **String.** A string is represented with a number of bytes. The first four bytes define the number of ASCII characters defined. For example, if there were four characters in the string then the first four bytes would be: 0, 0, 0, 4, followed by the four characters in the string. Note that this differs from the way that the C programming language represents strings, as C uses the NULL ASCII character to define the end of a string.

A single-precision floating-point value uses 32 bits, where the most-significant bit represents the sign bit (S), the next eight bits represents the exponent of the number in base 2, minus 127 (E). The final 23 bits represent the base-2 fractional part of the number's mantissa (F). The standard format is:

$$\text{Value} = -1^{S} \times 2^{(E-127)} \times 1.F$$

For example:

$$
\begin{aligned}
1.23 \quad &= \text{3F9D 70A4h} \\
&= 0\ 01111111\ 00111010111000010100100\text{b} \\
&= -1^{0} \times 2^{(127-127)} \times (1 + 2^{-3} + 2^{-4} + 2^{-5} + 2^{-9} + 2^{-10} + 2^{-11} + 2^{-16} + 2^{-18} + 2^{-21}) \\
-5.67 \quad &= \text{C0B5 70A4h} \\
&= 1\ 10000001\ 01101010111000010100100\text{b} \\
&= -1^{1} \times 2^{(129-127)} \times (1 + 2^{-2} + 2^{-3} + 2^{-5} + 2^{-7} + 2^{-9} + 2^{-10} + 2^{-11} + 2^{-15} + 2^{-17} + 2^{-20}) \\
100.442 \quad &= \text{42C8 E24Eh} \\
&= 0\ 10000101\ 10010001110001001001110\text{b} \\
&= -1^{0} \times 2^{(133-127)} \times (1 + 2^{-1} + 2^{-4} + 2^{-8} + 2^{-9} + 2^{-10} + 2^{-14} + 2^{-17} + 2^{-20} + 2^{-21} + 2^{-22})
\end{aligned}
$$

A single-precision floating-point value uses 64 bits, where the most-significant bit represents the sign bit (S), the next eight bits represents the exponent of the number in base 2, minus 1023 (E). The final 52 bits represent the base-2 fractional part of the number's mantissa (F). The complete listing of all the data formats represented in XDR is given in Section 22.8.

Integers can either be signed or unsigned. Unsigned integers do not have a sign. Signed integers use a 2's complement notation, where the binary digits have a '1' in the most significant bit column if the number is negative, else it is a '0'. To convert a decimal value into 2's complement notation, the magnitude of the negative number is represented in binary form. Next, all the bits are inverted and a '1' is added. The following example illustrates the 16-bit 2s complement representation of the decimal value −65.

+65	00000000 01000001
invert	11111111 10111110
add 1	11111111 10111111

Thus, −65 is 11111111 1011111 in 16-bit 2s complement notation. Table 7.1 shows that with 16 bits the range of values that can be represented in 2's complement is from −32767 to 32 768 (that is, 65536 values).

Table 7.1 16-bit 2s complement notation

Decimal	*2s complement*
−32 768	10000000 00000000
−32 767	10000000 00000001
::::	::::
−2	11111111 11111110
−1	11111111 11111111
0	00000000 00000000
1	00000000 00000001
2	00000000 00000010
::::	::
32 766	01111111 11111110
32 767	01111111 11111111

7.3 NFS

The Network File System (**NFS**) is defined in RFC1094 and allows computers to share the same files over a network. It was originally developed by Sun Microsystems, and has the great advantage that it is independent of the host operating system and can provide data sharing among different types of systems (heterogeneous systems). This is achieved using Remote Procedure Call (RPC), on top of XDR, which provides a standard method of representing data types. RPC is defined in RFC1057, and XDR is RFC1014.

NFS uses a client/server architecture where a computer can act as an NFS client, an NFS server or both. An NFS client makes requests to access data and files on servers; the server then makes that specific resource available to the client. NFS servers are passive and stateless. They wait for requests from clients and do not maintain any information on the client. One advantage of servers being stateless is that it is possible to reboot servers without adverse consequences to the client. Servers do not preserve the current status of any of their clients, which means that a client can simply retry a request from a server, if it fails to get a response (in the event of a failure of the network or the server). If the server was stateful, the client would have to know that a server had crashed or that the network connection had broken, so that it knew which state is should be in, when the connection was returned, or when the server came back on-line.

The server grants remote access privileges to a restricted set of clients, which allows clients to mount remote directory trees onto their local file system. The components of NFS are as follows:

- **NFS** remote file access may be accompanied by network information service (NIS).
- External data representation (**XDR**), which is a universal data representation, used by all nodes, and provides a common data representation if applications are to run transparently on a heterogeneous network or if data is to be shared among heterogeneous systems. Each node translates machine-dependent data formats to XDR format when sending and translating data. It is XDR that enables heterogeneous nodes and operating systems to communicate with each other over the network.
- Remote Procedure Call (**RPC**) allows clients to transparently execute procedures on remote systems of the network. NFS services run on top of the RPC, which corresponds to the session layer of the OSI model.

- Network lock manager (`rpc.lockd`) allows users to co-ordinate and control access to information on the network. It supports file locking and synchronizes access to shared files.

Figure 7.3 shows how the protocols fit into the OSI model.

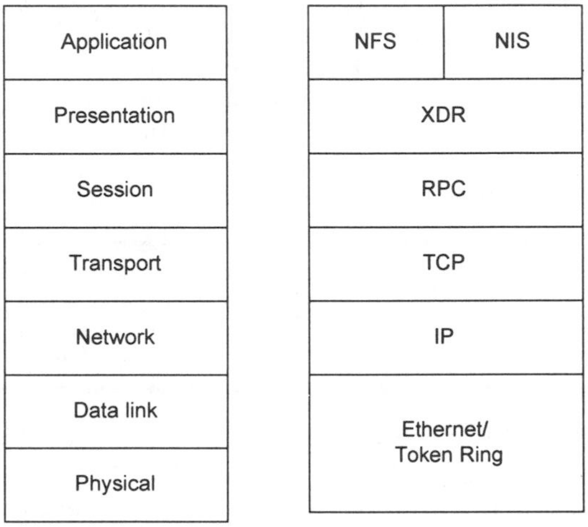

OSI model	NFS services
Application	NFS / NIS
Presentation	XDR
Session	RPC
Transport	TCP
Network	IP
Data link	Ethernet/ Token Ring
Physical	

Figure 7.3 NFS services protocol stack

7.4 NFS protocol

NFS assumes a hierarchical file structure. It can be used to mount file systems which map into a single tree (as the UNIX file system), or it can be used to add a file system as one of a forest of drives (as Microsoft Windows). NFS looks up one component of a pathname, at a time, as different file systems use different separators to identify a pathname (for example, UNIX uses '/' and Microsoft Windows uses '\', while others use periods).

The main NFS protocol is defined as a set of procedures with arguments and results defined using the RPC language. Each of the procedures is synchronous, and the client can assume that a response from a request completes the operation. The producers used are:

No.	Procedure	Name	Description
0	void NULL(void)	No operation	Used for server testing.
1	attrstat GETATTR(fhandle)	Get file attributes	Returns the attributes of the file specified by fhandle.

2	attrstat SETATTR(sattrargs)	Set file attributes	The sattrargs contains fields which are either –1 or are the new value for the file attributes. The sattargs structure is:
			struct sattrargs {
			fhandle file;
			sattr attributes;
			};
			where fhandle is the file handle that is passed between the client and the server. The structure of the sattr is:

			struct sattr {
			unsigned int mode; // file mode
			unsigned int uid; // user identification number
			unsigned int gid; // group identification number
			unsigned int size; // size of the file
			timeval atime; // time file last accessed
			timeval mtime; // time file was last modified
			};

6	readres READ(readargs)	Read from file	Reads a number of bytes of data (given by count), from a given file offset. The format of the readargs structure is:
			struct readargs {
			fhandle file; // used to represent file
			unsigned offset; // starting position
			unsigned count; // number of bytes to be read
			unsigned totalcount; // not used
			};

8	attrstat WRITE(writeargs)	Write to file	Writes data to a file starting at a given offset. The format of writeargs structure is:
			struct writeargs {
			fhandle file; // used to represent file
			unsigned beginoffset; // not used
			unsigned offset; // starting position
			unsigned totalcount; // not used
			nfsdata data; // data to be written
			};

9	diropres CREATE(createargs)	Create file	Creates a file in the given directory, with a given set of attributes. The format of the createargs structure is:
			struct createargs {
			diropargs where;
			sattr attributes;
			};
			The format of the diropags structure is:
			struct diropargs {
			fhandle dir; // used to represent directory
			filename name; // name of file
			};

10	stat REMOVE(diropargs)	Remove file	See above.

11	stat RENAME(renameargs)	Rename file	The format of the renameargs structure is:
			struct renameargs {
			diropargs from;
			diropargs to;
			};
			For example, from.name is changed to to.name.

12	stat LINK(linkargs)	Create link to file	The format of the linkargs structure is:
			struct linkargs {
			fhandle from;
			diropargs to;
			};

13	stat SYMLINK(symlinkargs)	Create symbolic link	The format of the aymlinkargs structure is:
			struct symlinkargs {
			diropargs from;
			path to;
			sattr attributes;
			};

14	diropres MKDIR(createargs)	Create directory	The format of the createargs structure is:
			struct createargs {
			diropargs where;
			sattr attributes;
			};

15	stat RMDIR(diropargs)	Remove directory	

16	readdirres READDIR(readdirargs)	Read from directory	The format of the readdirargs structure is:
			struct readdirargs {
			fhandle dir;
			nfscookie cookie; // Used to get the entries
			// starting at the beginning of the directory
			unsigned count;
			// maximum number of entries
			};

Figure 7.4 shows a client sending RPC procedures to the server, which responds back with the required data, parameters or with a status flag. Typical status flags are: NFS_OK (success), NFSERR_PERM (not owner), NFSERR_NOENT (no such file or directory), NFSERR_IO (some sort of hard error occurred), NFSERR_NXIO (no such device or address), NFSERR_ACCES (permission denied), NFSERR_EXIST (the file specified already exists), NFSERR_NODEV (no such device), NFSERR_NOTDIR (not a directory), NFSERR_ISDIR (the caller specified a directory in a non-directory operation), NFSERR_FBIG (file too large) and NFSERR_NAMETOOLONG (name too long).

7.4.1 Network Information Service (NIS)

As networks grow in size it becomes more difficult for the system administrator to maintain the security of the network. An important factor is the maintenance of a passwords file, where new users are added with the group, and any other information (such as their default home directory). In most networks a user should be able to log into any computer within a domain. Thus a global password and configuration files are required. This can be achieved with NIS, which is an optional network control program which maintains the network configuration files over a network. NIS allows the system manager to centralize the key configuration files on a single master server. If anyone wants to log into the network the master server is consulted (or one of its slave servers). Figure 7.5 illustrates some of the files that the server maintains; these include password (which contains the passwords for all the users within the domain), and groups (the group that the user is associated with). It is thus easy for the system administrator to add and delete users from the NIS server, and these changes will be reflected over the domain. A user cannot log into any of the clients, without the client checking with the server to see if they have a valid login and password.

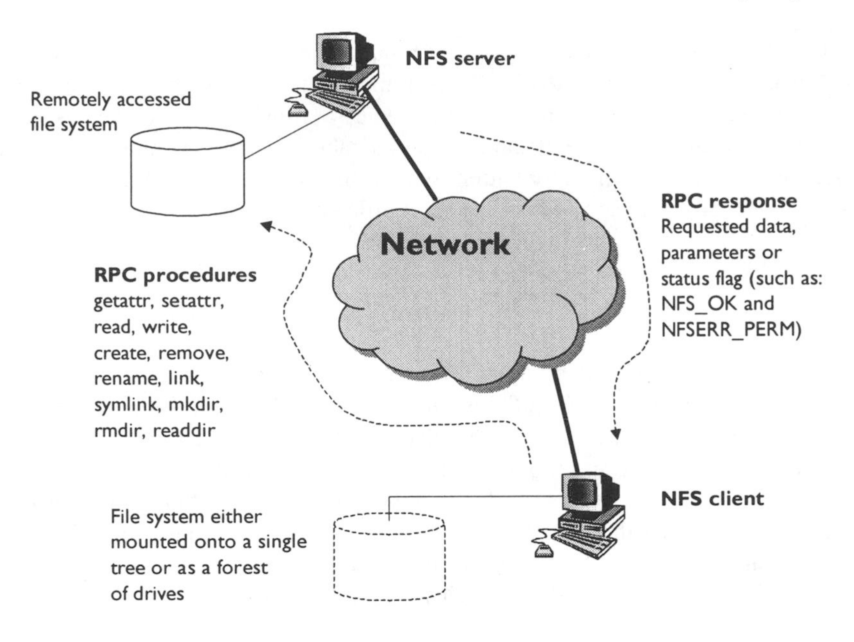

Figure 7.4 RPC procedures and responses

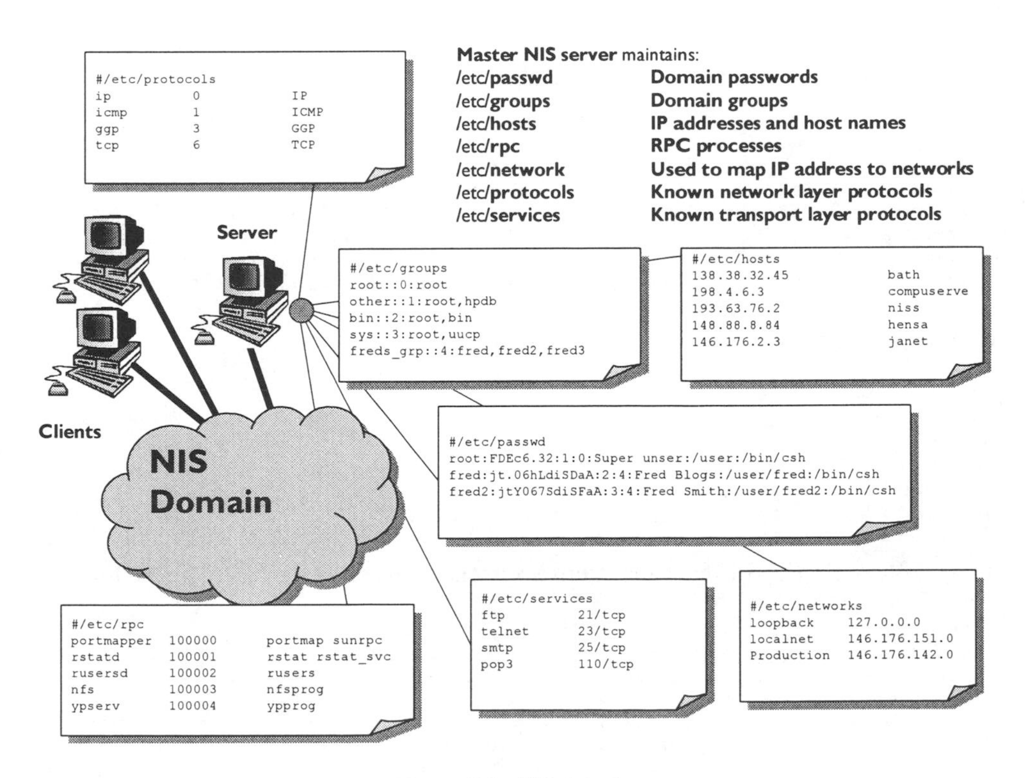

Figure 7.5 NIS domain

Previously NIS was named *Yellow Pages* (**YP**), but has changed its name as this is a registered trademark of the company British Telecommunications. NIS normally administers the network configuration files such as /etc/group (which defines the user groups), /etc/hosts (which defines the IP address and symbolic names of nodes on a network), /etc/passwd (which contains information, such as user names, encrypted passwords, home directories, and so on). An excerpt from a passwd file is:

```
root:FDEc6.32:1:0:Super user:/user:/bin/csh
fred:jt.06hLdiSDaA:2:4:Fred Blogs:/user/fred:/bin/csh
fred2:jtY067SdiSFaA:3:4:Fred Smith:/user/fred2:/bin/csh
```

This passwd file has three defined users; these are root, fred and fred2. The encrypted password is given in the second field (between the first and second colon), and the third field is a unique number that defines the user (in this case fred is 2 and fred2 is 3). The fourth field in this case defines the group number (which ties up with the /etc/groups file). An example of a groups file is given next. It can be seen from this file that group 4 is defined as freds_grp, and contains three users: fred, fred2 and fred3. The fifth field is simply a comment field and in this case it contains the user's names. In the next field each user's home directory is defined and the final field contains the initial UNIX shell (in this case it is the C-shell).

```
root::0:root
other::1:root,hpdb
bin::2:root,bin
sys::3:root,uucp
freds_grp::4:fred,fred2,fred3
```

A sample listing of a directory shows that a file owned by fred has the group name freds_grp.

```
> ls -l
-r-sr-xr-x   1 fred      freds_grp  24576    Apr 22  2000 file1
-r-xr-xr-x  13 fred      freds_grp  40       Apr 22  2000 file2
dr-xr-xr-x   2 fred      freds_grp  1024     Aug  5 14:01 myfile
-r-xr-sr-x   1 fred      freds_grp  24576    Apr 22  2000 text2.ps
-r-xr-xr-x   2 fred      freds_grp  16384    Apr 22  2000 temp1.txt
```

An excerpt from the /etc/hosts file is shown next.

```
138.38.32.45       bath
198.4.6.3          compuserve
193.63.76.2        niss
148.88.8.84        hensa
146.176.2.3        janet
146.176.151.51     sun
```

The /etc/protocols file contains information with known protocols used on the Internet.

```
# The form for each entry is:
# <official protocol name> <protocol number> <aliases>
# Internet (IP) protocols

ip      0  IP     # internet protocol, pseudo protocol number
icmp    1  ICMP   # internet control message protocol
ggp     3  GGP    # gateway-gateway protocol
tcp     6  TCP    # transmission control protocol
egp     8  EGP    # exterior gateway protocol
```

```
pup       12 PUP      # PARC universal packet protocol
udp       17 UDP      # user datagram protocol
hmp       20 HMP      # host monitoring protocol
xns-idp   22 XNS-IDP  # Xerox NS IDP
rdp       27 RDP      # "reliable datagram" protocol
```

The `/etc/netgroup` file defines network-wide groups used for permission checking when doing remote mounts, remote logins, and remote shells. Here is a sample file:

```
# The format for each entry is: groupname  member1  member2 ...
#     (hostname, username, domainname)
engineering hardware software (host3, mikey, hp)
hardware (hardwhost1, chm, hp)    (hardwhost2, dae, hp)
software (softwhost1, jad, hp)    (softwhost2, dds, hp)
```

NIS master server and slave server

With NIS, a single node on a network acts as the NIS master server, with a number of NIS slave servers, which receive their NIS information from the master server. The slaves are important in that they hold copies of the most up-to-date version of the NIS database, so if the master were to crash, or become uncontactable, the slaves could still provide password, group, and other NIS information to the clients in the domain. The slaves also relieve the workload on the master, as it may become busy responding to many NIS requests. When a client first starts up it sends out a broadcast to all NIS servers (master or slaves) on the network and waits for the first one to respond. The client then binds to the first that responds and addresses all NIS requests to that server. If this server becomes inoperative then an NIS client will automatically rebind to the first NIS server which responds to another broadcast. Figure 7.6 illustrates this.

Table 7.2 outlines the records which are used in the NIS database (or NIS map). This file consists of logical records with a search key and a related value for each record. For example, in the `passwd.byname` map, the users' login names are the keys and the matching lines from `/etc/passwd` are the values.

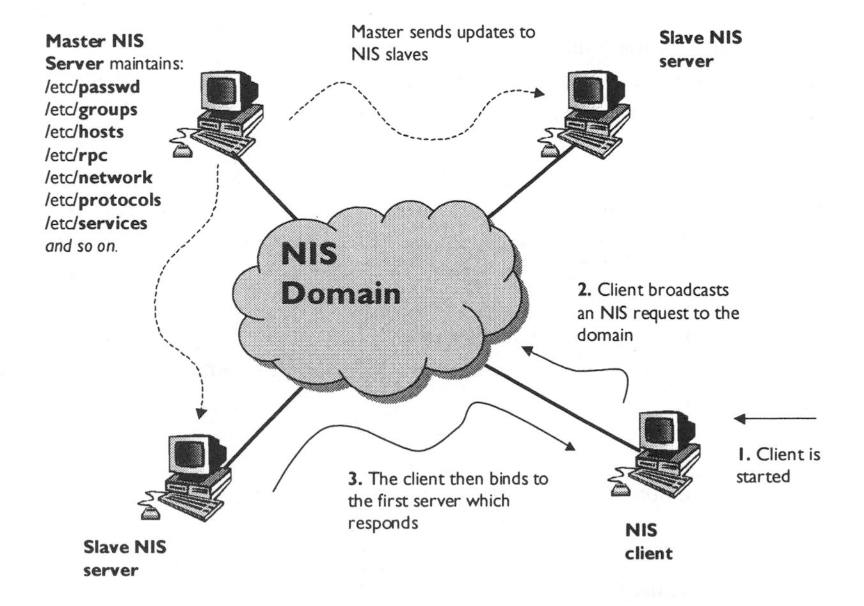

Figure 7.6 NIS domain

Table 7.2 NIS database components

NIS map	*File maintained*	*Description*
group.bygid group.byname	`/etc/group`	Maintains user groups.
hosts.byaddr hosts.byname	`/etc/hosts`	Maintains a list of IP addresses and symbolic names.
netgroup.byhost netgroup.byuser	`/etc/netgroup`	Contains a mapping of network group names to a set of node, user and NIS domain names.
networks.byaddr networks.byname	`/etc/network`	Defines network-wide groups used for permission checking when doing remote mounts, remote logins, and remote shells.
passwd.byname passwd.byuid	`/etc/passwd`	Contains details such as user names and encrypted passwords.
protocols.byname protocols.bynumber	`/etc/protocols`	Contains information with known protocols used on the Internet.
rpc.bynumber rpc.byname	`/etc/rpc`	Maps the RPC program names to the RPC program numbers and vice versa. This file is static; it is already correctly configured.
services.byname servi.bynp	`/etc/services`	
mail.byaddr mail.aliases	`/etc/aliases`	

NIS domain

An NIS domain is a logical grouping of the set of maps contained on NIS servers. The rules for NIS domains are:

- All nodes in an NIS domain have the same domain name.
- Only one master server exists on an NIS domain.
- Each NIS domain can have zero or more slave servers.

An NIS domain is a subdirectory of `/usr/etc/yp` on each NIS server, where the name of the subdirectory is the name of the NIS domain. All directories that appear under `/usr/etc/yp` are assumed to be domains that are served by an NIS server. Thus to remove a domain being served, the user deletes the domain's subdirectory name from `/etc/etc/yp` on all of its servers.

The start-up file on most UNIX systems is the `/etc/rc` file. This automatically calls the `/etc/netnfsrc` file which contains the default NIS domain name, and uses the program `domainname`.

7.4.2 *NFS remote file access*

To initially mount a remote directory (or file system) onto a local computer the superuser must do the following:

- On the server, export the directory to the client.
- On the client, mount (or import) the directory.

For example suppose the remote directory /user is to be mounted onto the host miranda as the directory /win. To achieve this operation the following are setup:

1. The superuser logs on to the remote server and edits the file /etc/exports adding the /user directory.
2. The superuser then runs the program exportfs to make the /user directory available to the client.

```
% exportfs -a
```

3. The superuser then logs into the client and creates a mount point /win (empty directory).

```
% mkdir /mnt
```

4. The remote directory can then be mounted with:

```
% mount miranda:/user /win
```

NFS maintains the file /etc/mnttab which contains a record of the mounted file systems. The general format is:

```
special_file_name  dir  type  opts  freq  passno  mount_time  cnode_id
```

where *mount_time* contains the time the file system was mounted using mount. Sample contents of /etc/mnttab could be:

```
/dev/dsk/c201d6s0    /                  hfs   defaults 0 1 850144122 1
/dev/dsk/c201d5s0    /win               hfs   defaults 1 2 850144127 1
castor:/win          /net/castor_win    nfs   rw,suid  0 0 850144231 0
miranda:/win         /net/miranda_win   nfs   rw,suid  0 0 850144291 0
spica:/usr/opt       /opt               nfs   rw,suid  0 0 850305936 0
triton:/win          /net/triton_win    nfs   rw,suid  0 0 850305936 0
```

In this case there are two local drivers (/dev/dsk/c201d6s0 is mounted as the root directory and /dev/dsk/c201d5s0 is mounted locally as /win). There are also four remote directories which are mounted from remote servers (castor, miranda, spica and triton). The directory mounted from castor is the /win directory and it is mounted locally as /net/castor_win. hfs defines a UNIX format disk and nfs defines that the disk is mounted over NFS.

A disk can be unmounted from a system using the umount command, e.g.

```
% umount miranda:/win
```

7.4.3 NIS commands

NIS commands allow the maintenance of network information. The main commands are as follows:

- domainname which displays or changes the current NIS domain name.
- ypcat which lists the specified NIS map contents.
- ypinit which, on a master server, builds a map using the networking files in /etc. On a slave server the map is built using the master server.
- ypmake which is a script that builds standard NIS maps from files such as /etc/passwd, /etc/groups, and so on.

- `ypmatch` which prints the specified NIS map data (values) associated with one or more keys.
- `yppasswd` which can be used to change (or install) a user's password in the NIS `passwd` map.
- `ypwhich` which is used to print the host name of the NIS server supplying NIS services to an NIS client.
- `ypxfr` which transfers the NIS map from one slave server to another.

For example the command:

```
ypcat group.byname
```

lists the group name, the group ID and the members of the group. Here is an example of changing a user's password for the NIS domain. In this case, the user `bill_b` changes the network-wide password on the master server `pollux`.

```
% yppasswd
Changing NIS password for bill_b...
Old NIS password: ********
New password: *******
Retype new password: *******
The NIS passwd has been changed on pollux, the master NIS passwd server.
```

The next example uses the `ypcat` program.

```
% ypcat group
students:*:200:msc01,msc02,msc03,msc04
nogroup:*:-2:
daemon::5:root,daemon
users::20:root,msc08
other::1:root,hpdb
root::0:root
```

The next example shows the `ypmake` command file which rebuilds the NIS database.

```
# ypmake
For NIS domain eece:
The passwd map(s) are up-to-date.
Building the group map(s)... group build complete.
   Pushing the group map(s):  group.bygid  group.byname
The hosts map(s) are up-to-date.
The networks map(s) are up-to-date.
The rpc map(s) are up-to-date.
The services map(s) are up-to-date.
The protocols map(s) are up-to-date.
The netgroup map(s) are up-to-date.
ypmake complete:
```

7.4.4 Network configuration files

The main files used to setup networking are as follows:

`/etc/checklist` is a list of directories or files that are automatically mounted at boot time.

`/etc/exports` contains a list of directories or files that clients may import.

`/etc/inetd.conf` contains information about servers started by `inetd` (the Internet daemon). Listing 7.1 shows an example of the inetd.conf file. It can be seen that it includes the service name, socket type (stream or datagram), the protocol (TCP or UDP), flags, the owner, the server path, and any other arguments. Lines which begin with the '#' character are ignored by `inetd`. It can be seen that many of the Internet-related programs, such as FTP and TELNET are started here, as well as the login program (LOGIN).

`/etc/netgroup` contains a mapping of network group names to a set of node, user, and NIS domain names.

`/etc/netnfsrc` is automatically started at run time and initiates the required daemons and servers, and defines the node as a client or server.

`/etc/rpc` maps the RPC program names to the RPC program numbers and vice versa.

`/usr/adm/inetd.sec` checks the Internet address of the host requesting a service against the list of hosts allowed to use the service.

Listing 7.1 (inetd.conf)

```
# <service_name> <sock_type> <proto> <flags> <user> <server_path> <args>
#
# Echo, discard and daytime are used primarily for testing.
echo     stream   tcp   nowait   root  internal
echo     dgram    udp   wait     root  internal
discard  stream   tcp   nowait   root  internal
discard  dgram    udp   wait     root  internal
daytime  stream   tcp   nowait   root  internal
daytime  dgram    udp   wait     root  internal
time     dgram    udp   wait     root  internal
#
# These are standard services.
ftp      stream   tcp   nowait   root  /usr/sbin/tcpd /usr/sbin/wu.ftpd
telnet   stream   tcp   nowait   root  /usr/sbin/tcpd /usr/sbin/in.telnetd
#
# Shell, login, exec and talk are BSD protocols.
shell    stream   tcp   nowait   root  /usr/sbin/tcpd /usr/sbin/in.rshd
login    stream   tcp   nowait   root  /usr/sbin/tcpd /usr/sbin/in.rlogind
#exec    stream   tcp   nowait   root  /usr/sbin/tcpd /usr/sbin/in.rexecd
talk     dgram    udp   wait     root  /usr/sbin/tcpd /usr/sbin/in.ntalkd
ntalk    dgram    udp   wait     root  /usr/sbin/tcpd /usr/sbin/in.ntalkd
#
# Pop mail servers
pop3     stream   tcp   nowait   root  /usr/sbin/tcpd /usr/sbin/in.pop3d
#
bootps   dgram    udp   wait     root  /usr/sbin/tcpd /usr/sbin/in.bootpd
#
# Finger, systat and netstat give out user information which may be
# valuable to potential "system crackers."  Many sites choose to disable
# some or all of these services to improve security.
finger   stream   tcp   nowait   daemon   /usr/sbin/tcpd /usr/sbin/in.fingerd
systat   stream   tcp   nowait   guest    /usr/sbin/tcpd /usr/bin/ps -auwwx
netstat  stream   tcp   nowait   guest    /usr/sbin/tcpd /bin/netstat -f inet
#
# Ident service is used for net authentication
auth     stream   tcp   nowait   root     /usr/sbin/in.identd    in.identd
```

Daemons

Networking programs normally initiate networking daemons which are background processes and are always running. Their main function is to wait for a request to perform a task. Typical daemons are:

`biod` which is asynchronous block I/O daemons for NFS clients.

`inetd` which is an Internet daemon that listens to service ports. It listens for service requests and calls the appropriate server. The server it calls depends on the contents of the `/etc/inetd.conf` file.

`nfsd` which is the NFS server daemon. It is used by the client for reading and writing to a remote directory and it sends a request to the remote server `nfsd` process.

`pcnfsd` which is a PC user authentication daemon.

`portmap` which is an RPC program to port number conversion daemon. When a client makes an RPC call to a given program number, it first contacts `portmap` on the server node to determine the port number where RPC requests should be sent.

Here is an extract from the processes that run a networked UNIX workstation:

```
 UID   PID  PPID  C    STIME TTY       TIME COMMAND
root   100     1  0  Dec  9  ?         0:00 /etc/portmap
root   138     1  0  Dec  9  ?         0:00 /etc/inetd
root    93     1  0  Dec  9  ?         0:00 /etc/rlbdaemon
root   104     1  0  Dec  9  ?         9:20 /usr/etc/ypserv
root   106     1  0  Dec  9  ?         0:00 /etc/ypbind
root   122   120  0  Dec  9  ?         0:00 /etc/nfsd 4
root   120     1  0  Dec  9  ?         0:00 /etc/nfsd 4
root   116     1  0  Dec  9  ?         0:00 /usr/etc/rpc.yppasswdd
root   123   120  0  Dec  9  ?         0:00 /etc/nfsd 4
root   128     1  0  Dec  9  ?         0:02 /etc/biod 4
root   131     1  0  Dec  9  ?         0:00 /etc/pcnfsd
root   133     1  0  Dec  9  ?         0:00 /usr/etc/rpc.statd
root   135     1  0  Dec  9  ?         0:00 /usr/etc/rpc.lockd
root  4652     1  0 14:33:15 ?         0:00 /etc/pcnfsd
root  4649     1  0 14:33:15 ?         0:00 /usr/etc/rpc.mountd
```

7.5 Exercises

The following questions are multiple choice. Please select from a–d.

7.5.1 What is the protocol that allows external drives to be mounted over a network:
(a) RPC (b) NIS
(c) XDR (d) NFS

7.5.2 Which is the term used when mounting a drive locally on a drive partition:
(a) Single tree (b) Forest of drives
(c) Mounted drives (d) Transparent drives

7.5.3 Which is the term used when mounting a drive locally onto a global file system:
(a) Single tree (b) Forest of drives
(c) Mounted drives (d) Transparent drives

7.5.4 What is the protocol that allows configuration files to be stored in a certain place:
(a) RPC (b) NIS
(c) XDR (d) NFS

7.5.5 Which operating system uses a single tree:
(a) UNIX (b) Novell NetWare
(c) DOS (d) Microsoft Windows

7.5.6 What is the protocol that allows computers to exchange data in a standard format:

(a)	RPC	(b)	NIS
(c)	XDR	(d)	NFS

7.5.7 Which operating system uses a single tree:

(a)	UNIX	(b)	Novell NetWare
(c)	DOS	(d)	Microsoft Windows

7.5.8 Which important files are maintained by the NIS server:

(a)	`passwd` and `group`	(b)	`passwords` and `group`
(c)	`password` and `group`	(d)	`passwords` and `groups`

7.5.9 In UNIX, which command is used to mount a remote drive:

(a)	`mount`	(b)	`export`
(c)	`fs`	(d)	`mnt`

7.5.10 In UNIX, which file stores the drives to be mounted:

(a)	`/etc/export`	(b)	`/etc/files`
(c)	`/etc/mnttab`	(d)	`/etc/tab`

7.5.11 Investigate the advantages and drawbacks of using a single tree system and a forest of drives. If possible, identify a network which uses one of these types. Which type do UNIX, Novell NetWare and Microsoft Windows use?

7.5.12 Show that 10.21 in decimal is represented by 0100 0001 0010 0011 0101 1100 0010 1001b in IEEE format.

7.5.13 Show that 999.9213 in decimal is represented by 0100 0100 0111 1001 1111 1010 1111 0111b in IEEE format.

7.5.14 Show that –32.59 in decimal is represented by 1100 0010 0000 0010 0101 1100 0010 1001b in IEEE format.

8 Data Communications

8.1 Introduction

The growth in data communications is creating one of the largest and most important industries in the world. It is a technology that brings benefits to virtually every individual in the world. Without it many organizations could not work efficiently. They are also creating industries that never existed before, such as digital TV, electronic commerce, electronic delivery of video and music, and, the best of them all, electronic mail. The trend for transmitting data is to transmit digital information, thus if the original information is in the form of an analogue signal, it must first be converted into a digital form. This can then be transmitted over a digital network. At one time computer type data was sent over a network which was matched to transmitting this type of data, such as Ethernet and Token Ring, and speech was sent over a telephone-type network. The future will see the total integration of both real-time (such as speech) and non-real-time (such as computer-type data) into a single integrated digital network (IDN), as illustrated in Figure 8.1. The true integrator is ATM, which will be covered in Chapter 13. Networking technologies such as Ethernet are likely to remain a standard network connection onto a network, as they have become de-facto standards.

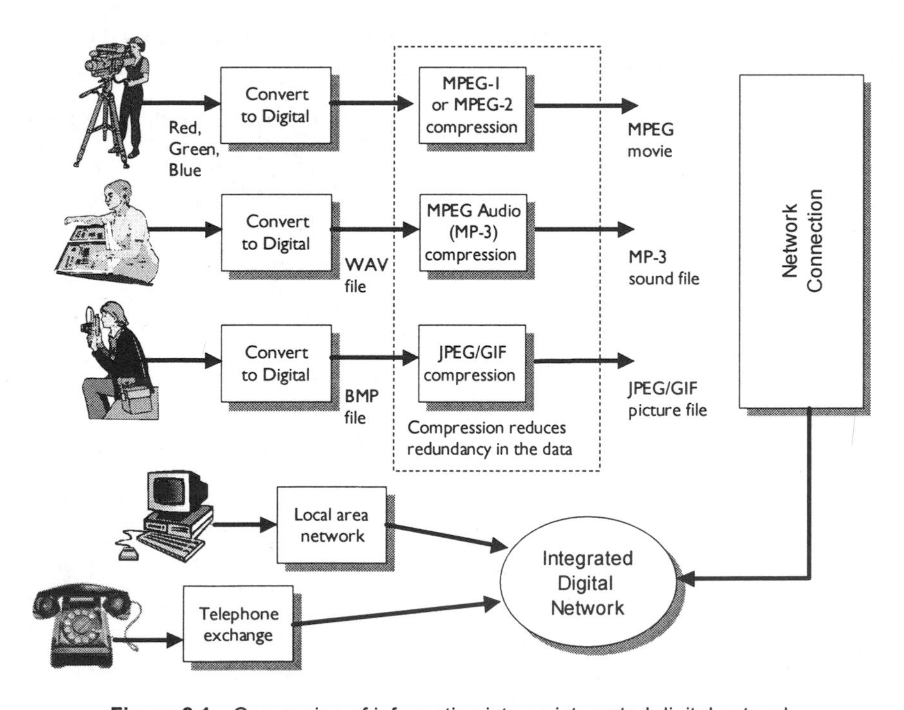

Figure 8.1 Conversion of information into an integrated digital network

The communications channel normally provides the main limitation on the amount of data that can be transmitted. This normally relates to its bandwidth, which can either be dedicated to the transmission of data from one source to a destination, or can be shared between more than one source, to more than one destination. Communications systems vary a great deal in the way they setup a connection, such as:

- **Bandwidth contention, bandwidth sharing or reserved bandwidth.** Some communication systems reserve bandwidth for a connection (such as ISDN and ATM), while others allow systems to contend for it (such as Ethernet). Normally the most efficient scheme is to allow systems to share the bandwidth. In this, some nodes can have more of the available bandwidth, if they require, while others can have a lesser share.
- **Virtual path, dedicated line or datagram.** Some communication systems allow for a virtual path to be setup between the two connected systems, while others support a dedicated line between the two systems. A dedicated line provides a guaranteed bandwidth for the length of the conversation, while a virtual path should support a certain amount of bandwidth, as a connection has been setup to support the data being transmitted. In a datagram-based system, there is no setup for the route and all of the transmitted datagrams (data packets) take an independent path from the source to the destination.
- **Global addressing, local addressing or no addressing.** An addressing structure provides for individual data packets to have an associated destination address. Each of the devices involved in the routing of the data read this address and send the data packet off on the optimal path. This type of addressed system normally uses datagrams. A typical global addressing structure is IP addressing, which is the standard addressing scheme for the Internet. In a non-addressed system the data is not tagged with the destination address, and only contains enough information to get it from one device to another. This is the technique that is typically used in setting up a virtual path. No addressing is used when circuit connection is setup, as all the data takes the same route.

> Shakespeare's networking sayings:
>
> - *To contend to not contend. That is the Ethernet question.*
> - *Whether it is nobler to Ethernet than to ATM.*
> - *A path, a path, my kingdom for a virtual path.*
> - *Oh poor, Token Ring. I knew him so well.*
>
> – W. Buchanan

An important point is that the bandwidth of a complete system is limited to the bandwidth of the element of the system which has the least bandwidth. This is similar to hi-fi systems where the performance of the system is limited by the worst element in the system, as illustrated in Figure 8.6.

8.2 Data transfer

Data is transferred from the source to the destination through a data path. This data can either be passed in a serial manner (one bit at a time) or in parallel (several bits at a time). In a parallel system, the bits are normally passed in a multiple of eight bits at a time. Typical parallel data transmissions are 8 bits, 16 bits, 32 bits, 64 bits or 128 bits wide.

Parallel transmission is normally faster than serial transmission (as it can transmit more bits in a single operation), but requires many more lines (thus requiring more wires in the cable). A parallel data transmission system normally requires extra data handshaking lines to synchronise the flow of data between devices. Serial data transmission normally uses a start and end bit sequence to define the start and end of transmission, as illustrated in Figure 8.2.

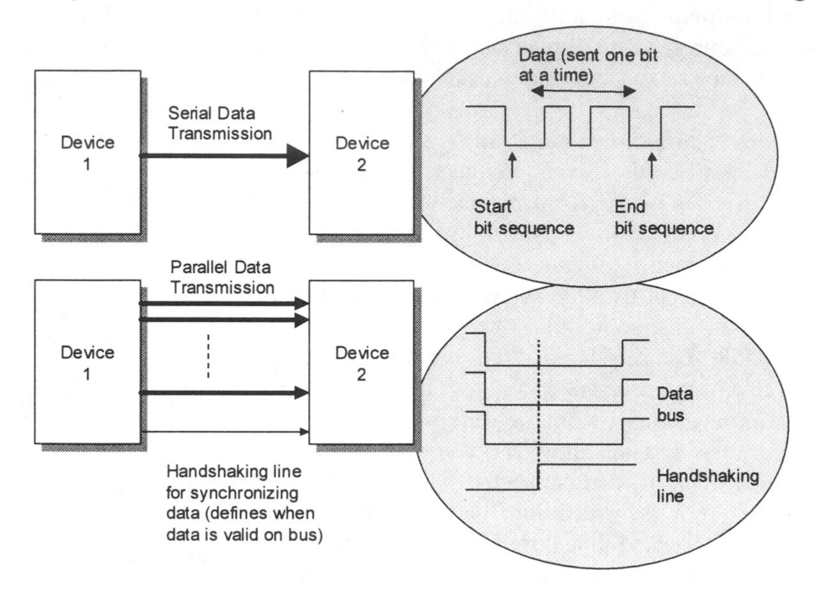

Figure 8.2 Serial and parallel data transmission

Parallel busses are typically used for local transmission systems, or where there are no problems with cables with a relatively large number of wires. Typically, parallel communication systems are SCSI and the parallel port, and typical serial transmission systems are RS-232 and Ethernet, and most communication systems.

Serial communications can operate at very high transmission rates; the main limiting factor is the transmission channel and the transmitter/receiver electronics. Gigabit Ethernet, for example, uses a transmission rate of 1 Gbps (125 MB/s) over high-quality twisted-pair copper cables, or over fiber-optic cables (although this is a theoretical rate as more than one bit is sent at a time). For a 32-bit parallel communication system, this would require a clocking rate of only 31.25 MHz (which requires much lower quality connectors and cables than the equivalent serial interface).

The main types of communication are:

- **Simplex communication.** Only one device can communicate with the other, and thus only requires handshaking lines for one direction.
- **Half-duplex communication.** This allows communications from one device to the other, in any direction, and thus requires handshaking for either direction.
- **Full-duplex communications.** This allows communication from one device to another, in either direction, at the same time. A good example of this is in a telephone system, where a caller can send and receive at the same time. This requires separate transmit and receive data lines, and separate handshaking for either direction.

Often the transmitter and receiver are operating at different speeds, where the transmitter can send faster than the receiver can receive the data, or vice-versa. To stop too much data being transmitted before it can be processed, there must be an orderly transfer of data. This is normally achieved with handshaking, either with special handshaking lines, or by using software methods (such as sending special data characters to start and stop the flow of data).

8.3 Data transfer rates

The amount of data that a system can transfer at a time is normally defined either in bits per second (bps) or bytes per second (B/s). The more bytes (or bits) that can be transferred the faster the transfer will be. Typically serial busses are defined in bps, whereas parallel busses use B/s.

The transfer of the data occurs at regular intervals, which are defined by the period of the transfer clock. This period is either defined as a time interval (in seconds), or as a frequency (in Hz). For example, if a clock operates at a rate of 1 000 000 cycles per second, its frequency is 1 MHz, and its time interval will be one millionth of a second (1×10^{-6} s).

In general, if f is the clock frequency (in Hz), then the clock period (in seconds) will be

$$T = \frac{1}{f} \text{ sec}$$

Conversion from clock frequency to clock time interval

For example, if the clock frequency is 8 MHz, then the clock period will be:

$$T = \frac{1}{8\times10^{6}} = 0.000000125 \text{ sec}$$
$$= 0.125\,\mu\text{s}$$

Example of a calculation of clock time interval from clock frequency

The data transfer rate (in bits/second) is defined as:

$$\text{Data transfer rate (bps)} = \frac{\text{Number of bits transmitted per operation (bits)}}{\text{Transfer time per operation (s)}}$$

If operated with a fixed clock frequency for each operation then the data transfer rate (in bits/second) will be

$$\text{Data transfer rate (bps)} = \text{Number of bits transmitted per operation (bits)} \times \text{Clocking rate (Hz)}$$

For example, the ISA bus uses an 8 MHz (8×10^{6} Hz) clocking frequency and can transfer 16 bits at a time. Thus the maximum data transfer rate (in bps) will be:

$$\text{Data transfer rate} = 16\times8\times10^{6} = 128\times10^{6} \text{ bps} = 128 \text{ Mbps}$$

Often it is required that the data rate is given in B/s, rather than bps. To convert from bps to B/s, the bps value is divided by eight. Thus to convert 128 Mbps to B/s

$$\text{Data transfer rate} = 128 \text{ Mbps}$$
$$= \frac{128}{8} \text{Mbps} = 16 \text{ MB/s}$$

Example conversion from bps to B/s

For serial communication, if the time to transmit a single bit is 104.167 μs then the maximum data rate will be

$$\text{Data transfer rate} = \frac{1}{104.167\times10^{-6}} = 9600 \text{ bps}$$

Example conversion to bps for a serial transmission with a given transfer time interval

8.4 Electrical signals

Any electrical signal can be analyzed either in the time-domain or in the frequency-domain. A time-varying signal contains a range of frequencies. If the signal is repetitive (that is, it repeats after a given time) then the frequencies contained in it will also be discrete.

The standard form of a single frequency signal is:

$$V(t) = V \sin(2\pi ft + \theta)$$

where $v(t)$ is the time varying voltage (V), V is the peak voltage (V), f the signal frequency (Hz) and θ its phase (°).

A signal in the time-domain is a time varying voltage. In the frequency domain it is voltage amplitude against frequency. Figure 8.3 shows how a single frequency is represented in the time-domain and the frequency domain. It shows that for a signal with a period T that the frequency of the signal is 1/T Hz. The signal frequency is represented in the frequency domain as a single vertical arrow at that frequency, where the amplitude of the arrow represents the amplitude of the signal.

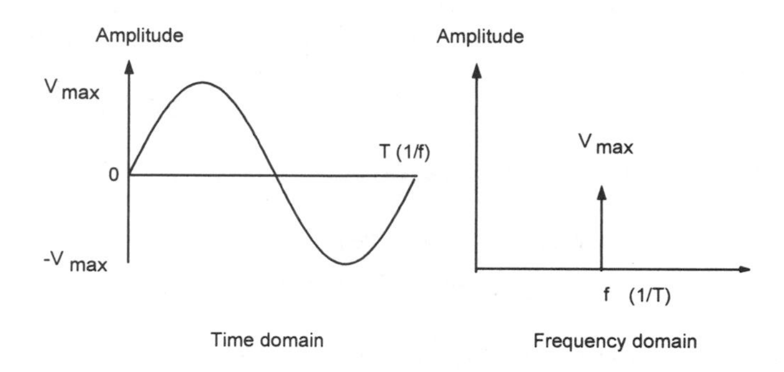

Figure 8.3 Representation of signal in frequency and time domain

8.5 Bandwidth

In general, in a communication system, bandwidth is defined as the range of frequencies contained in a signal. As an approximation it is the difference between the highest and the lowest signal frequency, as illustrated in Figure 8.4. For example, if a signal has an upper frequency of 100 MHz and a lower of 75 MHz then the signal bandwidth is 25 MHz. Normally, the larger the bandwidth the greater the information that can be sent. Unfortunately, normally, the larger the bandwidth the more noise that is added to the signal. The bandwidth of a signal is normally limited to reduce the amount of noise and to increase the number of signals transmitted. Table 8.1 shows typical bandwidths for different signals.

The two most significant limitations on a communication system performance are noise and bandwidth. In a data communications system the bandwidth is normally defined in terms for the maximum bit rate. As will be shown this can be approximated to twice the maximum frequency of transmission through the system.

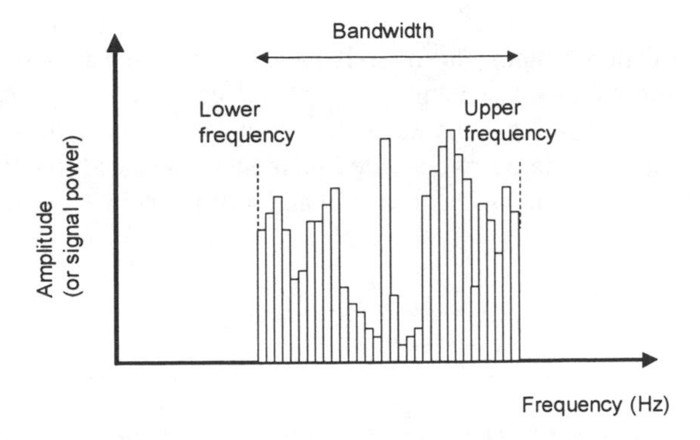

Figure 8.4 Signal bandwidth

Table 8.1 Typical signal bandwidths

Application	*Bandwidth*
Telephone speech	4 kHz
Hi-fi audio	20 kHz
FM radio	200 kHz
TV signals	6 MHz
Satellite communications	500 MHz

8.6 Signal frequency content

The greater the rate of change of an electronic signal the higher the frequencies that will be contained in its the frequency content. Figure 8.5 shows two repetitive signals. The upper signal has a DC component (zero frequency) and four frequencies, f_1 to f_4. The lower signal has a greater rate of change than the upper signal and it thus contains a higher frequency content, from f_1 to f_6.

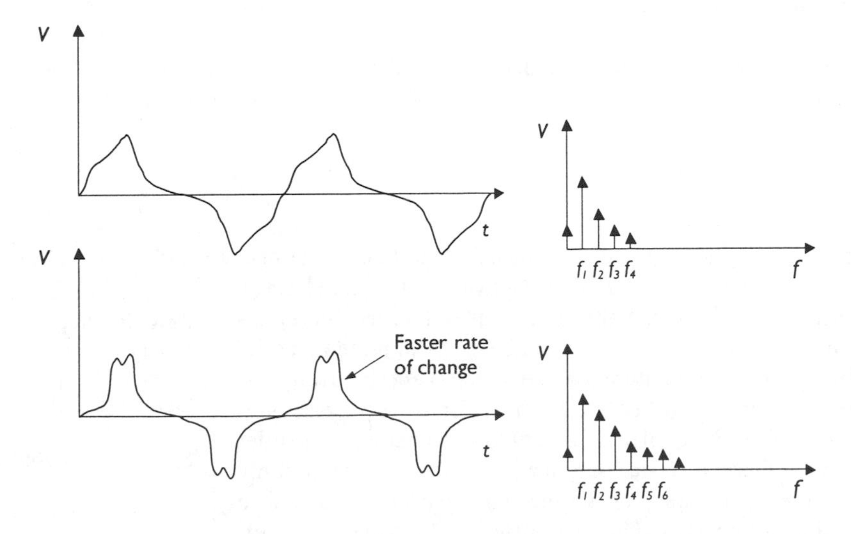

Figure 8.5 Frequency content of two repetitive signals

Typically, in a cascaded system, the overall bandwidth of the system is defined by the lowest bandwidth of the cascaded elements. Digital pulses have a very high rate of change around their edges. Thus, digital signals normally require a larger bandwidth than analogue signals. In a digital system made up of cascaded elements, each with its own bandwidth, the overall bandwidth will be given by the lowest bandwidth element (as defined in bps), as illustrated in Figure 8.6.

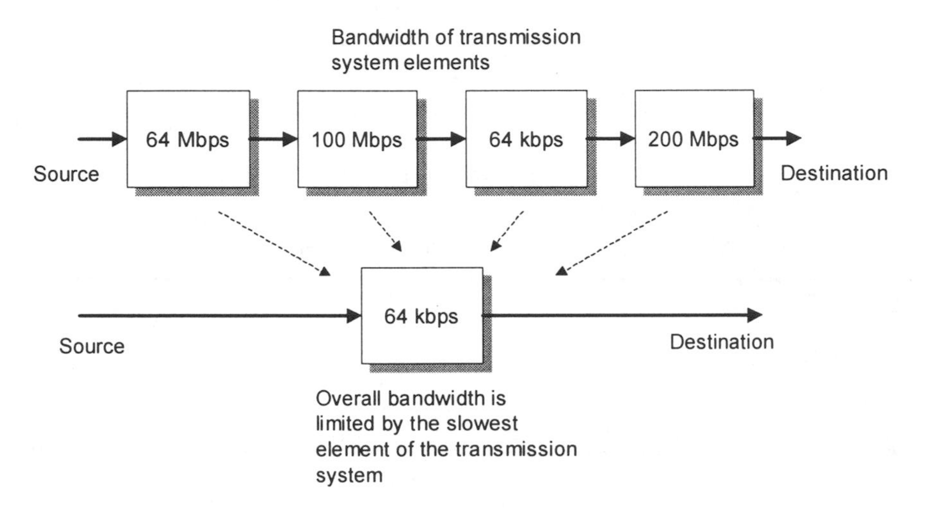

Figure 8.6 Overall bandwidth related to the system bandwidth elements

8.7 Noise and signal distortion

Noise is any unwanted signal added to information transmission. The main sources of noise on a communication system are:

- **Thermal noise.** Thermal noise occurs from the random movement of electrons in a conductor and is independent of frequency. The noise power can be predicted from the formula:

$$N = k\,T\,B$$

where N is the noise power in Watts, k is Boltzman's constant (1.38×10^{-23} J/K), T is the temperature (in K) and B the bandwidth of channel (Hz). Thermal noise is predictable and is spread across the bandwidth of the system. It is unavoidable but can be reduced by reducing the temperature of the components causing the thermal noise. Many receivers which detect very small signals require to be cooled to a very low temperature in order to reduce thermal noise. A typical example is in astronomy where the temperature of the receiving sensor is reduced to almost absolute zero. Thermal noise is a fundamental limiting factor of any communications system.

> Best file compression around:
>
> DEL *.* {DOS}
>
> or rm –r *.* {UNIX}
>
> gives 100% compression

- **Cross-talk.** Electrical signals propagate with an electric and a magnetic field. If two conductors are laid beside each other then the magnetic field from one couples into the other. This is known as crosstalk, where one signal interferes with another. Analogue systems tend to be affected more by crosstalk than digital ones.
- **Impulse noise.** Impulse noise is any unpredictable electromagnetic disturbance, such as from lightning or from energy radiated from an electric motor. It is normally characterized by a relatively high energy, short duration pulse. It is of little importance to an analogue transmission system as it can usually be filtered out at the receiver. However, impulse noise in a digital system can cause the corruption of a significant number of bits.

A signal can be distorted in many ways, especially due to the electrical characteristics of the transmitter and receiver and also the characteristics of the transmission media. An electrical cable contains inductance, capacitance and resistance. The inductance and capacitance have the effect of distorting the shape of the signal whereas resistance causes the amplitude of the signal to reduce (and also to lose power).

8.8 Capacity

The information-carrying capacity of a communications system is directly proportional to the bandwidth of the signals it carries. The greater the bandwidth, the greater the information-carrying capacity. An important parameter for determining the capacity of a channel is the *signal-to-noise ratio* (SNR). This is normally defined in decibels as the following:

$$\frac{S}{N}(dB) = 10\log_{10}\frac{Signal\ Power}{Noise\ Power}$$

In a digital system, Nyquist predicted that the maximum capacity, in bits/sec, of a channel subject to noise is given by:

$$Capacity = B.\log_2\left[1+\frac{S}{N}\right] \quad \text{bits/sec}$$

For example, if the signal power is 100 mW, and the noise power is 20 nW, then:

$$\frac{S}{N}(dB) = 10\log_{10}\frac{100\times10^{-3}}{20\times10^{-9}}\ \text{dB}$$

$$\frac{S}{N}(dB) = 10\times\log_{10}\left[5\times10^{6}\right]\text{dB}$$

$$\frac{S}{N}(dB) = 6.7\ \text{dB}$$

where B is the bandwidth of the system and S/N is the signal-to-noise ratio. For example if the signal-to-noise ratio is 10 000 and the bandwidth is 100 kHz, then the maximum capacity is:

$$\begin{aligned} Capacity &= 10^5.\log_2\left(1+10^4\right)\ \text{bits/sec} \\ &\approx 10^5.\frac{\log_{10}\left(10^4\right)}{\log_{10}(2)}\ \text{bits/sec} \\ &= 13.3\times10^5\ \text{bits/sec} \end{aligned}$$

$$\log_x(y) = \frac{\log_{10}(y)}{\log_{10}(x)}$$

Attenuation is the loss of signal power and is normally frequency dependent. A low-pass channel is one which attenuates, or reduces, the high frequency components of the signal more than the low frequency parts. A band-pass channel attenuates both high and low frequencies more than a band in the middle.

The bandwidth of a system is usually defined as the range of frequencies passed which are not attenuated by more than half their original power level. The end points in Figure 8.7 are marked at 3 dB (the –3 dB point) above the minimum signal attenuation.

Bandwidth is one of the most fundamental factors as it limits the amount of information which can be carried in a channel at a given time. It can be shown that the maximum possible symbol bit rate of a digital system on a noiseless, band-limited channel is twice the channel bandwidth, or:

Maximum symbol rate (symbols/sec)
= 2 × Bandwidth of channel

If a signal is transmitted over a channel which only passes a narrow range of frequencies than is contained in the signal then the signal will be distorted. This is illustrated in Figure 8.8, where the maximum frequency content occurs when the 10101... bit sequence occurs. The minimum frequency content of this bit pattern will be B Hz. The symbol bit rate will thus be twice the highest frequency of the channel. The reason that the rate is referred to as a symbol rate, and not a bit rate, is that the symbol rate can differ from the bit rate. This is because more than one bit can be sent for each symbol. This typically happens with modems, where more than one bit is sent for every symbol. For example, several amplitudes of symbols are to be sent, such as four amplitudes can be used to represent two bits. The symbol rate for speech limited channels will be 8,000 symbols per second (as the maximum frequency is 4 kHz). As four bits can be sent for every symbol, the bit rate will be 32 kbps.

If the frequency characteristics of the channel are known then the receiver can be given appropriate compensatory characteristics. For example, a receiving amplifier could boost higher frequency signals more than the lower frequencies. This is commonly done with telephone lines, where it is known as channel equalization.

Intel Processors that shook the world

4004: Nov 1971, 108 kHz, 4-bit data bus, 2300 transistors, 640 B of addressable memory.

8008: Apr 1972, 200 kHz, 8-bit data bus, 3500 transistors, 16 kB of addressable memory.

8080: Apr 1974, 2 MHz, 8-bit data bus, 6000 transistors, 64 kB of addressable memory.

8086: June 1978, 5/8/10 MHz, 16-bit data bus, 29000 transistors, 1 MB of addressable memory.

8088: June 1979, 5/8 MHz, 8-bit data bus, 29000 transistors, 1 MB of addressable memory.

80386: Oct 1985, 16/20/25/33 MHz, 32-bit data bus, 275000 transistors, 4 GB of addressable memory.

80486: Apr 1989, 25/33/50 MHz, 32-bit data bus, 1.2million transistors, 4 GB of addressable memory.

Pentium: Mar 1993, 60/66 MHz, 64-bit data bus, 3.1 million transistors, 4 GB of addressable memory.

Pentium II/III: Mar 1997, 200 MHz and on, 64-bit data bus, 7.5 million transistors, 64 GB of addressable memory.

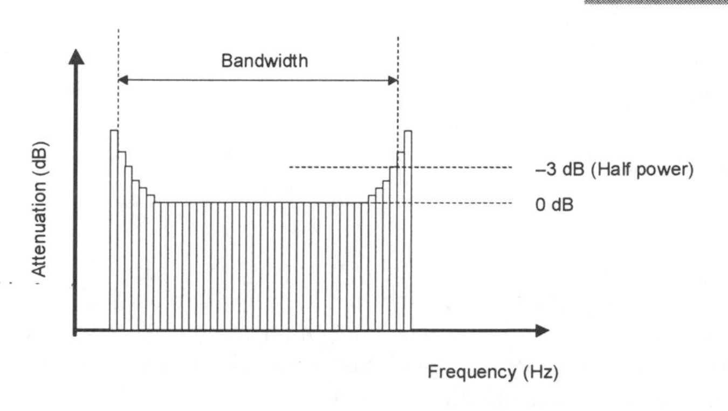

Figure 8.7 Bandwidth of a channel

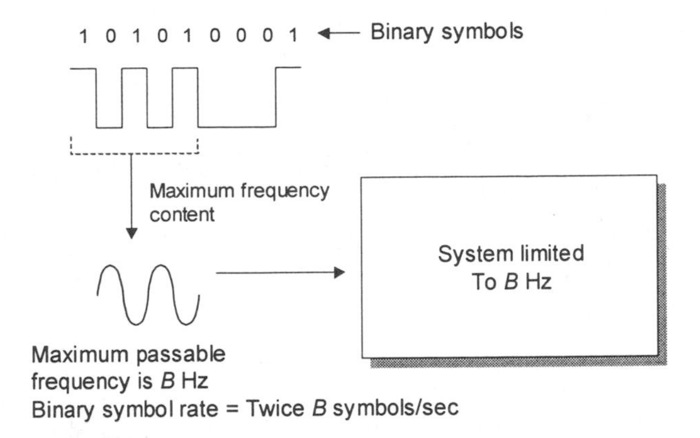

Figure 8.8 Maximum binary symbol rate is twice the frequency of the bandwidth

8.9 Modulation

Modulation allows the transmission of a signal through a transmission medium by carrying it on a carrier wave (which can propagate through a given media). It also adds extra information that allows the receiver to pick-up the signal (allowing the transmitted signal to be 'tuned-into'). In radio transmissions, each audio signal from the radio station could not be propagated through air without adding it to a high-frequency carrier wave. The different carrier frequencies also allows many radio stations to be transmitted simultaneously, and received by selecting the required carrier frequency.

> Do witches run spell checkers?
>
> A computer's attention span is as long as its power cord.

There are three main methods used to modulate: amplitude, frequency and phase modulation. With amplitude modulation (AM) the information signal varies the amplitude of a carrier wave. In frequency modulation (FM) it varies the frequency of the wave and with phase modulation (PM) it varies the phase.

8.9.1 Amplitude modulation (AM)

AM is the simplest form of modulation where the information signal modulates a higher frequency carrier. The modulation index, *m*, is the ratio of the signal amplitude to the carrier amplitude, and is always less than or equal to 1. It is given by:

$$m = \frac{V_{signal}}{V_{carrier}}$$

Figure 8.9 shows three differing modulation indices. In Figure 8.9 (a) the information signal has a relatively small amplitude compared with the carrier signal, given a relatively small modulation index. In Figure 8.9 (b) the signal amplitude is approximately half of the carrier amplitude, and in Figure 8.9 (c) the signal amplitude is almost equal to the carrier's amplitude (giving a modulation index of near unity).

AM is generally susceptible to noise and fading as it is dependent on the amplitude of the modulated wave. Binary information can be transmitted by assigning discrete amplitudes to bit patterns.

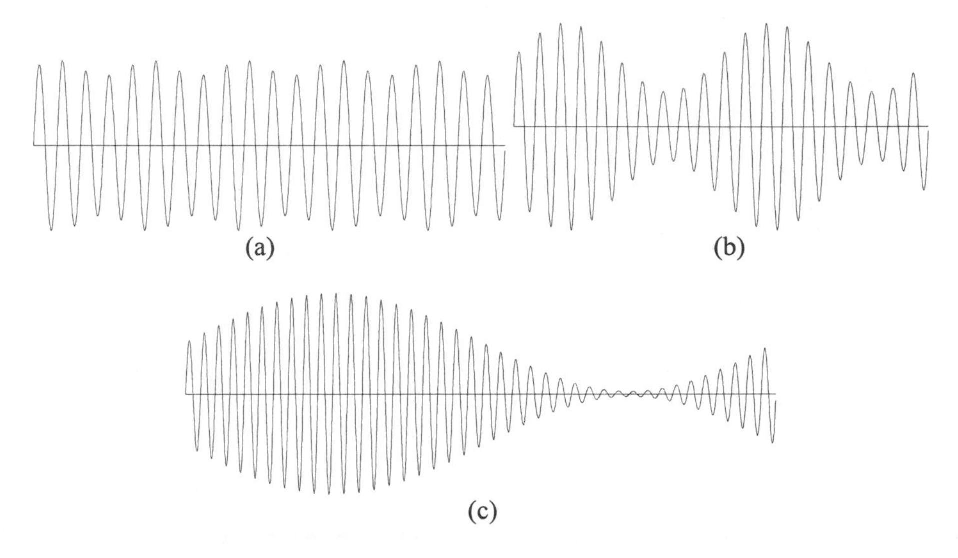

Figure 8.9 AM waveform

8.9.2 Frequency modulation (FM)

Frequency modulation involves the modulation of the frequency of a carrier. FM is preferable to AM as it is less affected by noise because the information is contained in the change of frequency and not the amplitude. Thus, the only noise that affects the signal is limited to a small band of frequencies contained in the carrier. The information in an AM waveform is contained in its amplitude which can be easily affected by noise.

Figure 8.10 shows a modulator/demodulator FM system. A typical device used in FM is a Phased-Locked Loop (PLL) which converts the received frequency-modulated signal into a signal voltage. It locks onto frequencies within a certain range (named the capture range) and follows the modulated signal within a given frequency band (named the lock range). Typically binary information can be sent by using two frequencies, the upper frequency representing a zero, and the lower frequency representing a one. Modems can transmit binary information by using different frequencies to represent bit patterns.

8.9.3 Phase modulation (PM)

Phase modulation involves the modulating the phase of the carrier. PM is less affected by noise than AM because the information is contained in the change of phase and, like FM, not in its amplitude. As with FM, binary information can be transmitted by assigning discrete phases to bit sequences. For example, a zero phase could represent a zero, and a 180° phase shift could represent a one (see Section 8.10).

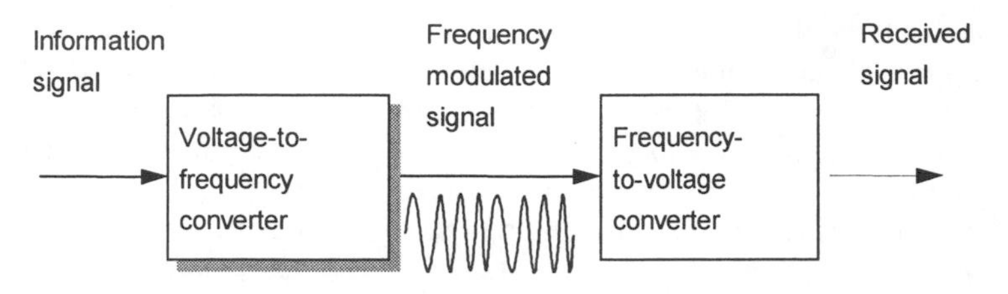

Figure 8.10 Frequency modulation

8.10 Digital modulation

Digital modulation changes the characteristic of a carrier according to binary information. With a sine wave carrier the amplitude, frequency or phase can be varied. Figure 8.11 illustrates the three basic types: amplitude-shift keying (ASK), frequency-shift keying (FSK) and phase-shift keying (PSK).

8.10.1 Frequency-shift keying (FSK)

FSK, in the most basic case, represents a 1 (a mark) by one frequency and a 0 (a space) by another. These frequencies lie within the bandwidth of the transmission channel. On a V.21, 300 bps, full-duplex modem the originator modem uses the frequency 980 Hz to represent a mark and 1180 Hz a space. The answering modem transmits with 1650 Hz for a mark and 1850 Hz for a space. These four frequencies allow the caller originator and the answering modem to communicate at the same time; that is, full-duplex communication.

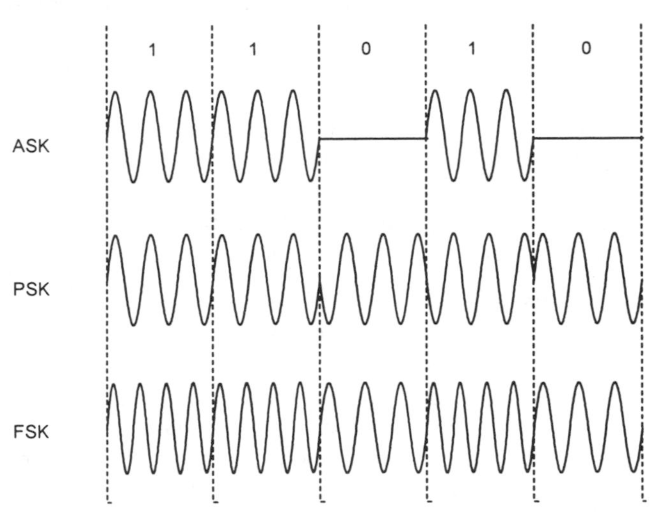

Figure 8.11 Waveforms for ASK, PSK and FSK

FSK modems are inefficient in their use of bandwidth, with the result that the maximum data rate over normal telephone lines is 1800 bps. Typically, for rates over 1200 bps, other modulation schemes are used.

8.10.2 Phase-shift keying (PSK)

In coherent PSK a carrier gets no phase shift for a 0 and a 180° phase shift for a 1, as given next:

$$0 \quad \Rightarrow \quad 0^\circ \qquad\qquad 1 \quad \Rightarrow \quad 180^\circ$$

Its main advantage over FSK is that as it uses a single frequency it uses much less bandwidth. It is thus less affected by noise, and has an advantage over ASK because its information is not contained in the amplitude of the carrier, thus again it is less affected by noise

8.10.3 M*-ary modulation*

With *M*-ary modulation a change in amplitude, phase or frequency represents one of *M* possible signals. It is possible to have *M*-ary FSK, *M*-ary PSK and *M*-ary ASK modulation schemes. This is where the baud rate differs from the bit rate. The bit rate is the true measure of the rate of the line, whereas the baud rate only indicates the signalling element rate, which might be a half or a quarter of the bit rate.

For four-phase differential phase-shift keying (DPSK) the bits are grouped into two and each group is assigned a certain phase shift. For two bits there are four combinations: a 00 is coded as 0°, 01 coded as 90°, and so on:

00 ⇒	0°	01 ⇒	90°
11 ⇒	180°	10 ⇒	270°

It is also possible to change a mixture of amplitude, phase or frequency. *M*-ary amplitude-phase keying (APK) varies both the amplitude and phase of a carrier to represent *M* possible bit patterns.

M-ary quadrature amplitude modulation (QAM) changes the amplitude and phase of the carrier. 16-QAM uses four amplitudes and four phase shifts, allowing it to code four bits at a time. In this case, the baud rate will be a quarter of the bit rate.

Typical technologies for modems are:

FSK	– used up to 1200 bps
Four-phase DPSK	– used at 2400 bps
Eight-phase DPSK	– used at 4800 bps
16-QAM	– used at 9600 bps

Most modern modems operate with V.90 (56kbps), V.22bis (2400 bps), V.32 (9600 bps), V.32bis (14 400 bps); some standards are outlined in Table 8.2. The V.32 and V.32bis modems can be enhanced with echo cancellation. They also typically have built-in compression using either the V.42bis standard or MNP level 5.

Table 8.2 Example modems

Type	*Bit rate(bps)*	*Modulation*
V.21	300	FSK
V.22	1 200	PSK
V.22bis	2 400	ASK/PSK
V.27ter	4 800	PSK
V.29	9 600	PSK
V.32	9 600	ASK/PSK
V.32bis	14 400	ASK/PSK
V.34	28 800	ASK/PSK

8.10.4 V.42bis and MNP compression

There are two main standards used in modems for compression. The V.42bis standard is defined by the ITU and the MNP (Microcom Networking Protocol) has been developed by a company named Microcom. Most modems will try to compress using V.42bis but if this fails they try MNP level 5. V.42bis uses the Lempel-Ziv algorithm, which builds dictionaries of code words for recurring characters in the data stream. These code words normally take up fewer bits than the uncoded bits. V.42bis is associated with the V.42 standard which covers error correction.

8.10.5 V.22bis modems

V.22bis modems allow transmission at up to 2400 bps. It uses four amplitudes and four phases. Figure 8.12 shows the 16 combinations of phase and amplitude for a V.22bis modem. It can be seen that there are 12 different phase shifts and four different amplitudes.

Each transmission is known as a symbol, thus each transmitted symbol contains 4 bits. The transmission rate for a symbol is 600 symbols per second (or 600 Baud), thus the bit rate will be 2 400 bps.

Trellis coding tries to ensure that consecutive symbols differ as much as possible.

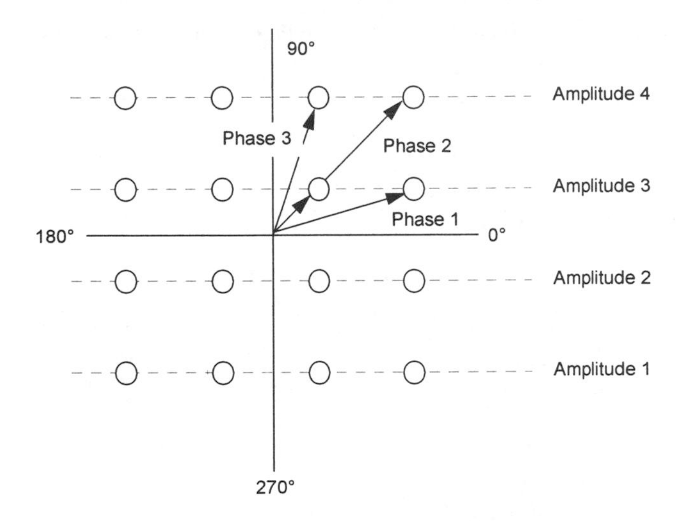

Figure 8.12 Phase and amplitude coding for V.32

8.10.6 V.32 modems

V.32 modems include echo cancellation which allows signals to be transmitted in both directions at the same time. Previous modems used different frequencies to transmit on different channels. Echo cancellation uses DSP (digital signal processing) to subtract the sending signal from the received signal.

V.32 modems use trellis encoding to enhance error detection and correction. They encode 32 signaling combinations of amplitude and phase. Each of the symbols contains four data bits and a single trellis bit (for error detection). The basic symbol rate is 2400 bps; thus the actual data rate will be 9600 bps. A V.32bis modem uses seven bits per symbol; thus the data rate will be 14 400 bps (2400 × 6).

8.11 Multiplexing

Multiplexing is a method of sending information from many sources over a single transmission media. For example, satellite communications and optical fibers allow many information channels to be transmitted simultaneously. There are two main methods of achieving this, either by separation in time with time-division multiplexing (TDM) or separation in frequency with frequency-division multiplexing (FDM).

8.11.1 Frequency-Division Multiplexing (FDM)

With FDM each channel uses a different frequency band. An example of this is FM radio and satellite communications. With FM radio, many channels share the same transmission media but are separated into different carrier frequencies. Satellite communication normally involves an earth station transmitting on one frequency (the up-link frequency) and the satellite relays this signal at a lower frequency (the down-link frequency).

Figure 8.13 shows an FDM radio system where each radio station is assigned a range of frequencies for their transmission. The receiver then tunes into the required carrier frequency.

8.11.2 Time-Division Multiplexing (TDM)

With TDM different sources have a time slot in which their information is transmitted. The most common type of modulation in TDM systems is pulsed code modulation (PCM). With PCM, analogue signals are sampled and converted into digital codes. These are then transmitted as binary digits.

In a PCM-TDM system, several voice-band channels are sampled and converted into PCM codes. Each channel gets a time slot and each time slot is built up into a frame. The complete frame has extra data added to it to allow synchronization. Figure 8.14 shows a PCM-TDM system with three sources.

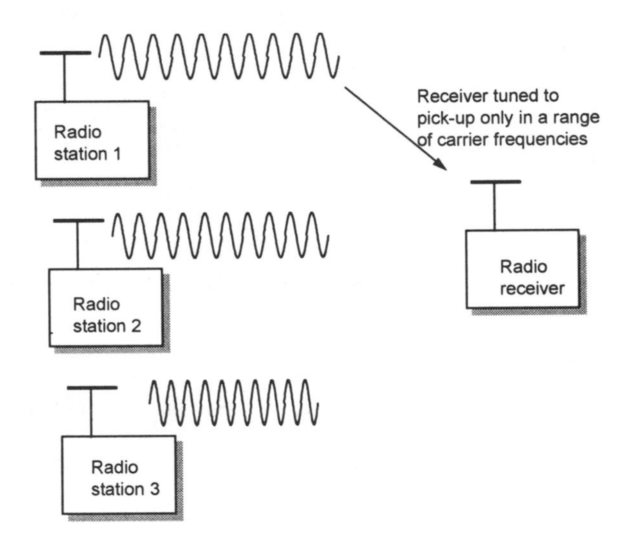

Figure 8.13 FDM radio system

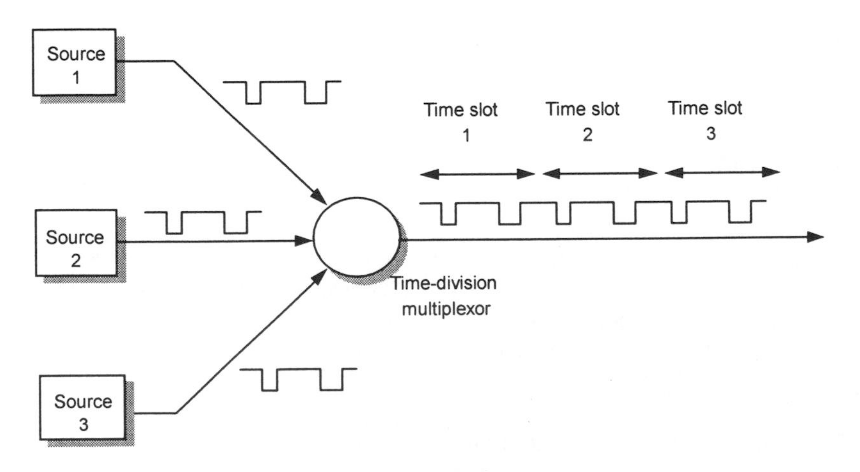

Figure 8.14 TDM system

8.12 Frequency carrier

Often a digital signal cannot be transmitted over channel without being carried on a carrier frequency. The frequency carrier of a signal is important and is chosen for several reasons, such as the:

- Signal bandwidth.
- Signal frequency spectrum.
- Transmission channel characteristics.

> In a coaxial cable the dielectric constant is approximately 9.
>
> Thus, the speed of propagation will be one-third the speed of light, that is:
>
> **100,000,000 m/s**
>
> In 1 ns (1×10^{-9}s), it will travel:
>
> Distance $= 1\times10^{8}\times10^{-9}$
> $= 0.1$ m = **10 cm**

Figure 8.15 shows the frequency spectrum of electromagnetic (EM) waves. The microwave spectrum is sometimes split into millimeter wave and microwaves and the radio spectrum splits into seven main bands from ELF (used for very long distance communications) to VHF (used for FM radio).

Normally, radio and lower frequency microwaves are specified as frequencies. Whereas, EM waves from high frequency (millimeter wavelength) microwaves upwards are specified as a wavelength.

The wavelength of a signal is the ratio of its speed of propagation (u) to its frequency (f). It is thus given by:

$$\lambda = \frac{u}{f}$$

In free space an electromagnetic wave propagates at the speed of light (300 000 000 m s^{-1} or 186 000 miles s^{-1}). For example, if the carrier frequency of an FM radio station is 97.3 MHz then its transmitted wavelength is 3.08 m. If an AM radio station transmits at 909 kHz then the carrier wavelength is 330 m. Typically, the length of radio antenna is designed to be half the wavelength of the received wavelength. This is the reason why FM aerials are normally between 1 and 2 m in length whereas in AM and LW aerials a long coil of wire is wrapped round a magnetic core. Note that a 50 Hz mains frequency propagates through space with a wavelength of 6 000 000 m.

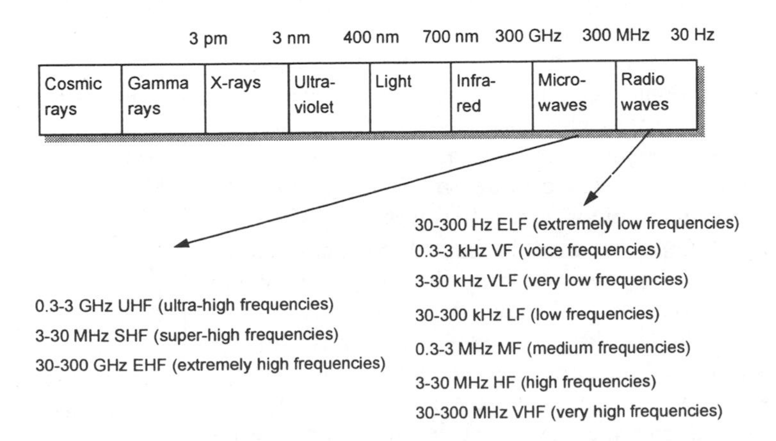

Figure 8.15 EM frequency spectrum

If an EM wave propagates through a dense material then its speed slows. In terms of the dielectric constant, ε_r of a material (which is related to density) then the speed of propagation is:

$$u = \frac{c}{\sqrt{\varepsilon_r}}$$

Each classification of EM waves has its own characteristics. The main classifications of EM waves used for communication are:

- **Radio waves.** The lower the frequency of a radio wave the more able it is to bend round objects. Defense applications use low frequency communications, as they can be transmitted over large distances, and over and round solid objects. The trade-off is that the lower the frequency the less the information that can be carried. LW (MF) and AM (HF) signals can propagate large distances, but FM (VHF) requires repeaters because they cannot bend round and over solid objects such as trees and hills. Long wave radio (LW) transmitters operate from approximately 100 to 300 kHz, medium wave (AM) from 0.5 to 2 MHz and VHF radio (FM) from 87 to 108 MHz.
- **Microwaves.** Microwaves have the advantage over optical waves (light, infra-red and ultra-violet) in that they can propagate well through water and thus can be transmitted thorough clouds, rain, and so on. If they are of a high enough frequency they can propagate through the ionosphere and out into outer space. This property is used in satellite communications where the transmitter bounces microwave energy off a satellite, which is then picked up at a receiving station. Radar and mobile radio applications use these properties. Their main disadvantage is that they will not bend round large objects, as their wavelength is too small. Included in this classification is UHF (used to transmit TV signals), SHF (satellite communications) and EHF waves.
- **Infrared.** Infrared is used in optical communications. When used as a carrier frequency the transmitted signal can have a very large bandwidth because the carrier frequency is high. It is extensively used for line-of-site communications especially in remote control applications. Infra-red radiation is basically the propagation of heat. Heat received from the sun propagates as infra-red radiation.

Radio waves:

- Low frequency: Propagate over long distances. Waves can bend around objects. Low bandwidth.
- Medium frequency. Long wave (LW) – long wave radio. High frequency (HF) – AM radio.
- High frequency. Normally used in line-of-sight communications. Very high frequency (VHF) – FM radio.

Microwaves:
Ultra-high frequency – UHF (TV).
Extra high frequency – EHF (Mobile radio).
Millimeter waves (Military communications).
Propagates well through clouds.

Medium bandwidth, such as one TV channel.

Infrared/Ultraviolet:
Does not propagate well through free-space. Normally used in light-of-sight communications. Very high bandwidth. For example, a 900 nm wave has a frequency of 3.33×10^{14} Hz, which allows for a bandwidth of 3.33×10^{13} Hz. It could thus provide for millions of TV channels, all at the same time.

Light spectrum:

Richard
Of
York
Gave
Battle
In
Vain.

(Infrared) **ROY.G.BIV** (Ultraviolet)

- **Light.** Light is the only part of the spectrum that humans can 'see'. It is a very small part of the spectrum and ranges from 300 to 900 nm. Colors contained are Red, Orange, Yellow, Green, Blue, Indigo and Violet.
- **Ultra-violet.** As with infrared it is used in optical communications. In high enough exposures it can cause skin cancer. The ozone layer blocks much of the ultra-violet radiation from the sun.

8.13 Routing of data

Two nodes can communicate over a physical distance. This distance could involve short distances, such as within a building, or worldwide. In order for data to be delivered to a recipient, a path must exist for it. Normally this route is setup by switching, which can be:

Intel inside

Few companies have ever managed to get their logo on other manufacturers products. The Intel Inside logo was one of these successes. Others include:

- Dolby noise reduction.
- Teflon non-stick material.
- Nutrasweet.

- **Circuit switching.** This type of switching uses a dedicated line to make the connection between the source and destination. The bit rate can vary as required, and possibly underutilized, but there tends to be little delay in sending the data.
- **Packet switching.** This type of switching involves the data being split into data packets. Each packet sent contains the data and also a packet header that has the information that is used to route the packet through the network. Each node on the path reads the data packet and sends it onto another node. The transport can either be:
    - Datagram. This is where the data packets travel from the source to the destination, and can take any path through the interconnected network. This technique has an advantage, over setting up a fixed path, as data packets can take alternative paths when there is heavy traffic on parts of the network. It also does not require a call setup.
    - Virtual Circuit. This is where all the data packets are routed along the same path. It differs from circuit switching in that there is no dedicated path for the data.
- **Multirate Circuit Switching.** Traditionally TDM (Time Division Multiplexing) is used to transmit data over a PSN (Public Switched Network). This uses a circuit switching technology with a fixed data rate. Multirate rates allow transmitters to transmit to different destination over a single physical connection. In ISDN, a node can transmit to two different destinations with a single connection (each of 64 kbps). The bit rate, though, is fixed at 64 kbps, and it is difficult to achieve a variable bit rate (VBR). Multirate circuit switching has fixed channels for the data.
- **Frame Relay.** This method is similar to packet switching, but the data packets have a variable length and not fixed in length. This allows for variable bit rates.
- **Cell Relay.** This method uses fixed packets (cells), and is a progression of the frame relay and multirate circuit switching. Cell relay allows for the definition of virtual channels with data rates dynamically defined. Using a small cell size allows almost constant data rate even though it uses packets.

A wide area network (WAN) connects one node to another over relatively large distances via an arbitrary graph of switching nodes. For the transmission of digital data, then data is either sent through a public data network (PDN) or through dedicated company connections. As shown in Figure 8.16, there are two main types of connection over the public telephone network, circuit switching and packet switching.

With circuit switched, a physical, or a reserved multiplexed, connection exists between two nodes, a typical example is the public-switched telephone network (PSTN). The connection must be made before transferring any data. In the past this connection took a relatively long time to set-up (typically over 10 seconds), but with the increase in digital exchanges it has reduced to less than a second. The usage of digital exchanges has also allowed the transmission of digital data, over PSTNs, at rates of 64 kbps and greater. This type of network is known as a circuit-switched digital network (CSDN). Its main disadvantage is that a permanent connection is set-up between the nodes. This is wasteful in time and can be costly. Another disadvantage is that the transmitting and receiving nodes must be operating at the same speed. A CSDN, also, does not perform any error detection or flow control.

> **First bug**
>
> The first computer bug was really a bug. Grace Murray Hopper, who also invented COBOL, coined the term bug when she traced an error in the Mark II computer to a moth trapped in a relay. She carefully removed the bug and taped it to her logbook.

Packet switching involves segmenting data into packets that propagate within a digital network. They either follow a pre-determined route or are routed individually to the receiving node via packet-switched exchanges (PSE) or routers. These examine the destination addresses and based on an internal routing directory pass it to the next PSE on the route. As with circuit switching, data can propagate over a fixed route. This differs from circuit switching in that the path is not an actual physical circuit (or a reserved multiplexed channel). As it is not a physical circuit it is normally defined as a virtual circuit. This virtual circuit is less wasteful on channel resources as other data can be sent when there are gaps in the data flow. Table 8.3 gives a comparison of the two types.

> **Pentium bug**
>
> The Pentium bug was found in October 1994 by Thomas Nicely, a mathematician from Lynchburgh College, Virginia. He found that one divided by 824,633,702,441 give the wrong answer. This bug eventually cost Intel over half a billion dollars, as they offered to replace bug-ridden Pentiums with correctly operating ones.

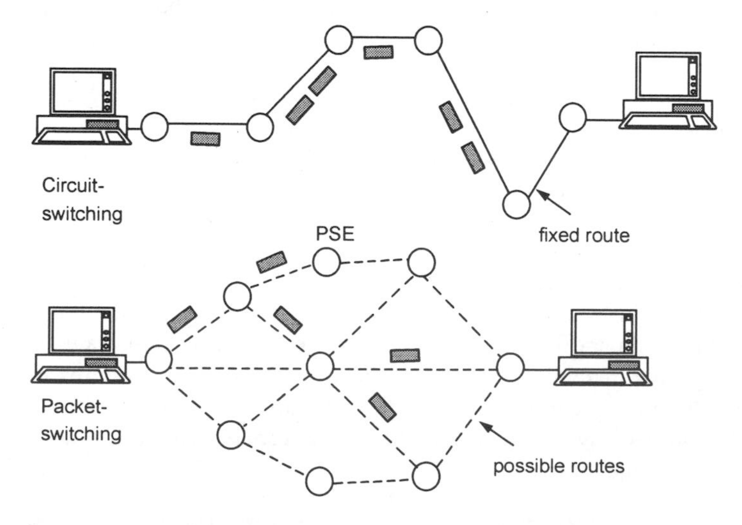

Figure 8.16 Circuit and packet switching

Table 8.3 Comparison of switching techniques

	Circuit-switching	*Packet-switching*
Investment in equipment	Minimal as it uses existing connections.	Expensive for initial investment
Error and flow control	None, this must be supplied by the end users.	Yes, using the FCS in the data link layer
Simultaneous transmissions and connections	No	Yes, nodes can communicate with many nodes at the same time and over many different routes
Allows for data to be sent without first setting up a connection	No	Yes, using datagrams
Response time	Once the link is set-up it provides a good reliable connection with little propagation delay	Response time depends on the size of the data packets and the traffic within the network

8.14 Exercises

8.14.1 The frequency response of a signal is given in the left-hand side of Figure 8.17, what is its bandwidth:

(a) 20 kHz (b) 80 kHz
(c) 10 Mbps (d) 120 kHz

8.14.2 Microwave EM waves have the advantage over optical waves in that they:

(a) Can carry more information
(b) Allow smaller transmitters to be designed
(c) They propagate faster
(d) Can propagate through water better

8.14.3 Radio stations transmit signals with which of the following:

(a) Frequency-division multiplexing (FDM)
(b) Time-division multiplexing (TDM)
(c) Frequency-hopping multiplexing (FHM)
(d) Time-hopping multiplexing (THM)

8.14.4 What is the most fundamental limitation on the maximum amount of data that can be transmitted over a channel:

(a) The application program (b) The type of data being transmitted
(c) Noise on the channel (d) The delay in transmission

8.14.5 What is the maximum bit rate for a digital signal if the bandwidth of the channel is 100 MHz:

(a) 10 Mbps (b) 100 Mbps
(c) 200 Mbps (d) 1000 Mbps

8.14.6 If the symbol rate is 100 symbols per second, and there are four bits transmitted for each symbol, what is the overall bit rate:

(a) 100 bps (b) 200 bps
(c) 400 bps (d) 1000 Mbps

8.14.7 If the symbol rate is 100 symbols per second, and each symbol has four different amplitudes, what is the overall bit rate:

(a) 100 bps (b) 200 bps
(c) 400 bps (d) 1000 Mbps

8.14.8 If the transfer rate is 400 Mbps, what time period for each bit:
(a) 2.5 ns (b) 4 ns
(c) 2.5 μs (d) 4 μs

8.14.9 If eight bits are transferred every 10 ns, what is the overall transfer rate:
(a) 8 Mbps (b) 80 Mbps
(c) 800 Mbps (d) 8000 Mbps

8.14.10 Referring to the right-hand side of Figure 8.17 which of the waveforms represent frequency-shift keying (FSK):
(a) A (b) B
(c) C (d) D

8.14.11 Referring to the right-hand side of Figure 8.17 which of the waveforms represent amplitude-shift keying (ASK):
(a) A (b) B
(c) C (d) D

8.14.12 Referring to the right-hand side of Figure 8.17 which of the waveforms represent phase-shift keying (PSK):

(a) A (b) B
(c) C (d) D

8.14.13 A digital system has a bandwidth of 100 kHz, and has a signal-to-noise ratio of 10,000. What is the capacity of the system:
(a) 400,004 bits/sec (b) 500,000 bits/sec
(c) 1,328,786 bits/sec (d) 1,000,000,000 bits/sec

8.14.14 If a system has a signal power of 10 mW, and a noise power of 1 μW, what is the signal-to-noise ratio (in dBs):
(a) 400,004 bits/sec (b) 500,000 bits/sec
(c) 1,328,786 bits/sec (d) 1,000,000,000 bits/sec

Amplitude (or power)
Signal bandwidth (f_{upper}-f_{lower})
Frequency
20 kHz
100 kHz
A
B
C
D

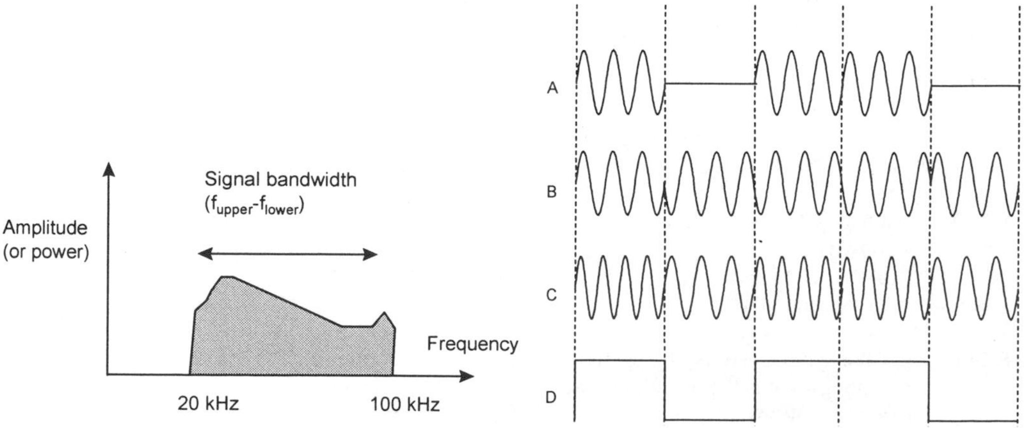

Figure 8.17 Diagrams

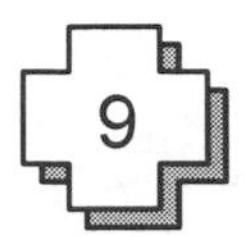

Compression

DEAR NET-ED

Question: *You say that high-quality audio uses 16 bits for each sample, but my CD player says that it uses **1- bit** conversion. Is this right?*

Yes. It is. It uses one bit at a time, as this is thought to give a smoother response. A major problem with CD recordings is that they sometimes lack warmth, and are a little sharp (as they are too perfect). One bit tracking tries to follow the movement of the audio signal. So your CD still uses 16-bit coding.

9.1 Introduction

Networks and the Internet have generally been used to transmit computer-type data, which is relatively easy to transmit as it is in a digital format, and, in most cases, does not have a large requirement in the amount of data transmitted from each user. The next great wave of usage of networks and the Internet will be the transmission of video, audio and speech. Each of these types of data, in their raw form, contains a great amount of data, and will generally swamp the rest of the network. Thus, a key to their acceptance will be compression, which tries to reduce the size of the stored/transmitted data, while still retaining the required information.

Information must be converted into a digital format before it can be stored on a computer or transmitted over a network. This is achieved by:

- **Images.** Images are converted into digital data by converting each of the pixels in the image into a digital value which represents the color of the pixel. The more colors that are used, the greater the amount of data that is required to store the image.
- **Motion video.** Motion video is basically a series of stored images, played back at a constant rate (scanning rate). This rate is 50 times per second for UK-type TV quality (PAL) or 60 times per second for US-type TV quality (NSTC).
- **Audio.** An audio file is converted by sampling the audio waveform at a constant rate (8000 times per second for telephone quality speech, and 44 100 times per second for hi-fi audio).

The basic conversion of this data into a digital form creates extremely large amount of digital data. For example:

Form	*Description*	*Storage requirement*
Bit-mapped image	1024×800, 65 536 colors	1.5 MB
Motion video	20 frames per second, 1024×800, 65 536 colors, 10 seconds	300 MB
Audio	60 minutes, hi-fi, stereo	600 MB

The raw form of image, video and audio contains massive amounts of redundant information, as there tends to be very little changes between one data sample and the next (Figure 9.1). One method used to compress the data is to store the changes between one data sample and the next. For example, in motion video, there are very few changes between one frame and the next (maybe just a few pixels). Thus, all that is required is to store the initial image, and then store the changes between the stored frame and the next, and so on. Occasionally it is important to store the complete frame, as the user may want to scan through the video, and start at any given point.

Motion video image compression relies on two facts:

- Images have a great deal of redundancy (repeated images, repetitive, superfluous, duplicated, exceeding what is necessary, and so on).
- The human eye and brain have limitations on what they can perceive.

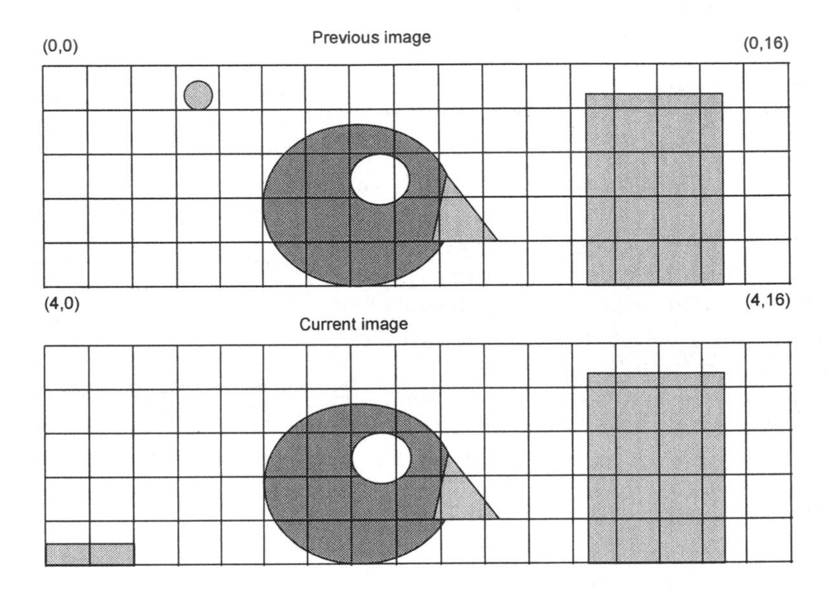

Figure 9.1 Information conversion into a digital form

The main forms of compression for images, video and audio are:

- **JPEG.** The JPEG (Joint Photographic Expert Group) compression technique is well matched to what the human eye and the brain perceive. For example, the brain is more susceptible to changes in luminance (brightness) and not so susceptible to color changes. More information on luminance can be stored, as apposed to information on color changes. JPEG can compress a photograph which is over 1 MB to less than 20 KB. Another typical image compression standard is GIF.
- **MPEG.** MPEG (Motion Picture Experts Group) uses many techniques to reduce the size of the motion video data. It uses the techniques that JPEG uses, to compress each of the images in the motion video. It also compresses between frames by only storing the information that is changing between frames. Typical compression rates are 130:1. Another typical video compression standard is AVI.
- **MPEG (MP-3).** The digital storage of audio allows for the data to be compressed.

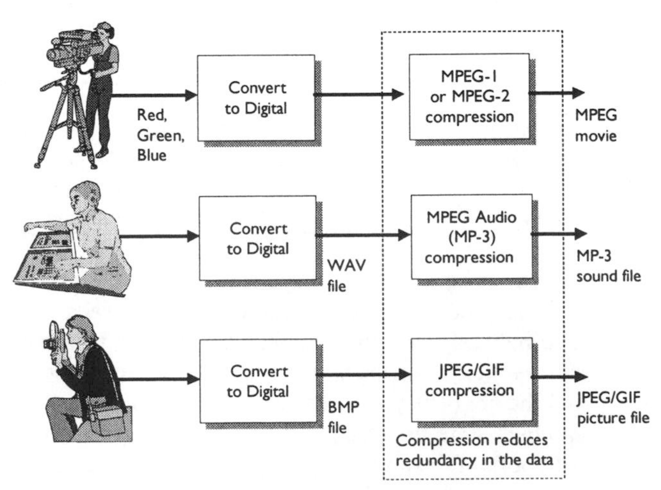

> **Monster or the Great Unifier?**
>
> With the growth in the Internet and global communications, we have created a global village, but have we also created an unregulated monster that is out of control? Well I think that Albert Einstein sums it best:
>
> *'Concern for man himself and his fate must always be the chief interest of all technical endeavors ... in order that the creations of our minds shall be a blessing and not a curse to mankind. Never forget that in the midst of your diagrams and equations.'*

Typically, on an audio CD, a stereo, hi-fi quality song uses about 10 MB for every minute of music (600 MB for 1 hour). The storage requirements are thus extremely large, as a few hours of music would fill many currently available hard disk drives. MP-3 audio uses a compression technique that understands the parts of the music that the human brain perceives, and retains this information, while discarding parts that it does not perceive. For example, the ear will generally only listen to loud instruments, and ignore instruments which are playing quietly. MP-3 is so successful that it can compress hi-fi quality audio into one-tenth of its normal, uncompressed, size. A standard, 60 min music CD can be compressed to around 60 MB. With the increasing size of electronic memory, it is now possible to store a whole music CD in electronic memory, rather than storing it on a CD or a hard disk. Another typical audio compression standard is Dolby AC-3.

9.2 Conversion to digital

Figure 9.2 outlines the conversion process for digital data (the upper diagram) and for analogue data (the lower diagram). When data is already in a digital form (such as text or animation) it is converted into a given data format (such as BMP, GIF, JPG, and so on). It can be further compressed before it is either stored, transmitted or processed. The lower diagram shows how an analogue signal (such as speech or video) is first sampled at regular intervals of time. These samples are then converted into a digital form with an ADC (analogue-to-digital converter). It can then be compressed and/or stored in a defined digital format (such as WAV, JPG, and so on). This digital form is then converted back into an analogue form with a DAC (digital-to-analogue converter).

9.3 Sampling theory

As a signal may be continually changing, a sample of it must be taken at given time intervals. The rate of sampling depends on its rate of change. For example, the temperature of the sea will not vary much over a short time but a video image of a sports match will. To encode a signal digitally it is normally sampled at fixed time intervals. Sufficient information is then extracted to allow the signal to be processed or reconstructed. Figure 9.3 shows a signal sampled every T_S seconds.

If a signal is to be reconstructed as the original signal it must be sampled at a rate defined by the Nyquist criterion. This states:

the sampling rate must be twice the highest frequency of the signal

For telephone speech channels, the maximum signal frequency is limited to 4 kHz and must thus be sampled at least 8000 times per second (8 kHz). This gives one sample every 125 μs.

Hi-fi quality audio has a maximum signal frequency of 20 kHz and must be sampled at least 40 000 times per second (many professional hi-fi sampling systems sample at 44.1 kHz). Video signals have a maximum frequency of 6 MHz, thus a video signal must be sampled at 12 MHz (or once every 83.3 ns).

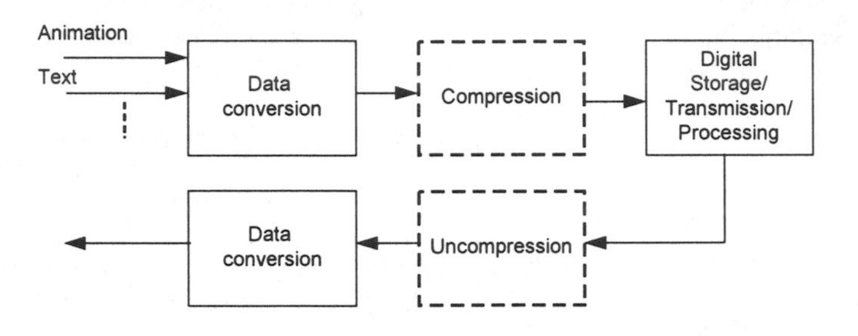

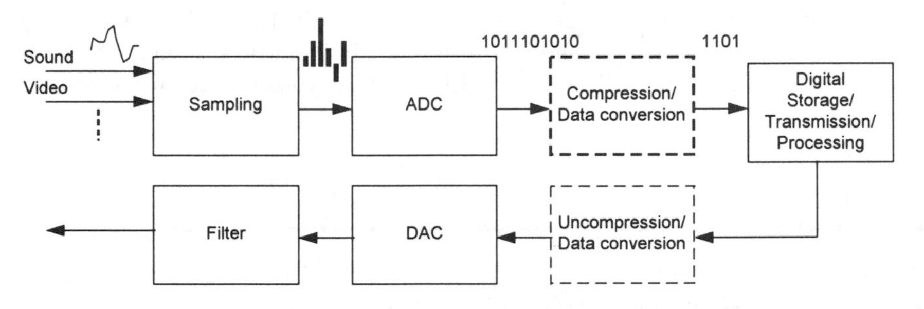

Figure 9.2 Information conversion into a digital form

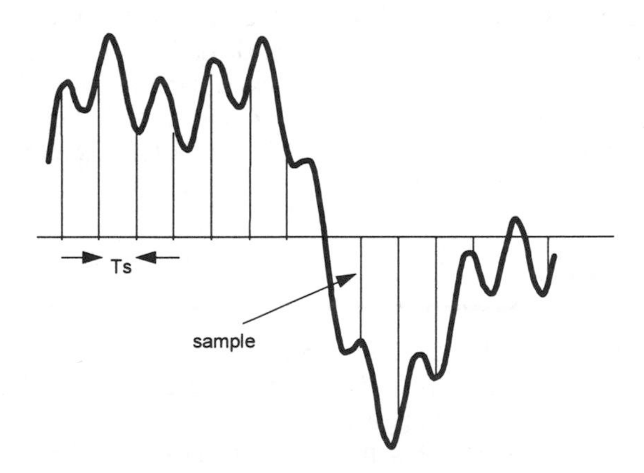

Figure 9.3 The sampling process

9.4 Quantization

Quantization involves converting an analogue level into a discrete quantized level. Figure 9.4 shows the conversion of an example waveform into a 4-bit digital code. In this case there are 16 discrete levels which are represented with the binary values 0000 to 1111. Value 1111 represents the maximum voltage level and value 0000 the minimum. It can be seen, in this case, that the digital codes for the four samples are 1011, 1011, 1001, 0111.

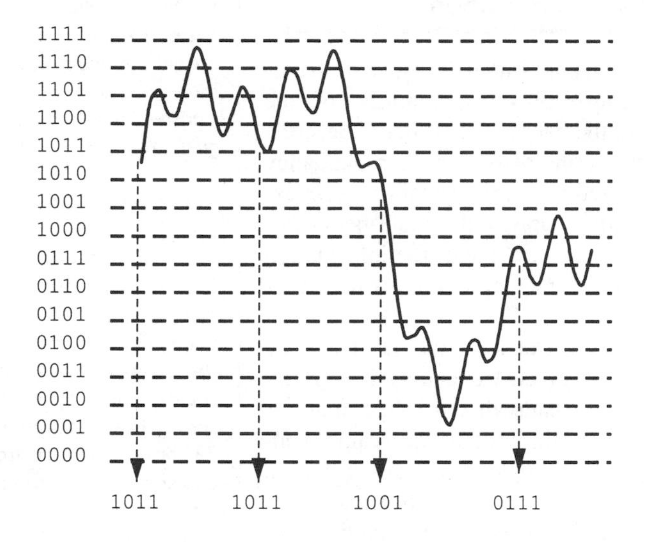

Figure 9.4 Converting an analogue waveform in a 4-bit digital form

The quantization process approximates the level of the analogue level to the nearest quantized level. This approximation leads to an error known as quantization error. The greater the number of levels the smaller the quantization error. Table 9.1 gives the number of levels for a given number of bits.

The maximum error between the original level and the quantized level occurs when the original level falls exactly halfway between two quantized levels. The maximum error will be a half of the smallest increment or

$$\text{Max error} = \pm\frac{1}{2}.\frac{\text{Full Scale}}{2^N}$$

If the range of a signal is between +5V and −5V, then a 12-bit convertor would give a maximum error of:

$$\text{Max error} = \pm\frac{1}{2}\frac{10}{2^{12}}$$
$$= 0.00122\ \text{V}$$

Table 9.1 states the quantization error (as a percentage) of a given number of bits. For example the maximum error with 8 bits is 0.2%, while for 16 bits it is only 0.000 76%.

Table 9.1 Number of quantization levels as a function of bits

Bits (N)	*Quantization levels*	*Accuracy (%)*
1	2	50
2	4	25
3	8	12.5
4	16	6.25
8	256	0.2
12	4 096	0.012
14	16 384	0.003
16	65 536	0.000 76

9.5 Compression methods

DEAR NET-ED

Question: *I can't understand it. I've just bought a brand-new, state-of-the-art 56 kbps modem, and all I ever get is a maximum transfer speed of 4.19 KB/s. Where am I going wrong, do I need a new ISP?*

No. Your ISP is providing an excellent service, as 56 kbps is split between sending and receiving. As users who access the Internet from modems typically need to receive more data than they send, the bandwidth for receiving is greater than the bandwidth for sending. You can thus receive at a faster rate than you can send. The maximum receiving rate is 33 kbps, which relates to a maximum transfer rate of 4.125 KB/s (there are 8 bits in a byte). If you need a higher-rate you should try ISDN which gives a total transfer rate of 128 kbps (16 KB/s). Otherwise consider ADSL (Asymmetric Digital Subscriber Line), which gives up to 9 Mbps receiving and 1.1 Mbps sending, over standard telephone lines.

Most transmission channels have a restricted bandwidth, either because of the limitations of the channel or because the bandwidth is shared between many users. Many forms of data have redundancy in it, thus if the redundant information was extracted the transmitted data would make better use of the bandwidth. This extraction of the information is normally achieved with compression, which takes many forms.

When compressing data it is important to take into account the type of data and how it is interpreted. For example pixels in an image may be distorted but it would still contain the required information. Whereas, a single erroneous bit in a computer data file can cause severe problems.

Video and sound images are normally compressed with a lossy compression whereas computer-type data has a lossless compression. The basic definitions are:

- **Lossless compression.** Where the information, once uncompressed, will be identical to the original uncompressed data. This will obviously be the case with computer-type data, such as data files, computer programs, and so on. Any loss of data will cause the file to be corrupted.
- **Lossy compression.** Where the information, once uncompressed, cannot be fully recovered. Lossy compression normally involves analyzing the data and determining which information has little effect on the resulting compressed data. For example, there is little difference, to the eye, between an image with 16.7 million colors (24-bit color information) and an image stored with 1024 colors (10-bit color information), but the storage will be reduced to 41.67% (many computer systems cannot even display 24-color information in certain resolutions). Compression of an image might also be used to reduce the resolution of the image. Again, the human eye might compensate for the loss of resolution.

Apart from lossy and lossless compression, data encoding is normally classified into two main areas: entropy encoding and source encoding. These are:

- **Entropy coding.** This does not take into account the characteristics of the data and treats all the bits in the same way; it produces lossless coding. Typically it uses:
  - Statistical encoding – where the coding analyses the statistical pattern of the data. For example if a source of text contains many more 'e' characters than 'z' characters then the character 'e' could be coded with very few bits and the character 'z' with many bits.
  - Suppressing repetitive sequences – many sources of information contain large amounts of receptive data. For example this page contains large amounts of 'white

space'. If the image of this page were to be stored, a special character sequence could represent long runs of 'white space'.

- **Source encoding.** This normally takes into account characteristics of the information. For example images normally contain many repetitive sequences, such as common pixel colors in neighboring pixels. This can be encoded as a special coding sequence. In video pictures, also, there are very few changes between one frame and the next. Thus typically the data encoded only stores the changes from one frame to the next.

9.6 Entropy encoding

Normally, general data compression does not take into account the type of data which is being compressed and is lossless. It can be applied to computer data files, documents, images, and so on. The two main techniques are statistical coding and repetitive sequence suppression. This section discusses two of the most widely used methods for general data compression: Huffman coding and Lempel-Ziv coding.

9.6.1 Huffman coding

Huffman coding uses a variable length code for each of the elements within the information. This normally involves analyzing the information to determine the probability of elements within the information. The most probable elements are coded with a few bits and the least probable coded with a greater number of bits.

The following example relates to characters. First, the textual information is scanned to determine the number of occurrences of a given letter. For example:

'b'	'c'	'e'	'i'	'o'	'p'
12	3	57	51	33	20

Next the characters are arranged in order of their number of occurrences, such as:

'e'	'i'	'o'	'p'	'b'	'c'
57	51	33	20	12	3

DEAR NET-ED

Question: *Everyone seems to be talking about MP-3, but what's so good about it?*

MP-3 audio is set to revolutionize the way that music is distributed and licensed. A typical audio track is sampled at 44 100 times per second, for two channels at 16 bits per sample. Thus the data rate is 1.411 Mbps (176 400 B/s), giving a total of 52 920 000 B (50.47 MB) for a five-minute song. As the storage of a CD is around 650 MB, it is possible to get 64 minutes from the CD.

Obviously it would take too long, with present bandwidths to download a five-minute audio file from the Internet in its raw form (over 3 hours with a 56 kbps modem). If the audio file was compressed with MP-3, it can be reduced to one-tenth of its original size, without losing much of its original content.

So, it is now possible, with MP-3, to get over 10 hours of hi-fi quality music on a CD. But the big change is likely to occur with songs being sampled, and downloaded over the Internet. Users would then pay for the license to play the music, and not for purchasing the CD.

Next the two least probable characters are assigned either a 0 or a 1. Figure 9.5 shows that the least probable ('c') has been assigned a 0 and the next least probable ('b') has been assigned a 1. The summation of the two occurrences is then taken to the next column and the occurrence values are again arranged in descending order (that is, 57, 51, 33, 20 and 15). As with the first column, the least probable occurrence is assigned a 0 and the next least probable occurrence is assigned a 1. This continues until the last column. The Huffman-coded values are then read from left to right and the bits are listed from right to left.

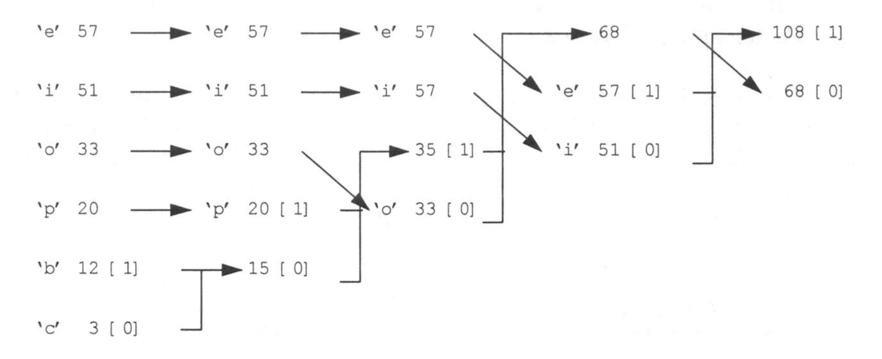

Figure 9.5 Huffman coding example

The final coding will be:

'e'	11	'i'	10
'o'	00	'p'	011
'b'	0101	'c'	0100

The great advantage of Huffman coding is that, although each character is coded with a different number of bits, the receiver will automatically determine the character whatever their order. For example if a 1 is followed by a 1 then the received character is an 'e'. If it is then followed by two 0s then it is an 'o'. Here is an example:

11000110100100110100

will be decoded as:

'e' 'o' 'p' 'c' 'i' 'p' 'c'

When transmitting or storing Huffman-coded data, the coding table needs to be stored with the data (if the table is generated dynamically). It is generally a good compression technique but it does not take into account higher order associations between characters. For example, the character 'q' is normally followed by the character 'u' (apart from words such as Iraq). An efficient coding scheme for text would be to encode a single character 'q' with a longer bit sequence than a 'qu' sequence.

Top 20 Computers of All Time

1. MITS Altair8800
2. Apple II
3. Commodore PET
4. Radio Shack TRS-80
5. Osborne I Portable
6. Xerox Star
7. IBM PC
8. Compaq Portable
9. Radio Shack TRS-80 Model 100
10. Apple Macintosh
11. IBM AT
12. Commodore Amiga 1000
13. Compaq Deskpro 386
14. Apple Macintosh II
15. Next Nextstation
16. NEC UltraLite
17. Sun SparcStation 1
18. IBM RS/6000
19. Apple Power Macintosh
20. IBM ThinkPad 701C

Byte, Sept 1995

9.6.2 Adaptive Huffman coding

Adaptive Huffman coding was first conceived by Faller and Gallager and then further refined by Knuth (so it is often called the FGK algorithm). It uses defined word schemes which determine the mapping from source messages to code words based upon a running estimate of the source message probabilities. The code is adaptive and changes so as to remain optimal for the current estimates. In this way, the adaptive Huffman codes respond to locality and the encoder thus learns the characteristics of the source data. The decoder must then learn along with the encoder by continually updating the Huffman tree so as to stay in synchronization with the encoder.

A second advantage of adaptive Huffman coding is that it only requires a single pass over the data. In many cases the adaptive Huffman method actually gives a better performance, in terms of number of bits transmitted, than static Huffman coding.

9.6.3 Lempel-Ziv coding

Around 1977, Abraham Lempel and Jacob Ziv developed the Lempel-Ziv class of adaptive dictionary data compression techniques (also known as LZ-77 coding). They are now some of the most popular compression techniques.

The LZ coding scheme takes into account repetition in phases, words or parts of words. These repeated parts could either be text or binary. A flag is normally used to identify coded and unencoded parts. An example piece of text could be:

'The receiver requires a receipt which is automatically sent when it is received.'

This has the repetitive sequence 'recei', and the encoded sequence could be modified with the flag sequence #*m*#*n* where *m* represents the number of characters to trace back to find the character sequence and *n* the number of replaced characters. Thus the encoded message could become:

'The receiver requires a #20#5pt which is automatically sent wh#6#2 it #30#2 #47#5ved.'

Normally a long sequence of text has many repeated words and phases, such as 'and', 'there', and so on. Note that in some cases this could lead to longer files if short sequences were replaced with codes that were longer than the actual sequence itself.

Top 20 most influential computer companies

1. Microsoft Corp.
2. Intel Corp.
3. IBM Corp.
4. Compaq Computer
5. Hewlett-Packard
6. Cisco Systems
7. Sun Microsystems
8. America Online
9. Dell Computer
10. Netscape
11. Oracle Corp.
12. 3Com Corp.
13. Yahoo! Inc.
14. Gateway Inc.
15. Adobe Systems
16. Apple Computer
17. Novell Inc.
18. Toshiba Corp.
19. Macromedia Inc.
20. Intuit Inc.

– *PC Magazine*, 1998.

9.6.4 Lempel-Ziv-Welsh coding

The Lempel-Ziv-Welsh (LZW) algorithm (also known LZ-78) builds a dictionary of frequently used groups of characters (or 8-bit binary values). Before the file is decoded, the compression dictionary must be sent (if transmitting data) or stored (if data is being stored). This method is good at compressing text files because text files contain ASCII characters (which are stored as 8-bit binary values) but not so good for graphics files, which may have repeating patterns of binary digits that might not be multiples of 8 bits.

A simple example is to use a six-character alphabet and a 16-entry dictionary, thus the resulting code word will have 4 bits. If the transmitted message is:

```
ababacdcdaaaaaaef
```

'But what ... is it good for?' Engineer at the Advanced Computing Systems Division of IBM, 1968, commenting on the microchip.

Then the transmitter and receiver would initially add the following to its dictionary:

```
0000        'a'   0001       2 'b'
0010        'c'   0011         'd'
0100        'e'   0101         'f'
0110–1111   empty
```

> 'I have traveled the length and breadth of this country and talked with the best people, and I can assure you that data processing is a fad that won't last out the year.'
> Editor, Prentice Hall, 1957

First the 'a' character is sent with 0000, next the 'b' character is sent and the transmitter checks to see that the 'ab' sequence has been stored in the dictionary. As it has not, it adds 'ab' to the dictionary, to give:

```
0000        'a'   0001         'b'
0010        'c'   0011         'd'
0100        'e'   0101         'f'
0110        'ab'  0111–1111    empty
```

The receiver will also add this to its table (thus the transmitter and receiver will always have the same tables). Next the transmitter reads the 'a' character and checks to see if the 'ba' sequence is in the code table. As it is not, it transmits the 'a' character as 0000, adds the 'ba' sequence to the dictionary, which will now contain:

```
0000        'a'
0001        'b'
0010        'c'
0011        'd'
0100        'e'
0101        'f'
0110        'ab'
0111        'ba'
1000–1111   empty
```

0000 0001 0000 0110 0010

'a' 'b' 'a' 'ba' 'c'

> **Most admired computer companies?**
>
> Microsoft (2)
> Dell Computer (3)
> Cisco Systems (4)
> Intel (8)
> Lucent Technologies (10)
>
> (Position in Fortune 500 Most Admired Companies list, 2000)

Next the transmitter reads the 'b' character and checks to see if the 'ba' sequence is in the table. As it is, it will transmit the code table address which identifies it, i.e. 0111. When this is received, the receiver detects that it is in its dictionary and it knows that the addressed sequence is 'ba'.

Next the transmitter reads a 'c' and checks for the character in its dictionary. As it is included, it transmits its address, i.e. 0010. When this is received, the receiver checks its dictionary and locates the character 'c'. This then continues with the transmitter and receiver maintaining identical copies of their dictionaries. A great deal of compression occurs when sending a sequence of one character, such as a long sequence of 'a'.

Typically, in a practical implementation of LZW, the dictionary size for LZW starts at 4K (4096). The dictionary then stores bytes from 0 to 255 and the addresses 256 to 4095 are used for strings (which can contain two or more characters). As there are 4096 entries then it is a 12-bit coding scheme (0 to 4096 gives 0 to 2^{12}–1 different addresses).

9.6.5 Statistical encoding

Statistical encoding is an entropy technique which identifies certain sequences within the data. These 'patterns' are then coded so that they are coded in fewer bits. Frequently used patterns are coded with fewer bits than less common patterns. For example, text files normally contain many more '*e*' characters than '*z*' characters. Thus the '*e*' character could be encoded with a few bits and the '*z*' with many bits. Statistical encoding is also known as arithmetic compression.

A typical statistical coding scheme is Huffman encoding. Initially the encoder scans through the file and generates a table of occurrences of each character. The codes are assigned to minimize the number of encoded bits, then stored in a codebook which must be transmitted with the data.

Table 9.2 shows a typical coding scheme for the characters '*a*' to '*z*'. It uses the same number of bits for each character. Morse code is an example of statistical encoding. It uses dots (a zero) and dashes (a one) to code characters, where a short space in time delimits each character. It uses short codes for the most probable letters and longer codes for less probable letters. In the form of zeros and ones it is stated in Table 9.3.

Thus the message:

```
this an
```

would be encoded as:

```
Message:      t       h       i       s               a       n
Simple code:  10011   00111   01000   10010   11010   00000   01101
Morse code:   1       0000    00      000     0011    01      10
```

This has reduced the number of bits used to represent the message from 35 (7×5) to 18.

Table 9.2 Simple coding scheme

a	00000	b	00001	c	00010	d	00011	e	00100
f	00101	g	00110	h	00111	i	01000	j	01001
k	01010	l	01011	m	01100	n	01101	o	01110
p	01111	q	10000	r	10001	s	10010	t	10011
u	10100	v	10101	w	10110	x	10111	y	11000
z	11001	SP	11010						

Table 9.3 Morse coding scheme

a	01	b	1000	c	1010	d	100	e	0
f	0010	g	110	h	0000	i	00	j	0111
k	101	l	0100	m	11	n	10	o	111
p	0110	q	1101	r	010	s	000	t	1
u	001	v	0001	w	011	x	1001	y	1011
z	1100	SP	0011						

9.6.6 Repetitive sequence suppression

Repetitive sequence suppression involves representing long runs of a certain bit sequence with a special character. A special bit sequence is then used to represent that character, followed by the number of times it appears in sequence. Typically 0s (zero) and ' ' (spaces) occur repetitively in text files. For example the data:

```
8.3200000000000
```

could be coded as:

```
8.32F11
```

where F represents the flag. In this case the number of stored characters has been reduced from 16 to 7. Many text sources have other characters which occur repetitively. Run-length encoding (RLE) uses this to encode any character sequence with a special flag followed by the number of characters and finally the character which is repeated. For example

> 'I think there is a world market for maybe five computers.'
> Thomas Watson, chairman of IBM, 1943

```
Fred       has     when.........
```

could be coded as:

```
FredF7 hasF7 whenF9.
```

where F represents the flag. In this case the number of stored characters has been reduced from 32 to 20. The 'F7 ' character code represents seven ' ' (spaces) and 'F9.' represents nine '.' characters.

9.7 Source compression

Source compression takes into account the type of information that is being compressed. Typically this is used with image, video and audio information, and these forms of data have information, which has little relevance on the image, images, or sound that is perceived. The main compression techniques are: JPEG, MEG and MP-3.

9.7.1 Image compression

> 'There is no reason anyone would want a computer in their home.'
> Ken Olson, president, chairman and founder of Digital Equipment Corp., 1977.

Data communication increasingly involves the transmission of still and moving images. Great savings in the number of bits can be made by compressing the images into a standard form. Some of these forms are outlined in Table 9.4. The main parameters in a graphics file are:

- The picture resolution. This is defined by the number of pixels in the *x*- and *y*-directions.
- The number of colors per pixel. If N bits are used for the bit color then the total number of displayable colors will be 2^N. For example an 8-bit color field defines 256 colors, a 24-bit color field gives 2^{24} or 16.7 M colors.
- Palette size. Some systems reduce the number of bits used to display a color by reducing the number of displayable colors for a given palette size.

Comparison of the different methods

This section uses example bitmapped images and shows how much the different techniques manage to compress them. The left-hand side of Figure 9.6 shows an image and Table 9.4 shows the resultant file size when it is saved in different formats. It can be seen that the BMP file format has the largest storage. The two main forms of BMP files are RGB (red, green, blue) encoded and RLE encoded. RGB coding saves the bit-map in an uncompressed form, whereas the RLE coding will reduce the total storage by compressing repetitive sequences. The GIF format manages to compress the file to around 40% of its original size and the TIF file achieves similar compression (mainly because both techniques use LZH compression). It can be seen that by far the best compression is achieved with JPEG which in both forms has compressed the file to under 10% of its original size.

Table 9.4 Typical standard compressed graphics formats

File	*Compression type*	*Max. resolution or colors*	
TIFF	Huffman RLE and/or LZW	48-bit color	TIFF (tagged image file format) is typically used to transfer graphics from one computer system to another. It allows high resolutions and colors of up to 48 bits (16 bits for red, green and blue).
GIF	LZW	65 536 × 65 536 (24-bit color, but only 256 display-able colors)	Standardized graphics file format which can be read by most graphics packages. It has similar graphics characteristics to PCX files and allows multiple images in a single file and interlaced graphics.
JPG	JPEG compression (DCT, Quantization and Huffman)	Depends on the compression	Excellent compression technique which produces lossy compression. It normally results in much greater compression than the methods outlined above.

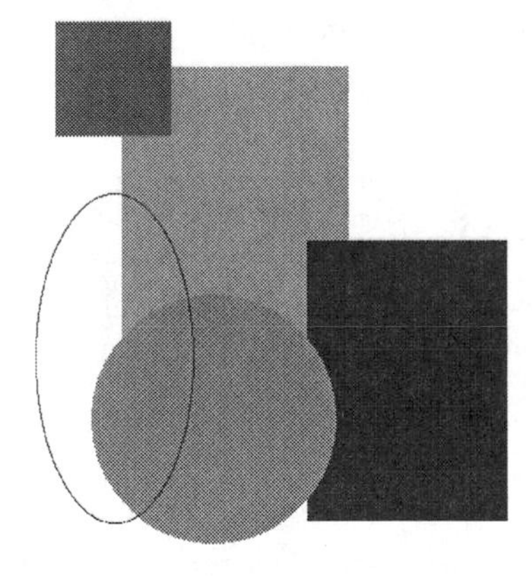

Figure 9.6 Sample graphics image

The reason that the compression ratios for GIF, TIF and BMP RLE are relatively high is that the image on the left-hand side of Figure 9.6 contains a lot of changing data. Most images will compress to less than 10% because they have large areas which do not change much. The right-hand side of Figure 9.7 shows a simple graphic of 500 × 500, 24-bit, which has large areas with identical colors. Table 9.6 shows that, in this case, the compression ratio is low. The RLE encoded BMP file is only 1% of the original as the graphic contains long runs of the same color. The GIF file has compressed to less than 1%. Note that the PCX, GIF and BMP RLE files have saved the image with only 256 colors. The JPG formats have the advantage that they have saved the image with the full 16.7M colors and give compression rates of around 2%.

Table 9.5 Compression on a graphics file

Type	*Size(B)*	*Compression (%)*	
BMP	308 278	100.0	BMP, RBG encoded (640 × 480, 256 colors)
BMP	301 584	97.8	BMP, RLE encoded
GIF	124 304	40.3	GIF, Version 89a, non-interlaced
GIF	127 849	41.5	GIF, Version 89a, interlaced
TIF	136 276	44.2	TIF, LZW compressed
TIF	81 106	26.3	TIF, CCITT Group 3, MONOCHROME
JPG	28 271	9.2	JPEG – JFIF Complaint (Standard coding)
JPG	26 511	8.6	JPEG – JFIF Complaint (Progressive coding)

Table 9.6 Compression on a graphics file with highly redundant data

Type	*Size (B)*	*Compression (%)*	
BMP	750 054	100.0	BMP, RBG encoded (500 × 500, 16.7 M colors)
BMP	7 832	1.0	BMP, RLE encoded (256 colors)
PCX	31 983	4.3	PCX, Version 5 (256 colors)
GIF	4 585	0.6	GIF, Version 89a, non-interlaced (256 colors)
TIF	26 072	3.5	TIF, LZW compressed (16.7 M colors)
JPG	15 800	2.1	JPEG (Standard coding, 16.7 M colors)
JPG	12 600	1.7	JPEG (Progressive coding, 16.7 M colors)

JPEG compression

The JPEG standard was developed by the Joint Photographic Expert Group (JPEG), a subcommittee of the ISO/IEC, and the standards produced can be summarized as follows:

It is a compression technique for gray-scale or color images and uses a combination of discrete cosine transform, quantization, run-length and Huffman coding.

JPEG is an excellent compression technique which produces lossy compression (although in one mode it is lossless). As seen from the previous section it has excellent compression ratio when applied to a color image. It has resulted from research into compression ratios and the resultant image quality. The main steps are:

- Data blocks — Generation of data blocks
- Source-encoding — Discrete cosine transform and quantization
- Entropy-encoding — Run-length encoding and Huffman encoding

Unfortunately, compared with GIF, TIFF and PCX, the compression process is relatively slow. It is also lossy in that some information is lost in the compression process. This information is perceived to have little effect on the decoded image.

GIF files typically take 24-bit color information (8 bits for red, 8 bits for green and 8 bits for blue) and convert it into an 8-bit color palette (thus reducing the number of bits stored to approximately one-third of the original). It then uses LZW compression to further reduce the storage. JPEG operates differently in that it stores changes in color. As the eye is very sensitive to brightness changes it is particularly sensitive to changes in brightness. If these changes are similar to the original then the eye will perceive the recovered image as very similar to the original.

DEAR NET-ED

Question: *Why, with video and images, do you convert from RGB into something else?*

Video cameras have sensors for Red, Green and Blue (the primary colors for video information). In TV, before color TV, these colors where converted into luminance (Y). When color TV arrived they had to hide the extra color information and then send it as U and V (Redness and Blueness). Thus for TV, RGB is converted into YUV. With images, the human eye is very sensitive to changes in brightness in any object, and not so sensitive to color changes. Thus color changes can be compressed more than the luminance. This is why RGB is converted in YC_bC_r. For example, 4:2:2 uses twice as many samples for luminance than redness and blueness, and 4:1:1 uses four times as many samples.

Color conversion and subsampling

In the first part of the JPEG compression, each color component (red, green and blue) is separated in luminance (brightness) and chrominance (color information) separately. JPEG allows more losses on the chrominance and less on the luminance. This is because the human eye is less sensitive to color changes than to brightness changes. In an RGB image, all three channels carry some brightness information but the green component has a stronger effect on brightness than the blue component.

A typical scheme for converting RGB into luminance and color is known as CCIR 601, which converts the components into Y (can be equated to brightness), C_b (blueness) and C_r (redness). The Y component can be used as a black and white version of the image. The components are computed from the RGB components:

$$Y = 0.299R + 0.587G + 0.114B$$
$$C_b = 0.1687R - 0.3313G + 0.5B$$
$$C_r = 0.5R - 0.4187G + 0.0813B$$

For the brightness it can be seen that green has the most effect and blue has the least effect. For the redness, the red color (of course) has the most effect and green the least. For the blueness, the blue color has the most effect and green the least. Note that the YC_bC_r components are often known as YUV.

A subsampling process is then conducted which samples the C_b and C_r components at a lower rate than the Y component. A typical sampling rate is four samples of the Y component to a single sample on the C_b and C_r component. This sampling rate is normally set with the compression parameters the lower the sampling, the smaller the compressed data and the shorter the compression time. The JPEG header contains all the information necessary to properly decode the JPEG data.

DCT coding

The DCT (discrete cosine transform) converts intensity data into frequency data, which can be used to tell how fast the intensities vary. In JPEG coding the image is segmented into 8×8 pixel rectangles, as illustrated in Figure 9.7. If the image contains several components (such as Y, C_b, C_r or R, G, B), then each of the components in the pixel blocks is operated on separately. If an image is subsampled, there will be more blocks of some components than of others. For example, for 2×2 sampling there will be four blocks of Y data for each block of C_b or C_r data.

The data points in the 8×8 pixel array start at the upper right at (0,0) and finish at the lower right at (7,7). At the point (*x*,*y*) the data value is *f*(*x*,*y*). The DCT produces a new 8×8 block ($u\times v$) of transformed data using the formula:

$$F(u,v)=\frac{1}{4}C(u)C(v)\left[\sum_{x=0}^{7}\sum_{y=0}^{7}f(x,y)\cos\frac{(2x+1)u\pi}{16}\cos\frac{(2y+1)v\pi}{16}\right]$$

where $C(z)=\frac{1}{\sqrt{2}}$ if $z=0$

or $=1$ if $z\neq 0$

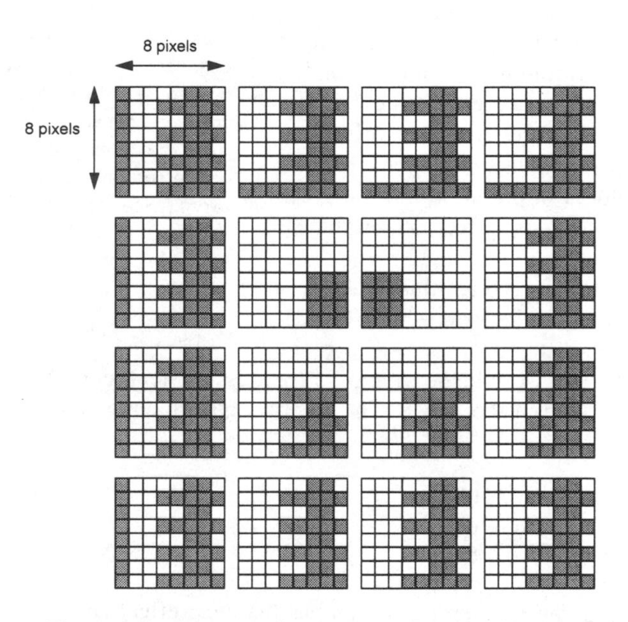

Figure 9.7 Segment of an image in 8x8 pixel blocks

This results in an array of space frequency $F(u,v)$ which gives the rate of change at a given point. These are normally 12-bit values which give a range of 0 to 1024. Each component specifies the degree to which the image changes over the sampled block. For example:

- $F(0,0)$ gives the average value of the 8×8 array.
- $F(1,0)$ gives the degree to which the values change slowly (low frequency).
- $F(7,7)$ gives the degree to which the values change most quickly in both directions (high frequency).

The coefficients are equivalent to representing changes of frequency within the data block. The value in the upper left block (0,0) is the DC or average value. The values to the right of a row have increasing horizontal frequency and the values to the bottom of a column have increasing vertical frequency. Many of the bands end up having zero or almost zero terms.

9.7.2 Motion video compression

Motion video contains massive amounts of redundant information. This is because each image has redundant information and also because there are very few changes from one image to the next.

Motion video image compression relies on two facts:

- Images have a great deal of redundancy (repeated images, repetitive, superfluous, duplicated, exceeding what is necessary, and so on).
- The human eye and brain have limitations on what they can perceive.

The Motion Picture Experts Group (MPEG) developed an international open standard for the compression of high-quality audio and video information. At the time, CD-ROM single-speed technology allowed a maximum bit rate of 1.2 Mbps and it is this rate that the standard was built around. These days, ×8 and ×10 CD-ROM bit rates are common.

MPEG's main aim was to provide good quality video and audio using hardware processors (and in some cases, on workstations with sufficient computing power, to perform the tasks using software). Figure 9.8 shows the main processing steps of encoding:

- Image conversion – normally involves converting images from RGB into YUV (or YC_rC_b) terms with optional color sub-sampling.
- Conversion into slices and macroblocks – a key part of MPEG's compression is the detection of movement within a frame. To detect motion a frame is subdivided into slices then each slice is divided into a number of macroblocks. Only the luminance component is then used for the motion calculations. In the subblock, luminance (Y) values use a 16×16 pixel macroblock, whereas the two chrominance components have 8×8 pixel macroblocks.
- Motion estimation – MPEG-1 uses a motion estimation algorithm to search for multiple blocks of pixels within a given search area and tries to track objects which move across the image.
- DCT conversion – as with JPEG, MPEG-1 uses the DCT method. This transform is used because it exploits the physiology of the human eye. It converts a block of pixels from the spatial domain into the frequency domain. This allows the higher-frequency terms to be reduced, as the human eye is less sensitive to high-frequency changes.
- Encoding – the final stages are run-length encoding and fixed Huffman coding to produce a variable-length code.

DEAR NET-ED

Question: *Why when I watch digital TV, or a DVD movie, does the screen sometimes display large rectangular blocks, or objects which seem to move incorrectly across the screen?*

MPEG splits images up into blocks. As part of the compression process, MPEG splits each frame into a series of blocks. These blocks are then transformed. To increase compression, MPEG sends the complete picture every so often, and then just sends updates in the differences between the frames. Thus if your reception is not very good then you may fail to get the complete update of the picture, and only receive parts for the update. Also MPEG tries to track moving objects, it will then group the moving object, and transmit how the object moves. Sometimes this has not been encoded very well, and the object seems to move incorrectly across the screen. Normally this is because there are not enough updates to the complete frame.

Question: *Why does MPEG have to send/store the complete picture every few frames. Would it not be possible to send/store one complete frame, and then just send/store the changes from frame to frame?*

This would work fine, and would give excellent compression, but the user would not be able to move quickly through the MPEG film, as the decoder would have to read the initial frame, and then all the updates to determine how the frames changed. Also if there were corrupt data, it would propagate through the whole film. Thus there is a compromise between the number of intermediate frames between each complete frame.

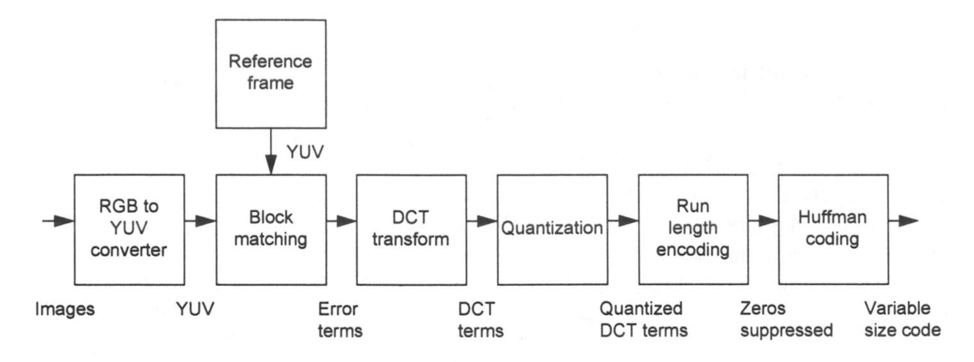

Figure 9.8 MPEG encoding with block matching

Color space conversion

The first stage of MPEG encoding is to convert a video image into the correct color space format. In most cases, the incoming data is in 24-bit RGB color format and is converted in 4:2:2 YC_rC_b (or YUV) form. Some information will obviously be lost but it results in some compression.

Slices and macroblocks

MPEG-1 compression tries to detect movement within a frame. This is done by subdividing a frame into slices and then subdividing each slice into a number of macroblocks. For example, a PAL format which has:

$$352 \times 288 \text{ pixel frame (101 376 pixels)}$$

can, when divided into 16×16 blocks, give a whole number of 396 macroblocks. Dividing 288 by 16 gives a whole number of 18 slices. Dividing 352 gives 22. Thus the image is split into 22 macroblocks in the *x*-direction and 18 in the *y*-direction, as illustrated in Figure 9.9.

Luminance (*Y*) values use a 16×16 pixel macroblock, whereas the two chrominance components have 8×8 pixel macroblocks. Note that only the luminance component is used for the motion calculations.

The Disappearing Internet

'Many people talk about the disappearing computer. Well, it has already disappeared. Computer chips may seem relatively large, but the actual silicon that they use covers a very small area. It's really just the casing and the pins that takes-up the space. If there were no interfaces to anything then the computer would be the size of this bookmark. Look at computer monitors, disk drives, and so on. All of these could blend themselves into our lives. Hard disks could be replaced by small pieces of silicon (Flash RAMs) and a computer monitor by a thin display. The most important thing though, will not be the processor, or the amount of memory, or whether it can play DVD movies, it will be its connection to the Internet. If you have access to the Internet, you are part of the world's biggest mainframe. Part of the large distributed system ever created (or should I say to evolve, as the Internet was never really created, or was it?). A living, breathing, digital world. The future is towards computers which will become a part of the fabric of our lives. So, what's going to disappear? The Internet, as it will become part of the fabrics of our lives. For too long our communications have been controlled by governments, large corporations and monopolistic telephone companies, but not any more.'

– W. Buchanan

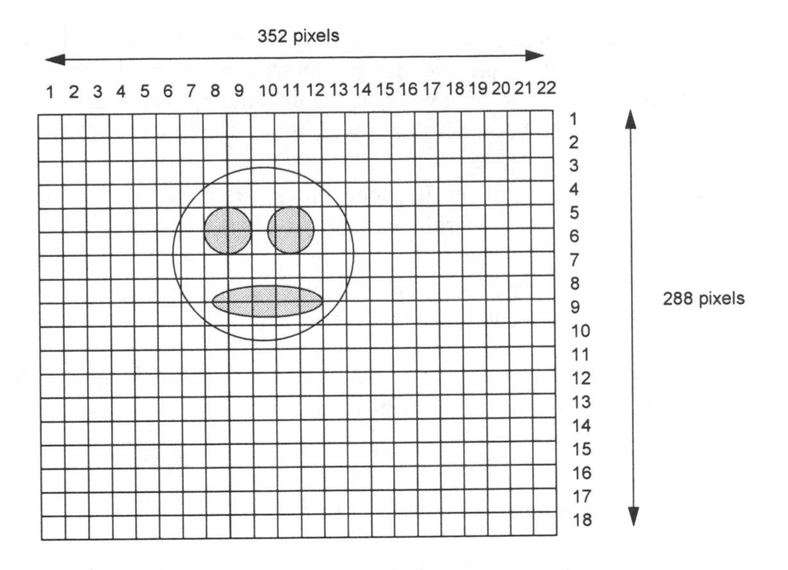

Figure 9.9 Segmentation of an image into subblocks

Motion estimation

MPEG-1 uses a motion estimation algorithm to search for multiple blocks of pixels within a given search area and tries to track objects which move across the image. Each luminance (Y) 16×16 macroblock is compared with other macroblocks within either a previous or future frame to find a close match. When a close match is found, a vector is used to describe where this block should be located, as well as any difference information from the compared block. As there tend to be very few changes from one frame to the next, it is far more efficient than using the original data.

Figure 9.1 shows two consecutive images of 2D luminance made up into 16×5 megablocks. Each of these blocks has 16×16 pixels. It can be seen that, in this example, there are very few differences between the two blocks. If the previous image is transmitted in its entirety then the current image can be transmitted with reference to the previous image. For example, the megablocks for $(0,1)$, $(0,2)$ and $(0,3)$ in the current block are the same as in the previous blocks. Thus they can be coded simply with a reference to the previous image. The $(0,4)$ megablock is different to the previous image, but the $(0,4)$ block is identical to the $(0,3)$ block of the previous image, thus a reference to this block is made. This can continue with most of the block in the image being identical to the previous image. The only other differences in the current image are at $(4,0)$ and $(4,1)$; these blocks can be stored in their entirety or specified with their differences to a previous similar block.

Each macroblock is compared mathematically with another block in a previous or future frame. The offset to find another block could be over a macroblock boundary or even a pixel boundary. The comparison then repeats until a match is found or the specified search area within the frame has been exhausted. If no match is available, the search process can be repeated using a different frame or the macroblock can be stored as a complete set of data. As previously stated, if a match is found, the vector information specifying where the matching macroblock is located is specified along with any difference information.

As the technique involves very many searches over a wide search area and there are many frames to be encoded, the encoder must normally be a high-powered workstation. This has several implications:

- An asymmetrical compression process is adopted, where a relatively large amount of computing power is required for the encoder and much less for the decoder. Normally the encoding is also done in non-real time whereas the decoder reads the data in real-time. As processing power and memory capacity increase, more computers will be able to compress video information in real-time.
- Encoders influence the quality of the decoded image dramatically. Encoding shortcuts, such as limited search areas and macroblock matching, can generate poor picture quality, irrespective of the quality of the encoder.
- The decoder normally requires a large amount of electronic memory to store past and future frames, which may be needed for motion estimation.

With the motion estimation completed, the raw data describing the frame can now be converted using the DCT algorithm ready for Huffman coding.

I, P and B-frames

MPEG video compression uses three main types of frames:

- **Intra frame (I-frame).** An intra frame, or I-frame, is a complete image and does not require any extra information to be added to it to make it complete. Thus no motion estimation processing has been performed on the I-frame. Mainly used to provide a known starting point, it is usually the first frame to be sent.
- **Predictive frame (P-frame).** The predictive frame, P-frame, uses the preceding I-frame as its reference and has motion estimation processing. Each macroblock in this frame is supplied either as a vector and difference with reference to the I-frame, or if no match was found, as a completely encoded macroblock (called an intracoded macroblock). The decoder must thus retain all I-frame information to allow the P-frame to be decoded.
- **Bidirectional frame (B-frame).** The bidirectional frame, B-frame, is similar to the P-frame except that its reference frames are to the nearest preceding I- or P-frame and the next future I- or P-frame. When compressing the data, the motion estimation works on the future frame first, followed by the past frame. If this does not give a good match, an average of the two frames is used. If all else fails, the macroblock can be intracoded. Needless to say, decoding B-frames requires that many I- and P-frames are retained in memory.

MPEG-1 allows frames to be ordered in any sequence. Unfortunately a large amount of reordering requires many frame buffers that must be stored until all dependencies are cleared.

The MPEG-1 format allows random access to a video sequence, thus the file must contain regular I-frames. It also allows enhanced modes such as fast forward, which means that an I-frame is required every 0.4 seconds, or 12 frames between each I-frame (at 30 fps).

At 30 fps, a typical sequence is a starting I-frame, followed by two B-frames, a P-frame, followed by two B-frames, and so on. This is known as a group of picture (GOP):

I⇒B⇒B⇒P⇒B⇒B⇒I⇒B⇒B⇒P⇒B⇒B⇒I⇒B⇒B⇒P⇒...

When decoding, the decoder must store the I-frame, the next two B-frames are also stored until the B-frame arrives. The next two B-frames have to be stored locally until the P-frame arrives. The P-frame can be decoded using the stored I-frame and the two B-frames can be decoded using the I- and P-frames. One solution of this is to reorder the frames so that the I- and P-frames are sent together followed by the two intermediate B-frames. Another more radical solution is not to send B-frames at all, simply to use I- and P-frames.

On computers with limited memory and limited processing power, the B-frames are difficult because:

- They increase the encoding computational load and memory storage. The inclusion of the previous and future I- and P-frames as well as the arithmetic average greatly increases the processing needed. The increased frame buffers to store frames allow the encode and decode processes to proceed. This argument is again less valid with the advent of large and high-density memories.
- They do not provide a direct reference in the same way that an I- or P-frame does.

The advantage of B-frames is that they lead to an improved signal-to-noise because of the averaging out of macroblocks between I- and P-frames. This averaging effectively reduces high-frequency random noise. It is particularly useful in lower bit rate applications, but is of less benefit with higher rates, which normally have improved signal-to-noise ratios.

Practical MPEG compression process

Most MPEG encoders can create MPEG-1 audio streams, MPEG-1 and MPEG-2 video streams, and MPEG-1 system streams. Each can be modified by a number of parameters. These are specified next.

Frame rate and data rate

The frame rate and data rate are the two main parameters which affect the quality of the encoded bitstream. The frame rate is normally set by the frame rate of the input format. Standard MPEG input frame rates are 23.976, 24, 25, 29.97, 30, 50, 59.94, and 60 frames/sec. Many encoders do not support all of these rates for the output so there are two modes which can be used to reduce or increase the frame rate, these are:

- Keep original number of frames. In this mode the frames are encoded frames as they are ordered in the input file, but MPEG players will play these files at the wrong speed.
- Keep original duration. In this mode the encoder either duplicates (to increase rate) or skip some input frames (to reduce the rate) to provide correct playback frame rate.

Most encoding systems will allow the user to specify the data rate of the encoded bitstream. The encode will then try to keep to this limit when it is encoding the input bitstream. For example, a single-speed CD-ROM requires a maximum data rate of 150 KB/sec. This rate is relatively low and there may be some degradation of quality to produce this. Reasonable quality requires about 300 to 600 KBs/sec.

Maximum motion vectors

MPEG uses motion estimation to reduce the encoded data rate. Motion estimation involves searching for the closest block of pixels on the previously encoded frame. It is a highly computing intensive task where each square block of pixels on the currently encoded frame is matched to the most similar block of pixels on the previously encoded frame. A pair of horizontal and vertical displacement values are then used to represent the pixel block, these values are named the motion vector.

Most MPEG encoders allow the user to select the maximum values of vertical and horizontal components of motion vectors. The larger these values then the greater the probability that a good match will be found to suit the output data rate. The two disadvantages of this are:

- The processing time increases, as there are many more pixel blocks to search through.
- The number of bits required to represent the motion vectors in the output file also increases. This increases the output data rate, but may not necessarily improve the quality of the output bitstream.

MPEG has three different types of encoded frames: I-, P-, and B-frames. Most encoding systems allow the user to specify the sequence of I-, P-, and B-frames. A typical sequence is to put several B-frames between each of the nearest P-frames, and several groups of P- and B-frames between the nearest I-frames, such as:

I⇒B⇒B⇒P⇒B⇒B⇒P⇒B⇒B⇒P⇒B⇒B⇒I ...

Thus the rules for the maximum motion vectors will be:

- P-frames. The motion vectors for these must be long enough to represent the motion of objects between the nearest P-frames.
- B-frames. The motion vectors must be long enough to represent the motion of objects between the current B-frame and each of (or at least one of) the nearest P-frames.

Thus, the motion vectors for P-frames are normally greater than for B-frames. When selecting the maximum motion vectors, the type of input video file should be analysed. For example, if there are faster horizontal movements than vertical movements, the maximum values of the horizontal motion vectors can be larger than the vertical vectors. Two examples are:

- Fast motion video and large frame sizes. P-frame maximum horizontal, vertical motion vector: 36, 30. B-frame maximum horizontal, vertical motion vector: 24, 16.
- Slow motion video and small frame sizes. P-frame maximum horizontal, vertical motion vector: 24, 16. B-frame maximum horizontal, vertical motion vector: 16, 8.

I-, P- and B-frames

Increasing the number of intermediate frames (such as P- and B-frames) can make significant savings. Each group of frames is coupled to the nearest I-frames and consists of a fixed number of subgroups, each of which has one P-frame and several B-frames.

The user can select the following:

- The number of B-frames between the nearest P-frames.
- The number of P-frames (subgroups comprising P- and B-frames) between the nearest I-frames.

I-frames are always larger that P or B-frames and P-frames are larger than B-frames. Increasing the number of P-frames reduces the number of P-frames, but too large a value of P-frames reduces the resolution of a backward/forward play system, as this requires I-frames. As P-frames require motion estimation, an increase in P-frames causes an increase in the encoding time, but reduces the overall bitrate. B-frames take slightly more time to encode than encoding P-frames as they require two types of motion estimation, one applied to the previous P- or I-frame and one applied to the next P- or I-frame. For example a typical setting is:

- I BBB P BBB P BBB I ... P-frames = 2, B-frames = 3.

- I BB P BB P BB P BB P BB P BB I ... P-frames = 5, B-frames = 2.
- I P P P P P P P P P P I... P-frames = 10, B-frames = 0.
- I I I I I I I I I I I... P-frames = 0, B-frames = 0.

DCT conversion

As with JPEG, MPEG-1 uses the DCT. It transforms macroblocks of luminance (16×16) and chrominance (8×8) into the frequency domain. This allows the higher-frequency terms to be reduced as the human eye is less sensitive to high-frequency changes. This type of coding is the same as used in JPEG still image conversion, that was described in the previous chapter.

Frames are broken up into slices 16 pixels high, and each slice is broken up into a vector of macroblocks having 16×16 pixels. Each macroblock contains luminance and chrominance components for each of four 8×8 pixel blocks. Color decimation can be applied to a macroblock, which yields four 8×8 blocks for luminance and two 8×8 blocks (C_b and C_r) of chrominance, using one chrominance value for each of the four luminance values. This is called the 4:2:0 format; two other formats are available (4:2:2 and 4:4:4, respectively known as two luminance per chrominance and one to one), which require higher data rates.

For each macroblock, a spatial offset difference between a macroblock in the predicted frame and the reference frame(s) is given if one exists (a motion vector), along with a luminance value and/or chrominance difference value (an error term) if needed. Macroblocks with no differences can be skipped except in intra frames. Blocks with differences are internally compressed, using a combination of a discrete cosine transform (DCT) algorithm on pixel blocks (or error blocks) and variable quantization on the resulting frequency coefficient (rounding off values to one of a limited set of values).

The DCT algorithm accepts signed, 9-bit pixel values and produces signed 12-bit coefficients. The DCT is applied to one block at a time, and works much as it does for JPEG, converting each 8×8 block into an 8×8 matrix of frequency coefficients. The variable quantization process divides each coefficient by a corresponding factor in a matching 8×8 matrix and rounds to an integer.

Quantization

As with JPEG the converted data is divided, or quantized, to remove higher-frequency components and to make more of the values zero. This results in numerous zero coefficients, particularly for high-frequency terms at the high end of the matrix. Accordingly, amplitudes are recorded in run-length form following a diagonal scan pattern from low frequency to high frequency.

Encoding

After the DCT and quantization state, the resultant data is then compressed using Huffman coding with a set of fixed tables. The Huffman code not only specifies the number of zeros, but also the value that ended the run of zeros. This is extremely efficient in compressing the zigzag DCT encoding method.

9.7.3 Audio compression

Audio signals normally use PCM (pulse coded modulation) codes which can be compressed to reduce the number of bits used to code the samples. For high-quality monochannel audio, the signal bandwidth is normally limited to 20 kHz, thus it is sampled at 44.1 kHz. If each sample is coded with 16 bits then the basic bit rate will be:

Digitized audio signal rate = 44.1×16 kbps = 705.6 kbps

For stereo signals the bit rate would be 1.4112 Mbps. Many digital audio systems add extra bits from error control and framing. This increases the bit rate.

Digital audio normally involves the processes of:

- Filtering the input signal.
- Sampling the signal at twice the highest frequency.
- Converting it into a digital form with an ADC (analogue-to-digital converter).
- Converting the parallel data into a serial form.
- Storing or transmitting the serial information.
- When reading (or receiving) the data the clock information is filtered out using a PLL (phase-locked loop).
- The recovered clock is then used with a SIPO (serial-in parallel-out) converter to convert the data back into a parallel form.
- Converting the digital data back into an analogue voltage.
- Filtering the analogue voltage.

These steps are illustrated in Figure 9.10. The clock recovery part is important; there is no need to save or transmit separate clock information because it can be embedded into the data. It also has the advantage that a clock becomes jittery when it is affected by noise, thus if the clock information is transmitted over relatively long distances it will be jittery.

DEAR NET-ED

Question: *All music seems to be becoming digital, but what's the great advantage when you* ***lose*** *something in the conversion?*

Yes. Something is lost in the conversion (the quantization error), but this stays constant, whereas the analogue value is likely to change. The benefits of converting to digital audio outweigh the drawbacks, such as:

- The quality of the digital audio system only depends on the conversion process, whereas the quality of an analogue audio system depends on the component parts of the system.
- Digital components tend to be easier and cheaper to produce than high-specification analogue components.
- Copying digital information is relatively easy and does not lead to a degradation of the signal.
- Digital storage tends to use less physical space than equivalent analogue forms.
- It is easier to transmit digital data.
- Information can be added to digital data so that errors can be corrected.
- Improved signal-to-noise ratios and dynamic ranges are possible with a digital audio system.

CD-quality stereo audio requires a bit rate of 1.411200 Mbps (2×16 bits $\times 44.1$ kHz). A single-speed CD-ROM can only transfer at a rate of 1.5 Mbps, and this rate must include both audio and video. Thus there is a great need for compression of both the video and audio data. The need to compress high-quality audio is also an increasing need as consumers expect higher-quality sound from TV systems and the increasing usage of digital audio radio.

A number of standards have been put forward for digital audio coding for TV/video systems. One of the first was MUSICAM, which is now part of the MPEG-1 coding system. The FCC Advisory Committee considered several audio systems for advanced television systems. There was generally no agreement on the best technology but finally it decided to conduct a side-by-side test. The winner was Dolby AC-3 followed closely by MPEG. Many cable and satellite TV systems now use either MPEG or Dolby AC-3 coding.

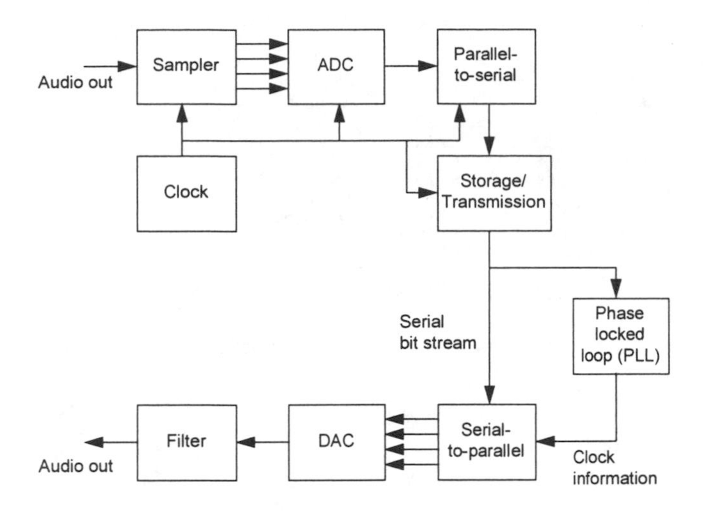

Figure 9.10 A digital audio system

Psycho-acoustic model

MPEG and Dolby AC-3 use the psycho-acoustic model to reduce the data rate, which exploits the characteristics of the human ear. This is similar to the method used in MPEG video compression which uses the fact that the human eye has a lack of sensitivity to the higher-frequency video components (that is, sharp changes of color or contrast). The psycho-acoustic model allows certain frequency components to be reduced in size without affecting the perceived audio quality as heard by the listener.

A well-known effect is the masking effect. This is where noise is only heard by a person when there are no other sounds to mask it. A typical example is high-frequency hiss from a compact cassette when there are quiet passages of music. When there are normal periods of music the louder music masks out the quieter hiss and they are not heard. In reality, the brain is masking out the part of the sound it wants to hear, even though the noise component is still there. When there is no music to mask the sound then the noise is heard.

Noise, itself, tends to occur across a wide range of frequencies, but the masking effect also occurs with sounds. A loud sound at a certain frequency masks out a quieter sound at a similar frequency. As a result the sound heard by the listener appears only to contain the loud sounds; the quieter sounds are masked out. The psycho-acoustic model tries to reduce the levels to those that would be perceived by the brain.

Figure 9.11 illustrates this psycho-acoustic process. In this case a masking level has been applied and all the amplitudes below this level have been reduced in size. Since these frequencies have been reduced in amplitude, then any noise associated with them is also significantly reduced. This basically has the effect of limiting the bandwidth of the signal to the key frequency ranges.

The psycho-acoustic model also takes into account non-linearities in the sensitivity of the ear. Its peak sensitivity is between 2 and 4 kHz (the range of the human voice) and it is least sensitive around the extremes of the frequency range (i.e. high and low frequencies). Any noise in the less sensitive frequency ranges is more easily masked, but it is important to minimize any noise in the peak range because it has a greater impact.

Masking can also be applied in the time domain, where it can be applied just before and after a strong sound (such as a change of between 30 and 40 dB). Typically, premasking occurs for about 2–5 ms before the sound is perceived by the listener and the post-masking effect lasts for about 100 ms after the end of the source.

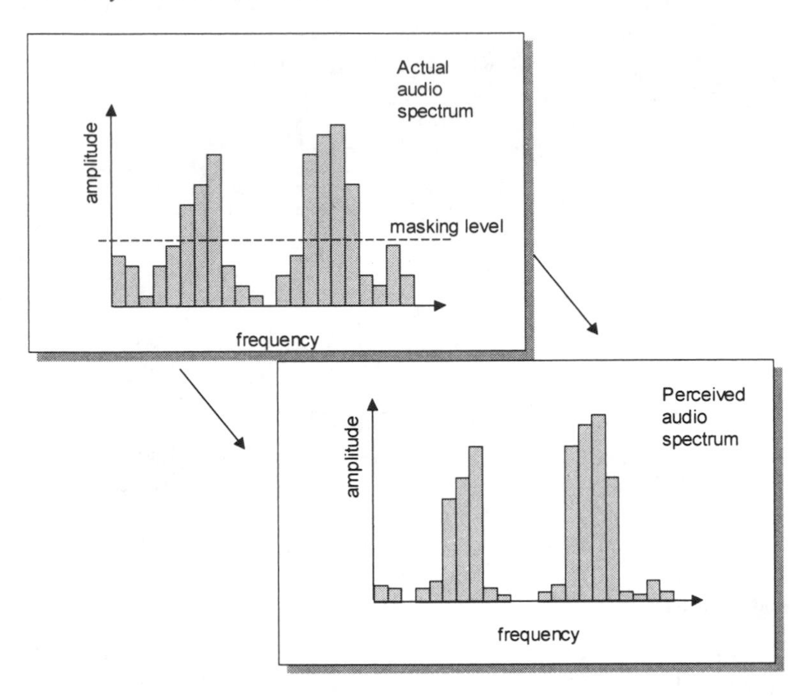

Figure 9.11 Difference between actual and perceived audio spectrum

MPEG basically has three different levels:

- MPEG-Audio Level I – uses a psycho-acoustic model to mask and reduce the sample size. It is basically a simplified version of MUSICAM and has a quality which is nearly equivalent to CD-quality audio. Its main advantage is that it allows the construction of simple encoders and decoders with medium performance and which will operate fairly well at 192 or 256 kbps.
- MPEG-Audio Level II – which is identical to the MUSICAM standard. It is also nearly equivalent to CD-quality audio and is optimized for a bit rate of 96 or 128 kbps per monophonic channel.
- MPEG-Audio Level III (MP-3) – which is a combination of the MUSICAM scheme and ASPEC, a sound compression scheme designed in Erlangen, Germany. Its main advantage is that it targets a bit rate of 64 kbps per audio channel. At that speed, the quality is very close to CD quality and produces a sound quality which is better than MPEG Level-II operating at 64 kbps.

The three levels are basically supersets of each other with Level III decoders being capable of decoding both Level I and Level II data. Level II is the most frequently used of the three standards. Level I is the simplest, while Level III gives the highest compression but is the most computational in coding.

The forward and backward compatible MPEG-2 system, following recommendations from SMPTE, EBU and others, has increased the audio capacity to five channels. Figure 9.12 shows an example of a 5-channel system; the key elements are:

- A center channel.
- Left and right surround channels.

- Left and right channels (as hi-fi stereo).

MPEG-2 also includes a low-frequency effects channel (called LFE, essentially a sub-woofer). This has a much lower bandwidth than the other channels. This type of system is often called a 5.1-channel system (5 main channels and LFE channel).

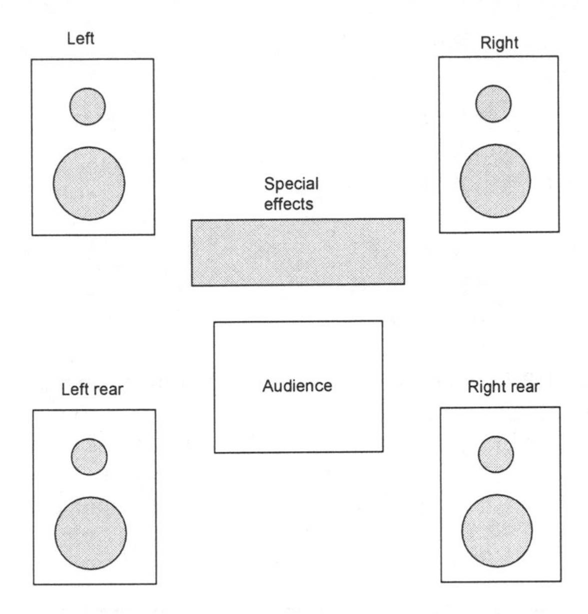

Figure 9.12 A 5.1 channel audio surround sound system

9.7.4 Speech

Speech and audio signals are normally converted into PCM, which can be stored or transmitted as a PCM code, or compressed to reduce the number of bits used to code the samples. Speech generally has a much smaller bandwidth than audio.

PCM parameters

Digital systems tend to be less affected by noise than analogue. The main source of noise is quantization noise which is caused by the finite number of quanitization levels converting to a digital code.

The main parameters in determining the quality of a PCM system are the dynamic range (DR) and the signal-to-noise ratio (SNR). The maximum error between the original level and the quantized level occurs when the original level falls exactly halfway between two quantized levels. This error will be half the smallest increment or

$$\text{Max error} = \pm\frac{1}{2}.\frac{\text{Full Scale}}{2^N}$$

The dynamic range is the ratio of the largest possible signal magnitude to the smallest possible signal magnitude. If the input signal uses the full range of the ADC then the maximum signal will be the full-scale voltage. The smallest signal which can be reproduced is one which toggles between one quantization level and the level above, or below. This signal amplitude, for an n-bit ADC, is the full-scale voltage divided by the number of quantization levels (that is, 2^n). Thus, for a linearly quantized signal:

$$\text{Dynamic range} = \frac{V_{\max}}{V_{\min}}$$

$$\text{Number of levels} = 2^n - 1$$

$$\text{Dynamic range} = 20\log\frac{V_{\max}}{V_{\max}/2^n - 1}\ \text{dB}$$

$$= 20\log(2^n - 1)\ \text{dB}$$

if 2^n is much greater that 1, then

$$\text{Dynamic range} \approx 20\log 2^n\ \text{dB}$$

$$= 20n\log 2\ \text{dB}$$

$$\approx 6.02n\ \text{dB}$$

Table 9.7 outlines the DR for a given number of bits. Normally the maximum number of bits is less than 20. The voltage ratio of a given number of bits is also given in square brackets [*ratio*]. For example an 8-bit system has a DR of 48.18 dB and the largest voltage amplitude is 256 times the smallest voltage amplitude. A 16-bit system has a DR of 96.33 dB and the largest voltage amplitude is 65 536 times the smallest voltage amplitude.

Table 9.7 Dynamic range of a digital system

Number of bits	*DR (dB)* [*ratio*]	*Number of bits*	*DR (dB)* [*ratio*]
1	6.02 [2]	11	66.23 [2 048]
2	12.04 [4]	12	72.25 [4 096]
3	18.06 [8]	13	78.27 [8 192]
4	24.08 [16]	14	84.29 [16 384]
6	36.12 [64]	16	96.33 [65 536]
7	42.14 [128]	17	102.35 [131 072]
8	48.16 [256]	18	108.37 [262 144]
10	60.21 [1 024]	20	120.41 [1 048 576]

Signal-to-noise ratio (SNR)
It can be shown that the SNR for a linearly quantized digital system is:

$$\text{SNR} = 1.76 + 6.02\,n\,\text{dB}$$

Table 9.8 outlines the SNR for a given number of bits. Normally the maximum number of bits is less than 20. The voltage ratio of a given number of bits is also given in square brackets [*ratio*]. For example an 8-bit system has an SNR of 49.92 dB and the largest rms voltage is 313.33 times the smallest rms voltage. A 16-bit system has an SNR of 96.33 dB and the largest rms voltage is 80 167.81 times the smallest rms voltage.

Differential encoding

Differential coding is a source-coding method which is used when there is a limited change from one value to the next. It is well suited to video and audio signals, especially audio, where the sampled values can only change within a given range. It is typically used in PCM (pulse code modulation) schemes to encode audio and video signals.

Table 9.8 Signal-to-noise ratio of a digital system

Number of bits	*SNR (dB) [ratio]*	*Number of bits*	*SNR (dB) [ratio]*
7	43.90 [156.68]	14	86.04 [20 044.72]
8	49.92 [313.33]	15	92.06 [40 086.67]
9	55.94 [626.61]	16	98.08 [80 167.81]
10	61.96 [1 253.14]	17	104.10 [160 324.5]
11	67.98 [2 506.11]	18	110.12 [320 626.9]
12	74.00 [5 011.87]	19	116.14 [641 209.6]
13	80.02 [10 023.05]	20	122.16 [1 282 331]

Delta modulation PCM

PCM coverts analogue samples into a digital code. Delta PCM uses a single-bit code to represent an analogue signal. With delta modulation a '1' is transmitted (or stored) if the analogue input is higher than the previous sample or a '0' if it is lower. It must obviously work at a higher rate than the Nyquist frequency, but because it uses only 1 bit, it normally uses a lower output bit rate. Figure 9.13 shows a delta modulation transmitter.

Initially the counter is set to zero. A sample is taken and if it is greater than the analogue value on the DAC output, the counter is incremented by 1, or it is decremented. This continues at a time interval given by the clock. Each time the present sample is greater than the previous sample, a '1' is transmitted; otherwise a '0' is transmitted. Figure 9.14 shows an example signal. The sampling frequency is chosen so that the tracking DAC can follow the input signal. This results in a higher sampling frequency, but because it only transmits one bit at a time, the output bit rate is normally reduced. Figure 9.15 shows that the receiver is almost identical to the transmitter except that it has no comparators.

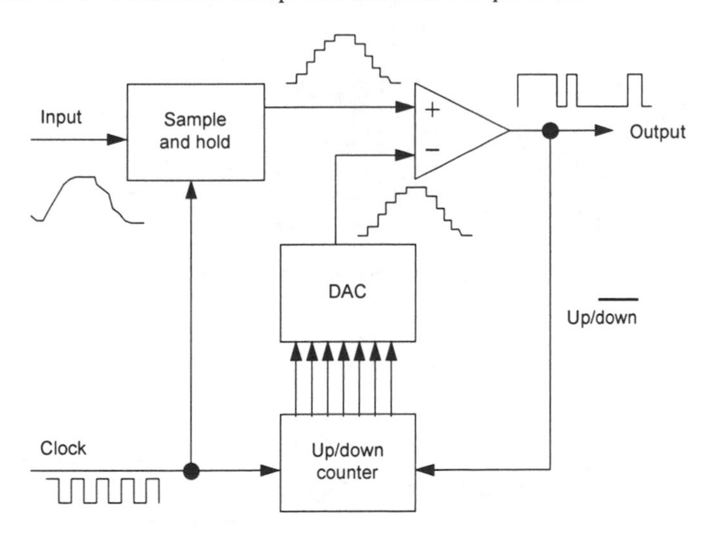

Figure 9.13 Delta modulation

Two problems with delta modulation are granular noise and slope overload:

- Slope overload. This occurs when the signal changes too fast for the modulator to keep up; see Figure 9.16 (left-hand side). It is possible to overcome this problem by increasing the clock frequency or increasing the step size.
- Granular noise. This occurs when the signal changes slowly in amplitude, as illustrated in Figure 9.16 (right-hand side). The reconstructed signal contains a noise which is not

present at the input. Granular noise is equivalent to quantization noise in a PCM system. It can be reduced by decreasing the step size, though there is a compromise between smaller step size and slope overload.

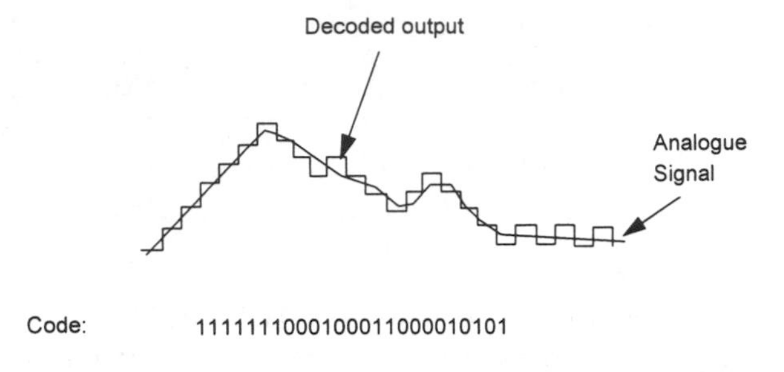

Figure 9.14 Delta modulator signal

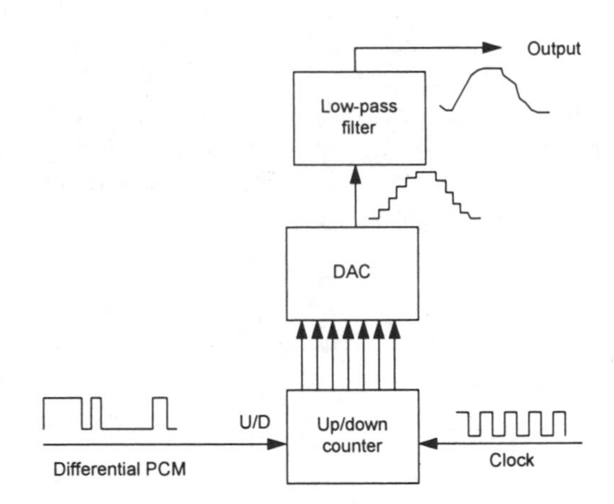

Figure 9.15 Delta modulator receiver

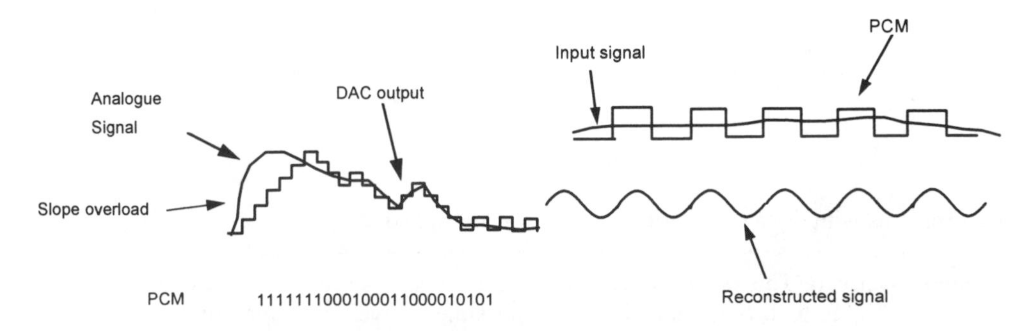

Figure 9.16 Slope overload and granular noise

Adaptive delta modulation PCM

Unfortunately delta modulation cannot react to very rapidly changing signals and will thus take a relatively long time to catch them up (known as slope overload). It also suffers when the signal does not change much as this ends up in a square wave signal (known as granular noise). One method of reducing granular noise and slope overload is to use adaptive delta PCM. With this method the step size is varied by the slope of the input signal. The larger the slope, the larger the step size; see Figure 9.17. The algorithms usually depend on the system and the characteristics of the signal. A typical algorithm is to start with a small step and increase it by a multiple until the required level is reached. The number of slopes will depend on the number of coded bits, such as 4 step sizes for 2 bits, 8 for 3 bits, and so on.

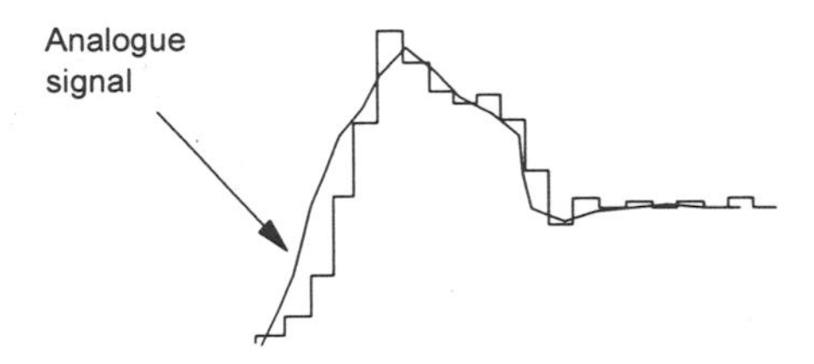

Figure 9.17 Variation of step size

Differential PCM (DPCM)

Speech signals tend not to change much between two samples. Thus similar codes are sent, which leads to a degree of redundancy. For example, in a certain sample it is likely the signal will only change within a range of voltages, as illustrated in Figure 9.18.

DPCM reduces the redundancy by transmitting the difference in the amplitude of two consecutive samples. Since the range of sample differences is typically less than the range of individual samples, fewer bits are required for DPCM than for conventional PCM.

Figure 9.19 shows a simplified transmitter and receiver. The input signal is filtered to half the sampling rate. This filter signal is then compared with the previous DPCM signal. The difference between them is then coded with the ADC.

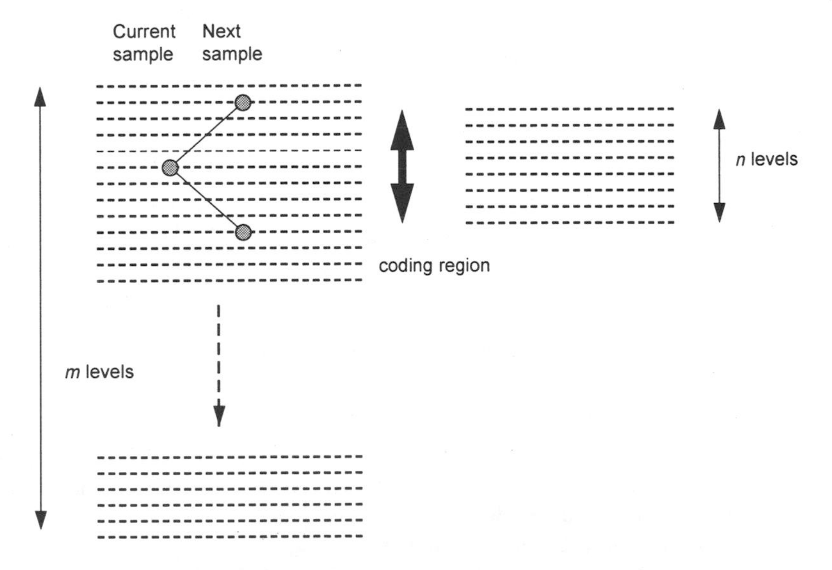

Figure 9.18 Normal and differential quantization

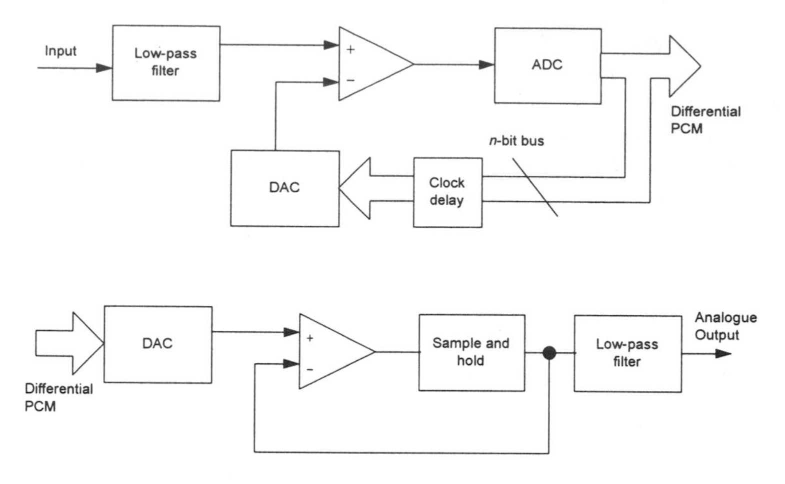

Figure 9.19 DPCM transmitter/receiver

Adaptive differential PCM (ADPCM)
ADPCM allows speech to be transmitted at 32 kbps with little noticeable loss of quality. As with differential PCM the quantizer operates on the difference between the current and previous samples. The adaptive quantizer uses a uniform quantization step *M*, but when the signal moves towards the limits of the quantization range, the step size *M* is increased. If it is around the center of the ranges, the step size is decreased. Within any other regions the step size hardly changes. Figure 9.20 illustrates this operation with a signal quantized to 16 levels. This results in 4-bit code.

The change of the quantization step is done by multiplying the quantization level, *M*, by a number slightly greater, or less, than 1 depending on the previously quantized level.

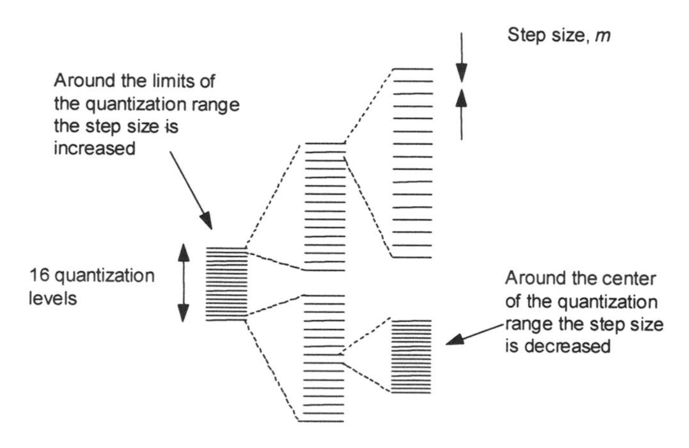

Figure 9.20 ADPCM quantization

Speech compression

Subjective and system tests have found that 12-bit coding is required to code speech signals, which give 4096 quantization levels. If linear quantization is applied then the quantization step is the same for quiet levels as for loud levels. Any quantization noise in the signal will be more noticeable at quiet levels than at loud levels. When the signal is loud, the signal itself swamps the quantization noise, as illustrated in Figure 9.21. Thus an improved coding mechanism is to use small quantization steps at low input levels and a higher one at high levels. This is achieved using non-linear compression.

The two most popular types of compression are A-Law (in European systems) and μ-Law (in the USA). These laws are similar and compress the 12-bit quantized speech code into an 8-bit compressed code. An example compression curve is shown in Figure 9.22.

As an approximation the two laws are split into 16 line segments. Starting from the origin and moving outwards, left and right, each segment has half the slope of the previous. Using an 8-bit compressed code at a sample rate of 8000 samples per second gives a bit rate of 64 kbps. ISDN uses this bit rate to transmit digitized speech. Figure 9.23 shows a basic transmission system.

A-Law and μ-Law companding

The companding and expansion encoding is normally implemented using either μ-Law or A-Law. A-Law is used in Europe and in many other countries, whereas μ-Law is used in North America and Japan. Both were defined by the CCITT in the G.711 recommendation and both use non-uniform quantization step sizes which increase logarithmically with signal level. μ-Law uses the compression characteristic of:

$$y = \frac{\log(1+\mu x)}{\log(1+x)} \quad \text{for } x \geq 0$$

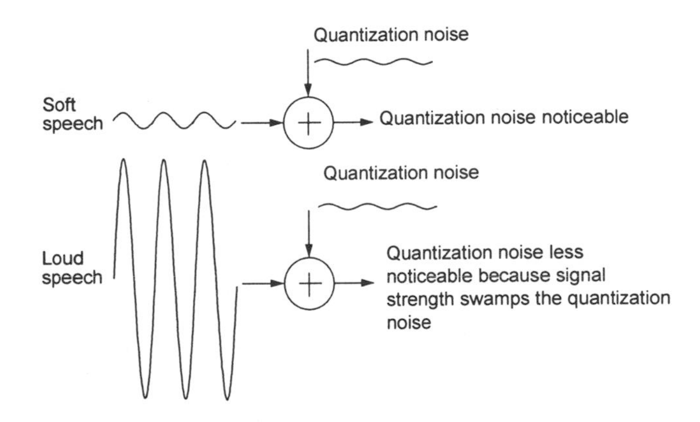

Figure 9.21 Quantization noise is more noticeable with low signal levels

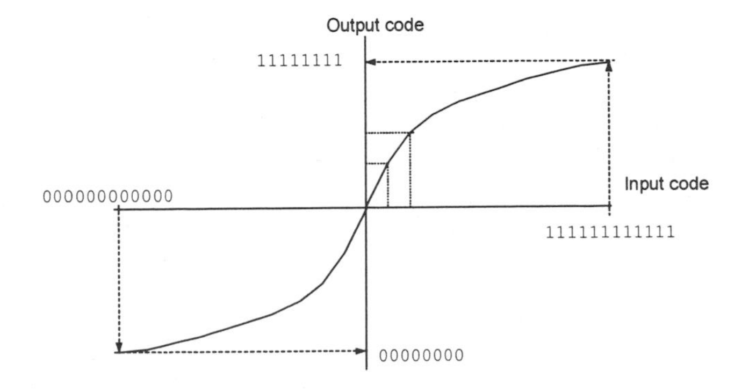

Figure 9.22 12-bit to 8-bit non-linear compression

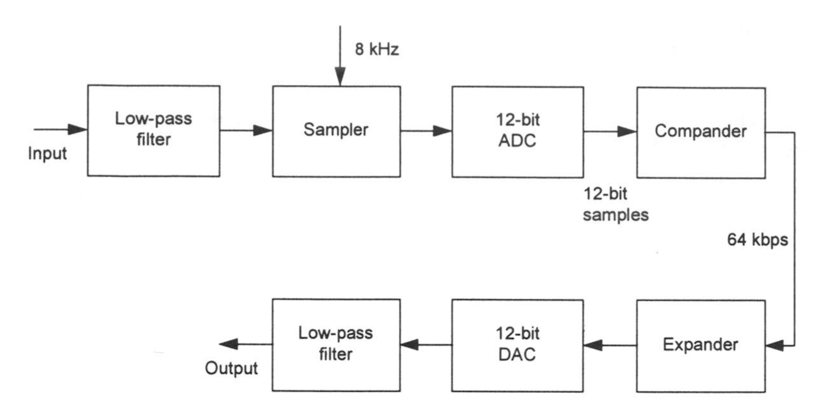

Figure 9.23 Typical PCM speech system

where y is the output magnitude
x is the input magnitude
μ is a positive factor which is chosen for the required compression characteristics

Figure 9.24 shows an example of μ-Law using $\mu=1$, $\mu=50$ and $\mu=255$. Using $\mu=0$ gives uniform conversion (linear quantization). Normally speech systems use $\mu=255$ as this characteristic is well matched to human hearing. An 8-bit implementation can achieve a small SNR and a dynamic range equivalent to that of a 12-bit uniform system.

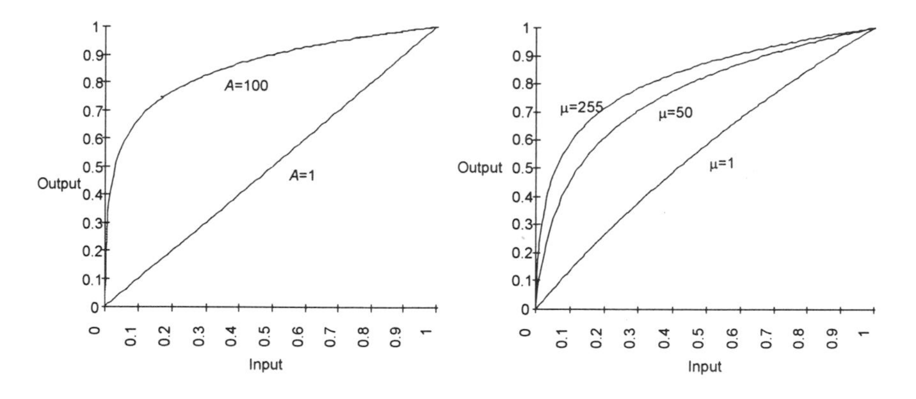

Figure 9.24 A-Law and μ-Law characteristics

The A-law also uses quantization characteristics that vary logarithmically. Figure 9.25 shows an example of A-Law using $A=1$ and $A=100$. Most A-Law speech systems use $A=87.56$. The compression characteristic is:

$$y = \begin{cases} \dfrac{Ax}{1+\log A} & \text{for } 0 \le |x| \le \dfrac{1}{A} \\ \dfrac{1+\log(Ax)}{1+\log A} & \text{for } \dfrac{1}{A} \le |x| \le 1 \end{cases}$$

where A is a positive integer.

Figure 9.25 shows two input waveforms, 1 V peak to peak and 0.1 V peak to peak. It can be seen that the companding process amplifies the lower amplitudes more than the large amplitudes. This causes low-amplitude speech signals to be boosted compared with loud speech. Also notice that the waveform has been distorted because the low amplitudes are amplified more than the large amplitudes.

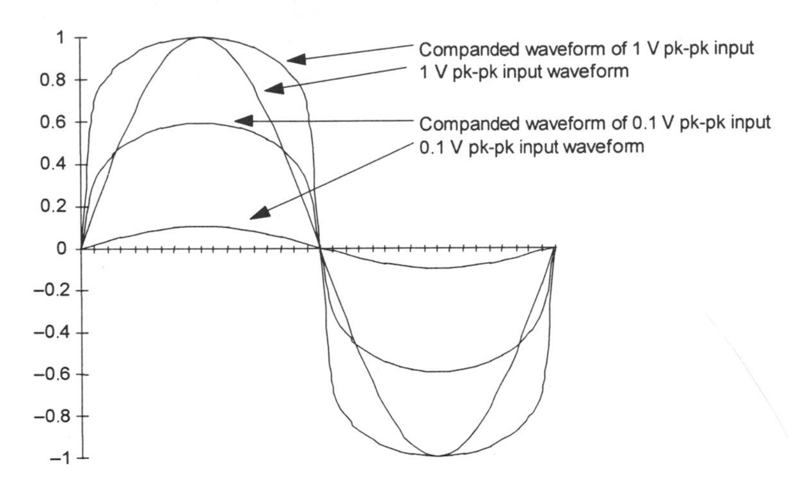

Figure 9.25 Effects of waveforms with μ-255 encoding

Digitally linearizable log-companding

The mathematical formulas for A-Law and μ-Law are normally approximated to a series of linear segments. This permits more precise control of the quantization characteristics. The chosen approximation used is to make the step sizes in consecutive segments change by a factor of 2. Figure 9.26 shows the characteristic of the piecewise linear conversion. It can be seen that the slope of each segment is twice the slope of the previous segment (although in A-Law 98.56, segment 0 and segment 1 have the same slope). Each segment has 16 quantization levels and there are 16 segments (8 for positive inputs and 8 for negative inputs). Thus 1 bit identifies the sign bit, 3 bits identify the segment (in the positive or negative part) and 4 bits identify the quantization level. The 8-bit companded values thus take the form:

```
SLLLQQQQ
```

where `S` is the sign bit, `LLL` is the segment number and `QQQQ` is the quantization level within the segment.

Speech sampling

With telephone-quality speech the signal bandwidth is normally limited to 4 kHz, thus it is sampled at 8 kHz. If each sample is coded with 8 bits then the basic bit rate will be:

Digitized speech signal rate = 8 × 8 kbps = 64 kbps

Table 9.9 outlines the main compression techniques for speech. The G.722 standard allows the best-quality signal. The maximum speech frequency is 7 kHz rather than 4 kHz in normal coding systems; this gives an equivalent of 14 coding bits. The G.728 allows extremely low bit rates (16 kbps).

Table 9.9 Speech compression standards

ITU standard	*Technology*	*Bit rate*	*Description*
G.711	PCM	64 kbps	Standard PCM
G.721	ADPCM	32 kbps	Adaptive delta PCM where each value is coded with 4 bits
G.722	SB-ADPCM	48, 56 and 64 kbps	Subband ADPCM allows for higher-quality audio signals with a sampling rate of 16 kHz
G.728	LD-CELP	16 kbps	Low-delay code excited linear prediction for low bit rates

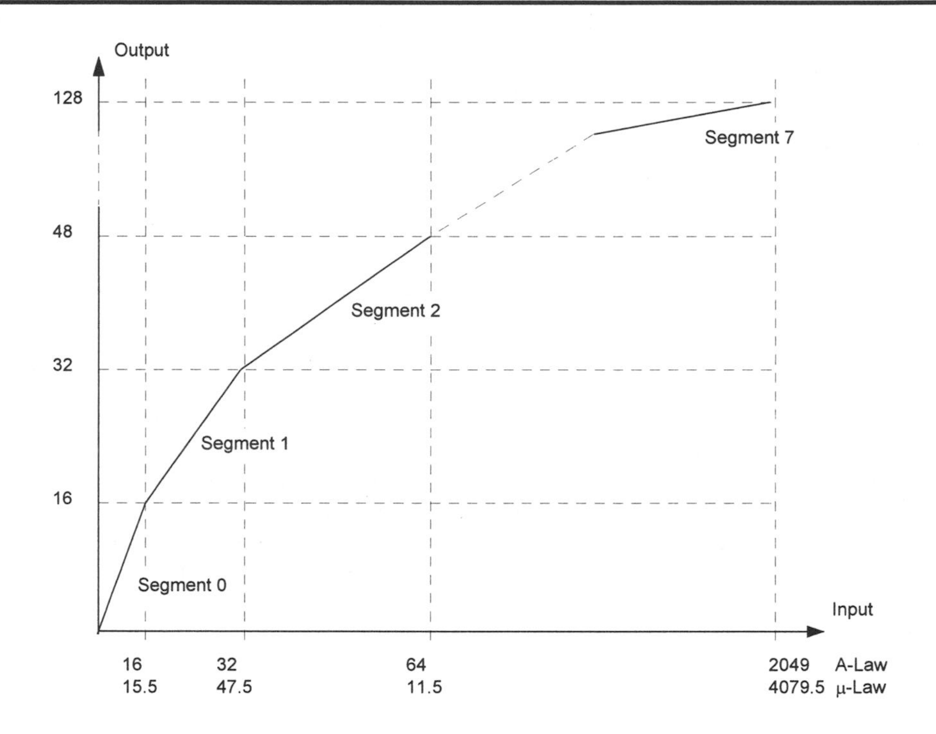

Figure 9.26 Piecewise linear compression for A-Law and μ-Law

PCM-TDM systems

Multiple channels of speech can be sent over a single line using time division multiplexing (TDM). In the UK a 30-channel PCM system is used, whereas the USA uses 24.

With a PCM-TDM system, several voice band channels are sampled, converted to PCM codes, these are then time division multiplexed onto a single transmission media. Each sampled channel is given a time slot and all the times slots are built up into a frame. The complete frame usually has extra data added to it such as synchronization data, and so on. Speech channels have a maximum frequency content of 4 kHz and are sampled at 8 kHz. This gives a sample time of 125 μs. In the UK a frame is built up with 32 time slots from TS0 to TS31. TS0 and TS16 provide extra frame and synchronization data. Each of the time slots has 8 bits, therefore the overall bit rate is:

Bits per time slot = 8
Number of time slots = 32
Time for frame = 125 μs

$$\text{Bit rate} = \frac{\text{No of bits}}{\text{Time}} = \frac{32 \times 8}{125 \times 10^{-6}} = 2048 \text{ kbps}$$

In the USA and Japan this bit rate is 1.544 Mbps. These bit rates are known as the primary rate multipliers. Further interleaving of several primary rate multipliers increases the rate to 6.312, 44.736 and 139.264 Mbps (for the USA) and 8.448, 34.368 and 139.264 Mbps (for the UK).

The UK multiframe format is given in Figure 9.27. In the UK format the multiframe has 16 frames. Each frame time slot 0 is used for synchronization and time slot 16 is used for signaling information. This information is sub-multiplexed over the 16 frames. During frame 0 a multiframe-alignment signal is transmitted in TS16 to identify the start of the multiframe structure. In the following frames, the eight binary digits available are shared by channels 1–15 and 16–30 for signaling purposes. TS16 is used as follows:

Frame 0 `0000XXXX`
Frames 1–15 `1234 5678`

where 1234 are the four signaling bits for channels 1,2,3, ..., 15 in consecutive frames, and 5678 are the four signaling bits for channels 16,17, 18,... 31 in consecutive frames. Thus in the first frame the `0000XXXX` code word is sent, in the next frame the first channel and the 16th channel appear in TS16, the next will contain the second and the 17th, and so on. Typical 4-bit signal information is:

`1111` – circuit idle/busy
`1101` – disconnection

TS0 contains a frame-alignment signal which enables the receiver to synchronize with the transmitter. The frame-alignment signal (`X0011011`) is transmitted in alternate frames. In the intermediate frames a signal known as a not-word is transmitted (`X10XXXXX`). The second binary digit is the complement of the corresponding binary digit in the frame-alignment signal. This reduces the possibility of demultiplexed misalignment to imitative frame-alignment signals.

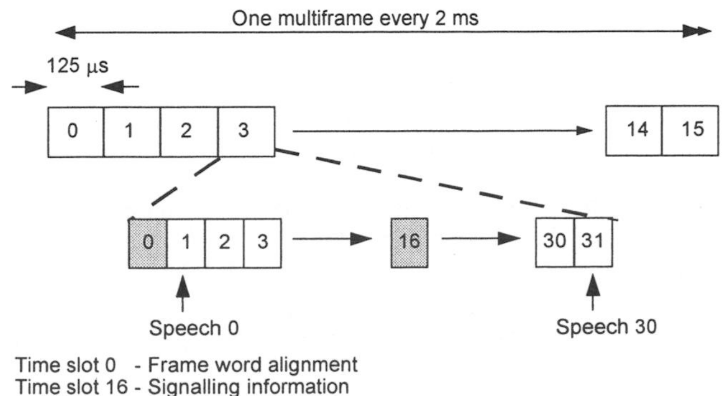

Figure 9.27 PCM-TDM multiframe format with 30 speech channels

Alternative frames:

TS0: `X0011011`
TS0: `X10XXXXX`

where `X` stands for don't care conditions.

9.8 Exercises

9.8.1 How much memory would an uncompressed 640×480 image with 256 colors take in memory:
(a) 300 KB (b) 2400 KB
(c) 76800 KB (d) 1 MB

9.8.2 For telephone-quality speech, what is the basic bit rate:
(a) 4 kbps (b) 64 kbps
(c) 80 kbps (d) 128 kbps

9.8.3 For 44.1 kHz, stereo, 16-bit sampled hi-fi, what is the basic bit rate:
(a) 150 kbps (b) 64 kbps
(c) 1 Mbps (d) 1.411 Mbps

9.8.4 How many analogue levels can be represented by a 12-bit analogue-to-digital convertor:
(a) 12 (b) 1024
(c) 4096 (d) 65536

9.8.5 If the sample rate is 25 MHz, what is the time between samples:
(a) 4 ns (b) 40 ns
(c) 4 μs (d) 40 μs

9.8.6 How many bits would be used to store 16 colors:
(a) 1 (b) 4
(c) 16 (d) 32

9.8.7 How many bits would be used to store 16K colors:
(a) 10 (b) 14
(c) 16 (d) 24

9.8.8 Which of the following does **not** use source compression:
(a) MPEG (b) MP-3
(c) LZ (d) JPG

9.8.9 For US/Japanese PCM-TDM systems, what is the basic bit rate:
(a) 1.411 Mbps (b) 1.544 Mbps
(c) 2.048 Mbps (d) 2.5 Mbps

9.8.10 For UK PCM-TDM systems, what is the basic bit rate:
(a) 1.411 Mbps (b) 1.544 Mbps
(c) 2.048 Mbps (d) 2.5 Mbps

9.8.11 Which of the following best describes Huffman coding:

(a) The number of bits allocated to each character depends on the number of occurrences
(b) Characters are back referenced to a previous occurrence
(c) It builds a data dictionary of character sequences
(d) Represents a special bit sequence for repetitive sequences

9.8.12 Which of the following best describes LZ coding:
(a) The number of bits allocated to each character depends on the number of occurrences
(b) Characters are back referenced to a previous occurrence
(c) It builds a data dictionary of character sequences
(d) Represents a special bit sequence for repetitive sequences

9.8.13 Which of the following best describes LZW coding:
(a) The number of bits allocated to each character depends on the number of occurrences
(b) Characters are back referenced to a previous occurrence
(c) It builds a data dictionary of character sequences
(d) Represents a special bit sequence for repetitive sequences

9.8.14 The two basic types of compression are lossless compression and lossy compression. Discuss the difference between the two types and give examples of where they would be used.

9.8.15 Explain the differences between source coding and entropy coding. Give examples of where they could possibly be used.

9.8.16 Determine the Huffman coding for the following characters:

'a'	[60]	'e'	[120]	'f'	[30]
'g'	[25]	'.'	[10]	'p'	[55]

Note that the number of occurrences of the character is given within brackets.

9.8.17 Explain how LZH coding operates with a 4 K dictionary size.

9.8.18 Explain why the compression rate varies for a GIF with the amount of changes in the graphic. Outline the differences in the compression rates given in Table 9.5 and Table 9.6.

9.8.19 Explain how RGB information is converted to YC_bC_r (YUV).

9.8.20 Using CCIR 601 color conversion complete Table 9.10.

9.8.21 Explain how subsampling reduces the amount of data stored.

9.8.22 Explain the main steps in MPEG coding.

9.8.23 Explain why a frame must be divisible by 16 in both the x- and y-directions. Also give an example of a frame split into a number of macroblocks.

9.8.24 Explain how I-, P- and B-frames might be used.

9.8.25 Show how the standard rate for PCM transmission is 64 kbps.

9.8.26 The three main hi-fi quality sampling rates are 32 kHz, 44.1 kHz and 48 kHz and the main classifications are broadcast quality, CD quality and professional quality. State which sampling rate normally goes with which classification.

Table 9.10 RGB to YC_bC_r and YC_bC_r to RGB

Red	*Green*	*Blue*	*Y*	C_b	C_r
1	1	1			
0.5	1	0			
1	0	0			
0	1	0			
0.5	0.5	0.5			
			0.6924	0.3	0.04878
			0.2	0.06748	0.03252
			0.369	0.05122	–0.12683
			0.3979	0.13557	0.37439

Sample exam questions

9.8.27 Explain each of the steps that an MPEG encoder uses to compress motion video. What effect, if any, do these steps have on the decoded video?

9.8.28 Explain why, to compress a color image, RGB is normally converted into luminance and chrominance. What techniques can be applied to the luminance and chrominance data to compress the image information?

9.8.29 Explain, using an example, how the Lempel-Ziv (LZ) compression algorithm operates. What type data does it compress best?

9.8.30 Explain, using an example, how the Lempel-Ziv-Welsh (LZW) algorithm operates.

9.8.31 Derive the basic bitrate transmission for single channel speech and also for stereo, hi-fi quality audio. State all assumptions made.

9.8.32 Explain the psycho-acoustic model and the masking effect. Also, contrast forward and backward adaptive methods in audio coding, giving their advantages and disadvantages.

10 Introduction to Networks

10.1 Introduction

Networking involves the interconnection of workstations, terminals and other networked devices. In most cases a network allows computers of different types to intercommunicate using a network protocol. The protocol that the computers use is thus more important to communication, than their actual make. Thus, in order for them to intercommunicate, computers on a network must have a common protocol.

Many of the first computers were standalone devices, and thus worked independently from other computers. This caused many problems, including:

- The difficulty to intercommunicate between computers.
- The difficulty in managing the configuration of the computers.
- The requirement for duplication of resources, as each computer required its own resource, such as a dedicated printer, a dedicated modem, and so on.

These problems were solved with local area networks (LANs), which connect computers and other devices within a single building. One of the great advantages of LANs was that they allowed the sharing of files and printers. They are also efficient in transferring files within an organization, but it was still difficult to transmit data over a large geographical area. This led to the development of WANs (Wide Area Networks), and MANs (Metropolitan Area Networks).

The order of size for networks are (in order of size):

- **Local area networks** (LANs), which connect over a relatively small geographical area, typically connecting computers within a single office or building. In most cases they connect to a common electronic connection – commonly known as a network backbone. LANs can connect to other networks either directly or through a WAN or MAN.
- **Metropolitan area networks** (MANs), which normally connect networks around a town or city. It is smaller than a WAN, but larger than a LAN. An example of a MAN is the EaStMAN (Edinburgh and Stirling MAN) network that connects universities and colleges in Edinburgh and Stirling, UK, as illustrated in Figure 10.1.
- **Wide area networks** (WANs), which connect networks over a large geographical area, such as between different buildings, towns or even countries.

The four main methods of connecting a network (or an independently connected computer) to another network are:

- Through a **modem** connection. A modem converts digital data into an analogue form that can be transmitted over a standard telephone line.
- Through an **ISDN** connection. An ISDN (integrated services digital network) connection uses the public telephone service and differs from a modem connection in that it sends data in a digital form.
- Through a **gateway**. A gateway connects one type of network to another type.

- Through a **bridge** or **router**. Bridges and routers normally connect one type of network to one of the same type. Normally, these days, gateways are routers.

The 1980s saw a growth in networks, but in many cases they became difficult to configure as new networking technologies used different types of hardware and software implementations, which caused incompatibilities with other types of networking equipment. Soon a number of specifications started to appear, each of which had difficulty in communicating with other types. To overcome this, in 1984, the International Organization for Standardization (ISO) developed a new model called the OSI Reference Model, which was based on research into network technologies such as:

- **SNA** (Systems Network Architecture). Developed by IBM in the 1970s. It is large and complex, and is similar in its approach to the developed OSI reference model, especially, as it also has seven layers.
- **DECNET.** Developed by DEC, and has been recently developed into DECnet/OSI which supports both OSI protocols and proprietary DEC protocols.
- **TCP/IP** (Transmission Control Protocol/Internet Protocol). Developed by US Department of Defense in the 1970s. Its main function was to allow the intercommunication of computers over a global network (the Internet). It is now used by all of the computers which connect to the Internet.

The developed model has since allowed:

- Compatibility – allows manufacturers to create networks which are compatible with each other.
- Interoperability – allows the transfer of data between different types of computers, no matter their architecture, their operating system, or their network connection type.

Along with the developed model, a number of standards have been defined, which are basically a set of rules or procedures that are either widely used or officially specified.

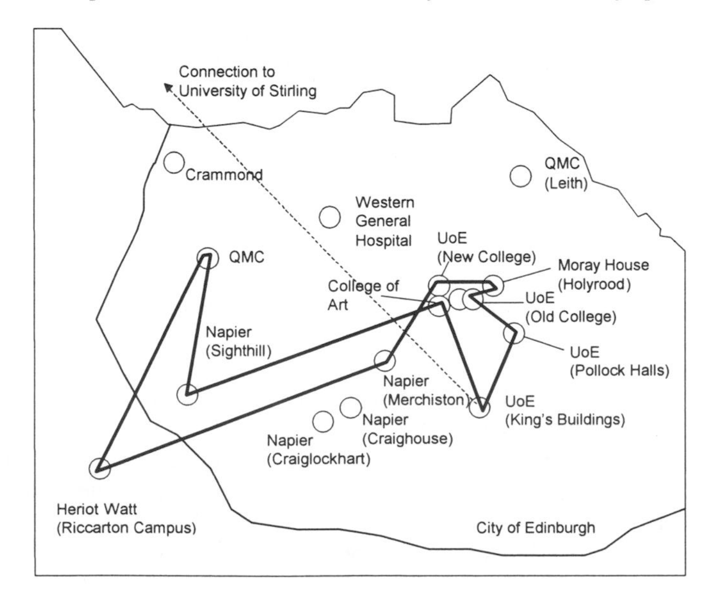

Figure 10.1 Layout of the EaStMAN network

10.2 Advantages and disadvantages of networks

Networks allow the orderly flow of information between connected nodes. Their main advantages are that:

- It is easier to set up new users and equipment.
- It allows the sharing of resources (see Figure 10.2).
- It is easier to administer users.
- It is easier to administer software licenses.
- It allows electronic mail to be sent between users.
- It allows simple electronic access to remote computers and sites.
- It allows the connection of different types of computers which can communicate with each other.

10.2.1 Sharing information

A major advantage of LANs is their ability to share information over a network. Normally, it is easier to store application programs at a single location and make them available to users rather than having copies individually installed on each computer (unless the application program requires special configurations or there are special licensing agreements). This saves on expensive disk space and increases the availability of common data and configurations. The disadvantage of this is that it increases the traffic on a network.

Most networks have a network manager, or network group, who manage the users and peripherals on a network. On a well-maintained network the network manager will:

- Control the users on the network, that is, who can and cannot login.
- Control which of the users are allowed to use which facilities.
- Control which of the users are allowed to run which application programs.
- Control the usage of software packages by limiting users to license agreements.
- Standardize the set up of application programs to a single source.
- Back-up important files on a regular basis onto a mass back-up system.
- Set up simple-to-use procedures to access programs, such as icons, menus, and so on.
- Possibly control PC (Personal Computer) viruses by running automatic scanning programs.
- Update application programs by modifying them at a single source.

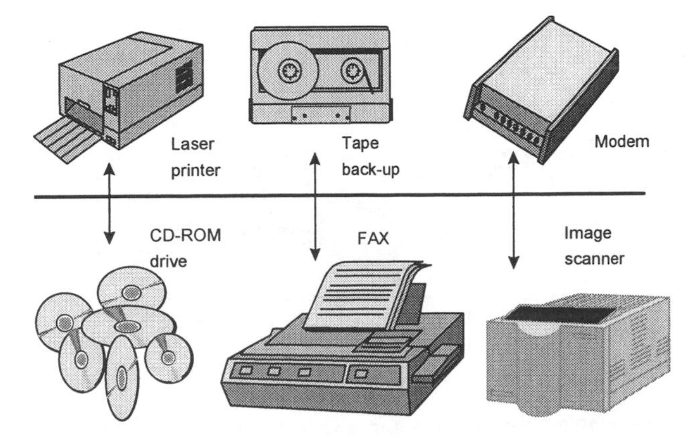

Figure 10.2 Local network with a range of facilities

10.2.2 Sharing disk resources (network file servers)

Many computer systems require access to a great deal of information and to run many application programs such as word processors, spreadsheets, compilers, presentation packages, computer-aided design (CAD) packages, and so on. Most local hard-disks could not store all the required data and application programs, thus a network allows users to access files and application programs on remote disks.

Some distributed, multi-tasking operating systems such as Unix and VMS allow all the hard disks on a network to be electronically linked as a single file system. Most PCs normally are networked to file servers, which provides networked file systems. A network file server thus allows users to access a central file system (for PCs) or a distributed file system (for Unix/VMS). This is illustrated in Figure 10.3.

10.2.3 Sharing resources

Computers not connected to a network may require extra peripherals such as printers, fax machines, modems, plotters, and so on. This may be resource inefficient, as other users cannot get access to them unless they are physically disconnected and connected to their own computer. Normally, it is more efficient to share resources over a network.

Access to networked peripherals is also likely to be simpler as the system manager can standardize configurations. Peripherals that are relatively difficult to set up such as plotters, fax machines and modems can be set up once and their configuration stored. The network manager can also bar certain users from using certain peripherals.

There is normally a trade-off between the usage of a peripheral and the number required. For example a single laser printer in a busy office may not be able to cope with the demand. A good network copes with this by segmentation, so that printers are assigned to different areas or users. The network may also allow for re-direction of printer data if a printer was to fail, or become busy.

10.2.4 Electronic mail

Electronic mail (e-mail) is one use of the Internet, which, according to most businesses, improves productivity. Traditional methods of sending mail within an office environment are inefficient, as it normally requires an individual requesting a secretary to type the letter. This must then be proofread and sent through the internal mail system, which is relatively slow and can be open to security breaches.

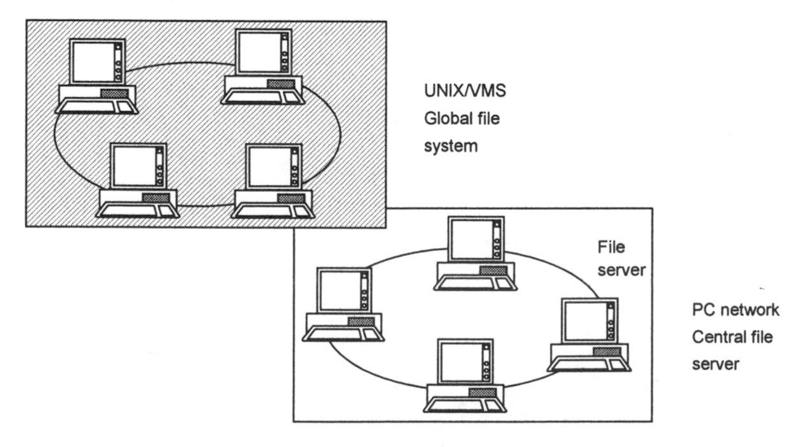

Figure 10.3 Sharing disk space with Unix/VMS and PC network

A faster method, and more secure method of sending information is to use electronic mail, where messages are sent almost in an instant. For example, a memo with 100 words can be sent within a fraction of a second. It is also simple to send to specific groups, various individuals, company-wide, and so on. Other types of data can also be sent with the mail message such as images, sound, and so on. It may also be possible to determine if a user has read the mail. The main advantages can be summarized as:

- It is normally much cheaper than using the telephone (although, as time equates to money for most companies, this relates any savings or costs to a user's typing speed).
- Many different types of data can be transmitted, such as images, documents, speech, and so on.
- It is much faster than the postal service.
- Users can filter incoming e-mail easier than incoming telephone calls.
- It normally cuts out the need for work to be typed, edited and printed by a secretary.
- It reduces the burden on the mailroom.
- It is normally more secure than traditional methods.
- It is relatively easy to send to groups of people (traditionally, either a circulation list was required or a copy to everyone in the group was required).
- It is usually possible to determine whether the recipient has actually read the message (the electronic mail system sends back an acknowledgement).

The main disadvantages are:

- It stops people using the telephone.
- It cannot be used as a legal document.
- Electronic mail messages can be sent impulsively and may be later regretted (sending by traditional methods normally allows for a rethink). In extreme cases messages can be sent to the wrong person (typically when replying to an e-mail message, where a message is sent to the entire mailing list [Reply to All] rather than the originator).
- It may be difficult to send to some remote sites. Some organizations have either no electronic mail or merely an intranet. Large companies are particularly wary of Internet connections and limit the amount of external traffic.
- Not everyone reads his or her electronic mail on a regular basis (although this is changing as more organizations adopt e-mail as the standard communications medium).

10.2.5 Peer-to-peer communication

A major problem with computers is to make them communicate with a different computer type or with another that possibly uses a different operating system. A local network allows different types of computers running different operating systems to share information over the network. This is named peer-to-peer exchange and is illustrated in Figure 10.4.

10.2.6 Remote login

A major advantage of networks is that they allow users to remotely log into other computers. The computer being logged into must be running a multi-tasking operating system, such as Unix. Figure 10.5 shows an example of three devices (a workstation, an X-windows terminal and a PC) logging into a powerful workstation. This method allows many less powerful computers to be linked to a few powerful machines.

10.2.7 Protecting information

Most computers have information which must be not be read or modified by certain users. It is difficult to protect information on a stand-alone computer, as typically all that is required is to wait until the user is not using the computer. On a network each user can be granted certain rights and privileges to stored information, which can be protected by a password.

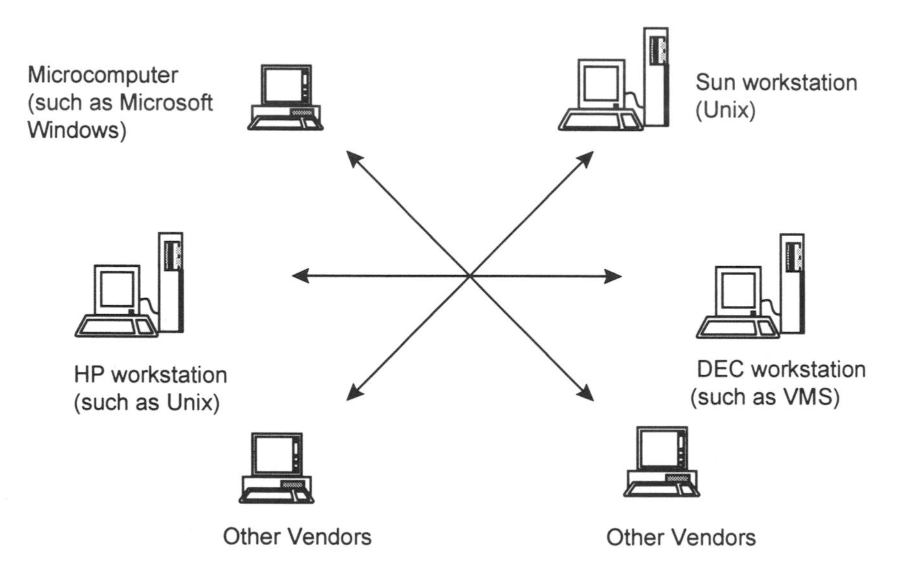

Figure 10.4 Peer-to-peer exchange over a network

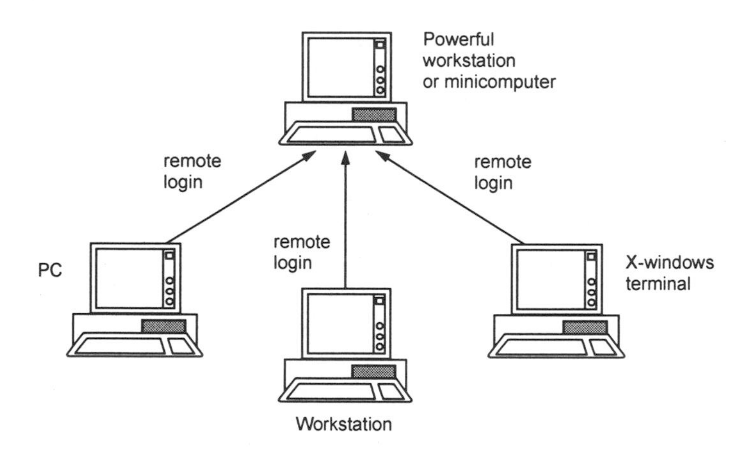

Figure 10.5 Remote login into other nodes

10.2.8 Centralized storage and backup of information

A particular problem with stand-alone computers is that when they crash the user can lose a lot of information, especially if they have not taken regular backups. As file sizes have increased it has also become more difficult for users to perform these backups as it normally involves spanning several floppy disks. Thus a better solution is to have a networked central storage and backup device. The network manager can then schedule backups at regular intervals (typically each day). If a network crash occurs on the central storage the manager can recover the previous backup, thus only losing a small amount of newly created data.

10.2.9 Disadvantages and potential pitfalls of networks

The main disadvantage of networks is that users become dependent upon them. For example, if a network file server develops a fault then many users may not be able to run application programs and get access to shared data. To overcome this a back-up server can be switched into action when the main server fails. A fault on a network may also stop users from being able to access peripherals such as printers and plotters. To minimize this, a network is normally segmented so that a failure in one part of it does not affect other parts.

Another major problem with networks is that their efficiency is very dependent on the skill of the system manager. A badly managed network may operate less efficiently than non-networked computers. Also, a badly run network may allow external users into it with little protection against them causing damage. Damage could also be caused by novices causing problems, such as deleting important files.

The main disadvantages are summarized:

- If a network file server develops a fault then users may not be able to run application programs.
- A fault on the network can cause users to lose data (especially if they have not saved the files they have recently been working with).
- If the network stops operating then it may not be possible to access various resources.
- Users work-throughput becomes dependent upon network and the skill of the system manager.
- It is difficult to make the system secure from hackers, novices or industrial espionage (again this depends on the skill of the system manager).
- Decisions on resource planning tend to become centralized, for example, what word processor is used, what printers are bought, and so on.
- Networks that have grown with little thought can be inefficient in the long term.
- As traffic increases on a network the performance degrades unless it is designed properly.
- Resources may be located too far away from some users.
- The larger the network becomes the more difficult it is to manage.

10.3 OSI model

The OSI reference model makes networks more manageable and then eases the problem of moving information between computers by divided the problem into seven smaller and more manageable tasks. A layer of the model solves each of the seven problem areas, these are: the physical layer, the data link layer, the network layer, the transport layer, the session layer, the presentation layer, and the application layer.

A major problem in the electronics industry is the interconnection of equipment and software compatibility. Other problems can occur in the connection of electronic equipment in one part of the world to another, in another part. For these reasons, the International Standards Organization (ISO) developed a model known as the OSI (open systems interconnection) model. Its main objects were to:

- Allow manufacturers of different systems to interconnect their equipment through standard interfaces.
- Allow software and hardware to integrate well and be portable on differing systems.
- Create a model which all the countries of the world use.

Figure 10.6 shows the OSI model. Data passes from the top layer of the sender to the bottom and then up from the bottom layer to the top on the recipient. Each layer on the sender, though, communicates directly to the recipient's corresponding layer, which creates a virtual data flow between layers.

The top layer (the application layer) initially gets data from an application and appends it with data that the recipient's application layer reads. This appended data passes to the next layer (the presentation layer). Again, it appends it with its own data, and so on, down to the physical layer. The physical layer is then responsible for transmitting the data to the recipient. The data sent can be termed as a data frame, whereas data sent by the network and the transport layers are typically referred to as a data packet and a data segment, respectively.

The basic function of each of the layers are:

OSI mnemonics

All
People
Seem
To
Need
Data
Processing

Please
Do
Not
Throw
Sausage
Pizza
Away

1. **Physical.** TRANSMISSION OF BINARY DATA. Defines the electrical characteristics of the communications channel and the transmitted signals, such as voltage levels, connector types, cabling, and so on.
2. **Data link.** MEDIA ACCESS. Ensures that the transmitted bits are received in a reliable way, such as adding extra bits to define the start and end of the data frame, adding extra error detection/correction bits and ensuring that multiple nodes do not try to access a common communication channel at the same time.
3. **Network.** ADDRESSING AND DETERMINING THE BEST PATH. Routes data frames through a network. If data packets require to go out of a network then the transport layer routes them through interconnected networks. Its task may involve, for example, splitting data for transmission and re-assembling it upon reception. The IP part of TCP/IP is involved with the network layer (or IPX in Novell NetWare).
4. **Transport.** END-TO-END CONNECTION RELIABILITY. Network transparent data transfer and transmission protocol, which supports the transmission of multiple streams from a single computer. The TCP part of TCP/IP is involved with the transport layer (or SPX in Novell NetWare).
5. **Session.** INTERHOST COMMUNICATION. Provides an open communications path with the other system. It involves the setting up, maintaining and closing down of a session. The communication channel and the internetworking of the data should be transparent to the session layer. A typical session protocol is telnet, which allows for remote login over a network.
6. **Presentation.** DATA REPRESENTATION and INTERPRETING. Uses a set of translations that allows the data to be interpreted properly. For example it may have to translate between two systems if they use different presentation standards, such as different character sets or differing character codes. The presentation layer can also add data encryption for security purposes.
7. **Application.** NETWORK SERVICES TO APPLICATION PROGRAMS. Provides network services to application programs, such as file transfer and electronic mail.

OSI model gives:

- Increased evolution.
- Modular engineering.
- Interoperable technology.
- Reduced complexity.
- Simplified teaching and learning.
- Standardized interfaces.

Figure 10.7 shows how typical networking systems fit into the OSI model. The data link and physical layers are covered by networking technologies such as Ethernet, Token Ring and FDDI. The networking layer is covered by IP (internet protocol) and transport by TCP (transport control protocol).

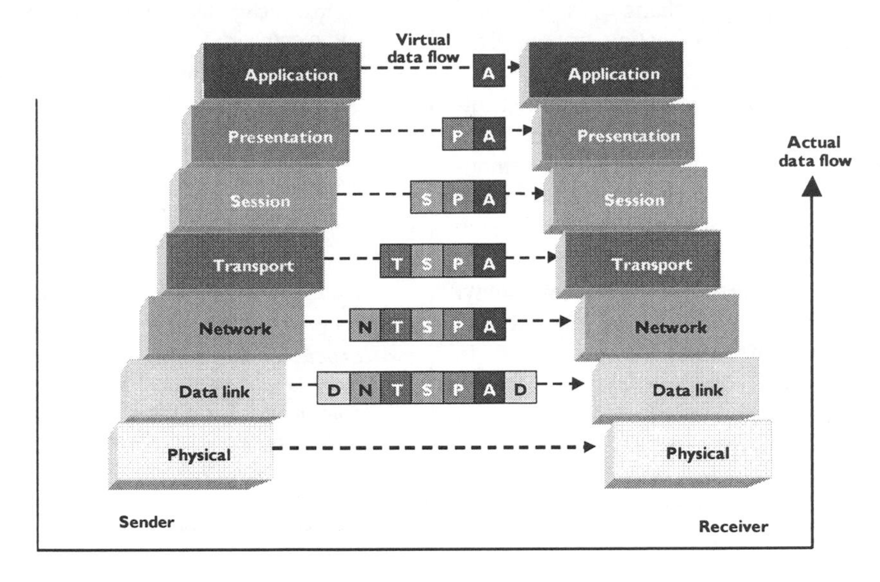

Figure 10.6 Seven-layer OSI model

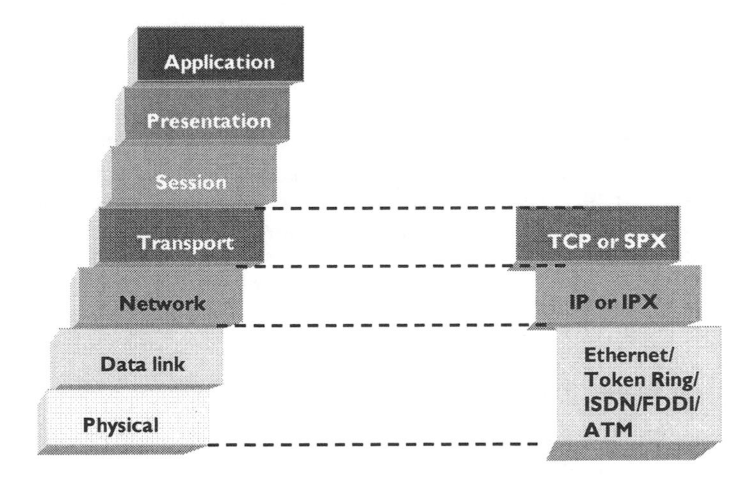

Figure 10.7 Typical technologies used in network communications

The layers can be grouped as follows:

- **Media layers.** This covers the physical and data link layers, as they control the physical delivery of messages over the network.
- **Host layers.** This covers the application, presentation, session and transport layers, as they provide for accurate delivery of data between computers on the network.

In general the OSI model has:

- **Increased evolution.** Systems are allowed to quickly change, as they still integrate well with existing systems. This speeds evolution.
- **Allows modular engineering.** This allows for systems to be designed in a modular way so that each of the components, whether they be hardware or software, can interface well with each other.
- **Guarantees interoperable technology.** This allows the transfer of data between computers of different types, either in their software, operating system, network hardware or computer hardware.
- **Reduced complexity.** The task of transmitting data from one application to another over a network is reduced in complexity as it is reduced to seven smaller tasks.
- **Simplifies teaching and learning.** The OSI model has been used as a standard method for teaching networking, and, as it is built up of layers, allows for easier learning of networking. Students can easily visualize the network in a given layer of abstraction.
- **Standardizes interfaces.** This allows for designers to design their products so that they can be easily plugged into one or more of the layers of the model. The actual implementation of the layer can be invisible to other layers.

OSI model layers:

- **A**pplication. Provides application programs, such as file transfer, print access and electronic mail.
- **P**resentation. Transforms the data into a form which the session layer and the application layer expect. It can perform encryption, translating character sets (such as converting binary values into text for transmitting a binary program over a text-based system), data compression and network redirections.
- **S**ession. Setting up, maintaining and closing down of a session. It should not depend on any specific transport or network layer, and should be able to communicate as if the session was created on a stand-alone computer (that is, the network is transparent to the session layer).
- **T**ransport. Provides for reliable end-to-end error and flow control. The network layer does not validate that the data packet has been successfully received, thus it is up to the transport layer to provide for error and flow control.
- **N**etwork. Defines the protocols that are responsible for delivering the data to the required destination.
- **D**ata link. Provides for the access to the network media and thus builds on the physical layer. It takes data packets from the upper level and frames it so that it can be transmitted from one node to another.
- **P**hysical. Provides for the actual transmission of the binary digits.

10.4 Foundations of the OSI model

The OSI reference model is purely an abstract model, and provides a conceptual framework which defines the network functions at each layer. It thus defines how data from the source (the network device that is sending data) is transmitted to the destination (the network device that is receiving data). This data is transmitted in the form of data packets. At the source the data is passed through all of the layers of the OSI model, with each layer adding its own information. The process of adding the extra information is known as encapsulation. The data packet is thus wrapped in a particular protocol header. For example, Ethernet networks require an Ethernet protocol header before transmitting onto the Ethernet network.

Figure 10.8 shows how the data link, network and transport layers are responsible for transporting data between applications (which basically covers the session, presentation and application layers). The data link layer delivers data between devices on a network segment, and the network layer is responsible for passing it between network segments and delivers the data at the destination (using routers). The transport layer concentrates (multiplexes) the data into a single data stream for transmission, and demultiplexes it at the destination.

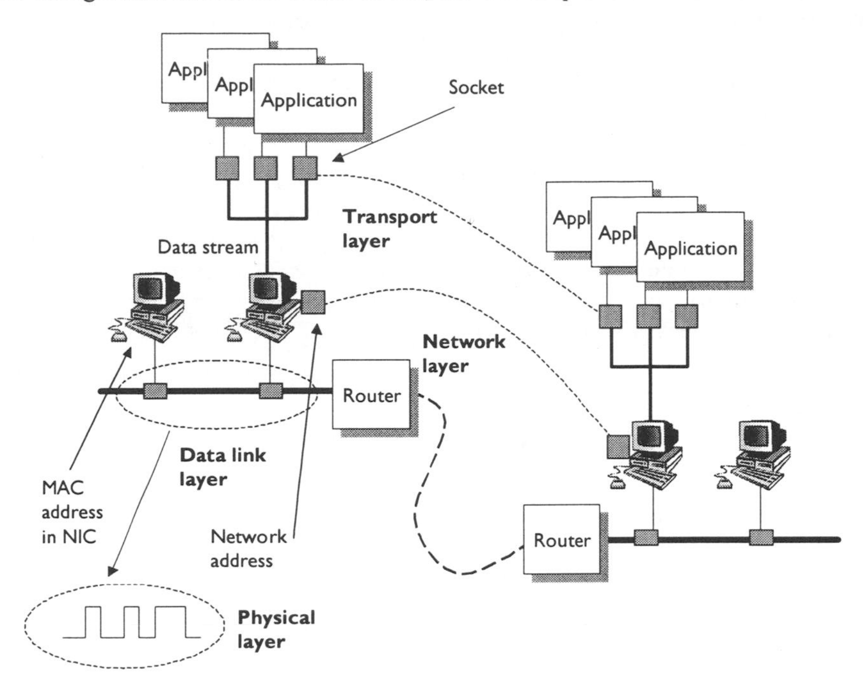

Figure 10.8 Networking showing lower-level layers

10.4.1 Physical layer – cables, voltages and connectors

Computers store information using digital digits (or bits), which have a level of a '0' or a '1'. The foundation level of the OSI model is the physical layer, and provides for the actual transmission of the binary digits. In most cases this is converted to an electrical voltage and sent over a copper cable, or converted in pulses of light and transmitted over a fiber-optic cable. The physical layer is thus responsible for the electrical, mechanical, procedural, and functional specifications for activating, maintaining, and deactivating the physical link between end systems (the end-user device on a network) and covers:

Choosing the right cable:

- Bandwidth (such as bit rate). Fiber-optic cable has the greatest bandwidth, followed by coaxial, and then by twisted-pair.
- Location (such as flammable environment, security).
- Expense. Fiber-optic cable is the most expensive, and twisted-pair the least.
- Long-term use (upgradeability over the years).

- **Cable.** Typically these are coaxial cable, twisted-pair cable or fiber-optic cable. Fiber-optic cable gives the best specification, and can be run for longer lengths than the other two types. Twisted-pair cable has the advantage over the other two in that it is relatively inexpensive to install, but does not have as good a specification as the others. The main decision for choosing a cable types relates to how much data that can be transmitted over it (the required bandwidth), its location (for example copper cables can cause electrical sparks, so they tend not to be used in flammable situations), its expense and its long-term usage. Typically cabling should have a lifetime of over ten years (which is much longer than most of the computers that connect to the network), thus they must have the potential to support future growth.

- **Electrical voltages, electrical currents, and intensities of light pulses.** Defines the levels of the voltages or light levels for transmission of the binary digits over the cable. Electrical impulses representing data are known as signals.
- **Connectors.** Defines the physical specifications of the connector, and the connections that are made.
- **Encoding.** Defines how the bits are represented by electrical or light signals. This normally involves matching the transmission of binary digits to the required transmission specification. Important considerations are to imbed the clock signal into the transmitted signals (which is important to properly recover the received data), or to try and reduce the average transmitted value to zero (as the average value does not contain any imbedded information). A typical encoding scheme is Manchester encoding which encodes the 0 and 1 as either a positive edge or a negative edge.

10.4.2 Data link layer – MAC addresses and NICs

Computers connect to the physical media using an NIC (network interface card). The data link layer provides for the access to the network media and thus builds on the physical layer. It takes data packets from the upper levels and frames them so that they can be transmitted from one node to another. The data link layer provides for:

- **Error control.** This provides for the addition of binary digits that can be used to identify if there has been an error in the transmission of one of more bits. If possible, there should be some mechanism for the destination to tell the source that it has received bits in error, and to request a retransmission.
- **Flow control.** This is where there is an orderly flow of transmitted data between the source and the destination, so that the source does not swamp the destination with data. Typically the destination sends back messages that indicate whether the destination can receive data, or not.
- **Line discipline.** This provides for the orderly access to the network media. If there was no orderly access, many nodes could try and get access to the network, at the same time, thus swamping the network. Typically only one node is allowed access to the network at a time. Techniques

OSI LAYERS

Physical (TRANSMISSION OF BINARY DATA):

- Cable.
- Electrical voltages, electrical currents and intensities of light pulses.
- Connectors.
- Encoding.

Data link (MEDIA ACCESS):

- Error control.
- Flow control.
- Line discipline.
- Network topology.
- Orderly delivery of frames.
- Physical addressing.

Network (ADDRESSING AND BEST PATH):

- Network addresses.
- Routing.

Transport (END-TO-END CONNECTION RELIABILITY):

- Connection type (connection-less/connection-oriented).
- Name service.

Session (INTERHOST COMMUNICATION):

- Setting up, maintaining and closing down of a session.

which allow an orderly access are collision detection (which detects when other nodes are trying to transmit at the same time) and token passing (which involves nodes passing an electronic token from one node to the next, nodes can only transmit when they capture the token).

- **Network topology.** Physical arrangement of network nodes and media within an enterprise networking structure.
- **Ordered delivery of frames.** This provides for sequencing of the data frames in the correct order, and allows the recipient to determine if there are any gaps in the sequence of the received data frames.
- **Physical addressing.** Each node on a network has a unique physical address (or hardware address), this is normally known as a MAC (Media Access Control) address. This address must be used if a node is to receive the transmitted data frame. The only other data frame that a node can receive is when the destination address is a broadcast (which is also received by all the nodes on the network). On Ethernet networks, the MAC address has six bytes, which is allocated by the IEEE.

Physical addresses and network addresses

The MAC address identifies the physical address of the NIC, and differs from the network address (which is also known as a protocol address) which is used by the network layer. An Ethernet address takes the form of a hexadecimal number, such as:

`0000.0E64.5432` or `00-00-0E-64-54-32`

and the network address, for IP, takes the form of a dot address, such as:

`146.176.151.130`

All computers that connect onto the Internet must have a unique IP address.

IPX addresses (for Novell NetWare) use an eight-digit hexadecimal address for the network address and the node portion is the 12-digit MAC address. From example:

`F5332B10:00000E645432`

Network address — Node address

Network addresses:

`146.176.151.130` (IP)
`F5332B10:00000E645432` (IPX)

MAC address:

`00-00-0E-64-54-32`

AppleTalk uses alphabet characters, such as:

`NewPrinter`

The physical address is physically setup in the NIC when it is manufactured and cannot be changed. It gives no information on the physical location of the NIC. The network address, on the other hand, is a software address, and gives some information on where a computer is logically located. The only way that a MAC address can be changed is to change the NIC card, whereas the network address is changed when the computer is moved from one network to another.

In Ethernet networks the following occurs:

- A data frame is transmitted onto the network with the destination MAC address.

- All the devices on the network read the destination MAC address to see if it matches their address.
- If it does not match the physical address, the device ignores the rest of the data frame, otherwise, the NIC card copies the data frame into its buffer, which is then read when the device is ready.

10.4.3 Network layer – protocols for reliable delivery

The network layer defines the protocols that are responsible for data delivery at the required destination, and requires:

- **Network addresses.** This identifies the actual logical location of the node (the network address), and the actual node (the node address). The form of the network address depends on the actual protocol. IP uses a dot address, such as 146.176.151.130 that identifies the network and the host. IPX address (for Novell NetWare) uses an eight-digit hexadecimal address to identify the network address and the node portion with a 12-digit MAC address, such as F5332B10: 00000E645432. Network addresses are setup in software and are loaded into the computer when it starts (assuming that it has some storage device to store its network address). This differs from the MAC address which is setup in the hardware. Like MAC addresses, no two computers on a network can have the same network address.
- **Routing.** This is passing of the data packets from one network segment to another, and involves routers. A router reads the network address and decides on which of its connections it should pass the data packet on to. Routing information is not static and must change as the conditions on the network change. Thus each route must maintain a routing table which is used to determine the route that the data packet takes. These routing tables are updated by each of the routers talking to each other using a routing protocol. Two typical routing protocols are Routing Information Protocol (RIP) and Open Shortest Path First (OSPF). RIP uses the least number of hops (which relates to the number of routers between the destination and the current router), whereas OSPF uses other metrics to determine the best route (such as latency and bandwidth capacity).

NETWORK LAYER

Network address (such as IP and IPX):

- NETWORK PART (146.176).
- HOST PART (151.130).

Network address changes when a computer is moved from one network to another.

Routing protocol (such as RIP, BGP and OSPF):

- Allows the communication between routers so that they can determine the optimal route to a destination.
- **RIP** is the most popular, and uses the number of hops that it takes to get to a destination (a hop is one router). It is limited to 16 hops.
- **OSPF** uses other metrics to determine the best route (known as route costs). It also allows for additional hierarchy, authentication of routing messages (using a password) and route load balancing.
- **BGP** (Border Gateway Protocol) is less complex than OSPF, but uses an Internet-like structure which assumes that the Internet is connected with a number of AANs (autonomously attached networks). These create boundaries around an organization, Internet service provider, and so on.

10.4.4 Transport layer – validates transmission and structures messages

The transport layer provides for reliable end-to-end error and flow control. This is required as the network layer does not validate that any data packets have been successfully received, thus it is up to the transport layer to provide for error and flow control. It involves:

- **Connection type.** This defines the method of handshaking of data between the source and the destination, and can be connection-oriented or connectionless. In a connectionless connection there are no acknowledgements and responses when the data is transmitted from the source to the destination. In a connection-oriented system, a virtual connection is set up, and data is acknowledged by the destination, by sending acknowledgement data from the destination to the source. The source will thus know if the data has been received correctly. In order to detect if data segments have been lost or are in error, each data segment has a sequence number. The destination sends back the acknowledgement with the data packet segment that it expects to receive from the source (thus acknowledging all previously transmitted data segment to the acknowledged data segment number). Figure 10.9 shows an example flow of information. Initially, for a connection-oriented connection, the transport layer creates a connection by negotiation, where both the source and destination pass the details of their connection, such as details of their socket (which is a unique number that defines the connection) and the data segment number that they will start to send on.
- **Name resolution.** This allows for the resolution of logical names to logical network addresses. It is often easier to access networked devices using a logical name, rather than their logical address, as these are easier to remember. A typical implementation on TCP/IP networks is the Domain Name Service (DNS) which resolves domain names to IP addresses. For example, a domain name of www.fredandco.com could be resolved to the network address of 11.22.33.44.

TRANSPORT LAYER

Connection type:

- **Connection-oriented** (or virtual circuit). The source and the destination set up a unique connection (typically known as a socket) and the destination acknowledges the successful receipt of data segments. Connection-oriented is thus reliable for system crashes on both the source and destination and data segment loss/error.
- **Connection-less** (or datagram). There is no virtual connection between the source and the destination, thus there is no guarantee that the data segments have been received correctly (or whether the destination is even there).

Name resolution:
Resolves logical names to network addresses. For example:

- **DNS**. Operates on the Internet and is used when a user uses a domain name (such as www.fred.co). DNS is thus used to return the IP address of the destination (such as 11.22.33.44). No data on the Internet can be transmitted unless the IP address is known.

10.4.5 Session layer

The session layer involves the setting up, maintaining and closing down of a session. This builds on the transport layer which provides the foundation for the connection over the network. The session provides a higher-level connection, such as a login procedure, or a remote connection. It is important that the session layer does not depend on any specific transport or network layer, and should be able to communicate as if the session was created on a standalone computer (that is, the network is transparent to the session layer). A typical session protocol is telnet, which allows for the remote login over a network.

10.4.6 Presentation layer

The presentation layer transforms the data into a form which the session layer and the application layer expect. It can perform encryption, translate character sets (such as converting binary values into text for transmitting a binary program over a text-based system), data compression and network redirections. An example protocol for the presentation layer is XDR.

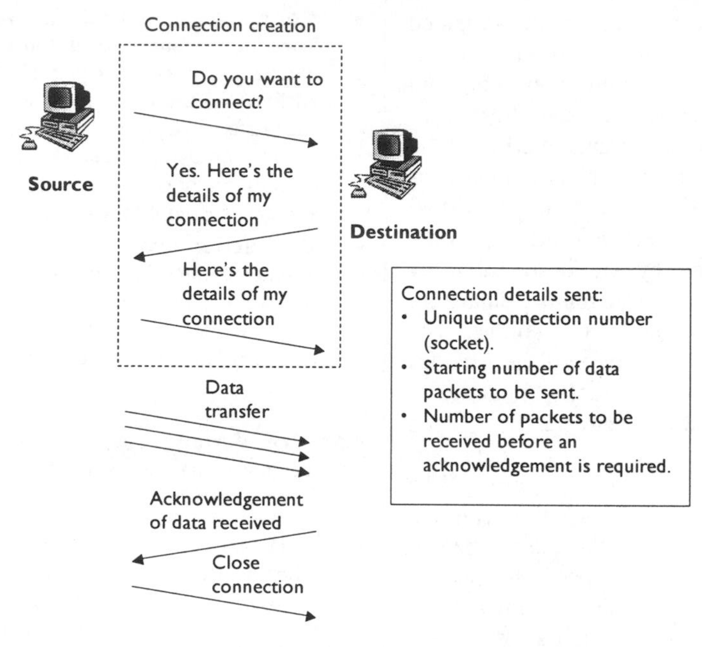

Figure 10.9 Basic transport layer connection-oriented protocol

10.4.7 Application layer

The application layer provides application programs, such as file transfer, print access and electronic mail.

10.5 Internetworking

Internetworking:

- Increases the number of nodes that can connect to a network.
- Extends the physical distance of a network.
- Localizes the traffic with network segments.
- Merges existing networks.
- Isolates network faults.

Networks can be constructed using a common connection for all the nodes that connect to the network. Unfortunately, the more devices that connect, the slower the network becomes. Thus there is a need for devices that split networks into segments, each of which contain locally attached nodes. Internetworking devices have many advantages, such as:

- They increase the number of nodes that can connect to the network than would be normally possible. Limitations on the number of nodes that connect to a network relate to the cable lengths and traffic constraints.
- They extend the physical distance of the network (the range of the network).
- They localize traffic within a network. Typically computers, which are geographically located close to each other, require to communicate with each other. Thus local communications should not have an effect on communications outside a given network segment.

- Merge existing networks. This allows connected networks to intercommunicate.
- Isolate network faults. This allows faults on one network to be contained within a given network, so that they do not affect other connected networks.

Typical internetworking devices are:

- **Repeater.** These operate at Layer 1 of the OSI model, and extend the physical length of a connection that would normally be possible with the cable type. They basically boost the electrical or light signals.
- **Bridges.** These pass data frames between networks using the MAC address (Layer 2 address).
- **Hubs.** These allow the interconnection of nodes and create a physically attached network.
- **Switches.** These allow simultaneous communication between two or more nodes, at a time.
- **Routers.** These pass data packets between connected networks, and operate on network addresses (Layer 3 address).

Internetworking devices:

- PHYSICAL LAYER: Repeaters.
- DATA LINK LAYER: Bridges, hubs and switches. Uses MAC address.
- NETWORK LAYER: Routers. Uses network address.

Networks connect to other networks through repeaters, bridges or routers. A repeater corresponds to the physical layer of the OSI model and routes data from one network segment to another. Bridges, on the other hand, route data using the data link layer (with the Media Access Control address – MAC address), whereas routers route data using the network layer (that is, using a network address, such as an IP address). Normally, at the data link layer, the transmitted data is known as a data frame, while at the network layer it is referred to as a data packet. Figure 10.10 illustrates the three interconnection types.

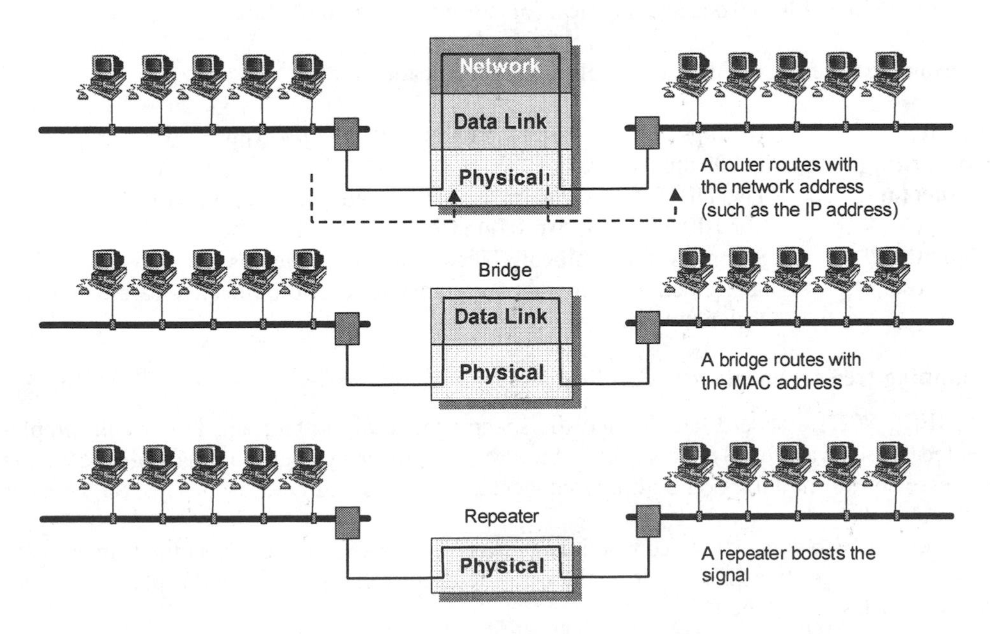

Figure 10.10 Repeaters, bridges and routers

10.5.1 Repeaters

All network connections suffer from a reduction in signal strength (attenuation) and digital pulse distortion. Thus, for a given cable specification and bit rate, each connection will have a maximum cable length that can be used to transmit the data reliably. Repeaters can be used to increase the maximum interconnection length, and may do the following:

- Reshape signal pulses.
- Pass all signals between attached segments.
- Boost signal power.
- Possibly translate between two different media types (such as between fiber-optic and twisted-pair cable).
- Transmit to more than one network. These are multiport repeaters and send data frames from any received segment to all the others. Multiport repeaters do not filter the traffic, as they blindly send received data frames to all the physically connected network segments.

10.5.2 Bridges

Bridges filter input and output traffic so that only data frames distended for another network segment are actually routed into that segment and only data frames destined for the outside are allowed out of the network segment.

The performance of a bridge is governed by two main factors:

- **The filtering rate.** A bridge reads the MAC address of the Ethernet/Token ring/FDDI node and then decides if it should forward the frames into the network. Filter rates for bridges range from around 5000 to 70 000 pps (packets per second).
- **The forward rate.** Once the bridge has decided to route the frame into the internetwork, the bridge must forward the frame onto the destination network. Forwarding rates range from 500 to 140 000 pps and a typical forwarding rate is 90 000 pps.

An example Ethernet bridge has the following specifications:

Bit rate: 10 Mbps **Filtering rate:** 17 500 pps
Forwarding rate: 11 000 pps
Connectors: Two DB15 AUI (female), one DB9 male console port, two BNC (for 10BASE2) or two RJ-45 (for 10BASE-T).
Algorithm: Spanning tree protocol. This automatically learns the addresses of all devices on both interconnected networks and builds a separate table for each network.

Spanning tree architecture (STA) bridges

The IEEE 802.1 standard has defined the spanning tree algorithm, and is normally implemented as software on STA-compliant bridges. On power-up they automatically learn the addresses of all the nodes on both interconnected networks and build up a separate table for each network.

They can also support two connections between two LANs so that when the primary path becomes disabled, the spanning tree algorithm re-enables the previously disabled redundant link, as illustrated in Figure 10.11.

Source route bridging

With source route bridging, a source device, not the bridge, is used to send special explorer packets. These are then used to determine the best path to the destination. Explorer packets are sent out from the source routing bridges until they reach their destination workstation. Each source routing bridge along the route enters its address in the routing information field (RIF) of the explorer packet. The destination node then sends back the completed RIF field to the source node. When the source device has determined the best path to the destination, it sends the data message along with the path instructions to the local bridge, which then forwards the data message according to the received path instructions.

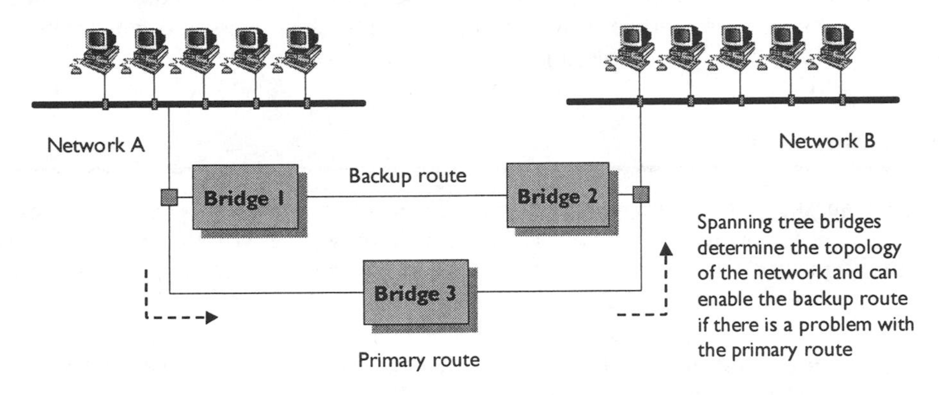

Figure 10.11 Spanning tree bridges

10.5.3 Routers

Routers examine the network address field and determine the best route for a data packet. They have the great advantage in that they normally support several different types of network layer protocols.

Routers need to communicate with other routers so that they can exchange routing information. Most network operating systems have associated routing protocols which support the transfer of routing information. Typical routing protocols for Internet communications are:

- BGP (border gateway protocol).
- EGP (exterior gateway protocol).
- OSPF (open shortest path first).
- RIP (routing information protocol).

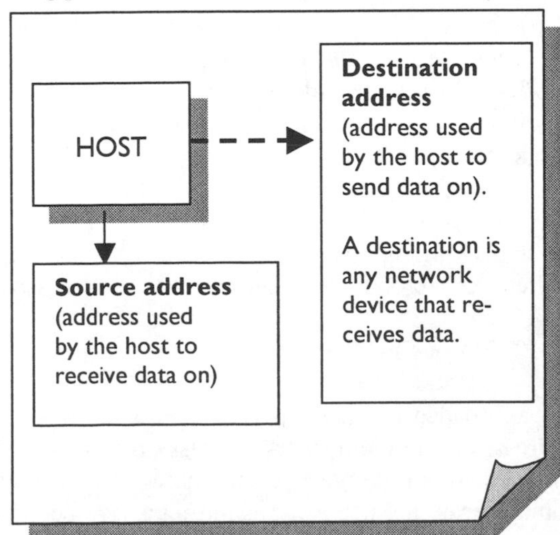

Most routers support RIP and EGP. In the past, RIP was the most popular router protocol standard, and its widespread use is due, in no small part, to the fact that it was distributed along with the Berkeley Software Distribution (BSD) of UNIX (from which most commercial versions of UNIX are derived). It suffers from several disadvantages and has been largely replaced by OSFP and EGB. These newer protocols have the advantage over RIP in that they can handle large internetworks, as well as reducing routing table update traffic.

RIP uses a distance-vector algorithm, which measures the number of network jumps (known as hops), up to a maximum of 16, to the destination router (a value of 16 identifies that the destination is not reachable). This has the disadvantage that the smallest number of hops may not be the best route from a source to a destination. The OSPF and EGB protocol uses a link state algorithm that can decide between multiple paths to the destination router. These are based, not only on hops, but also on other parameters such as delay (latency), capacity, reliability and throughput.

With distance-vector routing, each router maintains routing tables by communicating with neighboring routers. The number of hops in its own table are then computed as it knows the number of hops to local routers. Unfortunately, their routing tables can take some time to be updated when changes occur, because it takes time for all the routers to communicate with each other (known as slow convergence).

10.6 Broadcasts

A puzzling question that most people ask is how the host knows what the network address of the computer it is communicating with is, and how it knows the MAC address of the host that it communicates with. In order to determine these a host must send out a broadcast to all the hosts on its network segment. There are two main types of broadcasts:

- **Requests for a destination MAC address.** If a host does not know what the destination MAC address is, it sends out a broadcast request to all the hosts on the network segment. The host which has a matching network address responds back with its MAC address in the source MAC address field. The MAC and network addresses are then stored in the memory of the host, so that they can be used in future communications. This process is known as ARP (Address Resolution Protocol), and is illustrated in Figure 10.12.
- **Requests for a network address.** If a host does not know network address for a given MAC address, it sends out a request with the MAC address. A server on the network normally then responds back with the network address for the given MAC address. This process is known as RARP (Reverse Address Resolution Protocol), as it is only really used when a node requires to know what its own network address is (as it has no local storage to store it).

Most networking technologies have a special MAC address for a broadcast. Ethernet uses the address:

FF-FF-FF-FF-FF-FF

for a broadcast. There are also network broadcast addresses using the network address (known as multicast), where all the nodes on the network listen to the communication (such as transmitting a video conference to many nodes on a network, at the same time), but they are used for different purposes than with broadcast MAC addresses, which are used to get network information.

Bridges:

- Forward broadcasts.
- Forward traffic to unknown addresses.
- Do not modify data frame.
- Build tables of MAC addresses.
- Use the same network address for all of its ports.

Routers:

- Do not forward broadcasts.
- Do not forward traffic to unknown addresses.
- Modify data packet header.
- Build tables of network addresses.
- Use a different network address for each of its ports.

Bridges always forward broadcast addresses, but routers do not. Another advantage of routers over bridges is that a router will not forward to an unknown address, whereas a bridge will. The blocking of broadcasts is a great advantage in routers as it stops the broadcasts being sent to hosts on other networks, thus limiting the traffic on other networks.

10.7 Bits, frames, packets and segments

As was seen, each of the OSI layers communicates directly with the equivalent layer on the receiving host. The data that is transmitted in each of the lower layers is referred to in a different way. Data that passes from layer to layer is called protocol data units (PDUs). These PDUs are referred to in different ways in each of the layers. At the physical level they are referred to as bits, at the data link layer they are referred to as frames, at the network layer they are referred to as packets, and at the transport layer they are referred to as segments. This is illustrated in Figure 10.13.

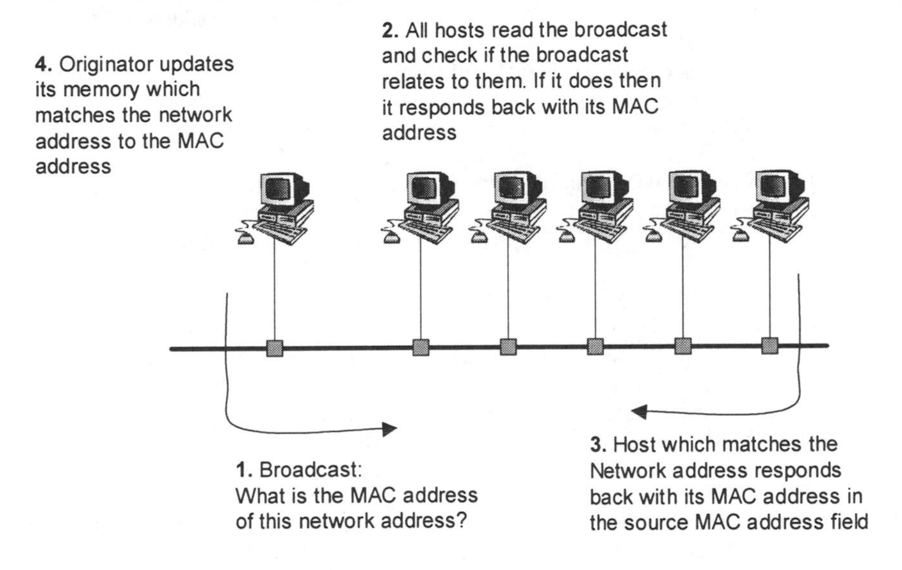

2. All hosts read the broadcast and check if the broadcast relates to them. If it does then it responds back with its MAC address
4. Originator updates its memory which matches the network address to the MAC address
1. Broadcast: What is the MAC address of this network address?
3. Host which matches the Network address responds back with its MAC address in the source MAC address field

Figure 10.12 Broadcasts for MAC address

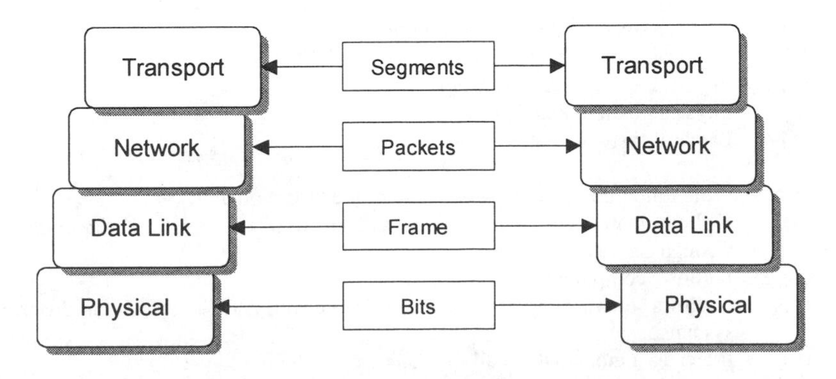

Figure 10.13 Bits, frames, packets and segments

10.8 Exercises

The following questions are multiple choice. Please select from a–d.

10.8.1 What must be common to computers if they are to communicate over a network:
(a) Use the same operating system (b) Use the same protocol.
(c) Manufactured by the same company (d) Use similar hardware

10.8.2 The standard networking protocol used on the Internet is:
(a) Ethernet (b) TCP/IP
(c) SPX/IPX (d) Token Ring

10.8.3 The default networking protocol used by Novell NetWare is:
(a) Ethernet (b) TCP/IP
(c) SPX/IPX (d) Token Ring

10.8.4 What address is used by computers connecting to the Internet:
(a) IP address (b) Node address
(c) IPX address (d) Router address

10.8.5 What network address is used by Novell NetWare :
(a) IP address (b) Node address
(c) IPX address (d) Router address

10.8.6 Which of the following is an IPX address:
(a) `0000.0E64.5432` (b) `146.176.151.130`
(c) `F5332B10:00000E645432` (d) `NETPRINTER`

10.8.7 Which of the following is an IP address:
(a) `0000.0E64.5432` (b) `146.176.151.130`
(c) `F5332B10:00000E645432` (d) `NETPRINTER`

10.8.8 Which organization originally developed the OSI model:
(a) DARPA (b) ISO
(c) CERN (d) IEEE

10.8.9 A network which connects computers within a single building is normally defined as:
(a) local area network (b) wide area network
(c) metropolitan area network (d) enterprise area network

10.8.10 The three lower layers of the OSI model are:
(a) Physical, data link and application
(b) Physical, network and transport
(c) Physical, data link and network
(d) Physical, network and session

10.8.11 Which of the following is **not** an advantage of the OSI model:
(a) It allows manufacturers of different systems to interconnect their equipment through standard interfaces
(b) It allows computers to operate faster
(c) It allows software and hardware to integrate well and be portable on differing systems
(d) It creates a model which all the countries of the world use

10.8.12 The function of the data link layer in the OSI model is:

(a) Control of session between two nodes
(b) Ensures reliable transmission of bits
(c) Routing, switching and flow control over a network
(d) Network transparent data transfer and transmission control

10.8.13 Which of the following is a MAC address:
(a) `0000.0E64.5432` (b) `146.176.151.130`
(c) `F5332B10:00000E645432` (d) `NETPRINTER`

10.8.14 Which best defines a WAN:
(a) Networks that connect users across a large geographic area.
(b) Networks that connect devices with a workgroup
(c) The interconnection of LANs with a building
(d) Networks that connect mixed protocol networks

10.8.15 Which of the following was solved with the use of networks:
(a) Loss of data
(b) Slow computer operation
(c) Inability to communicate and lack of management
(d) Lack of compatibility of application software packages

10.8.16 In early networks, why was connecting over a network difficult:
(a) Slow operation of the network
(b) New networking technologies used different hardware and software specifications
(c) Lack of compatibility of application software packages
(d) Lack of support from computer manufacturers

10.8.17 What is a source address:
(a) An address that the host receives data on
(b) An address that the host sends data on
(c) An address that changes as the host transmits data
(d) An address that changes as the host receives data

10.8.18 Which of the following is the correct order of the OSI model:
(a) Physical, data, network, session, transport, application, presentation
(b) Physical, network, data, transport, session, presentation, application
(c) Physical, data, network, transport, session, presentation, application
(d) Physical, data, network, transport, presentation, session, application

10.8.19 Which of the following is **not** a reason for using a layered network model:
(a) Standardizes interfaces of hardware and software
(b) Simplifies teaching and learning
(b) Reduces overlap
(d) Improve interoperability

10.8.20 Which of the following is **not** a purpose of the OSI model:
(a) Ensures that computers can operate as stand-alone devices
(b) Defines a standard way to describe how data travels on a network
(d) Replaces a number of competing standards with a single standard
(d) Simplifies the networking process by splitting it up into seven layers

10.8.21 Which of the following best describes the session layer:
(a) Defines data structures and negotiation data transfer syntax
(b) Ensures reliable transit of data across the physical layer
(c) Provides connectivity and path selection between two end systems
(d) Manages data exchange between presentation layer entities

10.8.22 Which of the following best describes the presentation layer:
- (a) Defines data structures and negotiation data transfer syntax
- (b) Ensures reliable transit of data across the physical layer
- (c) Provides connectivity and path selection between two end systems
- (d) Manages data exchange between presentation layer entities

10.8.23 What functions are the data link layer concerned with:
- (a) Physical addressing, network topology, and media access
- (b) Manages data exchange between presentation layer entities
- (c) Provides connectivity and path selection between two end systems
- (d) Establishment, maintenance, and termination of virtual circuits, transport fault detection, recovery, and information flow control

10.8.24 Which of the following best defines data encapsulation:
- (a) Segmenting data into a number of data packets
- (b) Encrypting data so that it can only be read by the destination user
- (c) Adding additional information that the data can be viewed properly
- (d) Wrapping the data in a particular protocol header

10.8.25 What is one function of the physical layer of the OSI model?
- (a) Physical addressing of computers on the network media
- (b) Data transmission across the network media
- (c) Defines the format of the data frame that is transmitted
- (d) Network services to applications

10.8.26 How is the function of the physical layer accomplished:
- (a) Using networking protocols
- (b) Using physical addressing
- (c) Conversion of signals into binary patterns.
- (d) Using wires, connectors, and voltages.

10.8.27 Which of the following best identifies networking media:
- (a) Encoding of data
- (b) Encryption of data
- (c) Physical connections through which transmission signals pass
- (d) Synchronization of data between nodes

10.8.28 Which are the most important considerations when considering network media:
- (a) Location, bandwidth requirements, and cable expense
- (b) Types of computers that will connect onto the network
- (c) Distance between nodes and their type
- (d) Availability of various media types

10.8.29 What does encoding achieve:
- (a) Represents data bits by using different voltages or light patterns
- (b) To physically connect one node to another
- (c) Converts data into binary bit patterns
- (d) Synchronization of data between nodes

10.8.30 Which of the following best defines a destination:
- (a) Destination network segment
- (b) Any terminal on a network
- (c) Network device that receives data
- (d) Destination router

10.8.31 What is a user device on a network usually referred to as:

(a) Application interface (b) Computer
(c) Source (d) End system

10.8.32 Which part of the host connects onto the network media:

(a) Media interface adaptor (b) Parallel port
(c) Local bus (d) NIC card

10.8.33 Which is the best definition of a network address:

(a) An address which uses a standardized network layer address for every device or port that connects onto a network
(b) A 32-bit address that defines the location of the network and the host
(c) A physical address which uses a standardized data link layer address for every device or port that connects onto a network
(d) A logical name that is assigned by a name server

10.8.34 Which is the best definition of a MAC address:

(a) An address which uses a standardized network layer address for every device or port that connects onto a network
(b) A 32-bit address that defines the location of the network and the host
(c) A physical address which uses a standardized data link layer address for every device or port that connects onto a network
(d) A logical name that is assigned by a name server

10.8.35 What method would be used to change the IP address of a computer:

(a) Move the computer to another network (b) Change the NIC
(c) Change the bridge (d) Change the router

10.8.36 What method would be used to change the MAC address of a computer:

(a) Move the computer to another network (b) Change the NIC
(c) Change the bridge (d) Change the router

10.8.37 When would the network address of the computer change:

(a) When moving the computer to another network
(b) After changing the NIC
(c) After changing the bridge
(d) After changing the router

10.8.38 Why are internetworking devices used:

(a) They allow a greater number of nodes to connect to a network, they extend the network distance, and merge separate networks.
(b) They support different physical media types
(c) They provide redundancy for additional paths
(d) They allow different protocols to used

10.8.39 What are electrical impulses representing data called?

(a) signals (b) nodes
(c) routes (d) repeaters

10.8.40 Which of the following forward broadcasts onto other connected networks:

(a) routers (b) bridges
(c) NIC card (d) file servers

10.8.41 What is name given to the data passed between the layers of the OSI model:

(a) protocol units (b) data identity
(c) data intervals (d) protocol data units

10.8.42 What is data passed at the data link layer referred to as:

(a)	bits	(b)	frames
(c)	packets	(d)	segments

10.8.43 What is data passed at the network layer referred to as:

(a)	bits	(b)	frames
(c)	packets	(d)	segments

10.9 Notes from the Author

Many of the great inventions/developments of our time were things that were not really predicted, such as CD-ROMs, RADAR, silicon transistors, fiber-optic cables, and, of course, the Internet. The Internet itself is basically an infrastructure of interconnected networks which run a common protocol. The nightmare of interfacing the many computer systems around the world was solved because of two simple protocols: TCP and IP. Without them the Internet would not have evolved so quickly and possibly would not have occurred at all. TCP and IP are excellent protocols as they are simple and can be run over any type of network, on any type of computer system.

The Internet is often confused with the World Wide Web (WWW), but the WWW is only one application of the Internet. Others include electronic mail (the No. 1 application), file transfer, remote login, and so on.

The amount of information transmitted over networks increases by a large factor every year. This is due to local area networks, wide area networks and traffic over the Internet. It is currently estimated that traffic on the Internet doubles every 100 days and that three people join the Internet every second. This means an eight-fold increase in traffic over a whole year. It is hard to imagine such growth in any other technological area. Imagine if cars were eight times faster each year, or could carry eight times the number of passengers each year (and of course roads and driveways would have to be eight times larger each year).

Networks have grown vastly since the 1970s, and most companies now have some form of network. At the beginning of the 1980s, PCs were relatively complex machines to use, and required application programs to be installed locally to their disk drives. Many modern computers now run their application programs over a network, which makes the administration of the application software much simpler, and also allows users to share their resources.

The topology of a network is all-important, as it can severely affect the performance of the network, and can also be used to find network faults. I have run a network for many years and know the problems that can occur if a network grows without any long-term strategy. Many users (especially managers) perceive that a network can be expanded to an infinite degree. Many also think that new users can simply be added to the network without a thought on the amount of traffic that they are likely to generate, and its effect on other users. It is thus important for network managers to have a short-term, a medium-term and a long-term plan for the network.

So, what are the basic elements of a network? I would say:

- *IP addresses/domain names (but only if the network connects to the Internet or uses TCP/IP).*
- *A network operating system (such as Microsoft Windows, Novell NetWare, UNIX and Linux). Many companies run more than one type of network operating system, which causes many problems, but has the advantage of being able to migrate from one network*

operating system to another. One type of network operating system can also have advantages over other types. For example, UNIX is a very robust networking operating system which has good network security and directly supports TCP/IP for all network traffic.

- *The cables (twisted-pair, fiber-optic or coaxial cables). These directly affect the bit rate of the network, its reliability and the ease of upgrade of the network.*
- *Network servers, client/server connections and peer-to-peer connections.*
- *Bridges, routers and repeaters. These help to isolate traffic from one network segment to another. Routers and bridges are always a good long-term investment and help to isolate network traffic and can also isolate segment faults.*

The networking topology of the future is likely to evolve around a client/server architecture. With this, server machines run special programs which wait for connections from client machines. These server programs typically respond to networked applications, such as electronic mail, WWW, file transfer, remote login, date/time servers, and so on.

Many application programs are currently run over local area networks, but in the future many could be run over wide area networks, or even over the Internet. This means that computers would require the minimum amount of configuration and allows the standardization of programs at a single point (this also helps with bug fixes and updates). There may also be a time when software licensing is charged by the amount of time that a user actually uses the package. This requires applications to be run from a central source (the server).

The Internet, networks and increased computing power will have great effects on all areas of life, whether they are in commerce, in industry or in home life. The standardization of networking technology has allowed for the standardization of systems, especially in electronic mail, and remote working. The key of this success is the worldwide acceptance of the TCP/IP protocol, which allows different computer systems over the world to communicate, no matter their type, their architecture, or their operating system.

The Internet is likely to have a great effect on how companies do business. It is likely in the coming years that many companies will become reliant on electronic commerce for much of their business, whether it is by direct sales over the Internet or the integration of their financial operation in an electronic form.

Electronic commerce involves customers using electronic communications to purchase goods, typically using the Internet. This will change the way that many businesses do business, and the way that consumers purchase their goods. Society is now moving from a cash based society to a cashless society. Most consumers now use ATMs (Automatic Telling Machines) for cash withdrawals, and debit and credit cards to purchase goods. The future is likely to see an increase in consumers using electronic methods to pay for their goods. An important key to the acceptance of Internet-based purchases is that they must be secure, and cannot be used by criminals to make false purchases, or criminals setting up companies which take payments for incorrect services.

Computer networks are a crucial part of many organizations and many users now even have a network connection in their own home. Without networks, there would be no electronic mail, no Internet access and no networked applications. It is one of the fastest growing technological areas and brings benefits to virtually every country in the world. With the interconnection of networks to the Internet, the world has truly become a Global Village. For many people, especially children, the first place to search for a given topic is the World Wide Web (WWW).

Who would believe the pace of technology over ten short years, such as:

- *From networks of tens of computers operating at speeds of thousands of bits per second to networks with thousands of computers operating at billions of bits per second.*

- *From organizations that passed paper documents back and forward, to the totally paperless organizations.*
- *From people who sent one letter each month to people who send tens of electronic mails every day.*
- *From sending letters around the world which would take days or weeks to arrive, to the transmission of information around the world within a fraction of a second.*
- *From businesses that relied on central operations to ones that can be distributed around the world, but can communicate as if they were next door.*
- *From the transmission of memos which could be viewed by people and organizations which were not meant to read the message, to the transmission of messages that can only be read by the intended destination (and maybe, by space aliens). Not even the CIA can decrypt these messages.*
- *From written signatures that can be easily forged, to digital signatures which are almost impossible to forge, and not only authenticate the sender but also all of the contents of a message.*

These days virtually every computer in a company is networked and networks are key to the effective working of an organization. Without them, few people could work effectively. They provide us with:

- *Electronic mail.*
- *Networked application software.*
- *Remote connections.*
- *Shared printers.*
- *Networked video conferencing.*
- *Remote control of remote equipment.*
- *Remote data acquisition.*
- *Shared disk resources.*

Networking Types and Cables

11.1 Introduction

Most computers in organizations connect to a network using a LAN (Local Area Network). These networks normally consist of a backbone, which is the common link to all the networks within the organization. This backbone allows users on different network segments to communicate and allows data into and out of the local network. Figure 11.1 shows a local area network which contains various segments: LAN A, LAN B, LAN C, LAN D, LAN E and LAN F. These are connected to the local network via the BACKBONE 1. Thus, if LAN A talks to LAN E then the data must travel out of LAN A, onto BACKBONE1, then into LAN D and through onto LAN E.

Networks are partitioned from other networks either using a bridge, a gateway or a router. A bridge links a network of one type to an identical type, such as Ethernet to Ethernet, or Token Ring to Token Ring. A gateway connects two dissimilar types of networks and routers operate in a similar way to gateways and can either connect to two similar or dissimilar networks. The essential operation of a gateway, bridge or router is that they only allow data traffic through that is intended for another network, which is outside the connected network. This filters traffic and stops traffic, not intended for the network, from clogging-up the backbone. Most modern bridges, gateways and routers are intelligent and can automatically determine the topology of the network. They do this by intercommunicating.

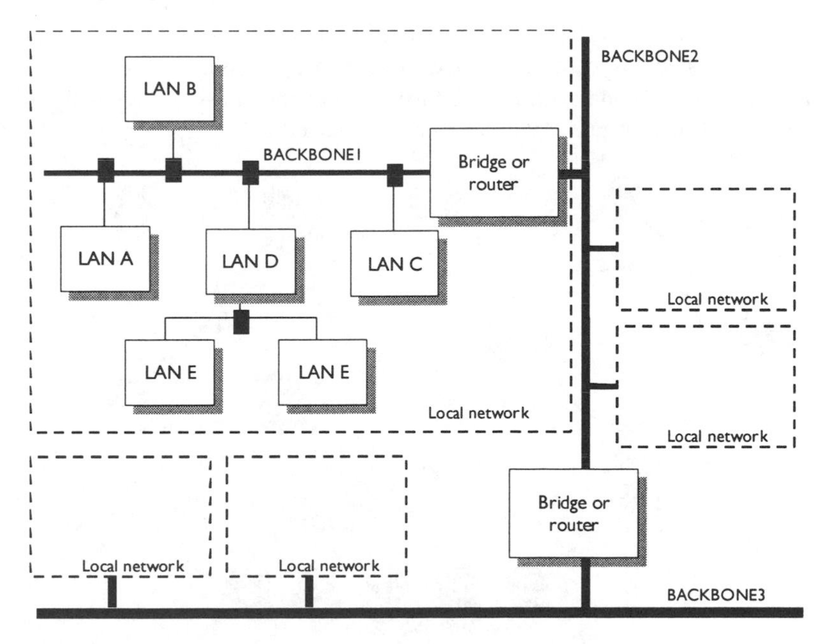

Figure 11.1 Interconnection of local networks

11.2 Network topologies

> **Enterprise network**
> An enterprise network is a network which binds together data, communication, computing and file services in a single organization. They typically require the services of a WAN to provide interconnection between sites, and a LAN to provide interconnection between computers and file servers.

There are three basic topologies for LANs, which are shown in Figure 11.2. These are:

- A **star** network. This type of network uses a central server to route data between clients.
- A **ring** network. This type of network uses a ring in which data is passed from node to node, either in a clockwise or an anti-clockwise direction. Normally a token is passed from node to node, and a node can only transmit when it gets the token.
- A **bus** network. In this type of network all the nodes on a network segment connect to the same physical cable. They must thus contend to get access to the network.

There are other topologies which are either a combination of two or more of the basic topologies or are derivatives of the main types. A typical topology is a **tree** topology, that is essentially a combined star and a bus network, as illustrated in Figure 11.3. A concentrator (or hub) is used to connect the nodes to the network.

11.2.1 Star network

In a star topology, a central server (or switching hub) switches data around the network (Figure 11.4). Data traffic between nodes and the server will thus be relatively low. Its main advantages are:

- Since the data rate is relatively low between central server and the node, a low-specification twisted-pair cable can be used to connect the nodes to the server.
- A fault on one of the nodes will not affect the rest of the network. Typically, mainframe computers use a central server with terminals connected to it.

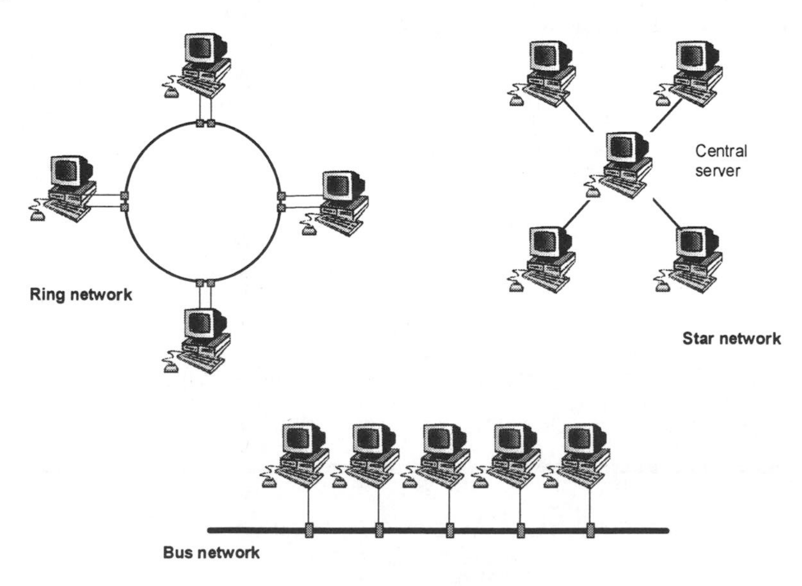

Figure 11.2 Network topologies

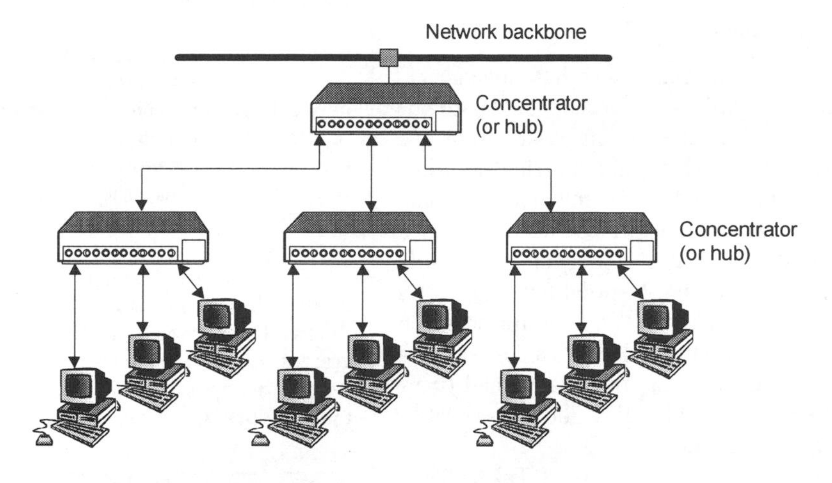
Network backbone
Concentrator (or hub)
Concentrator (or hub)

Figure 11.3 Tree topology

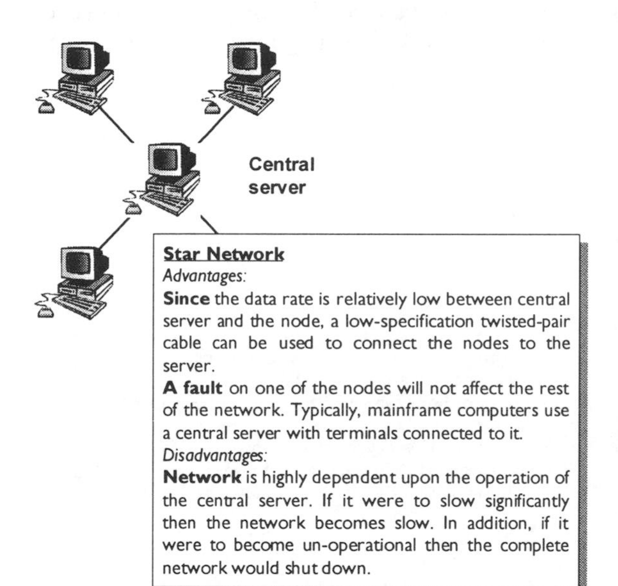
Central server
Star Network
Advantages:
Since the data rate is relatively low between central server and the node, a low-specification twisted-pair cable can be used to connect the nodes to the server.
A fault on one of the nodes will not affect the rest of the network. Typically, mainframe computers use a central server with terminals connected to it.
Disadvantages:
Network is highly dependent upon the operation of the central server. If it were to slow significantly then the network becomes slow. In addition, if it were to become un-operational then the complete network would shut down.

Figure 11.4 Star topology

The main disadvantage of this type of topology is that the network is highly dependent upon the operation of the central server. If it were to slow significantly then the network becomes slow. In addition, if it were to become un-operational then the complete network would be shut down.

An Ethernet hub acts as multiport repeaters (a concentrator). They can be either active or passive. An active hub connects to the network media, and also regenerates the signal, whereas a passive hub simply connects devices onto the networking media.

DEAR NET-ED

Question: *I've been told that I should not use copper cables to connect networks between two buildings. Why?*

Networks use digital signals. These digital signals are referenced to a local ground level (which eventually connects to the earth connection). The ground level can vary between different buildings (and can be large enough to give someone an electrical shock). Thus the ground connection between the two buildings must be broken. If possible for safety, and for reliable digital transmission, you should use a fiber-optic connection.

Also, copper cables can carry electrical surges (such as from lightning strikes), and airborne electrical noise. Electrical surges can cause great damage, and noise can cause the network performance to degrade (as it can cause bit errors).

If possible use fiber-optic cables for any long run of networking media. They tend to produce fewer problems, and allow for easy upgrades (as they have a much greater bandwidth than copper-based cables).

11.2.2 Ring network

In a ring network, computers link together to form a ring. To allow an orderly access to the ring, a single electronic token passes from one computer to the next around the ring, as illustrated in Figure 11.5. A computer can only transmit data when it captures the token. In a manner similar to the star network, each link between nodes is a point-to-point link and allows the usage of almost any type of transmission medium. Typically twisted-pair cables allow a bit rate of up to 16 Mbps, but coaxial and fiber-optic cables are normally used for extra reliability and higher data rates.

A typical ring network is IBM Token Ring. The main advantage of token ring networks is that all nodes on the network have an equal chance of transmitting data. Unfortunately, it suffers from several problems; the most severe is that if one of the nodes goes down then the whole network may go down (as it is not able to pass the token onto the next node).

Logical and physical star

Ethernet networks use a bus-type network, but when it connects to a hub the network can be seen as a physical star topology as the hub can be seen as a central point. If it were to fail, then the whole network may fail. Inside the hub the Ethernet connection still uses a bus network.

This is also the case for a ring network which uses MAU (Multistation Access Units) which is like a hub but creates a virtual ring. The MAU can be seen as a physical star, although the actual network is a ring topology.

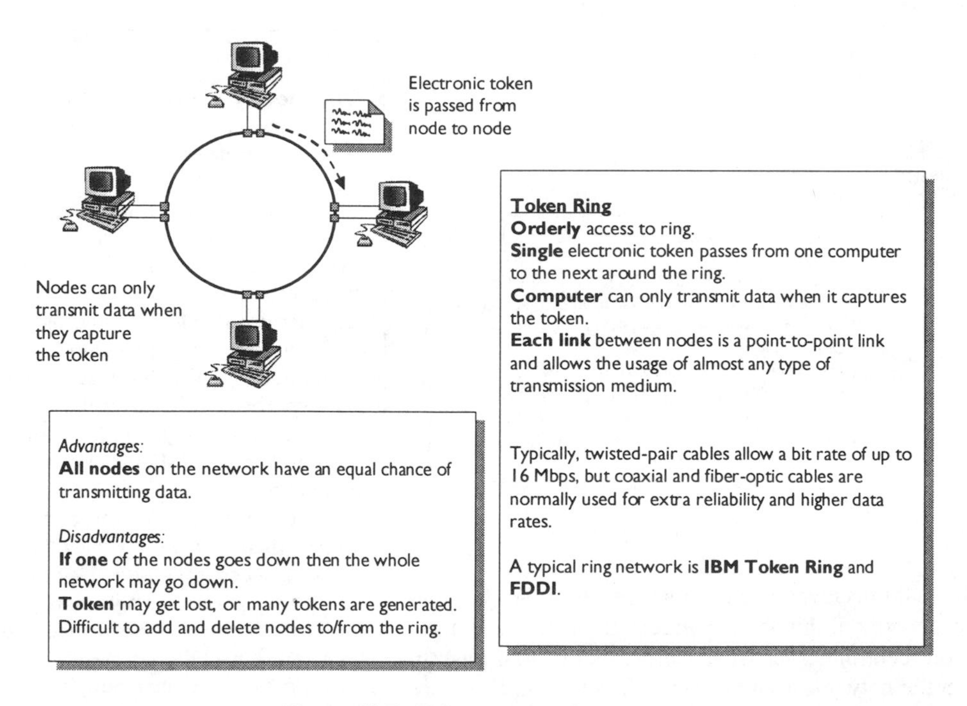

Figure 11.5 Token passing ring network

11.2.3 Bus network

A bus network uses a multi-drop transmission medium, as shown in Figure 11.6, where all nodes on the network share a common bus and thus share communications. This allows only one device to communicate at a time. A distributed medium access protocol determines which station is to transmit. As with the ring network, data frames contain source and destination addresses, where each station monitors the bus and copies frames addressed to itself.

Twisted-pair cables give data rates up to 100 Mbps, whereas, coaxial and fiber-optic cables give higher bit rates and longer transmission distances. A bus network is a good compromise over the other two topologies as it allows relatively high data rates. Also, if a node goes down, it does not normally affect the rest of the network. The main disadvantage of this topology is that it requires a network protocol to detect when two nodes are transmitting at the same time. It also does not cope well with heavy traffic rates. A typical bus network is Ethernet 2.0.

Bus networks require a termination at either end of the bus, as the signal requires to be absorbed at the end of the bus (else it would bounce off the end of the open-circuited bus). This prevents signals from bouncing back and being received again by workstations attached to the bus. Ring and star networks do not require termination as they are automatically terminated. With a star network the connected nodes automatically terminate the end of the connection.

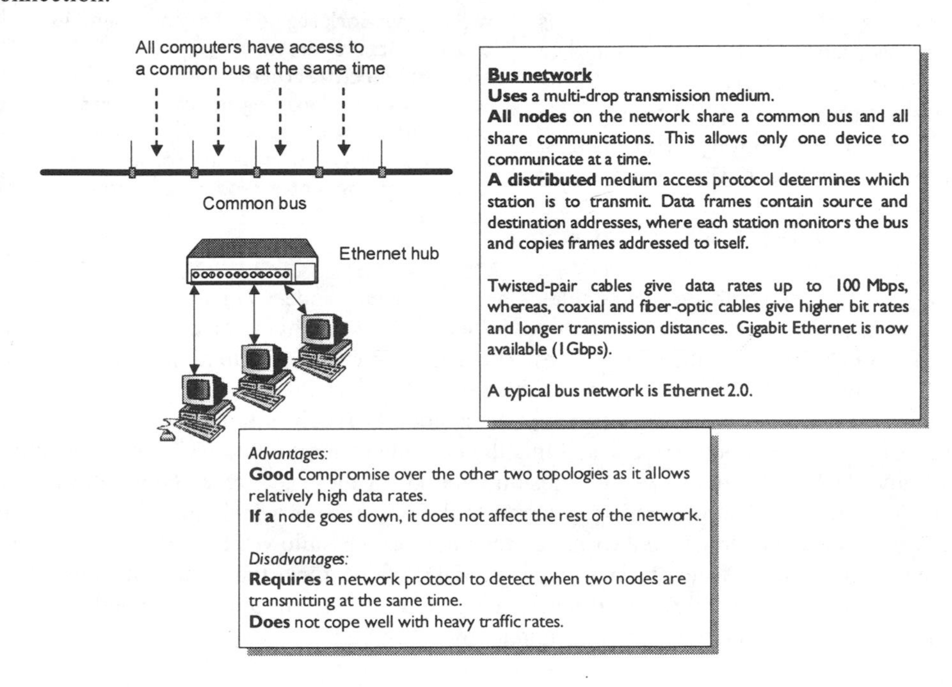

Figure 11.6 Bus topology

11.3 Token Ring

Token Ring networks were developed by several manufacturers, the most prevalent being the IBM Token Ring. Unlike Ethernet, they cope well with high network traffic loadings, and were at one time extremely popular but Ethernet has since overtaken their popularity. Token Ring networks have, in the past, suffered from network management problems and poor network fault tolerance.

11.3.1 Operation

A Token Ring network circulates an electronic token (named a control token) around a closed electronic loop. Each node on the network reads the token and repeats it to the next node. The control token circulates around the ring even when there is no data being transmitted.

Collision:
This is the result of two nodes transmitting at the same time. The frames from each node are damaged when they meet each other on the physical media.
Collision domain:
On an Ethernet network, when two or more nodes try and transmit at the same time the data frame is damaged, and the nodes must backoff from the network. The network area within which data packets originate and collide is called a collision domain.
Backoff:
The retransmission delay enforced when a collision occurs.
Broadcasts:
Data frames that are sent to all the nodes within a network segment. They are identified by a broadcast address.
Network architecture:
A combination of existing standards (rules or procedures that the network complies with) and protocols (set of rules and conventions that govern how networked nodes inter-communicate).

Nodes on a Token Ring network wishing to transmit must await a token. When they get it, they fill a frame with data and add the source and destination addresses then send it to the next node. The data frame then circulates around the ring until it reaches the destination node. It then reads the data into its local memory area (or buffer) and marks an acknowledgement on the data frame. This then circulates back to the source (or originating) node. When it receives the frame, it tests it to determine whether it contains an acknowledgement. If it does then the source node knows that the data frame was received correctly, else the node is not responding. If the source node has finished transmitting data then it transmits a new token, which can be used by other nodes on the ring.

Figure 11.7(a)–(d) shows a typical interchange between node B and node A. Initially, in (a), the control token circulates between all the nodes. This token does not contain any data and is only a few bytes long. When node B finally receives the token it then transmits a data frame, as illustrated in (b). This data frame is passed to node C, then to node D and finally onto A. Node A then reads the data in the data frame and returns an acknowledgement to node B, as illustrated in (c). After node B receives the acknowledgement, it passes a control token onto node C and this then circulates until a node wishes to transmit a data frame. No nodes are allowed to transmit data unless they have received a valid control token. A distributed control protocol determines the sequence in which nodes transmit. This gives each node equal access to the ring, as each node is only allowed to send one data frame (although some Token Ring systems, such as FDDI, allow for a time limit on transmitting data, before the token is released). It must then give up the token to the next node, and wait for the token to return before it can transmit another data frame.

11.3.2 Token Ring maintenance

A Token Ring system requires considerable maintenance; it must perform the following functions:

- Ring initialization – when the network is started, or after the ring has been broken, it must be reinitialized. A co-operative decentralized algorithm sorts out which node starts a new token, which goes next, and so on.
- Adding to the ring – if a new node is to be physically connected to the ring then the network must be shut down and reinitialized.
- Deletion from the ring – a node can disconnect itself from the ring by joining together its predecessor and its successor. Again, the network may have to be shut down and reinitialized.
- Fault management – typical Token Ring errors occur when two nodes think it is their turn to transmit or when the ring is broken as no node thinks that it is their turn.

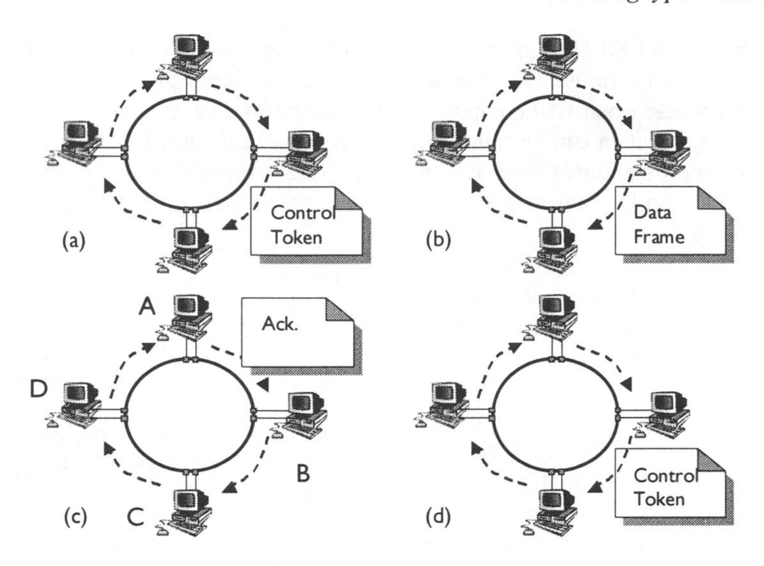

Figure 11.7 Example data exchange

11.3.3 Token Ring multistation access units (MAUs)

The problems of adding and deleting nodes to or from a ring network are significantly reduced with a multistation access unit (MAU). Normally, a MAU allows nodes to be switched in and out of a network using a changeover switch or by automatic electronic switching (known as auto-loopback). This has the advantage of not shutting down the network when nodes are added and deleted or when they develop faults.

11.4 Ethernet

Most of the computers in business now connect through a LAN and the most commonly used LAN is Ethernet. DEC, Intel and the Xerox Corporation initially developed Ethernet and the IEEE 802 committee has since defined standards for it, the most common of which are Ethernet 2.0 and IEEE 802.3.

In itself Ethernet cannot make a network and needs some other protocol such as TCP/IP to allow nodes to communicate. Unfortunately, Ethernet in its standard form does not cope well with heavy traffic, but this is offset by the following:

- Ethernet networks are easy to plan and cheap to install.
- Ethernet network components, such as network cards and connectors, are cheap and well supported.
- It is a well-proven technology, which is fairly robust and reliable.
- It is simple to add and delete computers on the network.
- It is supported by most software and hardware systems.

DEAR NET-ED

Question: *What do I need to create a basic network?*

All you really need is two computers, two Ethernet NICs, a hub, and some patch cables. The patch cables connect the computers to the hub, and the hub creates the network. The computers can then simply make a peer-to-peer connection with each.

UNIX/Linux will allow you to access one computer from another, using TELNET, FTP, NDS, and so on, but you would have to assign each computer a unique IP address. As long as you do not connect onto the Internet, you can choose any IP address.

Microsoft Windows uses its own protocol (NetBEUI) to make a peer-to-peer connection (with file/printer sharing).

A major problem with Ethernet is that, because computers must contend to get access to the network, there is no guarantee that they will get access within a given time. This contention also causes problems when two computers try to communicate at the same time, they must both back off and no data can be transmitted. In its standard form Ethernet allows a bit rate of 10 Mbps, but new standards for fast Ethernet systems minimize the problems of contention and also increase the bit rate to 100 Mbps (and even 1 Gbps). Ethernet uses coaxial, fiber-optic or twisted-pair cable.

Ethernet uses a shared-media, bus-type network topology where all nodes share a common bus. These nodes must then contend for access to the network as only one node can communicate at a time. Data is then transmitted in frames which contain the MAC (media access control) source and destination addresses of the sending and receiving node, respectively. The local shared media is known as a segment. Each node on the network monitors the segment and copies any frames addressed to it.

Ethernet uses carrier sense, multiple access with collision detection (CSMA/CD). On a CSMA/CD network, nodes monitor the bus (or Ether) to determine if it is busy. A node wishing to transmit data waits for an idle condition then transmits its message. Unfortunately, collisions can occur when two nodes transmit at the same time, thus nodes must monitor the cable when they transmit. When a collision occurs, both nodes stop transmitting frames and transmit a jamming signal. This informs all nodes on the network that a collision has occurred. Each of the nodes involved in the collision then waits a random period of time before attempting a re-transmission. As each node waits for a random delay time then there can be a prioritization of the nodes on the network, as illustrated in Figure 11.8.

Each node on the network must be able to detect collisions and be capable of transmitting and receiving simultaneously. These nodes either connect onto a common Ethernet connection or can connect to an Ethernet hub, as illustrated in Figure 11.8. Nodes thus contend for the network and are not guaranteed access to it. Collisions generally slow the network.

DEAR NET-ED

Question: *Everywhere I read, it says that Ethernet has so many problems, and isn't really a very good networking technique. So why is it so popular?*

Local area networks have evolved over the years. At one time the big contest was between Token Ring, and Ethernet. Which was best? Well Token Ring was always better at coping with network traffic than Ethernet, especially when the network traffic was heavy. But, remember these were the days before hubs. Thus most network connections were made from computer to computer with coaxial cable. The big problem with Token Ring was when there was a bad connection or when a computer was disconnected from the network, as this brought the whole network down. Ethernet (10BASE) proved much easier to add and delete computers to and from the network. Thus it triumphed over Token Ring. Soon Ethernet NICs cost much less than Token Ring cards, and were available from many sources (typically, these days, Token Ring cards will cost up to over five times as much as Ethernet ones).

Ethernet has coped well with the evolving networks, and the new hubs made it even easier to connect computers to a network. It faced a big problem, though, when the number of users of a network increased by a large factor. Its answer to this was 100BASE, which ramped up the bit rate by a factor of ten. This worked well, but it suffered when handling traffic over wide areas. Ethernet had a final trump card: 1000BASE, which gives a bit rate of 1 Gbps.

Thus, whatever we throw at Ethernet, it fights back by either ramping up the bit rate (from 10 Mbps to 100 Mbps to 1 Gbps) or it allows multiple simultaneous network connections (through Ethernet switches). So, don't dismiss the King, he's going to be around for a while yet.

1
Two nodes transmit at the same time
2
Nodes detect there has been a collision
3
Nodes transmit a jamming signal
4
Nodes wait a random period before retransmitting
CSMA/CD
Ethernet uses carrier sense, multiple access with collision detection (CSMA/CD).
Nodes monitor the bus (or Ether) to determine if it is busy. A node wishing to send data waits for an idle condition then transmits its message.
Collisions can occur when two nodes transmit at the same time, thus nodes must monitor the cable when they transmit.
When a collision occurs, both nodes stop transmitting frames and transmit a jamming signal.
This informs all nodes on the network that a collision has occurred.
Each of the nodes involved in the collision then waits a random period of time before attempting a re-transmission.
As each node has a random delay time then there can be a prioritization of the nodes on the network.
All computers have access to a common bus at the same time
Common bus

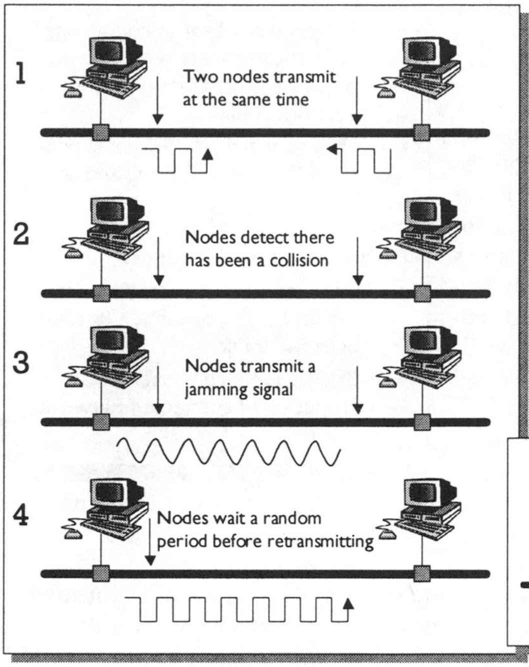

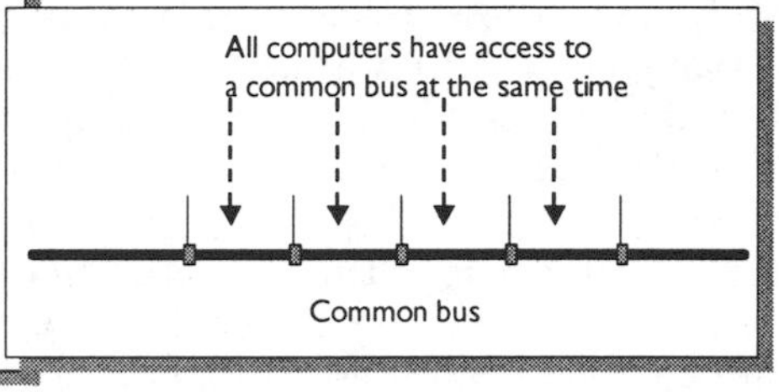

Figure 11.8 CSMA/CD

DEAR NET-ED

Question: *How do you connect a fiber-optic cable to a connector?*

It takes a little bit of skill, but basically it is just glued onto the end.

Question: *And, how do you get an RJ-45 connector onto twisted-pair cable?*

You strip about 0.5 inch of the outer jacket and fan-out the wires in the correct order. Next you push them fully into the RJ-45 connector, and finally use the special crimping tool to clamp the cable, and make the required contacts. No soldering is involved.

11.4.1 Hubs, bridges and routers

Repeaters operate at layer 1 of the OSI model, and are used to increase the number of nodes that can connect to a network segment, and the distance that it can cover. They do this by amplifying, retiming and reshaping the digital signals. A hub is a repeater with multiple ports, and can be thought of as being the center point of a star topology network. It is often known as a multiport repeater (or as a concentrator in Ethernet). Hubs generally:

- Amplify signals.
- Propagate the signal through the network.
- Do not filter traffic. This is a major disadvantage with hubs and repeaters as data arriving at any of the ports is automatically transmitted to all the other ports connected to the hub.
- Do not determine the path.
- Centralize the connection to the network. This is normally a major problem when using a star connected network, but hubs are normally reliable and can be easily interchanged if they do not operate properly.

Hubs do not filter traffic, so that collisions affect all the connected nodes within the collision domain. The more collisions there are the slower the network segment becomes. There are two main ways to overcome this:

- **Bridges.** These examine the destination MAC address (or station address) of the transmitted data frame, and will not retransmit data frames which are not destined for another network segment. They maintain a table with connected MAC addresses, and do not forward any data frames if the MAC address is on the network segment that originated it, else it forwards to all connected segments.
- **Router.** These examine the network (typically the IP or IPX address) of the transmitted data packet, and will only transmit it out of a network segment if it is destined for a node on another network.

Advantages of bridges:

- Segment networks.
- Reduce collision domains.
- Filter network traffic.
- Bridges work best in connecting network segments which do not require a great amount of internetwork traffic.
- They forward any high-level protocol (TCP/IP, SPX/IPX, and so on).

Disadvantages of bridges:

- If the internetwork traffic is large, the bridge can become a bottleneck for traffic.
- Bridges forward broadcast data frames to all connected networks.

Bridges and routers bound a network segment, whereas a repeater extends it, as illustrated in Figure 11.9. As they are outside the network segment, bridges and routers thus do not forward collisions. Broadcast data frames are sent out by a node if it does not know the MAC address of the destination. Bridges forward these broadcasts to all the connected network segments, and every device on the connected network segments must listen to these data frames. Broadcast storms result when too many broadcasts are sent out over the network, which can cause network time-outs, where the network slows down. Routers do not forward broadcasts and thus cope better with broadcast storms.

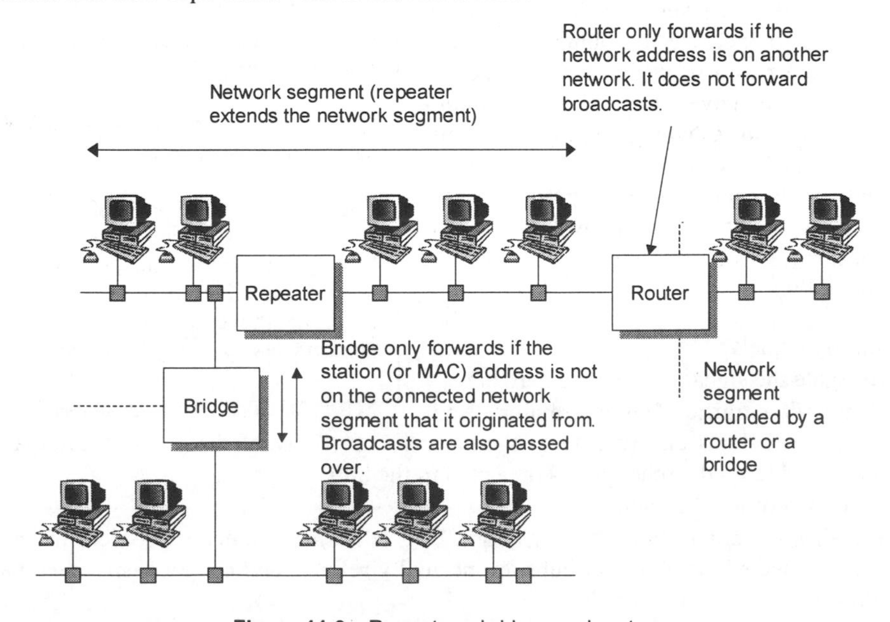

Figure 11.9 Repeaters, bridges and routers

11.5 LAN components

LANs are high-speed, low-error data networks that span a relatively small geographic area. They connect workstations, peripherals, terminals, and other devices in a single building or other geographically limited area. The required hardware is:

- **Workstations.** These are the devices that users use to gain access to the network. Typically they are PCs, and they run the application programs, thus important decisions in purchasing a computer are: whether they run the required application software packages; whether they run the required network operating system; and whether they can run as a stand-alone computer (when the network fails).
- **Networking media.** This is the media that connects the parts of the network. There are four main types of media: unshielded twisted-pair cable, shielded twisted-pair cable, coaxial cable, and fiber-optic cable. Network media is fundamentally important, as the rest of the network will not work without it. It is often said that a network is as reliable as its cabling, and that the networking media is the most important part of the whole network. It is the most limiting factor in the network, and sets the maximum limit for network traffic. In most cases it has a life of more than ten years, and thus must be chosen to support expansion over that time.
- **NIC cards.** This is the device that connects the workstation or file server to the network, and is where the physical or MAC address is located. Every workstation and file server has a NIC card, which typically plugs into one of the expansion slots on the motherboard. Its three main functions are:
  - To form data frames and send them out onto the networking media.
  - To receive data frames coming from the networking media and transform them into information that the workstation can understand.
  - To provide an orderly access to the shared networking media.
- **Hub or wiring center.** In networks with a file server it is not possible to connect every

Routers

Routers are the key components of the Internet. They communicate with each other and try and determine the best way to get to a remote network. As every computer which connects to the Internet must have an IP address, they use these addresses to route data around the Internet. Without routers we would not have an Internet. Routers are generally the best device to isolate traffic from one network and another, as they will only forward data packets if the destination is not on the current network.

Advantages of routers:

- Intelligently route data to find the best path using the network address. A bridge will route if the MAC address is not on the originating segment, whereas a router will intelligently decide whether to forward, or not.
- They do not forward broadcasts, thus they reduce the effect of broadcast storms.

Disadvantages of routers:

- Slower than bridges, as they must process the data packet at a higher level. The data frame is then forwarded in a modified form.
- They are network protocol dependent, whereas bridges will forward any high-level protocol as it is operating on the level 2 (as long as it connects two networks of the same types, such as Ethernet-to-Ethernet). Routers interpret the network level data using the required protocol, such as IP or IPX.

Hail the Router. Along with TCP/IP, the true uniter of the People of the World.

workstation to it, each with a separate NIC card within the server. Thus most LANs use a hub, which is a network device that serves as the center of a star-topology network, each of the connections being independent. On Ethernet networks, hubs are multiport repeaters and are used to provide for multiple connections.

Additionally a network may have one or more file servers, which are used to store important network information, and are a central storage facility for the network. Typically they contain networked application programs and have several general features:

- **Highly optimized**. As many users have access to them they must operate quickly and are thus generally built to a high specification with an optimized architecture. Normally they contain a large amount of memory and storage facilities.
- **Located in a convenient place** for the workstations that connect to it. This is important as most of the workstations which connect to the network require the file server to provide application software, and other networking facilities, such as being a print server, login validation, and so on. Thus the file server must be located in an optimum place so that each workstation can have fast access to it.
- **Reliable connection**. The file server is typically the most important computer on the network, and if it were to fail to communicate with the rest of the network it may cause a significant loss of service. Thus the file server is typically connected using reliable cabling with a robust network connection.

Coaxial cable is typically available in two main flavors:

- **Thicknet.** Thick and rigid cable which is difficult to bend and often difficult to install. It typically has a distinctive yellow outer cover.
- **Thinnet.** Thinner cable which is easier to work with (0.18 inch).

Advantages of coaxial cable

- Outer copper braid or metallic foil provides a shield to reduce the amount of interference.
- They can be run unboosted for longer distances than either shielded or unshielded twisted-pair cable.
- Less expensive than fiber-optic.
- Well-known technology. Coaxial cable has been used extensively in the past, especially in radio, TV and microwave applications.

Disadvantages of coaxial cable

- The cable is relatively thick, and the thicker the cable the more difficult it is to work with.
- More expensive than twisted-pair cable.
- In thinnet, the outer copper or metallic braid of the cable comprises half the electrical circuit. Thus special care must be taken to ensure that it is properly grounded, at both ends of the cable. If it is not properly grounded, it can result in electrical noise that interferes with transmittal of the signal on the networking media.

11.5.1 Cables

Cabling is one of the most important elements in a network, and is typically the limiting factor on the speed of the network. The four main types of networking media are:

- Shielded twisted-pair cable (STP).
- Unshielded twisted-pair cable (UTP).
- Coaxial cable.
- Fiber-optic cable.

The type of network media determines how fast the data travels along the media, and also the maximum data rate that can be carried. Twisted-pair and coaxial cable use copper wires to carry electrical signals, while fiber-optic cable carries light pulses. Fiber-optic cables generally support the fast data transfer rate.

All signals are affected by degradation when they are applied onto networking media. These are either internal or external, such as:

- **Internal.** In copper cables there are electrical parameters such as resistance (opposition to the flow of electrons), capacitance (the opposition to changes in voltage) and inductance (the opposition to changes in current) can cause signals to degrade. Resistance causes a loss of power (or signal attenuation), whereas capacitance and inductance cause the signals to lose their shape.
- **External.** These are external sources of electrical impulses that cause the electrical signals to change their shape. They are caused either by electromagnetic interference (EMI) and radio frequency interference (RFI), and are typically generated from lighting, electrical motors, and radio systems. In copper cables, each wire of the cable acts as an antenna, and absorbs electrical signals from other wires in the cable (know as crosstalk) and from EMI and RFI sources outside the cable. These sources are known as noise and can distort the electrical signals so that it is difficult to determine the original data.

Advantages of UTP:

- The cable is thin and easy to work with. This makes it easy to install.
- Less expensive than other types of networking media.
- When used with an RJ connector (RJ-45 or RJ-11), it provides a reliable connection.
- Data rates can be as fast as coaxial cables (as UTP cables now have an excellent specification).

Disadvantages of UTP:

- More prone to electrical noise and interference than other types of networking media (as there is no shield between the pairs).
- Can carry electrical surges.

Methods used to reduce signal attenuation, and coupled noise are:

- **Cancellation.** Electrical conductors produce a small circular magnetic field around themselves when an electrical current flows in them. If two wires are placed beside each other, and there is an opposite current flowing, then the magnetic fields will tend to cancel. This magnetic field can be reduced to almost zero by twisting the two opposite wires together. This technique is called twisted-pairs. The same goes for external magnetic fields coupling into the twisted-pairs, again they will cancel each other out. Thus twisted-pairs (or self-shielding) are useful for reducing external coupling of electromagnetic noise and crosstalk. The direction of these magnetic lines of force is determined by the direction in which current flows along the wire. If two wires are part of the same electrical circuit, electrons flow from the negative voltage source to the destination along one wire and from the destination to the positive voltage source along the other wire.
- **Shielding.** This combats EMI and RFI by wrapping a metal braid or foil around each wire pair or group of wire pairs, which acts as a barrier from external noise. This increases the size and cost of the cable, and is typically only used when there are large sources of external radiation, such as when placed near electrical motors. However, as with increasing the size of the conductors, using braid or foil covering increases the diameter of the cable, and it will increase the cost as well. Therefore, cancellation is the more commonly used technique to protect the wire from undesirable interference.
- **Match cables.** The characteristic impedance of a cable is important, and cables and connectors must always be chosen so that they have the same characteristic impedance. If they are not matched there can be a significant power loss or pulse reflections from the junction between the cable and the connection. For twisted-pair cables, this characteris-

tic impedance is typically 100 Ω, and for coaxial cable it is 50 Ω (for networking) and 75 Ω (for TV applications).
- **Improve the cable.** Increasing the thickness of the conductors reduces the electrical resistance, and increasing the thickness of the insulating material reduces the amount of crosstalk. These changes tend to be expensive and increase the size of the cable.

An important consideration when selecting a cable, especially in hazardous areas, is its jacket. Typically it is made from plastic, Teflon, or composite material. Problem areas are:

- **Carrying fire.** This is where the cable can carry fire from one part of a building to another. Typically it is where cables are installed between walls, in an elevator shaft or pass through an air-handling unit.
- **Producing toxic smoke when lit.** When burnt, plastic cable jackets can create toxic smoke.

To protect against these problems, network cables must always comply with fire codes, building codes, and safety standards. These are more important than other factors, such as cable size, speed, cost, and difficulty of installation.

Advantages of fiber optics:
- Excellent reliability, and are extensively used as network backbones.
- Immune from crosstalk, EMI and RFI.
- Can be run for longer distances than copper cables.
- They do not create grounding problems, thus they can be used to connect between two sites with a different ground potential.
- Very thin flat cable that can be easily run within confined spaces.
- Can be used in hazardous conditions, as it does not create electrical sparks.
- Immune from lightning strikes.

Disadvantages of fiber optics:
- More expensive and more difficult to install than any other networking media.
- Requires a trained installer to create a good cable connection.
- Too expensive in most situations to provide fiber connections to every workstation.

11.5.2 Unshielded twisted-pair cable

The most popular type of cabling is unshielded twisted-pair, which comprises of four-wire pair. Unshielded twisted-pair cable does not have a shield around each of the pairs, it thus relies on:

- **Cancellation effect.** The twists of each pair produces a cancellation effect which limits degradation caused by EMI and RFI.
- **Variation of twists.** With this the number of twists in the wire pairs varies from one to the other, which reduces the amount of crosstalk between the pairs. There are strict limits on the maximum number twists or braids per foot of cable.
- **Accurate characteristic impedance.** For this the characteristic impedance is around 100 Ω in order to produce a good match between the cable and any connection.

11.5.3 Shielded twisted-pair cable

STP cable is similar to UTP but has shielding on each of the pairs, thus reducing the effect of crosstalk, EMI and RFI. Unlike coaxial cable, the shielding does not act as part of the circuit, but it must be properly grounded at each end to enhance the shielding effect (as a non-ground shield will act like an antenna and pick-up electrical noise). Its only disadvantage is that it is more expensive than UTP, although it can suffer from the same problems of coaxial cable if either end of the shield is not grounded.

11.5.4 Fiber-optic cables

One of the greatest revolutions in data communications is the usage of light waves to transmit digital pulses through fiber-optic cables. A light carrying system has an almost unlimited information capacity, and theoretically, it has more than 200 000 times the capacity of a satellite TV system.

Optoelectronics is the branch of electronics which deals with light, and uses electronic devices that use light operate within the optical part of the electromagnetic frequency spectrum, as shown in Figure 11.10. There are three main bands in the optical frequency spectrum, these are:

- Infra-red – the band of light wavelengths that are too long to been seen by the human eye.
- Visible – the band of light wavelengths that the human eye responds to.
- Ultra-violet – band of light wavelengths that are too short for the human eye to see.

DEAR NET-ED

Question: *I live in Edinburgh, and my friend lives in London. How long does it take for a digital pulse to travel from Edinburgh to London?*

Well, there are a lot of assumptions to be made. First we'll assume that there are no intermediate devices in the cable that connects Edinburgh and London, and we'll assume that it is fiber-optic cable, which propagates light pulses at one-third the speed of light (10^8 m/s). Thus for a distance of 500 miles (804.65km,) the time will be:

$$T = \frac{Distance}{Speed} = \frac{804.65 \times 10^3}{1 \times 10^8} = 0.0080465$$

$$= 8.05\,\text{ms}$$

Wavelength and color

Wavelength is defined as the physical distance between two successive points of the same electrical phase. Figure 11.11 (a) shows a wave and its wavelength. The wavelength is dependent upon the frequency of the wave f, and the velocity of light, c (3×10^8 m/s) and is given by:

$$\lambda = \frac{c}{f}$$

The optical spectrum ranges from wavelengths of 0.005 mm to 4000 mm. In frequency terms these are extremely large values from 6×10^{16} Hz to 7.5×10^{10} Hz. It is thus much simpler to talk in terms of wavelengths rather than frequencies.

The human eye sees violet at one end of the color spectrum and red on the other. In-between, the eye sees blue, indigo, green, yellow and orange. Two beams of light that have the same wavelength are seen as the same color and the same colors usually have the same wavelength. Figure 11.11 (b) shows the color spectrum.

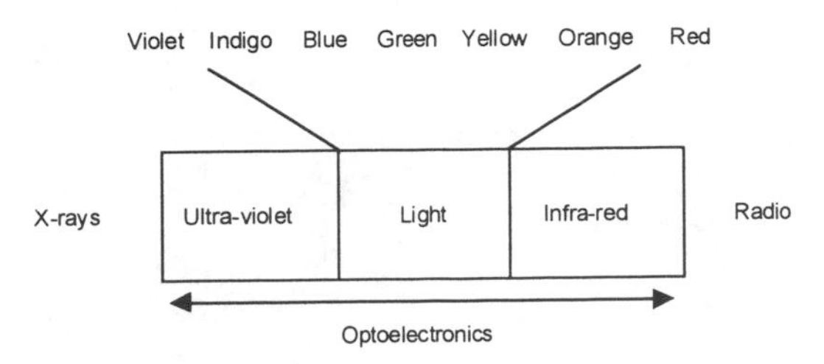

Figure 11.10 EM optoelectronics spectrum

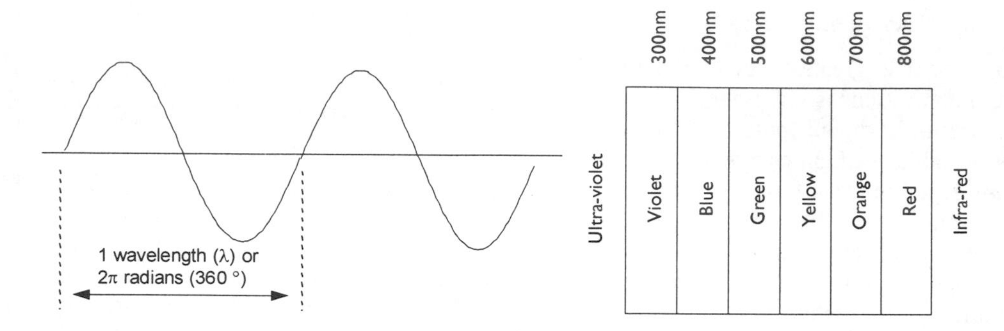

Figure 11.11 Wavelength of wave and color spectrum

Velocity of propagation and refractive index

In free space electromagnetic waves travel at approximately 300 000 000 m/sec (186 000 miles/sec). However, their velocity is lower when they travel through denser materials. When traveling from one material to another which is less dense then the light ray is refracted (or bent) away from the normal, as illustrated in Figure 11.12.

The amount of bending or refraction at the interface between two materials of different densities depends on the refractive index of the two materials. This index is the ratio of the velocity of propagation of a light ray in free space to the velocity of propagation of a light ray the material, as given by:

$$n = \frac{c}{v}$$

where c is speed of light in free space and v is the speed of light in a given medium. Typical refractive indexes are given in Table 11.1.

Optical fibers

Optical fibers are transparent, dielectric cylinders surrounded by a second transparent dielectric cylinder. Light is transported by a series of reflections from wall to wall from the interface between a core (inner cylinder) and its cladding (outer cylinder). A cross-section of a fiber is given in Figure 11.13.

Reflections occur because the core has a higher reflective index than the cladding (it thus has a higher density). Abrupt differences in the refractive index cause the light wave to bounce from the core/cladding interface back through the core to its opposite wall. Thus the light is transported from a light source to a light detector at the other end of the fiber.

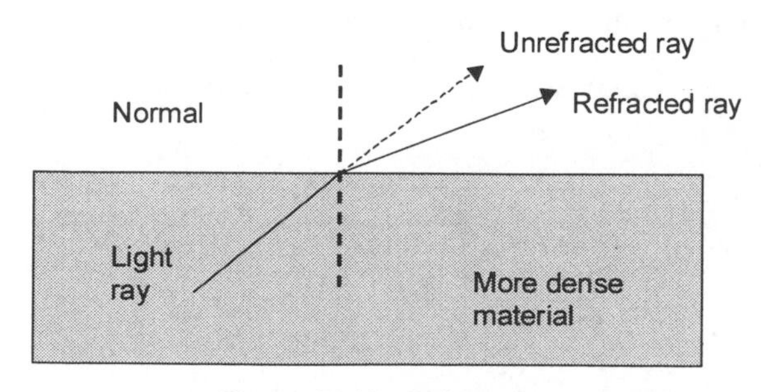

Figure 11.12 Refracted ray

Table 11.1 Refractive index of sample materials

Medium	*Refractive index*
Air	1.0003
Water	1.33
Glass fiber	1.5–1.9
Diamond	2.0–2.42
Gallium arsenide	3.6
Silicon	3.4

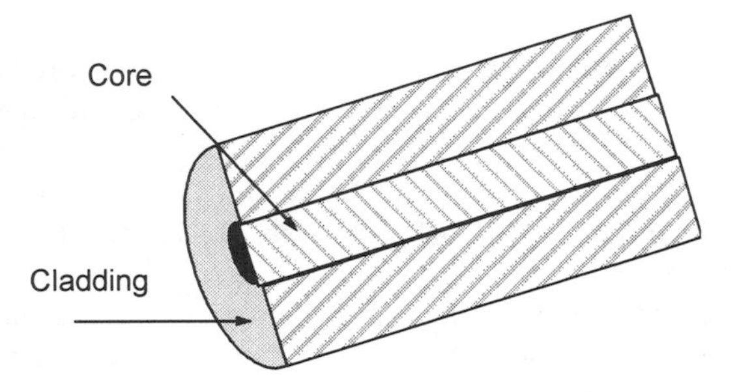

Figure 11.13 Cross-section of an optical fiber

Optical fibers transmit light by total internal reflection (TIR), where light rays passing between the boundaries of two optically transparent media of different densities experience refraction, as shown in Figure 11.14. This changed direction can be determined according to Snell's Law:

$$n_1 \sin\theta_1 = n_2 \sin\theta_2$$

Thus

$$\theta_2 = \sin^{-1}\left[\frac{n_1}{n_2}\sin\theta_1\right]$$

The angle at which the ray travels along the interface between the two materials is called the critical angle (θ_c). If the incident ray is greater than this angle, the ray will be totally reflected from the outer cladding. It then propagates along the fiber being reflected by the cladding on the way, as shown in Figure 11.15. The angle at which the reflection occurs is called the acceptance angle, and if the initial ray is entered at an angle of at least the acceptance angle, then the ray will bounce along the inner core.

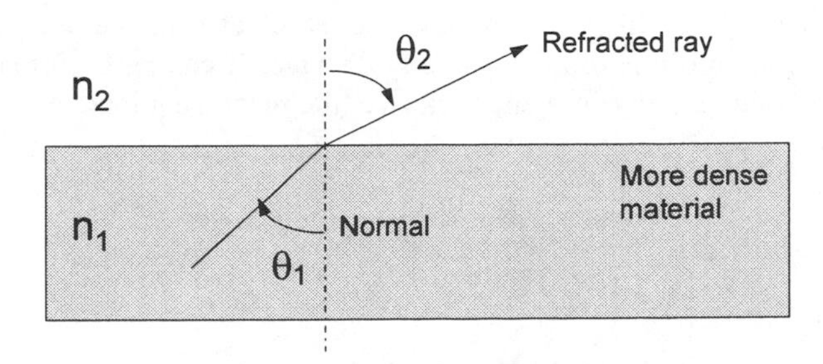

Figure 11.14 Refraction

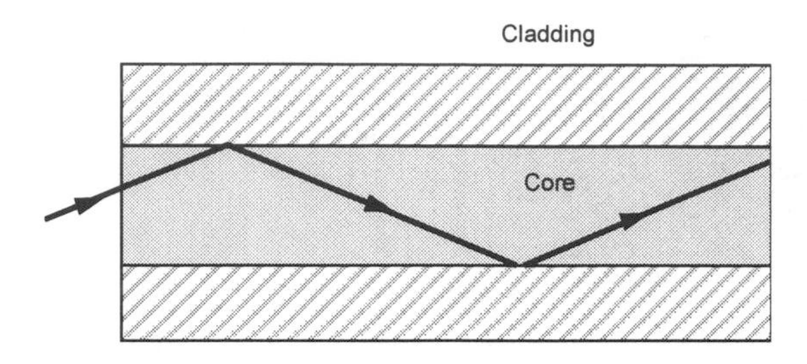

Figure 11.15 Light propagating in an optical fiber

Fiber-optic losses result in a lower transmitted light power, which reduces the system bandwidth, information transmission rate, efficiency and overall system capacity. The main losses are:

- **Absorption losses.** Impurities in the glass fiber cause the transmitted wave to be absorbed and converted into heat.
- **Material scattering.** Extremely small irregularities in the structure of the cable cause light to be diffracted. This causes the light to disperse or spread out in many directions. A greater loss occurs at visible wavelengths than at infrared.
- **Chromatic distortion.** Caused by each wavelength of light traveling at differing speeds. They thus arrive at the receiver at different times causing a distorted pulse shape. Monochromatic light reduces this type of distortion.
- **Radiation losses.** Caused by small bends and kinks in the fiber that scatters the wave.
- **Modal dispersion.** Caused by light taking different paths through the fiber. This will each have a different propagation time to travel along the fiber. These different paths are described as modes. Figure 11.16 shows two rays taking different paths. Ray 2 will take a longer time to get to the receiver than ray 1.
- **Coupling losses.** Caused by light being lost at mismatches at terminations between fiber/fiber, light source/fiber, and so on.

Optical fiber cables are either glass-based or plastic-based, and typically carry infra-red signals (it is thus important to never look directly into a fiber optic cable which is transmitting infra-red signals, as it can damage your eyes). Table 11.2 shows the characteristics of two typical fiber-optic cables, one using glass, and the other plastic. It can be seen that the inside core and the cladding diameters are relatively small, i.e. fractions of a millimeter. Normally the cladding is covered in a coating which is then covered in a jacket. These give the cable mechanical strength and also make it easier to work with. In the case of the 50/125 μm glass cable in Table 11.2 the outer diameter of the cable is 3.2 mm but the inner core diameter is just 50 μm. Normally glass fiber cables have better electrical characteristic over plastic equivalents, but are more prone to breakage and damage. It can be seen that the glass cable has improved bandwidth and lower attenuation over the plastic equivalent.

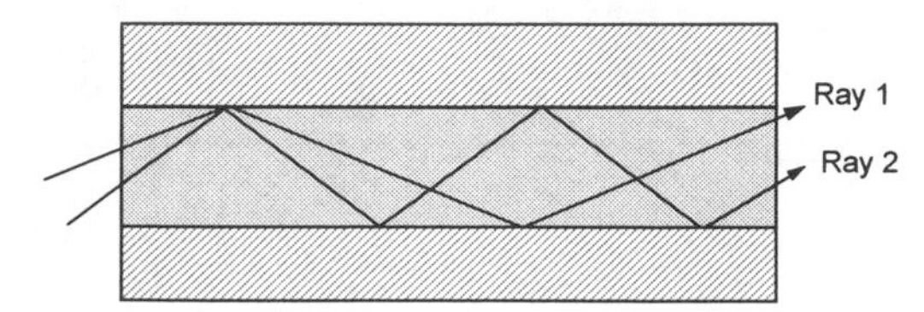

Figure 11.16 Light propagating in different modes

Table 11.2 Typical fiber-optic cables characteristics

	50/125 μm glass	*200 μm PCS*
Construction	Glass	Plastic coated silica (PCS)
Core diameter	50 μm	200 μm
Cladding diameter	125 μm	389 μm
Coating diameter	250 μm	600 μm
Jacket material	Polyethylene	PVC
Overall diameter	3.2 mm	4.8 mm
Connector	9 mm SMA	9 mm SMA
Bandwidth	400 MHz/km	25 MHz/km
Minimum bend radius	30 mm	50 mm
Temperature range	–15 °C to +60 °C	–10 °C to +50 °C
Attenuation @820 nm	3 dB/km	7 dB/km

There are many advantages in using fiber-optics, including:

- Fiber systems have a greater capacity due to the inherently larger bandwidths available with optical frequencies. Metallic cables contain capacitance and inductance along their conductors, which cause them to act like low-pass filters. This limits bandwidth and also the speed of propagation of the electrical pulse.
- Fiber systems are immune from cross-talk between cables caused by magnetic induction. Glass fibers are non-conductors of electricity and therefore do not have a magnetic field associated with them. In metallic cables, the primary cause of cross-talk is magnetic induction between conductors located near each other.
- Fiber cables do not suffer from static interference caused by lightning, electric motors, fluorescent lights, and other electrical noise sources. This immunity is because fibers are non-conductors of electricity.
- Fiber systems have greater electrical isolation thus allowing equipment greater protection from damage due to external sources. For example if a part of a network was hit by a lightning pulse then it may damage one of the optical receivers but a high voltage pulse cannot travel along the optical cable and damage sensitive equipment on other parts of the network. They also prevent electrical noise traveling from one part of a network to another, as illustrated in Figure 11.17.
- Fiber cables do not radiate energy and therefore cannot cause interference with other communications systems. This characteristic makes fiber systems ideal for military applications, where the effect of nuclear weapons (EMP-electromagnetic pulse interference) has a devastating effect on conventional communications systems.
- Fiber cables are more resistant to environmental extremes. They operate over a larger temperature variation than copper cables and are affected less by corrosive liquids and gases.
- Fiber cables are safer to install and maintain, as glass and plastic have no electrical currents or voltages associated with them. Optical fibers can be used around volatile liquids and gases without worrying about the risk of explosions or fires. They are also smaller and more lightweight than copper cables.
- Fiber cables are more secure than copper cables and are virtually impossible to tap into without users knowing about it.

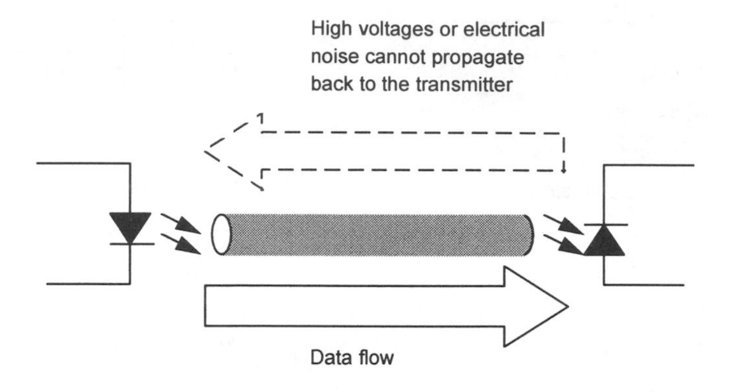
High voltages or electrical
noise cannot propagate
back to the transmitter
Data flow

Figure 11.17 Fiber-optic isolation

11.6 Cabling standards

The main standards agencies for network cabling are:

- Institute of Electrical and Electronic Engineers (IEEE). The IEEE defined cabling standards for Ethernet and Token Ring in their 802.3 and 802.5 standards, respectively.
- Underwriters Laboratories (UL). An independent agency in the United States that tests product safety. In networking it rates twisted-pair cables. The UL has also defined a cable marking system which identifies the cable and whether it is shielded or not. An example marking is 'IEEE 802.3 Coaxial Trunk (**UL**) Type CL2'. This is known as the UL marking system.
- Electrical Industries Association (EIA)/Telecommunications Industry Association (TIA). These organizations have developed many of the networking media standards, especially:
  - EIA/TIA-568. This standard defines UTP cabling, such as Category 1 (Cat-1), Category 2 (Cat-2), Category 3 (Cat-3), Category 4 (Cat-4) and Category 5 (Cat-5) cables. Cat-5 cables offer the best specification of the five types of cables, and are the most often used (although Cat-6 cabling is now become popular). Networking does not use Cat-1 or Cat-2 cable as these are used for voice circuits.
  - EIA/TIA-568A. Updated standard to EIA/TIA-568 which includes fiber-optic cable and link performance.
  - EIA/TIA-569. This standard defines cable interconnections and pathways, such as telecommunications closets, backbones, and so on.

The EIA/TIA have defined most of the important standards relating to networking media, and provide a foundation for multi-product and multi-vendor networks. The EIA/TIA-568A standard defines six main elements of cabling in a LAN (as illustrated in Figure 11.18):

- **Horizontal cabling.** This is defined as the cabling that runs from the telecommunications outlet to a horizontal cross-connect, and is basically the cable which runs from a wiring closet (a central point for cabling) to a workstation, and consists of:

  - *Horizontal cabling*: It may be up to 90m of 4- or 25-pair of Cat- 5 UTP cable.
  - *Telecommunications outlet*: The device in which the horizontal cable terminates at the work area.
  - *Cable terminations.*
  - *Cross-connections*: The device for interconnecting cable runs
  - *Patch Cords*: These are used in points were the network configuration will change frequently.
  - *Transition or consolidation point*: It connects standard horizontal cable to special flat cable designed to run under carpets.

- **Wiring closets.** This is where the horizontal distribution cables are terminated. All recognized types of horizontal cabling are terminated on compatible connecting hardware. Cross connection occurs with jumpers or patch cords to provide flexible connectivity for extending various services to users at the telecommunications outlets.
- **Backbone cabling.** This is defined as the interconnections between wiring closets, entrance facilities and between buildings that are part of the same LAN. It is also known as vertical cabling, and consists of the backbone cables, main and intermediate cross-connects, mechanical terminations, and patch cords or jumpers used for backbone-to-backbone cross-connection. This includes:

  - Vertical connections between floors (risers)
  - Cables between an equipment room and building entrance facilities.
  - Cables between buildings (inter-building).

- **Equipment rooms.** This is a centralized space for telecommunications equipment that serves users in the building. Equipment rooms usually house equipment of higher complexity than telecommunication closets. Any or all of a telecommunications closet may be provided by an equipment room.
- **Work areas.**
- **Entrance facilities.** This is the point where the outside cables and associated hardware are brought into the building interfacing and interfaces with the backbone cabling. The Entrance facility is typically in the basement (where access can be carefully controlled).

The EIA/TIA-568A specification defines for horizontal cabling:

- Four-pair unshielded 100 Ω cable.
- Two-pair shielded 150 Ω cable.
- Two fibers of 62.5/125 μm multimode cable. One cable for sending and the other for receiving.
- 50 Ω coaxial cable. This cable is not recommended for new installations, and is unlikely to be supported in future specifications.

The maximum lengths of horizontal cabling for unshielded twisted-pair cable are given in Figure 11.18. They are 90 m (295 feet) for a horizontal cable run, 3 m (9.8 feet) for a workstation cable and 6 m (20 feet) for a patch cord.

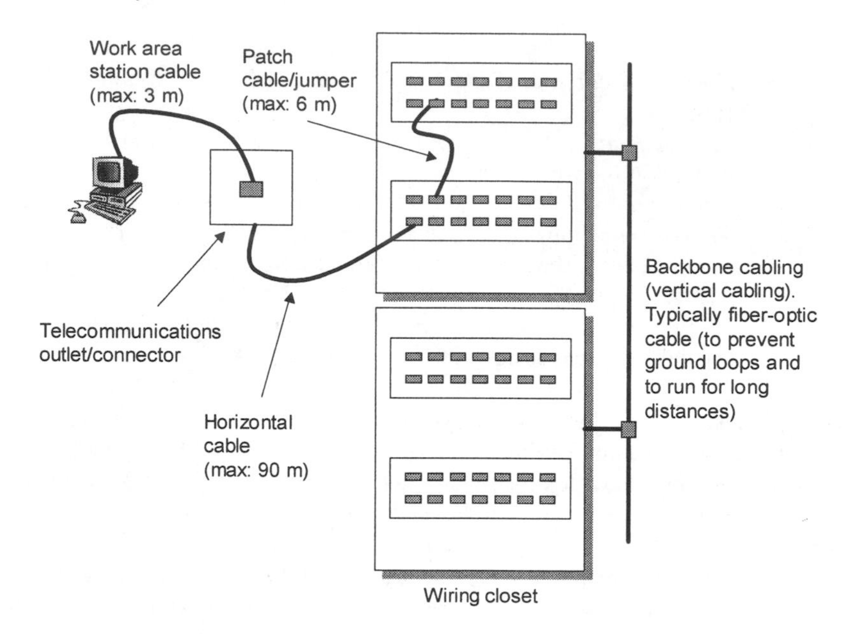

Figure 11.18 Cabling classifications

Most modern Ethernet and Token Ring networks connect onto a hub, which is basically a physically star connected network. As has been seen the maximum length of horizontal cabling is 90 m, along with a 6m patch cord and a workstation connect cable of 3 m, giving the maximum length of cabling from the hub to the workstations of 100 m, as illustrated in Figure 11.19. This gives a maximum coverage area of approximately 200 m by 200 m. It can be seen that workstations outside this area could not be connected using EIA/TIA-568A cabling. This is because attenuation and interference may cause unreliable reception of the data. One solution to this is to add a repeater if the connection is greater than 100 m. This leads to an extended star topology.

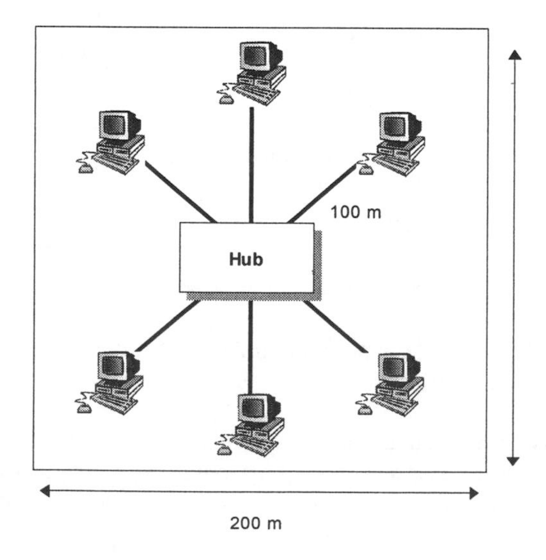

Figure 11.19 Maximum cabling area for a LAN for horizontal cabling runs

11.6.1 RJ-45 jacks

Terminals in an RJ-45 jack:
BLUE: Blue/White and White Blue.
GREEN: Green/White and White/Green.
BROWN: Brown/White and White/Brown.
ORANGE: Orange/White and White/Orange.

The RJ-45 connector is the standard connector used with the EIA/TIA-568A standard for connecting Cat-5 UTP at the telecommunications outlet. It has eight pins into which the four pairs connect into, each of which are color coded (a telephone-type connector only contains four pins), with blue, green, orange and brown terminals, which correspond to the wires found in each of the twisted-pairs used in Cat-5 UTP. The blue wire with a white strip and the white wire with the blue strip (the blue twisted-pair) are connected to the blue terminals, and so on. The wires are connected into the terminals, using a punch tool to force the wires into the terminals and also to strip the sheath away from each of the wires.

Normally the socket is mounted onto the wall so that the RJ-45 plug can connect to it. The two main types of socket are surface mounted and flush mounted. The surface mount technique is the easiest and fastest method of mounting an RJ-45 jack and does not involve cutting into the wall. They are permanently mounted onto the wall with an adhesive-backed box. Once mounted it cannot be moved. If it is required to be moved, the surface mounted RJ-45 jack can use a screw-mounted instead of an adhesive box. Surface-mounted jacks must be used where it is not possible to cut the wall, such as with concrete block walls. A flush mounted unit involves cutting into the wall.

The cable can either be mounted behind a wall (such as in plasterboard walls) or can be mounted to the wall, typically with a raceway, which is a wall-mounted channel with a removable cover. Raceways are made out of plastic or metal and are available in two forms:

- **Decorative.** These have a good-looking finish and are typically used within rooms, but are limited in the number of cables they can enclose (typically a maximum of two cables).
- **Gutter.** Not as good-looking as decorative raceways, but they can be used to hold many more cables. They are typically used in attics.

An important factor is never to run power cables alongside UTP cables, as power cables emit EMI which can corrupt data

DEAR NET-ED

Question: *What are the main rules that I should use when I install network cables?*
Well, the initial installation is important as well installed cable will reduce the likelihood of problems in the future. Cabling problems tend to be one of the top causes of network problems. The rules can be summarized as:

- **Untwisting cables.** The maximum amount of untwisted in a Cat-5 cable is ½ in; this is to maximize the cancellation effect.
- **Cable bend.** The maximum cable bend is 90°.
- **Staples** should never be used as these pierce the outer jacket of the cable. Attach cable ties to cables going on the same path, but never secure them too tightly. If possible secure the cable with cable ties, cable support bars, wire management panels and releasable Velcro straps.
- **Try and minimize outer cable twists and stretching the cable**, and never allow the cable to become kinked, as this changes the characteristic impedance of the cable. The cables within can untwist when stretched.
- **Leave enough cable at each end** so that it can be properly terminated. It is less expensive to add an extra few meters onto the length at either end, than it is to have to re-run the whole cable. Typically, the cable run will have an extra few meters hidden below in the floor, or above in the ceiling, in order to compensate for extra lengths.

signals. If possible cables should never be loose run along ceilings, but should be supported by ladder racks. Cable, which is laid through spaces where air is circulated, must be fire-rated in order to ensure that the cable does not carry the fire from one place to another. Also whenever someone is working within an attic, ceiling or wall, the electrical power must be isolated to prevent electrical shock.

11.6.2 Rules for installing cable

Once the cable is run, it is important for future reference to make a cut sheet, which is a diagram which shows the location of cable runs, and all the rooms and hallways within a building. The EIA/TIA-606 standard defines terminations, media, pathways, spaces, grounding of communications equipment in commercial buildings. It also defines the labels for the end of the cables and their termination (and compliance with the UL969 specification for legibility and adhesion). Labels should be long lasting, and should, if possible, reflect the actual location of the destination connection and be color coded to identify the usage of each of the cables. For example, all the cables that go into Production will be labeled with green labels, while all the cables that go to Sales would be labeled with blue labels. Figure 11.20 shows an example of a cut sheet diagram. In this case the rooms are 100, 101, 102, 103 and 104. Different cables which run to RJ-45 jacks in a single room are labeled with an A, B, C, and so on. In room 100 there are three cables, these are labeled 100A, 100B and 100C. These would be labeled on the hub, on either end of the cables, and also on the faceplate of the RJ-45 jack.

11.6.3 Patch panels and wiring closets

Wiring closets provide a convenient place where cables can be terminated, and is the center of the topology. Along with cables, it is typically the location for routers, bridges, patch panels and hubs. A patch panel is an interconnecting device which connects workstations to hubs and repeaters together using horizontal cabling. It is basically a switchboard with a collection of pin locations (the terminals at the back of the patch connector) and ports (the RJ-45 jack connections on the front of the patch panel). They are typically mounted into a 19 in rack, as illustrated in Figure 11.21. Some larger networks have more than one wiring closet, and have a main distribution facility (MDF) and other intermediate distribution facilities (IDFs), which connect to the MDF.

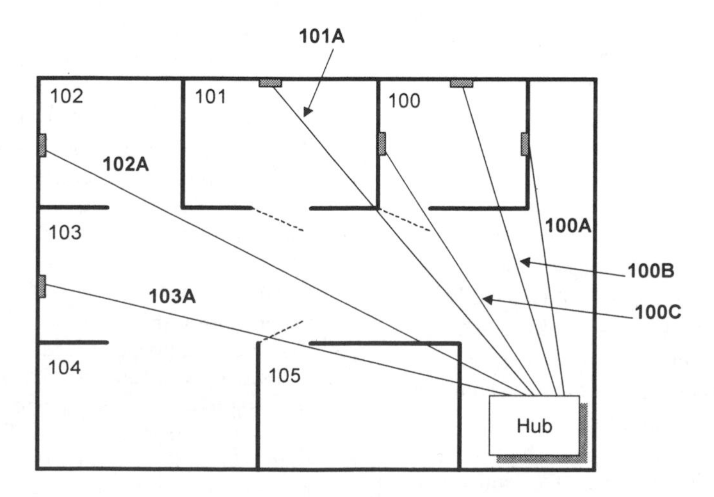

Figure 11.20 Cut sheets example

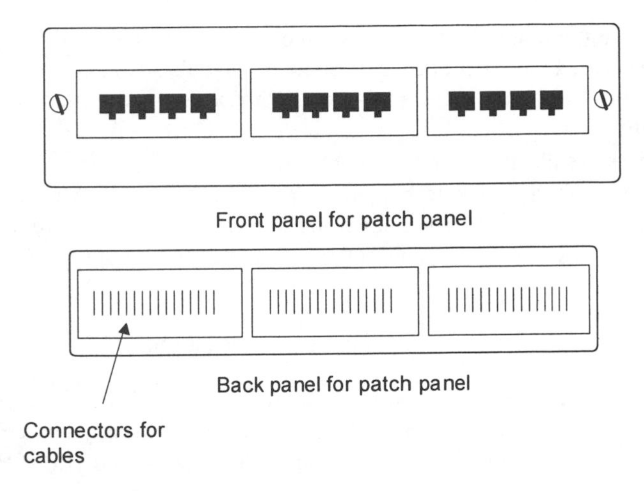

Figure 11.21 Patch panel front and back

11.6.4 Cable testing

Testing of network cables is important not only in detecting faults but also in determining if the cabling is conforming to the required specification. Both the IEEE and the EIA/TIA have tests for cabling after installation. These tests provide the baseline for the network. Along with a wire map (which shows the location of all the cable connections), an important device in testing the network is a cable tester, which determines:

- **Cable distance.** This is an important measure to verify that the connected cables are not too long for the given specification. Time domain reflectometers (TDRs) can measure distance, as they send a pulse down the cable which is then reflected from an open- or short-circuit connection at the far end of the cable. The TDR measures the time that the pulse takes to come back and as it can approximate the speed of propagation, it can determine the cable distance (speed of propagation multiplied by time). This device is typically accurate to within a meter. TDRs can also be used to determine the distance to a cable break (an open circuit) or a cable short (a short circuit).
- **Signal attenuation**. To measure attenuation a signal inject is applied at the far end of the cable, and the signal level is measured at the near end. From this, attenuation is calculated. Typically this is measured at various frequencies, such as 100 MHz for Cat-5 cable (as it must support up to 100 Mbps). The EIA/TIA-568A specification defines the maximum amount of attenuation on the cable.
- **Near-end crosstalk**. A typical problem which causes near-end crosstalk is when the pairs have become untwisted (such as when they have been pulled too tightly or have been untwisted too far where they are terminated), and the cable tester measures the crosstalk. On detecting large amounts of near-end crosstalk, the wires should be visually

DEAR NET-ED

Question: *I use a Dial-up connection from home, and an Ethernet connection at work. Is there any difference in the way that my applications operate?*

None at all (when you use TCP/IP communications). TCP/IP provides the interface between the networking technology and the application program, and have been designed so that the networking type is transparent to the application program, so, for example, it doesn't matter to a WWW browser that you connect to a modem or over a LAN.

inspected for any problems. A typical source of near-end crosstalk is split pairs. This is because the twisted-pairs do not carry opposite signals and thus the cancellation effect does not occur. The signal will thus interact with other pairs. Figure 11.22 shows an example of split pairs.

- **Crossed-pairs.** As illustrated in Figure 11.23.
- **External noise.** To detect external noise, all cables should be disconnected from the computer equipment and the noise level measured. If high levels are detected the source can be found by disconnecting each potential source, one by one, and then measuring the new level.

There are basically three main types of patch cable connection:

- **RJ-45 straight-through patch cable** (as shown on the left-hand side of Figure 11.24). This is the normal connection, where the cables connect directly from Pin 1 of the RJ-45 connector to Pin 1 of the other RJ-45 connector, and so on. In 10BASE-T only the orange and green pairs are used). The crossover between the transmit (TxData) and receive (RxData) occurs within the hub.
- **RJ-45 crossover patch cable** (as shown on the right-hand side of Figure 11.24). This is used when connecting between two workstations without the use of a hub or when connecting between hubs. It has a crossover between the orange (TxData) and the green (RxData).
- **RJ-11 connection.** An RJ-11 connector has six connectors, but only the middle four are used, and is typically used in telephone applications. In telephone applications, the colors which connect to these are BRGY (black-red-green-yellow) on one end, and YGRB on the other (the colors are reversed on one end of the cable). In networking applications the connections are: White/Orange (W/O), Blue (B), White/Blue (W/B) and Orange (O) and on the other side it is reversed. The blue pair make pair 1 and the orange pair make pair 2. White/Blue is a replacement for Green, Blue (/White) is a replacement for Red, White/Orange is a replacement for Black, and Orange (/White) is a replacement for Yellow.

The color codes for the RJ-45 connectors are defined in T-568A and T-568B standards (which are both part of the TIA/EIA-568A standard). A straight-through connection will work for either type (as Ethernet is color-blind), but the **T-568A** standard is recommended for new installations. The T-568A standard is an older standard (which was defined by AT&T and was previously known as 258A). Of course, a crossover cable can be made by wiring one end with T-568A and the other with T-568B.

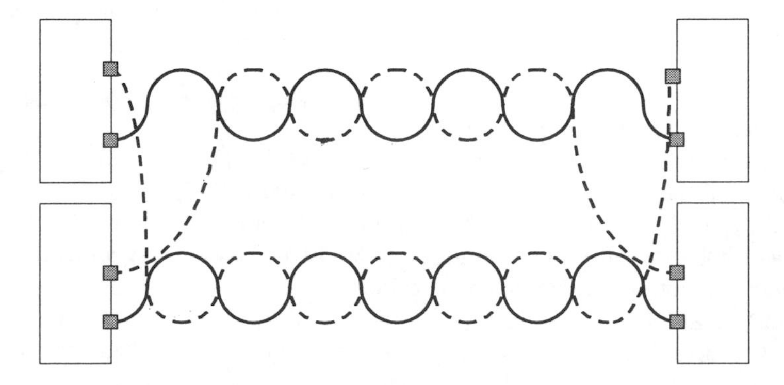

Figure 11.22 Split pairs

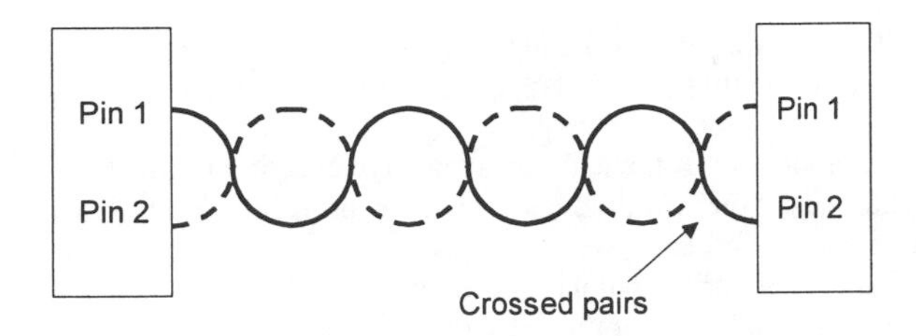

Figure 11.23 Crossed pairs

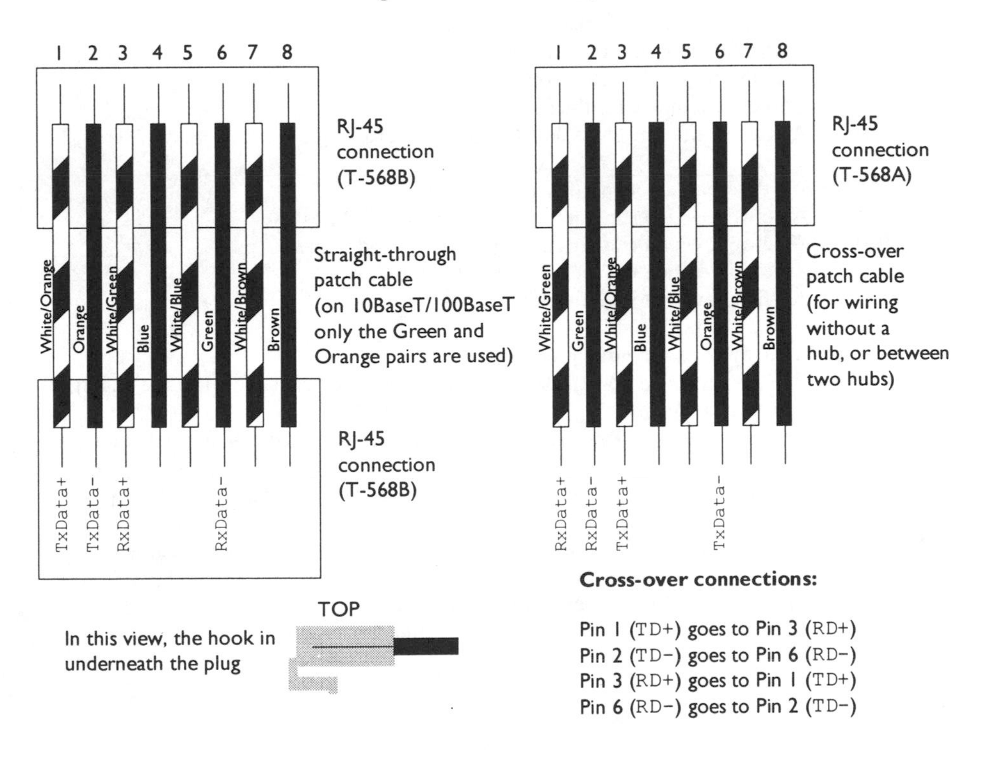

Figure 11.24 T-568A and T-568B connections

11.7 Exercises

The following questions are multiple choice. Please select from a–d.

11.7.1 The cable type which offers the highest bit rate is:
(a) Fiber-optic cable
(b) Twisted-pair cable
(c) Coaxial cable
(d) Untwisted-pair cable

11.7.2 Which of the following is the main disadvantage of a star network:
(a) That the data transmitted between the central server and the node is relatively high compared to other network topologies
(b) That the network is reliant on a central server

(c) All nodes compete for the network
(d) Nodes can only transmit data once they have a token

11.7.3 Which of the following is the main disadvantage of a ring network:
(a) That the data transmitted between the central server and the node is relatively high compared to other network topologies
(b) That the network is reliant on a central server
(c) All nodes compete for the network
(d) A break in the ring stops data from being transmitted

11.7.4 Which of the following is the main disadvantage of a bus network:
(a) Nodes can only transmit data once they have a token
(b) That the network is reliant on a central server
(c) All nodes compete for the network
(d) A break in the ring stops data from being transmitted

11.7.5 Which best describes a network segment?
(a) A network which contains a range of network addresses
(b) A network which contains a range of station addresses
(c) A network which has the same network topology
(d) A network which is bounded by routers or bridges

11.7.6 Which of the following is **not** a problem when using a bridge?
(a) Bridges become a bottleneck and actually slow down traffic when the data traffic becomes too great
(b) Bridges allow broadcast storms if too many broadcasts are sent out its multiple ports
(c) Bridges are not very efficient in large networks as they indiscriminately pass data frames from one network segment to all other segments
(d) Bridges reduce the number of collisions between interconnected segments

11.7.7 On a network which address does a bridge route with:
(a) IP address (b) Interrupt address
(c) MAC address (d) Source address

11.7.8 On a network which address does a router route with:
(a) IP address (b) Interrupt address
(c) MAC address (d) Source address

11.7.9 Which term is used for both multi-port repeaters and for the center of a star topology network:
(a) Bridge (b) Port
(c) Hub (d) File server

11.7.10 Which best describes a collision in an Ethernet network:
(a) The result of two nodes having the same IP address
(b) The result of two nodes transmitting simultaneously
(c) The result of two nodes having the same MAC address
(d) Two nodes with different network protocols

11.7.11 What is the name given to the retransmission delay after a collision:
(a) Retrans time (b) Collision timeout
(c) Interrupt time (d) Backoff

11.7.12 If a bridge detects that a destination address in a data frame is on the same network segment as the source:

(a) It passes the data frames between two network segments
(b) It forwards the data frame to all other network segments
(c) It stores the data frame for future transmission
(d) It does not forward the data frame to other network segments

11.7.13 Which of the following describes a broadcast storm:
(c) The area bounded by a network segment
(a) Data frames sent to all the nodes on a network segment
(b) An event where many broadcasts are sent simultaneously across the entire network.
(d) The area that defines the maximum propagation distance for a broadcast

11.7.14 Which of the following best describes a router:
(a) They are multiport repeater and at the center of a star topology network
(b) Amplifies the data signals
(c) Forwards data frame using the destination MAC addresses
(d) Forwards data packets using the destination network layer address

11.7.15 Which device solves excessive broadcast traffic:
(a) Bridge (b) Router
(c) Hub (d) File server

11.7.16 How does a router differ from a bridge:
(a) A bridge uses the network address, while a router uses the station address
(a) A bridge uses the station address, while a router uses the physical address
(c) Routers extend network segments, whereas bridges limit them
(d) Bridges modify the destination station address of the data frame, while routers do not

11.7.17 Which of the following is **not** true about bridges and routers?
(a) Routers allow access to the Internet, while bridges segment networks
(b) Routers operate at the network layer, while bridges operate at the data link layer
(c) Routers use IP addresses, whereas bridges use MAC addresses
(d) Routers extend network segment distances while bridges limit them

11.7.18 Which of the following should not be used when installing cables:
(a) Maximum cable bends is 90° (b) Use Velcro straps with cables
(c) Never overstretch cable
(d) Leave a maximum of 2 inches at either end

11.7.19 What type of cable should be used between buildings:
(a) UTP (b) STP
(c) Coaxial cable (d) Fiber-optic cable

11.7.20 What type of cable connects between wiring closets:
(a) Backbone cable (b) Horizontal cable
(c) Workstation connect (d) Patch cable

11.7.21 What type of cable connects between ports on a wiring closet:
(a) Backbone cable (b) Horizontal cable
(c) Workstation connect (d) Patch cable

11.7.22 What type of cable connects a computer to a wall mounted unit:
(a) Backbone cable (b) Horizontal cable
(c) Workstation connect (d) Patch cable

11.7.23 Which cable type is likely not to be recommended in future definitions of the EIA/TIA-568A specification for horizontal cabling:
- (a) Four-pair unshielded 100 Ω cable
- (b) Two-pair shielded 150 Ω cable
- (c) Two fibers of 62.5/125 μm multimode cable
- (d) 50 Ω coaxial cable.

11.7.24 When is a cross-connect patch cable used:
- (a) To connect a workstation to a hub
- (b) To connect over long distances
- (c) To connect workstations without the need for a hub
- (d) To connect to a router

11.7.25 Which two pairs are used in 10BaseT and 100BaseT:
- (a) Orange and blue
- (b) Orange and green
- (c) Brown and Red
- (d) Brown and blue

11.7.26 What is made when a cable uses T-568A at one end and T-568B at the other:
- (a) A cross-connect cable
- (b) A straight-through cable
- (c) A telephone cable
- (d) A backbone cable

11.7.27 State the main rules that should be used when installing a cable.

11.7.28 Clearly identify the main advantages that fiber-optic cables have over copper cables.

11.7.29 Show that the maximum cabling area for a LAN for horizontal cabling runs is approximately 200 m.

11.7.30 Locate some networking cables and determine if they contain the UL markings. If they do, determine the cable types (such as Cat-5). Also check the cable connector colors. Does it comply with T-568A or T-568B?

11.7.31 Locate a LAN, and determine the location of wall mounted units, cable types, wiring closets, hubs, and so on. What type of networking technology is used?

11.8 Note from the Author

Well, I hope you enjoyed our little delve in the practical world of networking. I apologize for confusing you a little, saying that an Ethernet hub is a physical star topology, but a logical bus. This, though, is an important concept as hubs provide for a central point of failure (as star networks do).

Twisted-pair cables have been a godsend for networks. They allow easy connect and deletion to/from a network. Their self-screen allows for reduced coupling with other nearby systems. This can be a serious problem in noisy environments, and may cause errors in the received bit pattern.

In this chapter we were introduced to some of the main characters in the Networking Hall of Fame. These are have made their way there because of their sheer usefulness. So here they are:

1. *TCP/IP*
2. *Ethernet*
3. *Electronic mail*
4. *Routers*

5. *Twisted-pair cable (and RJ-45 connectors)*
6. *Servers (especially WWW servers)*
7. *Fiber-optic cables*
8. *UNIX operating system (and its related networking protocols, such as NFS, TELNET, FTP, and so on).*
9. *Domain names*
10. *Hubs/switches/MAUs*

Ethernet

12.1 Introduction

The two main winners in networking technologies are likely to be Ethernet and ATM. Ethernet will be a winner because of its **popularity**, **reliability**, **compatibility**, **simplicity**, **ease-of-use** and **upgradeability**. These six simple words can overcome any great technological advancement, as Microsoft and Intel have found, in operating systems and in processors, respectively. Ethernet successfully beat off networking technologies, such as Token Ring, but its major weakness is that it does not cope well when the required bandwidth approaches the maximum bandwidth. This is due to the contentious nature of Ethernet, where nodes must contend to get access to the network. ATM and FDDI looked to be the solution for large-scale backbone-based networks, but Ethernet has a final trump card to play. Every time it looks like losing the battle, it improves itself ten-fold.

Another problem with Ethernet is that it cannot guarantee bandwidth, and thus does not cope well with real-time traffic, such as video and audio. These disadvantages have generally been overcome by simply increasing the bandwidth each time there is an increased requirement. It has thus gone from 10 Mbps to 100 Mbps, and now to 1 Gbps. The other trump card that Ethernet has is its popularity. It is estimated that Ethernet accounts for more than 85% of all installed network connections, which is well over 120 million computers. The other main types are: Token Ring, Fiber Distributed Data Interface (FDDI) and Asynchronous Transfer Mode (ATM). To make a network, a network protocol must sit above the networking technology. The most popular of these protocols are TCP/IP (for most Internet-based traffic, and on UNIX networks), SPX/IPX (on Novell NetWare networks) and NetBEUI (on Microsoft networks).

DEAR NET-ED

Question: *I would like some discipline in the design of my network. Thus, what are the main design steps?*

It is important to properly design your network, as incorrect planning can cause problems in the future. The basic steps are:

- **Analyse requirements.** This involves understanding and specifying the requirements of the network, especially its major uses. If possible future plans should be incorporated. One of the key features is the bandwidth requirements and the size of the network.
- **Develop LAN structure.** This step involves developing a LAN structure for these requirements. Typically in organizational networks this will be based on a star topology using Ethernet hubs/switches.
- **Set up addressing and routing.** The final step involves setting up IP addresses and subnets to add structure.

Several factors have contributed to making Ethernet the most popular network technology, these include:

- **Simplicity.** Easy to plan and cheap to install. The introduction of network hubs and twisted-pair cables has made Ethernet networks easy to connect to. It also has cheap and well-supported network components, such as network interface cards (NICs) and connectors (BNC and RJ-45).
- **Reliability.** Well-proven technology, which is fairly robust and reliable.

- **Ease-of-use.** It is simple to add and remove computers to/from the network.
- **Upgradeability and compatibility.** Ethernet has evolved from 10 Mbps, to 100 Mbps (Fast Ethernet, in 1985) and now to 1 Gbps (Gigabit Ethernet, in 1998). All three Ethernet speeds use the same basic data frame format (IEEE 802.3), have full-duplex operation and have the same flow control methods.
- **Popularity.** Supported by most software and hardware systems.

Ethernet was initially developed by DEC, Intel and the Xerox Corporation and has since been standardized by the IEEE 802 committee (as IEEE 802.3)

A major problem with Ethernet is that, because computers must contend to get access to the network, there is no guarantee that they will get access within a given time. This contention also causes problems when two computers try to communicate at the same time, they must both back off and no data can be transmitted. In its standard form Ethernet allows a bit rate of 10 Mbps, but new standards for fast Ethernet systems minimize the problems of contention and also increase the bit rate to 100 Mbps (and even 1 Gbps). Ethernet uses coaxial, fiber-optic or twisted-pair cable.

Ethernet uses a shared-media, bus-type network topology where all nodes share a common bus. These nodes must then contend for access to the network as only one node can communicate at a time. Data is then transmitted in frames which contain the MAC (media access control) source and destination addresses of the sending and receiving node, respectively. The local shared media is known as a segment, and each node on the network must monitor the segment and copy any frames addressed to it.

Ethernet uses carrier sense, multiple access with collision detection (CSMA/CD). On a

DEAR NET-ED

Question: *How do I go about gathering the data to define the requirements of a new network?*

The most important information that is required is the structure of the organization and how information flows between the units, as the designed network is likely to reflect this structure. The information will include:

- Understand the current network (if one exists) especially its strengths and weaknesses.
- Gather information on geographical locations.
- Determine current applications, and future plans for each site and for the organization.
- Develop organizational contacts. These will be the important people who will be involved in the development of the network. A mixture of technical and business skills always helps. Technical people tend to be driven by technology ('it should transfer files faster', 'it's easier to install', and so on), whereas business people tend to be driven by applications ('I just want access to a good spreadsheet', 'I want to be able to send e-mails to anyone in the company', and so on). It is also important to get someone involved who has experience of legal matters, and/or someone involved in Personnel matters.
- Determine the requirements for external network connections. This is an important decision as the security of the whole network may depend on the choices made on the external connections. Many large companies have a single point of connection to the external Internet as this allows organizations to properly manage internal and external connections to the Internet.
- Determine key objectives of the organization, especially related to mission-critical data and mission-critical operations. These should have top priority over other parts of the network. For example a hospital would declare its ambulance service as a mission-critical unit, whereas the cuts and bruises unit (if there was one) would not be.
- Determine who is in control of information services. This may be distributed over the organization or over centralized in an MIS (Management Information Service) unit.

CSMA/CD network, nodes monitor the bus (or Ether) to determine if it is busy. A node wishing to send data waits for an idle condition and then transmits its data frame. Unfortunately, collisions can occur when two nodes transmit at the same time, thus nodes must monitor the cable when they transmit. When a collision occurs, the nodes that caused the collision stop transmitting data frames and transmit a jamming signal. This informs all nodes on the network that a collision has occurred. Each of the nodes involved in the collision then wait a random period of time before attempting a re-transmission. As each node has a random delay time then there can be a prioritization of the nodes on the network.

Each node on the network must be able to detect collisions and be capable of transmitting and receiving simultaneously. These nodes either connect onto a common Ethernet connection or can connect to an Ethernet hub. Nodes thus contend for the network and are not guaranteed access to it, as illustrated in Figure 12.1. Collisions generally slow the network.

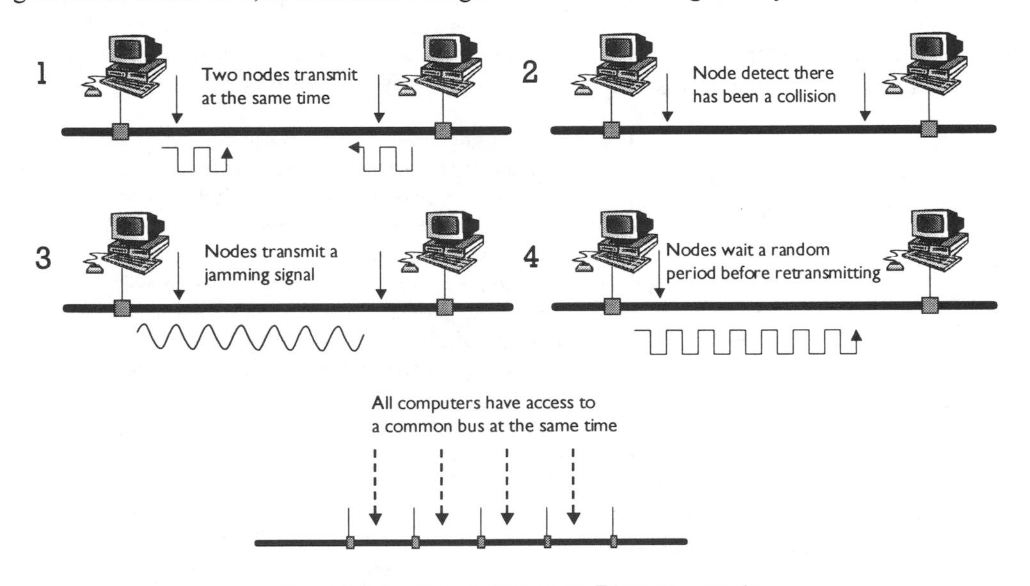

Figure 12.1 Connections to an Ethernet network

12.2 IEEE standards

The IEEE is the main standards organization for LANs and it refers to the standard for Ethernet as CSMA/CD. Figure 12.2 shows how the IEEE standards for CSMA/CD fit into the OSI model. The two layers of the IEEE standards correspond to the physical and data link layers of the OSI model. On Ethernet networks, most hardware complies with IEEE 802.3 standard. The MAC layer allows many nodes to share a single communication channel. It also adds the start and end frame delimiters, error detection bits, access control information, and source and destination addresses. Each frame also has an error detection scheme known as cyclic redundancy check (CRC).

12.2.1 Ethernet II

Most currently available systems implement either Ethernet II or IEEE 802.3 (although most networks are now defined as being IEEE 802.3 compliant). An Ethernet II frame is similar to the IEEE 802.3 frame; it consists of 8 bytes of preamble, 6 bytes of destination address, 6 bytes of source address, 2 bytes of frame type, between 46 and 1500 bytes of data, and 4 bytes of the frame check sequence field.

When the protocol is IPX/SPX the type field contains the bit pattern 1000 0001 0011 0111, but when the protocol is TCP/IP the type field contains 0000 1000 0000 0000.

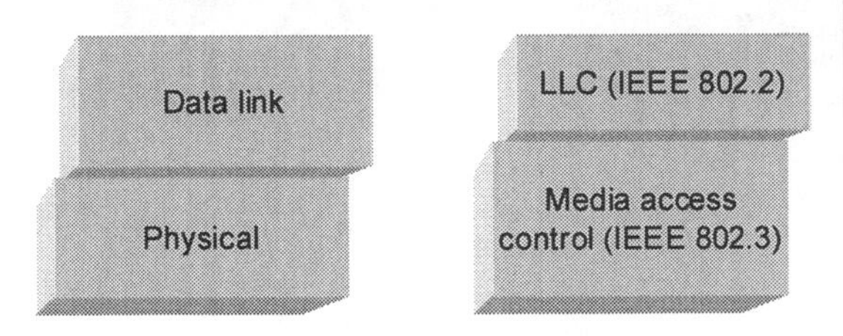

Figure 12.2 Standards for IEEE 802 LANs

DEAR NET-ED

Question: *In your previous reply you said that I should get someone involved with a legal or with a Personnel Department background. Why? Surely they will hold back the project.*

Not at all. A network design team which only involves technical people may not understand the complete requirements of the customer. In organizations, the customer will include the organization, itself, the workers, and the actual customers of the organization. Every organization must also meet its legal and moral responsibilities.

Networks and the Internet have caused all sorts of legal problems, which are extremely difficult to fully understand. A badly planned network which breaches civil liberties can be very costly in the future. A major decision is whether users are monitored in the applications they are running, and the files that they are accessing (and even how fast they typed). These could be used by management to determine how efficiently a worker is completing their tasks. Another major problem is users who access the Internet, either for mindless surfing, or for illegal downloads. Legal-minded individuals will be able to sort out these personal matters, especially when it relates to who is allowed access to information on certain users.

12.3 Ethernet – media access control (MAC) layer

When sending data the MAC layer takes the information from the LLC link layer. Figure 12.3 shows the IEEE 802.3 frame format. It contains 2 or 6 bytes for the source and destination addresses (16 or 48 bits each), 4 bytes for the CRC (32 bits) and 2 bytes for the LLC length (16 bits). The LLC part may be up to 1500 bytes long. The preamble and delay components define the start and end of the frame. The initial preamble and start delimiter are, in total, 8 bytes long and the delay component is a minimum of 96 bytes long.

A 7-byte preamble precedes the Ethernet 802.3 frame. Each byte of the preamble has a fixed binary pattern of `10101010` and each node on the network uses it to synchronize their clock and transmission timings. It also informs nodes that a frame is to be sent and for them to check the destination address in the frame.

At the end of the frame there is a 96-bit delay period, which provides the minimum delay between two frames. This slot time delay allows for the worst-case network propagation delay. The start delimiter field (SDF) is a single byte (or octet) of `10101011`, and follows the preamble and identifies that there is a valid frame being transmitted. Most Ethernet systems use a 48-bit MAC address for the sending and receiving node. Each Ethernet node has a unique MAC address, which is normally defined as hexadecimal digits, such as:

4C - 31 - 22 - 10 - F1 - 32 or 4C31 : 2210: F132.

A 48-bit address field allows 2^{48} different addresses (or approximately 281 474 976 710 000 different addresses). The LLC length field defines whether the frame contains information or it can be used to define the number of bytes in the logical link field. The logical link field can contain up to 1500 bytes of information and has a minimum of 46 bytes. If the information is greater than this upper limit then multiple frames are sent. Also, if the field is less than the lower limit then it is padded with extra redundant bits.

The 32-bit frame check sequence (or FCS) is an error detection scheme. It is used to determine transmission errors and is often referred to as a cyclic redundancy check (CRC) or simply as a checksum.

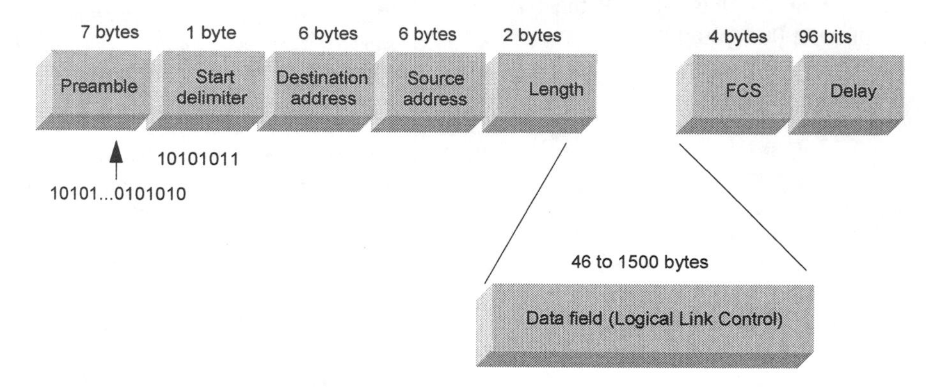

Figure 12.3 IEEE 802.3 frame format

If the transmission rate is 10 Mbps, the time for one bit to be transmitted will be:

$$T = \frac{1}{\text{bit rate}} = \frac{1}{10 \times 10^6} s = 100\,\text{ns}$$

Thus the maximum and minimum times to transmit a frame will be:

$$T_{\text{max}} = (7+1+6+6+2+1500+4+12) \times 8 \times 100\,\text{ns} = 1.2\,\text{ms}$$
$$T_{\text{min}} = (7+1+6+6+2+46+4+12) \times 8 \times 100\,\text{ns} = 67\mu\text{s}$$

It may be assumed that an electrical signal propagates at about half the speed of light ($c = 3 \times 10^8$ m/s). Thus, the time for a bit to propagate a distance of 500 m is:

$$T_{500m} = \frac{\text{dist}}{\text{speed}} = \frac{500}{1.5 \times 10^8} = 3.33\mu\text{s}$$

by which time, the number of bits transmitted will be:

$$\text{Number of bits transmitted} = \frac{T_{500m}}{T_{bit}} = \frac{3.33\mu\text{s}}{100\,\text{ns}} = 33.33$$

Thus, if two nodes are separated by 500 m then it will take more than 33 bits to be transmitted before a node can determine if there has been a collision on the line, as illustrated in Figure 12.4 (it will also take twice as long before a collision is detected by the transmitting node). If the propagation speed is less that this, it will take even longer. This shows the need for the preamble and the requirement for specifying a maximum segment length.

12.3.1 MAC address format

The MAC address is split into two parts. The first 24 bits identifies the manufacturer of the network card, and the second 24 bits identifies the serial number of the NIC. Example manufacturer codes are:

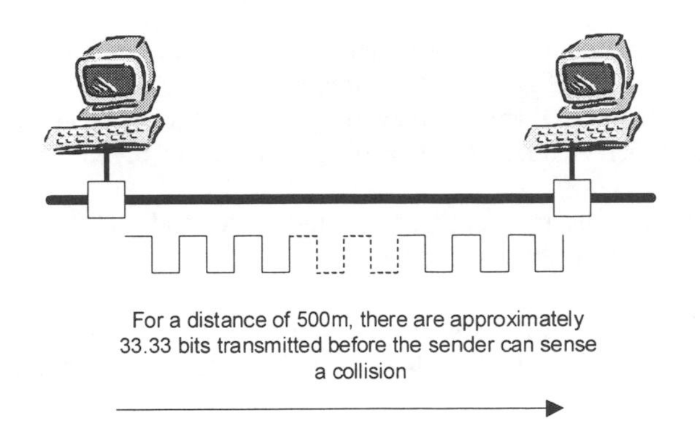

For a distance of 500m, there are approximately
33.33 bits transmitted before the sender can sense
a collision

Figure 12.4 Bits transmitted before a collision is detected

```
00000C  Cisco          00000E  Fujitsu            00000F  NeXT
00005A  S & Koch       00005E  IANA               000065  Network General
00AA00  Intel          020701  Racal InterLan     02608C  3Com
```

An example of an Ethernet card connected to a notebook (via the PCMCIA slot) is given next. It can be seen that the MAC address is 00-60-63-01-CF-15, and the IP address is 146.176.164.143. In Microsoft Windows, the command used to determine these settings is IPCONFIG (or WINIPCFG can be used, as given in Figure 12.29). The definition of the default gateway is important as this allows the node to communicate with nodes outside the current network segment.

```
C> ipconfig /all
Description . . . : Psion Combo 91C95 PCMCIA Ethernet Adapter
Physical Address. : 00-60-63-01-CF-15
IP Address. . . . : 146.176.164.143
Subnet Mask . . . : 255.255.255.0
Default Gateway . : 146.176.164.254
```

12.4 IEEE 802.2 and Ethernet SNAP

The LLC is embedded in the Ethernet frame and is defined by the IEEE 802.2 standard. Figure 12.5 illustrates how the LLC fields are inserted into the IEEE 802.3 frame. The DSAP and SSAP fields define the types of network protocol used. A SAP code of 1110 0000 identifies the network operating system layer as NetWare, whereas 0000 0110 identifies the TCP/IP protocol. The IEEE issues these SAP numbers. The control field is, among other things, for the sequencing of frames.

In some cases it was difficult to modify networks to be IEEE 802-compliant. Thus an alternative method was to identify the network protocol, known as Ethernet SNAP

DEAR NET-ED

Question: *What are the main requirements in designing a network?*

Business requirements.
Technical requirements. The main issues are media contention, reducing excessive broadcasts (routing tables, ARP requests, and so on), backbone requirements, support for real-time traffic and addressing issues.
Performance requirements. This is likely to involve a network load requirement analysis for the typical loading on the network, and also for the worst-case traffic loading. This will determine the requirement for client/server architectures. An analysis should also be made for the impact of new workstations being added to the network. It should also involve an analysis of the requirements for application software, especially in its bandwidth requirements. Multimedia applications tend to have a large bandwidth requirement, along with centralized database applications and file servers.
New application requirements.
Availability requirements. This defines the usefulness of the network, such as response time, resource availability, and so on.

(Subnetwork Access Protocol). This was defined to ease the transition to the IEEE 802.2 standard and is illustrated in Figure 12.6. It simply adds an extra two fields to the LLC field to define an organization ID and a network layer identifier. NetWare allows for either Ethernet SNAP or Ethernet 802.2 (as Novell used Ethernet SNAP to translate to Ethernet 802.2).

Non-compliant protocols are identified with the DSAP and SSAP code of 1010 1010, and a control code of 0000 0011. After these fields:

- Organization ID which indicates where the company that developed the embedded protocol belongs. If this field contains all zeros it indicates a non-company-specific generic Ethernet frame.
- EtherType field which defines the networking protocol. A TCP/IP protocol uses 0000 1000 0000 0000 for TCP/IP, while NetWare uses 1000 0001 0011 0111. NetWare frames adhering to this specification are known as NetWare 802.2 SNAP.

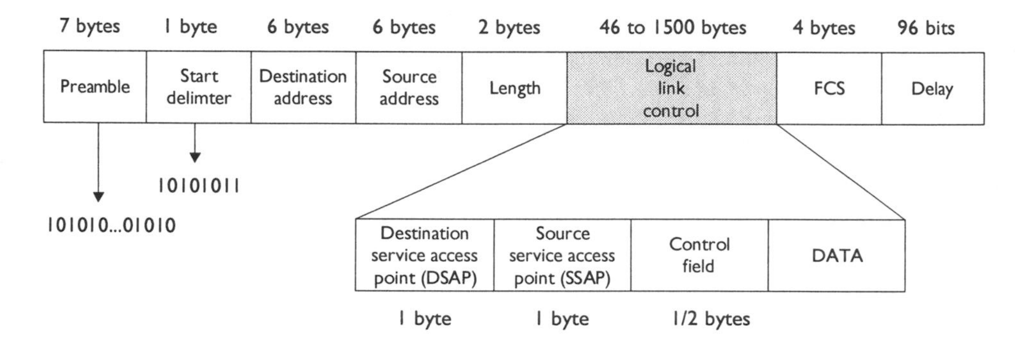

Figure 12.5 Ethernet IEEE 802.3 frame with LLC

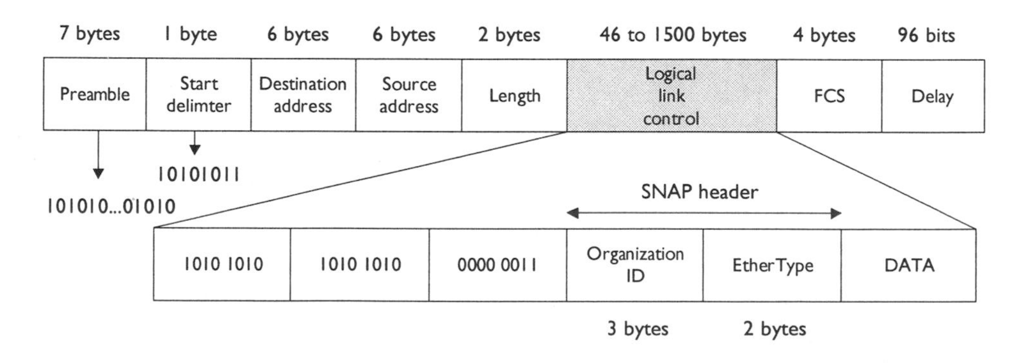

Figure 12.6 Ethernet IEEE 802.3 frame with LLC containing SNAP header

12.4.1 LLC protocol

The 802.3 frame provides some of the data link layer functions, such as node addressing (source and destination MAC addresses), the addition of framing bits (the preamble) and error control (the FCS). The rest of the functions of the data link layer are performed with the control field of the LLC field; these functions are:

- **Flow and error control.** Each data frame sent has a frame number. A control frame is sent from the destination to a source node informing it that it has or has not received the frames correctly.
- **Sequencing of data.** Large amounts of data are sliced and sent with frame numbers. The spliced data is then reassembled at the destination node.

DEAR NET-ED

Question: *How do I try to limit the number of collisions on an Ethernet segment?*

Ethernet collisions occur when two nodes try and transmit onto a network segment at the same time. When the transmitting nodes detect this, they transmit a jamming signal to the rest of the network. All the other nodes on the network detect this, and wait for one of the two colliding nodes to get access onto the network segment. These collisions reduce the overall bandwidth of the network segment. An important concept is the collision domain, which defines the physical distance by which a collision is propagated. Repeaters and hubs propagate collisions, but switches, bridges and routers do not. Thus if you want to reduce the amount of collision insert either a switch, a router or a bridge in a network segment.

Figure 12.7 shows the basic format of the LLC frame. There are three principal types of frame: information, supervisory and unnumbered. An information frame contains data, a supervisory frame is used for acknowledgment and flow control, and an unnumbered frame is used for control purposes. The first two bits of the control field determine which type of frame it is. If they are 0X (where X is a don't care) then it is an information frame, 10 specifies a supervisory frame and 11 specifies an unnumbered frame.

An information frame contains a send sequence number in the control field which ranges from 0 to 127. Each information frame has a consecutive number, N(S) (note that there is a roll-over from frame 127 to frame 0). The destination node acknowledges that it has received the frames by sending a supervisory frame. The 2-bit S-bit field specifies the function of the supervisory frame. This can either be set to Receiver Ready (RR), Receiver Not Ready (RNR) or Reject (REJ). If an RNR function is set then the destination node acknowledges that all frames up to the number stored in the receive sequence number N(R) field were received correctly. An RNR function also acknowledges the frames up to the number N(R), but informs the source node that the destination node wishes to stop communicating. The REJ function specifies that frame N(R) has been rejected and all other frames up to N(R) are acknowledged.

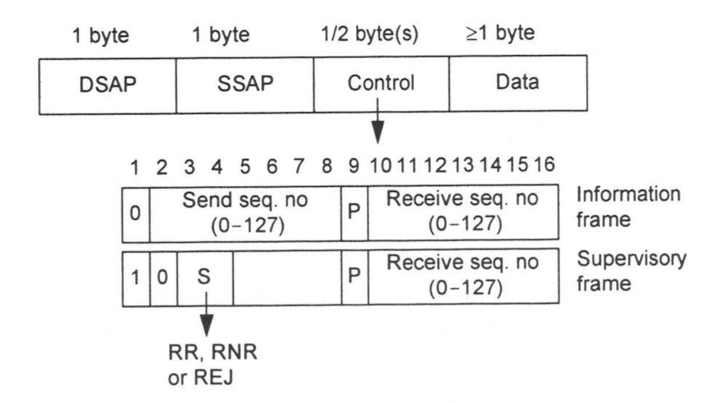

Figure 12.7 LLC frame format

12.5 OSI and the IEEE 802.3 standard

DEAR NET-ED

Question: *I've analysed the traffic on the network, and I've found that a large portion of the network traffic is related to broadcasts. How can I reduce their effect?*

Broadcasts are sent out when a node wants help from other nodes. Typically this happens when a node requires the MAC address for a known network address. The broadcast domain defines the physical distance by which a broadcast will be propagated. Hub, bridges and switches all propagate broadcasts, but routers do not. Thus, if you want to reduce the number of broadcasts on a network segment, insert a router, and it will intelligently route data packets into and out of a network segment without too many broadcasts (as the router handles external data routing).

Ethernet fits into the data link and the physical layer of the OSI model. These two layers only deal with the hardware of the network. The data link layer splits into two parts: the LLC and the MAC layer.

The IEEE 802.3 standard splits into three sub-layers:

- MAC (media access control).
- Physical signaling (PLS).
- Physical media attachment (PMA).

The interface between PLS and PMA is called the attachment unit interface (AUI) and the interface between PMA and the transmission media is called the media dependent interface (MDI). This grouping into modules allows Ethernet to be very flexible and to support a number of bit rates, signaling methods and media types. Figure 12.8 illustrates how the layers interconnect.

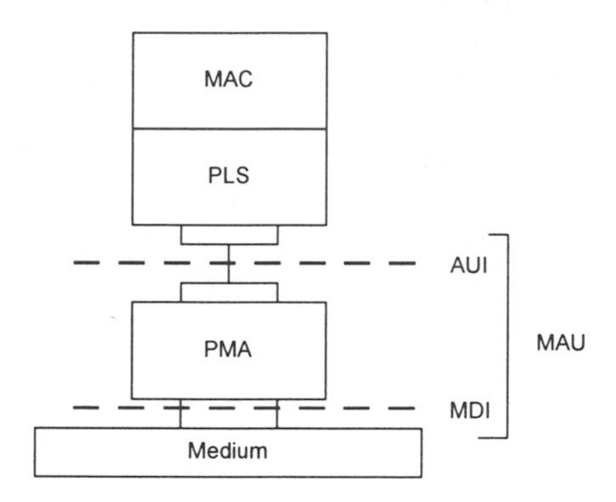

Figure 12.8 Organization of the IEEE 802.3 standard

12.5.1 Media access control (MAC)

CSMA/CD is implemented in the MAC layer. The functions of the MAC layer are:

- **When sending frames:** receive frames from LLC; control whether the data fills the LLC data field, if not add redundant bits; make the number of bytes an integer, and calculate the FCS; add the preamble, SFD and address fields to the frame; send the frame to the PLS in a serial bit stream.
- **When receiving frames:** receive one frame at a time from the PLS in a serial bit stream; check whether the destination address is the same as the local node; ensure the frame contains an integer number of bytes and the FCS is correct; remove the preamble, SFD, address fields, FCS and remove redundant bits from the LLC data field; send the data to the LLC.

- **Avoid collisions** when transmitting frames and keep the right distance between frames by not sending when another node is sending; when the medium is free, wait a specified period before starting to transmit.
- **Handle any collision** that appears by sending a jam signal; generate a random number and back off from sending during that random time.

12.5.2 Physical signaling (PLS) and physical medium attachment (PMA)

PLS defines transmission rates, types of encoding/decoding and signaling methods. In PMA a further definition of the transmission media is accomplished, such as coaxial, fiber or twisted-pair. PMA and MDI together form the media attachment unit (MAU), often known as the transceiver.

DEAR NET-ED

Question: *Can I use the OSI model to design my network?*

Yes. The OSI model can split the network up into identifiable areas. These are:

- **Physical layer.** Network media (typically Cat-5 cable or fiber-optic cable), hubs and repeaters. Cables are normally run conforming to the EIA/TIA-568A standard. This layer should allow for future expansion.
- **Data link layer.** Switches and bridges. These devices will define the size of the collision and broadcast domains.
- **Network layer.** Routers, addressing. This layer filters data packets between network segments.

12.6 Ethernet transceivers

Ethernet requires a minimal amount of hardware. The cables used to connect it are typically either unshielded twisted-pair cable (UTP) or coaxial cables. These cables must be terminated with their characteristic impedance, which is 50 Ω for coaxial cables and 100 Ω for UTP cables.

Each node has transmission and reception hardware to control access to the cable and also to monitor network traffic. The transmission/reception hardware is called a transceiver (short for *trans*mitter/re*ceiver*) and a controller builds up and strips down the frame. For 10 Mbps Ethernet, the transceiver builds the transmitted bits at a rate of 10 Mbps – thus the time for one bit is $1/10\times10^6$, which is 0.1 μs (100 ns).

The Ethernet transceiver transmits onto a single ether. When there are no nodes transmitting, the voltage on the line is +0.7 V. This provides a carrier sense signal for all nodes on the network, it is also known as the heartbeat. If a node detects this voltage then it knows that the network is active and there are no nodes currently transmitting.

Thus, when a node wishes to transmit a message it listens for a quiet period. Then, if two or more transmitters transmit at the same time, a collision results. When they detect a collision, each node transmits a 'jam' signal. The nodes involved in the collision then wait for a random period of time (ranging from 10 to 90 ms) before attempting to transmit again. Each node on the network also awaits a retransmission. Thus, collisions are inefficient in networks as they stop nodes from transmitting. Transceivers normally detect collisions by monitoring the DC (or average) voltage on the line.

When transmitting, a transceiver unit transmits the preamble of consecutive 1s and 0s. The coding used is a Manchester coding, which represents a 0 as a high to a low voltage transition and a 1 as a low to high voltage transition. A low voltage is −0.7 V and a high is +0.7 V. Thus, when the preamble is transmitted the voltage changes between +0.7 V and −0.7 V; as illustrated in Figure 12.9. If, after the transmission of the preamble, no collisions are detected then the rest of the frame is sent.

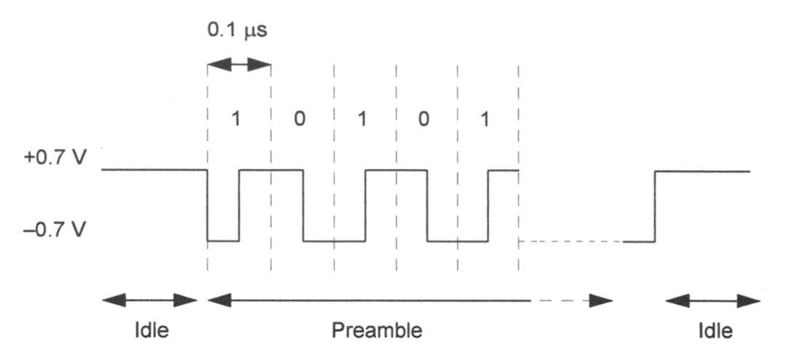

Figure 12.9 Ethernet digital signal

12.7 Ethernet types

The six main types of standard Ethernet are:

- Standard, or thick-wire, Ethernet (**10BASE5**).
- Thinnet, or thin-wire Ethernet, or Cheapernet (**10BASE2**).
- Twisted-pair Ethernet (**10BASE-T**).
- Optical fiber Ethernet (**10BASE-FL**).
- Fast Ethernet (**100BASE-TX** and 100VG-Any LAN).
- Gigabit Ethernet (1000BASE-SX, **1000BASE-T**, 1000BASE-LX and 1000BASE-CX).
- 10 Mbps broadband Ethernet using broadband coaxial cable (**10Broad36**). This has a distance limit of 3600 meters per segment.

DEAR NET-ED

Question: *I have a local Ethernet hub which I connect to. How far can I run a cable from the hub to my computer?*

If you use Cat-5 horizontal cable, you can get a maximum distance of 100 m (if you were to use fiber cable you could get up to 400 m). A hub can thus cover an area of 200 meters square (assuming that the hub is located in the center of the area).

Question: *Did you mention something about a cat?*

Yes. There are five categories of UTP cables defined in EIA/TIA-568A. Cat-1 is only suitable for telephone communications, Cat-2 supports up to 4 Mbps, Cat-3 supports up to 10 Mbps, Cat-4 supports up to 16 Mbps and Cat-5 supports up to 100 Mbps.

The thin- and thick-wire types connect directly to an Ethernet segment; these are shown in Figure 12.10 and Figure 12.11. Standard Ethernet, 10BASE5, uses a high specification cable (RG-50) and N-type plugs to connect the transceiver to the Ethernet segment. A node connects to the transceiver using a 9-pin D-type connector and a vampire (or bee-sting) connector can be used to clamp the transceiver to the backbone cable.

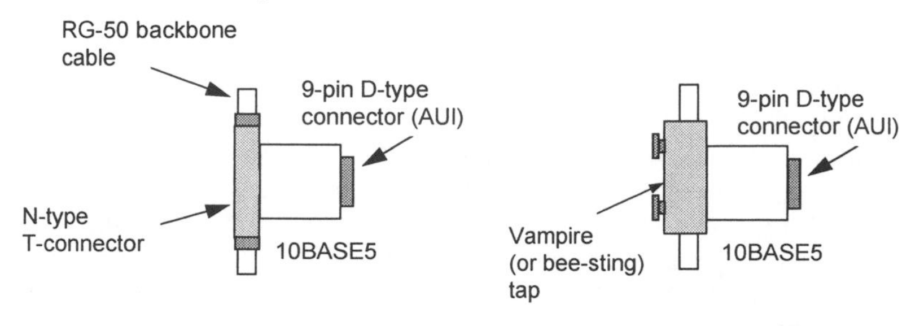

Figure 12.10 Ethernet connections for Thick Ethernet

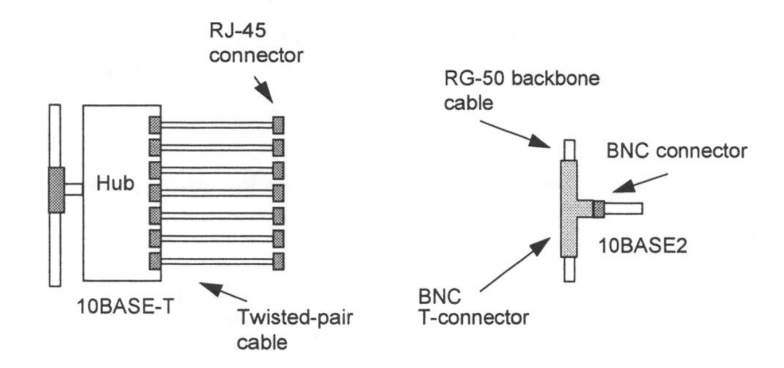

Figure 12.11 Ethernet connections for Thin Ethernet and 10BASE-T

Thin-wire, or Cheapernet, uses a lower specification cable (it has a lower inner conductor diameter). The cable connector required is also of a lower specification, that is, BNC rather than N-type connectors. In standard Ethernet the transceiver unit is connected directly onto the backbone tap. On a Cheapernet network the transceiver is integrated into the node.

Most modern Ethernet connections are to a 10BASE-T hub, which connects UTP cables to the Ethernet segment. An RJ-45 connector is used for 10BASE-T. The fiber-optic type, 10BASE-FL, allows long lengths of interconnected lines, typically up to 2 km. They use either SMA connectors or ST connectors. SMA connectors are screw-on types while ST connectors are push-on. Table 12.1 shows the basic specifications for the different types.

Table 12.1 10BASE network parameters

Parameter	*10BASE5*	*10BASE2*	*10BASE-T*
Common name	Standard or thick-wire	Thinnet or thin-wire	Twisted-pair
Data rate	10 Mbps	10 Mbps	10 Mbps
Maximum segment length	500 m	200 m	100 m
Maximum nodes on a segment	100	30	30
Maximum number of repeaters/nodes	2/1024	4/1024	4
Minimum node spacing	2.5 m	0.5 m	No limit
Location of transceiver electronics	Cable connection	In the node	In the node
Typical cable type	RG-50 (0.5 in diameter)	RG-6 (0.25 in diameter)	UTP cables
Connectors	N-type	BNC	RJ-45/Telco
Cable impedance	50 Ω	50 Ω	100 Ω

12.8 Twisted-pair hubs

Twisted-pair Ethernet (10BASE-T) nodes normally connect to the backbone using a hub, as illustrated in Figure 12.12. Connection to the twisted-pair cable is via an RJ-45 connector. The connection to the backbone can either be to thin- or thick-Ethernet. Hubs are also stackable, with one hub connected to another. This leads to concentrated area networks (CANs) and can be used to reduce the amount of traffic on the backbone. Twisted-pair hubs normally improve network performance.

10BASE-T uses two twisted-pair cables, one for transmit and one for receive. A collision occurs when the node (or hub) detects that it is receiving data when it is currently transmitting data.

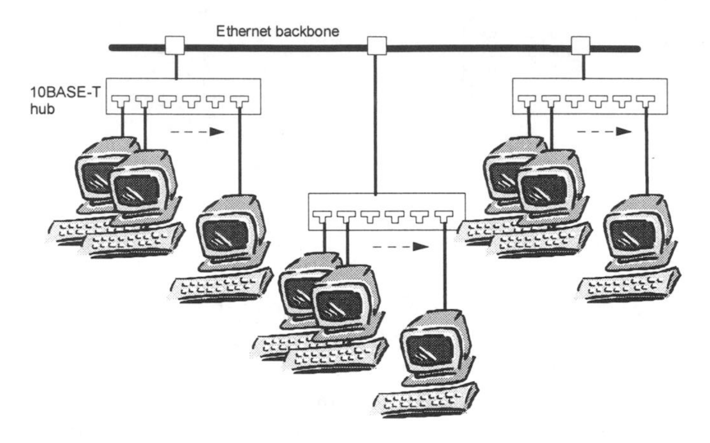

Figure 12.12 10BASE-T connection

12.9 100 Mbps Ethernet

Standard 10 Mbps Ethernet does not perform well when many users are running multimedia applications. Two improvements to the standard are Fast Ethernet and 100VG-AnyLAN. The IEEE has defined standards for both of them, IEEE 802.3u for Fast Ethernet and 802.12 for 100VG-AnyLAN. They are supported by many manufacturers and use bit rates of 100 Mbps, which gives at least 10 times the performance of standard Ethernet.

Standards relating to 100 Mbps Ethernet are:

- 100BASE-TX (twisted-pair) – which uses 100 Mbps over two pairs of Cat-5 UTP cable or two pairs of Type 1 STP cable.
- 100BASE-T4 (twisted-pair) – which is the physical layer standard for 100 Mbps over Cat-3, Cat-4 or Cat-5 UTP, and uses four pairs of UTP cable.
- 100VG-AnyLAN (twisted-pair) – which uses 100 Mbps over two pairs of Cat-5 UTP cable or two pairs of Type 1 STP cable.
- 100BASE-FX (fiber-optic cable) – which is the physical layer standard for 100 Mbps over fiber-optic cables.

Fast Ethernet, or 100BASE-T, is simply 10BASE-T running at 10 times the bit rate. It is a natural progression from standard Ethernet and thus allows existing Ethernet networks to be easily upgraded. Unfortunately, as with standard Ethernet, nodes contend for the network, reducing the network efficiency when there are high traffic rates. Also, as it uses collision detect, the maximum segment length is limited by the amount of time for the farthest nodes on a network to properly detect collisions. On a Fast Ethernet network with twisted-pair copper cables this distance is 100 m, and for a fiber-optic link, it is 400 m. Table 12.2 outlines the main network parameters for Fast Ethernet.

Since 100BASE-TX standards are compatible with 10BASE-TX networks then the network allows both 10 Mbps and 100 Mbps bit rates on the line. This makes upgrading simple, as the only additions to the network are dual-speed interface adapters. Nodes with the 100 Mbps capabilities can communicate at 100 Mbps, but they can also communicate with slower nodes, at 10 Mbps (initially a 10/100 Mbps NIC negotiates with the hub on the communicate speed).

Table 12.2 Fast Ethernet network parameters

	100BASE-TX	*100VG-AnyLAN*
Standard	IEEE 802.3	IEEE 802.12
Bit rate	100 Mbps	100 Mbps
Actual throughput	Up to 50 Mbps	Up to 96 Mbps
Maximum distance (hub to node)	100 m (twisted-pair, CAT-5) 400 m (fiber)	100 m (twisted-pair, CAT-3) 200 m (twisted-pair, CAT-5) 2 km (fiber)
Scaleability	None	Up to 400 Mbps
Advantages	Easy migration from 10BASE-T	Greater throughput, greater distance

The basic rules of a 100BASE-TX network are:

- The network topology is a star network and there must be no loops. The internals of the bus still connect the network as a bus network, but the hub can be seen as the central point of a star network, as when it becomes inoperative then the connected devices on the hub will not be able to communicate. This is illustrated in Figure 12.13.
- Cat-5 cable is used.
- Up to two hubs can be cascaded in a network.
- Each hub is the equivalent of 5 meters in latency.
- Segment length is limited to 100 meters.
- Network diameter must not exceed 200 meters. This is illustrated in Figure 11.19.

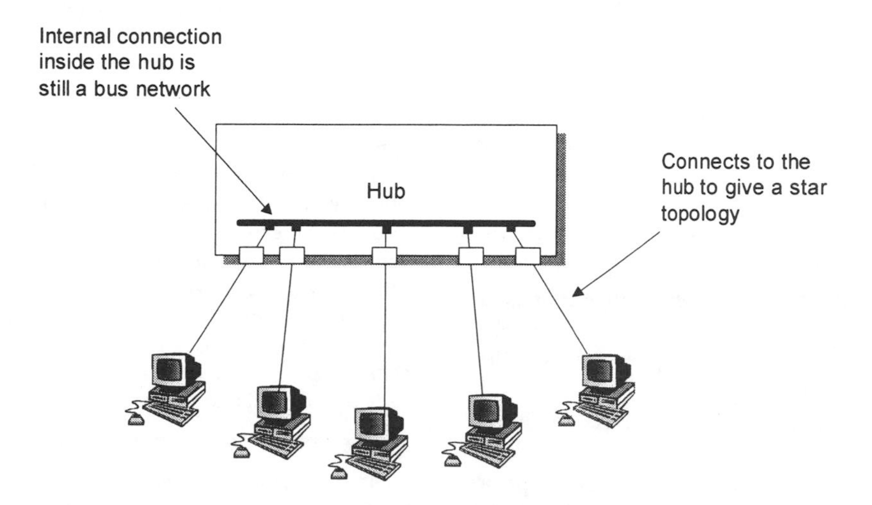

Figure 12.13 Hub connections

12.9.1 100BASE-T4

100BASE-T4 allows the use of standard Cat-3 cables, with eight wires made up of four twisted-pairs. 100BASE-T4 uses all of the pairs to transmit at 100 Mbps. This differs from 10BASE-T in that 10BASE-T uses only two pairs, one to transmit and one to receive. 100BASE-T allows compatibility with 10BASE-T in that the first two pairs (Pair 1 and Pair 2) are used in the same way as 10BASE-T connections. 100BASE-T4 then uses the other two pairs (Pair 3 and Pair 4) with half-duplex links between the hub and the node. The connections are illustrated in Figure 12.14.

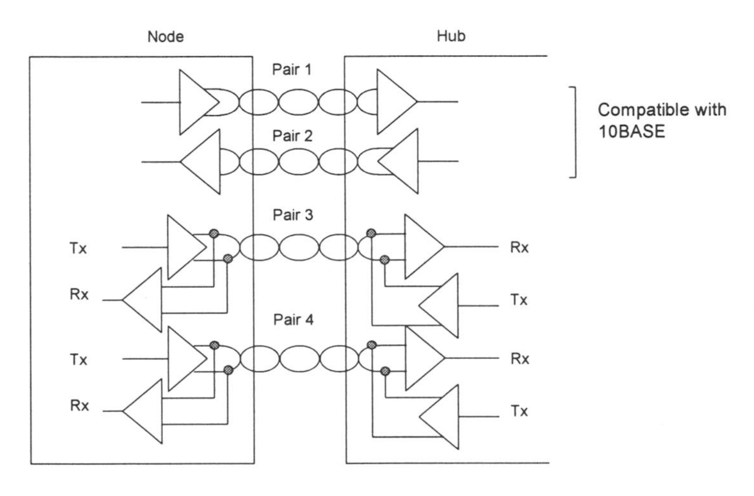

Figure 12.14 100BASE-T4 connections

12.9.2 100VG-AnyLAN

The 100VG-AnyLAN standard (IEEE 802.12) was developed mainly by Hewlett Packard and overcomes the contention problem by using a priority-based round-robin arbitration method, known as the demand priority access method (DPAM). Unlike Ethernet, nodes always connect to a hub which regularly scans its input ports to determine whether any nodes have requests pending.

100VG-AnyLAN has the great advantage that it supports both IEEE 802.3 (Ethernet) and IEEE 802.5 (Token Ring) frames and can thus integrate well with existing 10BaseT and Token Ring networks (allowing for gradual migration).

100VG-AnyLAN also has an in-built priority mechanism with two priority levels: a high-priority request and a normal-priority request. A normal-priority request is used for non-real-time data, such as data files, and so on. High-priority requests are used for real-time data, such as speech or video data. At present, there is limited usage of this feature and there is no support mechanism for this facility after the data has left the hub.

> DEAR NET-ED
>
> **Question:** *Which is best, enterprise servers or workgroup servers?*
>
> Well it all depends on your organization. Enterprise servers are typically used when all the users within an organization require access to a single resource, such as with electronic mail. Workgroup servers provide local access to data and application programs, and isolate traffic around these servers. Workgroup servers should be physically located where they are most required. Typically enterprise servers require to be more centralized in their location, and are more robust than workgroup servers, as the whole organization depends on them. Mirror servers (servers which have exact copies of the main enterprise server) can be used with an enterprise in order to reduce data traffic to the main server.

100VG-AnyLAN allows up to seven levels of hubs (i.e. one root and six cascaded hubs) with a maximum distance of 150 m between nodes. Unlike other forms of Ethernet, it allows any number of nodes to be connected to a segment (it is only limited to the speed of the hub).

12.9.3 Migration to Fast Ethernet

If an existing network is based on standard Ethernet then, in most cases, the best network upgrade is either to Fast Ethernet or 100VG-AnyLAN. Since the protocols and access methods are the same, there is no need to change any of the network management software or application programs. The upgrade path for Fast Ethernet is simple and could be:

- Upgrade high data rate nodes, such as servers or high-powered workstations to Fast Ethernet.
- Gradually upgrade NICs (network interface cards) on Ethernet segments to cards which support both 10BASE-T and 100BASE-T. These cards automatically detect the transmission rate to give either 10 or 100 Mbps.

The upgrade path to 100VG-AnyLAN is less easy as it relies on hubs and, unlike Fast Ethernet, most NICs have different network connectors, one for 10BASE-T and the other for 100VG-AnyLAN (although it is likely that more NICs will have automatic detection). A possible path could be:

- Upgrade high data rate nodes, such as servers or high-powered workstations to 100VG-AnyLAN.
- Install 100VG-AnyLAN hubs.
- Connect nodes to 100VG-AnyLAN hubs and change over connectors.

It is difficult to assess the performance differences between Fast Ethernet and 100VG-AnyLAN. Fast Ethernet uses a well-proven technology, but suffers from network contention. 100VG-AnyLAN is a relatively new technology and the handshaking with the hub increases delay time. The maximum data throughput of a 100BASE-TX network is limited to around 50 Mbps, whereas 100VG-AnyLAN allows rates up to 96 Mbps. 100VG-AnyLAN allows possible upgrades to 400 Mbps.

12.10 Switches and switching hubs

A switch is a very fast, low-latency, multiport bridge that is used to segment LANs. They are typically also used to increase communication rates between segments with multiple parallel conversations and also communication between differing networking technologies (such as between ATM and 100BASE-TX).

A 4-port switching hub is a repeater that contains four distinct network segments (as if there were four hubs in one device). Through software, any of the ports on the hub can connect directly to any of the four segments at any time. This allows for a maximum capacity of 40 Mbps in a single hub (using 10 Mbps for each network segment).

Ethernet switches overcome the contention problem on normal CSMA/CD networks, as they segment traffic by giving each connect a guaranteed bandwidth allocation. Figure 12.15 and Figure 12.16 show the two types of switches; their main features are:

- **Desktop switch** (or workgroup switch). These connect directly to nodes. They are economical with fixed configurations for end-node connections and are designed for standalone networks or distributed workgroups in a larger network.

- **Segment switch.** These connect both 10 Mbps workgroup switches and 100 Mbps interconnect (backbone) switches that are used to interconnect hubs and desktop switches. They are modular, high-performance switches for interconnecting workgroups in mid- to large-size networks.

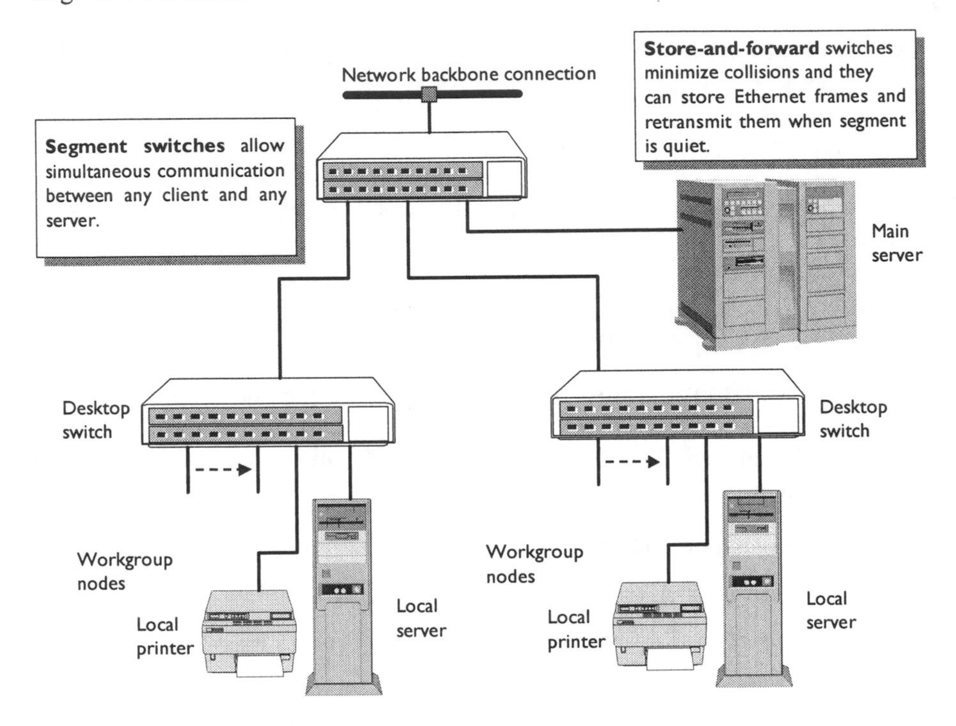

Figure 12.15 Desktop switch

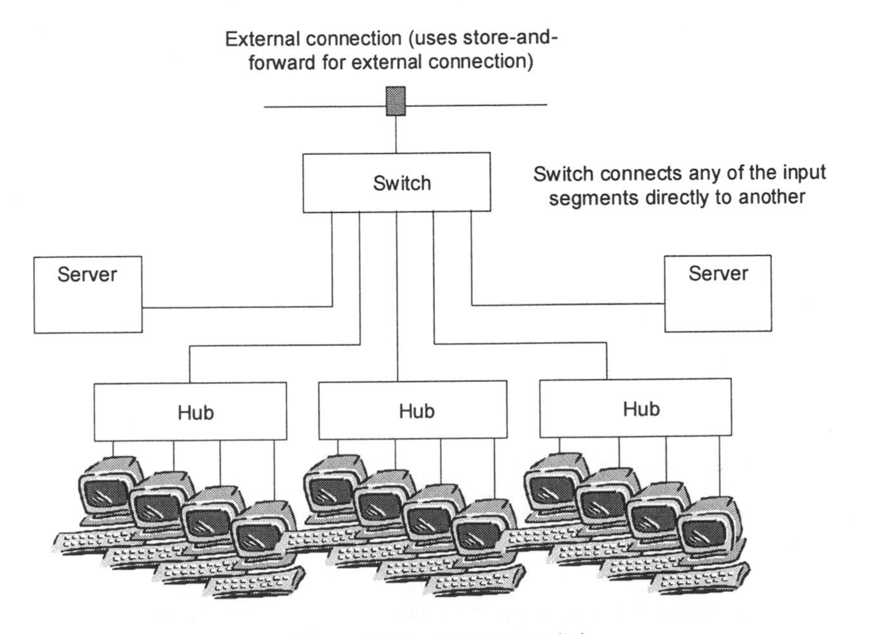

Figure 12.16 Segment switch

12.10.1 Segment switch

A segment switch allows simultaneous communication between any two nodes, and can simply replace existing Ethernet hubs. Figure 12.16 shows a switch with five ports each transmitting at 10 Mbps; this allows up to five simultaneous connections giving a maximum aggregated bandwidth of 50 Mbps. If the nodes support 100 Mbps communication then the maximum aggregated bandwidth will be 500 Mbps. To optimize the network, nodes should be connected to the switch that connects to the server with which it most often communicates. This allows for a direct connection with that server.

12.10.2 Desktop switch

A desktop switch can simply replace an existing 10BASE-T/100BASE-T hub. It has the advantage that any of the ports can connect directly to any other. In the network in Figure 12.15, any of the computers in the local workgroup can connect directly to any other, or to the printer, or a local server. This type of switch works well if there is a lot of local traffic, typically between a local server and local peripherals.

12.10.3 Asymmetric and symmetric switches

Switches can either be symmetric or asymmetric. A symmetric switch provides switched connections between ports with the same bandwidth, such as all 10 Mbps or all 100 Mbps ports. This gives an even distribution of network traffic across the switch. For example an 8-port, 10 Mbps switch will give a maximum throughput of 80 Mbps, as each of the ports can communicate at 10 Mbps.

Asymmetric switches provide differing bandwidths on each of the ports, typically either 10/100 Mbps (known as 10/100 switching) to 100 Mbps/1 Gbps. This type of switch is typically used in client/server applications, where the server requires a higher bandwidth than the client connections. Memory buffering is then used to store the faster bit rate port (such as 100 Mbps), and send it over the slower rate port (such as 10 Mbps). Figure 12.17 illustrates this.

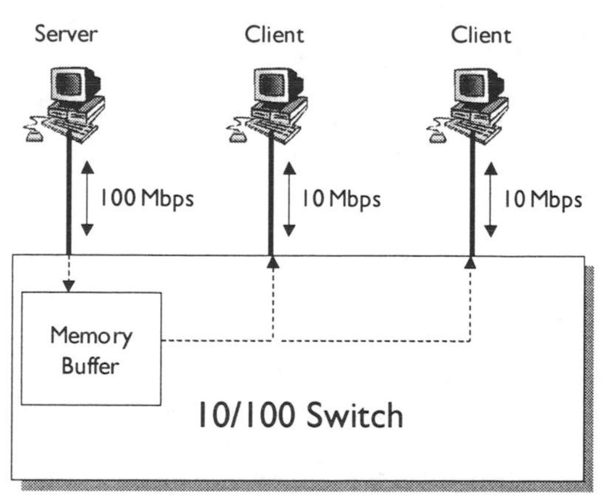

Figure 12.17 Asymmetric switching

12.10.4 Memory buffering

Memory buffers are important devices in a switch as they allow incoming data packets to be stored, before they are transmitted to the destination. The switch can use port based memory buffering or shared memory buffering. In a port-based system, data frames are stored in an area of memory that is associated with the incoming port. The data packet is then transmitted to the outgoing port only when all the packets ahead of it in the queue have been successfully transmitted. It is possible for a single packet to delay the transmission of all the packets in memory because of a busy destination port. This can stop other data packets which are destined for other unbusy ports from being delivered, as the packet at the top of the queue must be transmitted before they can.

With a shared memory technique, all the incoming packets are buffered in a common memory buffer, which is shared with all the switch ports. The packets in the buffer are then dynamically linked to the transmit port. This allows the packet to be received on one port and transmitted on another port without moving it into a different queue.

Cut-through switching

With cut-through switching, the switch reads the destination address before receiving the entire frame. The data frame is then forwarded before the entire frame arrives. This method has the advantage that there is less delay (latency) between the reception and transmission of a data packet, but has poor error detection, because it does not have a chance to detect any errors, before it has started to transmit the received data frame.

Cut-though switching is a packet switching in which data streams leave a destination before the end of the data stream has been fully received. This technique is also known as on-the-fly packet switching.

Store-and-forward switching

Store-and-forwarding techniques have been used extensively in bridges and routers, and are now used with switches. It involves reading the entire Ethernet frame, before forwarding it, with the required protocol and at the correct speed, to the destination port. This has the advantages of:

DEAR NET-ED

Question: *Will I do damage if I connect using incorrectly wired cable, also how do I know that I've connected everything correctly? For example, I have a fiber cable which has two connectors and both are the same, how do I get the TX to the RX, and vice-versa?*

It's unlikely that you will do any damage if you connect your cables round the wrong way, as all the inputs and outputs are electrically buffered. This allows them to sustain short-circuits, and incorrect wiring. The key of knowing if your connection is working is to look at the 'keep-alive' signal, which is typically a green LED on the NIC, hub, switch or router. If it is active, or flashing, you have made a proper connection.

With fiber-optic connections, the transceiver unit will activate two green LEDs when you have made a correct connection. If they are not active, swap the connections round and reconnect.

- Improved error checking. Bad frames are blocked from entering a network segment.
- Protocol filtering. Allows the switch to convert from one protocol to another.
- Speed matching. Typically, for Ethernet, reading at 10 Mbps or 100 Mbps and transmitting at 100 Mbps or 10 Mbps. Also, it can be used for matching between ATM (155 Mbps), FDDI (100 Mbps), Token Ring (4/16 Mbps) and Ethernet (10/100 Mbps).

The main disadvantage is system delay, as the frame must be totally read before it is transmitted thus there is a delay in the transmission. The improvement in error checking normally overcomes this disadvantage.

Figure 12.18 illustrates the two switching methods. With cut-though the data frame is forwarded to the destination before it is fully received. This technique operates on data frames, thus it is normally only used to transmit between networks of the same type (such as Ethernet), the same speed, and the same protocol. Store-and-forward techniques obviously have a greater latency but can be used to improve error detection, and interfacing between different network types (for example Ethernet to ATM), and different speeds.

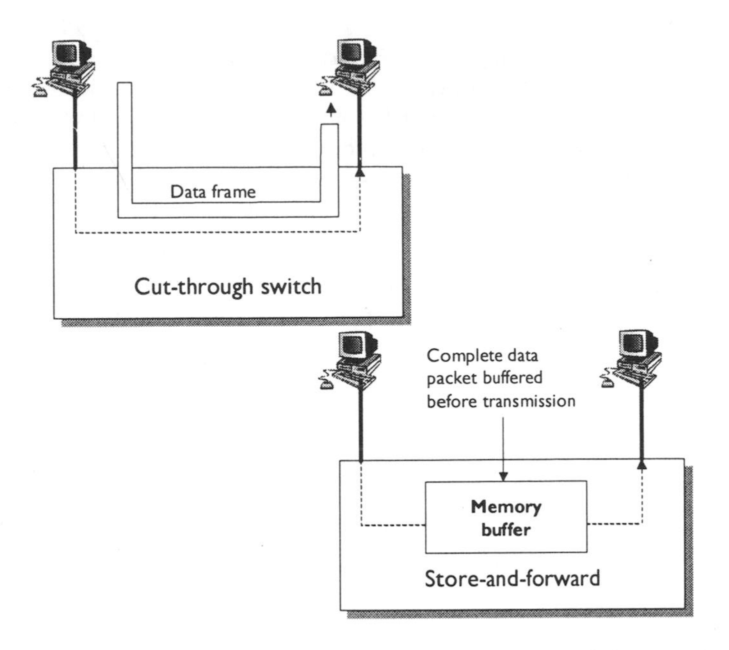

Figure 12.18 Cut-through and store-and-forward switches

12.10.5 Ethernet connections

Standard Ethernet (10BASE) uses a half-duplex connection, where the TX connects to the RX, and the RX to the TX, as illustrated in Figure 12.19. In full-duplex Ethernet uses switching to create a virtual circuit between two nodes (known as a FDES – full-duplex Ethernet switch). As there is a virtual point-to-point connection, nodes can transmit and receive at the same time. A wire is used to transmit and another to receive, thus the maximum total bandwidth for a single port operating at 10 Mbps is 20 Mbps. As it is a point-to-point connection there should be no collisions. This increases the actual bandwidth of a full-duplex Ethernet to nearly 100% of its capacity (while standard, half-duplex can only achieve a maximum of around 50% capacity).

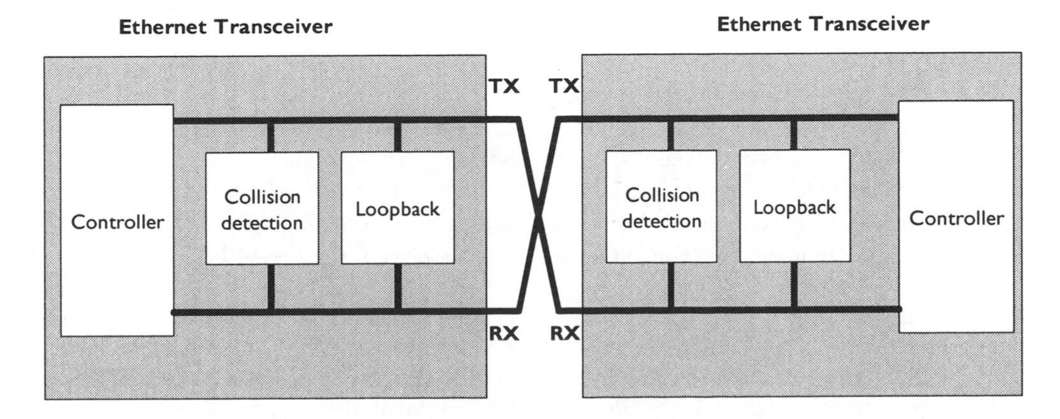

Figure 12.19 Ethernet transceiver

Switches learn the MAC address of devices by reading the source address of each data packet that they receive, and noting at which port the frame came from. This information is then added to its forwarding database (dynamic learning), which stores addresses in content addressable memory (CAM). Each address has a time stamp associated with it, and new references to the address cause the time stamp to be updated. Addresses which are not used for a given time, are aged-out.

12.11 Gigabit Ethernet

The IEEE 802.3 working group initiated the 802.3z Gigabit Ethernet task force to create the Gigabit Ethernet standard (which was finally defined in 1998). The Gigabit Ethernet Alliance (GEA) was founded in May 1996 and promotes Gigabit Ethernet collaboration between organizations. Companies, which were initially involved in the GEA, include: 3Com, Bay Networks, Cisco Systems, Compaq, Intel, LSI Logic, Sun and VLSI.

With Gigabit Ethernet, the amount of available bandwidth for a single segment is massive. For example, almost 125 million characters (125 MB) can be sent in a single second. A large reference book with over 1000 pages could be send over a network segment, ten times in a single second. Compare it also with a ×24, CD-ROM drive which transmits at a maximum rate of 3.6 MB/s (24×150 kB/sec). Gigabit Ethernet operates almost 35 times faster than this drive. With network switches, this bandwidth can be multiplied by a given factor, as they allow multiple simultaneous connections.

DEAR NET-ED

Question: *As usual, I'm confused. Sorry. You have said that Ethernet connections have a cross-over, but when I look at my patch cable, there isn't a cross-over, and pin 1 wires to pin 1, pin 2 to pin 2, and so on? Where's the crossover?*

You're totally correct, and so am I. The standard Ethernet connection must have a cross-over to connect the transmit to the receive, and vice-versa, but most hubs implement the cross-over inside the hub. Thus all you need is a straight-through cable. I've listed the standard cross-over connections in Section 12.18 (and Figure 11.24), but most of the time you do not need a cross-over when you're connecting to the front of a hub or a switch. It is only at the back of the hub that you may need a cross-over cable. If in doubt look at the 'keep-alive' LED. If it is off after you connect, it's likely that you've got the wrong cable (or the power isn't on, or you've not connected the other end, or the power isn't on the computer, and so on).

Often a cross-over connection is marked with:

Gigabit Ethernet is an excellent challenger for network backbones as it interconnects 10/100BASE-T switches, and also provides a high-bandwidth to high-performance servers. Initial aims were:

- Half/full-duplex operation at 1000 Mbps.
- Standard 802.3 Ethernet frame format. Gigabit Ethernet uses the same variable-length frame (64- to 1514-byte packets), and thus allows for easy upgrades.
- Standard CSMA/CD access method.
- Compatibility with existing 10BASE-T and 100BASE-T technologies.
- Development of an optional Gigabit Media Independent Interface (GMII).

The compatibility with existing 10/100BASE standards makes the upgrading to Gigabit Ethernet much easier, and considerably less risky than changing to other networking types, such as FDDI and ATM. It will happily interconnect with, and autosense, existing slower rated Ethernet devices. Figure 12.20 illustrates the functional elements of Gigabit Ethernet. Its main characteristics are:

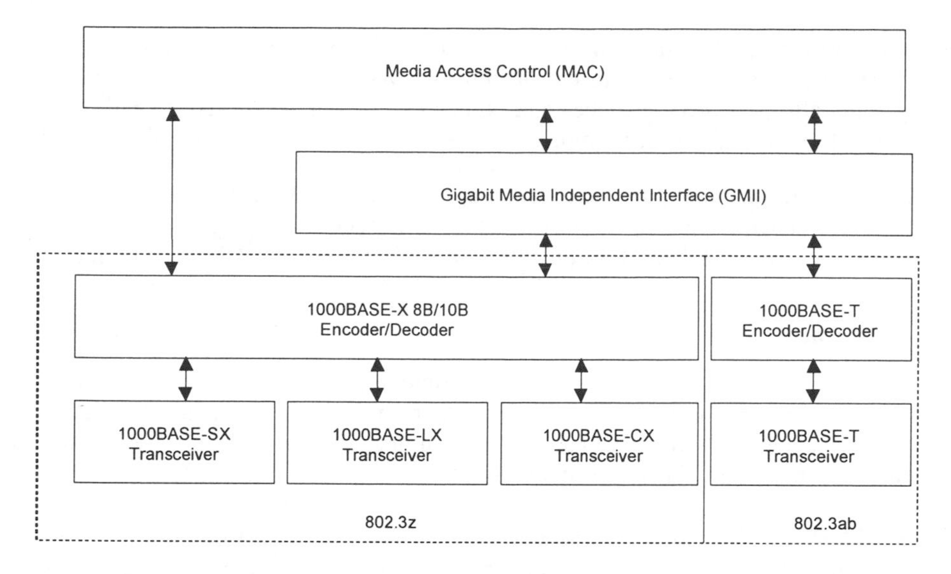

Figure 12.20 Gigabit Ethernet functional elements

- **Full-duplex communication.** As defined by the IEEE 802.3x specification, two nodes connected via a full-duplex, switched path can simultaneously send and receive frames (thus doubling the bandwidth). Gigabit Ethernet supports new full-duplex operating modes for switch-to-switch and switch-to-end-station connections, and half-duplex operating modes for shared connections using repeaters and the CSMA/CD access method.
- **Standard flow control.** Gigabit Ethernet uses standard Ethernet flow control to avoid congestion and overloading. When operating in half-duplex mode, Gigabit Ethernet adopts the same fundamental CSMA/CD access method to resolve contention for the shared media.
- **Enhanced CSMA/CD method.** This maintains a 200 m-collision diameter at gigabit speeds. Without this, small Ethernet packets could complete their transmission before the transmitting node could sense a collision, thereby violating the CSMA/CD method. To resolve this issue, both the minimum CSMA/CD carrier time and the Ethernet slot time (the time, measured in bits, required for a node to detect a collision) have been extended from 64 bytes (which is 51.2 μs for 10BASE and 5.12 μs for 100BASE) to 512 bytes (which is 4.1 μs for 1000BASE). The minimum frame length is still 64 bytes. Thus, frames smaller than 512 bytes have a new carrier extension field following the CRC field. Packets larger than 512 bytes are not extended.

DEAR NET-ED

Question: *Sorry to bother you again, but does it matter which port I connect my workstation to the hub with. Do I have to start from port 1, then port 2, and so on.*

No. Hubs and switches are autosensing and automatically use the port that you connect to. You should hopefully see an LED become active when you connect to the port. You can also connect to a cascaded hub/switch to any one of the ports.

Question: *I've got a dual 10/100 switching hub. Can I communicate at 100 Mbps, even though I only have a 10 Mbps networking card?*

No. The switching hub will automatically sense the speed of your networking card, and use that rate. The great advantage of buying a dual speed switch is that you can upgrade your network card over time.

- **Packet bursting.** The slot time changes affect the small-packet performance, but this has been offset by a new enhancement to the CSMA/CD algorithm, called packet bursting. This allows servers, switches and other devices to send bursts of small packets in order to fully utilize the bandwidth.

Devices operating in full-duplex mode (such as switches and buffered distributors) are not subject to the carrier extension, slot time extension or packet bursting changes. Full-duplex devices use the regular Ethernet 96-bit interframe gap (IFG) and 64-byte minimum frame size.

12.11.1 Ethernet transceiver

The IEEE 802.3z task force spent much of their time defining the Gigabit Ethernet standard for the transceiver (physical layer), which is responsible for the mechanical, electrical and procedural characteristics for establishing, maintaining and deactivating the physical link between network devices. The physical layers are:

- **1000BASE-SX** (Low cost, multi-mode fiber cables). These can be used for short interconnections and short backbone networks. The IEEE 802.3z task force has tried to integrate the new standard with existing cabling, whether it is twisted-pair cable, coaxial cable or fiber-optic cable. These tests involved firing lasers in long lengths of multi-mode fiber cables. From these tests it was found that a jitter component results which is caused by a phenomenon known as differential mode delay (DMD). The 1000BASE-SX standard has resolved this by carefully defining the shape of the laser signal, and enhanced conformance tests. Typical maximum lengths are: 62.5 µm, multi-mode fiber (up to 220 m) and 50 µm, multi-mode fiber (550 m).
- **1000BASE-LX** (Multi-mode/single mode-mode fiber cables). These can be used for longer runs, such as on backbones and campus networks. Single-mode fibers are covered by the long-wavelength standard, and provide for greater distances. External patch cords are used to reduce DMD. Typical lengths are: 62.5 µm, multi-mode fiber (up to 550 m), 50 µm, multi-mode fiber (up to 550 m) and 50 µm, single-mode fiber (up to 5 km).
- **1000BASE-CX** (Shielded Balanced Copper). This standard supports interconnection of equipment using a copper-based cable, typically up to 25 m. As with the 1000BASE-LX/SX standards, it uses the Fiber Channel-based 8B/10B coding to give a serial line rate of 1.25 Gbps. The 1000BASE-T is likely to supersede this standard, but it has been relatively easy to define, and to implement.
- **1000BASE-T** (UTP). This is a useful standard for connecting directly to workstations. The 802.3ab Task Force has been assigned the task of defining the 1000BASE-T physical layer standard for Gigabit Ethernet over four pairs of Cat-5 UTP cable, for cable distances of up to 100 m, or networks with a diameter of 200 m. As it can be used with existing cabling, and allows easy upgrades. Unfortunately, it requires new technology and new coding

DEAR NET-ED

Question: *Wow! One billion bits in a second from Ethernet using standard Cat-5 cable. Surely it isn't possible.*

It is, and the IEEE has standardized it. If the IEEE says that something works, then it works. In fact with a Gigabit Ethernet switch you can get even higher bandwidths.

Question: *So where will Gigabit Ethernet be used?*

It is unlikely that Gigabit Ethernet will be used for WANs, as it still suffers from too many problems, which get worse over large physical areas. It is most likely to be used for network backbones, and in campus area networks.

schemes in order to meet the potentially difficult and demanding parameters set by the previous Ethernet and Fast Ethernet standards.

12.11.2 Fiber Channel components

The IEEE 802.3 committee based much of the physical layer technology on the ANSI-backed X3.230 Fiber Channel project. This allowed many manufacturers to re-use physical-layer Fiber Channel components for new Gigabit Ethernet designs, and has allowed a faster development time than is normal, and increased the volume production of the components. These include optical components and high-speed 8B/10B encoders.

The 1000BASE-T standard uses enhanced DSP (Digital Signal Processing) and enhanced silicon technology to enable Gigabit Ethernet over UTP cabling. As Figure 12.20 shows, it does not use the 8B/10B encoding.

12.11.3 Buffered distributors

Along with repeaters, bridges and switches, a new device called a buffered distributor (or full-duplex repeater) has been developed for Gigabit Ethernet. It is a full-duplex, multiport, hub-like device that connects two or more Gigabit Ethernet segments. Unlike a bridge, and like a repeater, it forwards all the Ethernet frames from one segment to the others, but unlike a standard repeater, a buffered distributor buffers one, or more, incoming frames on each link before forwarding them. This reduces collisions on connected segments. The maximum bandwidth for a buffered distributor will still only be 1 Gbps, as opposed to Gigabit switches which allow multi-Gigabit bandwidths (but it reduces the number of collisions on a segment, as it buffers frames and waits until the segment is clear).

12.11.4 Quality of Service

Many, real-time, networked applications require a given Quality of Server (QoS), which might relate to bandwidth requirements, latency (network delays) and/or jitter. Unfortunately, there is nothing built into Ethernet that allows for a QoS, thus new techniques have been developed to overcome this. These include:

- **RSVP.** Allows nodes to request and guarantee a QoS, and works at a higher-level to Ethernet. For this, each network component in the chain must support RSVP and communicate appropriately. Unfortunately, this may require an extensive investment to totally support RSVP, thus many vendors have responded in implementing proprietary schemes, which may make parts of the network vendor-specific.
- **IEEE 802.1p and IEEE 802.1Q.** Allow a QoS over Ethernet by 'tagging' packets with an indication of the priority or class of service desired for the frames. These tags allow applications to communicate the priority of frames to internetworking devices. RSVP support can be achieved by mapping RSVP sessions into 802.1p service classes.
- **Routing.** Implemented at a higher layer.

12.11.5 Gigabit Ethernet migration

The greatest advantage of Gigabit Ethernet is that it is easy to upgrade existing Ethernet-based networks to higher bit rates. A typical migration might be:

- Switch-to-switch links. Involves upgrading the connections between switches to 1 Gbps. As 1000BASE switches support both 100BASE and 1000BASE then not all the switches require to be upgraded at the same time; this allows for gradual migration.
- Switch-to-Server links. Involves upgrading the connection between a switch and the server to 1 Gbps. The server requires an upgraded Gigabit Ethernet interface card.

- Switched Fast Ethernet Backbone. Involves upgrading a Fast Ethernet backbone switch to a 100/1000BASE switch. It thus supports 100BASE and 1000BASE switching, using existing cabling.
- Shared FDDI Backbone. Involves replacing FDDI attachments on the ring with Gigabit Ethernet switches or repeaters. Gigabit Ethernet uses the existing fiber-optic cable, and provides a greatly increased segment bandwidth.
- Upgrade NICs on nodes to 1 Gbps. It is unlikely that users will require 1 Gbps connections, but this facility is possible.

12.11.6 1000BASE-T

One of the greatest challenges of Gigabit Ethernet is to use existing Cat-5 cables, as this will allow fast upgrades. Two critical parameters, which are negligible at 10BASE speeds, are:

- Return loss. Defines the amount of signal energy that is reflected back towards the transmitter due to impedance mismatches in the link (typically from connector and cable bends).
- Far-End Crosstalk. Noise that is leaked from another cable pair. The higher the bit rate, the more crosstalk that is generated.

The 1000BASE-T Task Force estimates that less than 10% of the existing Cat-5 cables were improperly installed (as defined in ANSI/TIA/EIA568-A in 1995) and might not support 1000BASE-T (or even, 100BASE-TX). 100BASE-T uses two pairs, one for transmit and one for receive, and transmits at a symbol rate of 125Mbaud with a 3-level code. 1000BASE-T uses:

- All four pairs with a symbol rate of 125 MBaud (symbols/sec). One symbol contains two bits of information.
- Each transmitted pulse uses a 5-level PAM (Pulse Amplitude Modulation) line code, which allows two bits to be transmitted at a time.
- Simultaneous send and receive on each pair. Each connection uses a hybrid circuit to split the send and receive signals.
- Pulse shaping. Matches the characteristics of the transmitted signal to the channel so that the signal-to-noise ratio is minimized. It effectively reduces low frequency terms (which contain little data information, can cause distortion and cannot be passed over the transformer-coupled hybrid circuit), reduces high frequency terms (which increase crosstalk) and rejects any external high-frequency noise. It is thought that the transmitted signal spectrum for 1000BASE will be similar to 100BASE.
- Forward Error Correction (FEC). This provides a second level of coding that helps to recover the transmitted symbols in the presence of high noise and crosstalk. The FEC bit uses the fifth level of the 5-level PAM.

A 5-level code (–2, –1, 0, +1, +2) allows two bits to be sent at a time, if all four pairs are used then eight bits are sent at a time. If each pair transmits at a rate of 125 Mbaud (symbols/sec, giving 250 Mbps), the resulting bit rate will be 1 Gbps.

12.12 Bridges

Bridges are an excellent method of reducing traffic on network segments. A particular problem is when there is too much traffic on a network segment, as this can result in a great deal of collisions, and resultant backoff periods. The distance in which a collision can travel is called the collision domain. Collisions do not travel across a bridge and they thus reduce the impact of a collision on other network segments. Hubs, on the other hand, operate at the physical level and transmit the data to all the connected ports.

A bridge operates at the data link layer and uses the MAC address to forward data frames from one network segment to another. For this maintains a table which maps the MAC addresses of all the nodes on the network that it connects to. Ethernet networks normally use transparent bridges which automatically build up an address table from the nodes that connect to each of their ports, as illustrated in Figure 12.21. Token ring networks typically use source route bridges, where the entire route to a destination is predetermined, in real time, before sending of data to the destination.

A bridge determines if it should forward a data frame onto a network by examining the destination MAC address in the data frame. If the destination MAC address is not on the segment that originated the data frame, then the bridge forwards it to all the other connected segments. **Thus a bridge does not actually determine on which segment the destination is on, and blindly forwards data frames onto all other connected segments**.

The two major problems with bridges are:

- They work well when there is not too much intersegment traffic, but when the intersegment traffic becomes too heavy the bridge can actually become a bottleneck for traffic, and actually slow down communications.
- They spread and multiply broadcasts. A bridge forwards all broadcasts (which is identified with the FF-FF-FF-FF-FF-FF MAC address) to all other connected segments. If there are too many broadcasts, it can result in a broadcast storm, where broadcasts swamp transmitted data. The best way to overcome these storms is to use routers which do not forward broadcasts.

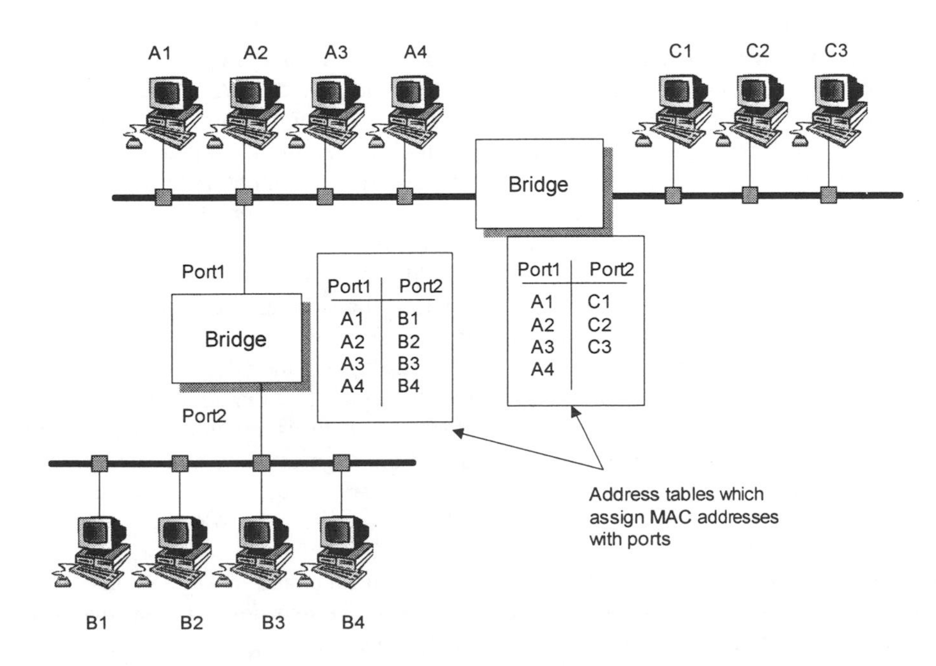

Figure 12.21 Transparent bridges with address tables

12.13 ARP

In order for data to be received by a node, the station address (typically the MAC address) and the network address (typically the IP address) must match. Thus how does a computer know the MAC address of the computer that it wants to communicate with? This is achieved with ARP (Address Resolution Protocol) which translates IP addresses to Ethernet addresses, and is used when IP packets are sent from a computer, and the Ethernet address is added to the Ethernet frame. A table look-up, called the ARP table, is used to translate the addresses. One column has the IP address and the other has the Ethernet address. The following is an example ARP table:

IP address	Ethernet address
146.176.150.2	00-80-C8-22-6B-E2
146.176.150.3	00-80-C8-22-CD-4E
146.176.150.4	00-80-C8-23-11-4C

The sequence of determining the Ethernet address is as follows:

1. An ARP request packet with a broadcast Ethernet address (`FF-FF-FF-FF-FF-FF`) is sent out on the network segment to every computer on the segment.
2. All the computers on the network segment read the broadcast Ethernet frame, and examine the Type field to determine if it is an ARP packet. If it is, then it is passed to the ARP module.
3. If the IP address of a receiving station matches the IP address in the IP data packet then it sends a response directly to the source Ethernet address.
4. The originator then receives the Ethernet frame and checks the Type field to determine if it is an ARP packet. If it is, then it adds the sender's IP address and Ethernet address to its ARP table. The IP packet can now be sent with the correct Ethernet address.

Each computer has a separate ARP table for each of its Ethernet interfaces, which is stored in the local memory (cache memory). The table is thus not static and is updated as the network changes. Figure 12.22 shows an example of an ARP request and an ARP reply. In this case the node with the IP address of 146.176.151.100 wants to communicate with a node with an IP address of 146.176.151.130, but does not know its MAC address. To determine it, it sends out an ARP request which is broadcast to all the nodes on the segment. All the nodes read the ARP request, and examine the destination IP address in the IP header. When the node whose IP address matches the destination IP address read the data frame, it sends back a data frame with its own IP address in the source IP address field and its MAC address in the data frame source MAC address. The destination addresses will be the same as the source address of the initial ARP request. Once the node that sent the request receives the reply it will update its local ARP table so that it can use the correct MAC address for the address node. Storing the ARP table thus reduces network traffic, as nodes only require to communicate with a destination node once in order to determine its MAC address. If this was not stored then each time a node wished to communicate with another node on the segment it would have to send an ARP request. All the nodes on the network listen to the ARP request and reply, even if they are not involved in passing their MAC address. When they hear the ARP reply they can all update their ARP tables.

ARP tables only remain current for a given amount of time, and must be updated in order to remain current. This is important as nodes can be added and deleted from the network. The process of deleting an ARP entry from the table is known as *ageing out.* With this, after a certain time period, nodes delete ARP entries.

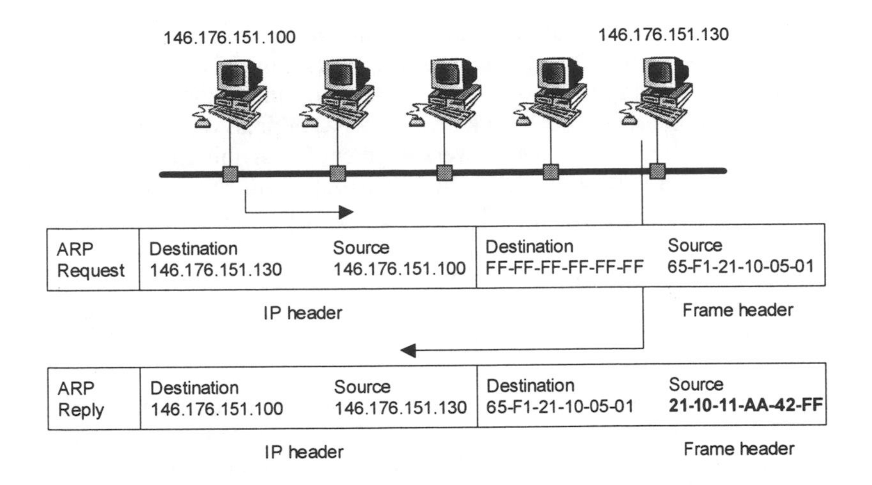

Figure 12.22 Example ARP request and reply

12.13.1 Routers and ARP

It has been shown how ARP allows a node to determine the MAC address of a node on the same network segment, but what happens when the destination is on a different network segment. This is done through routers, and shows the reason why a computer must know the IP address of the gateway node (normally a router).

Each router has a port (or interface) that connects to each of the network segments that it connects to. Each of these ports has an IP address, and each of them build-up ARP tables for each of its ports. The overall table maps IP addresses, MAC addresses and ports. For example:

Protocol	*Address*	*MAC address*	*Port*
IP	146.176.151.100	65-F1-21-10-05-01	Port_1
IP	146.176.151.100	23-EA-21-B8-F5-71	Port_1
IP	146.176.153.100	4F-DD-21-EE-05-22	Port_2
IP	146.176.153.100	21-F2-21-32-11-00	Port_2

In addition to IP addresses and MAC addresses of devices located on networks it is connected to, a router also possesses IP addresses and MAC addresses of other routers. It uses these addresses to direct data towards its final destination. If a router receives a data packet whose destination addresses are not in its routing table, it forwards the packet to the addresses of other routers which presumably do contain information about the destination host in their routing tables.

A node on a network cannot send an ARP request to a device on another network, as ARP requests are sent in broadcast mode and these are not forwarded by routers to other networks. If the node is on another network and it does not know its MAC address, it must seek out the services of a router. In this case the router is known as the default gateway, and the source node transmits a data frame with the MAC address of the router (the default gateway) and the IP address of the destination device (and not of the router). When the router reads the data frame it examines the destination IP address to determine if it requires to forward the data packet. For this it uses its routing tables. If it knows that the destination node is connected to one of its network segments it constructs a data frame and data packet with the required destination IP address and MAC address, and sends it onto the required network segment.

If the router does not detect that the destination node is not on one of its connected segments, it locates the MAC address of another router that is likely to know where it may be located and forwards the data to that router. This type of routing is known as indirect routing.

So what happens if the router does not know the MAC address of the addressed router (as ARP requests are confined to the one network segment)? For this the router sends out an ARP request on the local network segment. Another router, which is connected to the local network segment, then sends back an ARP reply which will contain the MAC address to the destination router.

Figure 12.23 shows an example of communication between two nodes on different network segments. In this case Node_A is communicating with Node_B. The sequence of operations will be:

- Node_A sends out a data frame with the MAC address of the gateway (00-80-55-43-FE-FF) and the destination node's IP address (146.176.120.2). The source MAC address of the data frame will be the MAC address of Node_A, and the source IP address will be the IP address of Node_A.
- Router_1 then sends a data frame with the MAC address of Router_2 (00-65-21-44-33-A1) and the destination node's IP address (146.176.120.2). The source MAC address will be the MAC address of Port_2 of Router_1 (which is 00-60-DD-E0-12-34), and the source IP address will be the IP address of Node_A.
- Router_2 then sends a data frame with the MAC address of Node_B (00-90-10-33-DE-EE) and the destination node's IP address (146.176.120.2). The data is thus received by Node_B. The source MAC address will be the MAC address of Port_1 of Router_2 (which is 00-10-32-11-BC-B1), and the source IP address will be the IP address of Node_A.

It should be noted that each port (or interface) of a router has a unique network address, and a unique MAC address.

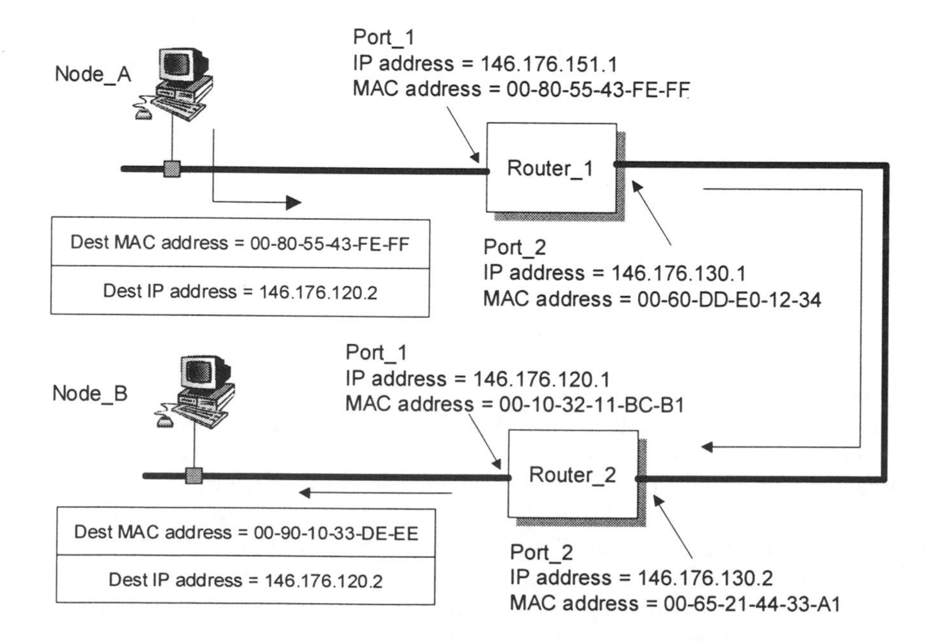

Figure 12.23 Example address over network segments

12.13.2 ARP definition

The RFC826 document defines the specification for ARP. The Ethernet frame allows for the definition of a 16-bit Ethernet Type field. This is set to ADDRESS_RESOLUTION for ARP and reverse ARP (RARP). Figure 12.24 shows the format of the ARP (and RARP) data packet format. The first two bytes define the hardware address space (ar$hrd). For Ethernet, this is set to ETHERNET (a value of 1). The next field defines the protocol address space (ar$pro). For IP, this is set to IP (a value of 2048). After this there are two fields (ar$hln and ar$pln) which define the length of the hardware (MAC address) and the protocol address (IP address). Typically this is six for MAC addresses, and four for IP addresses. After these two fields the opcode (ar$op) defines whether the packet is an ARP request or an ARP reply (a value of 1 for REQUEST, and 2 for REPLY). After these fields come the source MAC address and the source IP address (ar$sha and ar$spa), next the target MAC address and the target IP address (ar$tha and ar$tpa). If the data packet is an ARP request the target IP address will be unknown, as the source does not know what the target's MAC address is. Thus typical settings are:

ar$hrd = 1 [for Ethernet] ar$pro = 2048 [for IP]
ar$hln = 6 [for 48-bit MAC address] ar$pln = 4 [for 32-bit IP address]
ar$op = 1 | 2 [for REQUEST or REPLY]

Where the opcode is set to REQUEST, the hardware address of the source (ar$sha) and the target (ar$tha), and the source protocol address (ar$spa) will be defined, but the target protocol address (ar$tpa) will be undefined. If the opcode is a REPLY, then all the fields will be defined.

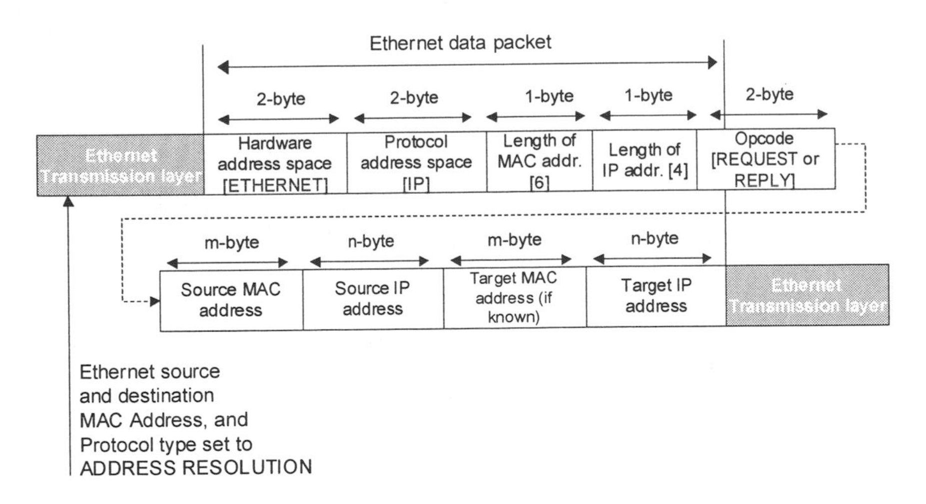

Figure 12.24 ARP data packet format

12.14 RARP

Reverse ARP (RARP) is much less common than ARP, and is the reverse of ARP. With RARP a node knows its MAC address but does not know what its IP address is. Most computers know what their IP address is as they have a hard disk to store it. Some computers, such as terminals or a diskless workstation, do not have local storage and thus must determine their IP address. To run a RARP service there must be a server which responds to

RARP requests. These servers have a table of IP and MAC addresses. In the transmission of RARP the network IP broadcast is used, which is the host part of the IP address set to all 1's (this will be covered in more detail in Chapter 14). Figure 12.25 shows an example RARP request and reply. Initially the RARP requester sends out a RARP request with an IP broadcast address (in this case 146.176.151.255), which is read by the RARP server. It then looks in its RARP table for the MAC address in the source MAC address field (in this case it is 65-F1-21-10-05-01). It then determines the IP address which corresponds to this MAC address (in this case it is 146.176.151.100), and sends back a RARP reply with the corresponding IP address, with the destination MAC address (in this case it is 65-F1-21-10-05-01).

The RFC903 document defines the specification for RARP. It defines that ARP treats all hosts (nodes on the network) as being equal, whether they be a client or a server, whereas RARP requires a server to maintain a database to map hardware addresses to IP addresses. It uses the same data packet as ARP (see Figure 12.24), but different opcode definitions:

ar$op = 3 | 4 [for REVERSE_REQUEST or REPLY_REVERSE]

Where the Opcode is set to REVERSE_REQUEST, the hardware address of the source (ar$sha) and the target (ar$tha) will be defined, whereas, the source protocol address (ar$spa) and the target protocol address (ar$tpa) will be undefined. If the source wishes to determine its protocol address then the target hardware address (ar$sha) will be set to the hardware address of the source.

Where the opcode is set to REPLY_REVERSE, the responder's hardware address (ar$sha) will be set to the hardware address of the responder, the responder's protocol address (ar$spa) will be set to the protocol address of the responder. The hardware address (ar$tha) will be set to the hardware address of the target (which is the same as the address in the request), and the target protocol address (ar$tpa) is set to the protocol address of the target (which is the main result of the reply).

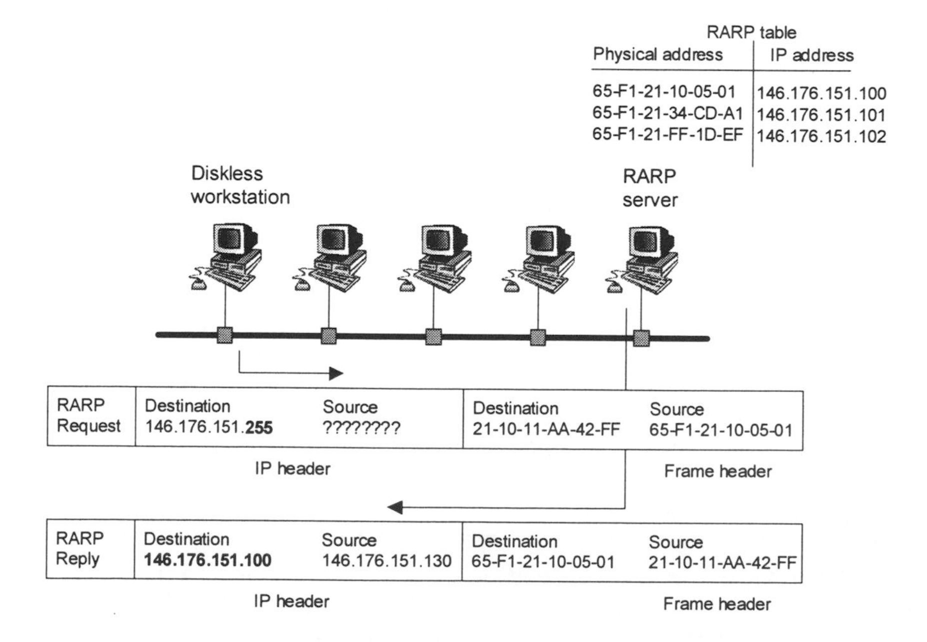

Figure 12.25 Example RARP request and reply

12.15 vLAN

vLANs are a new technology, which uses software to define a broadcast domain, rather than any physical connections. In a vLAN a message transmitted by one node is only received by other nodes with a certain criteria to be in the domain. It is made by logically grouping two or more nodes and a vLAN-initialized switching device, such as intelligent switches (which use the MAC address to forward data frames) or routers (which use the network address to route data packets). The important concept with vLANs is that the domain is defined by software, and not by physical connections.

There are two methods that can define the logical grouping of nodes within a vLAN:

- **Implicit tagging.** This uses a special tagging field which is inserted into the data frames or within data packets. It can be based upon the MAC address, a switch port number, protocol, or another parameter by which nodes can be logically grouped. The main problem with implicit tagging is that different vendors create different tags which make vendor interoperability difficult. This is known as frame filtering.
- **Explicit tagging.** This uses an additional field in the data frame or packet header. This can also lead to incompatibility problems, as different vendor equipment may not be able to read or process the additional field. This is known as frame identification.

It is thus difficult to create truly compatible vLANs until standards for implicit and explicit tags are standardized. One example of creating a vLAN is to map ports of a switch to create two or more virtual LANs. For example a switch could connect to two servers and 16 clients. The switch could be configured so that eight of the clients connected to one server through a vLAN, and the other eight onto the other server. This setup is configured in software, and not by the physical connection of the network. Figure 12.26 shows a possible implementation where nodes 1 to 8 create a vLAN through the switch with SERVER1, and nodes 9 to 16 create a vLAN with SERVER2. The switch would map ports to create the vLANs, where the two networks are now independent broadcast domains (network segments), and will only receive the broadcasts from each of their virtual LANs. Normally a switch would connect any one of its ports to another port, and allow simultaneous connection. In this case, the switch allows for multiple connections onto a segment. Now, with the vLAN, data frames transmitted on one network segment will stay within that segment and are not transmitted to the other vLAN.

The main advantages of using vLAN are:

- **Creation of virtual networks.** Just as many organizations build open-plan offices which can be changed when required, vLANs can be used to reconfigure the logical connections to a network without actually having to physically move any of the resources. This is especially useful in creating workgroups where users share the same resources, such as databases and disk storage.
- **Ease of administration.** vLANs allow networks to be easily configured, possibly at a distance from the configured networks. In the past reconfiguration has meant recabling and the movement of networked resources. With vLANs the resources can be configured with software to setup the required network connections.
- **Improved bandwidth usage.** Normally users who work in a similar area share resources. This is typically known as a workgroup. If workgroups can be isolated from other workgroups then traffic which stays within each of the workgroups does not affect other workgroups. A vLAN utilizes this concept by grouping users who share information and configuring the networked resources around them. This makes much better

usage of bandwidth than workgroup users who span network segments. The amount of broadcast traffic on the whole network is also reduced, as broadcasts can be isolated within each of the workgroups. A typical drain on network bandwidth is when network servers broadcast their services at regular intervals (in Novell NetWare this can be once every minute, and is known as the Service Advertising Protocol). With vLANs these broadcasts would be contained within each of the vLANs that the server is connected to.

- **Microsegmentation.** This involves dividing a network into smaller segments, which will increase the overall bandwidth available to networked devices.
- **Enhanced security.** vLANs help to isolate network traffic so that traffic which stays within a vLAN will not be transmitted outside it. Thus it is difficult for an external user to 'listen' to any of the data that is transmitted across the vLAN, unless they can get access to one of the ports of the vLAN device. This can be difficult as this would require a physical connection, and increases the chances of the external user being caught 'spying' on the network.
- **Relocate servers into secured locations.** vLANs allows for servers to be put in a physical location in which they cannot be tampered with. This will typically be in a secure room, which is under lock and key. The vLAN can be used to map hosts to servers.
- **Easy creation of IP subnets.** vLANs allow the creation of IP subnets, which are not dependent on the physical location of a node. Users can also remain part of a subnet, even if they move their computer.

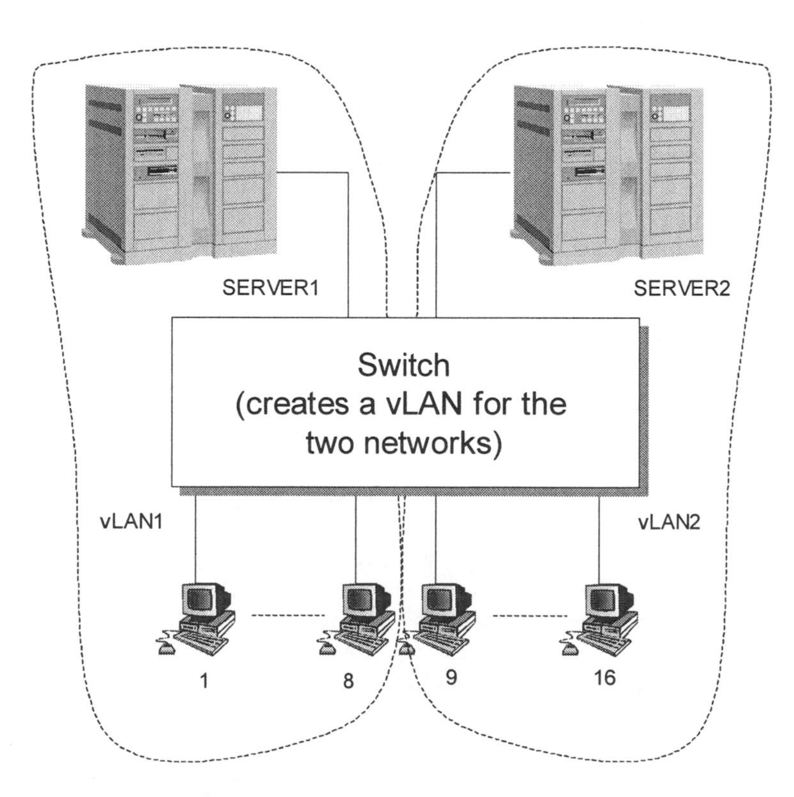

Figure 12.26 Creating a vLAN by mapping ports of a switch

There are also a number of key issues which need to be considered when installing a vLAN:

- **Multivendor compatibility.** This is an important issue as each vendor may try to implement a vLAN is a different way which may be incompatible with other equipment. Typically they differ in their tagging method, which is either implicit (such as tagging ports or MAC addresses) or explicit.
- **Interdomain communications.** In a vLAN frames or packets stay within the domain in which they are assigned (the broadcast domain). A problem occurs, though, when a host on one domain wants to communicate with another on another domain. One of the main objectives of a vLAN is to constrain traffic within a domain, and interdomain communications go against this objective.
- **Support upgradeability.** A major objective in any network is to allow an upgrade path for all the network components, as it is difficult to upgrade a network in a single stroke. Thus an important issue to install vLAN-aware equipment is the need to communicate with non-vLAN-aware (vLAN-unaware) equipment.
- **IP multicast address support.** The IP multicast address allows a single data packet to be received by many hosts. Thus an issue is how the vLAN deals with interdomain communications with IP multicast addresses.

A vLAN can be created by connecting workgroups by a common backbone, where broadcast frames are switched only between ports within the same vLAN. This requires port-mapping to establish the broadcast domain, which is based on a port ID, MAC address, protocol or application. Each frame is tagged with a VLAN ID. Figure 12.27 illustrates that switches are one of the core components of a VLAN. Each switch is intelligent enough to decide whether to forward data frame, based on VLAN metrics (such as port ID, MAC address or network address), and to communicate this information to other switches and routers within the network. The switching is based on frame filtering or frame identification.

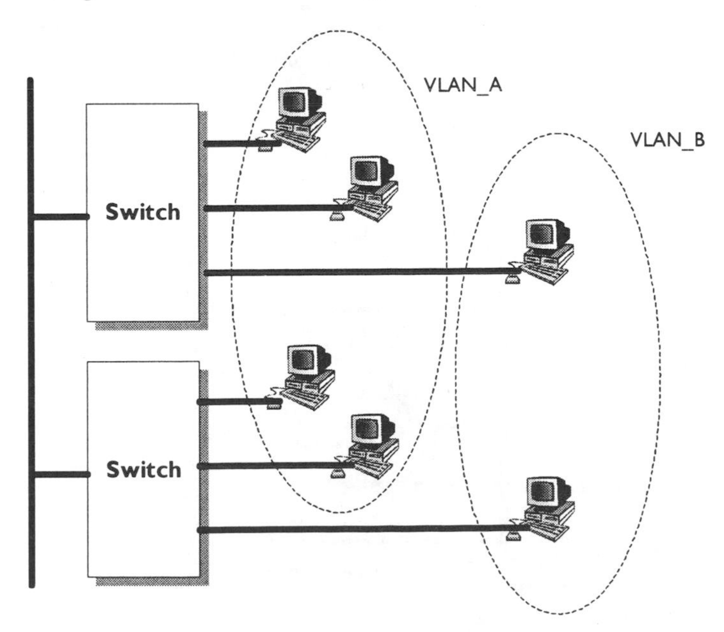

Figure 12.27 vLANs using a backbone and switches

Most early vLANs were based on frame filters, but the IEEE 802.1q vLAN standard is based on frame tagging, as this allows for scaleable networks. With frame tagging, each frame has a uniquely assigned user-defined ID. A unique identifier in the header of each frame is forwarded throughout the network backbone (vertical cabling), as illustrated in Figure 12.28. Each switch then reads the identifier, and if the frame is part of a network which it controls, the switch removes the identifier before the frame is transmitted to the target node (horizontal cabling). As the switching occurs at the data link layer, there is not a great processing time overhead.

vLANs rely on broadcasts to the virtual network, but they are constrained within the virtual network, and thus are not transmitted to other virtual networks. This should reduce the amount of overall network broadcasts (especially from broadcast storms). The broadcast domain can be reduced by limiting the number of switched ports which connect to a specific vLAN. The smaller the grouping, the lower the broadcast effect.

vLANs increase security of data as transmitted data is confined to the vLAN in which it is transmitted. These provide natural firewalls, in which external users cannot gain access to the data within a vLAN. This security occurs, as switch ports can be grouped based on the application type and access privileges. Restricted applications and resources can be placed in a secured VLAN group.

The two types of vLANs are:

- **Static vLANs.** These are ports on a switch that are statically assigned to a VLAN. These remain permanently assigned, until they are changed by the administrator. Static vLANs are secure and easy to configure, and are useful where vLANs are fairly well defined.
- **Dynamic VLANs.** These are ports on a switch which automatically determine their VLAN assignments. This is achieved with intelligent management software, using MAC addresses, logical addressing, or the protocol type of the data packets. Initially, where a node connects to the switch, the switch detects its MAC address entry in the VLAN management database and dynamically configures the port with the corresponding VLAN configuration. The advantage of dynamic vLANs is that they require less setup from the administrator (but the database must be initially created).

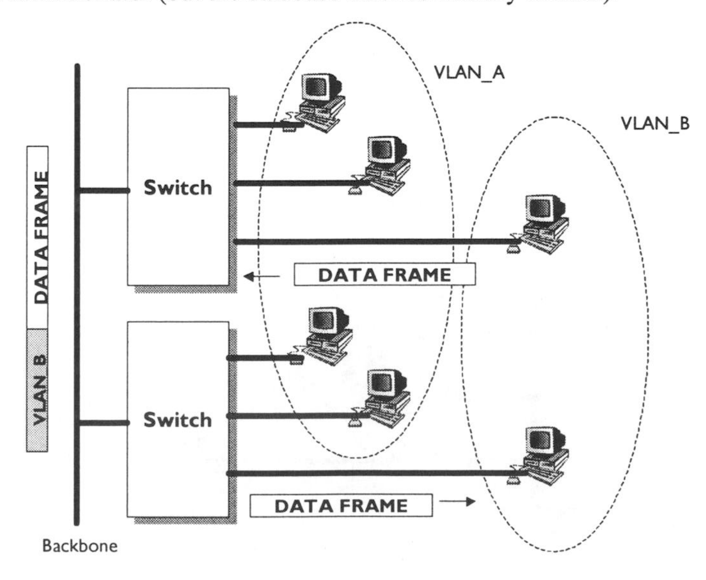

Figure 12.28 vLANs using frame tagging

12.16 Spanning-Tree Protocol

The spanning-tree protocol (STP) allows for redundant switched and bridge connections. Unicast frames have a specific destination MAC address, on which bridges and switches make their forwarding decisions. If they know that the MAC address is not on the source network segment, the bridge/switch floods the data frame onto all ports. The same thing will happen with frames with a broadcast address (FF-FF-FF-FF-FF-FF).

The best known STP is the Spanning Tree Algorithm (STA), which prevents loops by determining a stable spanning-tree network topology. Typically fault-tolerance networks are created with redundant paths. The STA is used to calculate a loop-free path using spanning-tree frames called bridge protocol data units (BPDUs). These are sent out by all STA-enabled bridges and switches are regular intervals, and are used to determine the spanning tree topology.

STP can quickly detect faults on a network connection and activates the stand-by network. The STP states are:

- Blocking. No frames forwarded, BPDUs heard.
- Listening. No frames forwarded, listening for frames.
- Learning. No frames forwarded, learning addresses.
- Forwarding. Frames forwarded, learning addresses.
- Disabled. No frames forwarded, no BPDUs heard.

12.17 Exercises

12.17.1 The base bit rate of standard Ethernet is:
(a) 1 kbps (b) 1 Mbps
(c) 10 Mbps (d) 100 Mbps

12.17.2 The base bit rate of Fast Ethernet is:
(a) 1 kbps (b) 1 Mbps
(c) 10 Mbps (d) 100 Mbps

12.17.3 Standard Ethernet (Thick-wire Ethernet) is also known as:
(a) 10BASE2 (b) 10BASE5
(c) 10BASE-T (d) 10BASE-FL

12.17.4 Thin-wire Ethernet (Cheapernet) is also known as:
(a) 10BASE2 (b) 10BASE5
(c) 10BASE-T (d) 10BASE-FL

12.17.5 Standard Ethernet (Thick-wire Ethernet) uses which type of cable:
(a) Twisted-pair cable (b) Coaxial cable
(c) Fiber-optic cable (d) Radio link

12.17.6 Thin-wire Ethernet (Cheapernet) uses which type of cable:
(a) Twisted-pair cable (b) Coaxial cable
(c) Fiber-optic cable (d) Radio link

12.17.7 Which cable type cannot be used for 100BASE networks:
(a) Cat-3 (b) Cat-5
(c) Coaxial cable (d) Fiber-optic

12.17.8 The IEEE standard for Ethernet is:
(a) IEEE 802.1 (b) IEEE 802.2
(c) IEEE 802.3 (d) IEEE 802.4

12.17.9 The main disadvantage of Ethernet is that:
(a) Computers must contend for the network
(b) It does not network well
(c) It is unreliable (d) It is not secure

12.17.10 A MAC address has how many bits:
(a) 8 bits (b) 24 bits
(c) 32 bits (d) 48 bits

12.17.11 Which MAC address used for a broadcast:
(a) 00-00-00-00-00-00 (b) FF-FF-FF-FF-FF-FF
(c) 12-34-56-78-9A-BC (d) 11-11-11-11-11-11

12.17.12 Which bit pattern identifies the start of an Ethernet frame:
(a) 11001100...1100 (b) 00000000...0000
(c) 11111111...1111 (d) 10101010...1010

12.17.13 The main standards relating to Ethernet networks are:
(a) IEEE 802.2 and IEEE 802.3
(b) IEEE 802.3 and IEEE 802.4
(c) ANSI X3T9.5 and IEEE 802.5
(d) EIA RS-422 and IEEE 802.3

12.17.14 Which layer in the Ethernet standard communicates with the OSI Network layer:
(a) the MAC layer (b) the LLC layer
(c) the Physical layer (d) the Protocol layer

12.17.15 Standard, or Thick-wire, Ethernet is also known as:
(a) 10BASE2 (b) 10BASE5
(c) 10BASE-T (d) 10BASE-F

12.17.16 Twisted-pair Ethernet is also known as:
(a) 10BASE2 (b) 10BASE5
(c) 10BASE-T (d) 10BASE-FL

12.17.17 Fiber-optic Ethernet is also known as:
(a) 10BASE2 (b) 10BASE5
(c) 10BASE-T (d) 10BASE-F

12.17.18 Which type of connector does twisted-pair Ethernet use when connecting to a network hub:
(a) N-type (b) BNC
(c) RJ-45 (d) SMA

12.17.19 Which type of connector does Cheapernet, or thin-wire Ethernet, use when connecting to the network backbone:
(a) N-type (b) BNC
(c) RJ-45 (d) SMA

12.17.20 What is the function of a repeater in an Ethernet network:
(a) It increases the bit rate
(b) It isolates network segments
(c) It prevents collisions
(d) It boosts the electrical signal

12.17.21 What devices do vLANs use:

(a) Switches (b) Hubs
(c) Routers (d) Repeaters

12.17.22 Clearly identify the requirements for ARP, and outline how a computer could determine the MAC address of another computer on a network segment.

12.17.23 Define the difference between a collision domain and a broadcast domain. Which devices can be used to limit the size of the collision domain, which can be to limit the broadcast domain?

12.17.24 Discuss the limitations of 10BASE5 and 10BASE2 Ethernet.

12.17.25 Discuss the main reasons for the preamble in an Ethernet frame.

12.17.26 Discuss 100 Mbps Ethernet technologies with respect to how they operate and their typical parameters.

12.17.27 Explain the usage of Ethernet SNAP.

12.17.28 State the main advantage of Manchester coding and show the line code for the bit sequence:

01111010101101010001011010

12.17.29 Explain the main functional differences between 100BASE-T, 100BASE-T4 and 100VG-AnyLAN.

12.17.30 Prove that the maximum length of segment that can be used with 10 Mbps Ethernet is 840 metres. Assume that the propagation speed is 1.5×10^8m/s and the length of the preamble is 56 bits. Note, a collision must be detected by the end of the transmission of the preamble. Also, why might the maximum length of the segment be less than this?

12.17.31 What problem might be encountered with Fast Ethernet, with respect to the maximum segment length?

12.17.32 **Examination question**

(a) Outline the main fields in the Ethernet IEEE 802.3 frame and the main objectives of preamble.

(b) Explain why the 3-level line code 8B6T is used in 100BASE-T4 and describe what is meant by DC wander.

(c) For Fast Ethernet:

(i) Determine the time taken for a single bit to propagate a distance of 500 m.
(ii) If two nodes are 500 m apart, estimate the minimum time to detect a collision.

(Assume the propagation speed is approximately half the speed of light $c=3\times10^8$ m/s)

12.17.33 Examination question

(a) Identify the usage of each of the fields of an IEEE 802.3 data frame.

(b) Show how a network could be setup using segment switches and network hubs. What advantages do switches have over conventional hubs? Also, explain the advantages of store-and-forward switching.

12.17.34 Examination question

(a) Outline the main 100 Mbps Ethernet technologies. Explain how 100BASE-T4 and 100VG-AnyLAN are able to use Cat-3 cable to transmit at 100 Mbps.

(b) Estimate the minimum and maximum time for the transmission of an Ethernet frame. Also, estimate the maximum number of bits that will be transmitted before a collision will be detected. Assume that the maximum segment length is 500 m and that the electrical signal propagates at half the speed of light. State all assumptions used.

12.18 Additional

The standard connections for 10BASE and 100BASE are:

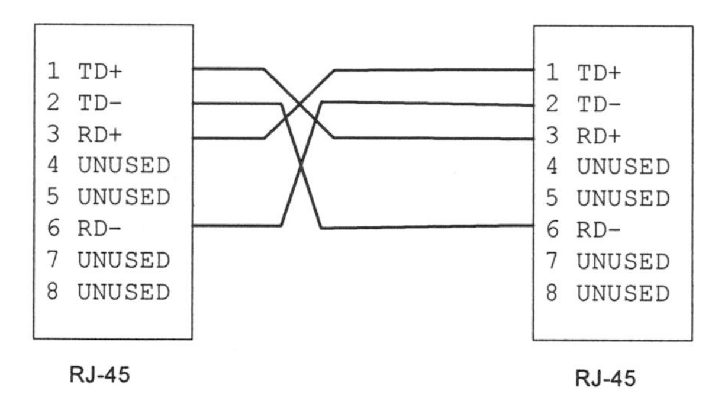

where RD is the receive signals (this is known as RECEIVE in 100BASE) and TD the transmit signals (TRANSMIT). These cable connections are difficult to setup and most connections use a straight through connection. Ports which have the cross-over connection internal in the port are marked with an 'X'.

The standard connections for 100BASE-T4 is given below:

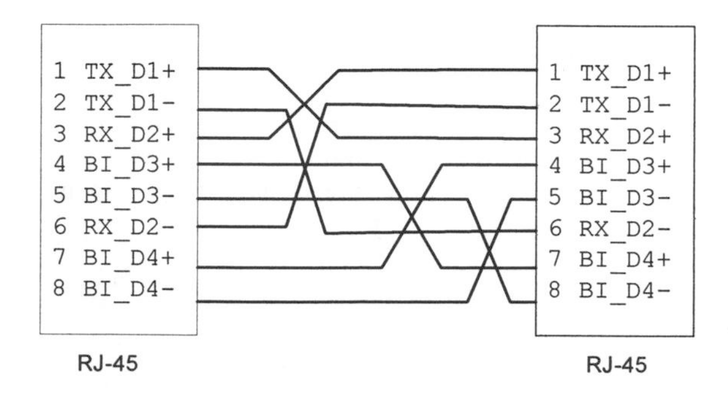

where BI represents the bi-directional transmission signals, TX the transmit signals and RX the receive signals. These cable connections are difficult to set-up and most connections use a straight through connection. Ports which have the cross-over connection internal in the port are marked with an 'X'.

Winipcfg is a useful program for determining networking settings. An example run is given in Figure 12.29. It can be seen that this program shows the MAC address, IP address, Host Name, and so on.

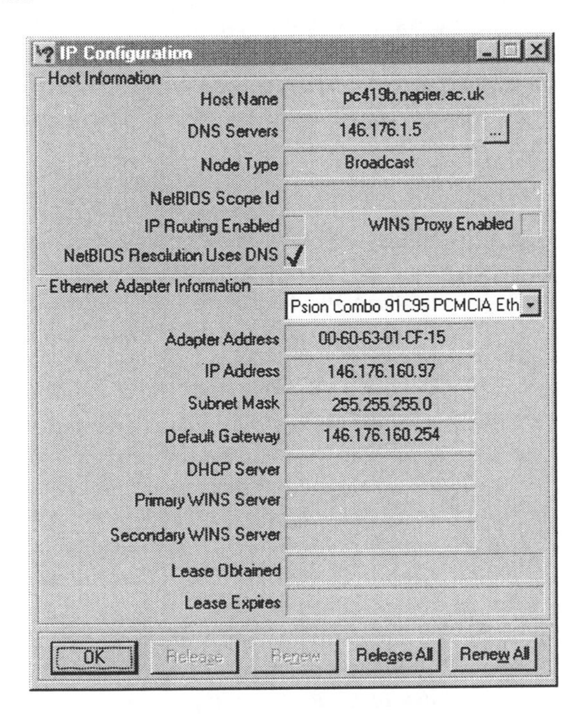

Figure 12.29 Winipcfg program

12.19 Note from the Author

TCP and IP have made the interconnection of the world possible, but Ethernet must be given a silver medal for building up the network from the ground up. Without Ethernet we would have never evolved organizational networks and the Internet so quickly. For anyone who has used a modem from home, and had to make a dial-up connection, will know how annoying this can be. But, Ethernet, plug in a cable from a hub to the computer, and it works seamlessly. No setting dial-ups, no screeching, no flashing lights, no tying up your phone line, no telephone bills, no bringing down the network while you connect or disconnect. Nothing. You don't even have to know what the physical address of the computer is. So how does it work. Well the key is ARP, as ARP allows a computer to broadcast a message to the rest of the network, asking for the MAC address of a given network address. Thus computers can quickly determine the physical addresses of all the devices on their network segment, simply by broadcasting an ARP request. So what if the destination is outside the network segment? Well with this the computer sends the data frame to the router (as it will know its address from a previous ARP request), but with the destination network address. The router detects that the destination address is outside the network segment, and that the data frame is addressed to itself. It will then forward to another router, or a network segment.

Until recently, it seemed unlikely that Ethernet would survive as a provider of network backbones and for campus networks, and its domain would stay, in the short-term, with connections to local computers. The world seemed destined for the global domination of ATM, the true integrator of real-time and non real-time data. This was due to Ethernet's lack of support for real-time traffic and that it does not cope well with traffic rates that approach the maximum bandwidth of a segment (as the number of collisions increases with the amount of traffic on a segment). ATM seemed to be the logical choice as it analyses the type of data being transmitted and reserves a route for the given quality of service. It looked as if ATM would migrate down from large-scale networks to the connection of computers, telephones, and all types of analogue/digital communications equipment. But, remember, not always the best technological solution wins the battle for the market – a specialist is normally always trumped by a good all-rounder.

Ethernet is the best poker player in town. It knows all the tricks. It's a heavyweight prize fighter. It'll slug it out with anyone, and win. It took Token Ring on, head to head, and thrashed it. So what would you choose for your corporate network? Would it be a technology that was cheap, and could give you 10 Mbps, 100 Mbps for your connections to workstations and server, and, possibly, 1 Gbps for your backbone. Ethernet always makes a sensible choice, as it's cheap and it's going to be around for a lot longer, yet. Any problems within an Ethernet network can be solved by segmenting the network, and by relocating servers. And for cable, it supports twisted-pair, coaxial and fiber. Who would have believed that you could get 1 Gbps down a standard Cat-5, twisted-pair cable. Amazing.

Ethernet also does not provide for quality of service and requires other higher-level protocols, such as IEEE 802.1p. These disadvantages are often outweighed by its simplicity, its upgradeability, its reliability and its compatibility. One way to overcome the contention problem is to provide a large enough bandwidth so that the network is not swamped by sources which burst data onto the network. For this, the gigabit Ethernet standard is likely to be the best solution for most networks.

A key method of increasing the bandwidth of a network is to replace hubs with switches, as switches allow simultaneous transmission between connected ports. Thus if the bandwidth of a single port on a switch is 100 Mbps, then a multi-port switch can give a throughput of several times this. But, switches have the potential of improving the configuration of networks.

Many workers are now used to open-plan offices, where the physical environment can be changed as workgroup evolve. This is a concept which is now appearing in networking, where virtual networks are created. With this computers connect to switches. The switch then tags data frames for destination virtual networks and puts the tagged data frame onto the backbone. Other switches then read the tag, and, if the destination is connected to one of their ports, they remove the data tag, and forward the data frame to the required port. This technique is now standardized with IEEE 802.1q, an important step in getting any networking technique accepted. Imagine if whole countries were setup like this. What we would have is a programmable network, where system administrators could connect any computer to any network. Presently we are constrained by the physical location of nodes.

Virtual networks will also bring enhanced security, where it will be possible to constrain the access to sensitive data. For example a server which contains data which must be kept secret can be located in a safe physical environment and only users which a valid MAC address would be allowed access to the data.

Hats off to the IEEE who have carefully developed the basic technology, after its initial conception by DEC, Intel and the Xerox Corporation.

13 ATM

13.1 Introduction

Most of the networking technologies discussed so far are good at carrying computer-type data and they provide a reliable connection between two nodes. Unfortunately, they are not as good at carrying real-time sampled data, such as digitized video or speech. Real-time data from speech and video requires constant sampling and these digitized samples must propagate through the network with the minimum of delay (latency). Any significant delay in transmission can cause the recovered signal to be severely distorted or for the connection to be lost. Ethernet, Token Ring and FDDI simply send the data into the network without first determining whether there is a communication channel for the data to be transported (and rely on the Transport layer to create a reliable connection).

Figure 13.1 shows some traffic profiles for sampled speech and computer-type data (a loading of 1 is the maximum loading). It can be seen that computer-type data tends to burst in periods of time. These bursts have a relatively heavy loading on the network. On the other hand, sampled speech has a relatively low loading on the network but requires a constant traffic throughput. It can be seen that if these traffic profiles were to be mixed onto the same network then the computer-type data would swamp the sampled speech data at various times.

Asynchronous transfer mode (ATM) overcomes the problems of transporting computer-type data and sampled real-time data by:

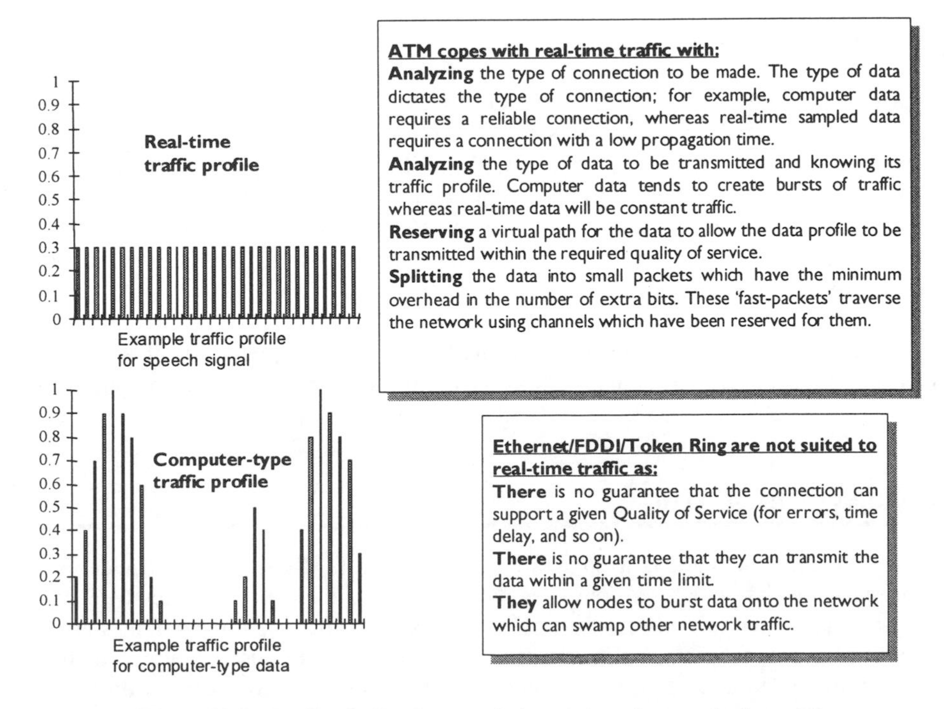

1
0.9
0.8
0.7
0.6
0.5
0.4
0.3
0.2
0.1
0
Real-time traffic profile
Example traffic profile for speech signal
Computer-type traffic profile
Example traffic profile for computer-type data
ATM copes with real-time traffic with:
Analyzing the type of connection to be made. The type of data dictates the type of connection; for example, computer data requires a reliable connection, whereas real-time sampled data requires a connection with a low propagation time.
Analyzing the type of data to be transmitted and knowing its traffic profile. Computer data tends to create bursts of traffic whereas real-time data will be constant traffic.
Reserving a virtual path for the data to allow the data profile to be transmitted within the required quality of service.
Splitting the data into small packets which have the minimum overhead in the number of extra bits. These 'fast-packets' traverse the network using channels which have been reserved for them.
Ethernet/FDDI/Token Ring are not suited to real-time traffic as:
There is no guarantee that the connection can support a given Quality of Service (for errors, time delay, and so on).
There is no guarantee that they can transmit the data within a given time limit.
They allow nodes to burst data onto the network which can swamp other network traffic.

Figure 13.1 Traffic profiles for sampled speech and computer-type data

- Analyzing the type of connection to be made. The type of data dictates the type of connection; for example, computer data requires a reliable connection, whereas real-time sampled data requires a connection with a low propagation delay.
- Analyzing the type of data to be transmitted and knowing its traffic profile. Computer data tends to create bursts of traffic whereas real-time data will be constant traffic.
- Reserving a virtual path for the data to allow the data profile to be transmitted within the required quality of service.
- Splitting the data into small packets which have the minimum overhead in the number of extra bits. These 'fast-packets' traverse the network using channels which have been reserved for them.

ATM has been developed mainly by the telecommunications companies. Unfortunately two standards currently exist. In the USA the ANSI T1S1 subcommittee have supported and investigated ATM and in Europe it has been investigated by ETSI. There are small differences between the two proposed standards, but they may converge into one common standard. The ITU-T has also dedicated study group XVIII to ATM-type systems with the objective of merging differences and creating one global standard for high-speed networks throughout the world. The main advantages are:

- Defined QoS (Quality of Service). This quality of service may relate to bandwidth, error rates, delays, peak bandwidth support, guaranteed path, and so on.
- Scaleable bandwidths. ATM integrates well with existing PCM-TDM systems and can fit into any layer of the existing telecommunication service.
- Connection-oriented. This involves the setup of a virtual circuit prior to data transfer. The sender of the data can thus be sure that the receiver is willing to accept the data, and that the required path has been setup.

Its disadvantages are that it:

- Is a complex networking technology. ATM is an extremely complex networking technology that requires a great deal of software processing.
- Requires an overlaying of existing protocols. ATM, itself, can provide the network layer of the OSI model, and does not required protocols such as IP. Unfortunately it would be too expensive to setup an infrastructure of ATM address servers, thus, IP addressing is typically used to route data between interconnected networks. This means that the network layer is implemented twice.

Note that ATM does not have any mechanisms for guaranteeing delivery of a cell, as it is assumed that higher-level protocols, such as TCP will be used to provide for acknowledgements.

13.2 Objectives of ATM

The major objective of ATM is to integrate real-time data (such as voice and video signals) and non-real-time data (such as computer data and file transfer). Computer-type data can typically be transferred in non-real-time but it is important that the connection is error free. In many application programs, a single bit error can cause serious damage. On the other hand, voice and video data require a constant sampling rate and low propagation delays, but are more tolerant to errors and any losses of small parts of the data.

An ATM network relies on user-supplied information to profile traffic flows so that the connection has the desired quality of service. Figure 13.2 gives four basic data types. These are further complicated by differing data types either sending data in a continually repeating fashion (such as telephone data) or with a variable frequency (such as interactive video). For example a high-resolution video image may need to be sent as several megabytes of data in a short time burst, but then nothing for a few seconds. For speech the signal must be sampled constantly at approximately 8 000 times per second.

Computer data will typically be sent in bursts, with either a high transfer rate (perhaps when running a computer package remotely over a network) or with a relatively slow transfer (such as when reading textural information). Conventional circuit-switched technology (such as ISDN and PCM-TDM) are thus wasteful in their connection because they either allocate a switched circuit (ISDN) or reserve a fixed time slot (PCM-TDM), no matter whether any data is being transmitted at that time. With circuit-switched technologies it may not be possible to service high burst rates by allocating either time slots or switched circuits when all of the other time slots are full, or because other switched circuits are being used.

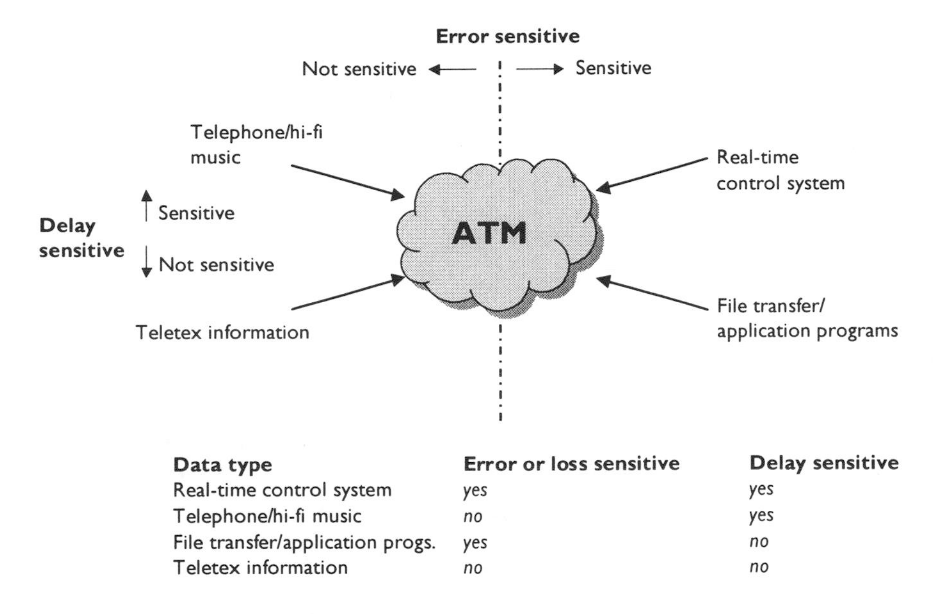

Data type	Error or loss sensitive	Delay sensitive
Real-time control system	*yes*	*yes*
Telephone/hi-fi music	*no*	*yes*
File transfer/application progs.	*yes*	*no*
Teletex information	*no*	*no*

Figure 13.2 Differing data types

13.3 ATM versus ISDN and PCM-TDM

ISDN and PCM-TDM use a synchronous transfer mode (STM) technique where a connection is made between two devices by circuit switching. For this the transmitting device is assigned a given time slot to transmit the data, which is fixed for the period of the transmission. The main problems with this type of transmission are:

- Not all the time slots are filled by data when there is light data traffic; this is wasteful in data transfer.
- When a connection is made between two endpoints a fixed time slot is assigned and data from that connection is always carried in that time slot. This is also wasteful because there may be no data being transmitted in certain time periods.

ATM overcomes these problems by splitting the data up into small fixed-length packets, known as cells. Each data cell is sent with its routing address and follows a fixed route through the network. The packets are small enough that, if they are lost, possibly due to congestion, they can either be requested (for high reliability) or cause little signal degradation (typically in voice and video traffic).

The address of devices on an ATM network are identified by a virtual circuit identifier (VCI), instead of by a time slot as in an STM network. The VCI is carried in the header portion of the fast packet.

13.4 ATM cells

The ATM cell, as specified by the ANSI T1S1 subcommittee, has 53 bytes, as shown in Figure 13.3. The first five bytes are the header and the remaining bytes are the information field which can hold up to 48 bytes of data. Optionally the data can contain a 4-byte ATM adaptation layer and 44 bytes of actual data. A bit in the control field of the header sets the data to either 44 or 48 bytes. The ATM adaptation layer field allows for fragmentation and reassembly of cells into larger packets at the source and destination, respectively. The control field also contains bits which specify whether it is a flow control cell or an ordinary data cell, and a bit to indicate whether this packet can be deleted in a congested network, and so on.

The ETSI definition of an ATM cell also contains 53 bytes with a 5-byte header and 48 bytes of data. The main differences are the number of bits in the VCI field, the number of bits in the header checksum, and the definitions and position of the control bits.

The IEEE 802.6 standard for the MAC layer of the metropolitan area network (MAN) DQDB (distributed queue dual bus) protocol is similar to the ATM cell.

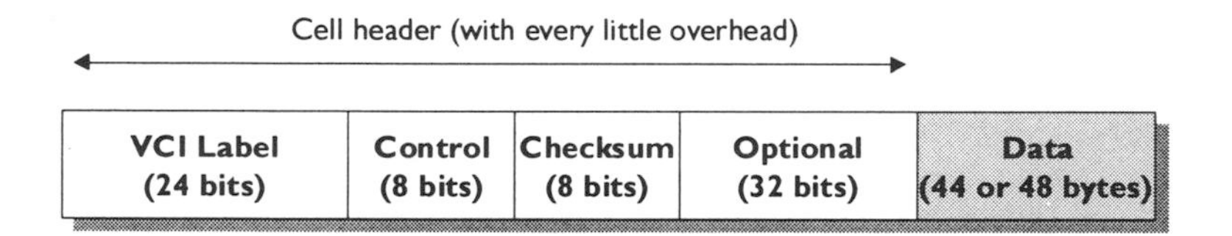

Figure 13.3 ATM cell

13.5 Routing cells within an ATM network

A user network interface (UNI) allows users to gain access to an ATM network. The UNI transmits data into the network with a set of agreed specifications and the network must then try to ensure the connection stays within those requirements. These requirements define the required quality of service for the entire duration of the connection.

In STM networks, data can change its position in each time slot in the interchanges over the global network. This can occur in ATM where the VCI label changes between intermediate nodes in the route. When a transmitting node wishes to communicate through the network it makes contact with the UNI and negotiates parameters such as destination, traffic type, peak and traffic requirements, delay and cell loss requirement, and so on. The UNI forwards this request to the network, from which the network computes a route based on the specified parameters and determines which links on each leg of the route can best support the requested quality of service and data traffic. It sends a connection set-up request to all the nodes in the path en route to the destination node.

Figure 13.4 shows an example of ATM routing. In this case, User 1 connects to Users 2 and 3. The virtual path set up between User 1 and User 2 is through the ATM switches 2, 3 and 4, whereas User 1 and User 3 connect through ATM switches 1, 5 and 6. A VCI number of 12 is assigned to the path between ATM switches 1 and 2, in the connection between User 1 and User 2. When ATM switch 2 receives a cell with a VCI number of 12 then it sends the cell to ATM switch 3 and gives it a new VCI number of 6. When it gets to ATM switch 3 it is routed to ATM switch 4 and given the VCI number of 22. The virtual circuit for User 1 to User 3 is through ATM switches 1, 5 and 6, and the VCI numbers used are 10 and 15. Once a connection is terminated the VCI labels assigned to the communications are used for other connections.

Certain users, or applications, can be assigned reserved VCI labels for special services that may be provided by the network. However, as the address field only has 24 bits it is unlikely that many of these requests would be granted. ATM does not provide for acknowledgements when the cells arrive at the destination (as it relies on the Transport layer for these services).

Note that as there is a virtual circuit set up between the transmitting and receiving node then cells are always delivered in the same order as they are transmitted. This is because cells cannot take alternative routes to the destination. Even if the cells are buffered at a node, they will still be transmitted in the correct sequence.

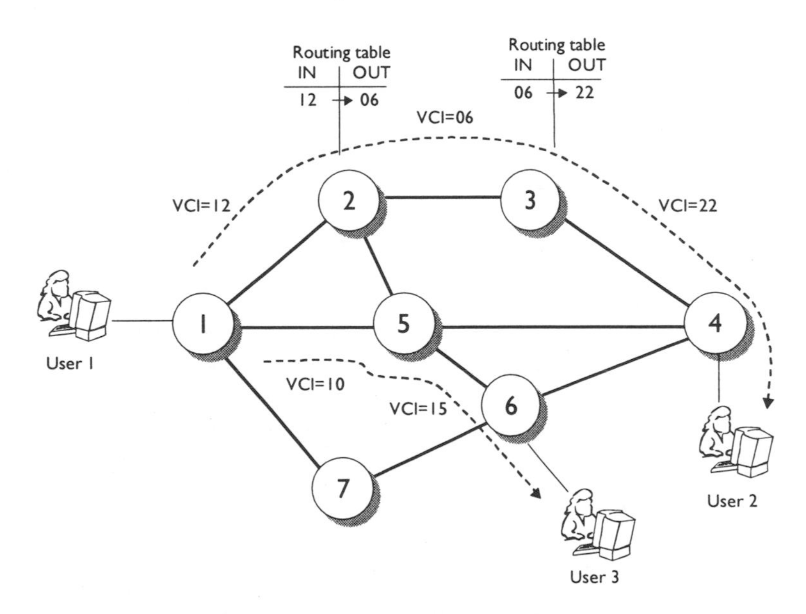

Figure 13.4 A virtual ATM virtual connection

13.5.1 VCI header

The 5-byte ATM user network cell header splits into six main fields:

- GFC (4 bits), which is the generic control bit, and is used only for local significance and is not transmitted from the sender to the receiver.
- VPI (8 bits), which is the path connection identifier (VPI). See next section for an explanation of virtual paths.
- VCI (16 bits) which is the virtual path/channel identifier (VCI). Its usage was described in the previous section. Each part of the route is allocated a VCI number.

- PT (3 bits), which is the payload type field. This is used to identify the higher-layer application or data type.
- CLP (1 bit), which is the cell loss priority, bit and indicates if a cell is expendable. When the network is busy an expendable cell may be deleted. If CLP=1, cells are dropped before any CLP=0 cells are dropped.
- HEC (8 bits), which is the header error control field. This is an 8-bit checksum for the header. It uses a cyclic code with a generating polynomial of $x^8 + x^2 + x + 1$. The first 4 bytes, written as a polynomial, are multiplied by x^8 and divided by the generating polynomial. Next 01010101 is added. The remainder is the HEC field, and transmitted to the receiver. At the receiver, 01010101 is first subtracted before further interpretation.

Note that the user-to-network cell differs from the network-to-network cell. A network-to-network cell, for a network-network interface (NNI), uses a 12-bit VPI field and does not have a GFC field. Otherwise, it is identical. Figure 13.5 shows the format of the UNI (User-network interface) cell, and the NNI cell.

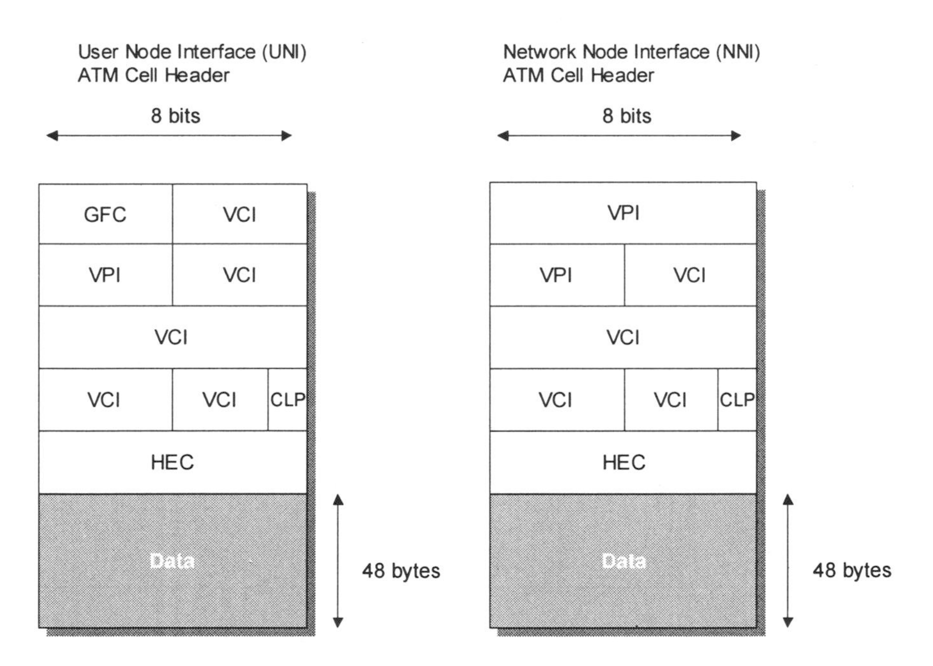

Figure 13.5 ATM cells for UNI and NNI

13.6 Virtual channels and virtual paths

Virtual circuits are set up between two users when a connection is made. Cells then travel over this fixed path through a reserved route. Often several virtual circuits take the same path, which can be grouped together to form a virtual path.

A virtual path is defined as a collection of virtual channels which have the same start and end points, as illustrated in Figure 13.6. These channels will take the same route. This makes the network administration easier and allows new virtual circuits, with the same route, to be easily set up.

Some of the advantages of virtual paths are:

- Network user groups or interconnected networks can be mapped to virtual paths and are thus easily administered.
- Simpler network architecture that consists of groups (virtual paths) with individual connections (virtual circuits).
- Less network administration and shorter connection times arise from fewer set-up connections.

Virtual circuits and virtual paths allow two levels of cell routing through the network. A VC switch routes virtual circuits and a VP switch routes virtual paths. Figure 13.7 shows a VP switch and a VC switch. In this case, the VP switch contains the routing table which maps VP1 to VP2, VP2 to VP3 and VP3 to VP1. This switch does not change the VCI number of the incoming virtual circuits (for example VC1 goes in as VC1 and exits as VC1).

The diagram shows the concepts between both types of switches. The VP switch on the left-hand side redirects the contents of a virtual path to a different virtual path. The virtual connections it contains are unchanged. This is similar to switching an input cable to a different physical cable. In a VC switch the virtual circuits are switched. In the case of Figure 13.7 the routing table will contain VC7 mapped to VC5, VC8 to VC6, and so on. The VC switch thus ignores the VP number and only routes the VC number. Thus the input and output VP number can change.

A connection is made by initially sending routing information cells through the network. When the connection is made, each switch in the route adds a link address for either a virtual path and or a virtual connection. The combination of VP and VC addressing allows for the support of any addressing scheme, including subscriber telephone numbering or IP addresses. Each of these address can be broken down in a chain of VPI/VCI addresses.

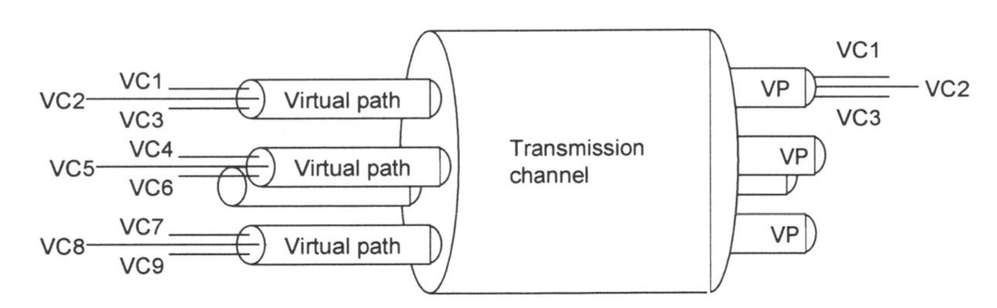

Figure 13.6 Virtual circuits and virtual paths

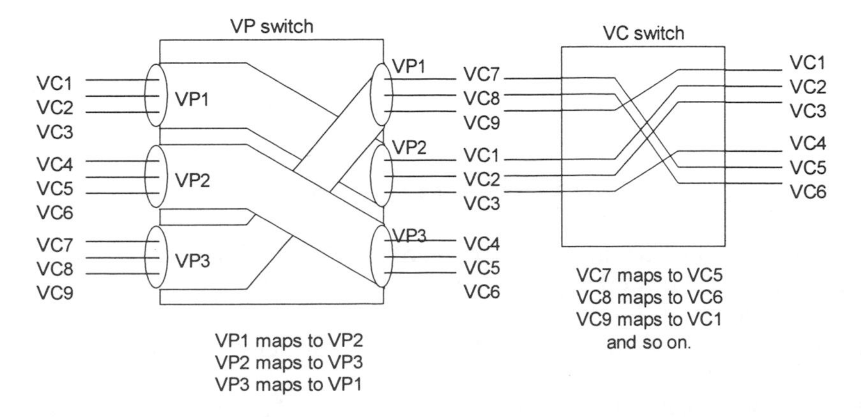

Figure 13.7 Virtual circuits and virtual paths

The two types of connections are thus:

- Permanent Virtual Connections (PVC). These are connections which are set up by some external mechanism, such as with network management software. Each of the switches on the route from the ATM source and destination ATM system are set up with the appropriate VPI/VCI values. Normally these are setup manually.
- Switched Virtual Connections (SVC). These are connections which are automatically set up using the signaling protocol. They are set up automatically, and require no manual interaction.

13.7 Statistical multiplexing

Fast packet switching attempts to solve the problem of unused time slots of STM. This is achieved by statistically multiplexing several connections on the same link based on their traffic characteristics. Applications, such as voice traffic, which requires a constant data transfer, are allowed safe routes through the network. Whereas several applications, which have bursts of traffic, may be assigned to the same link in the hope that statistically they will not all generate bursts of data at the same time. Even if some of them were to burst simultaneously, their data could be buffered and sent at a later time. This technique is called statistical multiplexing and allows the average traffic on the network to be evened out over a relatively short time period. This is impossible on an STM network.

13.8 ATM signaling and call set-up

ATM, as with most telecommunications systems, uses a single-pass approach to setting up a connection. Initially, the source connection (the source end-system) communicates a connection request to the destination connection (the destination end-point). The routing protocol manages the routing of the connection request and all subsequent data flow. The call is established with:

- A set-up message. This is initially sent, across the UNI, to the first ATM switch. It contains:

  - Destination end-system address.
  - Desired traffic.
  - Quality of service.
  - Information Elements (IE) defining particular desired higher-layer protocol bindings and so on.

- The initial ATM switch sends back a local call proceeding acknowledgement to the source end-system.
- The initial ATM switch invokes an ATM routing protocol, and propagates a signaling request across the network; it finally reaches the ATM switch connected to the destination end-system.
- The destination ATM switch connected to the destination end-system forwards the set-up message to the end-system, across its UNI.
- The destination end-system either accepts or rejects the connection. If necessary, the destination can negotiate the connection parameters. If the destination end-system rejects the connection request, it returns a release message. This is also sent back to the

source end-system and clears the connection, such as clearing any allocated VCI labels. A release message can also be used by any of the end-systems, or by the network, to clear an established connection.

- If the destination end-system accepts the call then the ATM switch, which connects to it, returns a connect message through the network, along the same path.
- When the source end-system receives and acknowledges the connection message, either node can then start transmitting data on the connection (the chain of allocated VCI labels defines the route).

The connections in an ATM network are:

- UNI (user-network interface). The UNI connects to ATM end-systems.
- NNI (network-node interface). The NNI is any physical or logical link across which two ATM switches exchange the NNI protocol.

13.8.1 Signaling packets

A connection is set up using signal information, which are sent using signaling packets. It uses a one-pass approach to setting up the virtual path, Along this path, each of the ATM switches are also preconfigured to receive any signaling packets sent across a connection, they pass them to the signaling process which will service them. In general, VCI labels from 0 to 31 for each VPI are reserved for control purposes. Thus all allocated VCI labels will be above 32. The signaling packet has the VCI label of **five**, and a VPI label of **zero**, and is passed from switch to switch. As it passes each switch, they will set up a connection identifier (VCI/VPI labels) for the connection. Two labels are used for each connection to identify the traffic flowing in one either of the two directions (such as from Switch *X* to Switch *Y*, and from Switch *Y* to Switch *X*). Figure 13.8 shows an example of a call set up from End system A to End system B. The phases of the connection will be:

- **Signaling request.** Initially a signal packet is initiated from End system A which requests to Switch 1 that it wants to connect with End system B, this is then passed onto Switch 2, then to Switch 3, and so on, until it reaches End system B. This is achieved with a Connect message.
- **Connection routed.** Each of the switches along the route will reserve VCI/VPI labels for data traveling in one of the two directions.
- **Connection accept/reject.** End system B will decide if it wishes to accept or reject the connection. If it accepts it sends back an Okay signal to all the switches along the route.
- **Data flow.** After the virtual circuit has been setup, the two end systems can transmit data using the VCI/VPI labels that have been allocated for the connection. If the end system rejects the connection, all the switches along the reserved route will delete the reserved VCI/VPI labels from their routing tables (using a Release message).
- **Connection teardown.** After the data has been transmitted a signaling packet is sent along the route, and all the switches delete the allocated VCI/VPI labels from their routing tables. This is achieved with a Release message.

The routing of the connection request, and the resultant virtual data path is defined by an ATM routing protocol, such as the P-NNI (Private-NNI) protocol. These protocols route the connection request using:

- Destination address.
- Traffic information. The ATM routing protocols can pass information on traffic flows,

so that connections can be setup so that they can avoid heavily used traffic routes. The ATM routing protocol will thus allow the network to made better use of the available bandwidth.

- QoS parameters. These are the parameters requested by the source end-system.

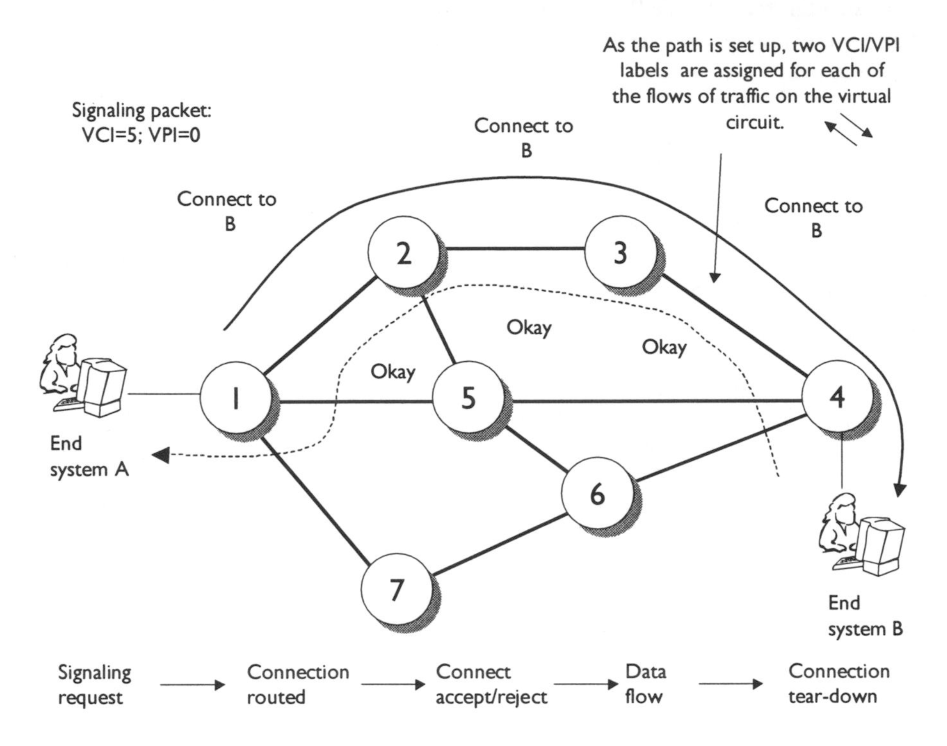

Figure 13.8 Connection set-up

The two fundamental types of ATM connections are:

- **Point-to-point connections.** This is a connection which connects two ATM end-systems, and data flow can either be unidirectional or bidirectional.
- **Point-to-multipoint connections.** This is a connection which connects a single source end-system (*the root node*) to many destination end-systems (*the leaves*). In this, an ATM switch replicates the cells for the multiply connected branches. This type of connection is unidirectional, and allows the root to transmit to the leaves, but not the leaves to transmit to the root, or to each other, on the same connection. This technique is similar to multicasting and/or broadcasting, which is typically used in Ethernet systems. The reason that these are unidirectional, is that if leaf nodes were to transmit cells they would be received by both the root node and all other leaf nodes.

13.8.2 ATM multicasting

As previously identified ATM does not support proper multicasting using bi-directional data flow. It is important, though, that ATM supports it, as most high-level protocols assume some form of multicast/broadcast. Broadcasting and multicasting, especially, is important in connection-less, shared medium networks, such as Ethernet. Several methods have been proposed to overcome this problem, these are:

- **Multicast server.** With this, all nodes wishing to transmit to a multicast group contact a multicast server through a point-to-point connection. Then, the multicast server connects to all nodes that are to receive the multicast packets through a point-to-multipoint connection, or point-to-point connections (with packet replication). The multicast server only sends out the received packets when the packets are serialized (that is, one packet is fully transmitted prior to the next being sent). This overcomes the problem of packet interleaving.
- **VP-multicasting.** With this, a multipoint-to-multipoint virtual path (VP) links all nodes in the multicast group. Each node is then given a unique VCI value within the VP, and the unique VCI label identifies interleaved packets. This technique requires a protocol which will uniquely allocate VCI values to nodes.
- **Overlaid point-to-multipoint connections.** With this, all the nodes in a multicast group set up a point-to-multipoint connection with each of the other nodes in the group. They then become leaves to all the other nodes, and all the nodes can both transmit/receive to/from all the other nodes.

The multicast server only requires two connections for each node, while the overlaid method requires *n* connections to connect to *n* transmitting nodes in the group. It has the disadvantage that it can be a bottleneck for group set up, and also a single-point-of-failure (as it is a centralized resequencer). The overlaid method has the advantage that there is no single point of failure, or bottlenecks, but has the disadvantage that it requires some mechanism for nodes to be informed of all the nodes that are currently in a group, and for all the existing nodes in a group to be told about a new member of the group.

13.9 ATM flow control

ATM cannot provide for a reactive end-to-end flow control because by the time a message is returned from the destination to the source, large amounts of data could have been sent along the ATM pipe, possibly making the congestion worse. The opposite can occur when the congestion on the network has cleared by the time the flow control message reaches the transmitter. The transmitter will thus reduce the data flow when there is little need. ATM tries to react to network congestion quickly, and it slowly reduces the input data flow to reduce congestion.

This rate-based scheme of flow control involves controlling the amount of data to a specified rate, agreed when the connection is made. It then automatically changes the rate based on the past history of the connection as well as the present congestion state of the network.

Data input is thus controlled by early detection of traffic congestion through closely monitoring the internal queues inside the ATM switches, as shown in Figure 13.8. The network then reacts gradually as the queues lengthen and reduces the traffic into the network from the transmitting UNI. This is an improvement over imposing a complete restriction on the data input when the route is totally congested. In summary, anticipation is better than desperation. A major objective of the flow control scheme is to try to affect only the streams which are causing the congestion, not the well-behaved streams.

The great advantage of ATM is that it guarantees a QoS, which involves setting up a suitable connection for the given QoS. It is thus important to understand how network traffic will vary over the time of the connection. The source node must inform the network about the traffic parameters and desired QoS for each direction of the requested connection upon initial set-up, such as bandwidth or delay (latency).

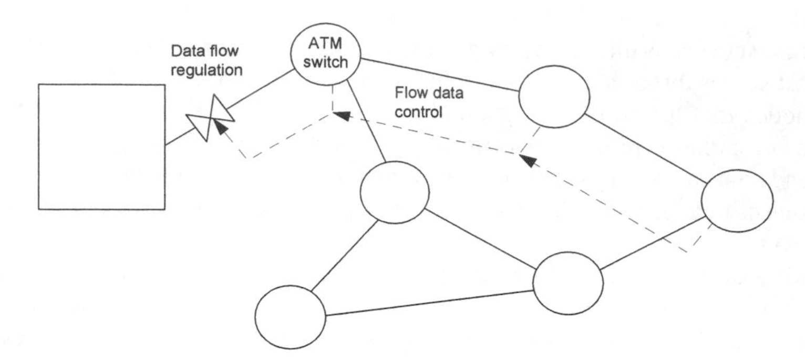

Figure 13.9 Flow control feedback from ATM switches

Current and proposed QoS levels are:

- **CBR** (Continuous Bit Rate). This is used to support constant bit rate traffic with a fixed timing relationship between data samples (such as with real-time speech).
- **VBR-RT** (Variable Bit Rate – Real Time). This is used to carry variable bit rate traffic that has a fixed time between samples. A typical application is variable bit rate video compression.
- **VBR-NRT** (Variable Bit Rate – Non-Real Time). This is used to carry variable bit rate traffic that has no timing relationship between data samples, but still requires a guaranteed QoS on bandwidth or latency. An example of this is in the interconnection of frame relay devices, where a guaranteed bandwidth is required for the connection.
- **ABR** (Available Bit Rate). This is used when the source and destination do not need to support a given bit rate, or any timing information. It is similar to VBR-NRT, but ABR service does not guarantee any bandwidth to the user, and the network will try to achieve the best service possible using feedback (flow control mechanisms, as shown in Figure 13.9) to increase the bandwidth available to the user, which is the Allowed Cell Rate (ACR). There may also be a Minimum Cell Rate (MCR), which will be the lowest value that the ACR can reach. It is likely this ABR will be supported with a rate-based system for ABR congestion control. With this Resource Management Cells (RMCs) or the explicit forward congestion indication (EFCI) bit within the ATM cell will be used to indicate to the source system that there is congestion in the network. The ACR is then controlled by a traffic-pacing algorithm, which will determine the traffic entering the network. The pacing will be either based on the number of received RM cells with a congestion indication or an explicit rate indication from the network. ABR is excellent at carrying LAN-type traffic, as the protocols that are used in LANs typically expect the maximum bandwidth that is available on the network, but will back-off or buffer data when the network becomes busy.
- **UBR** (Unspecified Bit Rate). This is used when there is no guarantee of QoS. Users can transmit as much data into the network as they want, but there is no guarantee on cell lost rate, delay, or delay variation. As UBR does not support any flow control mechanisms, it is important that there is cell buffering.

As has been seen there is no explicit priority field for an ATM connection. The only priority indicator in the cell is Cell Loss Priority (CLP) bit, which defines whether the cell can be dropped, or not. Cells with a CLP of 1 are dropped before cells with a CLP of 0). This bit can either be set by the end user, but is more likely to be set by the network. The basic parameters for defining traffic are:

- Burst Tolerance (BT). In VBR, the BT defines the maximum burst of transmitted contiguous cells.
- Cell Delay Variation Tolerance (CDVT).
- Minimum Cell Rate (MCR), if supported.
- Peak Cell Rate (PCR). In ABR, this defines the maximum value of ACR. The ACR will thus vary between MCR and PCR.
- Sustainable Cell Rate (SCR). In VBR, the SCR defines the long-term average cell rate.

These define the limits of a connection, but only some of these will be used for a given QoS.

13.10 ATM and the OSI model

The basic ATM cell fits roughly into the data link layer of the OSI model, but contains some network functions, such as end-to-end connection, flow control, and routing. It thus fits into layers 2 and 3 of the model, but is typically used to implement layers 2 and 3, with IP operating at layer 3, as shown in Figure 13.10. The layer 4 software layer, such as TCP (as covered in Chapter 15), can communicate directly with ATM, but typically IP is used as it provides for a global addressing structure.

The ATM network provides a virtual connection between two gateways and the IP protocol fragments IP packets into ATM cells at the transmitting UNI which are then reassembled back into the IP packets at the destination UNI.

With TCP/IP each host is assigned an IP address as it is the ATM gateway. Once the connection has been made then the cells are fragmented into the ATM network and follow a predetermined route through the network. At the receiver the cells are reassembled using the ATM adaptation layer. This reforms the original IP packet, which is then passed to the next layer.

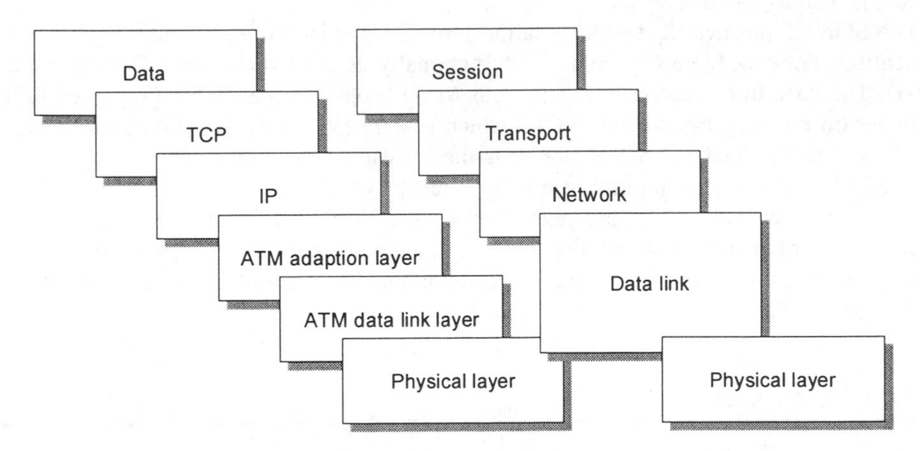

Figure 13.10 ATM and the OSI model

The functions of the three ATM layers are:

- ATM adaptation layer (AAL) – segmentation and reassembly of data into cells and vice-versa, such as convergence (CS) and segmentation (SAR). It is also involved with quality of service (QOS).

- ATM data link (ADL) – maintenance of cells and their routing through the network, such as generic flow control, cell VPI/VCI translation, and cell multiplex and demultiplex.
- ATM physical layer (PHY) – transmission and physical characteristics, such as cell rate decoupling, HEC header sequence generation/verification, cell delineation, transmission frame adaptation and transmission frame generation/recovery.

The overlay model corresponds to layer 2 of the OSI model (that is, the data link layer). But, ATM actually has most, if not all, of the characteristics of a layer 3 (this is, the network layer), such as having a hierarchical address space and a complex routing protocol. Unfortunately, the original OSI model did not support overlaid networks, where one network layer overlays on another. This has since been added to the model. Overlaid networks are useful when carrying a certain network protocol (such as TCP/IP or SPX/IPX) between two networks, transparently.

ATM can be used to build simpler networks as ATM addresses can be made to be hierarchical, whereas MAC addresses are flat addresses. A flat address is typically a randomly generated address, which does not give any clues about the physical location of the node. A MAC-level bridge does not require to be set up in the same way as routers are set, because they flood data frames through the network.

13.11 ATM physical layer

The physical layer is not an explicit part of the ATM definition, but is currently being considered by the standards organizations. T1S1 has standardized on SONET (Synchronous Optical NETwork) as the preferred physical layer, with STS-3c at 155.5 Mbps, STS-12 at 622 Mbps and STS-48 at 2.4 Gbps. The range is 800 m using a fiber-optic cable, or 100 m for coaxial cable (155/622 Mbps).

The SONET physical layer specification provides a standard worldwide digital telecommunications network hierarchy, known internationally as the Synchronous Digital Hierarchy (SDH). The base transmission rate, STS-1, is 51.84 Mbps. This is then multiplexed to make up higher bit rate streams, such as STS-3 which is 3 times STS-1, STS-12 which is 12 times STS-1, and so on. The 155 Mbps stream is the lowest bit rate for ATM traffic and is also called STM-1 (synchronous transport module – level 1).

The SDH specifies a standard method on how data is framed and transported synchronously across fiber-optic transmission links without requiring that all links and nodes have the same synchronized clock for data transmission and recovery. ATM is transported within the data part of the STM frame.

13.12 AAL service levels

The AAL layer uses cells to process data into a cell-based format and to provide information to configure the level of service required.

13.12.1 Processing data

The AAL performs two essential functions for processing data as shown: the higher-level protocols present a data unit with a specific format. This data frame is then converted using a convergence sublayer with the addition of a header and trailer that give information on how the data unit should be segmented into cells and reassembled at the destination. The data is then segmented into cells, together with the convergence subsystem information and other management data, and sent through the network.

13.12.2 AAL services

The AAL layer, as part of the process, also defines the level of service that the user wants from the connection. The following shows the four classes supported. For each class there is an associated AAL (as given in Table 13.1). Class A supports a constant bit rate with a connection and preserves timing information, and is typically used for voice transmission. Class B is similar to A but has a variable bit rate, and is typically it is used for video/audio data (such as, over the Internet). Class C also has a variable bit rate and is connection-oriented, but does not contain any information. This is typically used for non-real-time data, such as computer data. Class D is the same as class C, but is connectionless. This means there is no connection between the sender and the receiver before the data is transmitted.

There are four AAL services: AAL1 (for class A), AAL2 (for class B), AAL3/4 or AAL5 (for class C) and AAL3/4 (for class D).

Table 13.1 Four basic categories of data

	Timing information between source and destination	*Bit rate characteristics*	*Connection mode*
Class A	✓	Constant	Connection-oriented
Class B	✓	Variable	Connection-oriented
Class C	×	Variable	Connection-oriented
Class D	×	Variable	Connectionless

AAL1

The AAL1 supports class A and is intended for real-time voice traffic; it provides a constant bit rate and preserves timing information over a connection. The format of the 48 bytes of the data cell consists of 47 bytes of data, such as PCM or ADPCM code and a 1-byte header. Figure 13.11 shows the format of the cell, including the cell header. The 47-byte data field is described as the SAR-PDU (segmentation and reassembly protocol data units). The header consists of:

- SN (4 bits) which is a sequence number.
- SNP (4 bits) which is the sequence number protection.

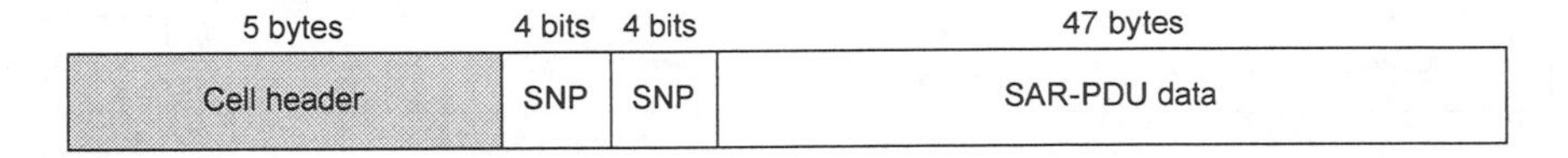

Figure 13.11 Cell format

AAL type 2

AAL type 2 is under further study.

AAL type 3/4

Type 3/4 is connection-oriented where the bit rate is variable and there is no need for timing information (Figure 13.12). It uses two main formats:

- SAR (segment and reassemble) which is segments of CPCS PDU with a SAR header and trailer. The extra SAR fields allow the data to be reassembled at the receiver. When the CPCS PDU data has been reassembled the header and trailer are discarded. The fields in the SAR are:

- Segment type (ST) identifies how the SAR has been segmented. Figure 13.13 shows how the CPCS PDU data has been segmented into five segments: one beginning segment, three continuation segments and one end segment. The ST field has 2 bits and can therefore contain one of four possible types:

  - SSM (single sequence message) identifies that the SAR contains the complete data.
  - BOM (beginning of message) identifies that it is the first SAR PDU in a sequence.
  - COM (continuation of message) identifies that it is an intermediate SAR PDU.
  - EOM (end of message) identifies that it is the last SAR PDU.

- Sequence number (SN) which is used to reassemble a SAR SDU and thus verify that all of the SAR PDUs have been received.
- Multiplexing indication (MI) which is a unique identifier associated with the set of SAR PDUs that carry a single SAR SDU.
- Length indication (LI) which defines the number of bytes in the SAR PDU. It can have a value between 4 and 44. The COM and BOM types will always have a value of 44. If the EOM field contains fewer than 44 bytes, it is padded to fill the remaining bytes. The LI then indicates the number of value bytes. For example if the LI is 20; there are only 20 value bytes in the SAR PDU, the other 24 are padding bytes.
- CRC which is a 10-bit CRC for the entire SAR PDU.

- CPCS (convergence protocol sublayer) takes data from the PDU. As this can be any length, the data is padded so it can be divided by four. A header and trailer are then added and the completed data stream is converted into one or more SAR PDU format cells. Figure 13.13 shows the format of the CPCS-PDU for AAL type 3/4.

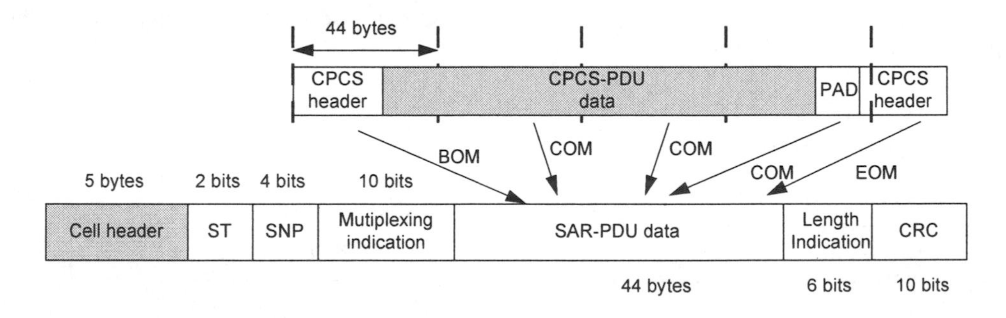

Figure 13.12 AAL type 3/4 cell format

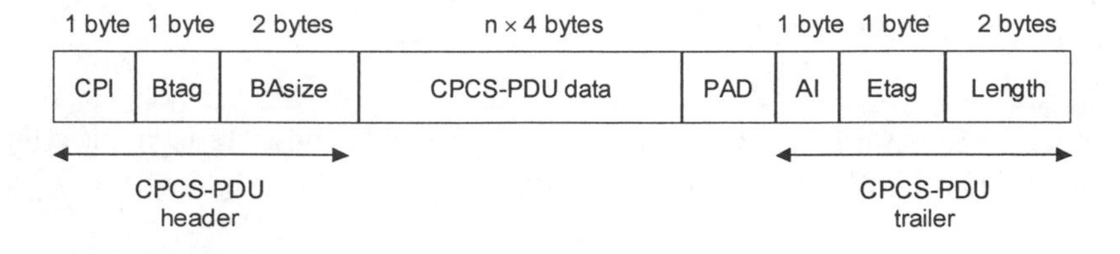

Figure 13.13 CPCS-PDU type 3/4 frame format

The fields in the CPCS-PDU are:

- CPI (common part indicator) which indicates how the remaining fields are interpreted (currently one version exists).
- Btag (beginning tag) which is a value associated with the CPCS-PDU data. The Etag has the same value as the Btag.
- BASize (buffer allocation size) which indicates the size of the buffer that must be reserved so that the completed message can be stored.
- AI (alignment) is a single byte which is added to make the trailer equal to 32 bits.
- Etag (end tag) which is the same as the Btag value.
- Length which gives the length of the CPCS PDU data field.

AAL type 5

AAL type 5 is a connectionless service; it has no timing information and can have a variable bit rate. It assumes that one of the levels above the AAL can establish and maintain a connection (such as TCP). Type 5 provides stronger error checking with a 32-bit CRC for the entire CPCS PDU, whereas type 3/4 only allows a 10-bit CRC which is error checking for each SAR PDU. The type 5 format is given in Figure 13.14. The fields are:

- CPCS-UU (CPCS user-to-user) indication.
- CPI (common part indicator) which indicates how the remaining fields are interpreted (currently one version exists).
- Length which gives the length of the CPCS-PDU data.
- CRC which is a 32-bit CRC field.

The type 5 CPCS-PDU is then segmented into groups of 44 bytes and the ATM cell header is added. Thus, type 5 does not have the overhead of the SAR-PDU header and trailer (i.e. it does not have ST, SN, MID, LI or CRC). This means it does not contain any sequence numbers. It is thus assumed that the cells will always be received in the correct order and none of the cells will be lost. Types 3/4 and 5 can be summarized as follows:

Type 3/4: SAR-PDU overhead is 4 bytes, CPCS-PDU overhead is 8 bytes.
Type 5: SAR-PDU overhead is 0 bytes, CPCS-PDU overhead is 8 bytes.

Type 5 can be characterized as:

- Strong error checking.
- Lack of sequence numbers.
- Reduced overhead of the SAR-PDU header and trailer.

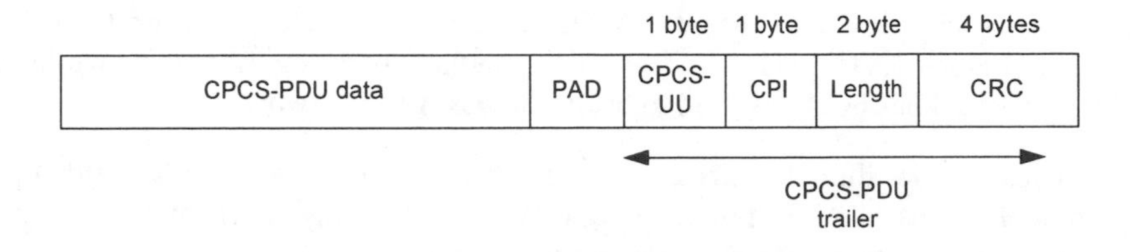

Figure 13.14 CPCS-PDU type 5 frame format

13.13 ATM LAN emulation

For ATM to get global acceptance it must support existing LANs and WANs, and must thus support existing network protocols, such as IP and IPX. There are two methods that ATM can use for this:

- **LANE (LAN Emulation).** With this, network layer packets are carried across the ATM network, so as to emulate a local area network. The LANE protocol defines the techniques to emulate IEEE 802.3 (Ethernet) and IEEE 802.5 (Token Ring) networks. LANE thus encapsulates the data in the appropriate LAN MAC packet format and sends it over the ATM network. The protocol does not, itself, try to emulate the actual media access control protocol of the specific LAN concerned (that is, CSMA/CD for Ethernet or token passing for 802.5). It makes the ATM network look like an Ethernet or Token Ring LAN. The main function of the LANE protocol is to resolve MAC addresses into ATM addresses. This is equivalent to a protocol for MAC bridging on ATM.
- **Native mode operation.** With this, address resolution techniques are used to map network layer addresses directly into ATM addresses. The network layer packets are then carried across the ATM network. Most network layer protocols can run directly over ATM networks, but only IP and IPX have been extensively implemented. Novell has outlined a protocol known as Connection Oriented IPX (CO-IPX), which allows IPX to be used over ATM networks. The LANE protocol does not fully use the Quality of Service (QoS) parameter and the capability for real-time multimedia applications. It is possible to run applications directly over ATM, without requiring the need for a network or a transport stack (no TCP/IP). This reduces the overhead of the IP and TCP protocols. Unfortunately, it limits the applications to running over purely ATM, which is difficult as long as there are non-ATM networks. In the future, with a totally integrated ATM network, from desktop-to-desktop, this will be possible. One of the main functions of the network layers is to provide network transparency, where the type of network is transparent to the application. If this layer is deleted it does not allow the applications to communicate over different networks.

13.13.1 LANE protocol

The LANE protocol allows for easy interfacing of ATM to existing IP or IPX-based networks, as it presents the same service interface of existing MAC protocols to network layer drivers (for example, an NDIS- or ODI-like driver interface), as illustrated in Figure 13.15.

The LANE protocol can be implement in two ways:

- ATM Network Interface Cards (NICs). The ATM NIC implements LANE, which will present the same software interface as a LAN NIC. This will increase the bandwidth of the LAN, as ATM offers high bit rates.
- Routers and switches. The routers or switches can also use the LANE protocol, as illustrated in Figure 13.16. This is an excellent method of bridging two networks, as the LANE protocol has been designed to bridge across ATM networks.

ATM switches are not directly affected by the LANE protocol, as LANE builds upon the overlay model. Thus, LANE protocols operate transparently over and through ATM switches, using only standard ATM signaling procedures.

13.13.2 IP over ATM

The transport of any network layer protocol over an overlaid ATM network involves: packet encapsulation and address resolution.

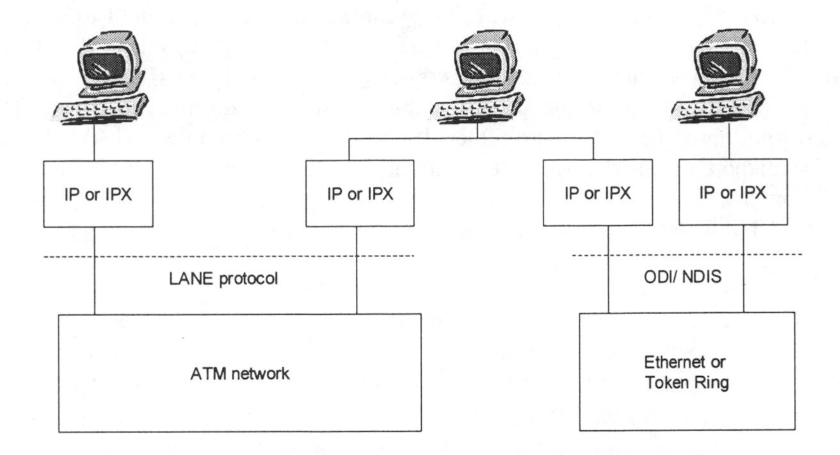

Figure 13.15 LANE protocol using NICs

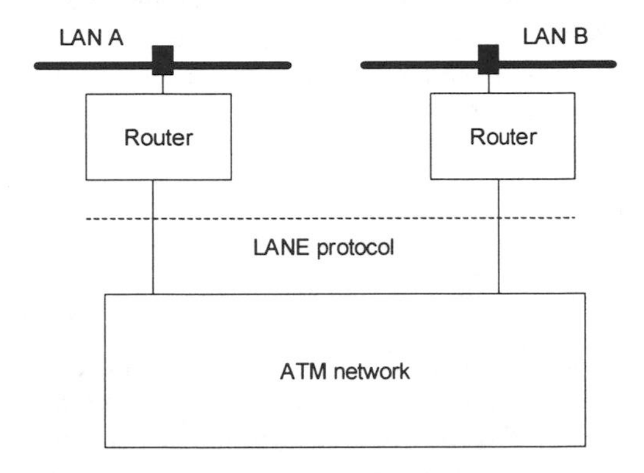

Figure 13.16 LAN protocol over routers

Packet encapsulation

A node must receive information about the kind of data packet that is contained within the encapsulated packet. The two main methods used are:

- LLC/SNAP encapsulation. With this, any type of protocol types can be carried across the connection. The LLC/SNAP header is then used from the encapsulated packet to identify the protocol type.
- VC multiplexing. With this, the protocol type is identified when the connection is set up, thus a single protocol is carried across the ATM connection. Therefore no multiplexing or packet type field is required within the packet, though the encapsulated packet may be prefixed with a pad field. The type of encapsulation used by LANE for data packets is actually a form of VC multiplexing. This type of encapsulation can be used where direct application-to-application ATM connectivity, bypassing lower level protocols, is desired. Such direct connectivity precludes the possibility of internetworking with nodes outside the ATM network.

Address resolution

Implementing IP over an ATM network requires a method of resolving IP addresses to ATM addresses. The address resolution table could be configured manually, but this is not a very scalable solution. The IP-Over-ATM working group has defined a protocol to support automatic address resolution of IP addresses in RFC 1577. This RFC introduces Logical IP Subnet (LIS), which, as a normal IP subnet, consists of a collection of IP nodes that make-up an ATM network and belong to the same IP subnet (just like a normal IP subnet). Then, for addresses resolution of the nodes within the LIS, each LIS has a single ATMARP server. All nodes (LIS clients) within the LIS are configured with the unique ATM address of the ATMARP server.

When an ATM nodes starts up, the following is implemented:

- The LIS client, using the configured address, connects to the ATMARP server.
- The ATMARP server transmits an Inverse ARP request to the LIS client, which requests the IP and ATM addresses of the node.
- ATMARP server adds the IP and ATM addresses of the client to its ATMARP table.
- Nodes then send an ATMARP request to the ATMARP server, which tries to resolve the destination IP address and return back its ATM address. If it cannot resolve it, it returns back an ATM_NAK response.
- On a successful return address, the LIS client has obtained the ATM address it can then communicate with the ATM node.
- The server periodically sends out Inverse ARP requests to client. If they do not respond, the server deletes their entry from its address table.

13.14 ATM addressing scheme

ATM, like any other network, requires a network address scheme which identifies the source and destination addresses. The ITU-T have developed a standardized, telephone-like, numbering system called E.164 for addressing public ATM networks. Unfortunately, E.164 addresses are public addresses and cannot typically be used within private networks. The ATM Forum has since extended ATM addressing to include private networks. For a private networking scheme in UNI 3.0/3.1 they evaluated two different models. These differ in the way that the ATM protocol layer is viewed in relation to existing protocol layers, such as IP and IPX layers, and are:

- Peer model. This model treats the ATM layer as a peer of existing network layers and uses the same addressing schemes within the ATM networks. Thus, ATM endpoints would be identified by their existing network layer address (such as an IP or an IPX address). The ATM signaling requests would then carry these addresses for the source and destination ATM switch. Also network layer routing protocols, such as RIP, and so on, can be used to route ATM signaling requests using existing network layer addresses. The peer model allows for simplified addressing.
- Overlay model. This model decouples the ATM layer from any existing network protocol and defines a new addressing structure and a new routing protocol. This, as with Ethernet, FDDI and Token Ring which use MAC addresses, allows all existing protocols to operate over an ATM network. For this reason, the model is known as the subnetwork or overlay model. Thus, all ATM switches need an ATM address, and possibly, also a network layer address (such as an IP or IPX address).

The disadvantage with the overlay model is that there needs to be an ATM address resolution protocol which maps network addresses (IP or IPX) to ATM addresses. The peer model does not need address resolution protocols, as it uses existing routing protocols.

The ATM Forum decided to implement the overlay model for UNI 3.0/3.1 signaling. This is mainly because the peer model would be difficult to implement as it must essentially act as a multiprotocol router and support address tables for all current protocols, as well as all of their existing routing protocols. In addition, currently available routing protocols for LANs and WANs do not map well into the QoS parameter.

The ATM Forum chose a private network-addressing scheme based on the OSI Network Service Access Point (NSAP) address. These are not true NSAP addresses, and are either ATM private network addresses or ATM end-point identifiers. They basically are subnetwork points of attachment.

An NSAP ATM address for a private network has 20 bytes, while a public network uses an E.164 address, as defined by the ITU-T. NSAP-based addresses have three main fields:

- Authority and Format Identifier (AFI). This defines the type and format of the Initial Domain Identifier (IDI).
- Initial Domain Identifier (IDI). This defines the address allocation and administration authority.
- Domain Specific Part (DSP). This defines the actual routing information.

There are three formats that private ATM addressing use for different definitions for the AFI and IDI parts. These are:

- NSAP Encoded E.164 format. The E.164 number is contained in the IDI.
- DCC Format. The IDI is a Data Country Code (DCC) which **identifies** the country, as specified in ISO 3166. These addresses are administered, in each country, by the ISO National Member Body.
- ICD Format. The IDI contains the International Code Designator (ICD). The ICD is allocated by the ISO 6523 registration authority. They identify particular international organizations.

13.14.1 NSAP

NSAP is defined in ISO/IEC 8348. It divides the address into two main parts: Initial Domain Part (IDP), which splits into the Authority and Format Identifier (AFI), and Initial Domain Identifier (IDI). The format is thus:

```
| IDP         | DSP
| AFI  | IDI  |
```

In the ISO/IEC 10589 specification, the DSP address includes an ID and SEL (1 byte selector) field which are used by level 1 routing. Typically, the ID part is taken from the ISO/IEC 8802 48-bit MAC address. The format is thus:

```
| IDP         | DSP                        |
| AFI  | IDI  |             | ID  | SEL    |
```

In the UNI-3.1 specification, the ID field is six bytes. It is also defined that the NSAP address format uses a maximum length of 20 bytes.

13.14.2 ICD Format

The format of an ICD scheme is:

```
AFI                      47                             (1 byte)
ICI                      xxxx                           (2 bytes)
```

```
Version              xx                            (1 byte)
Network              xxxxxx                        (3 bytes)
Tele traffic area    xx                            (1 byte)
Member identifier    xxxx                          (2 byte)
Member access point  xx                            (1 byte)
Area                 xx                            (1 byte)
Switch               xx                            (1 byte)
MAC-address          xxxxxxxxxxxx                  (6 bytes)
Nselector            xx                            (1 byte)
```

An example format is:

```
+--+--+--+--+--+--+--+--+--+--+--+--+--+--+--+--+--+--+--+--+
|47|00 23|00|00 00 03|xx|xx xx|xx|xx|xx| ESI MAC address |xx|
+--+--+--+--+--+--+--+--+--+--+--+--+--+--+--+--+--+--+--+--+
```

13.14.3 DCC Format

The DCC format is defined by the National Standards Organization 39528+1100. Its fields include:

```
AFI (39 or 38)       ISO DCC format                (1 byte)
IDI                                                (2 bytes)
CFI                  Country Format Identifier     (4 bits)
CDI                  Country Domain Identifier     (12 bits)
SFI                  SURFNet Format Identifier     (4 bits)
          0=CLN S in case of an organizational SDI
          1=in case of a network SDI
          2=ATM in case of an organization SDI
SDI                  SURFNet Domain Identifier     (2 bytes)
                     decimal encoded administrative numbers
ASDI                 Additional Domain Identifier  (4 bits)
NYU                  Not yet used                  (5 bytes)
ESI                  End System Identifier         (6 bytes)
SEL                  Selector                      (1 byte)
```

The EaStMAN network uses a 13-byte prefix, in the form:

```
39.826f.1107.16.7000.00.nnmm.ee.ff
```

where 39 is the AFI – ISO DCC, 826f is the IDI (indicating the UK), 1 is the CFI, 107 is the CDI, 1 for the SFI, 107 the SDI (country and domain), 16 for ASNI (for region), 00 (not yet used), `nnmm` for site code, `ee` for campus number and `ff` for switch number.

The `nn` part represents the institution, these are:

01	University of Edinburgh	02	Moray House
03	Queen Margaret College	04	Napier University
05	Heriot-Watt University	06	Edinburgh College of Art
07	University of Stirling		

and `mm` is the ring access point. These are assigned in a clockwise direction on the ring, starting from Kings Building (University of Edinburgh). In summary, the `nnmm` codes are assigned as follows:

University of Edinburgh:

Kings Buildings	0101	Pollock Halls	0102
Old College	0103	New College	0106

Moray House:			
MH-H	0204	MH-Cramond	0209
Queen Margaret College:			
QMC-Leith	0305	QMC-Corstorphine	030a
Napier University:			
Merchiston	0407	Sighthill	040b
Heriot-Watt:			
Riccarton campus	0508		
Edinburgh College of Art:			
ECA-L	060c	ECA-G	060d
University of Stirling:			
Stirling	070e		

The last two byte values (`ee` and `ff`) are allocated by the local institution. Typically, they can be used to identify the campus number (`ee`) and the switch number within the campus (`ff`).

13.14.4 E.164 Format (ATM Forum/95-0427R1)

The E.164 format provides a geographical scheme, but, as it is derived from the telephone system, the addresses are in short supply. Therefore, the E.164 NSAP format is recommended, which is extensible to ISDN. Its format is:

```
+--+--+--+--+--+--+--+--+--+--+--+--+--+--+--+--+--+--+--+--+
|45| Internat E.164 number |  HO-DSP   |xx xx xx xx xx xx|xx|
+--+--+--+--+--+--+--+--+--+--+--+--+--+--+--+--+--+--+--+--+
```

Where:

```
45       AFI for E.164 binary syntax
IDI      International E.164 number
HO-DSP   Extends the E.164 address to logically identify
         many devices in a single geographical location.
ESI      Similar to HO-DSP, extends the E.164 address.
SEL      Selector
```

13.15 ATM routing protocols

Routing protocols allow routing devices (such as IP routers and ATM switches) to intercommunicate and pass on information which can be used to find the optimum path for a connection. Routers typically use a crude technique called RIP which uses the number of hops (router counts) that it takes to get to a destination node. This type of routing protocol is not efficient, as it does not take into account traffic levels, delay, error levels, or reliability. ATM uses a connection-oriented technique, which initially involves setting up a connection.

The ATM Forum have defined a new protocol for private ATM network (that is, networks with NSAP format addresses), this is named the Private NNI (P-NNI) protocol. Public networks that use E.164 addressing use a different NNI protocol, such as the ITU-T B-ISUP signaling protocol and the ITU-T MTP routing protocol.

13.16 Exercises

13.16.1 The main advantage that ATM has over traditional LAN technologies is:
- (a) It is cheaper to implement
- (b) It properly supports real-time and non-real-time traffic
- (c) It is easier to interface to
- (d) It properly supports TCP/IP

13.16.2 What is the sampling rate for speech:

(a)	4 kHz	(b)	8 kHz
(c)	20 kHz	(d)	44.1 kHz

13.16.3 What is the sampling rate for hi-fi audio:

(a)	4 kHz	(b)	8 kHz
(c)	20 kHz	(d)	44.1 kHz

13.16.4 Discuss how ATM connections are more efficient in their transmission than ISDN and PCM-TDM. Also, determine the time between samples for speech (maximum frequency 4 kHz), audio (maximum frequency 20 kHz) and for video (maximum frequency 6 MHz).

13.16.5 Explain the fields in an ATM cell. Also explain how ATM cells are routed through a network, highlighting why the cells always take a fixed route.

13.16.6 Explain virtual channels, virtual paths and their uses.

13.16.7 Explain why computer-type data normally differs in its traffic profile from speech and video data.

13.16.8 Investigate a local WAN; determine its network topology and its network technology.

13.16.9 If there is access to an Internet connection, use the WWW to investigate the current status of the EaStMAN network (http://www.eastman.ac.uk).

13.16.10 **Exam question**

- (a) Explain, with the aid of a diagram, the format of an ATM cell. Discuss how the VCI label can be split into different fields.
- (b) Explain, with the aid of a diagram, how ATM uses the VCI virtual paths and circuits to route cells. What advantages do these types of routing have?

13.16.11 **Exam question**

- (a) Discuss how ATM deals with real-time and non-real-time traffic. Outline the advantages that ATM has over traditional network technologies, such as Ethernet.
- (b) Explain, with the use of an example, how ATM cells are routed. Explain also how the VCI virtual paths and circuits route cells.

IP

InterNIC (International Network Information Center) is the organization that serves the Internet community and supplies user assistance, documentation, training, and other services.

14.1 Introduction

Networking technologies such as Ethernet, Token Ring and FDDI provide a data link layer function that is, they allow a reliable connection between one node and another on the same network. They do not provide internetworking where data can be transferred from one network to another or from one network segment to another. For data to be transmitted across a network requires an addressing structure which is read by a gateway or a router. The interconnection of networks is known as internetworking (or an internet). Each part of an internet is a subnetwork (or subnet), and Transmission Control Protocol (TCP) and Internet Protocol (IP) are a pair of protocols that allow one subnet to communicate with another.

A protocol is a set of rules that allows the orderly exchange of information. The IP part corresponds to the network layer of the OSI model and the TCP part to the transport layer. Their operation is transparent to the physical and data link layers and can thus be used on Ethernet, FDDI or Token Ring networks. This is illustrated in Figure 14.1. The address of the Data Link layer corresponds to the physical address of the node, such as the MAC address (in Ethernet and Token Ring) or the telephone number (for a modem connection). The IP address is assigned to each node on the internet, and is used to identify the location of the network and any subnets.

TCP/IP was originally developed by the US Defense Advanced Research Projects Agency (DARPA), and its objective was to connect a number of universities and other research establishments to DARPA. The resultant internet is now known as the Internet. It has since outgrown this application and many commercial organizations now connect to the Internet. The Internet uses TCP/IP to transfer data, where each node on the Internet is assigned a unique network address, called an IP address. Note that any organization can have its own internets, but if it is to connect to the Internet then the addresses must conform to the Internet addressing format. Common applications that use TCP/IP communications are remote login and file transfer. Typical programs used in file transfer and login over TCP communication are `ftp` for file transfer program and `telnet` which allows remote log into another computer. The `ping` program determines if a node is responding to TCP/IP communications.

The ISO has adopted TCP/IP as the basis for the standards relating to the network and transport layers of the OSI model, and is known as ISO-IP. Most currently available systems conform to the IP addressing standard.

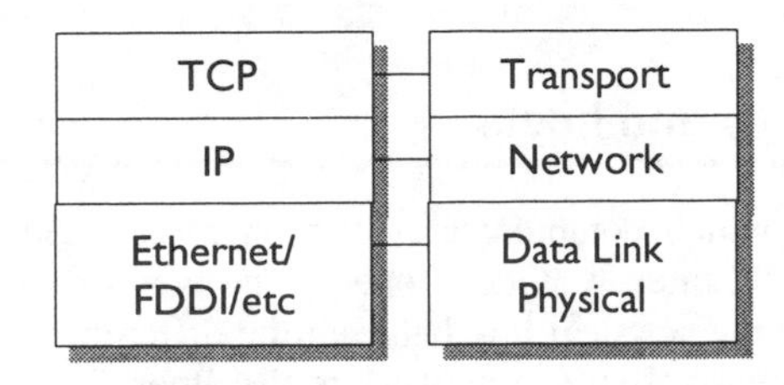

Figure 14.1 TCP/IP and the OSI model

14.2 Data encapsulation

The OSI model provides a clear model in understanding how data is transferred from one system to another. Each layer of the OSI model depends on the layers above and/or below it to provide a certain function. To achieve this data is encapsulated when passing from one layer to the next. Encapsulation happens when a lower layer gets a PDU (Protocol Data Unit) from an upper layer and uses this as a data field. It then adds to its own headers and trailers so that it can perform the required function. For example, as illustrated in Figure 14.2, if a user was sending an e-mail to another user the host computer would go through the following:

- **Step 1.** The application program makes contact with a network application program for electronic mail transfer [Application].
- **Step 2.** The computer converts the data into a form that can be transmitted over the network [Presentation].
- **Step 3.** The computer sets up a connection with the remote system and asks if it is willing to make a connection with it. Once accepted it will ask the remote system as to whether the user for whom the e-mail message is destined for actually exists on that system [Session].
- **Step 4.** The transport layer negotiates with the remote system about how it wants to receive the data, it then splits the data into a number of segments, each of which are numbered. Any problems with lost segments will be dealt with at this step [Transport]. DATA SEGMENT.
- **Step 5.** The network address of the destination and source are added to all the data segments [Network]. DATA PACKET.
- **Step 6.** The data packet is then framed in a format so that it can be transmitted over the local connection [Data link]. DATA FRAME.
- **Step 7.** The data frame is then taken and converted into a binary form and transmitted over a physical connection. BITS.

Media Layers

These layers control the physical delivery of messages of the network. They include:

Network (addressing and best path).
Data Link (access to media)
Physical layers (bit transmission).

Network: ROUTERS
Data link: BRIDGES/SWITCHES
Physical: HUBS/REPEATERS

Host layers

These layers provide accurate delivery of data between interconnected nodes. They are:

Application (provides network services to user applications, such as file transfer to an application program).
Presentation (provides data representation and encoding, and ensures that the data is in a form that can be used by the application or able to be transmitted over the network).
Session (establishes, maintains, and manages sessions between applications),
Transport (segments at the send and re-assembles the data into a data stream).

Bits→Frames→Packets→
Segments→Data (Bottom-Up)
Data→Segments→Packets→
Frames→Bits (Top-Down)

14.3 TCP/IP gateways and hosts

TCP/IP hosts are nodes which communicate over interconnected networks using TCP/IP communications. A TCP/IP gateway node connects one type of network to another. It contains hardware to provide the physical link between the different networks and the hardware and software to convert frames from one network to the other. Typically, it converts a Token Ring MAC layer to an equivalent Ethernet MAC layer, and vice versa.

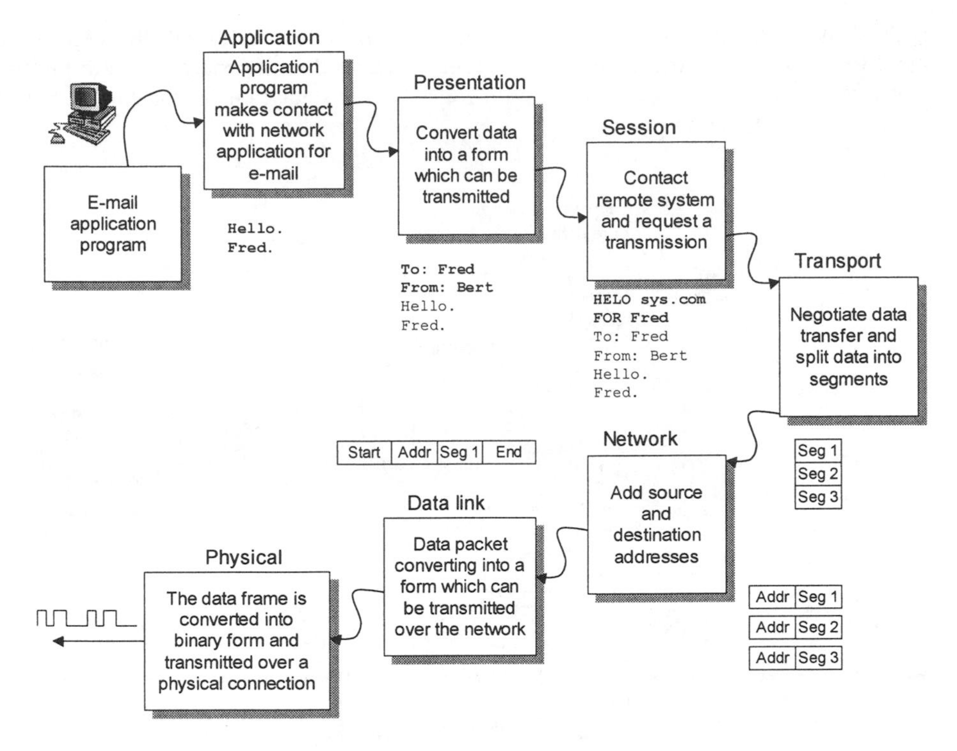

Figure 14.2 Data encapsulation

A router connects a network of a similar type to another of the same kind through a point-to-point link. The main operational difference between a gateway, a router, and a bridge is that for a Token Ring and Ethernet network, the bridge uses the 48-bit MAC address to route frames, whereas the gateway and router use the IP network address. As an analogy to the public telephone system, the MAC address would be equivalent to a randomly assigned telephone number, whereas the IP address would contain the information on where the telephone is logically located, such as which country, area code, and so on.

Figure 14.3 shows how a gateway (or router) routes information. It reads the data frame from the computer on network A, and reads the IP address contained in the frame and makes a decision whether it is routed out of network A to network B. If it does then it relays the frame to network B.

Transport layer: Segments

Network layer: Packets (Datagrams)

Data link layer: Frames

Physical layer: Bits

14.4 Function of the IP protocol

The main functions of the IP protocol are to:

- Route IP data packets – which are called internet datagrams – around an internet. The IP protocol program running on each node knows the location of the gateway on the network. The gateway must then be able to locate the interconnected network. Data then passes from node to gateway through the internet.
- Fragment the data into smaller units, if it is greater than a given amount (64 kB).
- Report errors. When a datagram is being routed or is being reassembled an error can occur. If this happens then the node that detects the error reports back to the source

node. Datagrams are deleted from the network if they travel through the network for more than a set time. Again, an error message is returned to the source node to inform it that the Internet routing could not find a route for the datagram or that the destination node, or network, does not exist.

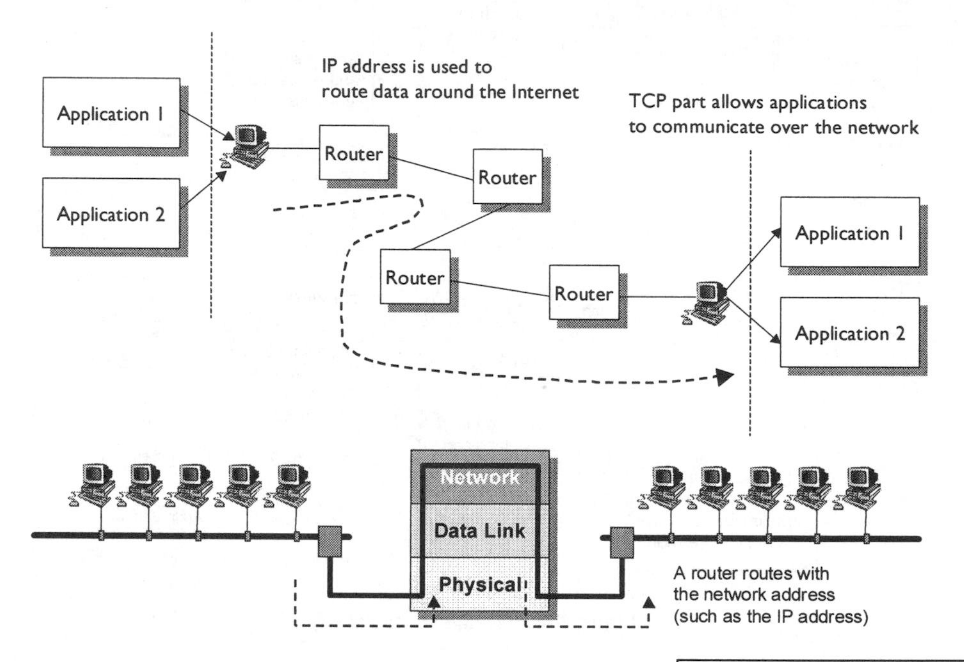

Figure 14.3 Internet gateway layers

APPLICATION LAYER

Computer Applications make use of network applications.

Computer applications:

- Spreadsheets.
- Word Processor.
- Database.
- WWW browser.

↓

Network applications (Application layer of OSI model):

- Electronic mail.
- File transfer.
- Remote access.
- WWW access.
- Network management.

14.5 Internet datagram

The IP protocol is an implementation of the network layer of the OSI model. It adds a data header onto the information passed from the transport layer, the resultant data packet is known as an internet datagram. The header contains information such as the destination and source IP addresses, the version number of the IP protocol and so on. Figure 14.4 shows its format.

The datagram can contain up to 65 536 bytes (64 KB) of data. If the data to be transmitted is less than, or equal to, 64 KB, then it is sent as one datagram. If it is more than this then the sender splits the data into fragments and sends multiple datagrams. When transmitted from the source each datagram is routed separately through the internet and the received fragments are finally reassembled at the destination.

The fields in the IP datagram are:

- **Version.** The TCP/IP `version number` helps gateways and nodes interpret the data unit correctly. Differing versions may have a different format. Most current implementations will have a version number of four (IPv4).

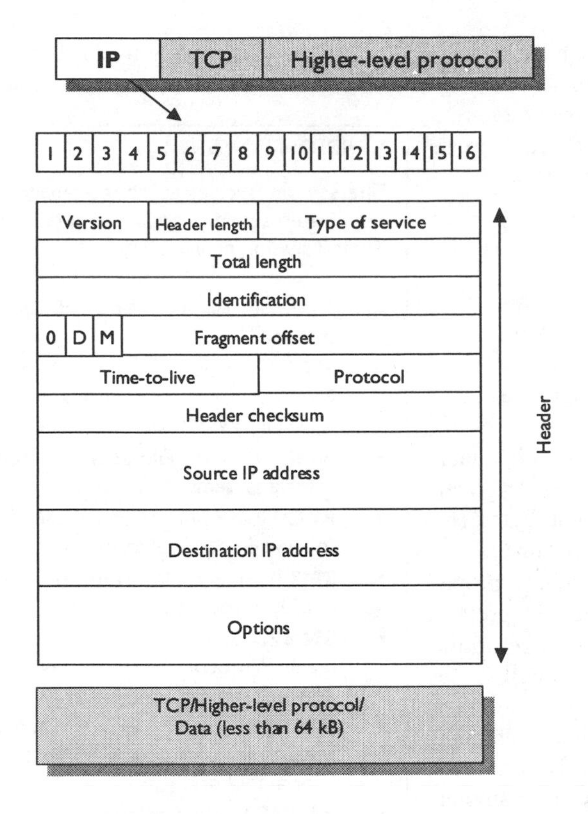

Figure 14.4 Internet datagram format and contents

- **Type of service.** The `type of service` bit field is an 8-bit bit pattern in the form `PPPDTRXX`, where `PPP` defines the priority of the datagram (from 0 to 7). The precedence levels are:

  111 (Network control) 110 (Internetwork control)
  101 (CRITIC/ECP) 100 (Flash override)
  011 (Flash) 010 (Immediate)
  001 (Priority) 000 (Routine)

  `D` sets a low delay service (0 – normal delay, 1 – low delay).
  `T` sets high throughput (0 – normal throughput, 1 – high throughput).
  `R` sets high reliability (0 – normal reliability, 1 – high reliability).

PRESENTATION LAYER

The presentation layer allows for code formatting and encoding. This ensures that the data is in a form that can be transmitted and received at the destination. A typical function is to translate from one system to another. Encryption is also another important function. Typical conversions and formats used at the presentation layer are:

- Text. ASCII, EBCDIC, Encryption, Binary-to-text, Character conversion.
- Graphics. JPEG, TIFF, GIF.
- Audio. MP-3, WAV, MIDI.
- Video. MPEG, QuickTime (AVI).

Typical encryption standards are:

- **RSA.** Rivest, Shamir and Aldeman (the creators of the RSA method). An encryption method where a user generates two keys: a public one and a private one. The public key is freely available to anyone who wants to send the user some encrypted data. It is used to encrypt the data distend for the user, and only the private key can decrypt the data.
- **PGP.** Pretty Good Privacy. Based on the RSA method, but high-optimized and fast.
- **DES.** Data Encryption Standard. A private-key standard where both the sender and the receiver use the same private key to encrypt and decrypt the data.
- **IDEA.** International Data Encryption Standard. Private-key method.

- **Header length** (4 bits). The `header length` defines the size of the data unit in multiples of four bytes (32 bits). The minimum length is five bytes and the maximum is 65 536 bytes. Padding bytes fill any unused spaces.
- **Identification** (16 bits). A value which is assigned by the sender to aid the assembly of the frames of a datagram.

- **D** and **M** bits. A gateway may route a datagram and split it into smaller fragments. The `D` bit informs the gateway that it should not fragment the data and thus it signifies that a receiving node should receive the data as a single unit or not at all. The `M` bit is the 'more fragments' bit and is used when data is split into fragments. The `fragment offset` contains the fragment number. The bit settings are:

  D – Don't fragment. 0 – may fragment, 1 – don't fragment.
  M – Last fragment. 0 – last fragment, 1 – more fragments.

- **Fragment offset** (13 bits). Indicates which datagram this fragment belongs to. The fragment offset is measured in units of eight bytes (64 bits). The first fragment has an offset of zero.
- **Time-to-live** (8 bits). A datagram could propagate through the internet indefinitely. To prevent this, the 8-bit `time-to-live` value is set to the maximum transit time in seconds and is set initially by the source IP. Each gateway then decrements this value by a defined amount. When it becomes zero the datagram is discarded. It can also be used define the maximum amount of time that a destination IP node should wait for the next datagram fragment.
- **Protocol** (8 bits). Different IP protocols can be used on the datagram. The 8-bit `protocol` field defines the type to be used. A full list is given in Table 14.7 (Section 14.18). Typical values are: 1 – ICMP and 6 – TCP.
- **Header checksum** (16 bits). The `header checksum` contains a 16-bit pattern for error detection. Since values within the header change from gateway to gateway (such as the time-to-live field), it must be recomputed every time the IP header is processed. The algorithm is:

  The 16-bit 1's complement of the 1's complement sum of all the 16-bit words in the header. When calculating the checksum the header checksum field is assumed to be set to a zero.

- **Source and destination IP addresses** (32 bits). The `source` and `destination IP addresses` are stored in the 32-bit source and destination IP address fields.
- **Options**. The `options` field contains information such as debugging, error control and routing information.

SESSION LAYER

The session layer establishes, manages and terminates sessions between applications. Typical session protocols are:

- **FTP.** File Transfer Protocol. Used to transfer files.
- **HTTP.** HyperText Transmission Protocol. Used to transfer files between a WWW server and a client.
- **NFS.** Network File Service. Used to link file systems.
- **RPC.** Remote-procedure call. Used to run remote applications.
- **SMTP.** Simple Mail Transport Protocol. Used to transmit e-mail.
- **SNMP.** Simple Network Management Protocol. Used to investigate network devices.
- **SQL.** Structured Query Language. Used to transfer database information over a network.
- **TELNET.** Used to remotely log into a remote computer.
- **DNS.** Domain Name Services. Used to convert logical names into network addresses.
- **BOOTP.** Boot Protocol. Used to assign network address (typically IP addresses), based on station addresses (typically, MAC addresses).
- **DHCP.** Dynamic Host Control Protocol. Used to assign network addresses.
- **WINS.** Used to assign network addresses.
- **NNTP.** Network News Transfer Protocol.
- **NTP.** Network Time Protocol.
- **FINGER.** Used to determine information on a user.

14.6 TCP/IP internets

Figure 14.5 illustrates a sample TCP/IP implementation. A gateway `MERCURY` provides a link between a Token Ring network (NETWORK A) and the Ethernet network (ETHER C). Another gateway `PLUTO` connects NETWORK B to ETHER C. The TCP/IP protocol allows a host on NETWORK A to communicate with `VAX01`.

> **Transport layer.** Segments and reassembles data into a data stream. It is also responsible for connection synchronization, flow control and error recovery.
>
> **Network layer.** Determine the best route for data to take from one place to another. Routers operate at this layer.

14.6.1 Selecting internet addresses

Each node using TCP/IP communications requires an IP address which is then matched to its Token Ring or Ethernet MAC address. The MAC address allows nodes on the same segment to communicate with each other. In order for nodes on a different network to communicate, each must be configured with an IP address.

Nodes on a TCP/IP network are either hosts or gateways. Any node that runs application software or are terminals are hosts. Any node that routes TCP/IP packets between networks is called a TCP/IP gateway node. This node must have the necessary network controller boards to physically interface to other networks it connects with.

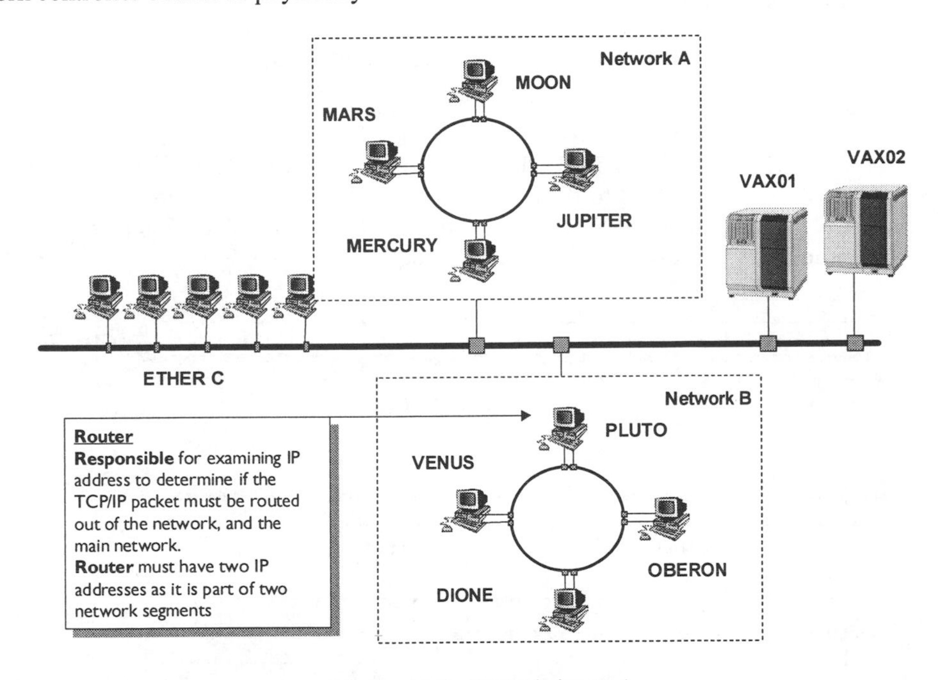
Network A
MOON
MARS
JUPITER
MERCURY
VAX01
VAX02
ETHER C
Network B
PLUTO
VENUS
OBERON
DIONE
Router
Responsible for examining IP address to determine if the TCP/IP packet must be routed out of the network, and the main network.
Router must have two IP addresses as it is part of two network segments

Figure 14.5 Example internet

14.6.2 Format of the IP address

A typical IP address consists of two fields: the left field (or the network number) identifies the network, and the right number (or the host number) identifies the particular host within that network. Figure 14.6 illustrates this. The IP address is 32 bits long and can address over four billion physical addresses (2^{32} or 4294967296 hosts). There are three main address formats and these are shown in Figure 14.7.

Each of these types is applicable to certain types of networks. Class A allows up to 128 (2^7) different networks and up to 16777216 (2^{24}) hosts on each network. Class B allows up to 16384 (2^{14}) networks and up to 65536 (2^{16}) hosts on each network. Class C allows up to 2097152 (2^{21}) networks each with up to 256 (2^8) hosts.

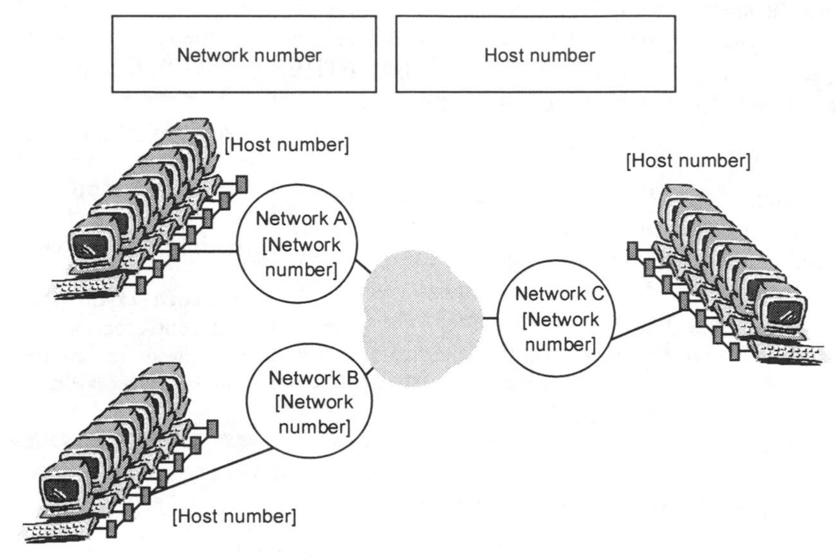

Figure 14.6 IP addressing over networks

The class A address is thus useful where there are a small number of networks with a large number of hosts connected to them. Class C is useful where there are many networks with a relatively small number of hosts connected to each network. Class B addressing gives a good compromise of networks and connected hosts.

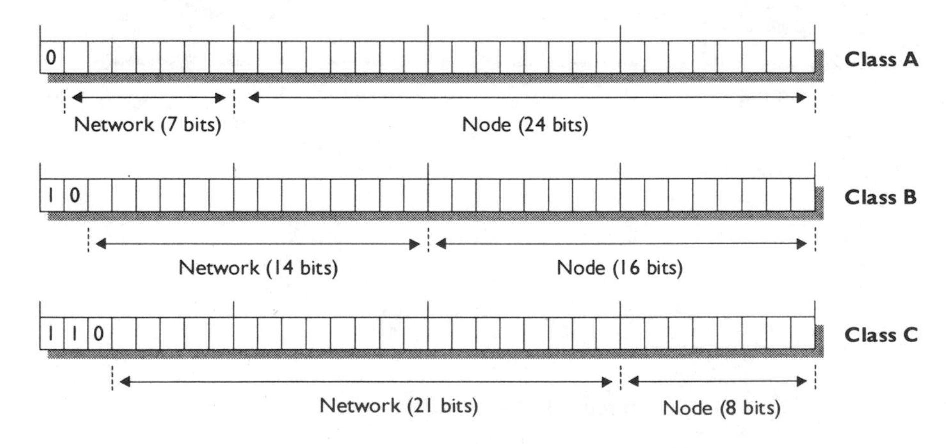

Figure 14.7 Type A, B and C IP address classes

When selecting internet addresses for the network, the address can be specified simply with decimal numbers within a specific range. The standard DARPA IP addressing format is of the form:

W.X.Y.Z

where W, X, Y and Z represent 1 byte of the IP address. As decimal numbers they range from 0 to 255. The 4 bytes together represent both the network and host address.

The valid range of the different IP addresses given in Table 14.1 defines the valid IP addresses. Thus for a class A type address there can be 127 networks and 16711680 (256×256×255) hosts. Class B can have 16320 (64×255) networks and class C can have 2088960 (32×256×255) networks and 255 hosts. Addresses above 223.255.254 are reserved, as are addresses with groups of zeros.

Table 14.1 Ranges of addresses for type A, B and C internet address

Type	*Network portion*	*Host portion*
A	1 – 126	0.0.1 – 255.255.254
B	128.1 – 191.254	0.1 – 255.254
C	192.0.1 – 223.255.254	1 – 254

The ping command allows a user to determine if a host is responding to TCP/IP communications. It also gives an indication of host's IP address.

ping www.mit.edu
Ping statistics for **18.181.0.31**:
Packets: Sent = 4, Received = 4, Lost = 0 (0% loss).

ping www.napier.ac.uk
Ping statistics for **146.176.1.8**:
Packets: Sent = 4, Received = 4, Lost = 0 (0% loss).

ping www.ja.net
Ping statistics for **194.82.140.71**:
Packets: Sent = 4, Received = 4, Lost = 0 (0% loss).

14.6.3 Range of IP addresses

The IP address splits into two main parts: the network part (which is assigned by InterNIC), and the host part (which is assigned by the local system administrator). The type of address that is allocated depends on the size of the organization and the number of hosts that it has on its network. Figure 14.8 shows how the binary notation can be represented in dotted notation, of which there are three commercial address classifications, these are:

- **Class A**. Large organizations with many nodes. MIT has a Class A address (18.0.0.0).
- **Class B.** Medium sized organizations with an average number of hosts. For example, Napier University has a Class B address (146.176.0.0).
- **Class C.** Small organizations or setups, with only a few hosts on each network.

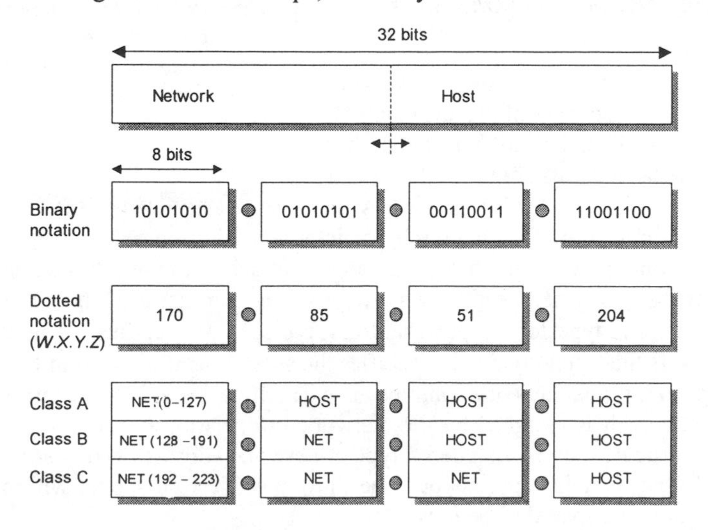

Figure 14.8 IP address format

The other two address classifications are used for multicast and research purposes. These are:

- **Class D.** Multicast. Single packets copied by the network and sent to a specific subset of network addresses. These addresses are specified in the destination address field.
- **Class E.** Research purposes.

Not all the IP addresses in the host part can be used to assign to a host. The two exceptions are:

- **All 0's in the host part** is reserved for the entire network. For example:

  32.0.0.0 (Class A network address).
  146.176.0.0 (Class B network address).
  199.20.30.0 (Class C network address).

- **All 1's in the host part** is reserved for the broadcast address, which is used to send a data packet to all the hosts on the network. For example:

  32.255.255.255 (Class A broadcast address for the network 32.0.0.0).
  146.176.255.255 (Class B broadcast address for the network 146.176.0.0).
  199.20.30.255 (Class C broadcast address for the network 199.20.30.0).

The other broadcast address is 255.255.255.255. Routers will not transmit broadcast addresses from one network segment to another.

TRANSPORT LAYER

The transport layer provides the required reliability in the delivery of data, and end-to-end services. It supports:

Segmentation and reassembly. This involves splitting data from one or more applications into segments, which are then transmitted onto a single data stream. The receiver must reassemble these segments back into the original data block.
Flow control. This allows the destination to control the amount of data but informing the source that it cannot receive any more data at the present.
Acknowledgement and retransmission. Acknowledgements are sent by the receiver after it has successfully received the transmitted data segments. This allows the transmitter to determine if the data has been received correctly, if not, it can retransmit the data.

Typical transport protocols are:

TCP. The standard protocol that is used on the Internet.
UDP. Similar to TCP, but does not have acknowledgement or flow control. It is thus unreliable, as the sender cannot guarantee that the data segments were received correctly.
SPX. The transport protocol that is used with Novell NetWare.

14.6.4 Creating IP addresses with subnet numbers

Besides selecting IP addresses of internets and host numbers, it is also possible to designate an intermediate number called a subnet number. Subnets extend the network field of the IP address beyond the limit defined by the type A, B, C scheme. They thus allow for a hierarchy of internets within a network. For example, it is possible to have one network number for a network attached to the internet, and various subnet numbers for each subnet within the network. This is illustrated in Figure 14.9. For an address `W.X.Y.Z` and type for a type A address, typically `W` specifies the network and `X` the subnet. For type B the `Y` field typically specifies the subnet, as illustrated in Figure 14.10.

To connect to a global network a number is normally assigned by a central authority. For the Internet network it is assigned by the Network Information Center (NIC). Typically, on the Internet an organization is assigned a type B network address. The first two fields of the address specify the organization network, the third specifies the subnet within the organization and the final value specifies the host.

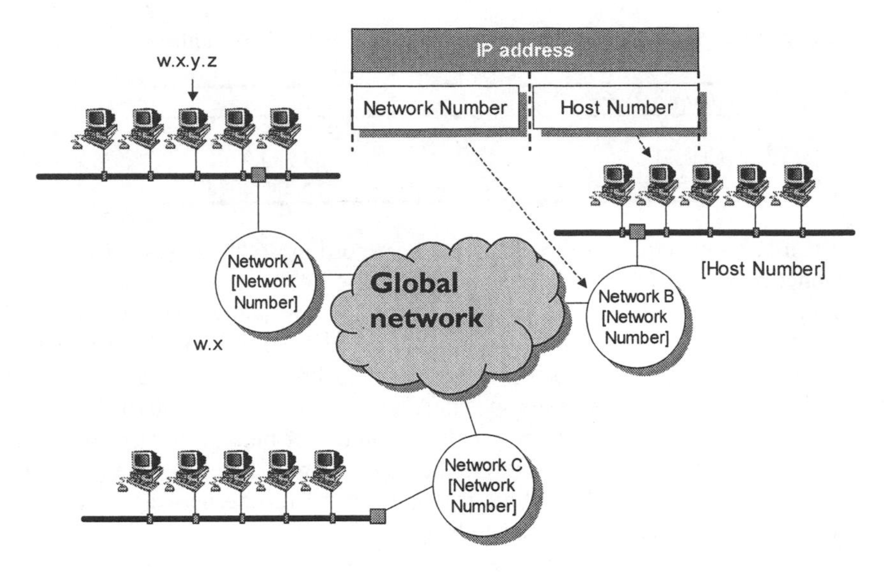

Figure 14.9 IP addresses with subnets

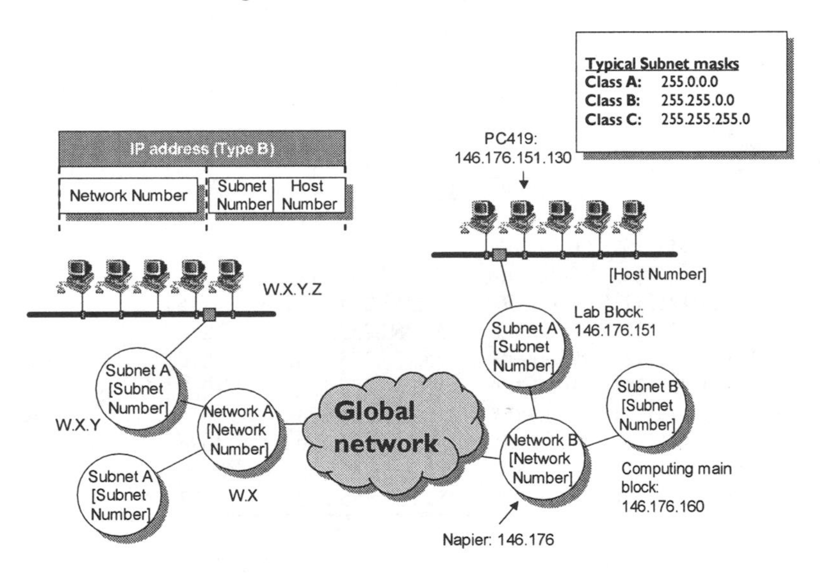

Figure 14.10 Internet addresses with subnets

14.6.5 Specifying subnet masks

If a subnet is used then a bit mask, or subnet mask, must be specified to show which part of the address is the network part and which is the host. The subnet mask is a 32-bit number that has 1's for bit positions specifying the network and subnet parts and 0's for the host part. A text file called `hosts` is normally used to set up the subnet mask. Table 14.2 shows example subnet masks. To set up the default mask the following line is added to the `hosts` file.

```
Hosts file
255.255.255.0  defaultmask
```

Table 14.2 Default subnet mask for type A, B and C IP addresses

Address Type	*Default mask*
Class A	255.0.0.0
Class B	255.255.0.0
Class C and Class B with a subnet	255.255.255.0

The subnet can use any number of bits from the host portion of the address. Table 14.3 outlines the subnet masks for Class B addresses, and Table 14.4 outlines the subnet masks for Class C addresses. The number of bits borrowed from the host address defines the maximum number of subnetworks. For example four bits borrowed from the host field will allow 14 different subnetworks (2^4-2, as 0000 is reserved for the whole network and 1111 is reserved for the broadcast address). The subnets will thus be defined by 0001, 0010, 0011 ... 1101 and 1110. The maximum number of hosts will be number of bits left in the host part, after the bits have been borrowed for the subnet, to the power of two, less two. The reason that the value is reduced by two is that the 0.0 address is reserved for the network, and the all 1's address is reserved for a broadcast address to that subnet.

Table 14.3 Subnet masks for a Class B address

Binary subnet address	*Dotted notation*	*Maximum number of subnets*	*Maximum number of hosts on each subnet*
11111111.11111111.11000000.00000000	255.255.192.0	2	16382
11111111.11111111.11100000.00000000	255.255.224.0	6	8190
11111111.11111111.11110000.00000000	255.255.240.0	14	4094
11111111.11111111.11111000.00000000	255.255.248.0	30	2046
11111111.11111111.10111100.00000000	255.255.252.0	62	1022
11111111.11111111.11111110.00000000	255.255.254.0	126	510
11111111.11111111.11111111.00000000	255.255.255.0	254	254
11111111.11111111.11111111.10000000	255.255.255.128	510	126
11111111.11111111.11111111.11000000	255.255.255.192	1022	62
11111111.11111111.11111111.11100000	255.255.255.224	2046	30
11111111.11111111.11111111.11110000	255.255.255.240	4094	14
11111111.11111111.11111111.11111000	255.255.255.248	8190	6
11111111.11111111.11111111.11111100	255.255.255.252	16382	2

Table 14.4 Subnet masks for a Class C address

Binary subnet address	*Dotted notation*	*Maximum number of subnets*	*Maximum number of hosts on each subnet*
11111111.11111111. 11111111.11000000	255.255.255.192	2	62
11111111.11111111. 11111111.11100000	255.255.255.224	6	30
11111111.11111111. 11111111.11110000	255.255.255.240	14	14
11111111.11111111. 11111111.11111000	255.255.255.248	30	6
11111111.11111111. 11111111.11111100	255.255.255.252	62	2

14.7 Internet naming structure

The Internet naming structure uses labels separated by periods; an example is `dcs.napier.ac.uk`. It uses a hierarchical structure where organizations are grouped into primary domain names, such as `com` (for commercial organizations), `edu` (for educational organizations), `gov` (for government organizations), `mil` (for military organizations), `net` (Internet network support centers) or `org` (other organizations). The primary domain name may also define the country in which the host is located, such as `uk` (United Kingdom), `fr` (France), and so on. All hosts on the Internet must be registered to one of these primary domain names.

The labels after the primary field describe the subnets within the network. For example in the address `eece.napier.ac.uk`, the `ac` label relates to an academic institution within the `uk`, `napier` to the name of the institution and `eece` the subnet within that organization. An example structure is illustrated in Figure 14.11.

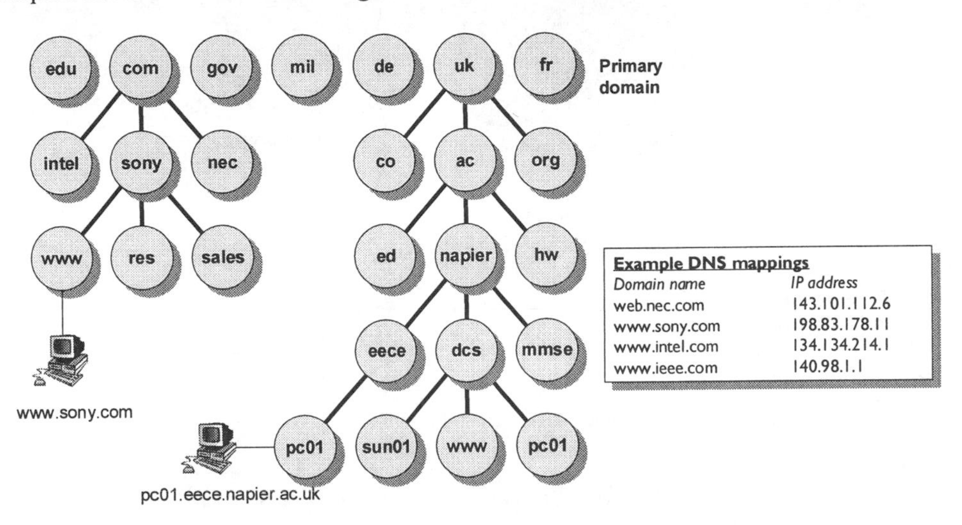

Figure 14.11 Example of domain naming

14.8 Domain name system

IP addresses are difficult to remember, thus Domain Name Services (DNS) are used to allow users to use symbolic names rather than IP addresses. DNS computers on the Internet determine the IP address of the named destination resource or application program. This dynamic mapping has the advantage that users and application programs can move around the Internet and are not fixed to an IP address. An analogy relates to the public telephone service. A telephone directory contains a list of subscribers and their associated telephone number. If someone looks for a telephone number, first the user name is looked up and their associated telephone number found. The telephone directory listing thus maps a user name (symbolic name) to an actual telephone number (the actual address). When a user enters a domain name (such as `www.fred.co.uk`) into the WWW browser, the local DNS server must try and resolve the domain name to an IP address, which can then be used to send the data to it. If it cannot resolve the IP address then the DNS server interrogates other servers to see if they know the required IP address, as illustrated in Figure 14.12. If it cannot be resolved then the WWW browser displays an error message.

Table 14.5 lists some Internet domain assignments for World Wide Web (WWW) servers. Note that domain assignments are not fixed and can change their corresponding IP addresses, if required. The binding between the symbolic name and its address can thus change at any time.

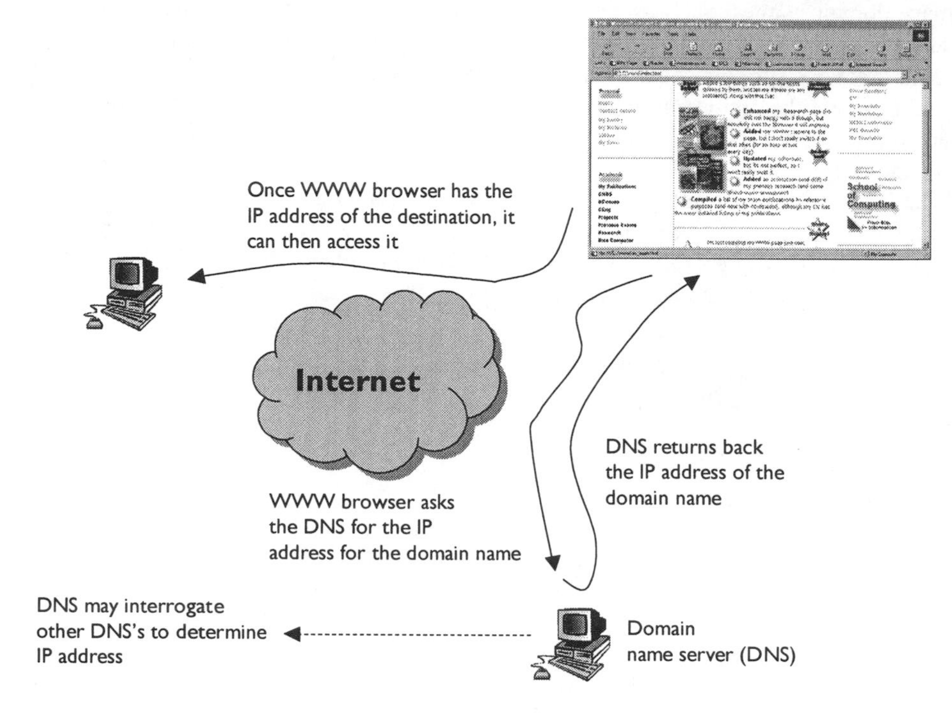
Once WWW browser has the
IP address of the destination, it
can then access it
Internet
DNS returns back
the IP address of the
domain name
WWW browser asks
the DNS for the IP
address for the domain name
DNS may interrogate
other DNS's to determine
IP address
Domain
name server (DNS)

Figure 14.12 Domain name server

Table 14.5 Internet domain assignments for web servers

Web server	*Internet domain name*	*Internet IP address*
NEC	web.nec.com	143.101.112.6
Sony	www.sony.com	198.83.178.11
Intel	www.intel.com	134.134.214.1
IEEE	www.ieee.com	140.98.1.1
University of Bath	www.bath.ac.uk	136.38.32.1
University of Edinburgh	www.ed.ac.uk	129.218.128.43
IEE	www.iee.org.uk	193.130.181.10
University of Manchester	www.man.ac.uk	130.88.203.16

14.9 Example network

A university network is shown in Figure 14.13. The connection to the outside global Internet is via the Janet gateway node and its IP address is `146.176.1.3`. Three subnets, `146.176.160`, `146.176.129` and `146.176.151`, connect the gateway to departmental bridges. The Computer Studies router address is `146.176.160.1` and the Electrical Department router has an address `146.176.151.254`.

The Electrical Department router links, through other routers, to the subnets `146.176.144`, `146.176.145`, `146.176.147`, `146.176.150` and `146.176.151`. The main bridge into the department connects to two Ethernet networks of PCs (subnets `146.176.150` and `146.176.151`) and to another router (`Router 1`). `Router 1` connects to the subnet `146.176.144`. Subnet `146.176.144` connects to workstations and X-terminals. It also connects to the gateway `Moon` that links the Token Ring subnet `146.176.145` with the Ethernet subnet `146.176.144`. The gateway `Oberon`, on the `146.176.145` subnet, connects to an Ethernet link `146.176.146`. This then connects to the gateway `Dione` that is also connected to the Token Ring subnet `146.176.147`.

The topology of the Electrical Department network is shown in Figure 14.14. Each node on the network is assigned an IP address. The *`hosts`* file for the set up in Figure 14.14 is shown next. For example the IP address of `Mimas` is `146.176.145.21` and for `miranda` it is `146.176.144.14`. Notice that the gateway nodes, `Oberon`, `Moon` and `Dione`, all have two IP addresses.

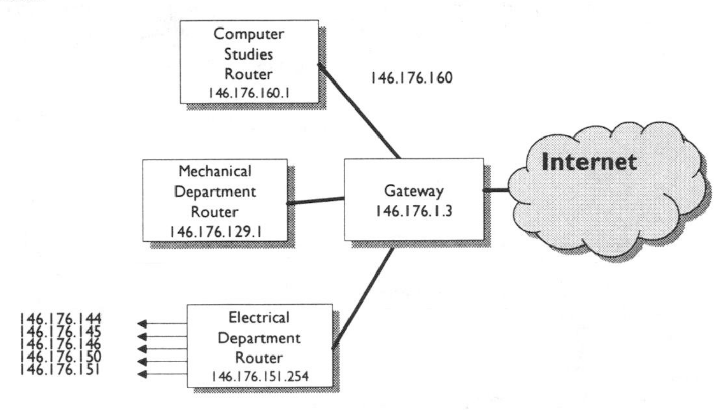

Figure 14.13 A university network

Contents of host file

```
146.176.1.3          janet
146.176.144.10       hp
146.176.145.21       mimas
146.176.144.11       mwave
146.176.144.13       vax
146.176.144.14       miranda
146.176.144.20       triton
146.176.146.23       oberon
146.176.145.23       oberon
146.176.145.24       moon
146.176.144.24       moon
146.176.147.25       uranus
146.176.146.30       dione
146.176.147.30       dione
146.176.147.31       saturn
146.176.147.32       mercury
146.176.147.33       earth
146.176.147.34       deimos
146.176.147.35       ariel
146.176.147.36       neptune
146.176.147.37       phobos
146.176.147.42       pluto
146.176.147.43       mars
146.176.147.22       jupiter
```

```
146.176.144.54      leda
146.176.144.55      castor
146.176.144.56      pollux
146.176.144.57      rigel
146.176.144.58      spica
146.176.151.254     cubridge
146.176.151.99      bridge_1
146.176.151.98      pc2
146.176.151.97      pc3
               :::::
146.176.151.71      pc29
146.176.151.70      pc30
146.176.151.99      ees99
146.176.150.61      eepc01
146.176.150.62      eepc02
255.255.255.0       defaultmask
```

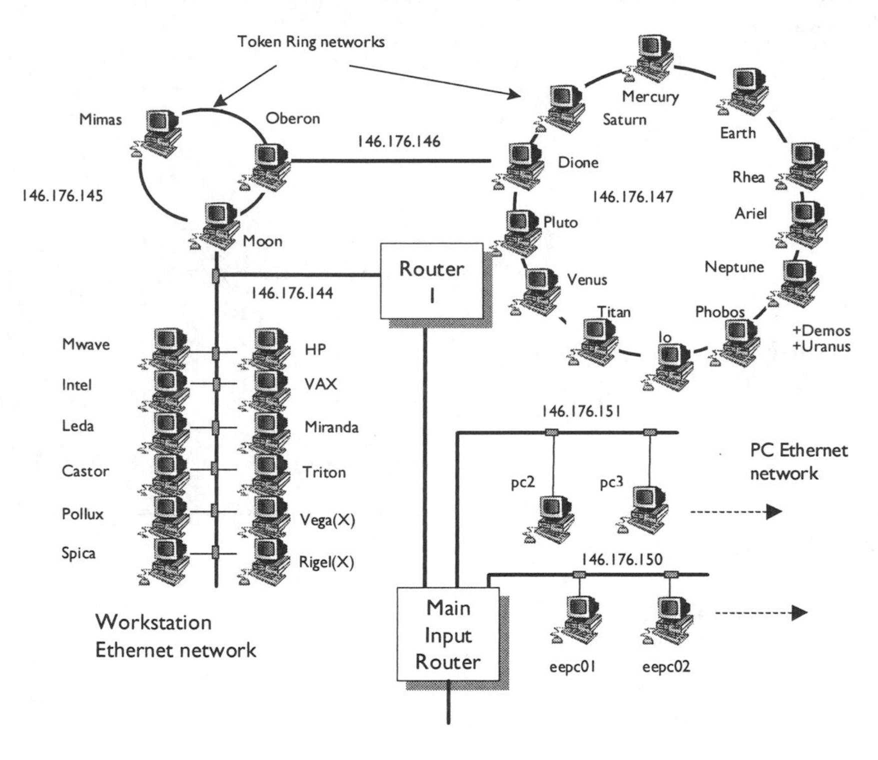

Figure 14.14 Network topology for the Electrical Department network

14.10 IP addresses for routers

Routers require an IP address for each of their ports, but they cannot use the zero host address as this is used to identify the whole network. For example, if there were five nodes on the 177.132.1 network (NETA), and six nodes on the 177.131.2 network (NETB), which are interconnected by a router (AB). Then the IP address could be assigned as:

Node NETA_1	177.131.1.2	Node NETA_2	177.131.1.3
Node NETA_3	177.131.1.4	Node NETA_4	177.131.1.5

Node NETA_5	177.131.1.6		
Node NETB_1	177.131.2.2	Node NETB_2	177.131.2.3
Node NETB_3	177.131.2.4	Node NETB_4	177.131.2.5
Node NETB_5	177.131.2.6	Node NETA_6	177.131.2.7
Router AB_E0	177.131.1.1	Router AB_E1	177.131.2.1

The router, in this case, has two ports. These have been defined with the IP address 177.131.1.1 (which connects to the 177.131.1 network), and 177.131.2.1 (which connects to the 177.131.2 network).

14.11 IP multicasting

Many applications of modern communications require the transmission of IP datagrams to multiple hosts. Typical applications are video conferencing, remote teaching, and so on. This is supported by IP multicasting, where a host group is identified by a single IP address. The main parameters of IP multicasting are:

- The group membership is dynamic.
- Hosts may join and leave the group at any time.
- There is also no limit to the location or number of members in a host group.
- A host may be a member of more than one group at a time.
- A host group may be permanent or transient. Permanent groups are well known and are administratively assigned a permanent IP address. The group is then dynamically associated with this IP address. IP multicast addresses that are not reserved to permanent groups are available for dynamic assignment to transient groups.
- Multicast routers forward IP multicast datagrams into the Internet.

ICMP messages

- Destination unreachable.
- Time-to-live exceeded.
- Parameter problem.
- Source quench.
- Redirect.
- Echo.
- Echo reply.
- Timestamp.
- Timestamp reply.
- Information request.
- Information reply.
- Address request.
- Address reply.

ICMP is used to send error and control messages.

14.11.1 Group addresses

A special group of addresses are assigned to multicasting. These are known as Class D addresses, and they begin with 1110 as their starting 4 bits (Class E addresses with the upper bits of 1111 are reserved for future uses). The Class D addresses thus range from:

224.0.0.0	(11100000 00000000 00000000 00000000)
239.255.255.255	(11101111 11111111 11111111 11111111)

The address 224.0.0.0 is reserved. 224.0.0.1 is also assigned to the permanent group of all IP hosts (including gateways), and is used to address all multicast hosts on the directly connected network. Reserved and allocated addresses are:

224.0.0.0	Reserved
224.0.0.1	All Systems on current subnet
224.0.0.2	All Routers on current subnet

224.0.0.3	Unassigned
224.0.0.5	OSPFIGP All Routers
224.0.0.6	OSPFIGP Designated Routers
224.0.0.9	RIP2 Routers
224.0.0.10–224.0.0.255	Unassigned
224.0.1.0	VMTP Managers Group
224.0.1.1	NTP Network Time Protocol
224.0.1.3	Rwhod
224.0.1.6	NSS – Name Service Server
224.0.1.7	AUDIONEWS
224.0.1.10–224.0.1.255	Unassigned
224.0.3.0–224.0.3.255	RFE Generic Service
224.0.4.0–224.0.4.255	RFE Individual Conferences
224.1.0.0–224.1.255.255	ST Multicast Groups
224.2.0.0–224.2.255.255	Multimedia Conference Calls
232.*x*.*x*.*x*	VMTP transient groups

All the above addresses are listed in the Domain Name Service under MCAST.NET and 224.IN-ADDR.ARPA. On an Ethernet or IEEE 802 network, the 23 low-order bits of the IP Multicast address are placed in the low-order 23 bits of the Ethernet or IEEE 802 net multicast address.

14.11.2 Conformance

There are three levels of conformance:

- Level 0. No IP multicasting support. In this, a Level 0 host ignores, or deletes, all Class D addressed datagrams.
- Level 1. Sending support, but not receiving. In this, a Level 1 host can send multicast datagrams, but cannot receive them.
- Level 2: Full multicasting support. In this, a Level 2 host can send and receive IP multicasting. It also requires the implementation of the Internet Group Management Protocol (IGMP).

14.12 IPv6

The IP header (IP Ver4) is added to higher-level data (as defined in RFC791). This header contains a 32-bit IP address of the destination node. Unfortunately, the standard 32-bit IP address is not large enough to support the growth in nodes connecting to the Internet. Thus a new standard, IP Version 6 (IP Ver6, aka, IP, The Next Generation, or IPng), has been developed to support a 128-bit address, as well as additional enhancements, such as authentication and data encryption.

DEAR NET-ED

Question: *If IP has been such a success, why do we need a new address scheme?*

IP has been a victim of its own success. No-one could have imagined how popular it would be. As it has a 32-bit address it can only support up to 4 billion addresses. Unfortunately not all these addresses can be used, as network addresses are allocated to organizations for their maximum requirement. Also, if an organization uses subnets, then it is unlikely that every subnet has its maximum capacity of hosts.

There are possibly enough IP addresses for all the computers in the world, but the next big wave is going to come from granting IP addresses to virtually every electronic device, such as mobile phones, faxes, printers, traffic lights, telephones, and so on. The stage after this is to grant every object in the world an IP address. This could include cars, trains, people, and even our pets.

The main techniques being investigated are:

- **TUBA** (TCP and UDP with bigger addresses).
- **CATNIP** (common architecture for the Internet). The main idea was to define a common packet format which was compatible with IP, CLNP (Connectionless Network Protocol) and IPX. CLNP was proposed by the OSI as a new protocol to replace IP, but it has never really been adopted (mainly because it was too inefficient).
- **SIPP** (Simple Internet protocol plus). This scheme increases the number of address bits from 32 to 64, and to get rid of unused fields in the IP header.

It is likely that none of these will provide the complete standard and the resulting standard will be a mixture of the three. The RFC1883 specification outlines the main changes as:

- **Expanded Addressing Capabilities.** The size of the IP address will be increased to 128 bits, rather than 32 bits. This will allow for more levels of addressing hierarchy, an increased number of addressable nodes and a simpler auto-configuration of addresses. With multicast routing, the scalability is improved by adding a scope field to the multicast addresses. As well as this, an anycast address has been added so that packets can be sent to any one of a group of nodes.
- **Improved IP Header Format.** This tidies the IPv4 header fields by dropping the least used options, or making them optional.
- **Improved Support for Extensions and Options.** These allow for different encodings of the IP header options, and thus allows for variable lengths and increased flexibility for new options.
- **Flow Labeling Capability.** A new capability is added to enable the labeling of packet belonging to particular traffic *flows* for which the sender requests special handling, such as non-default quality of service or *real-time* service.

DEAR NET-ED

Question: *Apart from increasing the number of IP addresses, why change the format, The Internet works, doesn't it, so why change it?*

Ah. Your perception of the Internet is based on what's available now. Few technologies have expanded so fast, and without virtually any inputs from the governments of the world. Look at the world-wide telephone system infrastructure, if it was based on the system that we had thirty years ago there's no way we could communicate as efficiently as we do. The Internet must do the same, if it is to keep pace with the increase in users, devices and the amount of information that can be transferred. At present, you possibly imagine that the Internet is an infrastructure of computers that have big boxes and sit on your desk, and are congregated around servers, and ISPs. In ten or twenty years this perception will change, and computers will almost become invisible, as will the Internet. To cope with this change we need a different infrastructure. To do this we need to identify its weaknesses:

- The Internet and its addressing structure was never really designed to be a global infrastructure and is constraining the access to resources and information.
- Information and databases tend to be static, and fixed to location.
- Difficult to group individual objects into larger objects.
- Difficult to add resources to the Internet (requires an ISP and a valid IP address).
- Search engines are not very good at gathering relevant information. On the WWW, typically users get pages of irrelevant information, which just happens to have the keyword which they are searching for.
- Resources are gathered around local servers.
- Resources are tied to locations with an IP address.
- IP addresses are not logically organized. The IP address given does not give any information about the geographical location of the destination. This then requires complex routing protocols in which routers pass on information about how to get to remote networks.

- Authentication and Privacy Capabilities. Extensions to support authentication, data integrity, and (optional) data confidentiality are specified for IPv6.

14.12.1 Autoconfiguration and multiple IP addresses

IPv4 requires a significant amount of human intervention to set up the address of each of the nodes. IPv6 improves this by supplying autoconfiguration renumbering facilities, which allows hosts to renumber without significant human intervention.

IPv4 has a stateful address structure, which either requires the user to manually set up the IP address of the computer or to use DHCP servers to provide IP addresses for a given MAC address. If a node moves from one subnet to another, the user must reconfigure the IP address, or request a new IP address from the DHCP. IPv6 supports a stateless autoconfiguration, where a host constructs its own IPv6. This occurs by adding its MAC address to a subnet prefix. The host automatically learns which subnet it is on by communicating from the router which is connected to the network that the host is connected to.

IPv6 supports multiple IP addresses for each host. These addresses can be either *valid*, *deprecated* or *invalid*. A valid address would be used for new and existing communications. A deprecated address could be used only for the existing communications (as they perhaps migrated to the new address). An invalid address would not be used for any communications. When renumbering, a host would deprecate the existing IP address, and set the new IP address as valid. All new communications would use the new IP address, but connections to the previous address would still operate. This allows a node to graduate migrate from one IP address to another.

14.12.2 IPv6 header format

Figure 14.15 shows the basic format of the IPv6 header. The main fields are:

- Version number (4 bits) – contains the version number, such as 6 for IP Ver6. It is used to differentiate between IPv4 and IPv6.
- Priority (4 bits) – indicates the priority of the datagram, and gives 16 levels of priority (0 to 15). The first eight values (0 to 7) are used where the source is providing congestion control (which is traffic that backs-off when congestion occurs), these are:

DEAR NET-ED

Question: *Can devices have more than one IP address?*

Yes. Many devices have more than one IP address. In fact each port that connects to a network must have an IP address. A good example of this is with routers, as they connect to two or more networks. Each of the ports of the router must have an IP address which relates to the network to which it connects to. For example if a router connects to three networks of:

146.176.151.0
146.176.152.0
146.176.140.0

then one IP address from each of the networks must be assigned to the router. Thus it could be assigned the following addresses for its ports:

146.176.151.1
146.176.152.1
146.176.140.1

Question: *Can these addresses be used again for one of the hosts on the connected networks?*

No way. No two ports on the Internet can have the same address.

Question: *Okay, sorry I asked. So what addresses cannot be used for the ports, or the hosts?*

All zeros in the host field, as this identifies the network, and all 1's in the host field as this identifies the broadcast address. Thus in the example above, 146.176.151.0 and 146.176.151.255 could not be used (these addresses use a Class B address with a subnet in the third field).

- 0 defines no priority.
- 1 defines background traffic (such as netnews).
- 2 defines unattended transfer (such as e-mail), 3 (reserved).
- 4 defines attended bulk transfer (FTP, NFS), 5 (reserved).
- 6 defines interactive traffic (such as telnet, X-windows).
- 7 defines control traffic (such as routing protocols, SNMP).

The other values are used for traffic that will not backoff in response to congestion (such as real-time traffic). The lowest priority for this is 8 (traffic which is the most willing to be discarded) and the highest is 15 (traffic which is the least willing to be discarded).

DEAR NET-ED

Question: *Sometimes when I connect to the Internet everything seems fine, but I cannot access WWW sites, and it seems to load pages from a WWW cache?*

This is a common problem, and it is likely that you are connected to the Internet, but the Domain Name Server is not reachable. This means that you cannot resolve domain names into IP addresses. The way to check this is to use the IP address in the URL. For example:

http://www.mypage.com/index.html

could be accessed with:

http://199.199.140.10/index.html

If you can get access with this, you should investigate your DNS. Remember you can normally specify several DNS's, thus find out the address of a remote DNS, just in case your local one goes off-line.

- Flow label (24 bits) – still experimental, but will be used to identify different data flow characteristics. It is assigned by the source and can be used to label data packets which require special handling by IPv6 routers, such as defined QoS (Quality of Service) or real-time services.
- Payload length (16 bits) – defines the total size of the IP datagram (and includes the IP header attached data).
- Next header – this field indicates which header follows the IP header (it uses the same IPv4). For example: 0 defines IP information; 1 defines ICMP information; 6 defines TCP information and 80 defines ISO-IP.
- Hop limit – defines the maximum number of hops that the datagram takes as it traverses the network. Each router decrements the hop limit by 1; when it reaches 0 it is deleted. This has been renamed from IPv4, where it was called Time-to-live, as it better describes the parameter.
- IP addresses (128 bits) – defines IP address. There will be three main groups of IP addresses: unicast, multicast and anycast. A unicast address identifies a particular host, a multicast address enables the hosts within a particular group to receive the same packet, and the anycast address will be addressed to a number of interfaces on a single multicast address.

There are various assigned values for the IP version label. These are:

Value	*Keyword*	*Description*
0		Reserved
4	IP	Internet Protocol (RFC791)
5	ST	ST Datagram Mode (RFC1190)
6	SIP	Simple Internet Protocol
7	TP/IX	TP/IX: The Next Internet
8	PIP	The P Internet Protocol
9	TUBA	TUBA
10–14		Unassigned
15		Reserved

IPv6 has a simple header, which can be extended if required. These are:

- Routing header.
- Authentication header.
- Destinations options header.
- Fragment header.
- Encrypted security payload.

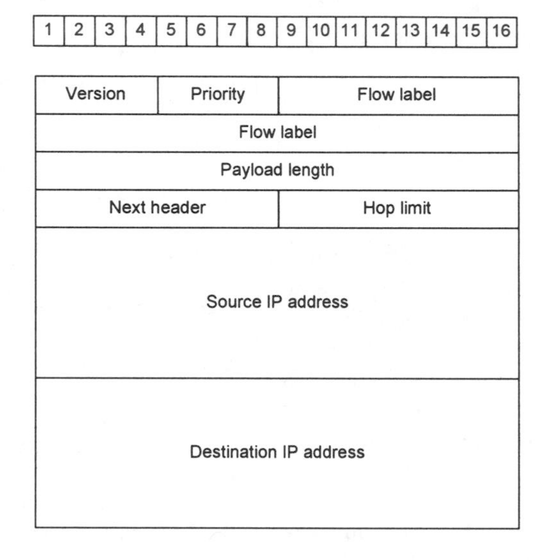

Figure 14.15 IP Ver6 header format

14.12.3 Flow labels

A flow is defined as a sequence of packets sent from a particular source to a particular (unicast or multicast) destination, which requires special handling by intervening routers. There can possibly be many active flows from a source to a destination, as well as other, non-associated traffic. Flows are uniquely identified by the combination of their source address and a non-zero flow label (1h to FFFFFFh). Packets that are not part of a flow are given a zero label.

All packets in the same flow have the same flow label, as well as having the same source and destination addresses, and priority level. The flow-handling lifetime must be set up as part of the set-up mechanism. A label is initially assigned pseudo-randomly and then is assigned sequentially. A source should not reuse a flow label for a new flow within the lifetime of any flow-handling state that could have been established in recent history of the lifetime of a flow. Thus a host must be careful when it restarts from a crash that it is not using a previously defined flow label that is still within its lifetime. Typically, this is overcome by storing a table of previously assigned flow labels.

14.13 Allocating IP addresses

IP addresses can either be allocated statically or dynamically. A static address is permanently assigned to a node, whereas a dynamically allocated address is assigned to a host when it requires connecting to the Internet. Dynamically assigned addresses have the following advantages over static addresses:

- **Limiting access to the Internet.** IP addresses can be mapped to MAC addresses. A node which requires an IP address will ask the IP granting server for an IP address. The server then checks the host's MAC address to determine if it is allowed to access the Internet. If it is not, the server does not return an IP address. The system administrator can thus set up a table which only includes the hosts which are required to connect to the Internet.
- **Authenticating nodes.** A typical hacking method is to steal an IP address and use it for the time of a connection. This can be overcome by making all of the nodes on the network ask the IP granting server for their IP address. It is thus not possible to steal an address, as the IP granting server will check the MAC address of the host.
- **Allocating from a pool of IP addresses.** An organization may be granted a limited range of IP addresses which is not enough to allocate to all the nodes in the organization. The IP granting server can thus be set up to allocate IP addresses to nodes as they require them. When all the IP addresses have been allocated, no more IP addresses can be given out. When a node is finished with its IP address, the IP address that was granted to it can be put back in the pool when it is finished with it.
- **Centralized configuration of IP addresses.** The system manager can easily setup IP addresses to nodes from the central IP granting server.
- **Barring computers from connecting to a network.** Some networks are set up so that they must get a valid IP address before they can connect to the network (typically in Unix-type networks). The IP granting server will check the MAC address of the requester, if it is not allowed the server will not grant it an IP address.

DEAR NET-ED

Question: *When I connect to an ISP, what is my IP address, and my domain name? Can I have the same IP address each time, and the same domain name?*

When you connect to your ISP you will be granted an IP address from a pool of assigned IP addresses. There is no guarantee that this will be the same each time you connect. Your domain name will also change, as it is bound to the IP address. It is possible to be allocated a static IP address, but you would have to pay some money to your ISP for the privilege. The advantage of this is that remote computers could connect to you when you connected via your ISP.

You can determine your current IP address if you use the command WINIPCFG (or IPCONFIG). This is particularly useful if you are playing games over the Internet.

The two main protocols which are used to dynamically allocate IP addresses are DHCP (Dynamic Host Configuration Program) and bootp (bootstrap protocol). DHCP is typically used by Microsoft Windows to get IP addresses, while bootp is sometimes used in Unix environments. The main disadvantage of dynamically assigned IP addresses is that the network is centralized on the single DHCP server. If this were to crash, no IP addresses can be assigned.

14.13.1 Bootp protocol

The bootp protocol allocates IP addresses to computers based on a table of network card MAC addresses. When a computer is first booted, the bootp server interrogates its MAC address and then looks up the bootp table for its entry. It then grants the corresponding IP address to the computer, which the computer then uses it for connections.

The bootp program is typically run on a Linux-based PC with the `bootp` program. The following shows that the `bootp` daemon program (bootpd) is currently running on a computer:

```
$ ps -ax

   PID TTY STAT  TIME COMMAND
     1 con S     0:06 init
    31 con S     0:01 /usr/sbin/inetd
 14142 con S     0:00 bootpd -d 1
    35 con S     0:00 /usr/sbin/lpd
    49 p 3 S     0:00 /sbin/agetty 38400 tty3
 14155 pp0 R     0:00 ps -ax
 10762 con S     0:18 /usr/sbin/named -b /usr/local/adm/named/named.boot
```

For the bootp system to operate then a table is required that reconciles the MAC addresses of the card to an IP address. In the previous example this table is contained in the `bootptab` file which is located in the `/etc` directory. The following file gives an example `bootptab`:

Contents of bootptab file

```
# /etc/bootptab: database for bootp server
# Legend:
#       first field -- hostname
#       hd -- home directory
#       bf -- bootfile
#       cs -- cookie servers
#       ds -- domain name servers
#       gw -- gateways
#       ha -- hardware address
#       ht -- hardware type
#       im -- impress servers
#       ip -- host IP address
#       rl -- resource location protocol servers
#       sm -- subnet mask
#       tc -- template host (points to similar host entry)
#       to -- time offset (seconds)
#       ts -- time servers
#hostname:ht=1:ha=ether_addr_in_hex:ip=ip_addr_in_dec:tc=allhost:
.default150:\
        :hd=/tmp:bf=null:\
        :ds=146.176.151.99 146.176.150.62 146.176.1.5:\
        :sm=255.255.255.0:gw=146.176.150.253:\
        :hn:vm=auto:to=0:
.default151:\
        :hd=/tmp:bf=null:\
        :ds=146.176.151.99 146.176.150.62 146.176.1.5:\
        :sm=255.255.255.0:gw=146.176.151.254:\
        :hn:vm=auto:to=0:
pc345: ht=ethernet: ha=0080C8226BE2:  ip=146.176.150.2: tc=.default150:
pc307: ht=ethernet: ha=0080C822CD4E:  ip=146.176.150.3: tc=.default150:
pc320: ht=ethernet: ha=0080C823114C:  ip=146.176.150.4: tc=.default150:
pc331: ht=ethernet: ha=0080C823124B:  ip=146.176.150.5: tc=.default150:
   :       :
pc460: ht=ethernet: ha=0000E8C7BB63:  ip=146.176.151.142: tc=.default151:
pc414: ht=ethernet: ha=0080C8246A84:  ip=146.176.151.143: tc=.default151:
pc405: ht=ethernet: ha=0080C82382EE:  ip=146.176.151.145: tc=.default151:
```

The format of the file is:

```
#hostname:ht=1:ha=ether_addr_in_hex:     ip=ip_addr_in_dec:tc=allhost:
```

where `hostname` is the hostname, the value defined after `ha=` is the Ethernet MAC address, the value after `ip=` is the IP address and the name after the `tc=` field defines the host information script. For example:

```
pc345:    ht=ethernet:    ha=0080C8226BE2:  ip=146.176.150.2: tc=.default150:
```

defines the hostname of `pc345`, `ethernet` indicates it is on an Ethernet network, and shows its IP address is `146.176.150.2`. The MAC address of the computer is `00:80:C8:22:6B:E2` and it is defined by the script `.default150`. This file defines a subnet of 255.255.255.0 and has associated DNS of

```
146.176.151.99 146.176.150.62 146.176.1.5
```

and uses the gateway at: `146.176.150.253`

14.14 Domain name server

Each institution on the Internet has a host that runs a process called the domain name server (DNS). The DNS maintains a database called the directory information base (DIB) which contains directory information for that institution. When a new host is added, the system manager adds its name and its IP address. It can then access the Internet.

14.14.1 DNS program

The DNS program is typically run on a Linux-based PC with a program called `named` (located in `/usr/sbin`) with an information file of `named.boot`. To run the program the following is used:

```
/usr/bin/named -b   /usr/local/adm/named/named.boot
```

The following shows that the DNS program is currently running.

```
$ ps -ax
  PID TTY STAT  TIME COMMAND
  295 con S     0:00 bootpd
   35 con S     0:00 /usr/sbin/lpd
  272 con S     0:00 /usr/sbin/named -b /usr/local/adm/named/named.boot
  264 p 1 S     0:01 bash
  306 pp0 R     0:00 ps -ax
```

In this case the data file `named.boot` is located in the `/usr/local/adm/named` directory. A sample `named.boot` file is:

```
/usr/local/adm/named - soabasefile
             eece.napier.ac.uk -main record of computer names
             net/net144   -reverse look-up database
             net/net145      "     "
             net/net146      "     "
             net/net147      "     "
             net/net150      "     "
             net/net151      "     "
```

This file specifies that the reverse look-up information on computers on the subnets 144, 145, 146, 147, 150 and 151 is contained in the `net144`, `net145`, `net146`, `net147`, `net150` and `net151` files, respectively. These are stored in the `net` subdirectory. The main file which contains the DNS information is, in this case, `eece.napier.ac.uk`.

Whenever a new computer is added onto a network, in this case, the `eece.napier.ac.uk` file and the `net/net1**` (where ** is the relevant subnet name) are updated to reflect the changes. Finally, the serial number at the top of these data files is updated to reflect the current date, such as 19970321 (for 21 March 1997).

The DNS program can then be tested using `nslookup`. For example:

```
$ nslookup
Default Server:  ees99.eece.napier.ac.uk
Address:  146.176.151.99
> src.doc.ic.ac.uk
Server:  ees99.eece.napier.ac.uk
Address:  146.176.151.99
Non-authoritative answer:
Name:    swallow.doc.ic.ac.uk
Address:  193.63.255.4
Aliases:  src.doc.ic.ac.uk
```

DEAR NET-ED

Question: *If I move my computer from one network to another, does the IP and MAC address stay the same, and what do I need to change?*

The MAC address will not change as the network card stays with the computer. If the computer is moved to a different subnet or onto a completely different network, the IP address must change, or the data will be routed back to the wrong network. Data would leave the relocated computer, and would arrive at the destination, but any data coming back would be routed to the previously attached network (and thus get lost). Another thing that is likely to change is the gateway. Nodes cannot communicate with the hosts outside their network if they do not know the IP address of the gateway (normally a router), thus if the network changes then the gateway is likely to be different.

The user may also have to set a new Domain Name Server (although a host can have several DNS entries). The first one listed in the DNS entries should be the one that is the most reliable and, possibly, the fastest.

Other changes may be to change the subnet mask (on a Class B network, with a subnet this is typically 255.255.255.0).

Question: *So why do you only have to specify the IP address of the gateway?*

Because the host uses an ARP request to determine the MAC address of the gateway.

14.15 DHCP

Dynamic Host Configuration Protocol (DHCP) allows for the transmission of configuration information over a TCP/IP network. Microsoft implemented DHCP on its Microsoft Windows operating system and many other vendors are incorporating it into their systems. It is based on the Bootstrap Protocol (BOOTP) and adds additional services, such as:

- Automatic allocation of reusable IP network addresses.
- Additional TCP/IP configuration options.

It has two components:

- A protocol for delivering host-specific configuration parameters from a DHCP server to a host.
- A mechanism for allocation of network addresses to hosts.

DHCP has been fully defined in the following RFCs:

RFC1533. DCHP options and BOOTP vendor extensions.
RFC1534. Interoperation between DHCP and BOOTP.
RFC1541. DHCP. RFC1542. Clarifications and Extensions for BOOTP.
RFC2131. DHCP. RFC2240. DHCP for Novell.

DHCP uses a client-server architecture, where the designated DHCP server hosts (servers) allocate network addresses and deliver configuration parameters to dynamically configured hosts (clients).

The three techniques that DHCP uses to assign IP addresses are:

- **Automatic** allocation. DHCP assigns a permanent IP address to a client.
- **Dynamic** allocation. DHCP assigns an IP address to a client for a limited period of time or when the client releases the address. It allows for automatic reuse of IP addresses that are no longer used by clients. It is typically used when there is a limited pool of IP addresses (which is less than the number of hosts) so that a host can only connect when it can get one of the IP addresses from the pool.
- **Manual** allocation. DHCP is used to convey an IP address which has been assigned by the network administrator. This allows DHCP to be used to eliminate assigning an IP address to a host through its operating system.

Networks can use several of these techniques.

DHCP messages are based on BOOTP messages, which allows DHCP to listen to a BOOTP relay agent and to allow integration of BOOTP clients and DHCP servers. A BOOTP relay agent is an Internet host or router that passes DHCP messages between DHCP clients and DHCP servers. DHCP uses the same relay agent behavior as the BOOTP protocol specification. BOOTP relay agents are useful because they eliminate the need for having a DHCP server on each physical network segment.

Some of the objectives of DHCP are:

- DHCP should be a mechanism rather than a policy. DHCP must allow local system administrators control over configuration parameters where desired.
- No requirements for manual configuration of clients.
- DHCP does not require a server on each subnet and should communicate with routers and BOOTP relay agents and clients.
- Ensure that the same IP address cannot be used by more than one DHCP client at a time.
- Restore DHCP client configuration when the client is rebooted.
- Provide automatic configuration for new clients.
- Support fixed or permanent allocation of configuration parameters.

14.15.1 Host configuration parameters

The host configuration parameters are:

Be a router	on/off
Non-local source routing	on/off
Policy filters for non-local source routing	(list)
Maximum reassembly size	integer
Default TTL	integer
PMTU ageing timeout	integer
MTU plateau table	(list)
IP-layer parameters:	
IP address	(address)
Subnet mask	(address mask)
MTU	integer
All-subnets-MTU	on/off
Broadcast address flavour	00000000h/FFFFFFFFh
Perform mask discovery	on/off
Be a mask supplier	on/off
Perform router discovery	on/off

Router solicitation address	(address)
Default routers, list of:	
router address	(address)
preference level	integer
Static routes, list of:	
destination	(host/subnet/net)
destination mask	(address mask)
type-of-service	integer
first-hop router	(address)
ignore redirects	on/off
PMTU	integer
perform PMTU discovery	on/off

Link-layer parameters, per interface:

Trailers	on/off
ARP cache timeout	integer
Ethernet encapsulation	(RFC 894/RFC 1042)

TCP parameters, per host:

TTL	integer
Keep-alive interval	integer
Keep-alive data size	0/1

Where MTU is Path MTU Discovery

14.15.2 Protocol outline

Figure 14.16 defines the format of a DHCP message. The numbers in parentheses indicate the size of each field in octets.

- Op. Defines message code/op code (1 for BOOTREQUEST and 2 for BOOTREPLY). 1 byte.
- Htype. Defines hardware address type, such as 1 for 10 Mbps Ethernet. 1 byte.
- Hlen. Hardware address length such as 6 for Ethernet. 1 byte.
- Hops. Client sets to 0, optionally used by relay agents when booting through a relay agent. 1 byte.
- Xid. Transaction ID which is a random number chosen by the client and used by the client and the server to associate messages and responses. 4 bytes.
- Secs. Set by client for the number of seconds elapsed since client began address acquisition. 2 bytes.
- Flags. Flags, the format is BRRR...R where R bits are reserved for future use (and must always be zero) and B defines the broadcast bit which overcomes the problem where some clients cannot accept IP unicast datagrams before the TCP/IP software is configured. 2 bytes.
- Ciaddr. Defines the client's IP address. It is only addressed when the client is in the BOUND, RENEW or REBINDING state and can respond to ARP requests. 4 bytes.
- Yiaddr. Clients IP address. 4 bytes.
- Siaddr. IP address of next server to use in bootstrap; returned in DHCPOFFER, DHCPACK by server. 4 bytes.
- Giaddr. Relay agent IP address and is used in booting through a relay agent. 4 bytes.
- Chaddr. Clients MAC address. 16 bytes.

- Sname. Optional null-terminated server host name string. 64 bytes.
- File. Boot null-terminated file name string. 128 bytes.
- Options. Optional parameters field. Variable number of bytes.

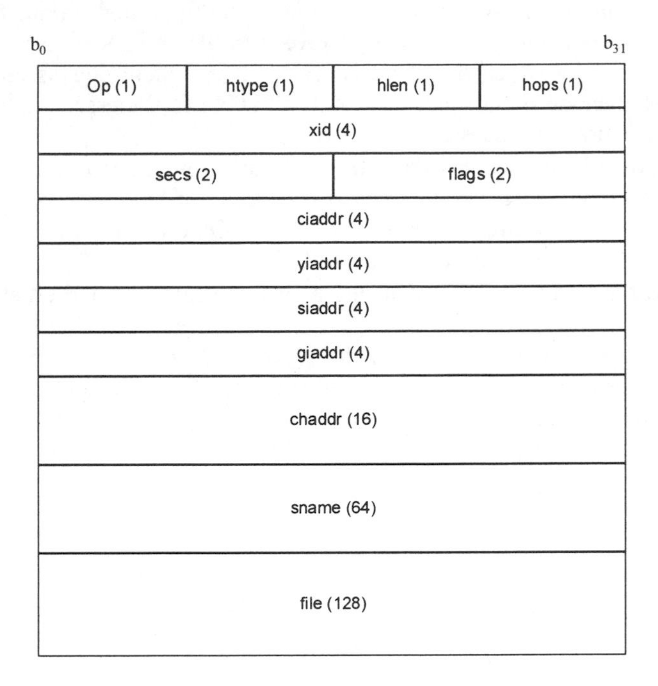

Figure 14.16 DHCP message format

14.15.3 Allocating a network address

The protocol between clients and server is defined by DHCP messages. These are:

- DHCPDISCOVER. The client broadcasts this to locate available servers. When a client initially is started it binds with an address of 0.0.0.0. It then sends out the DHCPDISCOVER message in a UDP packet to port 67 (which is the DHCP/BOOTP server port).
- DHCPOFFER. After a DHCPDISCOVER message, a server may respond to a client with a DHCPOFFER message that includes an available network address in the 'yiaddr' field. When allocating a new address, servers normally check that the offered network address is not already in use. Sending out an ICMP Echo Request to the new address can do this. If the client receives one or more DHCPOFFER messages from one or more servers then the client may choose to wait for multiple responses. It chooses the server based on the configuration parameters offered in the DHCPOFFER messages. The server sends out the DHCPOFFER message in a UDP packet to port 68 (which is the DHCP/BOOTP client port). This is sent as a broadcast as the client does not currently have an IP address.
- DHCPREQUEST. Sent by clients to servers when either requesting offered parameters from one server and implicitly declining offers from all others, confirming parameter allocation (such as after a system boot), or extending the time on a particular network address. The server selected in the DHCPREQUEST message responds with a DHCPACK message containing the configuration parameters for the requesting client. The 'client identifier' ('chaddr') and assigned network address define a unique identifier for the cli-

ents and are used by both the client and server to identify the lease. Servers not selected by the DHCPREQUEST message use the message as notification that the client has declined that server's offer.

- DHCPACK. Sent by the server to a client with configuration parameters, including a committed network address. The client receives the DHCPACK message with configuration parameters, after which the client is set up. If the client detects that the address is already in use then the client sends back a DHCPDECLINE message to the server and restarts the configuration process.
- DHCPNAK. Sent by the server to the client indicating that the client's network address is wrong (for example, the client has moved and does not have the correct subnet) or the time allocation for its network address has expired. On receiving a DHCPNAK message, the client restarts the configuration process.
- DHCPDECLINE. Sent by a client to the server indicating that the network address is already in use.
- DHCPRELEASE. Sent by a client to the server relinquishing the network address. It identifies the lease to be released with its 'client identifier' (chaddr) and network address in the DHCPRELEASE message.
- DHCPINFORM. Sent by a client to the server, asking only for local configuration parameters.

14.15.4 Time allocations

The client acquires the lease of a network time for specified time (either finite or infinite time). The units of time are unsigned integer values in seconds, although the value FFFFFFFFh is used to represent infinity. This gives a range of between 0 and approximately 100 years.

When a client cannot contact the local DHCP server and has knowledge of a previous network address then it may continue to use the previous assigned network address until the lease expires. If the lease expires before the client can contact a DHCP server then the client immediately stops using the previous network address and informs local users of the problem.

14.16 WINS

The Windows Internet Naming Service (WINS) is an excellent companion to DHCP. WINS provides a name registration and resolution on TCP/IP. It extends the function of DNS which will only map static IP addresses to TCP/IP host names. WINS is designed to resolve NetBIOS names on TCP/IP to dynamic network addresses assigned by DHCP. As it resolves NetBIOS names it is obviously aimed at Microsoft Windows-based (and DOS) networks.

14.16.1 NetBIOS name types

Microsoft networks identify computers by their NetBIOS name. Each is 16 characters long, and the 16th character represents the purpose of the name. An example list of a WINS database is:

Name		Type	Status
FRED	<00>	UNIQUE	Registered
BERT	<00>	UNIQUE	Registered
STAFF	<1C>	GROUP	Registered
STAFF	<1E>	GROUP	Registered

The values for the 16th byte are:

00	Workstation	03	Message service
06	RAS server service	1B	Domain master browser
1C	Domain group name	1D	Master browser name
1E	Normal group name (workgroup)	1F	NetDDE service
20	Server service	21	RAS client
BE	Network Monitor Agent	BF	Network Monitor Utility

On Microsoft Windows, the names in the WINS database can be shown with the `nbstat` command.

14.16.2 Microsoft Windows TCP/IP setup

Microsoft Windows uses the files `LMHOSTS`, `HOSTS` and `NETWORKS` to map TCP/IP names and network addresses. These are stored in the *<winNT root>*`\SYSTEM32\DRIVERS\ETC`. `LMHOSTS` maps IP addresses to a computer name. An example format is:

```
#IP-address          host-name
146.176.1.3          bills_pc
146.176.144.10       fred_pc       #DOM:STAFF
```

where comments have a preceding '#' symbol. To preserve compatibility with previous versions of Microsoft LAN Manager, special commands have been included after the comment symbol. These include:

```
#PRE
#DOM:domain
#INCLUDE fname
#BEGIN_ALTERNATE
#END_ALTERNATE
```

where

`#PRE` specifies the name that is preloaded into the memory of the computer and no further references to the `LMHOSTS` file will be made.

`#DOM:`*domain* specifies the name of the domain that the node belongs to.

`#BEGIN_ALTERNATE and #END_ALTERNATE` are used to group multiple `#include`'s

`#include` *fname* specifies other `LMHOST` files to include.

The `HOSTS` file format is IP address followed by the fully qualified name (FQDN) and then any aliases. Comments have a preceding '#' symbol. For example:

```
#IP Address         FQDN           Aliases
146.176.1.3         superjanet     janet
146.176.144.10      hp
146.176.145.21      mimas
146.176.144.11      mwave
146.176.144.13      vax
146.176.146.23      oberon
146.176.145.23      oberon
```

14.16.3 Name registration

The WINS server stores a WINS database, which maps IP addresses to NetBIOS names. The operation is as follows:

1. **Startup.** The WINS client sends a Name Registration Request in a UDP packet to the WINS server to register its NetBIOS name and IP address. When the WINS server receives it then it checks its stored database to make sure that the requested name is not already in use on the network.
2. **Unsuccessful registration.** When a client tries to register a name that is currently in use then the server sends the client a Denial message. The user of the client will then be informed that the computer's name is already in use on that network.
3. **Successful registration.** On successful name registration, the server sends a Name Registration Acknowledgement to the client. It fills the Time-to-Live (TTL) field to define the amount of time that the name registration will be active, after this time the server will cancel it.
4. **Initially re-registering name.** The WINS client must send a Name Refresh Request after a given time interval so that its name will not expire. The first request is made after one-eighth of the TTL time, and then if unsuccessful after periods of one-eighth of the TTL time. If it has not been able to contact the primary WINS server after half the time defined in the TTL field, then the client tries to contact the secondary WINS server (if there is one).
5. **Re-registering a name.** After the first registration, the following registration will be made 50 percent of the TTL time (instead of one-eighth of the time).
6. **Client shutdown.** When a WINS client shuts down, it sends a Name Release Request to the WINS server, releasing its name from the WINS database.

14.16.4 Name resolution and WINS proxy agents

A client that requires the resolution of a NetBIOS name to an IP will go through the following:

1. Checks the name is actually on the local computer; then looks in its own name resolution cache for a match.
2. Sends a required request for a directed name lookup to the WINS server. If it finds one the WINS server sends its IP address to the client.
3. If the WINS server does not find a match then the client broadcasts to the network for help.
4. If there is no response from a broadcast then the client looks into its local LMHOSTS file, else it will look in its local HOSTS file.

Many older Windows systems do not support WINS clients. To allow them to communicate with a WINS server a WINS proxy agent can be used. This agent listens to the network for clients broadcasting for NetBIOS names resolution. The WINS proxy agent then redirects them to the WINS server, which will then pass the IP address resolution to the proxy agent, which in turn will pass it onto the client.

14.17 ICMP

Messages, such as control data, information data and error recovery data, are carried between Internet hosts using the Internet Control Message Protocol (ICMP). These messages are sent with a standard IP header. Typical messages are:

- Destination unreachable (message type 3) – which is sent by a host on the network to say that a host is unreachable. The message can also include the reason the host cannot be reached.

- Echo request/echo reply (message type 8 or 0) – which is used to check the connectivity between two hosts. The `ping` command uses this message, where it sends an ICMP 'echo request' message to the target host and waits for the destination host to reply with an 'echo reply' message.
- Redirection (message type 5) – which is sent by a router to a host that is requesting its routing services. This helps to find the shortest path to a desired host.
- Source quelch (message type 4) – which is used when a host cannot receive anymore IP packets at the present (or reduce the flow).

An ICMP message is sent within an IP header, with the Version field, Source and Destination IP Addresses, and so on. The Type of Service field is set to a 0 and the Protocol field is set to a 1 (which identifies ICMP). After the IP header, follows the ICMP message, which starts with three fields, as shown in Figure 14.17. The message type has eight bits and identifies the type of message; as Table 14.6. The code fields are also eight bits long and a checksum field is 16 bits long. The checksum is the 1's complement of the 1's complement sum of all 16-bit words in the header (the checksum field is assumed to be zero in the addition).

The information after this field depends on the type of message, such as:

- For echo request and reply, the message header is followed by an 8-bit identifier, then an 8-bit sequence number followed by the original IP header.
- For destination unreachable, source quelch and time, the message header is followed by 32 bits which are unused and then the original IP header.
- For timestamp request, the message header is followed by a 16-bit identifier, then by a 16-bit sequence number, followed by a 32-bit originating timestamp.

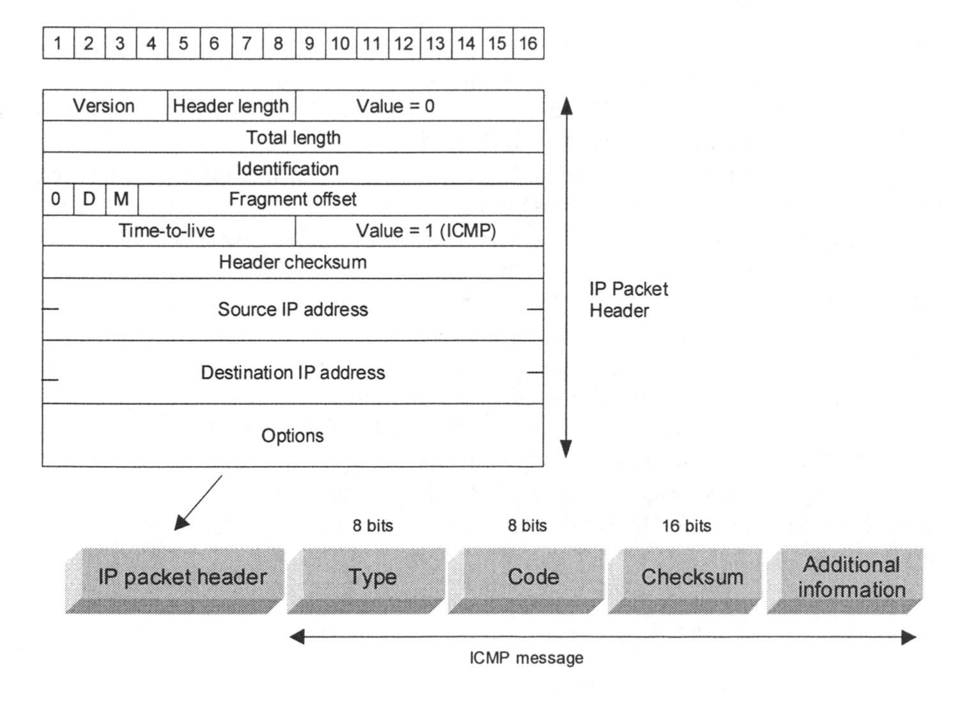

Figure 14.17 ICMP message format

Where:

- Pointer (8-bit). Identifies the byte location of the parameter error in the original IP header. For example, a value of 9 would identify the Protocol field, and 12 would identify the Source IP address field.
- Identifier (16-bit). Helps the matching of requests and replies (possibly set to zero). It can be used to identify a unique connection.
- Sequence Number (16-bit). Helps in matching request and replies (possibly set to zero).
- Timestamps (32-bit) – This is the time in milliseconds since midnight UT (Universal Time). If this is not possible then it is anytime, as long as the high-order bit of the timestamp is set to a 1 to indicate that it is non-standard time.
- Gateway address (32-bit). The address of the gateway to which network traffic specified in the original datagram should be sent to.
- Internet Header + 64 bits of Data Datagram. This is the original IP header and the first 64 byte of the data part. It is used by the host to match the match to the required high-level application (such as TCP port values).

The descriptions of the messages and replies are:

- Source Quelch Message (4). Sent by a gateway or a destination host when it discards a datagram (possibly through lack of buffer memory), and identifies that the sender should reduce the flow of traffic transmission. The host should then reduce the flow, and gradually increase it, as long as it does not receive any more Source Quelch messages.
- Time Exceeded Message (11). This is sent either by a gateway when a datagram has a Time-to-Live field which is zero and has been deleted, or when a host cannot reassemble a fragmented datagram due to missing fragments, within a certain time limit.
- Parameter Problem Message (12). Sent by a gateway or a host when they encounter a problem with one of the parameters in an IP header.
- Destination Unreachable Message (3). Sent by a gateway to identify that a host cannot be reached or a TCP port process does not exist.
- Redirected Message (5). Sent by a gateway to inform other gateways that there is a better route to a given network destination address.
- Information Reply Message (15). Sent in reply to an Information Request. See Information Request (16) for a typical usage.
- Information Request (16). This request can be sent with a fully specified Source IP address, and a zero Destination IP address. The replying IP gateway then replies with an Information Reply Message with its fully specified IP address. In this way the host can determine the network address that it is connected to.
- Echo Message (8). Requests an echo. See Echo Reply Message (0).
- Echo Reply Message (0). The data received in the Echo Message (8) must be returned in this message.

Table 14.6 Message type field value

Value	*Description*	*Code field*	*Additional information*
0	Echo Reply Message	0	16-bit Identifier 16-bit Sequence Number
3	Destination Unreachable	0 – net unreachable 1 – host unreachable 2 – protocol unreachable 3 – port unreachable 4 – fragmentation needed and D bit set 5 – source route failed	32 bits unused Internet Header + 64 bits of Original Data Datagram
4	Source Quench Message	0	32 bits unused Internet Header + 64 bits of Original Data Datagram
5	Redirect Message	0 – redirect datagram for the network 1 – redirect datagram for the host 2 – redirect datagram for the type of service and network 3 – redirect datagram for the type of service and host	32 bits Gateway address Internet Header + 64 bits of Original Data Datagram
8	Echo Request	0	
11	Time-to-live Exceeded	0 – time-to-live exceeded in transit 1 – fragment reassembly time exceeded	32 bits unused Internet Header + 64 bits of Original Data Datagram
12	Parameter Problem	0 – pointer indicates the error	8-bit Pointer 24 bits unused Internet Header + 64 bits of Original Data Datagram
13	Timestamp Request	0	16-bit Identifier 16-bit Sequence Number 32-bit Originate Timestamp 32-bit Receive Timestamp 32-bit Transmit Timestamp
14	Timestamp Reply	0	As above
15	Information Request	0	16-bit Identifier 16-bit Sequence Number
16	Information Reply	0	As above

14.18 Additional

Table 14.7 outlines the values that are used in the protocol field of the IP header.

Table 14.7 Assigned Internet protocol numbers

Value	*Protocol*	*Value*	*Protocol*
0	Reserved	18	Multiplexing
1	ICMP	19	DCN
2	IGMP (Internet group management)	20	TAC monitoring
3	Gateway-to-gateway	21–62	
4	CMCC gateway monitoring message	63	Any local network
5	ST	64	SATNET and backroom EXPAK
6	TCP	65	MIT subnet support
7	UCL	66–68	Unassigned
8	EGP (exterior gateway protocol)	69	SATNET monitoring
9	Secure	70	Unassigned
10	BBN RCC monitoring	71	Internet Packet core utility
11	NVP	72–75	Unassigned
12	PUP	76	Backroom SATNET monitoring
13	Pluribus	77	Unassigned
14	Telenet	78	WIDEBAND monitoring
15	XNET	79	WIDEBAND EXPAK
16	Chaos	80–254	Unassigned
17	User datagram	255	Reserved

14.19 Exercises

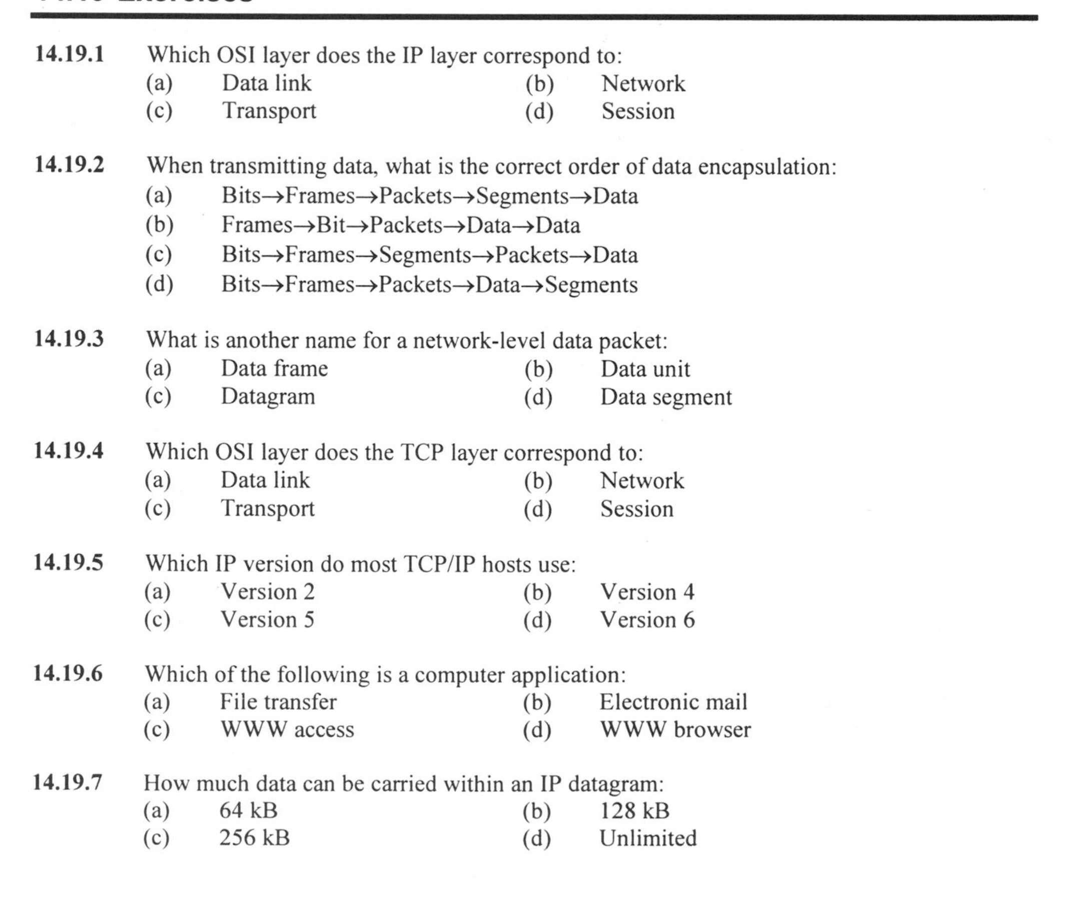

14.19.1 Which OSI layer does the IP layer correspond to:
(a) Data link (b) Network
(c) Transport (d) Session

14.19.2 When transmitting data, what is the correct order of data encapsulation:
(a) Bits→Frames→Packets→Segments→Data
(b) Frames→Bit→Packets→Data→Data
(c) Bits→Frames→Segments→Packets→Data
(d) Bits→Frames→Packets→Data→Segments

14.19.3 What is another name for a network-level data packet:
(a) Data frame (b) Data unit
(c) Datagram (d) Data segment

14.19.4 Which OSI layer does the TCP layer correspond to:
(a) Data link (b) Network
(c) Transport (d) Session

14.19.5 Which IP version do most TCP/IP hosts use:
(a) Version 2 (b) Version 4
(c) Version 5 (d) Version 6

14.19.6 Which of the following is a computer application:
(a) File transfer (b) Electronic mail
(c) WWW access (d) WWW browser

14.19.7 How much data can be carried within an IP datagram:
(a) 64 kB (b) 128 kB
(c) 256 kB (d) Unlimited

14.19.8 How many IP addresses are possible:

(a)	1 048 576	(b)	16 777 216
(c)	4 294 967 296	(d)	$3.402823669 \times 10^{38}$

14.19.9 How are IP datagrams deleted from the network:

(a) They are deleted when the Time-to-live field becomes zero
(b) They are never deleted, and will always be delivered
(c) They are buffered on intermediate systems, and then deleted after a given time
(d) They are returned to the originator if they are not deleted, and the originator either resends them or deletes them

14.19.10 Which is the function of the session layer:

(a) Formats the data into a data frame which can be sent over the network
(b) Segments the data into data segments
(c) Establishes, manages and terminates communication between applications.
(d) Routes data packets around the network

14.19.11 Electronic mail, remote access and file transfer are functions at which layer of the OSI model:

(a)	Data link	(b)	Transport
(c)	Application	(d)	Presentation

14.19.12 Which layer of the OSI model is responsible for how graphics and sound files are displayed:

(a)	Data link	(b)	Transport
(c)	Application	(d)	Presentation

14.19.13 Which of the following is a network application:

(a)	Word Processor	(b)	WWW browser
(c)	Spreadsheet	(d)	Remote access

14.19.14 Which are typical formats that are used at the presentation layer:

(a)	GIF, JPEG, MPEG	(b)	DOC, XLS, MDB
(c)	802.3, 802.5	(d)	HTTP, SMTP, FTP

14.19.15 Which layer of the OSI model encrypts and decrypts data:

(a)	Data link	(b)	Transport
(c)	Application	(d)	Presentation

14.19.16 Which of the following is a Class A IP address:

(a)	12.1.14.12	(b)	146.176.151.130
(c)	194.50.100.1	(d)	224.50.50.1

14.19.17 Which of the following is a Class D IP address:

(a)	12.1.14.12	(b)	146.176.151.130
(c)	194.50.100.1	(d)	224.50.50.1

14.19.18 What are Class D IP addresses used for:

(a)	Dynamic IP addressing	(b)	Testing networks
(c)	Static IP addressing	(d)	Multicasting

14.19.19 Which of the following is the country domain for Germany:

(a)	ge	(b)	de
(c)	dr	(d)	gy

14.19.20 Which service allows hosts to determine the IP address for a given domain name:

(a) TCP (b) ICMP
(c) ARP (d) DNS

14.19.21 Which protocol is used by a node to determine the Ethernet address to a host with a given IP address:

(a) TCP (b) ICMP
(c) ARP (d) DNS

14.19.22 Which Ethernet address is used for broadcast messages:

(a) FF-FF-FF-FF-FF-FF (b) 11-11-11-11-11-11-11
(c) 00-00-00-00-00-00 (d) AA-AA-AA-AA-AA-AA

14.19.23 How do computer applications differ from network applications:

(a) Computer applications only use the local computer for their operation, while network applications require a networking element
(b) Computer applications use a file server for their operation, while network applications require a networking element
(c) Computer applications only require some networking for their operation, while network applications do not require a networking element
(d) Computer applications run on a LAN, while networked applications run over a WAN

14.19.24 Outline how ARP uses the broadcast address and the Type field to identify that an ARP request is being transmitted. Also, discuss a typical ARP conversation.

14.19.25 Outline how the protocol is identified in the IP header. Discuss how the format of the data after the header differs with different protocols (such as TCP and ICMP).

14.19.26 Explain how ICMP and the Options field would be used to determine the following information:

(i) Whether a destination node is responding to TCP/IP communications
(ii) The route to a destination node
(iii) The route to a destination node, with the time delay between each gateway

14.19.27 Explain how the Options field can be used to set the route that a datagram can take.

14.19.28 Determine the IP addresses, and their type (i.e. class A, B or C), of the following 32-bit addresses:

(i) `10001100.01110001.00000001.00001001`
(ii) `01000000.01111101.01000001.11101001`
(iii) `10101110.01110001.00011101.00111001`

14.19.29 Determine the countries which use the following primary domain names:

(a) `de` (b) `nl` (c) `it` (d) `se` (e) `dk` (f) `sg`
(g) `ca` (h) `ch` (i) `tr` (j) `jp` (k) `au`

Determine some other domain names.

14.19.30 For a known TCP/IP network determine the names of the nodes and their Internet addresses. Also determine how the DNS is implemented and how IP addresses are granted.

14.19.31 If a subnet mask on a Class B network is 255.255.240.0, show that there can be 16 connected networks, each with 4095 nodes on a Class B network.

14.19.32 Exam questions

(a) Outline the format of the IP header and how its main fields are used.

(b) What are the main functions of the IP field, and how does this match with the OSI 7-layered model

(c) Explain the method that IP uses to stop data packets from transversing around the Internet, forever.

(d) Explain how different high-level protocols are defined in the IP header. What is the usage of ICMP?

(e) Derive the range of addresses for a Class A and a Class B IP network address, and the maximum number of nodes that can connect to each address allocation.

(f) Derive the maximum number of IP addresses available in IPv4 and also for IPv6.

(g) If a TCP/IP network has a subnet mask of 255.255.240.0, determine the maximum number of hosts that can be connected to the subnet.

(h) Explain the method that IP uses to differentiate between different IP addressing groups, such as Class A–E. To what type of organization might these classes be allocated to?

(i) Determine the IP addresses, and their type (i.e. class A, B or C), of the following 32-bit binary addresses:

(i) `10001100.01110001.00000001.00001001`
(ii) `01000000.01111101.01000001.11101001`
(iii) `10101110.01110001.00011101.00111001`

14.20 Note from the Author

Which two people have done more for world unity than anyone else? Well, Prof. TCP and Dr. IP must be somewhere in the Top 10. They have done more to unify the world than all the diplomats in the world have. They do not respect national borders, time zones, cultures, industrial conglomerates or anything like that. They allow the sharing of information around the world, and are totally open for anyone to use. Top marks to Prof. TCP and Dr. IP, the true champions of freedom and democracy.

Many of the great inventions/developments of our time were things that were not really predicted, such as CD-ROMs, RADAR, silicon transistors, fiber-optic cables, and, of course, the Internet. The Internet itself is basically an infrastructure of interconnected networks which run a common protocol. The nightmare of interfacing the many computer systems around the world was solved because of two simple protocols: TCP and IP. Without them the Internet would not have evolved so quickly and possibly would not have occurred at all. TCP and IP are excellent protocols as they are simple and can be run over any type of network, on any type of computer system.

The Internet is often confused with the World Wide Web (WWW), but the WWW is only one application of the Internet. Others include electronic mail (the No.1 application), file transfer, remote login and so on.

The amount of information transmitted over networks increases by a large factor every year. This is due to local area networks, wide area networks and of course, traffic over the Internet. It is currently estimated that traffic on the Internet doubles every 100 days and that three people join the Internet every second. This means an eightfold increase in traffic over a whole year. It is hard to imagine such growth in any other technological area. Imagine if cars were eight times faster each year, or could carry eight times the number of passengers each year (and of course roads and driveways would have to be eight times larger each year).

TCP/UDP

RFCs

RFC (Request For Comment) documents are published by the IAB (Internet Advisory Board) and are a quick way to define new standards, in which anyone can comment on. They are continually updated, but various documents define the main standards used on the Internet; these are:

RFC768	UDP
RFC791	IP
RFC792	ICMP
RFC793	TCP
RFC821	SMTP
RFC822	Format of e-mail messages
RFC854	Telnet
RFC959	FTP
RFC1034	Domain names
RFC1058	RIP
RFC1157	SNMP
RFC1521	MIME Pt. 1
RFC1522	MIME Pt. 2
RFC1939	POP Version 3

15.1 Introduction

The transport layer is important as it segments the data from the upper layers, and passes these onto the network layer, which adds the network address to the segments (these are then called packets). The network layer allows for delivery of these segments at the receiver. Once delivered the transport layer then takes over and reassembles the data segments back into a form that can be delivered to the layer above. These services are often known as end-to-end services, as the transport layer provides a logical connection between two end points on a network.

The network layer cannot be considered reliable, as the transmitter has no idea that the data packets have been received correctly. It is thus up to the transport layer to provide this, using:

- **Synchronization and acknowledgement.** Initially, when the transmitter makes contact with the receiver it makes a unique connection. The transmitter thus knows that the receiver is on-line, and willing to receive data.
- **Acknowledgements and retransmissions.** This allows the receiver to send back acknowledgements which tell the transmitter that the data segments have been received correctly. If no acknowledgements have been received, the transmitter can either resend the data, or can assume that the receiver has crashed and that the connection is to be terminated.
- **Flow control.** This allows the receiver to tell the transmitter that it cannot receive any more data at present. This typically happens when the receiver has filled-up its receiving buffer.
- **Windowing.** This is where the transmitter and the receiver agree on a window size when the connection is initially made. The window then defines the number of data segments that can be sent before the transmitter must wait for an acknowledgement from the receiver.
- **Multiple connections onto a single data stream.** The transport layer takes data from one or more applications; it then marks them with a unique connection number and segment number. At the receiver these can be demultiplexed to the correct application program.
- **Reordering of data segments.** All the data segments that are transmitted are marked with a sequence number. Thus if any are delivered in the incorrect order, or if any of them are missing, the receiver can easily reorder them or discard segments if one or more are missing.

The most important transport protocol is TCP, which is typically uses IP as an addressing scheme. TCP and IP (TCP/IP) are the standard protocols that are used on the Internet and also on Unix networks (where TCP/IP grew-up, but have since dwarfed their parent). TCP works well because it is simple, but reliable, and it is an open system where no single vendor has control over its specification and its development.

An important concept of TCP/IP communications is the usage of ports and sockets. A port identifies a process type (such as FTP, TELNET, and so on) and the socket identifies a unique connection number. In this way, TCP/IP can support multiple simultaneous connections of applications over a network, as illustrated in Figure 15.1.

The IP header is added to higher-level data, and contains the 32-bit IP address of the destination node. Unfortunately, the standard 32-bit IP address is not large enough to support the growth in nodes connecting to the Internet. Thus a new standard, IP Version 6, has been developed to support 128-bit addresses, as well as additional enhancements.

The transport layer really provides a foundation for the application program to create a data stream from one application program to another, just as if the two programs were running on the same computer, thus the data transfer operation should be transparent to the application program. It is thus important that the network protocol and the network type are invisible to the layers about the transport layer, as illustrated in Figure 15.2. This is important as application programmers can write programs which will run on any type of network protocol (such as SPX/IPX or TCP/IP) or any network type (such as, over ISDN, ATM or Ethernet). The protocol stack is the software that is used to provide the transport and network layers.

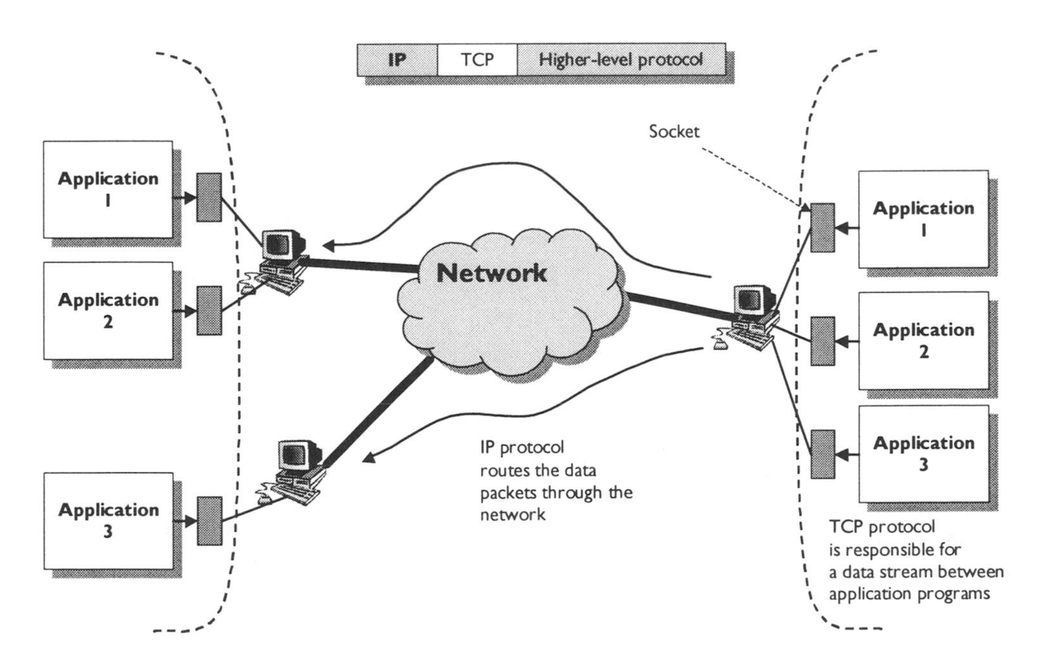

Figure 15.1 Roles of TCP and IP

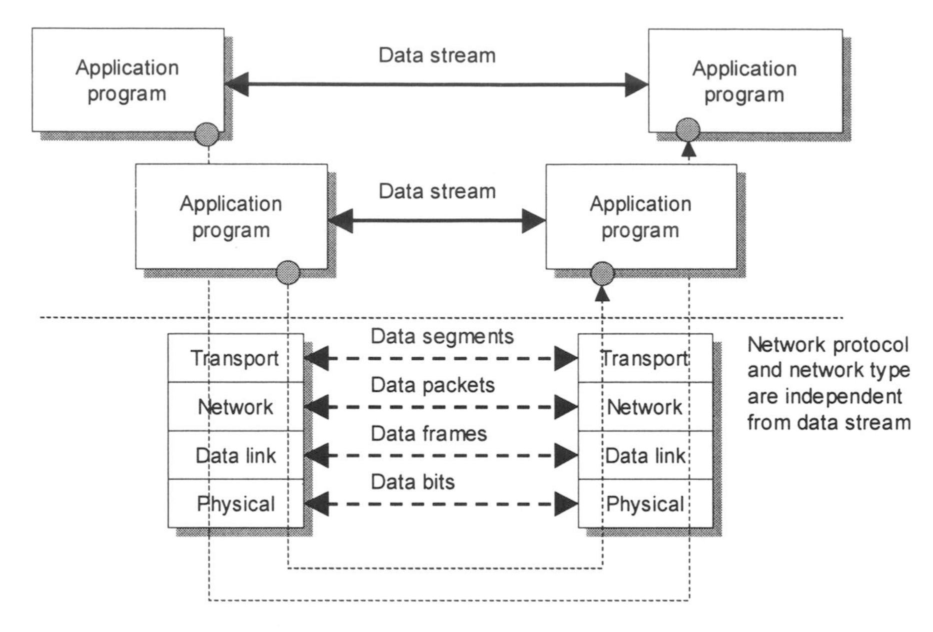

Figure 15.2 Data streams

15.2 Functions of the transport layer

The great trick of the transport level is to allow multiple applications to communicate over a network, at the same time. This is achieved by identifying each connection with a unique value (a socket number) and adding sequence numbers on each of the segments. Segments can also be sent to many different destinations (using the network layer to identify the destination).

> **TCP/IP protocol stack**
>
> - Protocols to support file transfer, e-mail, remote login, and other applications
> - Reliable and unreliable transports (TCP/UDP).
> - Connectionless datagram delivered at the network layer.
> - Connection-oriented with TCP, connection-less with UDP.

15.2.1 Establishing connections

An important concept in the transport layer is to make a reliable connection with a destination node. This requires that the transport layer on one host makes contact with the transport layer on the destination host, and create a connection-oriented session with the peer system. Before data can be transmitted the sender and the receiver negotiate the connection and agree the parameters for the data to be sent, such as how many data segments that can be sent before an acknowledgement, the unique connection number, and so on. This process is called synchronization. Once the connection has been agreed, the data can pass between the transmitter and receiver on a datastream.

Figure 15.3 shows an example of the negotiation between a transmitter and a receiver. Initially the transmitter requests synchronization, after which the transmitter and the receiver negotiate the connection parameters. If acceptable the receiver sends a synchronization signal, after which the transmitter acknowledges this with an acknowledgement. The connection is now made, and a data stream can flow between the two systems.

15.2.2 Flow control

Data that is received is typically buffered in memory, as the processor cannot deal with it immediately. If there is too much data arriving to be processed, the data buffer can often overflow, and all newly arriving data will be discarded (as there is no place to store it). Another problem can occur when several hosts are transmitting to a single host. There is thus a need for a mechanism which can tell hosts to stop sending data segments, and to wait until the data has been properly processed.

The transport layer copes with these problems by issuing a Not Ready indicator, which tells a transmitter not to send any more data, until the host sends a Ready indicator. After this the transmitter can send data. This process is illustrated in Figure 15.4.

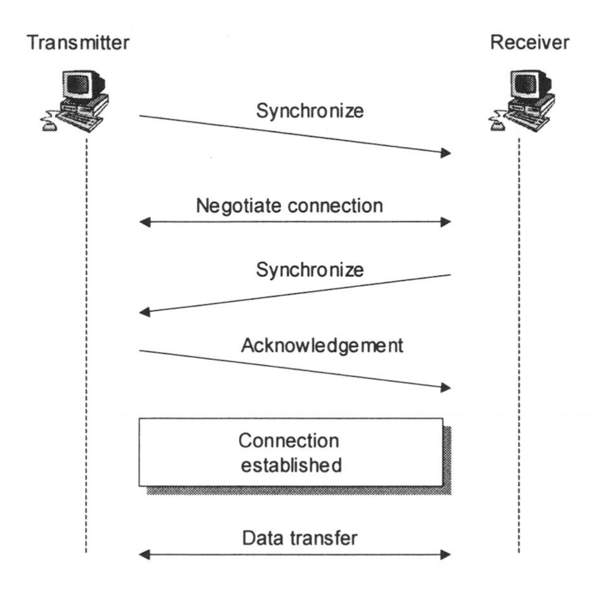

Figure 15.3 Synchronization and acknowledgement

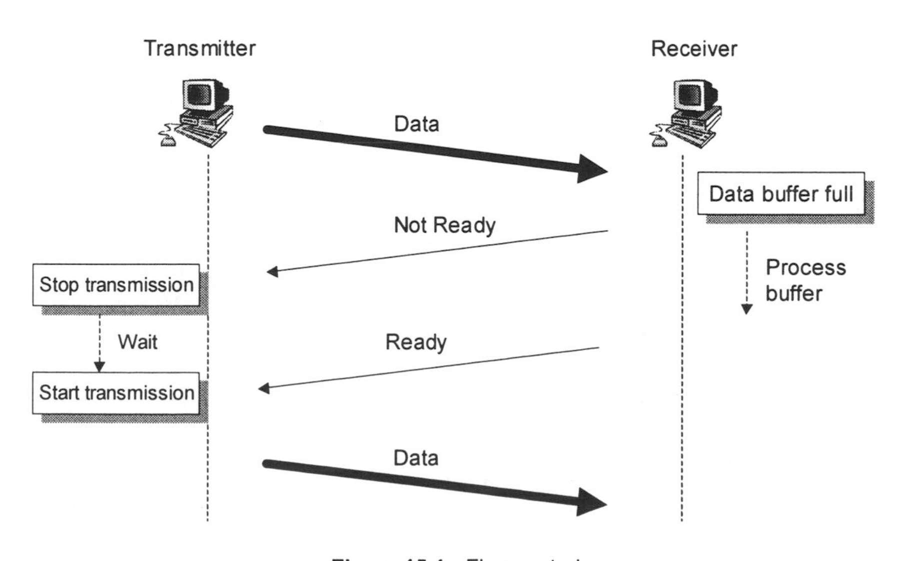

Figure 15.4 Flow control

15.2.3 Acknowledgement and windowing

The transport layer uses a technique called positive acknowledgement which only acknowledges correctly received data segments. The transmitter thus assumes that, after a certain period of time, if no acknowledgement has been received then the data has not been received correctly, and will thus retransmit the unacknowledged data segments. These will be sent with the same segment numbers as the original data segments so that the receiver does not end up with identical data segments. Thus two or more received data segments with the same sequence number will be deleted so that only a single copy is left.

> **DARPA**
>
> The Defense Advanced Research Projects Agency, which is a US government agency that has funded research relating to the Internet. Now known as ARPA.
>
> **TCP/IP**
>
> Two of the protocols developed by the US DoD in the 1970s to support the interconnection of networks.

Windowing provides for a method where the transmitter is forced to wait for an acknowledgement for the data segments that it has sent. The number of data segments that it is allowed to be sent before an acknowledgement, is set up when the connection is made. If the transmitter does not receive an acknowledgement within a given time limit, it will assume that the data did not arrive at the destination, and will then retransmit the data segments which were sent after the last acknowledged data segment.

If the window were set to unity, every data segment would have to be acknowledged. This, of course, would be inefficient and slow, as the transmitter would have to wait for every segment to be acknowledged. Thus the window allows for a number of segments to be transmitted before the requiring acknowledgements.

Figure 15.5 shows the transmission of data segments with a window size of three. Initially the transmitter sends data segments with a send sequence number (S) of 1–3. It then waits for the receiver to send an acknowledgement, which it does by informing the transmitter that it expects to receive a data segment with a value of 4, next (R=4). If no acknowledgement was received from the receiver, the transmitter would resend the data segments 1 to 3. Note that the sequence numbers can either relate to the packet number (such as 1, 2, 3, …) or to the byte number of the data being transmitted (this is the case in TCP communications). The reason that there is a different send (S) and receive (R) number is that both nodes may be transmitting and receiving data, thus they must both keep track of the data segments that are being sent and received.

OSI model	**Internet model**
Application *Presentation* *Session*	*Application*
Transport	*Transport (TCP)*
Network	*Internet (IP, ARP, ICMP, RARP)*
Data link *Physical*	*Network Interface*

> TCP/IP supports most existing networking technologies, including Ethernet, ATM, FDDI, ISDN, and so on.
>
> Internet applications:
>
> - File transfer (FTP, NFS).
> - E-mail (SMTP).
> - Remote login (TELNET).
> - Network management (SNMP).
> - Domain naming (DNS).

Often the start sequence number does not start at zero or one, as previous connections could be confused with new connections, thus a time-based initial value is used to define the start sequence number. Both the transmitter and the receiver agree on the start sequence number when the connection is negotiated.

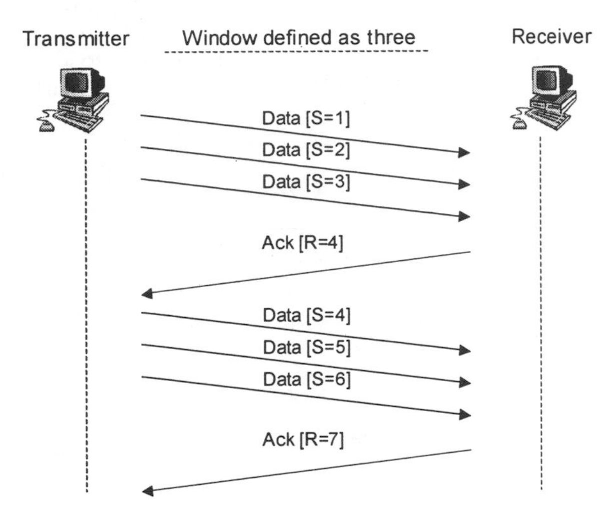

Figure 15.5 Windowing with a window of three

Top 10 TCP ports

21	FTP
23	TELNET
25	SMTP (E-mail)
37	Time
53	DNS (Naming)
69	TFTP
79	FINGER
110	POP-3
161	SNMP
520	RIP

15.3 TCP/IP model

No networking technology fits into the OSI model, as many of the existing networking standards were developed before the model was developed. Also the OSI model was meant to be an abstract method of viewing a network from its physical connection, through its hardware/software interface, right up to the application program. The key element that allows computers over the world to intercommunicate, no matter their operating system, their hardware, their network connection, or their application program is the networking protocol. At one time networking protocols were tied to specific systems, such as DECNET (for DEC-based network), DLC (for IBM-based networks), NetBEUI (for Microsoft networks) and SPX/IPX (for Novell NetWare networks). While most of these networking protocols are still used for local area networks, the most common protocol for worldwide communications is TCP/IP.

TCP/IP does not quite fit into the OSI model, as illustrated in Figure 15.6. The OSI model uses seven layers where the TCP/IP model uses four layers, which are:

- **Network access layer.** Specifies the procedures for transmitting data across the network, including how to access the physical medium, such as Ethernet and FDDI.
- **Internet layer.** Responsible for data addressing, transmission, and packet fragmentation and reassembly (IP protocol).
- **Transport layer.** Manages all aspects of data routing and delivery including session initiation, error control and sequence checking (TCP/UDP protocols). This includes part of the session layer of the OSI model.
- **Application layer.** Responsible for everything else. Applications must be responsible for all the presentation and part of the session layer.

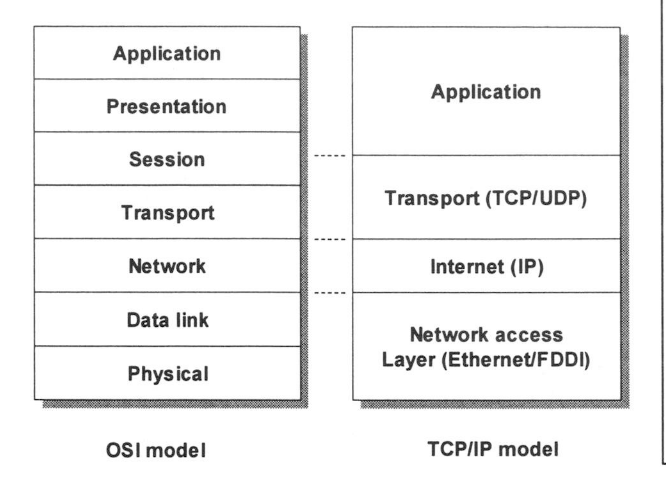

TCP port values:

Used to identify the application (such as 21 for FTP). The ranges are:

- 0–255. Public applications.
- 255–1023. Assigned to companies for commercial products.
- Above 1023. Unregulated, and can be used by application programmers, or dynamically assigned by an application.

Figure 15.6 OSI and TCP/IP model

15.4 Transmission control protocol

In the OSI model, TCP fits into the transport layer and IP fits into the network layer. TCP thus sits above IP, which means that the IP header is added onto the higher-level information (such as transport, session, presentation and application). The main function of TCP is to provide a robust and reliable transport protocol. It is characterized as a reliable, connection-oriented, acknowledged and datastream-oriented server service. IP, itself, does not support the connection of two nodes, whereas TCP does. With TCP, a connection is initially established and is then maintained for the length of the transmission.

The main aspects of TCP are:

- **Data Transfer.** Data is transmitted between two applications by packaging the data within TCP segments. This data is buffered and forwarded whenever necessary. A push function can be used when the data is required to be sent immediately.
- **Reliability.** TCP uses sequence numbers and positive acknowledgements (ACK) to keep track of transmitted segments. Thus, it can recover from data that is damaged, lost, duplicated, or delivered out of order, such as:
  - **Time-outs.** The transmitter waits for a given time (the timeout interval), and if it does not receive an ACK, the data is retransmitted.
  - **Sequence numbers.** The sequence numbers are used at the receiver to correctly order the packets and to delete duplicates.
  - **Error detection and recovery.** Each packet has a checksum, which is checked by the receiver. If it is incorrect the receiver discards it, and can use the acknowledgements to indicate the retransmission of the packets.

Window size. Defines the number of data segments that can be sent before an acknowledgement is required. The destination sends back an acknowledgement number for the data packet it expects to receive **NEXT**. Thus, if data segments 1, 2, 3 and 4 have been received the recipient sends back an acknowledgement value of 5.

Window size negotiation. This occurs when the connection is first made. The two hosts define the window in either direction (known as sliding window).

- **Flow Control.** TCP returns a window with every ACK. This window indicates a range of acceptable sequence numbers beyond the last segment successfully received. It also indicates the number of bytes that the sender can transmit before receiving further acknowledgements.
- **Multiplexing.** To support multiple connections to a single host, TCP provides a set of ports within each host. This, along with the IP addresses of the source and destination, makes a socket, and a pair of sockets uniquely identifies each connection. Ports are normally associated with various services and allow server programs to listen for defined port numbers.
- **Connections.** A connection is defined by the sockets, sequence numbers and window sizes. Each host must maintain this information for the length of the connection. When the connection is closed, all associated resources are freed. As TCP connections can be made with unreliable hosts and over unreliable communication channels, TCP uses a handshake mechanism with clock-based sequence numbers to avoid inaccurate connection initialization.
- **Precedence and Security.** TCP allows for different security and precedence levels.

TCP information contains simple acknowledgement messages and a set of sequential numbers. It also supports multiple simultaneous connections using destination and source port numbers, and manages them for both transmission and reception. As with IP, it supports data fragmentation and reassembly, and data multiplexing/demultiplexing.

The set-up and operation of TCP is as follows:

1. When a host wishes to make a connection, TCP sends out a request message to the destination machine that contains unique numbers called a socket number, and a port number. The port number has a value which is associated with the application (for example a TELNET connection has the port number 23 and an FTP connection has the port number 21). The message is then passed to the IP layer, which assembles a datagram for transmission to the destination.
2. When the destination host receives the connection request, it returns a message containing its own unique socket number and a port number. The socket number and port number thus identify the virtual connection between the two hosts.
3. After the connection has been made the data can flow between the two hosts (called a data stream).

UDP

UDP has no sequence numbers, and thus does not use windowing or acknowledgements. Thus the application layer must provide reliability. UDP is designed for applications which do not send data segment sequences, and do not require to make a virtual connection with the other side.

Applications:

- **DNS.** Supports domain name mapping to IP addresses.
- **NFS.** Supports a distributed file system.
- **SNMP.** Support network management for network devices.
- **TFTP** (Trivial FTP). A simplified version of FTP (typically used to update computers/routers with firmware updates).

After TCP receives the stream of data, it assembles the data into packets, called TCP segments. After the segment has been constructed, TCP adds a header (called the protocol data unit) to the front of the segment. This header contains information such as a checksum, the port number, the destination and source socket numbers, the socket number of both machines and segment sequence numbers. The TCP layer then sends the packaged segment down to the IP layer, which encapsulates it and sends it over the network as a datagram.

15.4.1 Ports and sockets

As previously mentioned, TCP adds a port number and socket number for each host. The port number identifies the required service, whereas the socket number is a unique number for that connection. Thus, a node can have several TELNET connections with the same port number but each connection will have a different socket number. A port number can be any value but there is a standard convention that most systems adopt. Table 15.1 defines some of the most common values. Standard applications normally use port values from 0 to 255, while unspecified applications can use values above 255. Section 15.11 outlines the main ports.

Table 15.1 Typical TCP port numbers

Port	*Process name*	*Notes*
20	FTP-DATA	File Transfer Protocol – data
21	FTP	File Transfer Protocol – control
23	TELNET	Telnet
25	SMTP	Simple Mail Transfer Protocol
49	LOGIN	Login Protocol
53	DOMAIN	Domain Name Server
79	FINGER	Finger
161	SNMP	SNMP

15.4.2 TCP header format

The sender's TCP layer communicates with the receiver's TCP layer using the TCP protocol data unit. It defines parameters such as the source port, destination port, and so on, and is illustrated in Figure 15.7.

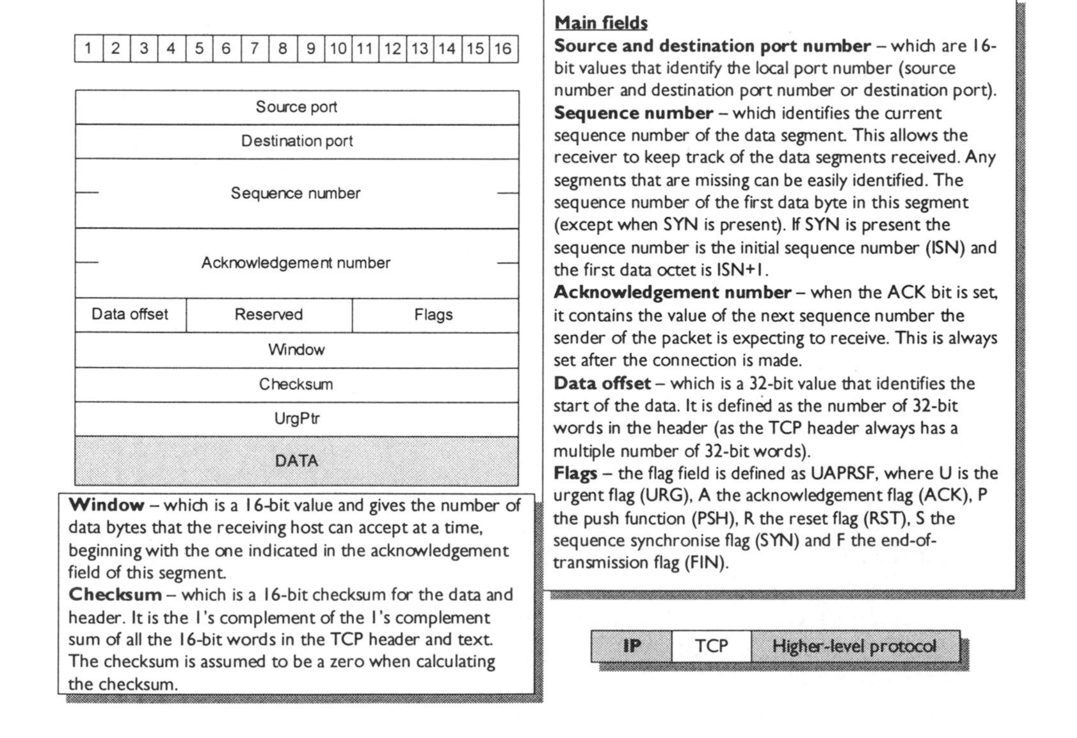
1 2 3 4 5 6 7 8 9 10 11 12 13 14 15 16
Source port
Destination port
Sequence number
Acknowledgement number
Data offset
Reserved
Flags
Window
Checksum
UrgPtr
DATA
Window – which is a 16-bit value and gives the number of data bytes that the receiving host can accept at a time, beginning with the one indicated in the acknowledgement field of this segment.
Checksum – which is a 16-bit checksum for the data and header. It is the 1's complement of the 1's complement sum of all the 16-bit words in the TCP header and text. The checksum is assumed to be a zero when calculating the checksum.
Main fields
Source and destination port number – which are 16-bit values that identify the local port number (source number and destination port number or destination port).
Sequence number – which identifies the current sequence number of the data segment. This allows the receiver to keep track of the data segments received. Any segments that are missing can be easily identified. The sequence number of the first data byte in this segment (except when SYN is present). If SYN is present the sequence number is the initial sequence number (ISN) and the first data octet is ISN+1.
Acknowledgement number – when the ACK bit is set, it contains the value of the next sequence number the sender of the packet is expecting to receive. This is always set after the connection is made.
Data offset – which is a 32-bit value that identifies the start of the data. It is defined as the number of 32-bit words in the header (as the TCP header always has a multiple number of 32-bit words).
Flags – the flag field is defined as UAPRSF, where U is the urgent flag (URG), A the acknowledgement flag (ACK), P the push function (PSH), R the reset flag (RST), S the sequence synchronise flag (SYN) and F the end-of-transmission flag (FIN).
IP
TCP
Higher-level protocol

Figure 15.7 TCP header format

The fields are:

- **Source and destination port number** – which are 16-bit values that identify the local port number (source number and destination port number or destination port).
- **Sequence number** – which identifies the current sequence number of the data segment. This allows the receiver to keep track of the data segments received. Any segments that are missing can be easily identified. The sequence number is the first data byte of the DATA segment (except when SYN is present). If SYN is present the sequence number is the initial sequence number (ISN) and the first data octet is ISN+1.
- **Acknowledgement number** – when the ACK bit is set, it contains the value of the next sequence number the sender of the packet is expecting to receive. This is always set after the connection is made.
- **Data offset** – which is a 32-bit value that identifies the start of the data. It is defined as the number of 32-bit words in the header (as the TCP header always has a multiple number of 32-bit words).
- **Flags** – the flag field is defined as `UAPRSF`, where `U` is the urgent flag (URG), `A` the acknowledgement flag (ACK), `P` the push function (PSH), `R` the reset flag (RST), `S` the sequence synchronize flag (SYN) and `F` the end-of-transmission flag (FIN).
- **Window** – which is a 16-bit value and gives the number of data bytes that the receiving host can accept at a time, beginning with the one indicated in the acknowledgement field of this segment.
- **Checksum** – which is a 16-bit checksum for the data and header. It is the 1's complement of the 1's complement sum of all the 16-bit words in the TCP header and data. The checksum is assumed to be a zero when calculating the checksum.
- **UrgPtr** – which is the urgent pointer and is used to identify an important area of data (most systems do not support this facility). It is only used when the URG bit is set. This field communicates the current value of the urgent pointer as a positive offset from the sequence number in this segment.
- **Padding** (variable) – The TCP header padding is used to ensure that the TCP header ends and data begins on a 32-bit boundary. The padding is composed of zeros.

Question. *If a computer has no permanent storage, how does it know its own IP address?*

Diskless hosts use the RARP protocol, which broadcasts a message to a RARP server. The RARP server looks-up the MAC address in the source address field in the data frame and sends back its IP address in a reply to the host.

Question. *How is it possible to simply connect a computer to an Ethernet network, and all the computers on the network are able to communicate with it, and how do they know when a computer has been disconnected?*

Computers use the ARP protocol, which allows nodes to determine the MAC address of computers on the network, from given IP addresses. Once they discover the destination MAC address, they update their ARP cache. After a given time, the entries in the table are updated (known as aging the entry).

Question: *How does a node broadcast to the network?*

There are two types of broadcasts. The first is a flooded broadcast, which has 1's in all parts of the IP address (255.255.255.255). The other it a directed broadcast, which has all 1's in the host part of a IP address. For example, to broadcast to the 146.176.151.0 network, the broadcast address is 146.176.151. 255, as all 1's in the host part of the address specifies a broadcast. Routers forward directed broadcasts, but not flooded addresses (as these are local). All hosts and routers must thus know what the subnet mask is.

- **Options** – Possibly added to segment header.

In TCP, a packet is termed as the complete TCP unit; that is, the header and the data. A segment is a logical unit of data, which is transferred between two TCP hosts. Thus a packet is made up of a header and a segment.

Top 5 Protocol numbers in the IP header:

Protocol number defines the transport protocol that follows. This tends to be either IP, ICMP (for testing networks) or a routing protocol (such as EGP).

1 ICMP
6 TCP
8 EGP (routing protocol)
17 UDP
88 IGRP (routing protocol)

15.3 UDP

TCP allows for a reliable connection-based transfer of data. The User Datagram Protocol (UDP) is an unreliable connection-less approach, where datagrams are sent into the network without any acknowledgements or connections (and thus relies on high-level protocols to provide for these). It is defined in RFC768 and uses IP as its underlying protocol. Its main advantage over TCP is that it has a minimal protocol mechanism, but does not guarantee delivery of any of the data. Figure 15.8 shows its format, which shows that the Protocol field in the IP header is set to 17 to identify UDP.

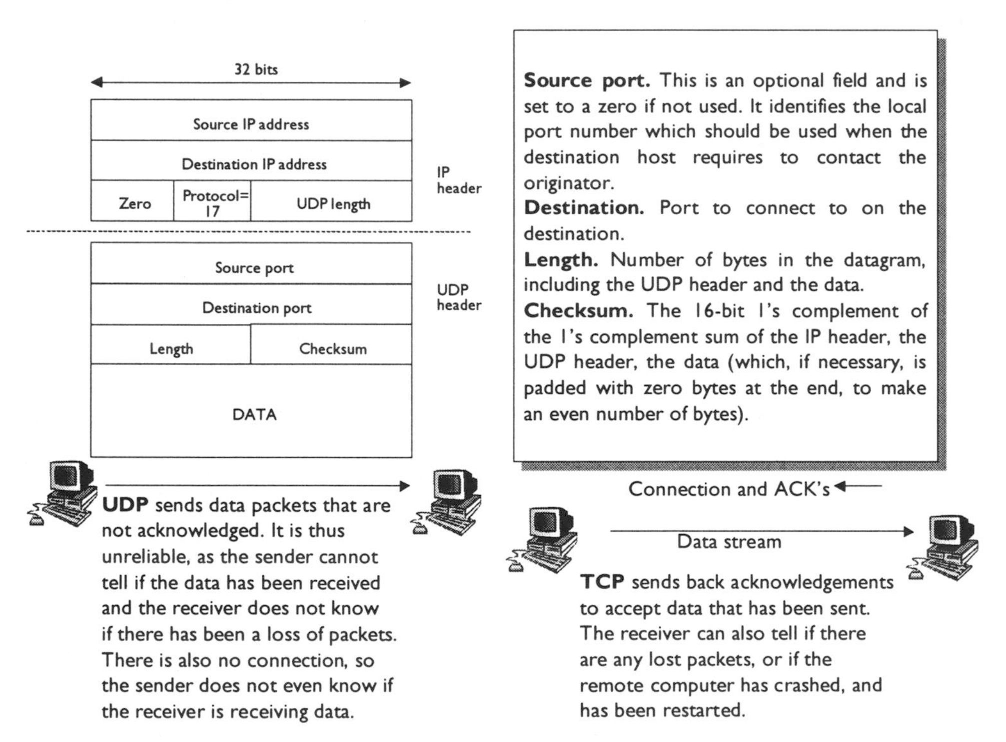

Figure 15.8 UDP header format

The fields are:

- **Source port.** This is an optional field and is set to a zero if not used. It identifies the local port number which should be used when the destination host requires to contact the originator.
- **Destination.** Port to connect to on the destination.

- **Length.** Number of bytes in the datagram, including the UDP header and the data.
- **Checksum.** The 16-bit 1's complement of the 1's complement sum of the IP header, the UDP header, the data (which, if necessary, is padded with zero bytes at the end, to make an even number of bytes).

UDP is used when hosts do not require to make a connection with the other side, and where reliability is built into a high-layer protocol. It is also used when there is no segmentation of data (as there are no segment values). Some applications are solely TCP or solely UDP, whereas the rest can use either. For example:

- TCP applications. FTP, TELNET, SMTP, FINGER, DNS and LOGIN.
- UDP applications. RIP, TFTP, NFS and SNMP.
- TCP or UDP applications. HTTP, POP-3 and ECHO.

Question. *Okay. I understand that both the MAC address and the IP address need to be specified for a node to receive data, but how does a node know the MAC address of the remote destination?*

1. A host looks up its local ARP cache (which is in its own RAM, and not stored to the permanent storage) to see if it knows the MAC address for a known IP address.
2. If it does not find the MAC address, it transmits an ARP request to the whole of the network (ARP requests do not travel over routers). The host who matches the transmitted IP address then responds with an ARP reply with its own MAC address in the source address field in the data frame. This is received by the originator of the request, which updates its local ARP cache, and then transmits with the required MAC address.

Question. *Oh, yes. I think I see it now, but what if the destination is on another network, possibly in another country, how does it determine the address of the destination?*

1. The host knows the IP address of the gateway for the network (normally a router). It then uses the MAC address of the gateway, but with the destination IP address of the host that the data is destined for. The gateway senses that the data frame is addressed to itself, and forwards it to the next gateway, and so on.
2. If the node does not know the MAC address of the gateway it will send out an ARP request to the network with the IP address of the gateway.

15.4 TCP specification

TCP is made reliable with the following:

- **Sequence numbers.** Each TCP packet is sent with a sequence number (which identifies byte numbers). Theoretically, each data byte is assigned a sequence number. The sequence number of the first data byte in the segment is transmitted with that segment and is called the segment sequence number (SSN).
- **Acknowledgements.** Packets contain an acknowledgement number, which is the sequence number of the next expected transmitted data byte in the reverse direction. On sending, a host stores the transmitted data in a storage buffer, and starts a timer. If the packet is acknowledged then this data is deleted, else, if no acknowledgement is received before the timer runs out, the packet is retransmitted.
- **Window.** With this, a host sends a window value which specifies the number of bytes, starting with the acknowledgement number, that the host can receive.

15.4.1 Connection establishment, clearing and data transmission

The main interfaces in TCP are shown in Figure 15.9. The calls from the application program to TCP include:

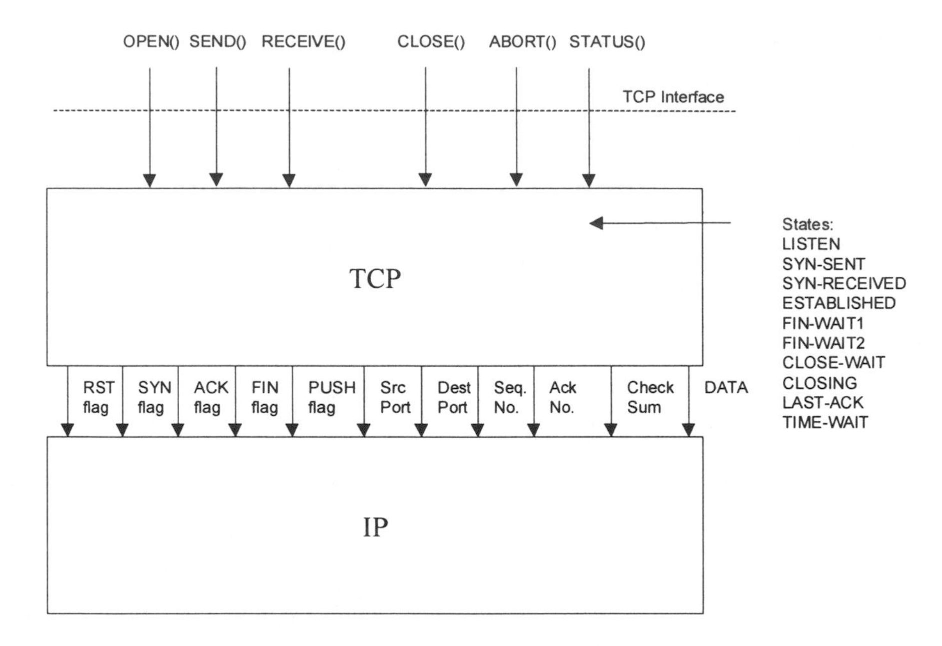

Figure 15.9 TCP interface

- OPEN and CLOSE. To open and close a connection.
- SEND and RECEIVE. To send and receive.
- STATUS. To receive status information.

The OPEN call initiates a connection with a local port and foreign socket arguments. A Transmission Control Block (TCB) stores the information on the connection. After a successful connection, TCP adds a local connection name by which the application program refers to the connection in subsequent calls.

The OPEN call supports two different types of call, as illustrated in Figure 15.10. These are:

- **Passive OPEN.** TCP waits for a connection from a foreign host, such as from an active OPEN. In this case, the foreign socket is defined by a zero. This is typically used by servers, such as TELNET and FTP servers. The connection can either be from a fully specified or an unspecified socket.
- **Active OPEN.** TCP actively connects to a foreign host, typically a server (which is opened with a passive OPEN) or with a peer-to-peer connection (with two active OPEN calls, at the same time, on each computer).

A connect is established with the transmission of TCP packets with the SYN control flag set and uses a three-way handshake (see Section 15.7). A connection is cleared by the exchange of packets with the FIN control flag set. Data flows in a stream using the SEND call to send data and RECEIVE to receive data.

Top-down testing of a network

1. Test application layer. Use TELNET or application layer protocol to test the connection.
2. PING. If Step 1 fails, test if the node is responding to TCP/IP communications.
3. TRACEROUTE. If Step 2 fails, test the route to the node, to determine where the problem occurs.

The PUSH flag is used to send data in the SEND immediately to the recipient. This is required as a sending TCP is allowed to collect data from the sending application program and sends the data in segments when convenient. Thus, the PUSH flag forces it to be sent. When the receiving TCP sees the PUSH flag, it does not wait for any more data from the sending TCP before passing the data to the receiving process.

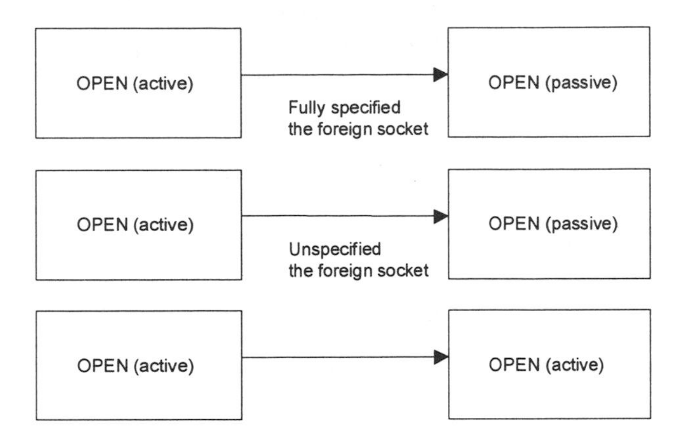

Figure 15.10 TCP connections

15.5 TCB parameters

Ethernet (n):
something used to catch the etherbunny

Table 15.2 outlines the send and receive packet parameters, as well as the current segment parameter, which are stored in the TCB. Along with this, the local and remote port number require to be stored.

Table 15.2 TCB parameters

Send Sequence Variables	*Receive Sequence Variables*	*Current Packet Variable*
SND.UNA Send unacknowledged	RCV.NXT Receive next	SEG.SEQ segment sequence number
SND.NXT Send next	RCV.WND Receive window	SEG.ACK segment acknowledgement number
SND.WND Send window	RCV.UP Receive urgent pointer	SEG.LEN segment length
SND.UP Send urgent pointer	IRS Initial receive sequence number	SEG.WND segment window
SND.WL1 Segment sequence number used for last window update		SEG.UP segment urgent pointer
SND.WL2 Segment acknowledgement number used for last window update		SEG.PRC segment precedence value
ISS Initial send sequence number		

15.6 Connection states

'This "telephone" has too many shortcomings to be seriously considered as a means of communication. The device is inherently of no value to us.'

– Western Union internal memo, 1876.

Figure 15.11 outlines the states which the connection goes into, and the events which cause them. The events from applications programs are: OPEN, SEND, RECEIVE, CLOSE, ABORT, and STATUS, and the events from the incoming TCP packets include the SYN, ACK, RST and FIN flags. The definition of each of the connection states are:

- **LISTEN.** This is the state in which TCP is waiting for a remote connection on a given port.
- **SYN-SENT.** This is the state where TCP is waiting for a matching connection request after it has sent a connection request.
- **SYN-RECEIVED.** This is the state where TCP is waiting for a confirming connection request acknowledgement after having both received and sent a connection request.
- **ESTABLISHED.** This is the state that represents an open connection. Any data received can be delivered to the application program. This is the normal state for data to be transmitted.
- **FIN-WAIT-1.** This is the state in which TCP is waiting for a connection termination request, or an acknowledgement of a connection termination, from the remote TCP.
- **FIN-WAIT-2.** This is the state in which TCP is waiting for a connection termination request from the remote TCP.
- **CLOSE-WAIT.** This is the state where TCP is waiting for a connection termination request from the local application.
- **CLOSING.** This is the state where TCP is waiting for a connection termination request acknowledgement from the remote TCP.
- **LAST-ACK.** This is the state where TCP is waiting for an acknowledgement of the connection termination request previously sent to the remote TCP.
- **TIME-WAIT.** This is the state in which TCP is waiting for enough time to pass to be sure the remote TCP received the acknowledgement of its connection termination request.
- **CLOSED.** This is the notational state, which occurs after the connection has been closed.

Question: *What's the difference between a data segment and a data packet?*

The transport layer uses data segments, whereas the network layer uses data packets. Data segments allow two or more applications to share the same transport connection. These segments are then split into data packets which have a given maximum size (typically for IP packets this is 64 KB) and each are tagged with a source and destination network address. Different applications can send data segments on a first-come, first-served basis.

The following shows a sample session (using the netstat command). The local address is defined with the node address followed by the port number (for example for the first entry, artemis is the local address and the local port is 1023). Ports that are known (such as login, shell and imap) are given names, while non-assigned ports are just defined with a port number. It can be seen that the local host has three current logins with the hosts aphrodite and leto. These connect to the remote ports of 1018, 1023 and 1019. The send and receive windows do not vary on each of the connections, and are set at 8760. An important field is Send-Q, for which a nonzero value indicates that the network for that particular host is severely congested, as it defines the number of unacknowledged segments.

```
TCP
   Local Address        Remote Address     Swind Send-Q Rwind Recv-Q  State
-------------------- -------------------- ----- ------ ----- ------ -------
artemis.1023         poseidon.shell        8760      0  8760      0 FIN_WAIT_2
artemis.1022         poseidon.1022         8760      0  8760      0 ESTABLISHED
artemis.1021         poseidon.shell        8760      0  8760      0 FIN_WAIT_2
artemis.1020         poseidon.1021         8760      0  8760      0 ESTABLISHED
artemis.43939        hades.701             8760      0  8760      0 CLOSE_WAIT
artemis.login        aphrodite.1018        8760      0  8760      0 ESTABLISHED
artemis.login        leto.1023             8760      0  8760      0 ESTABLISHED
artemis.login        aphrodite.1019        8760      0  8760      0 ESTABLISHED
artemis.50925        poseidon.imap         8760      0  8760      0 CLOSE_WAIT
```

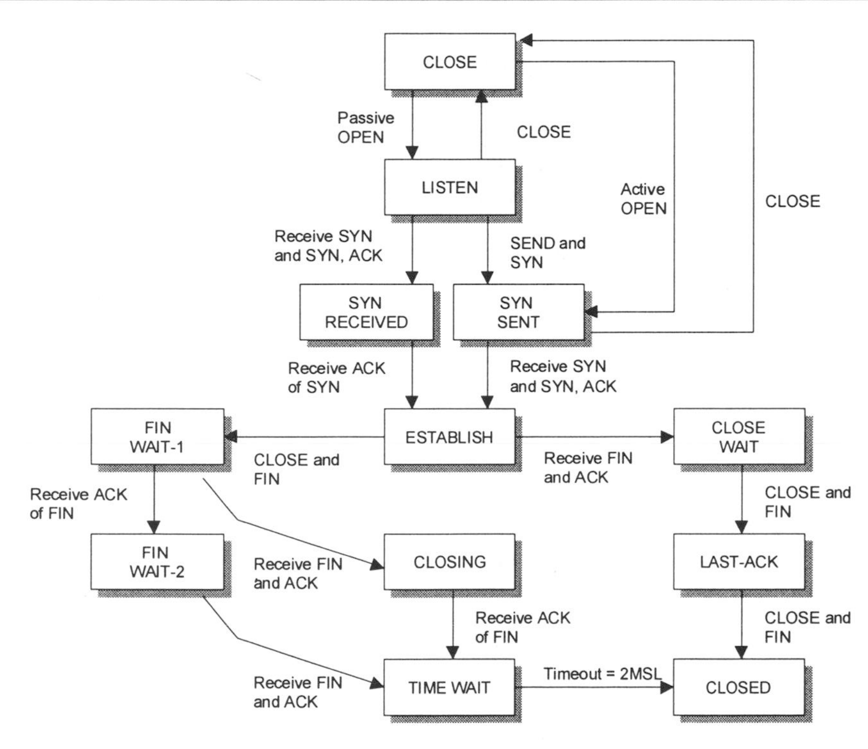

Figure 15.11 TCP connection states

15.6.1 Sequence numbers

TCP packets contain a 32-bit sequence number (0 to 4294967295), which relates to every byte sent. It uses a cumulative acknowledgement scheme, where an acknowledgement with a value of VAL, validates all bytes up to, but not including, byte VAL. Each byte at which the packet starts is numbered consecutively, after the first byte.

When sending data, TCP should receive acknowledgements for the transmitted data. The required TCB parameters will be:

SND.UNA	Oldest unacknowledged sequence number.
SND.NXT	Next sequence number to send.
SEG.ACK	Acknowledgement from the receiving TCP (next sequence number expected by the receiving TCP).
SEG.SEQ	First sequence number of a segment.
SEG.LEN	Number of bytes in the TCP packet.

SEG.SEQ+SEG.LEN–1	Last sequence number of a segment.

On receiving data, the following TCB parameters are required:

RCV.NXT	Next sequence number expected on an incoming segment, and is the left or lower edge of the receive window.
RCV.NXT+RCV.WND–1	Last sequence number expected on an incoming segment, and is the right or upper edge of the receive window.
SEG.SEQ	First sequence number occupied by the incoming segment.
SEG.SEQ+SEG.LEN–1	Last sequence number occupied by the incoming segment.

15.6.2 ISN selection

The Initial Sequence Number (ISN) is selected so that previous sockets are not confused with new sockets. Typically, this can happen when a host application crashes and then quickly re-establishes the connection before the other side can time-out the connection. To avoid this a 32-bit initial sequence number (ISN) generator is created when the connection is made, which is a number generated by a 32-bit clock, and is incremented approximately every 4 μs (giving an ISN cycle of 4.55 hours). Thus, within 4.55 hours, each ISN will be unique.

As each connection has a send and receive sequence number, these are an initial send sequence number (ISS) and an initial receive sequence number (IRS). When establishing a connection, the two TCPs synchronize their initial sequence numbers. This is done by exchanging connection establishing packets, with the SYN bit set and with the initial sequence numbers (these packets are typically called SYNs). Thus four packets must be initially exchanged:

- A sends to B. SYN with A_{SEQ}.
- B sends to A. ACK of the sequence number (A_{SEQ}).] Can be merged into a single state.
- B sends to A. SYN with B_{SEQ}.
- A sends to B. ACK of the sequence number (B_{SEQ}).

Note that the two intermediate steps can be combined into a single message, which is sometimes known as a three-way handshake. This handshake is necessary as the sequence numbers are not tied to a global clock, only to local clocks, and has many advantages, including the fact that old packets will be discarded as they occurred in a previous time.

To make sure that a sequence number is not duplicated, a host must wait for a maximum segment lifetime (MSL) before starting to retransmit packets (segments) after start-up or when recovering from a crash. An example MSL is 2 minutes. However, if it is recovering, and it has a memory of the previous sequence numbers, it may not need to wait for the MSL, as it can use sequence numbers which are much greater than the previously used sequence numbers.

15.7 Opening and closing a connection

Figure 15.12 shows a basic three-way handshake. The steps are:

1. The initial state on the initiator is CLOSED and, on the recipient, it is LISTEN (the recipient is waiting for a connection).
2. The initiator goes into the SYN-SENT state and sends a packet with the SYN bit set and then indicates that the starting sequence number will be 999 (the current sequence number, thus the next number sent will be 1000). When this is received the recipient goes into the SYN-RECEIVED state.

3. The recipient sends back a TCP packet with the SYN and ACK bits set (which identifies that it is a SYN packet and also that it is acknowledging the previous SYN packet). In this case, the recipient tells the originator that it will start transmitting at a sequence number of 100. The acknowledgement number is 1000, which is the sequence number that the recipient expects to receive next. When this is received, the originator goes into the ESTABLISHED state.
4. The originator sends back a TCP packet with the SYN and ACK bits set and the acknowledgement number is 101, which is the sequence number it expects to see next.
5. The originator transmits data with the sequence number of 1000.

	Originator				**Recipient**
1.	CLOSED				LISTEN
2.	SYN-SENT	→	<SEQ=999><CTL=SYN>		SYN-RECEIVED
3.	ESTABLISHED		<SEQ=100><ACK=1000><CTL=SYN,ACK>	←	SYN-RECEIVED
4.	ESTABLISHED	→	<SEQ=1000><ACK=101> <CTL=ACK>		ESTABLISHED
5.	ESTABLISHED	→	<SEQ=1000><ACK=101> <CTL=ACK><DATA>		ESTABLISHED

Figure 15.12 TCP connection

Note that the acknowledgement number acknowledges every sequence number up to but not including the acknowledgement number.

Figure 15.13 shows how the three-way handshake prevents old duplicate connection initiations from causing confusion. In state 3, a duplicate SYN has been received, which is from a previous connection. The recipient sends back an acknowledgement for this (4), but when this is received by the originator, the originator sends back a RST (reset) packet. This causes the recipient to go back into a LISTEN state. It will then receive the SYN packet sent in 2, and after acknowledging it, a connection is made.

TCP connections are half-open if one of the TCPs has closed or aborted, and the other end is still connected. Half-open connections can also occur if the two connections have become desynchronized because of a system crash. This connection is automatically reset if data is sent in either direction. This is because the sequence numbers will be incorrect, otherwise the connection will time-out.

A connection is normally closed with the CLOSE call. A host who has closed cannot continue to send, but can continue to RECEIVE until it is told to close by the other side. Figure 15.14 shows a typical sequence for closing a connection. Normally the application program sends a CLOSE call for the given connection. Next, a TCP packet is sent with the FIN bit set, the originator enters into the FIN-WAIT-1 state. When the other TCP has acknowledged the FIN and sent a FIN of its own, the first TCP can ACK this FIN.

	Originator				**Recipient**
1.	CLOSED				LISTEN
2.	SYN-SENT	→	<SEQ=999><CTL=SYN>		
3.	(duplicate)	→	<SEQ=900><CTL=SYN>		
4.	SYN-SENT		<SEQ=100><ACK=901> <CTL=SYN,ACK>	←	SYN-RECEIVED
5.	SYN-SENT	→	<SEQ=901><CTL=RST>		LISTEN
			(packet 2 received) →		
7.	SYN-SENT		<SEQ=100><ACK=1000><CTL=SYN,ACK>	←	SYN-RECEIVED
8.	ESTABLISHED	→	<SEQ=1000><ACK=101><CTL=ACK><DATA>		ESTABLISHED

Figure 15.13 TCP connection with duplicate connections

	Originator				Recipient
1.	ESTABLISHED (*CLOSE call*)				ESTABLISHED
2.	FIN-WAIT-1	→	<SEQ=1000><ACK=99> <CTL=SFIN,ACK>		CLOSE-WAIT
3.	FIN-WAIT-2		<SEQ=99><ACK=1001> <CTL=ACK>	←	CLOSE-WAIT
4.	TIME-WAIT		<SEQ=99><ACK=101><CTL=FIN,ACK>	←	LAST-ACK
5.	TIME-WAIT	→	<SEQ=1001><ACK=102><CTL=ACK>		CLOSED

Figure 15.14 TCP close connection

15.8 TCP user commands

The commands in this section characterize the interface between TCP and the application program. Their actual implementation depends on the operating system. Appendix C discusses practical WinSock implementations.

15.8.1 OPEN

The OPEN call initiates an active or a passive TCP connection. The basic parameters passed and returned from the call are given next. Parameters in brackets are optional.

Parameters passed:	local port, foreign socket, active/passive [, timeout] [, precedence] [, security/compartment] [, options])
Parameters returned:	local connection name

These parameters are defined as:

- Local port. The local port to be used.
- Foreign socket. The definition of the foreign socket.
- Active/passive. A passive flag causes TCP to LISTEN, else it will actively seek a connection.
- Timeout. If present, this parameter allows the caller to set up a timeout for all data submitted to TCP. If the data is not transmitted successfully within the timeout period the connection is aborted.
- Security/compartment. Specifies the security of the connection.
- Local connection name. A unique connection name is returned which identifies the socket.

15.8.2 SEND

The SEND call causes the data in the output buffer to be sent to the indicated connection. Most implementations return immediately from the SEND call, even if the data has not been sent, although some implementation will not return until either there is a timeout or the data has been sent. The basic parameters passed and returned from the call are given next. Parameters in brackets are optional.

Parameters passed:	local connection name, buffer address, byte count, PUSH flag, URGENT flag [,timeout]

These parameters are defined as:

- Local connection name. A unique connection name which identifies the socket.
- Buffer address. Address of data buffer.
- Byte count. Number of bytes in the buffer.
- PUSH flag. If this flag is set then the data will be transmitted immediately, else the TCP may wait until it has enough data.
- URGENT flag. Sets the urgent pointer.
- Timeout. Sets a new timeout for the connection.

15.8.3 RECEIVE

The RECEIVE call allocates a receiving buffer for the specified connection. Most implementations return immediately from the RECEIVE call, even if the data has not been received, although some implementation will not return until either there is a timeout or the data has been received. The basic parameters passed and returned from the call are given next. Parameters in brackets are optional.

Parameters passed:	local connection name, buffer address, byte count
Parameters returned:	byte count, URGENT flag, PUSH flag

These parameters are defined as:

- Local connection name. A unique connection name which identifies the socket.
- Buffer address. Address of the receive data buffer.
- Byte count. Number of bytes received in the buffer.
- PUSH flag. If this flag is set then the PUSH flag has been set on the received data.
- URGENT flag. If this flag is set then the URGENT flag has been set on the received data.

15.8.4 CLOSE

The CLOSE call closes the connections and releases associated resources. All pending SENDs will be transmitted, but after the CLOSE call has been implemented, no further SENDs can occur. RECEIVEs can occur until the other host has also closed the connection. The basic parameters passed and returned from the call are given next.

Parameters passed:	local connection name

15.8.5 STATUS

The STATUS call determines the current status of a connection, typically listing the TCBs. The basic parameters passed and returned from the call are given next.

Parameters passed:	local connection name
Parameters returned:	status data

The returned information should include status information on the following:

- local socket, foreign socket, local connection name;
- receive window, send window, connection state;
- number of buffers awaiting acknowledgement, number of buffers pending receipt;
- urgent state, precedence, security/compartment;
- transmission timeout.

15.8.6 ABORT

The ABORT call causes all pending SENDs and RECEIVEs to be aborted. All TCBs are also removed and a RESET message sent to the other TCP. The basic parameters passed and returned from the call are given next. Parameters in brackets are optional.

Parameters passed: local connection name

15.9 Exercises

15.9.1 Which of the following is not part of a TCP header (select one or more):
(a) Host IP address (b) Time-to-live field
(c) Host port number (d) Acknowledgement number

15.9.2 Which port does a TELNET server listen to:
(a) 21 (b) 25
(c) 25 (d) 80

15.9.3 Which of the following best describes the function of the transport layer:
(a) Segment and reassemble segments into a data stream
(b) Establishes, maintains and terminates sessions between applications
(c) Routes data packets around interconnected networks
(d) Formats data in a form which can be transmitted over the network

15.9.4 Which of the following best describes the function of the transport layer:
(a) Segment and reassemble segments into a data stream
(b) Establishes, maintains and terminates sessions between applications
(c) Routes data packets around interconnected networks
(d) Formats data in a form which can be transmitted over the network

15.9.5 Which port does an e-mail server (using SMTP) listen to:
(a) 21 (b) 25
(c) 25 (d) 80

15.9.6 Which port does a WWW server (using HTTP) listen to:
(a) 21 (b) 25
(c) 25 (d) 80

15.9.7 Which port does an FTP server listen to:
(a) 21 (b) 25
(c) 25 (d) 80

15.9.8 Which of the following best describes segmentation:
(a) Allows for many applications to share processor time
(b) Splits the data into smaller segments for faster transmission
(c) Allowing applications to share bandwidth on the network
(d) Forwards traffic from one network to another

15.9.9 What is the main difference between UDP and TCP:
(a) TCP uses sequence numbers, makes connections and uses acknowledgements
(b) They use different addressing schemes
(c) They use different port allocations
(d) UDP only supports one-way traffic, while TCP supports multiplexed traffic

15.9.10 What is the main method that TCP uses to create a reliable connection:

(a) Enhanced error correction
(b) Specially coded data
(c) Encrypted data
(d) Sequence numbers and acknowledgements

15.9.11 What TCP method controls the amount of data segments that can be transmitted before an acknowledgement must be sent:
(a) Segmentation (b) Windowing
(c) Flow control (d) Routing

15.9.12 Which TCP method stops the receiver's buffer from overflowing:
(a) Segmentation (b) Windowing
(c) Flow control (d) Routing

15.9.13 What are the initial states when making a TCP connection:
(a) Segmentation and recovery
(b) Acknowledgements and retransmissions
(c) Flow control
(d) Synchronization and acknowledgements between the hosts

15.9.14 If the TCP window size is five, how many data segments can be sent before an acknowledgement is required:
(a) 1 (b) 5
(c) 6 (d) Any number

15.9.15 Why do hosts use TCP flow control:
(a) To monitor network traffic
(b) To avoid their data buffers from overflowing
(c) To interrupt the processor when the computer is busy
(d) To create a constant flow of data

15.9.16 How is the initial sequence number of a TCP packet generated:
(a) Randomly
(b) From a 32-bit clock which is updated every 4 μs
(c) From a universal Internet-based clock
(d) From the system clock

15.9.17 How many packets are exchanged in setting up an established TCP connection:
(a) 1 (b) 2
(c) 3 (d) 4

15.9.18 Outline the operation of the three-way handshaking.

15.9.19 What advantages does TCP have over UDP. Investigate server applications which use UDP.

15.9.20 **Exam question**
(a) Outline the main features of TCP.
(b) Explain how TCP uses ports and sockets to set up a data stream.
(c) Discuss the fields within the TCP header and outline how two applications can communicate over a network using TCP.
(d) Outline why TCP is reliable, and how it copes with system crashes and lost data packets.
(e) Outline the operation of the three-way handshaking in TCP communications.
(f) Explain how the IP header is used to differentiate between TCP and UDP. Outline the format of the UDP header.

(g) The TCP protocol is described as a connection-oriented, reliable protocol, whereas the IP is described as a connectionless, unreliable protocol. Discuss, with reference to the TCP and IP headers, these descriptions.

15.10 Note from the Author

So what really makes the Internet work? Why does the WWW work so well? How can we run so many applications over the Internet at the same time? How do we know that our data has been received? How does the data actually know how to get to a certain destination? Well, it's to do with TCP, IP and routing protocols. These three parts make the whole of the Internet work, and work reliably. The IP part is responsible for getting the data packets from the source to the destination (using the IP address), the TCP part is then responsible for sending the data to the required application program (using TCP sockets and sequence numbers), and the routing protocols are responsible for passing on information about how to get to destinations (using protocols such as RIP). Isn't it wonderful how a user can run a few WWW browsers, a TELNET session, an FTP session, a video conferencing session, and it all works, seamlessly, even if there are multiple destinations. Well it's really Mr IP who's sending the data to the right place, and deciding if the data on the network is for them, and Dr TCP who either tags all the outgoing data and clearly identifies the virtual connection, or reorders and passes the received data to the required application. But, Dr TCP is no egghead who lives in the realms of academia, he makes sure that all the data is properly received, and makes sure that anything that he sends, he gets a signed receipt for. If he doesn't get a receipt, he'll resend his data. But, what if the place that he's sending the data has blown-up, or there's a postal strike. Well Dr TCP has that side covered also, he sends the data again, and then waits for a time-out period. If no data is received, he gives up.

TCP and IP unite the world and allow everyone in the world to communicate, no matter which computer they use, which operating system they are running, which language they speak, or which network they use. It fits onto virtually every type of networking technology. It'll work with Ethernet, ATM (although with some difficulty), FDDI, ISDN, Modems, RS-232, blah, blah, blah. So, whom should we thank for giving us these two great protocols. Of course DARPA should be congratulated for conceiving it, but the main award must go to the shy, but dependable workhorse of the computing industry: UNIX. Through UNIX, TCP, UDP and IP have been allowed to blossom, and show their full potential.

UDP transmission can be likened to sending electronic mail. In most electronic mail packages the user can request that a receipt is sent back to the originator when the electronic mail has been opened. This is equivalent to TCP, where data is acknowledged after a certain amount of data has been sent. If the user does not receive a receipt for its electronic mail then it will send another one, until it is receipted or until there is a reply. UDP is equivalent to a user sending an electronic mail without asking for a receipt, thus the originator has no idea if the data has been received, or not.

TCP/IP is an excellent method for networked communications, as IP provides the routing of the data, and TCP allows acknowledgements for the data. Thus, the data can always be guaranteed to be correct. Unfortunately there is an overhead in the connection of the TCP socket, where the two communicating stations must exchange parameters before the connection is made, then they must maintain and acknowledge received TCP packets. UDP has the advantage that it is connectionless. So there is no need for a connection to be made, and data is simply thrown in the network, without the requirement for acknowledgements. Thus UDP packets are much less reliable in their operation, and a sending station cannot guarantee that the data is going to be received. UDP is thus useful for remote data acquisition where data can be simply transmitted without it being requested or without a TCP/IP connection being made.

The concept of ports and sockets is important in TCP/IP. Servers wait and listen on a given port number. They only read packets which have the correct port number. For example, a WWW server listens for data on port 80, and an FTP server listens for port 21. Thus a properly set up communication network requires a knowledge of the ports which are accessed. An excellent method for virus writers and hackers to get into a network is to install a program which responds to a given port which the hacker uses to connect to. Once into the system they can do a great deal of damage. Programming languages such as Java have built-in security to reduce this problem.

So why does TCP have a PhD, and IP doesn't. Well TCP operates at a higher layer and allows the whole system to operate reliably. It does an excellent job, whereas IP is a child that has grown up too quickly for its own use. It's excellent the way that IP has created a world-wide addressing structure, but it's limited. Why is it that my IP address is 146.176.151.130, while the address of a computer in the same street, that uses the same Internet connection has the address of 138.154.33.100. Well it's because the IP address gives no indication about the location of a node. Thus, we need complex routing protocols, in which routers use to pass information about the best way to get to a node. Some of these routing protocols, like IP, have grown up too quickly, and have outgrown their usage. The worst offender is RIP, which basically defines the number of hops (the number of routers in the path to the destination) that it takes to get to a destination. Unfortunately the maximum number of hops is defined at 15, thus if a destination is more that 16, the destination is not reachable. The other problem with RIP is that it's a bit lazy, and basically doesn't try too hard. It doesn't want to know about the bandwidth of a connection, or reliability, or its cost. Hop count is hardly very taxing on computing the best way to get to a destination. Just imagine if you were a car driver, and when you looked at a map you choose the route which has the minimum number of junctions. This could take you around country roads, or through congested roads. Normally we would pick the route that allows the highest average speed (the highest bandwidth), rather than the one with the minimum number of junctions.

So the Internet we have is the Internet we have. TCP, IP and RIP are there, and they're not going to be moved for a long time. But the great thing about the Internet is that it's not going to go away either. It also allows for migration. The key to this is found in IP which can exist on its own with any other transport protocol above it, and it allows for different version, which will allow the Internet to migrate away from the current setup towards a more sensible structure (such as one based on area codes). IP addresses are also very static. If you move your computer from one network to another you must change your IP address. This normally requires skilled operators to make it work. Why can the network do it for you? In the next chapter, let's have a bit of fun, and look at how the Internet could look in the future.

In this chapter, and Appendix C, I have presented the two opposite ends of code development for TCP/IP communications. The C++ code is complex, but very powerful, and allows for a great deal of flexibility. On the other hand, the Visual Basic code is simple to implement but is difficult to implement for non-typical applications. Thus, the code used tends to reflect the type of application. In many cases Visual Basic gives an easy-to-implement package, with the required functionality. I've seen many a student wilt at the prospect of implementing a Microsoft Windows program in C++. 'Where do I start', is always the first comment, and then 'How do I do text input', and so on. Visual Basic, on the other hand, has matured into an excellent development system which hides much of the complexity of Microsoft Windows away from the developer. So, don't worry about computer language snobbery. Pick the best language to implement the specification.

15.11 TCP/IP services reference

Port	*Service*	*Comment*	*Port*	*Service*	*Comment*
1	TCPmux		7	echo	
9	discard	Null	11	systat	Users
13	daytime		15	netstat	
17	qotd	Quote	18	msp	Message send protocol
19	chargen	ttytst source	21	ftp	
23	telnet		25	smtp	Mail
37	time	Timserver	39	rlp	Resource location
42	nameserver	IEN 116	43	whois	Nicname
53	domain	DNS	57	mtp	Deprecated
67	bootps	BOOTP server	67	bootps	
68	bootpc	BOOTP client	69	tftp	
70	gopher	Internet Gopher	77	rje	Netrjs
79	finger		80	www	WWW HTTP
87	link	Ttylink	88	kerberos	Kerberos v5
95	supdup		101	hostnames	
102	iso-tsap	ISODE	105	csnet-ns	CSO name server
107	rtelnet	Remote Telnet	109	pop2	POP version 2
110	pop3	POP version 3	111	sunrpc	
113	auth	Rap ID	115	sftp	
117	uucp-path		119	nntp	USENET
123	ntp	Network Timel	137	netbios-ns	NETBIOS Name Service
138	netbios-dgm	NETBIOS	139	netbios-ssn	NETBIOS session
143	imap2		161	snmp	SNMP
162	snmp-trap	SNMP trap	163	cmip-man	ISO management over IP
164	cmip-agent		177	xdmcp	X Display Manager
178	nextstep	NeXTStep	179	bgp	BGP
191	prospero		194	irc	Internet Relay Chat
199	smux	SNMP Multiplexer	201	at-rtmp	AppleTalk routing
202	at-nbp	AppleTalk name binding	204	at-echo	AppleTalk echo
206	at-zis	AppleTalk zone information	210	z3950	NISO Z39.50 database
213	ipx	IPX	220	imap3	Interactive Mail Access
372	ulistserv	UNIX Listserv	512	exec	Comsat 513 login
513	who	Whod	514	shell	No passwords used
514	syslog		515	printer	Line printer spooler
517	talk		518	ntalk	
520	route	RIP	525	timed	Timeserver
526	tempo	Newdate	530	courier	Rpc
531	conference	Chat	532	netnews	Readnews

16 Routing Protocols

RFCs for Routing Protocols

RFC827	EGP (updated by RFC904)
RFC1771	BGP-4
RFC1267	BGP-3
RFC1163	BGP
RFC1058	RIP (updated by RFC1388)
RFC1583	OSPF Version 2 (updated by RFC2328)

16.1 Introduction

Routers filter network traffic so that the only internetwork traffic flows into and out of a network. In many cases there are several possible routes that can be taken between two nodes on different networks. Consider the network in Figure 16.1. In this case the upper network shows the connection between two nodes A and B through routers 1 to 6. It can be seen from the lower diagram that there are four routes that the data can take. To stop traffic taking a long route or even one that does not exist, each router must maintain a routing table so that it knows where the data must be sent when it receives data destined for a remote node.

For routers to find the best route they must communicate with their neighbors to find the best way through the network. This measure can be defined in a number of ways, such as the number of router hops to the remote node, the bandwidth on each link, latency, average error rates, current network traffic, and so on. Many routers use the number of router hops to determine the route, which is not always the best measure, as it may include a congested route. As with road traffic, it is often better to take the freeway (which is equivalent to a high-bandwidth route) than it is to take a route which has lower speed limits, or has a great deal of traffic congestion and/or traffic lights.

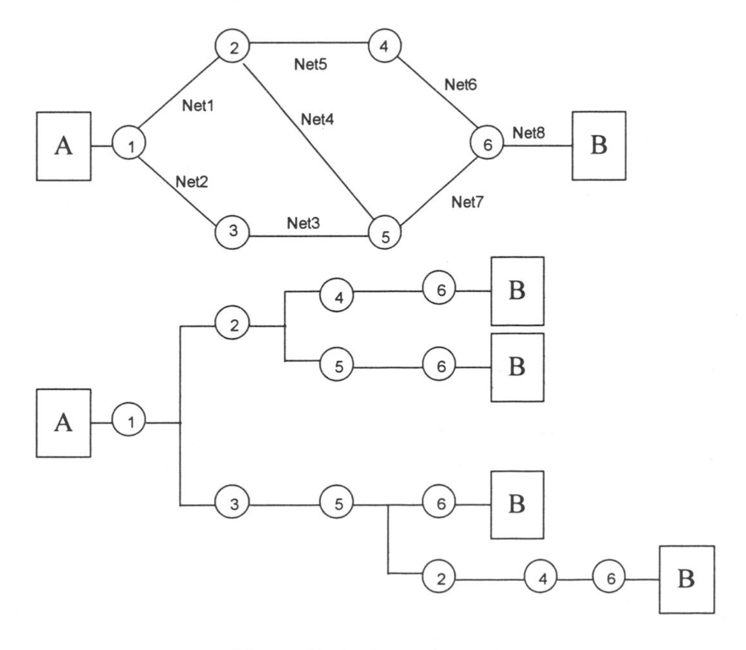

Figure 16.1 Example routing

Each router communicates with its neighbors to build-up a routing table. For example in Figure 16.1 the routing table for router 1 could be:

Destination	*Distance (hops)*	*Next router*	*Output port*
Net5	1	2	(Net1)
Net7	2	3	(Net2)
Net8	3	2	(Net1)
Net3	1	3	(Net2)
	And so on.		

It can be seen that the best route (measured by hops) from node A to Net8 is to go via Net1. This is the only information that the router needs to store. When the data gets to router 2 it has the choice of whether to send it to Net4 or Net5, as both routes get to Net8 in two hops.

A better method of determining the best route is to have some measure of the delay. For this routers pass delay information about their neighbors. For example, if the relative delay in Net5 was 1.5 and in Net6 it was 1.25, and the relative delay in Net4 was 1.1 and in Net7 was 1.3, then the relative delay between router 2 and router 6 can be calculated as:

$$\text{Route}(2,4,6) = 1.5 + 1.25 = 2.75 \qquad \text{Route}(2,5,6) = 1.1 + 1.3 = 2.4$$

Thus the best route is via router 5.

Another technique used to determine the best route is error probability. In this case the probability of an error is multiplied to give the total probability. The route with the lowest error probability will be the most reliable route. For example:

$$P_e(2-5)=0.01 \qquad P_e(5-6)=0.15$$
$$P_e(2-4)=0.05 \qquad P_e(4-6)=0.1$$

Thus,

$$P_{noerror}(2,5,6) = (1-0.01)\times(1-0.15) = 0.8415$$
$$P_{noerror}(2,4,6) = (1-0.05)\times(1-0.1) = 0.855$$

Thus, the route via router 4 is the most reliable.

DEAR NET-ED

Question: *How does a router know that it is getting routing information, and not an IP data packet?*

A routing packet is identified in the protocol field in the IP header. For example the OSPF routing protocol is defined by an 89 in the IP protocol field of the IP header (TCP is defined as 6, and UDP as 17). Example protocol numbers are:

1	Internet Control Message (ICMP)
3	Gateway-to-Gateway (GGP)
8	Exterior Gateway Protocol (EGP)
9	Any private interior gateway (IGP)
45	Inter-Domain Routing Protocol (IDRP)
86	Dissimilar Gateway Protocol (DGP)
88	Interior GRP (IGRP)
89	OSPF

16.2 Routing fundamentals

Layer 3 of the OSI model covers the network layer. There are two types of protocol at this level. These are:

- **Routing protocols.** A routing protocol provides a mechanism for routers to share routing information. These protocols allow routers to pass information between themselves, and update their routing tables. Examples of routing protocols are Routing Information Protocol (RIP), Interior

Gateway Routing Protocol (IGRP), Enhanced Interior Gateway Routing Protocol (EIGRP), and Open Shortest Path First (OSPF).

- **Routed protocols.** These protocols are any network layer protocol that allows for the addressing of a host and a destination on a network, such as IP and IPX. Routers are responsible for passing a data packet onto the next router in, if possible, an optimal way, based on the destination network address. The definition of an optimal way depends on many things, especially its reachability. With IP, routers on the path between a source and a destination, examine the network part of the IP address to achieve their routing. Only the last router, which is connected to the destination node network, examines the host part of the IP address.

A route can either use static or dynamic routing. These are:

- **Dynamic routing.** In dynamic routing, the routers monitor the network, and can change their routing tables based on the current network conditions. The network thus adapts to changing conditions. Unfortunately, this method tends to reveal everything known about an internetwork to the rest of the network. This may be inappropriate for security reasons.
- **Static routing.** In static routing, a system administrator sets up a manual route when there is only one route to get to a network (a stub network). This type of configuring reduces the overhead of dynamic routing. Static routing also allows the internetwork administrator to specify the information that is advertised about restricted parts of a network.
- **Default routing.** These are manually defined by the system administrator and define the path that is taken if there is not a known route for the destination.

In order to achieve dynamic routing, each router uses a metric for a route to a destination. The route with the lowest metric wins and the router sends the data packet onto the next router in the best path. There are many ways to define the best route, these include:

- **Bandwidth.** The data capacity of a link, which is typically defined in bps.
- **Delay.** The amount of time that is required to send a packet from the source to a destination.
- **Load.** A measure of the amount of activity on a route.
- **Reliability.** Relates to the error rate of the link.
- **Hop count.** Defined by the number of routers that it takes between the current router and the destination.
- **Ticks.** Defines the delay of a link by a number of ticks of a clock.
- **Cost.** An arbitrary value which defines the cost of a link, such as financial expense, bandwidth, and so on.

Dynamic routing

Important definitions for a dynamic routing protocol are:

- How and when updates are sent. The shorter the time between updates the faster that the network will adapt to changes in the network, but will obviously use more bandwidth than slower updated protocols.
- The knowledge that is contained in the updates.
- How to determine the locations of recipients of updates.

Routers can use a single metric (such as hop count with RIP) or multiple metrics, where the updates in routing information can be sent by:

- **Broadcast.** In broadcast, routers transmit their information to other routers at regular intervals. A typical broadcast routing protocol is RIP, in which routers send their complete routing table once every few minutes, to all of their neighbors. This technique tends to be wasteful in bandwidth, as changes in the route do not vary much over short amounts of time.
- **Event-driven**. In event-driven routing protocols, routing information is only sent when there is a change in the topology or state of the network. This technique tends to be more efficient than broadcast, as it does not use up as much bandwidth.

Routing protocols suffer from several problems. The main problem with dynamic routing protocols is the amount of time that a network will take to change its routing to take into account changes in topology, whether it is due to failure, growth or reconfiguration. The knowledge that is passed between routers must be accurate and represent the true nature of any changes. This is known as convergence, and occurs when all of the routers on an internetwork have the same knowledge. An efficient network should have fast convergence, as it reduces the time that routers use outdated information, which would be used to send data packets on an incorrect route, or is sent over an inefficient route.

16.3 Routing protocol techniques

There are three main types of routing protocols (as illustrated in Figure 16.2):

- **Distance-vector.** Distance-vector routing uses a distance-vector algorithm (such as the Bellman-Ford routing algorithm), which uses a direction (vector) and distance to any link in the internetwork to determine the best route. Each router periodically sends information to each of its neighbors on the cost that it takes to get to a distance node. Typically this cost relates to the hop count (as with RIP). The main problem with distance-vector is that updates to the network are step-by-step, and it has high bandwidth requirements as each router sends its complete routing table to all of its neighbors at regular intervals.
- **Link-state.** Link-state involves each router building up the complete topology of the entire internetwork (or at least of the partition on which the router is situated), thus each router contains the same information. With this method, routers only send information to all of the other routers when there is a change in the topology of the network. Link-state is also known as shortest path first. Typical link-state protocols are OSPF, BGP and EGP. With OSPF, each router builds a hierarchical topology of the internetwork, with itself at the top of the tree. The main problem with link-state is that routers require much more processing power to update the database, and more memory as routers require to build a database with details of all the routers on the network.
- **Hybrid.** A mixture of distance-vector and link-state. Typical hybrid routing protocols are IS-IS and Enhanced IGRP.

An outline of the main techniques used in routing protocols is outlined in Figure 16.2.

16.3.1 Distance-vector

In the distance-vector routing protocol, each router initially identifies each of its neighbors. Next it defines each of its ports which connect to a network as having a distance of zero. The discovery process continues by communicating with each of its neighbors. A typical distance-vector algorithm is RIP, which uses a hop count as a metric, but other algorithms can use different cost metrics (such as bandwidth or delay).

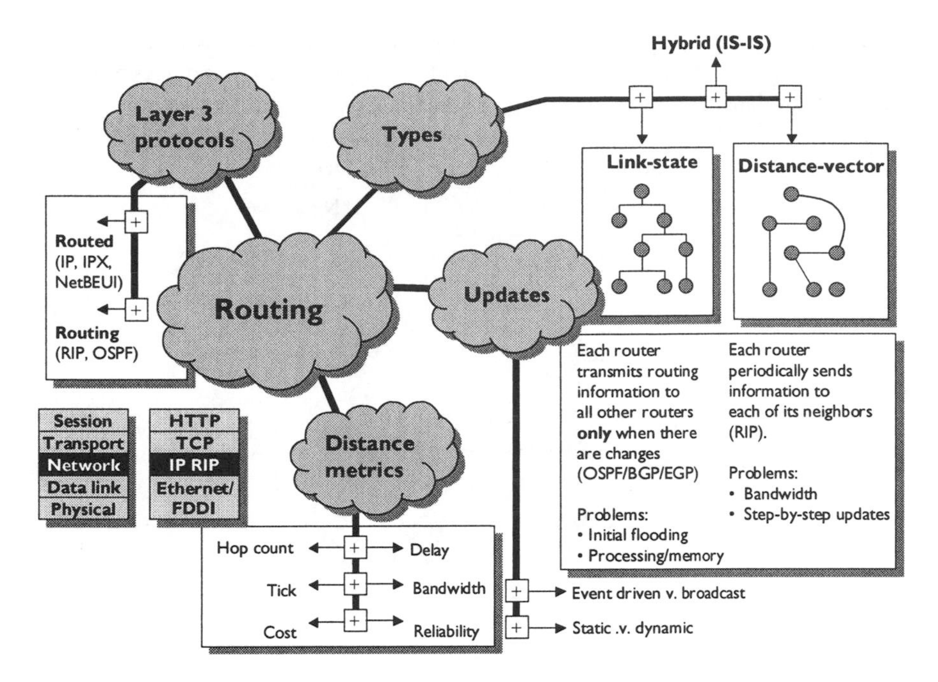

Figure 16.2 Example routing

An example network is shown in Figure 16.3. In this case Router Z communicates with Router W and Router Y and receives their routing tables. Each router will maintain a routing table which defines the number of hops that it takes to get to a destination and the port of the router that the router should use. An example sequence may be:

- All routers communicate to determine their neighbors. For example, Router X knows that its neighbors are Router Y and Router W.
- All routers set the 'count to network' that they connect to, at zero.
- Router X then tells Router Y and Router W that it takes zero hops to get to Network A. Router Y will add one hop for the destination to Network A (hop count to A is one), and Router W will add one hop for the destination to Network A (hop count to A is one).
- Router Y then tells Router X and Router Z that it takes zero hops to get to Network B. Router X will add one hop for the destination to Network B (hop count to B is one), and Router Z will add one hop for the destination to Network B (hop count to B is one).
- Router Z then tells Router W and Router Y that it takes zero hops to get to Network C. Router W will add one hop for the destination to Network C (hop count to C is one), and Router Y will add one hop for the destination to Network C (hop count to C is one).
- Router X sends its new updated routing table to W and Y, and informs them that it takes one hop to get to Network B. Router Y will not change its entry for Network B as it currently has a hop count of zero, but Router W will update its routing table for Network B to two hops.

... and so on. After the convergence time, the routers will then build up the tables shown in Figure 16.3. If a router gets information that it can reach the route from more than one path, it will take the lowest hop count for its table entry. In the example in Figure 16.3, there are two equal paths to get to Network B from Router W, each with a hop count of two. The router can thus use any one of these.

It should be noted that distance-vector is inefficient, as each router will send out its entire routing table at regular intervals. This is inefficient when there are no updates in the topology of the network. Updates also take some time, as they must be passed from router-to-router.

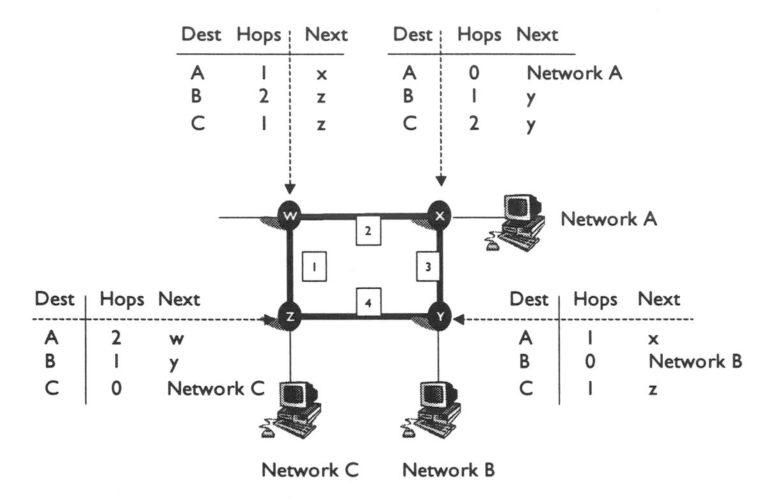

Figure 16.3 Example routing

Problems with distance-vector routing protocols

The major problems with distance-vector routing protocols (such as RIP) are:

- **Routing Loops.** These occur when slow convergence causes inconsistent routing entities when a new configuration occurs (Figure 16.4). In this case, Network A becomes unavailable. Router V will report this to Router Y, which will then report to Router Z and Router X. Both Routers X and Z will stop routing to Network A. Unfortunately Router W still thinks it can reach Network A with 3 hops, thus Router Z will receive information that says that Router W can get to Network A in 3 hops, and that it is unreachable from Router Y. Thus Router Z updates its routing table so that Network A is reachable in 4 hops, and that the next router to the destination is Router W. Router Z will then send its updated information to Router Y which informs it that there is a path to Network A from Router Z to Router W, and so on. Router Y will then inform Router X, and so on. Thus any data packet which is destined for Network A will now loop around the loop from Router Z to Router W to Router X to Router Y to Router Z, and so on.

DEAR NET-ED

Question: *Some people talk about gateways, and I've got a gateway option in my settings for my network connection. So what's the difference between a gateway and a router?*

A gateway is an old fashioned way of defining the entry and exit point of a network. Most of the time a gateway is a router. You need to define the IP address of the gateway (the router) before you can communicate with external networks.

Question: *So isn't a gateway a better definition for it?*

Well I suppose it is, but it's a router, really. In the past computers were sometimes set up to run a routing protocol and had two or more network cards. These systems acted as gateways.

- **Counting to Infinity.** As has been seen in Figure 16.4, data packets can loop around forever, because of incorrect routing information. In this loop, the distance-vector of the hop count will increment each time the packet goes through a router.

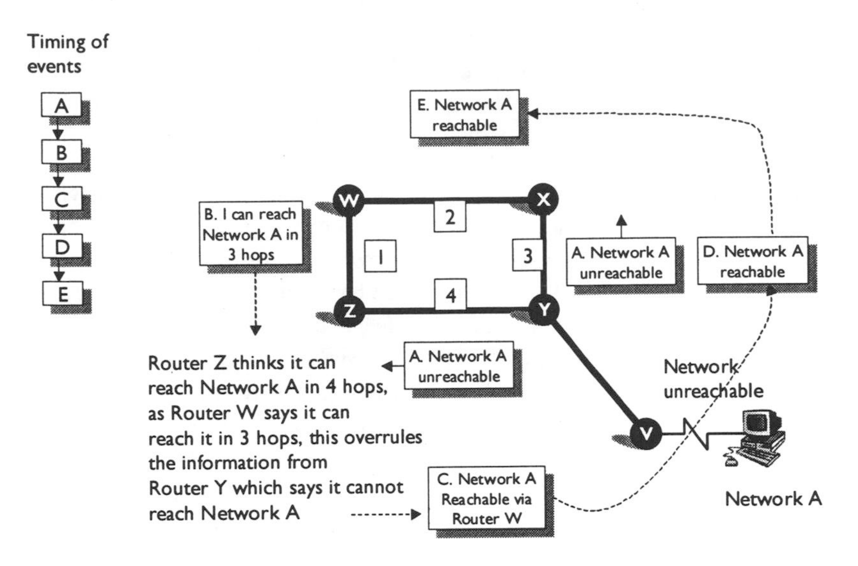

Figure 16.4 Routing loops

There are several solutions to the count-to-infinity and routing loop problems. These are:

- **Setting infinity values.** The count-to-infinity will eventually resolve itself when the routers have counted to infinity (as infinity will be constrained with the maximum definable value), but while the network is counting to this value, the routing information will be incorrect. To reduce the time that it takes to get to this maximum, a maximum value is normally defined. In RIP this value is set at 16 hops for hop-count distance-vectors, thus the maximum number of hops that can occur is **15**. This leads to a problem in that a destination which has a distance of more than 15 hops is unreachable, as a value of 16 or more defines that the network is unreachable.
- **Split horizon.** This method tries to overcome routing loops. With this routers do not update their routing table with information on a destination if they know that the network is already connected to the router (that is, the router knows more about the state of the network than any other router, as it connects to it). Thus in Figure 16.3, Router Z and Router X will not send routing information on Network B to Router Y, as they know that Network B is connected to Router Y.
- **Hold-Down Timers.** This method overcomes the count-to-infinity problem. With a hold-time time, a router starts a hold-time timer when it receives an update from a neighbor indicating that a previously accessible network is now inaccessible. It also marks the route as inaccessible. There are then three possible situations:

    - If, at any time before the hold-down timer expires, an update is sent from the same neighbor which alerted the initial problem saying that it is now accessible, the router marks the network as accessible and removes the hold-down timer.

- If an update arrives from a different neighboring router with a better metric than the original metric, the router marks the network as accessible and removes the hold-down timer.
- If, at any time before the hold-down timer expires, an update is sent from a different neighbor which alerted the initial problem saying that it is accessible, but has a poorer metric than the previously recorded metric, the update is ignored. Obviously after the timer has expired the network will still be prone to looping routes, but the timer allows for a longer time for the network to settle down and recover the correct information.

16.3.2 Link-state concepts

The link-state algorithm (known as shortest path first) maintains a complex database on the topology of a network. Each router thus has the complete picture of the whole of the network and has knowledge of distant routers, and how they interconnect. The distance-vector, on the other hand, has non-specific information on distant networks, and no knowledge of distant routers.

The link-state algorithm uses link-state advertisements (LSAs) to advertise routing information from routers. From this each router build-up a topological database with themselves at the top of the tree, and uses a shortest path first (SPF) algorithm to determine the best route to get to a destination. A typical implementation is Open Shortest Path First (OSPF), which is defined in RFC1583, and uses Dijkstra's algorithm to determine the best path. An outline of it is given in Figure 16.5. Initially, on start-up, each router must discover the routes which it connects to. Next each router advertises its connection to all the other routers on the network. From this it builds up a topology database with the information. There may be multiple routes to

DEAR NET-ED

Question: *Most of the systems I have worked with use the RIP routing protocol. If it is so popular, why should I use anything else?*

RIP is a distance-vector routing protocol which uses a metric to determine the best route to a network. A metric-based system is not really a problem, but RIP uses a very simple method to define the metric: the hop count. This in no way defines the bandwidth on any of the inter-connected networks, or the delay, or really anything, apart from the number of routers that it encounters. Another major problem is that the maximum hop count is set at 15, thus if a destination is further than 15 routers away, it cannot be reached.

I could go on all day talking about the problems of RIP, but I will not because it's what makes a lot of networks work. A major problem, though, is that, unlike link-state protocols, each router transmits the complete contents of its routing table to all of its neighbors (even if there have been no changes to its connected networks). This occurs every 30 seconds, and is thus wasteful of bandwidth.

Routing loops can also occur, but these can be overcome with hold-down timers, which do not allow any updates to the metric for a network which is known to be down, for a given time (the hold-down time).

If you really must have a distance-vector approach, choose IGRP, as it better defines the best route, as it uses things like bandwidth, delay, and so on, to define the metric. It also only has an update time period of once every 90 seconds, rather that once every 30 seconds for RIP.

Typically routers can run one of many routing protocols, and you can choose the one that fits your network.

get to a destination, thus the SPF algorithm is used to determine the best route. It is this route that is then used in the topological database. All the routers on the network will thus have the same information. The distance-vector approach uses constant updates of routing tables between neighbors, whereas the link-state approach only sends routing information when there is a change in the topology of the network. This information floods to all the routers on the network (the discovery process), thus all routers are updated with the new information.

In order for the link-state algorithm to work, a router must keep communicating with its neighbors to determine if they are still responding to communications, and if they have any changes in their link metrics. Each router only uses the most up-to-date information from LSA packets to build up the topological database.

The main problem with link-state, as apposed to distance-vector, is that the algorithm requires much more processing power, and also increased amount of memory to store up-to-date LSAs and the complete topological database. With Dijkstra's algorithm the processing task is proportional to the number of links in the internetwork multiplied by the number of routers in the internetwork.

Link-state routing is much more efficient in bandwidth, though, as the routing information is only passed when there is a change in the topology of the internetwork. Initially, though, there is a high requirement for bandwidth, as each of the routers must flood routing information through the internetwork. After the internetwork has converged, there is a reduced requirement for bandwidth, as apposed to the distance-vector method which has a constant demand on bandwidth.

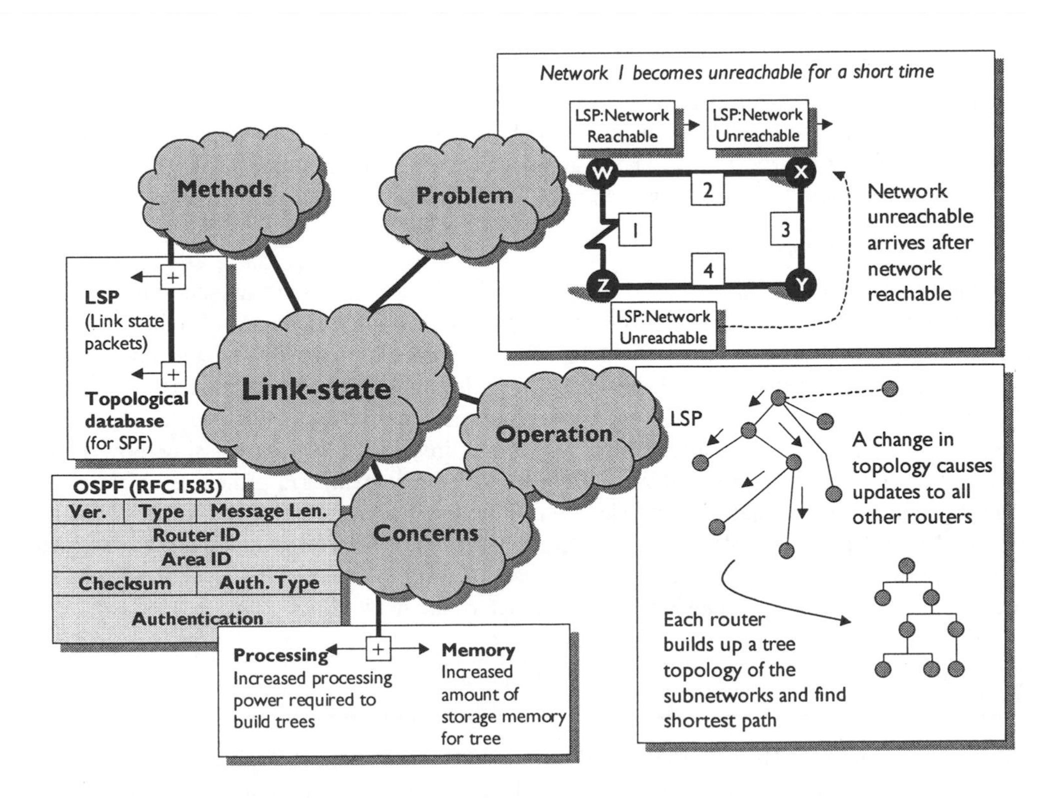

Figure 16.5 Link-state overview

The main problems with link-state are:

* **Link-state updates.** This problem is illustrated in Figure 16.5. In this case, Network link 1 becomes unavailable for a short time. Thus Router W and Router Z transmit the information that this link is unavailable to the rest of the network. In this case the information will be received by Router X and Router Y. If the network then becomes available from Router W, Router W will send out Network Reachable advertisement. If Router X receives this before it receives the Network Unreachable then Router X thinks that the network is still unavailable, even though it is available via Router W. This problem can cause whole sections of a network to become unavailable.
* **Scaling.** A problem with link-state occurs when scaling-up large internetworks when one network comes up before other parts of the network. This causes a timing problem where differing reachability information is sent between routers, thus routers might learn about different versions of the topology before they construct their SPF trees and routing tables. On a large internetwork, parts that update more quickly can cause problems for parts that update more slowly.

DEAR NET-ED

Question: *So which is better distance-vector or link-state?*

Well there are advantages and disadvantages with both types. With a distance-vector approach each router sends their complete routing table to their neighbors, at given time intervals. If the network interconnections are not varying this can be wasteful of bandwidth. Another problem is that updates to the network is done on a step-by-step basis (ripple effect), and networks may take some time to converge (that is, to have a consistent view of the complete network).

Link-state routing protocols have the advantage in that they only transmit updates to the rest of the interconnected network when they sense a change in the interconnected parameters. These changes are then broadcast to the rest of the interconnected network. This is thus more efficient in its use of bandwidth, but suffers from initial flooding when the network is first switched on. The convergence is faster than distance-vector, as each router should have the same routing table.

16.3.3 Hybrid routing

An important third classification of routing algorithm is hybrid routing, which is a combination of both distance-vector and link-state routing, and is named balanced-hybrid routing. These routing protocols use distance-vectors for more accurate metrics to determine the best paths to destination networks, and report routing information only when there is a change in the topology of the network. The event-driven nature of hybrid routing allows for rapid convergence (as with link-state protocols), but it requires much less processing power and memory than a link-state protocol would require. Examples of hybrid protocols are Intermediate System-to-Intermediate System (IS-IS) routing and Enhanced Interior Gateway Routing Protocol (EIGRP). The IS-IS routing protocol has been defined OSI link-state hierarchical routing protocol which is based on DECnet Phase V routing whereby ISs (routers) exchange routing information based on a single metric to determine network topology.

16.4 RIP

Most routers support RIP and EGP. In the past, RIP was the most popular router protocol standard. Its widespread use is due, in no small part, to the fact that it was distributed along with the Berkeley Software Distribution (BSD) of UNIX (from which most commercial versions of UNIX are derived), and was originally defined in RFC 1058. Unfortunately it suffers from several disadvantages and has been largely replaced by OSFP and EGB, which have the advantage over RIP in that they can handle large internetworks, as well as reducing routing table update traffic. By default, in RIP, each router transmits its complete routing table to their neighbor once every 30 seconds (although this time is configurable in most routers).

RIP uses a distance-vector algorithm which measures the number of network jumps (known as hops), up to a maximum of 16, to the destination router. This has the disadvantage that the smallest number of hops may not be the best route from source to destination. The OSPF and EGB protocols use a link state algorithm that can decide between multiple paths to the destination router, which are are based, not only on hops, but also on other parameters such as delay, capacity, reliability and throughput.

With distance-vector routing each router maintains tables by communicating with neighboring routers. The number of hops in its own table is then computed, as it knows the number of hops to local routers. Unfortunately, the routing table can take some time to be updated when changes occur, because it takes time for all the routers to communicate with each other (known as slow convergence).

RIP packets add to the general network traffic as each router broadcasts its entire routing table every 30–60 seconds. Figure 16.6 outlines the RIP packet format. The fields are:

- **Operation** (2 bytes) – this field gives an indication that the RIP packet is either a request or a response. The first 8 bits of the field give the command/request name and the next 8 bits give the version number.
- **Network number** (4 bytes of IP addresses) – this field defines the assigned network address number to which the routing information applies (note that, although 4 bytes are shown, there are in fact 14 bytes reserved for the address. In RIP version 1 (RIPv1), with

Question: *So, apart from a World War, what relatively uninsured event could trigger a world-wide panic for over a week and end up costing over $15 billion of damage?*

A simple worm, which poked its head above the surface on the fourth day of May 2000. It was the day that will go down in history as the day that the world said 'I Love You'. Unfortunately it was not a message which would bring world peace, nor was it a sign of affection. Never before had such a global threat grown so quickly and attacked so many systems. So why was it able to spread so quickly and cause so much damage? The reasons are three-fold:

- **Internet transparency.** The Internet is of course, as it should be, a vast complex infrastructure that requires to have many different paths from one computer to another.
- **The openness and interconnection of global e-mail system.** Electronic mail is obviously an easy target as it is one of the most useful applications of the Internet. Thus it is a portal that allows external threats into an organization, but its usefulness typically outweighs its potential threats.
- **Microsoft's scripting tools.** This was caused by Microsoft's openness in allowing users to script system events and write simple programs which integrate into their office tools (normally known as Macros) and also perform powerful operating system commands, such as deleting files, and sending information over the Internet. Unfortunately, while they are extremely useful for some users, few users actually use the full power of their facilities, and would prefer that they did not exist, as they are a potential source of damage.

IP traffic, 10 of the bytes were unused; RIPv2 uses the 14-byte address field for other purposes, such as subnet masks.

- **Number of router hops** (2 bytes) – this field indicates the number of routers that a packet must go through in order to reach the required destination. Each router adds a single hop, the minimum number is 1 and the maximum is 16. The maximum number of hops to a destination is thus limited to 15.
- **Number of ticks** (2 bytes) – this field indicates the amount of time (in 1/18 second) it will take for a packet to reach a given destination. Note that a route which has the fewest hops may not necessarily be the fastest route.

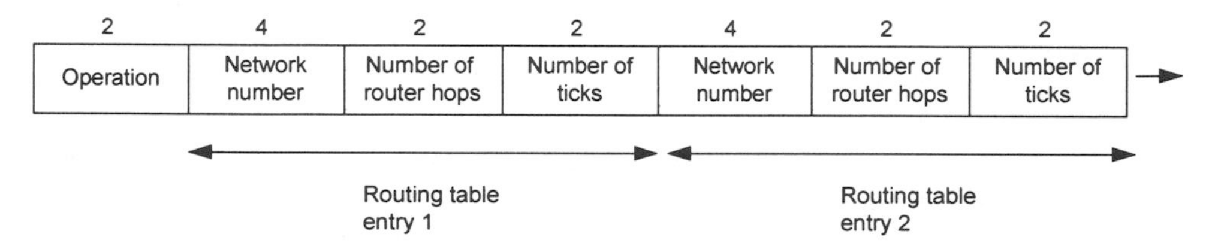

Figure 16.6 RIP packet format

16.5 OSPF

The OSPF (Open Shortest Path First) is an open, non-proprietary standard which was created by the IETF (Internet Engineering Task Force), a task force of the IAB. It is a link state routing protocol and is thus able to maintain a complete and more current view of the total internetwork, than distance-vector routing protocols. Link state routing protocols have these features:

- They use link state packets (**LSPs**) which are special datagrams that determine the names of and the cost or distance to any neighboring routers and associated networks.
- Any information learned about the network is then passed to all known routers, and not just neighboring routers, using LSPs. Thus all routers have a fuller knowledge of the entire internetwork than the view of only the immediate neighbors (as with distance-vector routing).

OSPF adds to these features with:

- **Additional hierarchy.** OSPF allows the global network to be split into areas. Thus a router in a domain does not necessarily have to know how to reach all the networks with a domain, it simply has to send to the right area.
- **Authentication of routing messages** using an 8-byte password. This length is not long enough to stop unauthorized users from causing damage. Its main purpose is to reduce the traffic from misconfigured routers. Typically a misconfigured router will inform the network that it can reach all nodes with no overhead.
- **Load balancing.** OSPF allows multiple routes to the same place to be assigned the same cost and will cause traffic to be distributed evenly over those routes.

Figure 16.7 shows the OSPF header. The fields in the header are:

- **A version number** (1 byte) which, in current implementations, has the version number of 2.
- **The type field** (1 byte) which can range from 1 to 5. Type 1 is the Hello message and the others are to request, send and acknowledge the receipt of link state messages. Nodes, to convince their neighbors that they are alive and reachable, use hello messages. If a router fails to receive these messages from one of its neighbors for a period of time, it assumes that the node is no longer directly reachable and updates its link state information accordingly.
- **Router ID** (4 bytes) identifies the sender of the message.
- **Area ID** (4 bytes) is an identifier to the area in which the node is located.
- **Authentication field** can either be set to 0 (none) or 1. If it is set to 1 then the authentication contains an 8-byte password.

The Hello packet is used to establish and maintain a connection. It is used to determine the routers that are connected to the current router. The connected routers then agree on HelloInterval and RouterDeadInterval values. The HelloIntervalue defines the number of seconds between Hello packets. The smaller the value, the faster the detection of topological changes. For example, X.25 typically uses 30 sec and LANs use 10 sec. The RouterDeadInterval defines the number of seconds before a router assumes that a route is down. It should be a multiple of HelloInterval (such as four times).

When a router thinks that it does not have the correct information on a part of a route it sends Link-state Request, which request parts of a neighbor's database. The requested neighbor then sends back a Database Description which describes the requested part of its database.

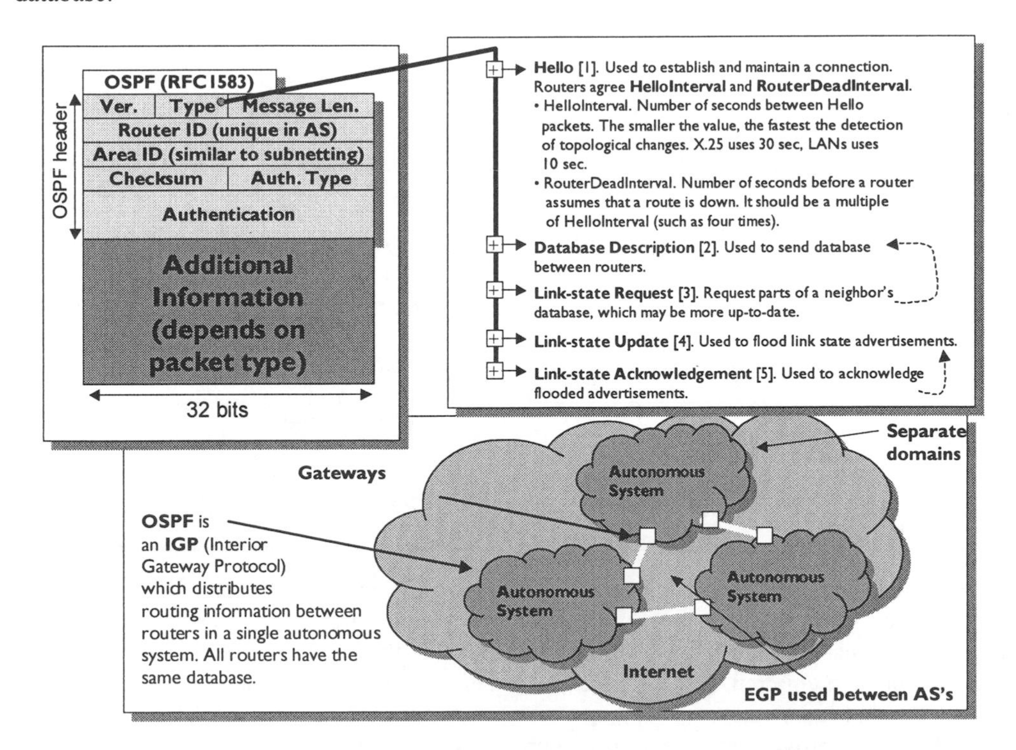

Figure 16.7 OSPF overview

When a router detects a change in its connections it sends a Link-state Update message, which is then flooded to all the routers on the internetwork. Routers return back a Link-state Acknowledgement to acknowledge the flooded advertisements.

It would of course be impossible for every router in the world to know about every other router and every link, thus each internetwork is segmented into Automomous Systems (ASs), which are bounded by a gateway. In these each router knows the complete topology of the AS. An interior routing protocol (such as OSPF) is used to transmit routing information within the AS, and an exterior routing protocol (such as EGP) is used to route between ASs, as illustrated in Figure 16.7. The Network Information Center (NIC) assigns a unique 16-bit number to enterprises for ASs.

The usage of ASs provides for a similar architecture for the Internet, where data packets are routed from one domain into another. For example, all the universities in France could define one domain. Anyone communicating with them will be routed in a defined domain through a designated gateway. The routers within the domain would then have a complete picture of all the internetworks within the domain. ASs also help to hide the architecture of the interior network from other routers outside the AS.

16.6 IGRP

IGRP (Interior Gateway Routing Protocol) is a distance-vector routing protocol (as RIP), which is used as an interior routing within an AS. Like RIP, it transmits routing information at regular intervals, but unlike RIP it uses a much better measure of the metric, these include:

- Bandwidth.
- Delay.
- Load.
- Reliability.
- Maximum transmission unit (MTU), which defines the maximum data packet size that an interface can handle.

It advertises routing information every 90 seconds. The key points of IGRP are:

- Handles complex networks, as the metric can define problems (rather than the basic hop count used in RIP).
- It allows for more efficient routing, as the network can cope with different delays and bandwidths.
- Scaleable for very large networks.

16.7 EGP/BGP

The two main interdomain routing protocols in recent history are **EGP** and **BGP**. EGP suffers from several limitations, and its principal one is that it treats the Internet as a tree-like structure, as illustrated in Figure 16.8. This assumes that the structure of the Internet is made up of parents and children, with a single backbone. A more typical topology for the Internet is illustrated in Figure 16.9.

BGP is an improvement on EGP (the fourth version of BGP is known as BGP-4). Unfortunately it is more complex than EGP, but not as complex as OSPF. BGP assumes that the

Internet is made up of an arbitrarily interconnected set of nodes. It then assumes the Internet connects to a number of **AAN**s (autonomously attached networks), as illustrated in Figure 16.10, which create boundaries around organizations, Internet service providers, and so on. It then assumes that, once they are in the AAN, the packets will be properly routed.

Most routing algorithms try to find the quickest way through the network, whereas BGP tries to find any path through the network. Thus the main goal is reachability instead of the number of hops to the destination. So finding a path which is nearly optimal is a good achievement. The AAN administrator selects at least one node to be a BGP speaker and also one or more border gateways. These gateways simply route the packet into and out of the AAN. The border gateways are the routers through which packets reach the AAN.

The speaker on the AAN broadcasts its reachability information to all the networks within its AAN. This information states only whether a destination AAN can be reached; it does not describe any other metrics. An important point is that BGP is not a distance-vector or link state protocol because it transmits complete routing information instead of partial information.

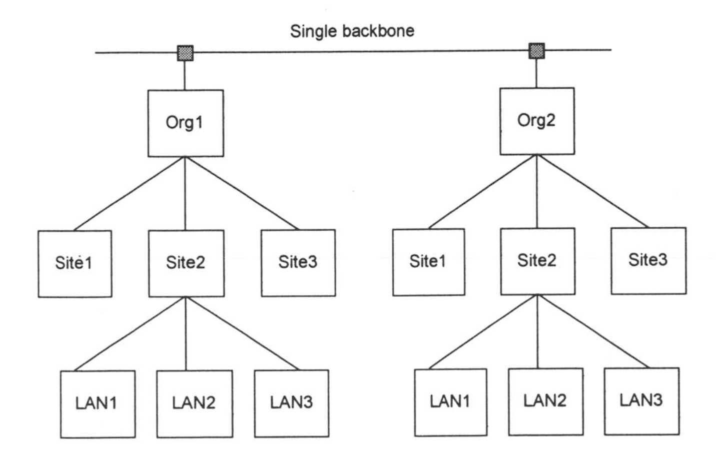

Figure 16.8 Tree-like topology

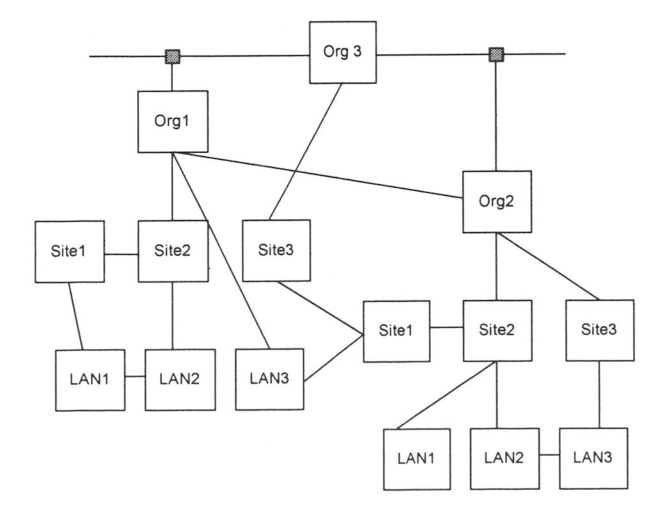

Figure 16.9 Network with multiple backbones

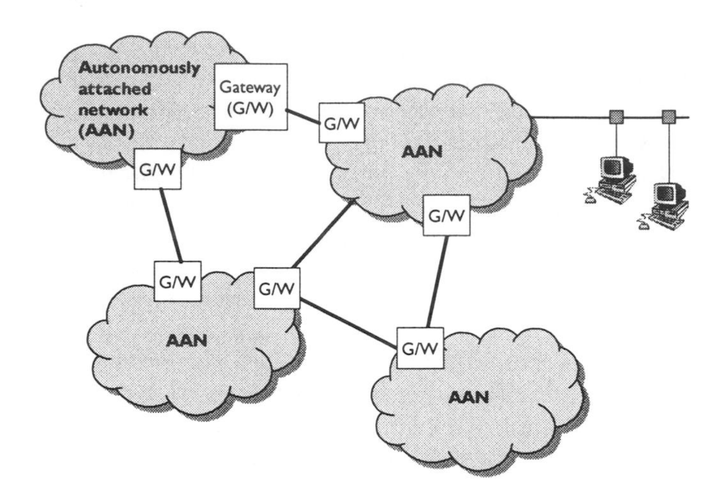

Figure 16.10 Autonomously attached networks

The BGP update packet also contains information on routes which cannot be reached (withdrawn routes), and the content of the BGP-4 update packet is:

- **Unfeasible routes length** (2 bytes). **Withdrawn routes** (variable length).
- **Total path attribute length** (2 bytes). **Path attributes** (variable length).
- **Network layer reachability information** (variable length). This can contain extra information, such as 'use AAN 1 in preference to AAN 2'.

16.8 BGP specification

Border Gateway Protocol (**BGP**) is an inter-Autonomous System routing protocol (exterior routing protocol), which builds on EGP. The main function of a BGP-based system is to communicate network reachability information with other BGP systems. Initially two systems exchange messages to open and confirm the connection parameters, and then transmit the entire BGP routing table. After this, incremental updates are sent as the routing tables change.

Each message has a fixed-size header and may or may not be followed a data portion. The fields are:

- **Marker.** Contains a value that the receiver of the message can predict. It can be used to detect a loss of synchronization between a pair of BGP peers, and to authenticate incoming BGP messages. 16 bytes.
- **Length.** Indicates the total length, in bytes, of the message, including the header. It must always be greater than 18 and no greater than 4096. 2 bytes.
- **Type.** Indicates the type of message, such as 1 – OPEN, 2 – UPDATE, 3 – NOTIFICATION and 4 – KEEPALIVE.

16.8.1 OPEN message

The OPEN message is the first message sent after a connection has been made. A KEEPALIVE message is sent back confirming the OPEN message. After this the UPDATE, KEEPALIVE, and NOTIFICATION messages can be exchanged.

Figure 16.11 shows the extra information added to the fixed-size BGP header. It has the following fields:

- **Version.** Indicates the protocol version number of the message. Typical values are 2, 3 or 4. 1 byte.
- **My Autonomous System.** Identifies the sender's Autonomous System number. 2 bytes.
- **Hold Time.** Indicates the maximum number of seconds that can elapse between the receipt of successive KEEPALIVE and/or UPDATE and/or NOTIFICATION messages. 2 bytes.
- **Authentication Code.** Indicates the authentication mechanism being used. This should define the form and meaning of the Authentication Data and the algorithm for computing values of Marker fields.
- **Authentication Data.** The form and meaning of this field is a variable-length field which depends on the Authentication Code.

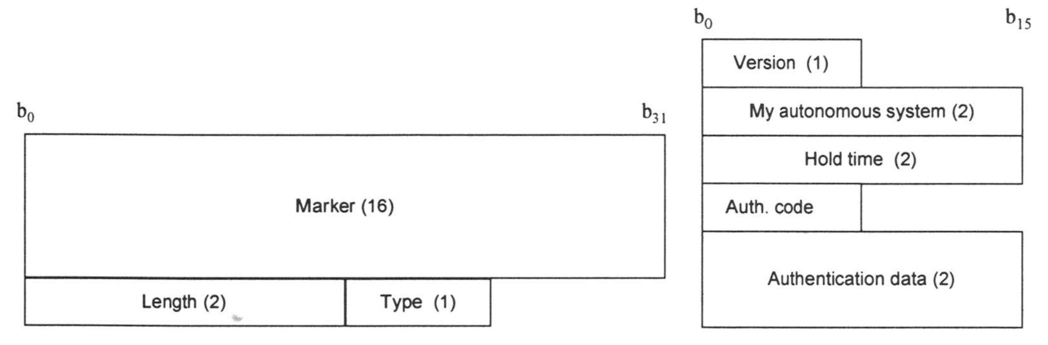

Figure 16.11 BGP message header and BGP OPEN message data

16.8.2 UPDATE message format

The UPDATE message is used to transfer routing information between BGP peers. This information is used to construct a graph describing the relationships of the various Autonomous Systems. The extra information added to the fixed-size BGP header is as follows:

- **Total Path Attribute Length.** Indicates the length of the Path Attributes field in bytes. Its value must allow the number of Network fields to be determined as specified below. 2 bytes.
- **Path Attributes.** A variable length sequence of path attributes. Each path contains the attributes for attribute type, attribute length and attribute value. These are:
    - Attribute Type. Consists of the Attribute Flags byte (b_0–b_7) followed by the Attribute Type Code byte (b_8–b_{15}). 2 bytes. The format of the Attribute flags are:

DEAR NET-ED

Question: *What's an autonomous system and how does it help with routing?*

An autonomous system (AS) simplifies the structure of the Internet, and is a logical grouping of routers within one or more organizations. InterNIC assigns unique 16-bit AS addresses to organizations. A typical protocol that uses ASs is IGRP (Interior Gateway Routing Protocol).

Question: *So what's the difference between interior and exterior routing protocols?*

Exterior routing protocols route between AS's, whereas interior routing protocols route with a single AS. Typical interior routing protocols are IGRP (distance-vector), Enhanced IGRP (balanced hybrid), RIP (distance-vector) and OSPF (link state).

b_0 Optional bit. Defines whether the attribute is optional (if set to 1) or well-known (if set to 0).

b_1 Transitive bit. Defines whether an optional attribute is transitive (if set to 1) or non-transitive (if set to 0).

b_2 Partial bit. Defines whether the information contained in the optional transitive attribute is partial (if set to 1) or complete (if set to 0). For well-known attributes and for optional non-transitive attributes the Partial bit must be set to 0.

b_3 Extended Length bit. Defines whether the Attribute Length is one byte (if set to 0) or two bytes (if set to 1).

b_4–b_7 Unused and set to zero.

- Attribute Type Code. Contains the Attribute Type Code.

- **Network.** Each Internet network number indicates one network whose Inter-Autonomous System routing is described by the Path Attributes. 4-byte.

The total number of Network fields in the UPDATE message can be determined by the formula:

$$\text{Message\_Length} = 19 + \text{Total\_Path\_Attribute\_Length} + 4 \times \text{No\_of\_networks}$$

16.8.3 KEEPALIVE and NOTIFICATION message format

The KEEPALIVE message consists of only the message header (and is thus only 19 bytes long) and is used to determine if peers are reachable. Unlike other routing protocols, such as RIP, BGP does not continually poll its peers to determine if they are still reachable, instead peers exchange KEEPALIVE messages (which must be less than the hold time of the OPEN message). A typical maximum time between KEEPALIVE messages is one-third of the hold time period.

The NOTIFICATION message is sent when an error condition occurs and the BGP connection is immediately closed after sending it. The extra information added to the fixed-size BGP header is as follows:

- **Error code.** Indicates the type of NOTIFICATION. 1 byte. Defined error codes are:
  1 Message Header Error 2 OPEN Message Error
  3 UPDATE Message Error 4 Hold Timer Expired
  5 Finite State Machine Error 6 Cease

- **Error subcode.** Provides more specific information about the error. 1 byte. Each Error Code can have one or more Error Subcodes associated with it. Error subcodes are:

  - Message Header Error subcodes:
    1 – Connection Not Synchronized. 2 – Bad Message Length.
    3 – Bad Message Type.

  - OPEN Message Error subcodes:
    1 – Unsupported Version Number. 2 – Bad Peer AS.
    3 – Unsupported Authentication Code. 4 – Authentication Failure.

- UPDATE Message Error subcodes:
  1 – Malformed Attribute List.
  2 – Unrecognized Well-known Attribute.
  3 – Missing Well-known Attribute.
  4 – Attribute Flags Error.
  5 – Attribute Length Error.
  6 – Invalid ORIGIN Attribute
  7 – AS Routing Loop.
  8 – Invalid NEXT_HOP Attribute.
  9 – Optional Attribute Error.
  10 – Invalid Network Field.

- Data. Diagnostic data which identifies the reason for the NOTIFICATION message. Its contents depends on the Error Code and Error Subcode. Variable-length.

The message length can be determined by:

Message_Length = 21 + Data_Length

16.8.4 Path attributes

Path attributes in the UPDATE message fall into four separate categories:

- **Well-known mandatory.**
- **Well-known discretionary.**
- **Optional transitive.**
- **Optional non-transitive.**

Attributes which are well known must be recognized by all BGP implementations. If they are mandatory they must be included in every UPDATE message. Discretionary attributes may or may not be sent in an UPDATE message. Table 16.1 defines the well-known attributes.

Table 16.1 Well-known attributes

Attribute Name	*Type code*	*Length*	*Attribute category*	*Description*
ORIGIN	1	1	Well-known, mandatory	Defines the origin of the path information. Values are: 0 IGP – network(s) are interior to the originating AS. 1 EGP – network(s) learned via EGP. 2 INCOMPLETE – network(s) learned by some other means.
AS_PATH	2	variable	Well-known, mandatory	AS_PATH attribute enumerates the ASs that must be traversed to reach the networks listed in the UPDATE message.
NEXT_HOP	3	4	Well-known, mandatory	Defines the IP address of the border router that should be used as the next hop to the networks listed in the UPDATE message.
UNREACHABLE	4	0	Well-known, discretionary	Used to notify a BGP peer that some of the previously advertised routes have become unreachable.
INTER-AS METRIC	5	2	Optional, non-transitive	May be used on external (inter-AS) links to discriminate between multiple exit or entry points to the same neighboring AS.

16.8.5 BGP state transitions and actions

BGP states are: 1 – Idle; 2 – Connect; 3 – Active; 4 – OpenSent; 5 – OpenConfirm; 6 – Established. The events are: 1 – BGP Start; 2 – BGP Stop; 3 – BGP Transport connection open; 4 – BGP Transport connection closed; 5 – BGP Transport connection open failed; 6 – BGP Transport fatal error; 7 – ConnectRetry timer expired; 8 – Holdtime timer expired; 9 – KeepAlive timer expired; 10 – Receive OPEN message; 11 – Receive KEEPALIVE message; 12 – Receive UPDATE messages; 13 – Receive NOTIFICATION message. The following defines the state transitions of the BGP FSM and the actions that occur.

EVENT	ACTIONS	MESSAGE_SENT	NEXT_STATE
Idle (1)			
1	Initialize resources	none	2
	Start ConnectRetry timer		
	Initiate a transport connection		
Connect (2)			
1	none	none	2
3	Complete initialization	OPEN	4
	Clear ConnectRetry timer		
5	Restart ConnectRetry timer	none	3
7	Restart ConnectRetry timer	none	2
	Initiate a transport connection		
Active (3)			
1	none	none	3
3	Complete initialization	OPEN	4
	Clear ConnectRetry timer		
5	Close connection		3
	Restart ConnectRetry timer		
7	Restart ConnectRetry timer	none	2
	Initiate a transport connection		
OpenSent (4)			
1	none	none	4
4	Close transport connection	none	3
	Restart ConnectRetry timer		
6	Release resources	none	1
10	Process OPEN is OK	KEEPALIVE	5
	Process OPEN failed	NOTIFICATION	1
OpenConfirm (5)			
1	none	none	5
4	Release resources	none	1
6	Release resources	none	1
9	Restart KeepAlive timer	KEEPALIVE	5
11	Complete initialization	none	6
	Restart Holdtime timer		
13	Close transport connection		1
	Release resources		
Established (6)			
1	none	none	6
4	Release resources	none	1
6	Release resources	none	1
9	Restart KeepAlive timer	KEEPALIVE	6
11	Restart Holdtime timer	KEEPALIVE	6
12	Process UPDATE is OK	UPDATE	6
	Process UPDATE failed	NOTIFICATION	1

16.9 Exercises

16.9.1 RIP is an acronym for:
(a) Routing Internet Protocol (b) Routing Information Protocol
(c) Routable Internet Protocol (d) Routable Information Protocol

16.9.2 OSPF is an acronym for:
(a) Open System Path Format (b) Open Shortest Path First
(c) Open System Protocol Formation
(d) Open Shortest Path Format

16.9.3 What is the maximum number of hops for the RIP to a destination:
(a) 5 (b) 15
(c) 128 (d) No maximum

16.9.4 Which of the following is not a routed protocol:
(a) IP (b) IPX
(c) AppleTalk (d) RIP

16.9.5 Which of the following is not a routing protocol:
(a) IP (b) OSPF
(c) EGP (d) EGP

16.9.6 How is a routing packet defined:
(a) Value in IP protocol field in the IP header
(b) Special header at the start of the packet
(c) Sent on a different channel than data packets
(d) It has no IP header

16.9.7 Which of the following is not a metric that can be used in a routing protocol:
(a) Memory requirements (b) Reliability
(c) Bandwidth (d) Delay

16.9.8 How does RIP send routing information updates:
(a) Each router sends routing packets to determine the time it takes to get to a distant node
(b) Each router continuously sends its routing table to all the routers on the network
(c) Each router only sends their routing information when there is a change in the topology of the internetwork
(d) Each router sends its routing tables to each of the neighbors at periodic intervals

16.9.9 Which metric does RIP use to determine the best route:
(a) Delay (b) Bandwidth
(c) Reliability (d) Hop count

16.9.10 How does OSPF send routing information updates:
(a) Each router sends routing packets to determine the time it takes to get to a distant node
(b) Each router continuously sends its routing table to all the routers on the network
(c) Each router only sends its routing information when there is a change in the topology of the internetwork
(d) Each router sends its routing tables to each of the neighbors at periodic intervals

16.9.11 What is a routing loop:

(a) Where no nodes are connected to a route
(b) Where data packets are returned to the sender when they are not delivered
(c) Where data packets continuously loop around a series of routers
(d) Where there is no end connection from a router

16.9.12 What causes count-to-infinity:
(a) Where no nodes are connected to a route
(b) Where data packets are returned to the sender when they are not delivered
(c) Where data packets continuously loop around a series of routers
(d) Where there is no end connection from a router

16.9.13 What type of routing protocol is used outside an AS:
(a) Interior Routed Protocol (b) Interior Gateway Protocol
(c) Exterior Gateway Protocol (d) Interior Routed Protocol

16.9.14 What type of routing protocol is used inside an AS:
(a) Interior Routed Protocol (b) Interior Gateway Protocol
(c) Exterior Gateway Protocol (d) Interior Routed Protocol

16.9.15 In OSPF, how does a router keep in contact with its neighbors:
(a) Hello packets (b) Special bit sequence
(c) I'm Alive packet (d) Uses an interrupt line

16.9.16 In RIP, by default, what is the time between routers sending their routing tables to their neighbors:
(a) 1 second (b) 30 sec
(c) 1 min (d) 1 hr

16.9.17 In OSPF, when are updates sent to other routers on the internetwork:
(a) Periodically
(b) Random periods
(c) When there's a change in the topology of the internetwork
(d) Only when they receive information from other routers

16.9.18 In OSPF, which of the following is most true:
(a) All routers on an internetwork have the same routing information
(b) All routers on an internetwork have different routing information
(c) All routers on an internetwork contain only part of the routing information for the internetwork
(d) Routers only see the network from their neighbors point-of-view

16.9.19 In OSPF, how are topology updates transmitted to the rest of the internetwork:
(a) Passed from one router to the next
(b) Other routers request the updated routing information
(c) Other routers map their memory to all other routers
(d) Routing information is flooding into the network

16.9.20 Contrast link-state and distance-vector routing protocols, giving an example of each.

16.9.21 Discuss the main drawbacks with RIP, and how they are overcome.

16.10 Note from the Author

Okay. That was routing protocols. This chapter, as much as any other, should give you a more in-depth knowledge of how the Internet works. So, as a bit of fun, let's do a little bit of crystal ball looking, and try and predict where the Internet will evolve in 10 or 20 years in the future.

Most people are well aware of the disappearing computer, well, in fact, it has almost disappeared. The only reason that we still see computers, is that they still require large screens, and keyboards, and floppy disks, and so on. The actual physical size of the electronics is small. For example a high-performance microprocessor has a silicon size of less than one inch square. The next step from the disappearing computer is the disappearing Internet, where the interface to the Internet is invisible to the user.

So what are the things that are holding back this development:

- *The Internet and its addressing structure were never really designed to be a global infrastructure and is constraining the access to resources and information.*
- *Information and databases tend to be static, and fixed to location.*
- *Difficult to group individual objects into larger objects.*
- *Difficult to add resources to the Internet (requires an ISP and a valid IP address).*
- *Search engines are not very good at gathering relevant information. On the WWW, typically users get pages of irrelevant information, which just happens to have the keyword which they are searching for.*
- *Resources are gathered around local servers.*
- *Resources are tied to locations with an IP address.*
- *IP addresses are not logically organized.*
- *Infrastructure of the Internet requires complex routing.*

So where are we now, and where will we be in the future:

Now	Future
Computers use IP addresses, which require complex routing	Virtually every object in the world can be addressed
Location of resources is tied to networks	Resources and objects can move
Movement of resources requires specialist configuration	Automatic tracking of resources
Difficult to add objects onto the Internet (normally requires an IP address and a network adaptor)	Easy to add objects onto the Internet (done automatically)
Difficult to address resources, which tend to be fixed in their location	Objects become real, and do not have a fixed location

In the future many objects will have their own address, and can thus be addressed over the Internet. This will require that each object has a unique ID, that will map to a network address (which is built up from a geographical addressing structure). This is similar to the way that each network adaptor has a MAC address, which is mapped to a network address (its IP address). The MAC address does not contain any information on the location of the object, but the network address does.

TCP/IP Commands

17.1 Introduction

There are several standard programs and protocols available over TCP/IP connections, these include:

- FTP (File Transfer Protocol) – transfers file between computers.
- PING – determines if a node is responding to TCP/IP communications. The PING command is probably the easiest way to determine if a node is alive, and responding to communications.
- TRACEROUTE (or tracert) – determines the route to a remote host.
- NSLOOKUP – determines the IP address of a remote host for a given host name.
- TELNET – allows remote login using TCP/IP.
- HTTP (Hypertext Transfer Protocol) – which is the protocol used in the World Wide Web (WWW) and can be used for client-server applications involving hypertext.
- MIME (Multipurpose Internet Mail Extension) – gives enhanced electronic mail facilities over TCP/IP.
- SMTP (Simple Mail Management Protocol) – gives simple electronic mail facilities.

An interface to a host has two components: the physical (the hardware) and the logical (the software). The hardware is the actual connection between devices, and the software controls the messages that are passed between adjacent devices. When a fault develops on a network, the administrator will use a number of methods to test the link, normally using the PING, TRACEROUTE and TELNET programs. The first steps might be as given in Figure 17.1. These are:

- **PING.** This is the simplest method in determining if a node is reachable by TCP/IP communications. It also gives a measure of the delay in reaching the remote host. If the host is reachable by PING it is likely that it has not crashed and that the network connection to it is reliable. With ping, the host sends out a packet to the destination host and then waits for a reply packet from that host. The results give path-to-host reliability, delays over the path, and whether the host can be reached or is functioning. If possible the user should test the ping command with both the domain name and the IP address of the host, as this will test to see if the domain name server is operating correctly (or reachable). If the ping command works with the IP address, and not with the domain name, then either the domain name has been specified wrongly or the domain name server is not operating correctly.
- **TELNET.** If the remote host is a server, and the user needs to test if the server is responding to server applications, a TELNET session is typically initiated (or often an FTP session is also used). Another method of testing a server is to determine if it is responding to the HTTP protocol (if it has been setup to run a WWW server). The Telnet connection has the advantage of testing the application layer of the OSI model.
- **TRACEROUTE.** The traceroute command can be used to determine the route that data takes to reach a remote host. This will show if the data is being dropped at any point in

the connection. It is similar to the ping command, except that instead of testing end-to-end connectivity, traceroute tests each step along the way. The traceroute command takes advantage of the error messages generated by routers when a packet exceeds its Time-To-Live (TTL) value. It sends several packets and displays the round-trip time for each. The main advantage of the traceroute command is that it shows which router in the path was the last one to be reached (fault isolation). An administrator can then determine the parts of the network that are operating properly.

If the administrator determines that there is a local software problem with a host, it is likely that the computer would be given a hardware reboot. If there is a local problem with a router, bridge or hub, the administrator would check the equipment by asking the following questions:

- Is the carrier detect signal active? Most transmission lines transmit a signal even when there are no data frames being transmitted at a time. This is called carrier detect, and many devices will either give a software indication on whether a link is active, or activate an LED to show an active link.
- Are the keepalive messages being received? Many devices send keepalive signals between each other to show that they are still responding to communications (normally a good network connection will be identified with an active green LED).
- Can data packets be sent across the physical link?
- Is there a good physical link between devices? This involves inspecting the cable that connects to a device. A typical problem is when a user disconnects a cable to connect another device.

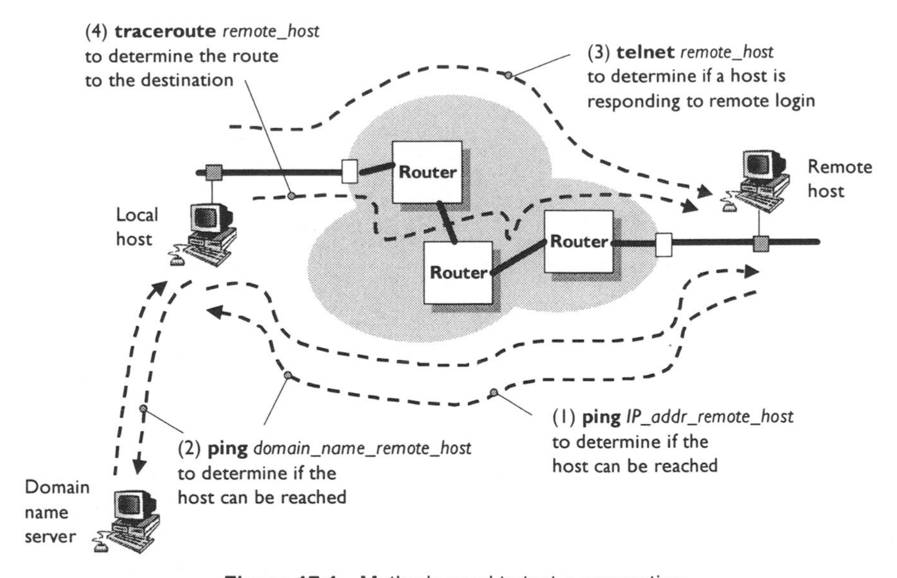

Figure 17.1 Methods used to test a connection

17.2 ping

The `ping` program (Packet Internet Gopher) determines whether a node is responding to TCP/IP communication. It is typically used to trace problems in networks and uses the Internet Control Message Protocol (ICMP) to send a response request from the target node.

Sample run 17.1 shows that `miranda` is active and `ariel` isn't.

Sample run 17.1: Using PING command

```
C:\WINDOWS>ping miranda
miranda (146.176.144.14) is alive
C:\WINDOWS>ping ariel
no reply from ariel (146.176.147.35)
```

The `ping` program can also be used to determine the delay between one host and another, and also if there are any IP packet losses. In Sample run 17.2 the local host is `pc419.eece.napier.ac.uk` (which is on the `146.176.151` segment); the host `miranda` is tested (which is on the `146.176.144` segment). It can be seen that, on average, the delay is only 1 ms and there is no loss of packets.

In Sample run 17.3 the destination node (`www.napier.ac.uk`) is located within the same building but is on a different IP segment (`147.176.2`). It is also routed through a router. It can be seen that the packet delay has increased to between 9 and 10 ms. Again, there is no packet loss.

Sample run 17.2: Using PING command

```
225 % ping miranda
PING miranda.eece.napier.ac.uk: 64 byte packets
64 bytes from 146.176.144.14: icmp_seq=0. time=1. ms
64 bytes from 146.176.144.14: icmp_seq=1. time=1. ms
64 bytes from 146.176.144.14: icmp_seq=2. time=1. ms
3 packets transmitted, 3 packets received, 0% packet loss
round-trip (ms)  min/avg/max = 1/1/1
```

Sample run 17.3: Using PING command

```
226 % ping www.napier.ac.uk
PING central.napier.ac.uk: 64 byte packets
64 bytes from 146.176.2.3: icmp_seq=0. time=9. ms
64 bytes from 146.176.2.3: icmp_seq=1. time=9. ms
64 bytes from 146.176.2.3: icmp_seq=2. time=10. ms
3 packets transmitted, 3 packets received, 0% packet loss
round-trip (ms)  min/avg/max = 9/9/10
```

Sample run 17.4 shows a connection between Edinburgh and Bath in the UK (`www.bath.ac.uk` has an IP address of `138.38.32.5`). This is a distance of approximately 500 miles and it can be seen that the delay is now between 30 and 49 ms. This time there is 25% packet loss.

Finally, in Sample run 17.5 the `ping` program tests a link between Edinburgh, UK, and a WWW server in the USA (`home.microsoft.com`, which has the IP address of `207.68.137.51`). It can be seen that in this case, the delay is between 447 and 468 ms, and the loss is 60%.

Sample run 17.4: Using PING command

```
222 % ping www.bath.ac.uk
PING jess.bath.ac.uk: 64 byte packets
64 bytes from 138.38.32.5: icmp_seq=0. time=49. ms
64 bytes from 138.38.32.5: icmp_seq=2. time=35. ms
64 bytes from 138.38.32.5: icmp_seq=3. time=30. ms
4 packets transmitted, 3 packets received, 25% packet loss
round-trip (ms)  min/avg/max = 30/38/49
```

A similar utility program to `ping` is `spray` which uses Remote Procedure Call (RPC) to send a continuous stream of ICMP messages. It is useful when testing a network connection for its burst characteristics. This differs from `ping`, which waits for a predetermined amount of time between messages.

Sample run 17.5: Ping command with packet loss

```
224 % ping home.microsoft.com
PING home.microsoft.com: 64 byte packets
64 bytes from 207.68.137.51: icmp_seq=2. time=447. ms
64 bytes from 207.68.137.51: icmp_seq=3. time=468. ms
5 packets transmitted, 2 packets received, 60% packet loss
```

17.3 ftp (file transfer protocol)

The `ftp` program uses the TCP/IP protocol to transfer files to and from remote nodes. If necessary, it reads the *hosts* file to determine the IP address. Once the user has logged into the remote node, the commands that can be used are similar to DOS commands such as `cd` (change directory), `dir` (list directory), `open` (open node), `close` (close node), `pwd` (present working directory). The `get` command copies a file from the remote node and the `put` command copies it to the remote node.

The type of file to be transferred must also be specified. This file can be ASCII text (the command `ascii`) or binary (the command `binary`). The FTP protocol is defined in RFC959, and a full set of commands and responses are detailed in Appendix B.

Another typical file transfer protocol is Trivial File Transfer Protocol, which is a simplified version of FTP that allows files to be transferred over a network. It uses UDP communications rather than TCP.

17.4 traceroute

The `traceroute` program traces the route of an IP packet through the Internet. It uses the IP protocol time-to-live field and attempts to get an ICMP TIME_EXCEEDED response from each gateway along the path to a defined host. The default probe datagram length is 38 bytes (although the sample runs use 40 byte packets by default). Sample run 17.6 shows an example of `traceroute` from a PC (`pc419.eece.napier.ac.uk`). It can be seen that initially it goes through a bridge (`pcbridge.eece.napier.ac.uk`) and then to the destination (`miranda.eece.napier.ac.uk`).

Sample run 17.7 shows the route from a PC (`pc419.eece.napier.ac.uk`) to a destination node (`www.bath.ac.uk`). Initially, from the originator, the route goes through a gateway (`146.176.151.254`) and then goes through a routing switch (`146.176.1.27`) and onto EaStMAN ring via `146.176.3.1`. The route then goes round the EaStMAN to a gateway at the University of Edinburgh (`smds-gw.ed.ja.net`). It is then routed onto the SuperJanet network and reaches a gateway at the University of Bath (`smds-gw.bath.ja.net`). It then goes to another gateway (`jips-gw.bath.ac.uk`) and finally to its destination (`jess.bath.ac.uk`). Figure 17.2 shows the route the packet takes (refer to Appendix C.2).

Note that gateways 4 and 8 hops away either don't send ICMP 'time exceeded' messages or send them with time-to-live values that are too small to be returned to the originator.

Sample run 17.6: Example traceroute

```
www:~/www$ traceroute miranda
traceroute to miranda.eece.napier.ac.uk (146.176.144.14), 30 hops max,
     40 byte packets
1  pcbridge.eece.napier.ac.uk (146.176.151.252)  2.684 ms  1.762 ms 1.725 ms
2  miranda.eece.napier.ac.uk (146.176.144.14)  2.451 ms  2.554 ms   2.357 ms
```

Sample run 17.8 shows an example route from a local host at Napier University, UK, to the USA. As before, it goes through the local gateway (`146.176.151.254`) and then goes through three other gateways to get onto the SMDS SuperJANET connection. The data packet then travels down this connection to University College London (`gw5.ulcc.ja.net`). It then goes onto high speed connections to the USA and arrives at a US gateway (`mcinet-2.sprintnap.net`). Next, it travels to `core2-hssi2-0.WestOrange.mci.net` before reaching the Microsoft Corporation gateway in Seattle (`microsoft.Seattle.mci.net`). It finally finds its way to the destination (`207.68.145.53`). The total journey time is just less than half a second.

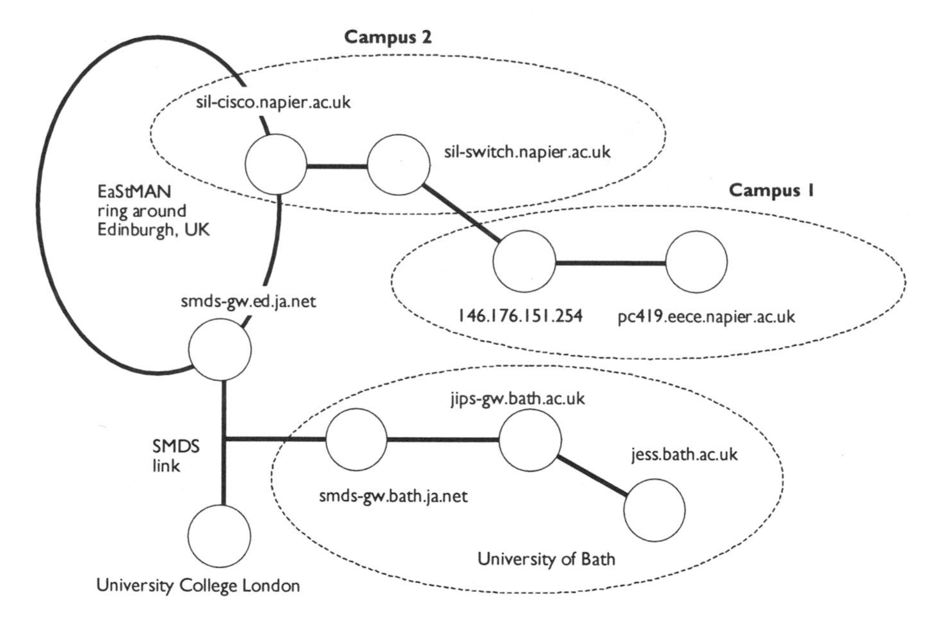

Figure 17.2 Route between local host and the University of Bath

Sample run 17.7: Example traceroute

```
www:~/www$ traceroute www.bath.ac.uk
traceroute to jess.bath.ac.uk (138.38.32.5), 30 hops max, 40 byte packets
1  146.176.151.254 (146.176.151.254)  2.806 ms  2.76 ms  2.491 ms
2  sil-switch.napier.ac.uk (146.176.1.27)  19.315 ms  11.29 ms  6.285 ms
3  sil-cisco.napier.ac.uk (146.176.3.1)  6.427 ms  8.407 ms  8.872 ms
4  * * *
5  smds-gw.ed.ja.net (193.63.106.129)  8.98 ms  30.308 ms  398.623 ms
6  smds-gw.bath.ja.net (193.63.203.68)  39.104 ms  46.833 ms  38.036 ms
7  jips-gw.bath.ac.uk (146.97.104.2)  32.908 ms  41.336 ms  42.429 ms
8  * * *
9  jess.bath.ac.uk (138.38.32.5)  41.045 ms *  41.93 ms
```

Sample run 17.8: Example traceroute

```
> traceroute home.microsoft.com
1  146.176.151.254 (146.176.151.254)  2.931 ms  2.68 ms  2.658 ms
2  sil-switch.napier.ac.uk (146.176.1.27)  6.216 ms  8.818 ms  5.885 ms
3  sil-cisco.napier.ac.uk (146.176.3.1)  6.502 ms  6.638 ms  10.218 ms
4  * * *
5  smds-gw.ed.ja.net (193.63.106.129)  18.367 ms  9.242 ms  15.145 ms
6  smds-gw.ulcc.ja.net (193.63.203.33)  42.644 ms  36.794 ms  34.555 ms
7  gw5.ulcc.ja.net (128.86.1.80)  31.906 ms  30.053 ms  39.151 ms
8  icm-london-1.icp.net (193.63.175.53)  29.368 ms  25.42 ms  31.347 ms
9  198.67.131.193 (198.67.131.193)  119.195 ms  120.482 ms  67.479 ms
10  icm-pen-1-H2/0-T3.icp.net (198.67.131.25)  115.314 ms  126.152 ms
     149.982 ms
11  icm-pen-10-P4/0-OC3C.icp.net (198.67.142.69)  139.27 ms  197.953 ms
     195.722 ms
12  mcinet-2.sprintnap.net (192.157.69.48)  199.267 ms  267.446 ms 287.834 ms
13  core2-hssi2-0.WestOrange.mci.net (204.70.1.49)  216.006 ms  688.139 ms
     228.968 ms
14  microsoft.Seattle.mci.net (166.48.209.250)  310.447 ms  282.882 ms
     313.619 ms
15  * microsoft.Seattle.mci.net (166.48.209.250)  324.797 ms  309.518 ms
16  * 207.68.145.53 (207.68.145.53)  435.195 ms *
```

17.5 nslookup

The `nslookup` program interrogates the local `hosts` file or a DNS server to determine the IP address of an Internet node. If it cannot find it in the local file then it communicates with gateways outside its own network to see if they know the address. Sample run 17.9 shows that the IP address of `www.intel.com` is `134.134.214.1`.

Sample run 17.9: Example of nslookup

```
C:\> nslookup
Default Server:  ees99.eece.napier.ac.uk
Address:  146.176.151.99
> www.intel.com
Server:  ees99.eece.napier.ac.uk
Address:  146.176.151.99
Name:    web.jf.intel.com
Address:  134.134.214.1
Aliases:  www.intel.com
230 % nslookup home.microsoft.com
Non-authoritative answer:
Name:    home.microsoft.com
Addresses:  207.68.137.69, 207.68.156.11, 207.68.156.14, 207.68.156.56
207.68.137.48, 207.68.137.51
```

17.6 Windows programs

Microsoft Windows has seven main programs which can be used to diagnose network problems:

arp Address resolution protocol (ARP) is used to determine the MAC address for a given IP address.

ipconfig Displays the current TCP/IP settings for the host.

winipcfg	As ipconfig, but a Windows version.
netstat	Displays the status of all TCP/IP connections on the host.
nslookup	Determines the IP address for a given domain name (Windows NT only).
ping	Determines if a remote host is communicating using TCP/IP.
tracert	Traces the route from the current host to a remote host.

17.6.1 Ping

The ping command is a standard program that can be used to determine if a host is responding to TCP/IP communications. Its format is:

```
Usage: ping [ -t] [ -a] [ -n count] [ -l size] [ -f] [ -i TTL] [ -v TOS]
            [ -r count] [ -s count] [[ -j host-list] | [ -k host-list]]
            [ -w timeout] destination-list

Options:
    -t             Ping the specified host until stopped.
    -a             Resolve addresses to hostnames.
    -n count       Number of echo requests to send.
    -l size        Send buffer size.
    -f             Set Don't Fragment flag in packet.
    -i TTL         Time To Live.
    -v TOS         Type Of Service.
    -r count       Record route for count hops.
    -s count       Timestamp for count hops.
    -w timeout     Timeout in milliseconds to wait for each reply.
```

Figure 17.3 shows that ping displays the IP address of the host for the entered domain name. In this case the `www.napier.ac.uk` host is contacted. It can be seen that its IP address is `146.176.1.8`. Ping also displays the time delay for each time that it contacts the remote host. In the example in Figure 17.3 the delays are: 153 ms; 137 ms; 148 ms and 152 ms. It also displays the value of the TTL (time-to-live) field, which, in this case, is 115.

The ping command is run from a Command window, which is typically selected by either:

- Clicking on MS-DOS prompt on the Desktop.
- Selecting the MS-DOS prompt option in the Start menu.
- Typing `command.com` in the Start→Run option, as shown next.

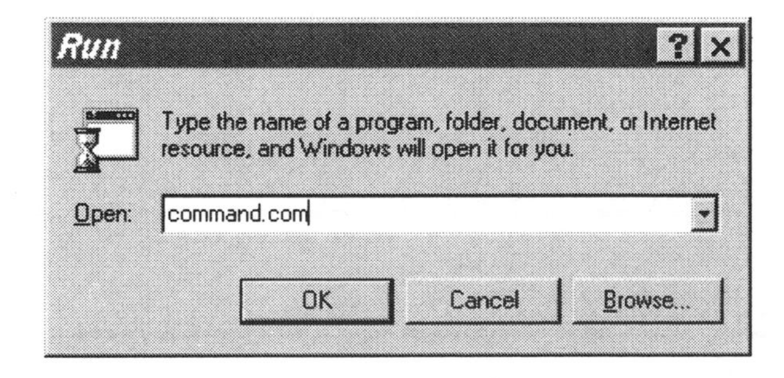

17.6.2 Tracert

The traceroute (`tracert`) program can be used to determine the route from the local host to a remote host. Its usage is:

```
Usage: tracert [ -d] [ -h maximum_hops] [ -j host-list] [ -w timeout] target_name
```

```
Options:
    -d                   Do not resolve addresses to hostnames.
    -h maximum_hops      Maximum number of hops to search for target.
    -j host-list         Loose source route along host-list.
    -w timeout           Wait timeout milliseconds for each reply.
```

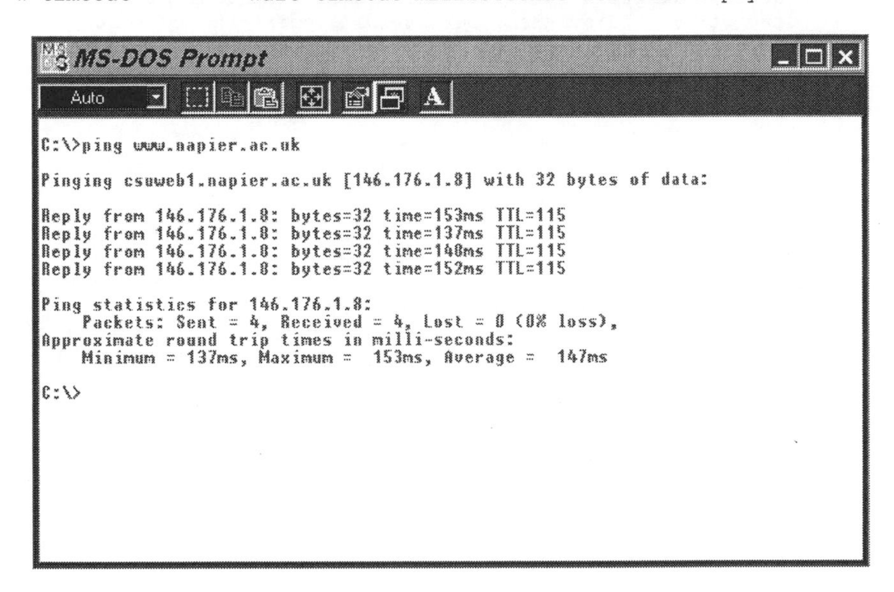

Figure 17.3 Example of the ping command

For example to trace the route between a local host and the `www.bath.ac.uk` host:

```
C:\WINDOWS>tracert www.bath.ac.uk
Tracing route to jess.bath.ac.uk [138.38.32.5]
  1     1 ms     1 ms     2 ms  146.176.151.254
  2     2 ms     1 ms     2 ms  146.176.7.250
  3     4 ms     4 ms     4 ms  146.176.9.1
  4     4 ms     2 ms     2 ms  oc3-f1-0.net.ed.ac.uk [194.81.56.65]
  5     3 ms     4 ms     6 ms  scot-x-gw2-a1-0-8.eastman.net.uk [194.81.56.30]
  6     4 ms     4 ms    28 ms  scot-pop-east.ja.net [146.97.250.53]
  7     8 ms     8 ms     7 ms  scot-pop-east.leeds-core.ja.net [146.97.254.41]
  8    13 ms    13 ms    13 ms  leeds.bweman.site.ja.net [146.97.254.102]
  9    15 ms    14 ms    15 ms  194.82.125.210
 10    16 ms    15 ms    16 ms  bath-gw-1.bwe.net.uk [194.82.125.198]
 11     *        *        *     Request timed out.
 12    48 ms    18 ms    20 ms  jess.bath.ac.uk [138.38.32.5]
Trace complete.
```

17.6.3 Netstat/IPConfig

The netstat (`netstat`) program can be used to determine TCP/IP protocol statistics and the current TCP/IP network connections. Its format is:

```
NETSTAT [-a] [-e] [-n] [-s] [-p proto] [-r] [interval]
  -a        Displays all connections and listening ports.
  -e        Displays Ethernet statistics.
  -n        Displays addresses and port numbers in numerical form.
  -p proto  Shows connections for the protocol specified by proto; proto
            may be TCP or UDP.  If used with the -s option to display
            per-protocol statistics, proto may be TCP, UDP, or IP.
  -r        Displays the routing table.
```

```
  -s        Displays per-protocol statistics.  By default, statistics
            are shown for TCP, UDP and IP; the -p option may be used to
            specify a subset of the default.
  Interval  Redisplays selected statistics, pausing interval second
            between each display.  Press CTRL+C to stop redisplaying
            statistics.  If omitted, netstat will print the current
            configuration information once.
```

Examples of its usage are:

```
C:\WINDOWS>netstat -a
Active Connections
  Proto  Local Address      Foreign Address                State
  TCP    bill-s:1060        csunt1.napier.ac.uk:80         CLOSE_WAIT
  TCP    bill-s:1061        csunt1.napier.ac.uk:80         CLOSE_WAIT
  TCP    bill-s:1062        csunt1.napier.ac.uk:80         CLOSE_WAIT
  TCP    bill-s:1063        csunt1.napier.ac.uk:80         CLOSE_WAIT
  TCP    bill-s:1097        artemis.dcs.napier.ac.uk:ftp   ESTABLISHED
  TCP    bill-s:1106        www.eece.napier.ac.uk:telnet   CLOSE_WAIT

C:\WINDOWS>netstat -s
Active Connections
  Proto  Local Address          Foreign Address        State
  TCP    62.136.29.56:1060      146.176.1.24:80        CLOSE_WAIT
  TCP    62.136.29.56:1061      146.176.1.24:80        CLOSE_WAIT
  TCP    62.136.29.56:1062      146.176.1.24:80        CLOSE_WAIT
  TCP    62.136.29.56:1063      146.176.1.24:80        CLOSE_WAIT
  TCP    62.136.29.56:1097      146.176.161.5:21       ESTABLISHED
  TCP    62.136.29.56:1106      146.176.151.139:23     CLOSE_WAIT
```

The Ipconfig program can be used to determine the IP configuration (otherwise WINIPCFG can be used). For example, to determine the IP address of the host computer:

```
C:\WINDOWS>ipconfig
0 Ethernet adapter :
       IP Address. . . . . . . . . : 62.136.29.56
       Subnet Mask . . . . . . . . : 255.0.0.0
       Default Gateway . . . . . . : 62.136.29.56

C:\WINDOWS>ipconfig /all
       Host Name . . . . . . . . . : BILL'S
       DNS Servers . . . . . . . . : 195.92.193.8
                                     194.152.64.35
       Node Type . . . . . . . . . : Broadcast
       NetBIOS Scope ID. . . . . . :
       IP Routing Enabled. . . . . : No
       WINS Proxy Enabled. . . . . : No
       NetBIOS Resolution Uses DNS : No

0 Ethernet adapter :
       Description . . . . . . . . : PPP Adapter.
       Physical Address. . . . . . : 44-45-53-54-00-00
       DHCP Enabled. . . . . . . . : Yes
       IP Address. . . . . . . . . : 62.136.29.56
       Subnet Mask . . . . . . . . : 255.0.0.0
       Default Gateway . . . . . . : 62.136.29.56
       DHCP Server . . . . . . . . : 255.255.255.255
       Primary WINS Server . . . . :
       Secondary WINS Server . . . :
       Lease Obtained. . . . . . . : 01 01 80 00:00:00
       Lease Expires . . . . . . . : 01 01 80 00:00:00
```

17.7 Exercises

17.7.1 Which TCP/IP command allows a user to test to see if a node is responding to TCP/IP communications:

(a) nslookup (b) telnet
(c) ping (d) ftp

17.7.2 Which TCP/IP command allows a user to determine the IP address of a host address:

(a) nslookup (b) telnet
(c) ping (d) ftp

17.7.3 What is the size of the IP address in IPv4:

(a) 32 bits (b) 64 bits
(c) 128 bits (d) Any size

17.7.4 What is the size of the IP address in IPv6:

(a) 32 bits (b) 64 bits
(c) 128 bits (d) Any size

17.7.5 Using the `ping` program determine if the following nodes are responding:

(i) `www.eece.napier.ac.uk` (ii) `home.microsoft.com`
(iii) `www.intel.com`

17.7.6 Using the `traceroute` program determine the route from your local host to the following destinations:

(i) `www.napier.ac.uk` (ii) `home.microsoft.com`
(iii) `www.intel.com`

Identify each part of the route and note the timing information.

17.7.7 Outline how IPv6 uses priority levels to differentiate between traffic that can back-off and that which is not willing to back-off. How is the priority level of real-time traffic dealt with?

17.7.8 Describe how IPv6 uses flow labels to route traffic.

17.7.9 An example trace for a computer on the `146.176.150` segment to a computer at `hotmail.com` gives:

```
146.176.150.253     eece_1.eece.napier.ac.uk
146.176.151.254     Unavailable
146.176.7.250       Unavailable
146.176.9.1         Unavailable
194.81.56.126       scot-x-gw.eastman.net.uk
146.97.253.33       scot-pop.ja.net
146.97.254.41       leeds-core.ja.net
146.97.254.54       external-gw.ja.net
128.86.1.80         tglobe.gw2.ja.net
193.62.157.10       ppt-gw.ja.net
207.45.215.165      gin-ppt-bb2.Teleglobe.net
207.45.223.42       gin-mtt-core1.Teleglobe.net
207.45.223.109      gin-nyy-core1.Teleglobe.net
207.45.215.165      bbr05-p1-1.jrcy01.exodus.net
207.45.215.165      bbr02-p6-1.jrcy01.exodus.net
207.45.215.165      bbr02-p0-1.sntc01.exodus.net
```

Determine the routers on the local network, the EASTMAN networks, the JANET network and the connection between the UK and the USA. If possible, trace the route to the `scot-x-gw.eastman.net.net` router.

17.7.10 Using IPConfig, determine the following settings for your local computer and another computer (the first column contains a sample session):

	Example connection	Local computer	Another computer (please ask someone for their settings)
IP address	`62.136.29.56`		
Host name	`BILL'S`		
Subnet mask	`255.0.0.0`		
Gateway	`62.136.29.56`		
DNS server(s)	`195.92.193.8` `194.152.64.35`		
Ethernet physical address	`44-45-53-54-00-00`		
Full domain name of computer	`Bill-s.freeserve.co.uk`		
DHCP server IP address (if used). Note that a DHCP server will only be used if the computer does not have a static IP address.	`255.255.255.255`		

17.7.11 Start two different WWW sessions (with different sites), then run `netstat`. From this determine the local port that is used for the connection, and complete the table given next (an example is given in the first row).

Connection type (TCP/UDP)	*Local hostname*	*Local port value*	*Destination hostname*	*Destination port value*	*State*
TCP	bill-s	1060	csunt1.napier.ac.uk	80	CLOSE_WAIT

17.8 Note from the Author

This chapter has discussed some of the practical aspects of networking. As someone who has run a network for many years, I have found that the ping and traceroute commands are invaluable for tracing faults. One network that I used to look after was a 10BASE2-based Ethernet network. Unfortunately, as it was based on coaxial cable, whenever someone disconnected their computer, it would bring the whole network segment down, or it would slow it down until it was hardly useable. One method that we used to bring the network back up was to disconnect all the computers from the segment. A node nearest the correctly operating segment would be connected, and the other terminated with a terminator. The node was then tested to see if it could log into the network. The nodes were then put back on-line one at a time, until the fault was traced (as illustrated in Figure 17.4).

A Digital Voltmeter (DVM) was a useful as it could be used to test for the heartbeat voltage. In these days of network hubs and switches, the best method of determining if a node is operating is to view the green LED on the network card and the hub/switch.

The greatest problem that I had was when I looked after a Token Ring network, which worked well, and was very efficient. Unfortunately, a network fault caused terrible problems, as the whole network would go down. To improve fault detection we disconnected each node in turn, and inserted a by-pass cable. The problem from there was getting the ring back online, and for it to generate a new token. As the network grew in size, the more difficult it became to find faults. Thus, Token Ring is an excellent networking topology but, at the time, suffered from too many problems. The way was thus clear for Ethernet to carve a massive niche for LANs.

One method to reduce the number of nodes which are affected by a fault is to segment the network, using bridges, switches or routers. A router offers the best segmentation as it intelligently routes data in to and out of a network segment. Routers have many advantages over bridges, as they do not forward broadcasts (which are typically used with ARP for nodes to find-out the MAC address for a given IP address), and not to forward traffic with an unknown address (as a router will only forward is they know that the destination is outside the network segment).

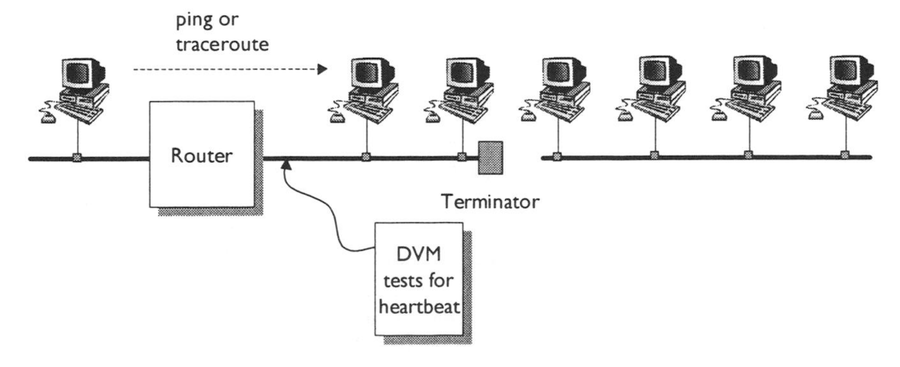

Figure 17.4 Fault-finding on a bus network

18

Routers

18.1 Introduction

Routers are key elements of the Internet, and without routers there would be no Internet. They are highly secure and complex systems which run their own operating systems, and if operated correctly they can significantly improve the performance of interconnected networks. Their main function is to examine the network address of the data packets and decide the route that the data packet should take. To be able to determine the best route that the data packets should take they require routing protocols, such as RIP and EGP. These routing protocols allow routers to intercommunicate with other routers so that they can determine the structure of the interconnected networks. A routed protocol is one that is routed through interconnected networks, such as IP.

Each router has a number of interfaces; a data packet received on one interface is examined to determine if the destination address is on another subnet. If it is then it is routed to one of the other interfaces, if not then the data packet will not be routed onto another port. Figure 18.1 shows a router with four interfaces: Ether0, Ether1, Serial0 and Serial1. Serial interfaces are typically used for WANs, whereas Ethernet-type interfaces are typically used in LANs. Typically, also, the Ethernet interface is via a 15-pin D-type connector which interfaces to a transceiver unit onto either a twisted-pair or a coaxial connector.

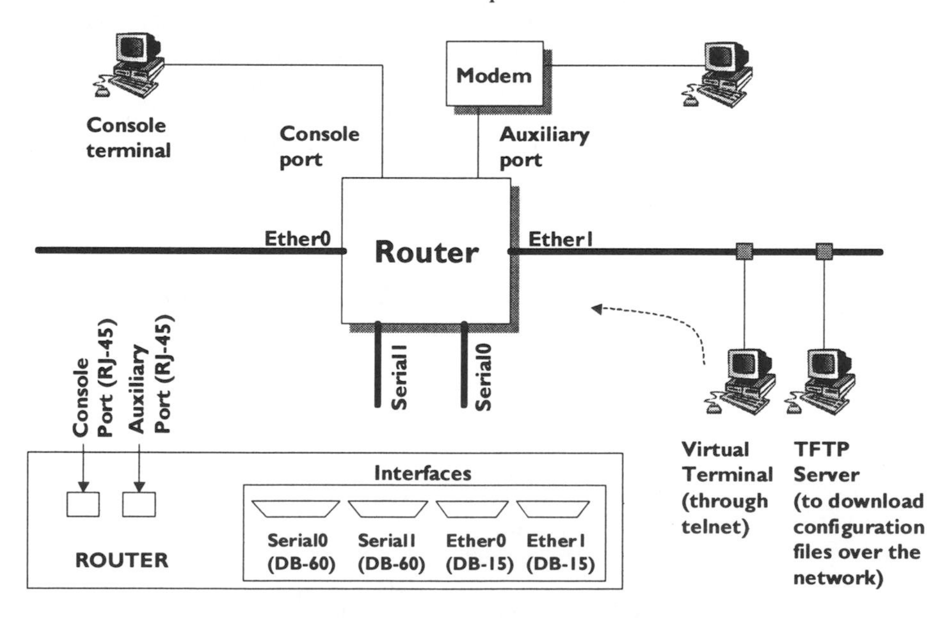

Figure 18.1 Router interfaces and configuration methods

The router must obviously be configured for the IP addresses of its connected networks, the protocols it should use, and so on. The router itself can be configured in a number of ways, such as:

- **Console terminal.** The console terminal connects to the router via the console port. This is a standard RS-232 connection and typically uses an RJ-45 connector on the router, and a 9-pin D-type connection on the console (typically a PC). When the router is initially started it must be configured from the console terminal, which can then initiate configuration from one of the following methods.
- **Via a modem.** A modem can connect to the router via the auxiliary port on the router. This allows an administrator to log into the router remotely via a telephone line. This is useful when the network connection to the router becomes inoperable, and only a telephone connection is available.
- **Virtual terminals.** The console terminal can set up a number of virtual terminals which can connect into the router from one of its interfaces. These typically use a telnet connection to change the configuration of the router.
- **From a TFTP server.** The configuration files for the router can be downloaded from a TFTP server.

This chapter includes practical setups of routers, and Figure 18.2 shows the setup of the system on which these setups are based on. The network uses five routers (LAB-A, LAB-B, LAB-C, LAB-D and LAB-E), a switch and two hubs. Each link that the routers connect to must be assigned a unique IP address. As these are Class C network addresses they all end with a zero. For example, LAB-A connects to 205.7.5.0 (from its Ether1 interface), 192.5.5.0 (from its Ether0 interface) and 201.100.11.0 (from its Serial0 interface). Each of the ports of the routers must be given a unique IP address. Thus the ports on LAB-A and LAB-B are assigned as follows:

Router	*Port*	*IP address*	*Router*	*Port*	*IP address*
LAB-A	Ether1	205.7.5.1	LAB-A	Ether0	192.5.5.1
LAB-A	Serial0	201.100.11.1	LAB-B	Serial1	201.100.11.2
LAB-B	Serial0	199.6.13.1	LAB-B	Ether0	219.17.100.1

Each of these IP addresses relates to the networks that it connects to, such as Ether1 connects to the 205.7.5.0 network. All the computers which connect to the switch on this network will have the same network part of their address, and will range from 205.7.5.2 to 205.7.5.254 (as 205.7.5.255 is used for a broadcast address to the 205.7.5.0 network). As all of the addresses are Class C addresses, the subnet mask will be 255.255.255.0 in all cases.

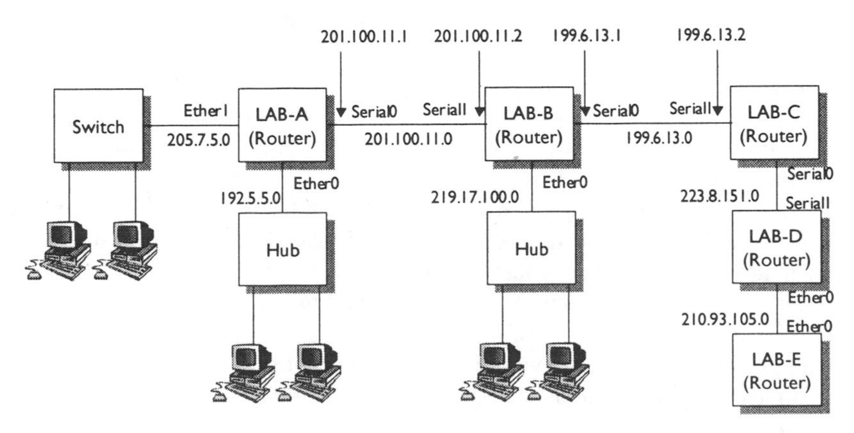

Figure 18.2 Sample interconnect network

18.2 Router configuration and startup

The router must be started in an orderly manner, so that it can operate correctly. The first step is to run a power-up, self-test. After this is successful the router goes through the following steps:

- **The bootstrap loader** is loaded from ROM and run on the processor.
- **The operation system** (Cisco IOS – Internetwork Operating System) is then loaded from the boot field of a configuration register (which specifies either boot from flash memory, boot from the network or manual boot). The lower four bits of the configuration register define the boot field. A value of 0000 defines ROM monitor mode, where the user can manually boot the router with a b command at the prompt, a value of 0001b defines that the boot should occur automatically from ROM, while any other value (from 0010b to 1111b) defines that the system is booted from the configuration defined in the NVRAM (Non-volatile RAM). Normally these binary values are defined in hexadecimal, thus 0x00 defines a manual boot, 0x01 defines an automatic boot from ROM and 0x02 defines that the system should boot from the commands given in NVRAM.
- The operating system is then **booted**, and it determines the hardware and the software on the system, and displays these to the console terminal.
- The operating system then loads the **configuration file** from NVRAM and executes it one line at a time. These lines start different processes, and define addresses and protocol types.
- If there is no configuration file in NVRAM, the router automatically goes into user **setup mode**, where the router asks the user questions about the router configuration. Once these have been specified the router saves these to NVRAM, so that the settings will be saved. Once saved, the router should automatically boot, without going into user setup mode. As much as possible the router tries to discover its environment, and tries to minimize the settings that the user has to add. Typically values are given in squared brackets, which are defaults that the user can choose if the return key is pressed at the option.

Each router has memory which stores information on the setup of the router, and also the operating system of the router. The configuration file stores important information on the setup of the interfaces that is specific to the router. The configuration and other processes are stored in memory. The different types and uses of memory are:

- **NVRAM.** This type of memory does not lose its contents when the power is withdrawn, but can be written to. It is used to store the router's backup/startup configuration file. One of the options in the configuration is where the operating system image is loaded from, typically either from flash memory, or from a TFTP server.
- **Flash.** This is erasable, reprogrammable ROM, which keeps its contents when the power is taken away. It is used in the router to contain one or more copies of the operating system image and microcode. Flash memory allows for easy updates to the operating system software, without having to replace any parts of the hardware.
- **ROM.** This is a permanent type of memory, which cannot be changed, and does not lose its contents when the power is withdrawn. On the router it contains power-on diagnostics, a bootstrap program, and operating system software. Upgrades to ROM require a change of a ROM integrated circuit.
- **RAM.** This is the main memory of the router and stores running programs and the current running configuration file. Along with this the RAM stores routing tables, ARP

cache, packet buffering and packet hold queues. The contents of the RAM are lost when the power is withdrawn.

The configuration file stores information on the global configuration, processes and interface information for the router and its interface ports. Initially, when the system is booted, a bootstrap program is executed from the ROM, which then loads the operating system image in the RAM, as illustrated in Figure 18.3. The operating system then starts a number of routines which operate on the different protocols that are associated with the ports. This includes updating routing tables, buffering data packets and moving data, as well as executing user commands.

To be able to login into a router the user must have a password. This gets the user into the User EXEC mode, which allows the user to view the configuration of the router, and perform simple operations. As the router could be breached by an external hacker, a further password is required to enter into the Privileged EXEC mode, in which the user can actually change the configuration of the router, and perform extensive operations. Figure 18.3 shows the other modes that the router can be put into. For example, the RXBOOT mode allows the user to recover corrupted passwords, and a setup mode allows the router to be initially configured when it is first powered-on. The global configuration mode allows for simple, one-line, configuration tasks.

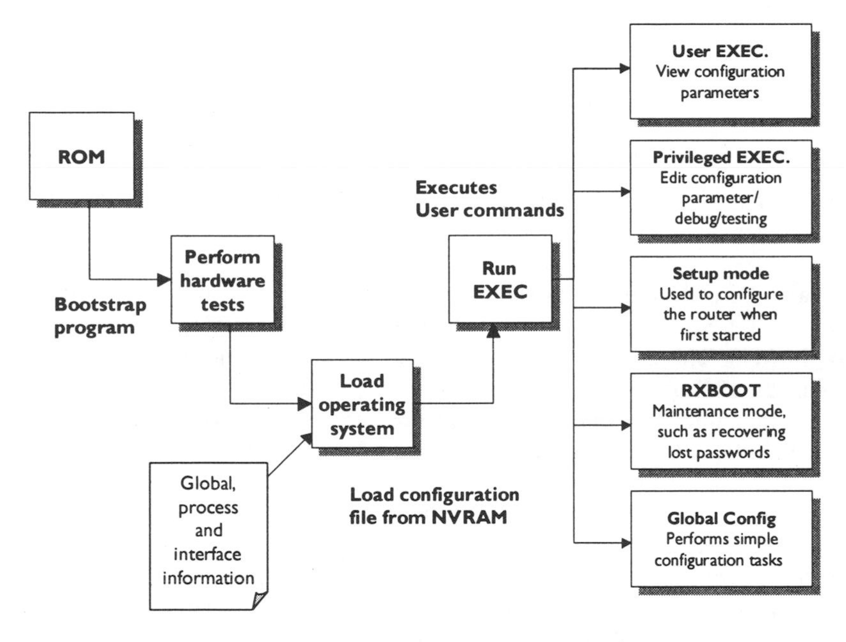

Figure 18.3 Router startup and router modes

18.3 Router commands

The router has a whole series of commands which can be used to determine the status of the router, and to change its configuration. There are two main modes that can be used: User Exec mode and Privileged Exec mode. Initially the user logs into the User Exec mode, and then uses a further secret password to get into the Privileged Exec mode. The enable command is used to go from User Exec mode into Privileged Exec mode. The prompt for User Exec is > and the prompt for Privileged Exec mode is #. Session run 18.1 shows an example of a user logging into a router and then logging into the Privileged mode. The ? character is used to give help on commands (with ? on its own) or with subcommands (such as show ?, as illustrated in Session run 18.17). A complete list of User Exec and Privileged Exec commands is given in Session run 18.15 and Session run 18.16.

Session run 18.1

```
LAB-A con0 is now available
Press RETURN to get started.
User Access Verification
Password:  *******
LAB-A>   ?
Exec commands:
  access-enable     Create a temporary Access-List entry
  access-profile    Apply user-profile to interface
  clear             Reset functions
  connect           Open a terminal connection
      ::::::::
LAB-A>   enable
Password:  **********
LAB-A#   ?
Exec commands:
  access-enable     Create a temporary Access-List entry
  access-profile    Apply user-profile to interface
  access-template   Create a temporary Access-List entry
  cd                Change current directory
      ::::::::
```

18.3.1 Router status commands

The show command in user or privileged mode can be used to display the status of the router. These include:

- **show arp.** Displays the current status of router's ARP tables, which map IP addresses to MAC address. Session run 18.6 gives an example of this command.
- **show buffers.** This command shows detailed statistics on the buffers within the router. Session run 18.2 gives an example of this command. In this case, the memory buffers split into small buffers (104 bytes), middle buffers (600 bytes), big buffers (1524 bytes), very big buffers (4520 bytes), large buffers (5024 bytes) and huge buffers (18024 bytes).
- **show flash.** This command displays information on the data stored in the flash memory. An example is given in Session run 18.8.
- **show hosts.** This command displays a lists of connected hosts and their IP addresses.
- **show interfaces.** This command displays statistics for all interfaces configured on the router. Session run 18.10 shows an example.
- **show mem.** This command displays the usage of the routers memory. Session run 18.3 shows an example.
- **show processes.** This command shows the active processes.
- **show protocols.** This command displays the status of currently running protocols (such as IP, IPX, AppleTalk and DECnet). It can be seen from Session run 18.9 that there are three active interfaces (Ethernet0, Serial0 and Serial1), and that each of the interfaces is operating ('they are up'). For example the IP address of Ethernet0 interface is 219.17.100.1/24, which specifies that it has an IP address of 219.17.100.1 and that 24 bits are used to define the network part of the address (as expected as it is a Class C address).

- **show running-config.** This command displays the active configuration file.
- **show stacks.** This command displays the usage of stacks within the router. Session run 18.4 shows an example.
- **show startup.** Displays the startup configuration file.
- **show version.** This command display information on the hardware, software version, configuration file name, and the boot image. Session run 18.7 shows an example for a Cisco 2500-series router.

Session run 18.2

```
LAB-A#   show buffers
Buffer elements:
     500 in free list (500 max allowed)
     2026 hits, 0 misses, 0 created
Public buffer pools:
Small buffers, 104 bytes (total 50, permanent 50):
     49 in free list (20 min, 150 max allowed)
     669 hits, 0 misses, 0 trims, 0 created
      ::::::::
Huge buffers, 18024 bytes (total 0, permanent 0):
     0 in free list (0 min, 4 max allowed)
     0 hits, 0 misses, 0 trims, 0 created
     0 failures (0 no memory)
Interface buffer pools:
Ethernet0 buffers, 1524 bytes (total 32, permanent 32):
     8 in free list (0 min, 32 max allowed)
     24 hits, 0 fallbacks
     8 max cache size, 8 in cache
      ::::::::
Serial0 buffers, 1524 bytes (total 32, permanent 32):
     7 in free list (0 min, 32 max allowed)
     102 hits, 0 fallbacks
     8 max cache size, 8 in cache
```

Session run 18.3

```
LAB-A#   show memory
                Head    Total(b)     Used(b)     Free(b)  Lowest(b) Largest(b)
Processor      84D10    1549040     1537528       11512          0       4560
      I/O     200000    2097152      437652     1659500    1463544    1463544
         Processor memory
 Address  Bytes Prev.     Next    Ref   PrevF    NextF    Alloc PC  What
84D10      1064 0         85164   1                      31A6070    List Elements
85164      2864 84D10     85CC0   1                      31A6070    List Headers
85CC0      3992 85164     86C84   1                      3148B48    TTY data
```

Session run 18.4

```
LAB-A#   show stacks
Minimum process stacks:
 Free/Size   Name
 3528/4000   Router Init
 2380/4000   Init
 3468/4000   RADIUS INITCONFIG
 3380/4000   DHCP Client

Interrupt level stacks:
Level    Called Unused/Size  Name
  3           3   2772/3000  Serial interface state change interrupt
  4        1440   2612/3000  Network interfaces
  5       32703   2888/3000  Console Uart
```

18.3.2 Show hosts

The `show hosts` command displays the hosts that are connected to the local router and their IP addresses. Session 18.5 shows an example session.

Session run 18.5

```
LAB-A>   show hosts
Default domain is not set
Name/address lookup uses domain service
Name servers are 255.255.255.255

Host                     Flags       Age Type    Address(es)
LAB-B                    (perm, OK)  17   IP     201.100.11.2  219.17.100.1
                                                 199.6.13.1
LAB-C                    (perm, OK)  18   IP     199.6.13.2  223.8.151.1
                                                 204.204.7.1
LAB-D                    (perm, OK)  19   IP     204.204.7.2  210.93.105.1
LAB-E                    (perm, OK)  18   IP     210.93.105.2
LAB-A                    (perm, OK)  19   IP     192.5.5.1  205.7.5.1
                                                 201.100.11.1
```

18.3.3 Show arp

Arp tables are very important in that they allow a host to determine the MAC address of a host with a given IP address. To complete the table a host sends out ARP requests on the network and the host which has the IP address defined in the ARP request replies back with its MAC address. Session run 18.6 shows an example of an ARP table, which has four entries in it. In this case (see Figure 18.2), router LAB-A has four connected Ethernet adaptors, two of its own (205.7.5.1 and 192.5.5.1), a switch (205.7.5.254) and a hub (192.5.5.12). Figure 18.4 illustrates the layout of the connections. It can be seen that each Ethernet interface port has an associated IP address, and a unique IP address.

Session run 18.6

```
LAB-A#   show arp
Protocol  Address          Age (min)  Hardware Addr   Type   Interface
Internet  205.7.5.254          108    0030.8071.9f40  ARPA   Ethernet1
Internet  192.5.5.1              -    0010.7b81.1d72  ARPA   Ethernet0
Internet  192.5.5.12             1    0000.b430.b332  ARPA   Ethernet0
Internet  205.7.5.1              -    0010.7b81.1d73  ARPA   Ethernet1
```

18.3.4 Show version

The `show version` command gives information on the operating system version, the hardware and the amount of on-board memory. It can be seen from Session run 18.7 that the amount of on-board flash memory is 8 MB, with 32 KB of NVRAM, and there are four interfaces (Serial0, Serial1, Ethernet0 and Ethernet1).

Flash memory (EEPROM) contains the operating system image, which can be upgraded electronically with an operating system update. In this case the current flash file is c2500-d-l.120-4 (which defines that the image is for a Cisco 2500-series router-special features, and so on). The NVRAM is updated with a startup configuration file, and can be updated by the user at any time. It should be noted that the show version command also displays the 32-bit configuration register value, which, in this case, is 0x2102. The last four bits define the boot location, which, in this case, is 0010b (which causes the boot to occur from the location defined in NVRAM).

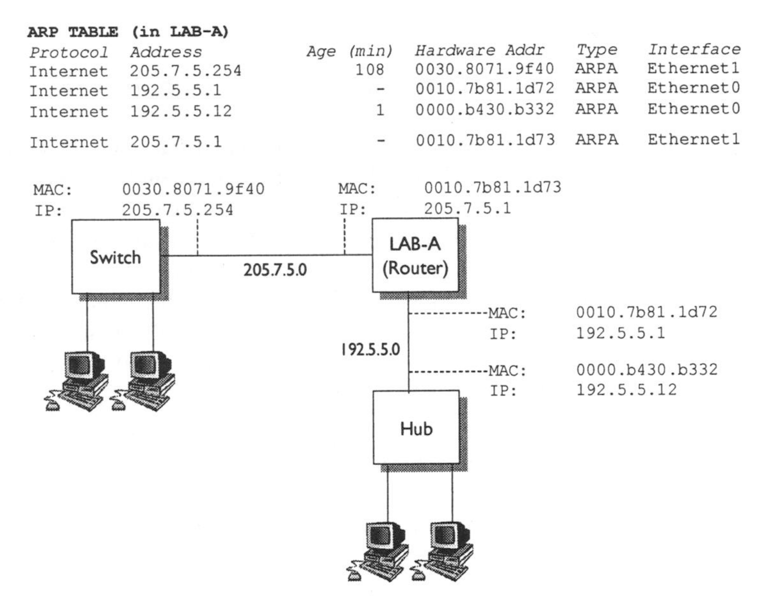

ARP TABLE (in LAB-A)

Protocol	Address	Age (min)	Hardware Addr	Type	Interface
Internet	205.7.5.254	108	0030.8071.9f40	ARPA	Ethernet1
Internet	192.5.5.1	-	0010.7b81.1d72	ARPA	Ethernet0
Internet	192.5.5.12	1	0000.b430.b332	ARPA	Ethernet0
Internet	205.7.5.1	-	0010.7b81.1d73	ARPA	Ethernet1

Figure 18.4 Example of an ARP table

Session run 18.7

```
LAB-A#   show version
Cisco Internetwork Operating System Software
IOS (tm) 2500 Software (C2500-D-L), Version 12.0(4), RELEASE SOFTWARE (fc1)
Copyright (c) 1986-1999 by cisco Systems, Inc.
Compiled Wed 14-Apr-99 21:21 by ccai
Image text-base: 0x03037C88, data-base: 0x00001000
ROM: System Bootstrap, Version 11.0(10c)XB2, PLATFORM SPECIFIC RELEASE SOFTWARE
(fc1)
BOOTFLASH: 3000 Bootstrap Software (IGS-BOOT-R), Version 11.0(10c)XB2, PLATFORM
SPECIFIC RELEASE SOFTWARE (fc1)

LAB-A uptime is 5 minutes
System restarted by reload
System image file is "flash:c2500-d-l.120-4"
cisco 2500 (68030) processor (revision L) with 2048K/2048K bytes of memory.
Processor board ID 13583483, with hardware revision 00000000
Bridging software.
X.25 software, Version 3.0.0.
2 Ethernet/IEEE 802.3 interface(s)
2 Serial network interface(s)
32K bytes of non-volatile configuration memory.
8192K bytes of processor board System flash (Read ONLY)
Configuration register is 0x2102
```

18.3.5 Show flash

The `show flash` command displays important configuration information. It can be seen that, in Session run 18.8, the hostname is LAB-A, and that it connects to four interfaces (Ethernet0, Ethernet1, Serial0 and Serial1). Each of these interfaces has an associated IP address (such as 192.5.5.1 for Ethernet0), and a subnet mask (255.255.255.0). It can also be seen that the IP addresses of each of the interfaces of the other routers has been defined. For example:

```
ip host LAB-B 201.100.11.2 219.17.100.1 199.6.13.1
```

defines that LAB-B connects has three interface ports, which have IP addresses of 201.100.11.2, 219.17.100.1 and 199.6.13.1.

It can be seen from the initial information from the command that there is a total of 8 MB of memory in the flash memory (8192 KB), of which over 6 MB of memory has been used. If there is enough memory, more than one operating system image can be stored. In this case there is possibly not enough memory for another image. If more than one image was to be stored, the user would have to upgrade the flash memory. When downloading a new image from a TFTP server, the following should be done: check the size, name and pathname of the new image on the TFTP server (using DIR or ls commands on an FTP session) and if there is enough flash memory for the new image. Note also that there can be different passwords setup for each of the configuration sources, these are defined from:

- Line con 0. The console terminal. In the example, the password is Cisco.
- Line aux. The auxiliary terminal.
- Line vty 0 4. Defines a password from incoming Telnet sessions.

Session run 18.8

```
LAB-A#   show flash
System flash directory:
File  Length   Name/status
  1   6788464  c2500-d-l.120-4
[ 6788528 bytes used, 1600080 available, 8388608 total]
8192K bytes of processor board System flash (Read ONLY)
Current configuration:
!
version 12.0
service timestamps debug uptime
service timestamps log uptime
no service password-encryption
!
hostname LAB-A
!                                              <-- Encrypted secret password
enable secret 5 $1$EHcO$ro5NY5aMGtbyViNE/m7VG1
enable password cisco
!
ip subnet-zero
ip host LAB-B 201.100.11.2 219.17.100.1 199.6.13.1
ip host LAB-C 199.6.13.2 223.8.151.1 204.204.7.1
ip host LAB-D 204.204.7.2 210.93.105.1
ip host LAB-E 210.93.105.2
ip host LAB-A 192.5.5.1 205.7.5.1 201.100.11.1
!
interface Ethernet0
 ip address 192.5.5.1 255.255.255.0
 no ip directed-broadcast
 no mop enabled
!
interface Ethernet1
 ip address 205.7.5.1 255.255.255.0
 no ip directed-broadcast
!
interface Serial0
 ip address 201.100.11.1 255.255.255.224
 no ip directed-broadcast
 no ip mroute-cache
 no fair-queue
```

```
 clockrate 56000
!
interface Serial1
 no ip address
 no ip directed-broadcast
 shutdown
!
router rip
 network 192.5.5.0
 network 201.100.11.0
 no ip directed-broadcast
 shutdown
!
router rip
 network 192.5.5.0
 network 201.100.11.0
 network 205.7.5.0
!
no ip classless
!
snmp-server community public RO
!
line con 0
 password cisco
 login
 transport input none
line aux 0
line vty 0 4
 password cisco
 login
!
end
```

In the privileged mode, the user can copy the contents of the flash memory to another file or to a TFTP server. This can be used to create a backup copy of the operating system before it is overwritten by a new version, or in case it is corrupted. The command to copy the flash memory to a TFTP server is (the default options are given in square brackets):

```
LAB-A# copy flash tftp
IP address of remote TFTP server [ 201.10.30.3] ?
Source filename [ c2500-d-1.120-4] ?
Writing c2500-d-1.120-4  !!!!!!!!!!!!!!!!!!!!!!!!
successful tftp write
```

to copy the file back from the TFTP requires the following command:

```
LAB-A# copy tftp flash tftp
IP address of remote TFTP server [ 201.10.30.3] ?
Source filename [ ] ? c2500-d-1.120-5
Copy c2500-d-1.120-4 from 201.10.30.3 into flash memory? [ confirm]
erase before writing? [ confirm]
Clearing and initializing flash memory
[ Ok]
```

18.3.6 Show protocols

The `show protocols` command is important as it shows the interfaces that are connected, and their status. In Session run 18.9 it can be seen that the LAB-B router has three interfaces (Ethernet0, Serial0 and Serial1). It shows that each of these interfaces is currently connected, and that their protocol is operating (in this case the protocol used is IP).

Session run 18.9

```
Lab-B>   show protocols
Global values:
  Internet Protocol routing is enabled
Ethernet0 is up, line protocol is up
  Internet address is 219.17.100.1/24
Serial0 is up, line protocol is up
  Internet address is 199.6.13.1/24
Serial1 is up, line protocol is up
  Internet address is 201.100.11.2/24
```

Shows that the IP address uses 24 bits for the network portion (Class C).

18.3.7 Show interfaces

The `show protocols` shows some basic information on the interfaces that the router uses, whereas the `show interfaces` command in Session run 18.10 gives a detailed view of any, or all, of the interfaces on the router. It defines whether its link is currently operating, the IP and MAC address of the Ethernet adaptors (for example, Ethernet0 has a MAC address of 0010.7b81.1d72 and an IP address of 192.5.5.1). The counters can be cleared with the clear counters command. By resetting the counters the user can get a better idea of the current status of the network. The `show interfaces` command also displays:

- Bit rate on interface. It can be seen that the bit rate for Ethernet0 is 10 Mbps (BW 10000 Kbit).
- Error rate. Along with all this information, the show interfaces gives information on the number of errors that have occurred on an interface. In this case there are no errors, and defines the number of data packets that have been received.
- Time since last reboot. In this case the amount of up time is 5 minutes.
- Delay. In this case the delay is 1 ms (1000 usec).
- Keep-alive time. In this case it is once every 10 seconds.
- Number of collisions. In this case there have been no collisions.
- Input/output rate. In packets per second and bits per second.

Session run 18.10

```
LAB-A#   show interfaces
Ethernet0 is up, line protocol is up
  Hardware is Lance, address is 0010.7b81.1d72 (bia 0010.7b81.1d72)
  Internet address is 192.5.5.1/24
  MTU 1500 bytes, BW 10000 Kbit, DLY 1000 usec, rely 255/255, load 1/255
  Encapsulation ARPA, loopback not set, keepalive set (10 sec)
  ARP type: ARPA, ARP Timeout 04:00:00
  Last input 00:00:43, output 00:00:00, output hang never
  Last clearing of "show interface" counters never
  Queueing strategy: fifo
  Output queue 0/40, 0 drops; input queue 0/75, 0 drops
  5 minute input rate 0 bits/sec, 0 packets/sec
  5 minute output rate 0 bits/sec, 0 packets/sec
     62 packets input, 6425 bytes, 0 no buffer
     Received 50 broadcasts, 0 runts, 0 giants, 0 throttles
     0 input errors, 0 CRC, 0 frame, 0 overrun, 0 ignored, 0 abort
     0 input packets with dribble condition detected
     261 packets output, 31216 bytes, 0 underruns
     0 output errors, 0 collisions, 2 interface resets
     0 lost carrier, 0 no carrier
Ethernet1 is up, line protocol is up
  Hardware is Lance, address is 0010.7b81.1d73 (bia 0010.7b81.1d73)
  Internet address is 205.7.5.1/24   ... and so on.
```

18.3.8 Running-config and startup-config

Two important commands are `show running-config` and `show startup-config`, as these show the current settings (with running-config), and the settings which are used at boot (startup-config). These settings can differ if the user has made a change to the settings of the router. Session run 18.11 show an example configuration from the Lab-B router. It can be seen that the Lab-B router has three interfaces (Ethernet0, Serial0 and Serial1). It uses the `show running-config` (the `show startup-config` can be used to display the backup configuration). It can be seen from Session run 18.11, that the hostname is Lab-B (hostname Lab-B), and that the networks it connects to are: 199.6.13.0, 201.100.11.0 and 219.17.100.0.

The `copy running-config startup-config` command is useful in storing any changes that a user has made to the setup of the router, so that the router will use the updated settings when the router has been rebooted. This should only be done when the user is sure that the changes that have been made are correct.

The `erase startup-config` command deletes the backup configuration file in NVRAM, and the reload command causes the system to reload the startup-configuration and reconfigure the system, as it was when the system was booted. This allows the user to undo any updates to the running configurations. Note that these commands can only be run from Privileged EXEC mode (which requires the secret password).

The running-config file can be transferred to the TFTP server with `copy tftp running-config` and copied back with `copy running-config tftp`.

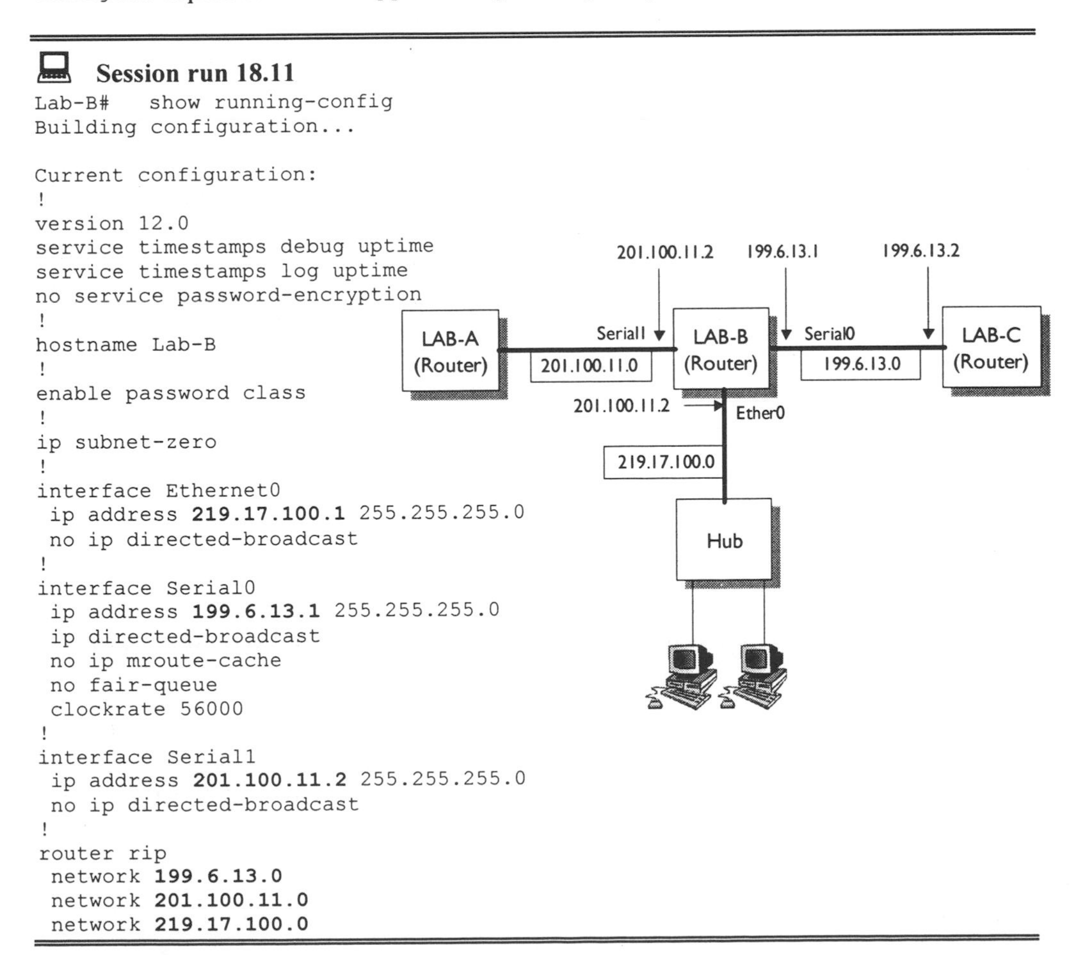

Session run 18.11

```
Lab-B#   show running-config
Building configuration...

Current configuration:
!
version 12.0
service timestamps debug uptime
service timestamps log uptime
no service password-encryption
!
hostname Lab-B
!
enable password class
!
ip subnet-zero
!
interface Ethernet0
 ip address 219.17.100.1 255.255.255.0
 no ip directed-broadcast
!
interface Serial0
 ip address 199.6.13.1 255.255.255.0
 ip directed-broadcast
 no ip mroute-cache
 no fair-queue
 clockrate 56000
!
interface Serial1
 ip address 201.100.11.2 255.255.255.0
 no ip directed-broadcast
!
router rip
 network 199.6.13.0
 network 201.100.11.0
 network 219.17.100.0
```

```
!
no ip classless
!
!
line con 0
 password cisco
 login
 transport input none
line aux 0
line vty 0 4
 password cisco
 login
!
end
```

18.4 Cisco discovery protocol

Cisco routers are a foundation part of the Internet. To aid the testing and configuration of routing systems, Cisco have developed a protocol called the Cisco Discovery Protocol (CDP). It operates above the data link layer, and below the upper network-layer protocols, which allows CDP devices (running Cisco IOS Release 10.3 or later) to intercommunicate with each other and automatically discover other neighboring Cisco devices. The great advantage of this protocol is that it can support any network and transport layer protocols, and not just TCP and IP, such as SPX/IPX and AppleTalk.

The `show cdp` command allows a user to view the configuration of other neighboring routers. Session run 18.12 shows an example `show cdp` from the Lab-A router. It displays the device ID (in this case, it shows that the neighbors are Lab-B and 003080718F40), the type of device (in this case, Lab-B is a router (R) and 003080719F0 is an Ethernet switch T/S), the port which it connects to (Lab-B connects via Ser 1 port, and 003080719F40 connects from the AUI port), the port which it connects from (Lab-B connects from Ser 0 port, and 003080719F40 connects from the Eth 1 port) and the type of devices it is (in this case, the router connects to a Cisco 2500-series and a Cisco Catalyst 1900-series switch).

Session run 18.12

```
LAB-A>   show cdp neighbors

Capability Codes: R - Router, T - Trans Bridge, B - Source Route Bridge
                  S - Switch, H - Host, I - IGMP, r - Repeater

Device ID       Local Intrfce     Holdtme    Capability  Platform  Port ID
Lab-B              Ser 0          141           R         2500     Ser 1
003080719F40       Eth 1          157          T S        1900     AUI
```

It is also possible to get detailed information on neighbors with the `show cdp neighbors detail` command, as outlined in Session run 18.13. There are default values for the time between CDP updates and for aging CDP entries. Typically these are set at 60 seconds and 180 seconds, respectively. Old updates are automatically deleted when new updates arrive. Holdtime defines the amount of time (in seconds) since a CDP frame arrived with the required information.

Session run 18.13

```
LAB-A>   show cdp neighbors detail
-------------------------
Device ID: Lab-B
Entry address(es):
  IP address: 201.100.11.2
Platform: cisco 2500,  Capabilities: Router
Interface: Serial0,  Port ID (outgoing port): Serial1
Holdtime : 171 sec

Version :
Cisco Internetwork Operating System Software
IOS (tm) 2500 Software (C2500-D-L), Version 12.0(4), RELEASE SOFTWARE (fc1)
Copyright (c) 1986-1999 by cisco Systems, Inc.
Compiled Wed 14-Apr-99 21:21 by ccai

-------------------------
Device ID: 003080719F40
Entry address(es):
  IP address: 205.7.5.254
Platform: cisco 1900,  Capabilities: Trans-Bridge Switch
Interface: Ethernet1,  Port ID (outgoing port): AUI
Holdtime : 127 sec

Version : V8.01
```

To check other neighbors, and their connections, the Telnet command is used to log into the neighboring devices. Session run 18.14 shows an example of login into Lab-B and the switch.

Session run 18.14

```
Lab-A>   telnet 201.100.11.2


Lab-B>   show cdp neighbors
Capability Codes: R - Router, T - Trans Bridge, B - Source Route Bridge
                  S - Switch, H - Host, I - IGMP, r - Repeater

Device ID      Local Intrfce     Holdtme    Capability  Platform   Port ID
LAB-C             Ser 0          148           R        2500       Ser 1
LAB-A             Ser 1          144           R        2500       Ser 0
Lab-B>  telnet 205.7.5.254
        Catalyst 1900 - IP Configuration
        Ethernet Address:  00-30-80-71-9F-40
    ----------------------- Settings -----------------------------------
     [I] IP address                                205.7.5.254
     [S] Subnet mask                               255.255.255.0
     [G] Default gateway                           205.7.5.1
     [M] IP address of DNS server 1                0.0.0.0
     [N] IP address of DNS server 2                0.0.0.0
     [D] Domain name                               dcs.napier.ac.uk
     [R] Use Routing Information Protocol          Enabled
    ----------------------- Actions ------------------------------------
     [P] Ping
     [C] Clear cached DNS entries
     [X] Exit to previous menu
Enter Selection:
```

18.5 Exercises

18.5.1 What is found in NVRAM:
(a) Operating system (IOS) (b) Configuration file
(c) Bootstrap program (d) Routing tables

18.5.2 What is found in flash memory:
(a) Operating system (IOS) (b) Configuration file
(c) Bootstrap program (d) Routing tables

18.5.3 What is found in ROM:
(a) Operating system (IOS) (b) Configuration file
(c) Bootstrap program (d) Routing tables

18.5.4 To upgrade the ROM, which is required:
(a) Change the ROM IC (b) Update flash memory
(c) Download IOS to ROM (d) Save to NVRAM

18.5.5 To upgrade the IOS, which is required:
(a) Change the ROM IC (b) Update flash memory
(c) Download IOS to ROM (d) Save to NVRAM

18.5.6 On a router, which command allows the user to get into the Privileged Exec mode:
(a) `su` (b) `privileged`
(c) `exec` (d) `enable`

18.5.7 On a router, which command can be used to determine the IP addresses of connected hosts:
(a) `show hosts` (b) `show ip`
(c) `show domain` (d) `show neighbors`

18.5.8 Which command shows the current running parameters for a router:
(a) `show running` (b) `show config`
(c) `show running-config` (d) `show startup-config`

18.5.9 Which command shows the parameters that will be used when the router is booted:
(a) `show running` (b) `show config`
(c) `show running-config` (d) `show startup-config`

18.5.10 On a router, which command can be used to determine if any of the ports are not enabled:
(a) `show no-startup` (b) `show startup-config`
(c) `show protocols` (d) `show ports`

18.5.11 Which is the correct order for a Cisco router startup:
(a) load bootstrap, load configuration file, test hardware and load operating system
(b) test hardware, load bootstrap, load configuration file and load operating system
(c) test hardware, load bootstrap, load operating system and load configuration file
(d) load operating system, load bootstrap, test hardware and load configuration file

18.6 Cisco router commands

The User Exec commands include:

Session run 18.15

```
access-enable       Create a temporary Access-List entry
access-profile      Apply user-profile to interface
clear               Reset functions
connect             Open a terminal connection
disable             Turn off privileged commands
disconnect          Disconnect an existing network connection
enable              Turn on privileged commands
exit                Exit from the EXEC
help                Description of the interactive help system
lock                Lock the terminal
login               Log in as a particular user
logout              Exit from the EXEC
mrinfo              Request neighbor and version information from a
                    multicast router
mstat               Show statistics after multiple multicast traceroutes
mtrace              Trace reverse multicast path from destination to source
name-connection     Name an existing network connection
pad                 Open a X.29 PAD connection
ping                Send echo messages
ppp                 Start IETF Point-to-Point Protocol (PPP)
resume              Resume an active network connection
```

The Privileged EXEC commands include:

Session run 18.16

```
access-enable       Create a temporary Access-List entry
access-profile      Apply user-profile to interface
access-template     Create a temporary Access-List entry
bfe                 For manual emergency modes setting
cd                  Change current directory
clear               Reset functions
clock               Manage the system clock
configure           Enter configuration mode
connect             Open a terminal connection
copy                Copy from one file to another
debug               Debugging functions (see also 'undebug')
delete              Delete a file
dir                 List files on a filesystem
disable             Turn off privileged commands
disconnect          Disconnect an existing network connection
enable              Turn on privileged commands
erase               Erase a filesystem
exit                Exit from the EXEC
help                Description of the interactive help system
lock                Lock the terminal
login               Log in as a particular user
logout              Exit from the EXEC
more                Display the contents of a file
mstat               Show statistics after multiple multicast traceroutes
mtrace              Trace reverse multicast path from destination to source
name-connection     Name an existing network connection
no                  Disable debugging functions
pad                 Open a X.29 PAD connection
ping                Send echo messages
ppp                 Start IETF Point-to-Point Protocol (PPP)
pwd                 Display current working directory
reload              Halt and perform a cold restart
resume              Resume an active network connection
rlogin              Open an rlogin connection
rsh                 Execute a remote command
```

```
send                Send a message to other tty lines
set                 Set system parameter (not config)
setup               Run the SETUP command facility
show                Show running system information
slip                Start Serial-line IP (SLIP)
systat              Display information about terminal lines
telnet              Open a telnet connection
terminal            Set terminal line parameters
test                Test subsystems, memory, and interfaces
traceroute          Trace route to destination
tunnel              Open a tunnel connection
undebug             Disable debugging functions (see also 'debug')
undelete            Undelete a file
verify              Verify a file
where               List active connections
write               Write running configuration to memory, network, or
                    terminal
```

Session run 18.17

```
LAB-A#   show ?
access-expression  List access expression
access-lists          List access lists
accounting        Accounting data for active sessions
aliases           Display alias commands
appletalk         AppleTalk information
arap              Show Appletalk Remote Access statistics
arp               ARP table
async             Information on terminal lines used as router interfaces
backup            Backup status
bridge            Bridge Forwarding/Filtering Database [ verbose]
buffers           Buffer pool statistics
cdp               CDP information
clock             Display the system clock
compress          Show compression statistics
configuration     Contents of Non-Volatile memory
controllers       Interface controller status
debugging         State of each debugging option
decnet            DECnet information
dhcp              Dynamic Host Configuration Protocol status
dialer            Dialer parameters and statistics
dnsix             Shows Dnsix/DMDP information
entry             Queued terminal entries
file              Show filesystem information
flash:            display information about flash: file system
flh-log           Flash Load Helper log buffer
frame-relay       Frame-Relay information
history           Display the session command history
hosts             IP domain-name, nameservers, and host table
interfaces        Interface status and configuration
ip                IP information
ipx               Novell IPX information
key               Key information
line              TTY line information
location          Display the system location
logging           Show the contents of logging buffers
memory            Memory statistics
modemcap          Show Modem Capabilities database
ntp               Network time protocol
ppp               PPP parameters and statistics
printers          Show LPD printer information
privilege         Show current privilege level
processes         Active process statistics
```

```
protocols        Active network routing protocols
queue            Show queue contents
queueing         Show queueing configuration
registry         Function registry information
reload           Scheduled reload information
rhosts           Remote-host+user equivalences
rif              RIF cache entries
rmon             rmon statistics
route-map        route-map information
running-config   Current operating configuration
sessions         Information about Telnet connections
smds             SMDS information
smf              Software MAC filter
smrp             Simple Multicast Routing Protocol (SMRP) information
snapshot         Snapshot parameters and statistics
snmp             snmp statistics
spanning-tree    Spanning tree topology
stacks           Process stack utilization
standby          Hot standby protocol information
startup-config   Contents of startup configuration
subscriber-policy  Subscriber policy
subsys           Show subsystem information
tacacs           Shows tacacs+ server statistics
tcp              Status of TCP connections
terminal         Display terminal configuration parameters
traffic-shape    traffic rate shaping configuration
users            Display information about terminal lines
version          System hardware and software status
whoami           Info on current tty line
x25              X.25 information
```

THE FIRST OF MANY

The world's first large electronic computer (1946), containing 19000 values was built at the University of Pennsylvania by John Eckert during World War II. It was called ENIAC (Electronic Numerical Integrator and Computer) and it ceased operation in 1957. By today's standards, it was a lumbering dinosaur and by the time it was dismantled it weighed over 30 tons and spread itself over 1500 square feet. Amazingly, it also consumed over 25 kW of electrical power (equivalent to the power of over 400, 60 W light bulbs), but could perform over 100000 calculations per second (which is reasonable, even by today's standards). Unfortunately, it was unreliable, and would only work for a few hours, on average, before a valve needed to be replaced.

19 Electronic Mail

19.1 Introduction

Electronic mail (e-mail) is one use of the Internet which, according to most businesses, improves productivity. Traditional methods of sending mail within an office environment are inefficient, as it normally requires an individual requesting a secretary to type the letter. This must then be proof-read and sent through the internal mail system, which is relatively slow and can be open to security breaches.

A faster method, and more secure method of sending information is to use electronic mail, where messages are sent almost in an instant. For example, a memo with 100 words will be sent in a fraction of a second. It is also simple to send to specific groups, various individuals, company-wide, and so on. Other types of data can also be sent with the mail message such as images, sound, and so on. It may also be possible to determine if a user has read the mail. The main advantages are:

- It is normally much cheaper than using the telephone (although, as time equates to money for most companies, this relates any savings or costs to a user's typing speed).
- Many different types of data can be transmitted, such as images, documents, speech, and so on.
- It is much faster than the postal service.
- Users can filter incoming e-mail easier than incoming telephone calls.
- It normally cuts out the need for work to be typed, edited and printed by a secretary.
- It reduces the burden on the mailroom.
- It is normally more secure than traditional methods.
- It is relatively easy to send to groups of people (traditionally, either a circulation list was required or a copy to everyone in the group was required).
- It is usually possible to determine whether the recipient has actually read the message (the electronic mail system sends back an acknowledgement).

The main disadvantages are:

- It stops people using the telephone.
- It cannot be used as a legal document.
- Electronic mail messages can be sent impulsively and may be later regretted (sending by traditional methods normally allows for a rethink). In extreme cases messages can be sent

E-mail standards:

- **RFC821.** SMTP. Simple Mail Transfer Protocol. Defines the transfer of an e-mail message from one system to another.
- **RFC822.** Defines the format of an e-mail message, with heading to define the sender and the recipient.
- **RFC1521/1522.** MIME. Multipurpose Internet Mail Extensions. Defines a mechanism for supporting multiple attachments and for differing content types.
- **RFC1939.** POP-3. Post Office Protocol – Version 3. Creation of a standard and simple mechanism to access e-mail messages on a mail server. As it is simple it allows for many different types of programs to download e-mail messages.

to the wrong person (typically when replying to an e-mail message, where a message is sent to the mailing list rather than the originator).

- It may be difficult to send to some remote sites. Many organizations have either no electronic mail or merely an intranet. Large companies are particularly wary of Internet connections and limit the amount of external traffic.
- Not everyone reads their electronic mail on a regular basis (although this is changing as more organizations adopt e-mail as the standard communications medium).

The main standards that relate to the protocols of e-mail transmission and reception are:

- Simple Mail Transfer Protocol (SMTP) – which is used with the TCP/IP protocol suite. It has traditionally been limited to the text-based electronic messages.
- Multipurpose Internet Mail Extension (MIME) – which allows the transmission and reception of mail that contains various types of data, such as speech, images and motion video. It is a newer standard than SMTP and uses much of its basic protocol.
- S/MIME (Secure MIME). RSA Data Security created S/MIME which supports encrypted e-mail transfers and digitally signed electronic mail.

19.2 Shared-file approach versus client/server approach

An e-mail system can use either a shared-file approach or a client/server approach. In a shared-file system the source mail client sends the mail message to the local post office. This post office then transfers control to a message transfer agent which then stores the message for a short time before sending it to the destination post office. The destination mail client periodically checks its own post office to determine if it has mail for it. This arrangement is often known as store and forward, and the process is illustrated in Figure 19.1. Most PC-based e-mail systems use this type of mechanism.

A client/server approach involves the source client setting up a real-time remote connection with the local post office, which then sets up a real-time connection with the destination, which in turn sets up a remote connection with the destination client. The message will thus arrive at the destination when all the connections are complete.

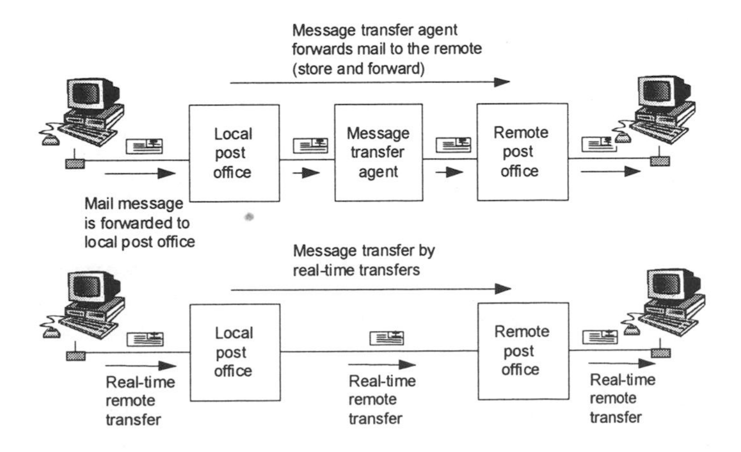

Figure 19.1 Shared-file versus client/server

19.3 Electronic mail overview

Figure 19.2 shows a typical e-mail architecture. It contains four main elements:

1. Post offices – where outgoing messages are temporally buffered (stored) before transmission and where incoming messages are stored. The post office runs the server software capable of routing messages (a message transfer agent) and maintaining the post office database.
2. Message transfer agents – for forwarding messages between post offices and to the destination clients. This software can either reside on the local post office or on a physically separate server.
3. Gateways – which provide part of the message transfer agent functionality. They translate between different e-mail systems, different e-mail addressing schemes and messaging protocols.
4. E-mail clients – normally the computer which connects to the post office. It contains three parts:

    - E-mail Application Program Interface (API), such as MAPI, VIM, MHS and CMC.
    - Messaging protocol. The main messaging protocols are SMTP or X.400. SMTP is defined in RFC 822 and RFC 821, whereas X.400 is an OSI-defined e-mail message delivery standard.
    - Network transport protocol, such as Ethernet, FDDI, and so on.

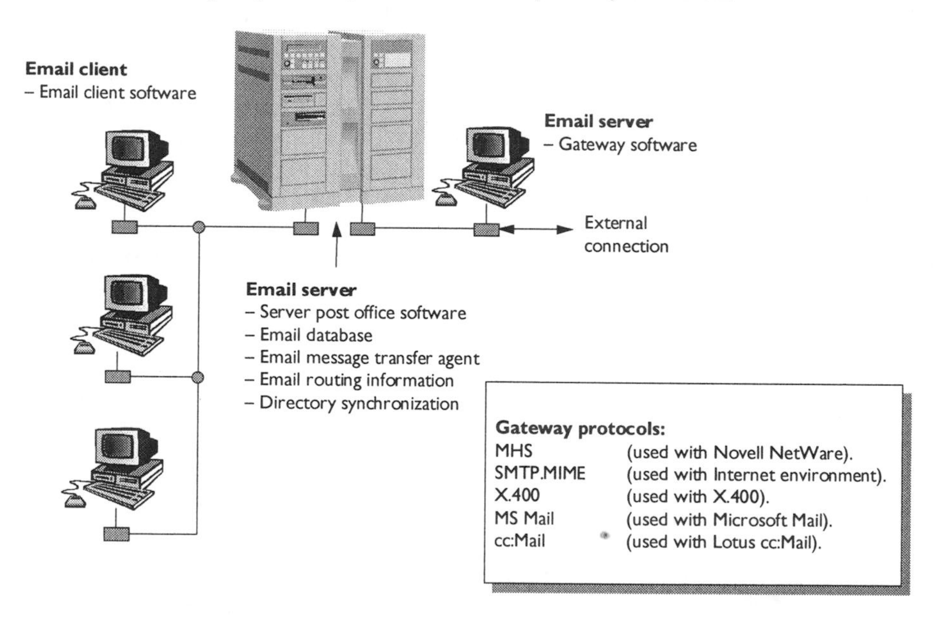

Figure 19.2 E-mail architecture

The main APIs are:

- MAPI (messaging API) – Microsoft part of Windows Operation Services Architecture.
- VIM (vendor-independent messaging) – Lotus, Apple, Novell and Borland derived e-mail API.
- MHS (message handling service) – Novell network interface which is often used as an e-mail gateway protocol.

- CMC (common mail call) – E-mail API associated with the X.400 native messaging protocol.

Gateways translate the e-mail message from one system to another, such as from Lotus cc:Mail to Microsoft Mail. Typical gateway protocols are:

- MHS (used with Novell NetWare).
- X.400 (used with X.400).
- cc:Mail (used with Lotus cc:Mail).
- SMTP.MIME (used with Internet environment).
- MS Mail (used with Microsoft Mail).

The Internet e-mail address is in the form of a name (such as `f.bloggs`), followed by an '`@`' and then the domain name (such as `anytown.ac.uk`). For example:

```
f.bloggs@anytown.ac.uk
```

19.4 SMTP

The IAB has defined the protocol SMTP in RFC821. This section discusses the protocol for transferring mail between hosts using the TCP/IP protocol. As SMTP is a transmission and reception protocol it does not actually define the format or contents of the transmitted message except that the data has 7-bit ASCII characters and that extra log information is added to the start of the delivered message to indicate the path the message took. The protocol itself is only concerned in reading the address header of the message.

19.4.1 SMTP operation

SMTP defines the conversation that takes place between an SMTP sender and an SMTP receiver. Its main functions are the transfer of messages and the provision of ancillary functions for mail destination verification and handling.

Initially the message is created by the user and a header is added which includes the recipient's e-mail address and other information. This message is then queued by the mail server, and when it has time, the mail server attempts to transmit it.

Each mail may have the following requirements:

- Each e-mail can have a list of destinations; the e-mail program makes copies of the messages and passes them onto the mail server.
- The user may maintain a mailing list, and the e-mail program must remove duplicates and replace mnemonic names with actual e-mail addresses.
- It allows for normal message provision, e.g. blind carbon copies (BCCs).

An SMTP mail server processes e-mail messages from an outgoing mail queue and then transmits them using one or more TCP connections with the destination. If the mail message is transmitted to the required host then the SMTP sender deletes the destination from the message's destination list. After all the destinations have been sent to, the sender then deletes the message from the queue.

If there are several recipients for a message on the same host, the SMTP protocol allows a single message to be sent to the specified recipients. Also, if there are several messages to be sent to a single host, the server can simply open a single TCP connection and all the messages can be transmitted in a single transfer (there is thus no need to set up a connection for each message).

SMTP also allows for efficient transfer with error messages. Typical errors include:

- Destination host is unreachable. A likely cause is that the destination host address is incorrect. For example, `f.bloggs@toy.ac.uk` might actually be `f.bloggs@toytown.ac.uk`.
- Destination host is out of operation. A likely cause is that the destination host has developed a fault or has been shut down.
- Mail recipient is not available on the host. Perhaps the recipient does not exist on that host, the recipient name is incorrect or the recipient has moved. For example, `fred.bloggs@toytown.ac.uk` might actually be `f.bloggs@toytown.ac.uk`. To overcome the problem of user names which are similar to a user's name then some systems allow for certain aliases for recipients, such as `f.bloggs`, `fred.bloggs` and `freddy.bloggs`, but there is a limit to the number of aliases that a user can have. If a user has moved then some systems allow for a redirection of the e-mail address. UNIX systems use the `.forward` file in the user's home directory for redirection. For example on a UNIX system, if the user has moved to `fred.bloggs@toytown.com` then this address is simply added to the `.forward` file.
- TCP connection failed on the transfer of the mail. A likely cause is that there was a timeout error on the connection (maybe due to the receiver or sender being busy or there was a fault in the connection).

SMTP senders have the responsibility for a message up to the point where the SMTP receiver indicates that the transfer is complete. This only indicates that the message has arrived at the SMTP receiver; it does not indicate that:

- The message has been delivered to the recipient's mailbox.
- The recipient has read the message.

Thus, SMTP does not guarantee to recover from lost messages and gives no end-to-end acknowledgement on successful receipt (normally this is achieved by an acknowledgement message being returned). Nor are error indications guaranteed. However, TCP connections are normally fairly reliable.

If an error occurs in reception, a message will normally be sent back to the sender to explain the problem. The user can then attempt to determine the problem with the message. SMTP receivers accept an arriving message and either place it in a user's mailbox or, if that user is located at another host, copies it to the local outgoing mail queue for forwarding. Most transmitted messages go from the sender's machine to the host over a single TCP connection. But sometimes the connection will be made over multiple TCP connections over multiple hosts. The sender specifying a route to the destination in the form of a sequence of servers can achieve this.

19.4.2 SMTP overview

An SMTP sender initiates a TCP connection. When this is successful it sends a series of commands to the receiver, and the receiver returns a single reply for each command. All commands and responses are sent with ASCII characters and are terminated with the carriage return (CR) and line feed (LF) characters (often known as CRLF).

Each command consists of a single line of text; beginning with a four-letter command code followed by in some cases an argument field. Most replies are a single line, although multiple-line replies are possible. Table 19.1 gives some sample commands.

Table 19.1 SMTP commands

Command	*Description*
HELO *domain*	Sends an identification of the domain
MAIL FROM: *sender-address*	Sends identification of the originator (sender-address)
RCPT FROM: *receiver-address*	Sends identification of the recipient (receiver-address)
DATA	Transfer text message
RSEY	Abort current mail transfer
QUIT	Shut down TCP connection
EXPN *mailing-list*	Send back membership of mailing list
SEND FROM: *sender-address*	Send mail message to the terminal
SOML FROM: *sender-address*	If possible, send mail message to the terminal, otherwise send to mailbox
VRFY username	Verify user name (username)

SMTP replies with a three-digit code and possibly other information. Some of the responses are listed in Table 19.2. The first digit gives the category of the reply, such as 2*xx* (a positive completion reply), 3*xx* (a positive intermediate reply), 4*xx* (a transient negative completion reply) and 5*xx* (a permanent negative completion reply). A positive reply indicates that the requested action has been accepted, and a negative reply indicates that the action was not accepted.

Positive completion reply indicates that the action has been successful, and a positive intermediate reply indicates that the action has been accepted but the receiver is waiting for some other action before it can give a positive completion reply. A transient negative completion reply indicates that there is a temporary error condition which can be cleared by other actions and a permanent negative completion reply indicates that the action was not accepted and no action was taken.

19.4.3 SMTP transfer

Figure 19.4 shows a successful e-mail transmission. For example if:

```
f.bloggs@toytown.ac.uk
```

is sending a message to:

```
a.person@place.ac.de
```

Then a possible sequence of events is:

- Set up TCP connection with receiver host.
- If the connection is successful, the receiver replies back with a 220 code (server ready). If it is unsuccessful, it returns back with a 421 code.
- Sender sends a `HELO` command to the hostname (such as `HELO toytown.ac.uk`).
- If the sender accepts the incoming mail message then the receiver returns a `250 OK` code. If it is unsuccessful then it returns a 421, 451, 452, 500, 501 or 552 code.
- Sender sends a `MAIL FROM: sender` command (such as `MAIL FROM: f.bloggs@toytown.ac.uk`).
- If the receiver accepts the incoming mail message from the sender then it returns a `250 OK` code. If it is unsuccessful then it returns codes such as 251, 450, 451, 452, 500, 501, 503, 550, 551, 552 or 553 code.
- Sender sends an `RCPT TO: receiver` command (such as `RCPT TO: a.person@place.ac.de`).

- If the receiver accepts the incoming mail message from the sender then it returns a `250 OK` code.
- Sender sends a `DATA` command.
- If the receiver accepts the incoming mail message from the sender then it returns a `354` code (start transmission of mail message).
- The sender then transmits the e-mail message.
- The end of the e-mail message is sent as two LF, CR characters.
- If the reception has been successful then the receiver sends back a `250 OK` code. If it is unsuccessful then it returns a 451, 452, 552 or 554 code.
- Sender starts the connection shutdown by sending a `QUIT` command.
- Finally the sender closes the TCP connection.

Table 19.2 SMTP responses

CMD	*Description*	*CMD*	*Description*
211	System status	500	Command unrecognized due to a syntax error
214	Help message	501	Invalid parameters or arguments
220	Service ready	502	Command not currently implemented
221	Service closing transmission channel	503	Bad sequence of commands
250	Request mail action completed successfully	504	Command parameter not currently implemented
251	Addressed user does not exist on system but will forward to receiver-address	550	Mailbox unavailable, request action not taken
354	Indicate to the sender that the mail message can now be sent. The end of the message is identified by two CR, LF characters	551	The addressed user is not local, please try receiver-address
421	Service is not available	552	Exceeded storage allocation, requested mail action aborted
450	Mailbox unavailable and the requested mail action was not taken	553	Mailbox name not allowed, requested action not taken
451	Local processing error, requested action aborted	554	Transaction failed
452	Insufficient storage, requested action not taken		

The following text shows some of the handshaking that is used in transmitting an electronic mail message (the UNIX mail –v command is used to show the handshaking. The commands sent to the mail server are highlighted in bold. It can be seen that the responses are 220 (Service ready), 221 (Request mail action completed successfully), 250 (Request mail action completed successfully) and 354 (Indicate to the sender that the mail message can now be sent).

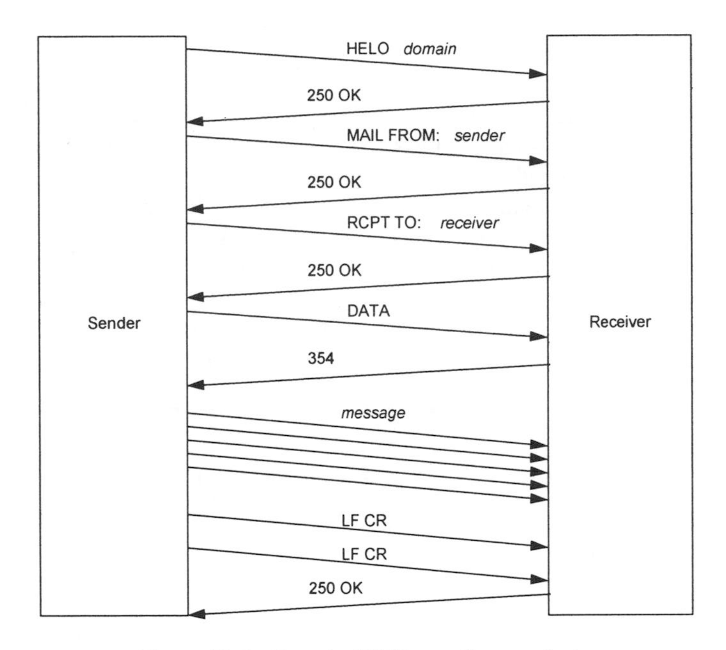

Figure 19.3 Sample SMTP e-mail transmission

```
> mail -v w.buchanan@napier.ac.uk
Subject: Test
This is a test message. Hello, how are you.
Fred.
EOT

w.buchanan@napier.ac.uk... Connecting to central.napier.ac.uk. (smtp)...
220 central.napier.ac.uk ESMTP Sendmail 8.9.1/8.9.1; Fri, 18 Dec 1998 15:55:45
GMT
>>> HELO www.eece.napier.ac.uk
250 central.napier.ac.uk Hello bill_b@www.eece.napier.ac.uk [ 146.176.151.139] ,
pleased to meet you
>>> MAIL From:<bill_b@www.eece.napier.ac.uk>
250 <bill_b@www.eece.napier.ac.uk>... Sender ok
>>> RCPT To:<w.buchanan@napier.ac.uk>
250 <w.buchanan@napier.ac.uk>... Recipient ok
>>> DATA
354 Enter mail, end with "." on a line by itself
>>> .
250 PAA24767 Message accepted for delivery
w.buchanan@napier.ac.uk... Sent (PAA24767 Message accepted for delivery)
Closing connection to central.napier.ac.uk.
>>> QUIT
221 central.napier.ac.uk closing connection
```

19.4.4 RFC 822

SMTP uses RFC 822, which defines the format of the transmitted message. RFC 822 contains two main parts:

- A header – which is basically the mail header and contains information for the successful transmission and delivery of a message. This typically contains the e-mail addresses for sender and receiver, the time the message was sent and received. Any computer involved in the transmission can added to the header.

- The contents.

Normally the e-mail-reading program will read the header and format the information to the screen to show the sender's e-mail address; it splits off the content of the message and displays it separately from the header.

An RFC 822 message contains a number of lines of text in the form of a memo (such as To:, From:, Bcc:, and so on). A header line usually has a keyword followed by a colon and then followed by keyword arguments. The specification also allows for a long line to be broken up into several lines.

Here is an RFC 822 message with the header shown in italics and the message body in bold. Table 19.3 explains some of the RFC 822 items in the header.

```
From FREDB@ACOMP.CO.UK Wed Jul  5 12:36:49 1995
Received: from ACOMP.CO.UK ([154.220.12.27]) by central.napier.ac.uk
(8.6.10/8.6.10) with SMTP id MAA16064 for <w.buchanan@central.napier.ac.uk>;

Wed, 5 Jul 1995 12:36:43 +0100

Received: from WPOAWUK-Message_Server by ACOMP.CO.UK
      with Novell_GroupWise; Wed, 05 Jul 1995 12:35:51 +0000

Message-Id: <sffa8725.082@ACOMP.CO.UK >

X-Mailer: Novell GroupWise 4.1

Date: Wed, 05 Jul 1995 12:35:07 +0000

From: Fred Bloggs <FREDB@ACOMP.CO.UK>

To: w.buchanan@central.napier.ac.uk
Subject:  Technical Question
Status: REO
Dear Bill
      I have a big problem. Please help.
Fred
```

Table 19.3 Header line descriptions

Header line	*Description*
From FREDB@ACOMP.CO.UK Wed Jul 5 12:36:49 1995	Sender of the e-mail is FREDB@ ACOM.CO.UK
Received: from ACOMP.CO.UK ([154.220.12.27]) by central.napier.ac.uk (8.6.10/8.6.10) with SMTP id MAA16064 for <w.buchanan@central.napier.ac.uk>; Wed, 5 Jul 1995 12:36:43 +0100	It was received by CENTRAL.NAPIER.AC.UK at 12:36 on 5 July 1995
Message-Id: <sffa8725.082@ACOMP.CO.UK >	Unique message ID
X-Mailer: Novell GroupWise 4.1	Gateway system
Date: Wed, 05 Jul 1995 12:35:07 +0000	Date of original message
From: Fred Bloggs <FREDB@ACOMP.CO.UK>	Sender's e-mail address and full name
To: w.buchanan@central.napier.ac.uk	Recipient's e-mail address
Subject: Technical Question	Mail subject

19.5 MIME

SMTP suffers from several drawbacks, such as:

- SMTP can only transmit ASCII characters and thus cannot transmit executable files or other binary objects.
- SMTP does not allow the attachment of files, such as images and audio.
- SMTP can only transmit 7-bit ASCII character thus it does not support an extended ASCII character set.

A new standard, Multipurpose Internet Mail Extension (MIME), has been defined for this purpose, which is compatible with existing RFC 822 implementations. It is defined in the specifications RFC 1521 and 1522. Its enhancements include the following:

- Five new message header fields in the RFC 822 header, which provide extra information about the body of the message.
- Use of various content formats to support multimedia electronic mail.
- Defined transfer encodings for transforming attached files.

The five new header fields defined in MIME are:

- MIME-version – a message that conforms to RFC 1521 or 1522 is MIME-version 1.0.
- Content-type – this field defines the type of data attached.
- Content-transfer-encoding – this field indicates the type of transformation necessary to represent the body in a format which can be transmitted as a message.
- Content-id – this field is used to uniquely identify MIME multiple attachments in the e-mail message.
- Content-description – this field is a plain-text description of the object with the body. It can be used by the user to determine the data type.

These fields can appear in a normal RFC 822 header. Figure 19.7 shows an example e-mail message. It can be seen that the API has split the message into two parts: the message part and the RFC 822 part. The RFC 822 part is shown in Figure 19.7. It can be seen that, in this case, the extra MIME messages are:

```
From: "Bill Buchanan" <w.buchanan@napier.ac.uk>
To: <f.bloggs@napier.ac.uk>
Subject: ECBS 2000 Referee Database
Date: Wed, 24 Nov 1999 00:09:22 -0000
MIME-Version: 1.0
```

This defines it as MIME Version 1.0; the data that the e-mail was sent, the person who sent the mail message, and so on.

RFC 822 example file listing (refer to Figure 19.8)

```
Received: from central.napier.ac.uk (146.176.2.3) by csumail1.napier.ac.uk
with SMTP  (IMA Internet Exchange 2.12 Enterprise) id 00232062; Wed, 24 Nov 99
04:35:35 +0000
Received: from mail6.svr.pol.co.uk (mail6.svr.pol.co.uk [195.92.193.212])
      by central.napier.ac.uk (8.9.1/8.9.1) with ESMTP id EAA03895;
      Wed, 24 Nov 1999 04:31:40 GMT
```

```
Received:from modem-111.doxycycline.dialup.pol.co.uk
([ 62.136.63.239] helo=bills)
      by mail6.svr.pol.co.uk with smtp (Exim 3.03 #0)
      id 11qQ0t-00069D-00; Wed, 24 Nov 1999 00:10:25 +0000
Reply-To: <w.buchanan@napier.ac.uk>
From: "Bill Buchanan" <w.buchanan@napier.ac.uk>
To: <w.buchanan@napier.ac.uk>
Subject: ECBS 2000 Referee Database
Date: Wed, 24 Nov 1999 00:09:22 -0000
Message-ID: <NDBBJGAHLDCDDONHKJEOMEBFCIAA.w.buchanan@napier.ac.uk>
MIME-Version: 1.0
Content-Type: multipart/mixed;   boundary="----
```

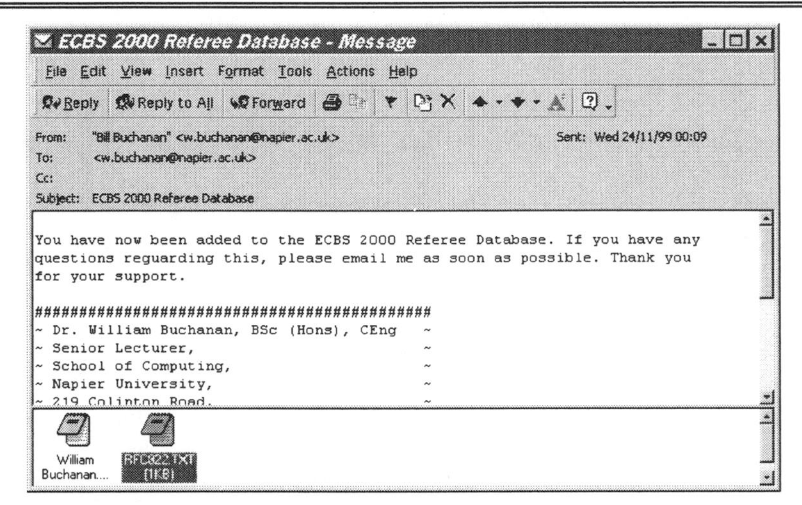

Figure 19.4 Sample e-mail message showing message and RFC822 part

```
Rfc822.txt - Notepad
File  Edit  Search  Help
Received: from central.napier.ac.uk (146.176.2.3) by csumail1.napier.ac.u
SMTP
  (IMA Internet Exchange 2.12 Enterprise) id 00232062; Wed, 24 Nov 99 04:
+0000
Received: from mail6.svr.pol.co.uk (mail6.svr.pol.co.uk [195.92.193.212])
        by central.napier.ac.uk (8.9.1/8.9.1) with ESMTP id EAA03895;
        Wed, 24 Nov 1999 04:31:40 GMT
Received: from modem-111.doxycycline.dialup.pol.co.uk ([62.136.63.239]
helo=bills)
        by mail6.svr.pol.co.uk with smtp (Exim 3.03 #0)
        id 11qQ0t-00069D-00; Wed, 24 Nov 1999 00:10:25 +0000
Reply-To: <w.buchanan@napier.ac.uk>
From: "Bill Buchanan" <w.buchanan@napier.ac.uk>
To: <w.buchanan@napier.ac.uk>
Subject: ECBS 2000 Referee Database
Date: Wed, 24 Nov 1999 00:09:22 -0000
Message-ID: <NDBBJGAHLDCDDONHKJEOMEBFCIAA.w.buchanan@napier.ac.uk>
MIME-Version: 1.0
Content-Type: multipart/mixed;
        boundary="----=_NextPart_000_0009_01BF3610.28FFE9A0"
```

Figure 19.5 RFC 822 part

19.5.1 MIME content types

Content types define the format of the attached files. There are a total of 16 different content types in seven major content groups. If the text body is pure text then no special transformation is required. RFC 1521 defines only one subtype, text/plain; this gives a standard ASCII character set.

A MIME-encoded e-mail can contain multiple attachments. The content-type header field includes a boundary which defines the delimiter between multiple attachments. A boundary always starts on a new line and has the format:

```
-- boundary name
```

The final boundary is:

```
-- boundary name --
```

For example, the following message contains two parts:

Example MIME file with 2 parts

```
From: Dr William Buchanan <w.buchanan@napier.ac.uk>
MIME-Version: 1.0
To: w.buchanan@napier.ac.uk
Subject: Any subject
Content-Type: multipart/mixed; boundary="boundary name"
This part of the message will be ignored.
-- boundary name
Content-Type: multipart/mixed; boundary="boundary name"
This is the first mail message part.
-- boundary name
And this is the second mail message part.
-- boundary name --
```

Table 19.4 MIME content-types

Content type	*Description*
text/plain	Unformatted text, such as ASCII
text/richtext	Rich text format which is similar to HTML
multipart/mixed	Each attachment is independent from the rest and all should be presented to the user in their initial ordering
multipart/parallel	Each attachment is independent from the others but the order is unimportant
multipart/alternative	Each attachment is a different version of the original data
multipart/digest	This is similar to multipart/mixed but each part is message/rfc822
message/rfc822	Contains the RFC 822 text
message/partial	Used in fragmented mail messages
message/external-body	Used to define a pointer to an external object (such as an ftp link)
image/jpeg	Defines a JPEG image using JFIF file format
image/gif	Defines GIF image
video/mpeg	Defines MPEG format
audio/basic	Defines 8-bit μ-Law encoding at 8 kHz sampling rate
application/postscript	Defines postscript format
application/octet-stream	Defines binary format which consists of 8-bit bytes

The part of the message after the initial header and before the first boundary can be used to add a comment. This is typically used to inform users that do not have a MIME-compatible program about the method used to encode the received file. A typical method for converting binary data into ASCII characters is to use the programs UUENCODE (to encode a binary file into text) or UUDECODE (to decode a uuencoded file).

The four subtypes of multipart type can be used to sequence the attachments; the main subtypes are:

- multipart/mixed subtype – which is used when attachments are independent but need to be arranged in a particular order.
- multipart/parallel subtype – which is used when the attachments should be present at the same time; a typical example is to present an animated file along with an audio attachment.
- multipart/alternative subtype – which is used to represent an attachment in a number of different formats.

19.5.2 Example MIME

The following file listing shows the message part of a MIME-encoded e-mail message (i.e. it excludes the RFC 822 header part). It can be seen that the sending e-mail system has added the comment about the MIME encoding. In this case the MIME boundaries have been defined by:

```
-- IMA.Boundary.760275638
```

Example MIME file

```
This is a Mime message, which your current mail reader
may not understand. Parts of the message will appear as
text. To process the remainder, you will need to use a Mime
compatible mail reader. Contact your vendor for details.

--IMA.Boundary.760275638

Content-Type: text/plain; charset=US-ASCII
Content-Transfer-Encoding: 7bit
Content-Description: cc:Mail note part

This is the original message .....

--IMA.Boundary.760275638--
```

19.5.3 Mail fragments

A mail message can be fragmented using the content-type field of message/partial and then reassembled back at the source. The standard format is:

```
Content-type: message/partial;
   id="idname"; number=x; total=y
```

where *idname* is the message identification (such as `xyz@hostname`, *x* is the number of the fragment out of a total of *y* fragments. For example, if a message had three fragments, they could be sent as:

Example MIME file with 3 fragments (first part)

```
From: Fred Bloggs <f.bloggs@toytown.ac.uk>
MIME-Version: 1.0
To: a.body@anytown.ac.uk
Subject: Any subject
Content-Type: message/partial;
      id="xyz@toytown.ac.uk"; number=1; total=3
Content=type: video/mpeg

First part of MPEG file
```

Example MIME file with 3 fragments (second part)

```
From: Fred Bloggs <f.bloggs@toytown.ac.uk>
MIME-Version: 1.0
To: a.body@anytown.ac.uk
Subject: Any subject
Content-Type: message/partial;
      id="xyz@toytown.ac.uk"; number=2; total=3
Content=type: video/mpeg

Second part of MPEG file
```

Example MIME file with 3 fragments (third part)

```
From: Fred Bloggs <f.bloggs@toytown.ac.uk>
MIME-Version: 1.0
To: a.body@anytown.ac.uk
Subject: Any subject
Content-Type: message/partial;
      id="xyz@toytown.ac.uk"; number=3; total=3
Content=type: video/mpeg

Third part of MPEG file
```

19.5.4 Transfer encodings

MIME allows for different transfer encodings within the message body:

- 7bit – no encoding, and all of the characters are 7-bit ASCII characters.
- 8bit – no encoding, and extended 8-bit ASCII characters are used.
- quoted-printable – encodes the data so that non-printing ASCII characters (such as line feeds and carriage returns) are displayed in a readable form.
- base64 – encodes by mapping 6-bit blocks of input to 8-bit blocks of output, all of which are printable ASCII characters.

When the transfer encoding is:

```
Content-transfer-encoding: quoted-printable
```

then the message has been encoded so that all non-printing characters have been converted to printable characters. A typical transform is to insert =*xx* where *xx* is the hexadecimal equivalent for the ASCII character. A form feed (FF) would be encoded with '=0C',

A transfer encoding of base64 is used to map 6-bit characters to a printable character. It is a useful method in disguising text in an encrypted form and also for converting binary data into a text format. It takes the input bitstream and reads it six bits at a time, then maps this to an 8-bit printable character. Table 19.5 shows the mapping.

Thus if a binary file had the bit sequence:

```
10100010101010001010101010
```

It would first be split into groups of 6 bits, as follows:

```
101000     101010  100010  101010  000000
```

This would be converted into the ASCII sequence:

```
YsSqA
```

which is in a transmittable form.

Thus the 7-bit ASCII sequence 'FRED' would use the bit pattern:

```
1000110 1010010 1000101 1000100
```

which would be split into groups of 6 bits as:

```
100011 010100 101000 101100 010000
```

which would be encoded as:

```
jUosQ
```

Table 19.5 MIME base64 encoding

Bit value	*Encoded character*	*Bit value*	*Encoded character*	*Bit value*	*Encoded character*	*Bit value*	*Encoded character*
0	A	16	Q	32	g	48	w
1	B	17	R	33	h	49	x
2	C	18	S	34	i	50	y
3	D	19	T	35	j	51	z
4	E	20	U	36	k	52	0
5	F	21	V	37	l	53	1
6	G	22	W	38	m	54	2
7	H	23	X	39	n	55	3
8	I	24	Y	40	o	56	4
9	J	25	Z	41	p	57	5
10	K	26	a	42	q	58	6
11	L	27	b	43	r	59	7
12	M	28	c	44	s	60	8
13	N	29	d	45	t	61	9
14	O	30	e	46	u	62	+
15	P	31	f	47	v	63	/

19.5.5 Example

The following parts of the RFC 822 messages.

(a)

```
Received: from publish.co.uk by ccmail1.publish.co.uk (SMTPLINK V2.11.01)
Return-Path: <FredB@local.exnet.com>
Received: from mailgate.exnet.com ([ 204.137.193.226] ) by zeus.publish.co.uk
with SMTP id <17025>; Wed, 2 Jul 1997 08:33:29 +0100
Received: from exnet.com (assam.exnet.com) by mailgate.exnet.com with SMTP id
AA09732 (5.67a/IDA-1.4.4 for m.smith@publish.co.uk); Wed, 2 Jul 1997 08:34:22
+0100
Received: from maildrop.exnet.com (ceylon.exnet.com) by exnet.com with SMTP id
AA10740 (5.67a/IDA-1.4.4 for <m.smith@publish.co.uk>); Wed,. 2 Jul 1997
08:34:10 +0100
Received: from local.exnet.com by maildrop.exnet.com (4.1/client-1.2DHD)
    id AA22007; Wed, 2 Jul 97 08:25:21 BST
From: FredB@local.exnet.com (Arthur Chapman)
Reply-To: FredB@local.exnet.com
To: b.smith@publish.co.uk
Subject: New proposal
Date: Wed, 2 Jul 1997 09:36:17 +0100
Message-Id: <66322430.1380704@local.exnet.com>
Organization: Local College
```

(b)

```
Received: from central.napier.ac.uk by ccmailgate.napier.ac.uk (SMTPLINK
V2.11.01) Return-Path: <fred@singnetw.com.sg>
Received: from server.singnetw.com.sg (server.singnetw.com.sg [165.21.1.15])
by central.napier.ac.uk (8.6.10/8.6.10) with ESMTP id DAA18783 for
<w.buchanan@napier.ac.uk>; Sun, 29 Jun 1997 03:15:27 GMT
Received: from si7410352.ntu.ac.sg (ts900-1908.singnet.com.sg [165.21.158.60])
    by melati.singnet.com.sg (8.8.5/8.8.5) with SMTP id KAA08773
    for <w.buchanan@napier.ac.uk.>; Sun, 29 Jun 1997 10:14:59 +0800 (SST)
Message-ID: <33B5C33B.6CCC@singnetw.com.sg>
Date: Sun, 29 Jun 1997 10:06:51 +0800
From: Fred Smith <fred@singnetw.com.sg>
X-Mailer: Mozilla 2.0 (Win95; I)
MIME-Version: 1.0
To: w.buchanan@napier.ac.uk
Subject: Chapter 15
Content-Type: text/plain; charset=us-ascii
Content-Transfer-Encoding: 7bit
```

(c)

```
Received: from central.napier.ac.uk by ccmailgate.napier.ac.uk (SMTPLINK
V2.11.01)
Return-Path: <bertb@scms.scotuni.ac.uk>
Received: from master.scms.scotuni.ac.uk ([193.62.32.5]) by
central.napier.ac.uk (8.6.10/8.6.10) with ESMTP id MAA20373 for
<w.buchanan@napier.ac.uk>; Tue, 1 Jul 1997 12:25:38 GMT
Received: from cerberus.scms.scotuni.ac.uk (cerberus.scms.scotuni.ac.uk
[193.62.32.46]) by master.scms.scotuni.ac.uk (8.6.9/8.6.9) with ESMTP id
MAA10056 for <w.buchanan@napier.ac.uk>; Tue, 1 Jul 1997 12:24:32 +0100
From: David Davidson <bertb@scms.scotuni.ac.uk>
Received: by cerberus.scms.scotuni.ac.uk (SMI-8.6/Dumb)
    id MAA03334; Tue, 1 Jul 1997 12:23:17 +0100
Date: Tue, 1 Jul 1997 12:23:17 +0100
Message-Id: <199707011123.MAA03334@cerberus.scms.scotuni.ac.uk>
To: w.buchanan@napier.ac.uk
Subject: Advert
Mime-Version: 1.0
Content-Type: text/plain; charset=us-ascii
Content-Transfer-Encoding: 7bit
Content-MD5: TzKyk+NON+vy6Cm6uqy9Cg==
```

19.6 Post Office Protocol (POP)

The Post Office Protocol was first defined in RFC918, but has since been replaced with POP-3, which is defined in RFC1939. The objective of POP is to create a standard method for users to access a mail server. E-mail messages are uploaded onto a mail server using SMTP, and then downloaded using POP. With POP the server listens for a connection, and when this occurs the server sends a greeting message, and waits for commands. The standard port reserved for POP transactions is 110. Like SMTP, it consists of case-insensitive commands with one or more arguments, followed by the Carriage Return (CR) and Line Feed (LF) characters, typically represented by CRLF. These keywords are either three or four characters long.

The client opens the connection by sending a USER and a PASS command, for the user name and password, respectively. If successful this will give access to the mailbox on the server. The client can then read the messages with the following commands:

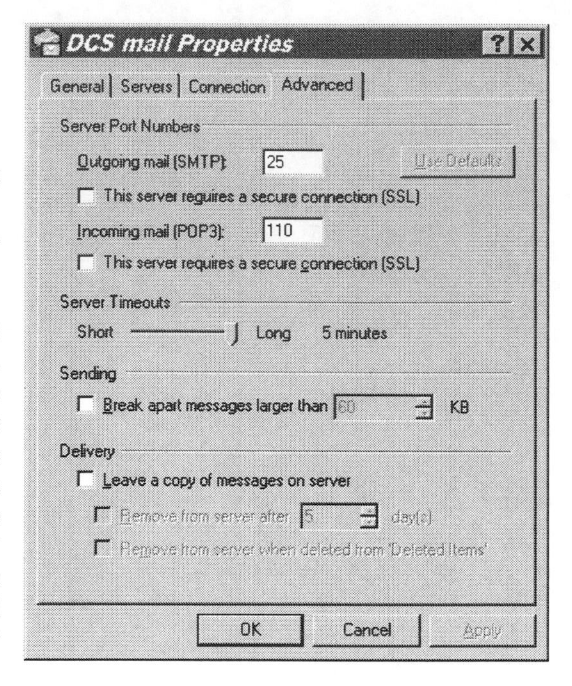

- RDEL. Reads and deletes all the messages from the mailbox.
- RETR. Reads the messages from the mailbox, and keeps them on the server.

When transferring, the e-mail server locks the mailbox, and waits for the client to respond. The client then uses the RECV command to receive data from the mailbox. When complete the client sends a RCVD command. If the RDEL was initially sent, the server will delete the messages from the mailbox, otherwise the messages will stay on the server. A QUIT command terminates the session. After each command the e-mail server must respond back. The commands and responses for POP can be summarized by:

Command	*Description*	*Possible responses*
USER *name*	Defines the name of the user.	"+OK", "-ERR"
PASS *password*	Defines the password for the user	"+OK", "-ERR"
RETR *mailbox*	Begins a mail reading transaction, but does not delete the messages once they have been transferred.	"+*val*", "-ERR"
RDEL *mailbox*	Begins a mail reading transaction, and deletes the messages once they have been transferred. The messages are not deleted until a RCEV command.	"+*val*", "-ERR"
RVEC	Acknowledges the reception of the mail messages.	"+OK", or aborted connection
RCVD	Confirms that client has received the mail messages.	"+OK", "-ERR"
QUIT	Client wishes to end the session.	"+OK" then close
NOOP	No operation but prompts the mail server for an OK response.	"+OK"
RSET	Sent by the client to inform the server to abort the current transaction.	"+OK"

An important objective of POP is that messages are normally downloaded from the e-mail server, and then deleted. POP is a simple protocol, and allows for many different programs to access e-mail servers, without having to implement a complex protocol. Typically WWW browsers can be setup to access the POP-3 server.

19.7 Exercises

19.7.1 The main drawback of SMTP is:

(a) It is incompatible with most e-mail systems
(b) It can only be used for text-based e-mail
(c) It is slow when transferring an e-mail
(d) It is only used on UNIX systems

19.7.2 The advantage that MIME has over SMTP is:
(a) That it allows the attachment of other data types (such as speech and images).
(b) It is faster when transferring
(c) It is compatible with more systems
(d) It is only used on UNIX systems

19.7.3 The share-file (store and forward) e-mail approach involves:
(a) Sending the mail message to multiple sites
(b) Sending the mail message in fragments
(c) Using real-time remote transfer
(d) Using a message transfer agent

19.7.4 The client/server e-mail approach involves:
(a) Sending the mail message to multiple sites
(b) Sending the mail message in fragments
(c) Using real-time remote transfer
(d) Using a message transfer agent

19.7.5 The main function of an e-mail post office is to:
(a) Forward messages to clients
(b) Store incoming messages and temporarily store outgoing messages
(c) Translate messages between different systems
(d) Provide the user interface program

19.7.6 The main function of an e-mail gateway is to:
(a) Forward messages to clients
(b) Store incoming messages and temporarily store outgoing messages
(c) Translate messages between different systems
(d) Provide the user interface program

19.7.7 The main function of an e-mail message transfer agent is to:
(a) Forward messages to clients
(b) Store incoming messages and temporarily store outgoing messages
(c) Translate messages between different systems
(d) Provide the user interface program

19.7.8 The standard text format used for e-mail messages is:
(a) ISO characters (b) ANSI characters
(c) EBCDIC characters (d) 7-bit ASCII characters

19.7.9 The SMTP transmission protocol has been defined in which IAB standard:
(a) RFC 822 (b) RFC 821
(c) IEEE 802.2 (d) RFC 802

19.7.10 The SMTP message format has been defined in which IAB standard:
(a) RFC 822 (b) RFC 821
(c) IEEE 802.2 (d) RFC 802

19.7.11 How is routing information added to an SMTP message:
(a) It is encoded with special codes
(b) It is added to the end of the message
(c) It is sent as a separate file
(d) It is added to the header of the message

19.7.12 The likely reason that the X.400 standard has never been universally accepted is:

(a) It is too complex (b) It is too simple
(c) It is difficult to incorporate into e-mail systems
(d) It is not an international standard

19.7.13 The main disadvantage of SMTP is:
(a) It is not suited to client/server applications
(b) It does not support file attachments
(c) It is too slow
(d) It is incompatible with many systems

19.7.14 Which TCP port does SMTP normally use:
(a) 21 (b) 23
(c) 25 (d) 110

19.7.15 Which TCP port does POP-3 normally use:
(a) 21 (b) 23
(c) 25 (d) 110

19.7.16 Which is the following is normally TRUE:
(a) E-mail messages are sent using SMTP and received using POP-3
(b) E-mail messages are sent using POP-3 and received using SMTP
(c) E-mail messages are sent and received using POP-3
(d) The POP-3 protocol is incompatible with many systems

19.7.17 How are attachments delimited in MIME encoded messages:
(a) `++ boundary name` (b) `-- boundary name`
(c) `<< boundary name` (d) `>> boundary name`

19.7.18 A typical format used to convert binary files into ASCII text is:
(a) LHA (b) PKZIP
(c) UUENCODE (d) ZIP

19.7.19 How is the type of attachment defined:
(a) It is defined in the address field
(b) It is sent as separate message
(c) Automatically with the type of encoding
(d) With the content type

19.7.20 Identify the main functional differences between SMTP and MIME.

19.7.21 Contrast shared-file and client/server approach for electronic mail.

19.7.22 Give an example set of messages between the sender and a recipient for a successful SMTP transfer.

19.7.23 Give an example set of messages between the sender and a recipient for an unsuccessful SMTP transfer.

19.7.24 If you have access to e-mail read an e-mail and identify each part of the header.

19.7.25 Explain how base64 encoding can be used to attach binary information.

19.7.26 Encode the following bit stream with base64 encoding:

(i) `0111000000000101011010100000001111110111110`
(ii) `1111100011101101011100100111100110101111111`

19.8 Note from the Author

Electronic mail is my favorite application of the Internet. My favorite question to a new set of students used to be: 'What's the difference between the World Wide Web and the Internet?' It shows the ignorance of the media that very few students could even differentiate the two. By now you should realize that the World Wide Web (the Web, Cyber Space, and so on) and the Internet are two different things. The World Wide Web is really just one of the applications of the Internet. Others include remote access, remote diagnostics, and, of course, electronic mail. For most the best application of the Internet is electronic mail. It has really enhanced business communications, and, like the telephone, has changed the way people operate.

One of the most amazing things about electronic mail is how quickly it was adopted as a world standard, and how well it has changed from being the domain of computer specialists to become usable by virtually every person who owns, or uses a computer. The keys to this success have been the adoption of TCP/IP as a standard transport mechanism, and, of course, the adoption of the RFC821 and RFC822 standards. These two simple protocols have allowed for electronic e-mail to be transferred between different systems, and different e-mail clients. Without them we may have been forced to adopt industry-driven standards. But, as we know, the Internet is the greatest open system, ever. It thrives on its openness, and worldwide standards. No one owns the Internet, and no one ever will. If anyone tries to dominate, they will be immediately shouted down, either by governments or by the users of the Internet.

Given that RFC821 and RFC822 were based on text-type messages, no one would have really expected that they would eventually support file attachments, and, even graphics and binary programs. MIME has done this successfully, but it has done it in the way that still makes the e-mail compatible with all previous standards. For a while, electronic mails became a bit messy, as not all e-mail systems were quite compatible with the new MIME standards, but with new versions the integration is almost seamless. Now, e-mail messages can be sent as HTML documents, with hypertext links, colored backgrounds, and colored text. To be totally compatible it is often best to send text messages in text format, as you can never guarantee that the recipient is fully compatible with the message format that you are sending. POP-3 has since enhanced the reception of e-mail, as it supports a simple message transfer system which allows many different types of programs to access POP-3 initialized servers. Typically WWW browsers are now used to download e-mail messages.

Electronic mail requires a different culture than traditional methods of communications. The key difference is that there is no immediate feedback (as there would be with a telephone call), and that messages can be sent within a relatively short time. These two things cause considerable problems, and most people have sent e-mails which have either been construed the wrong way or have sent messages which they quickly regret ('acting-on-the-spot'). So, as words of advice, try not to send e-mails without first thinking about their consequences (typically, allow yourself a cooling-off period before you blast users with an e-mail that you may later regret), and carefully read what you have written, so that it cannot be construed in the wrong way.

One way to lighten-up an e-mail, and to show that you are not being too serious is to use smilies, some of which are given next. For example, someone sending a lighthearted e-mail about not getting a job might say:

```
Well, thanks for your advice, but it didn't help because I didn't get
the job. So I don't think I'll be asking you again.

-- Fred Smith.
```

Might get construed as abusive, as the person sounds as if they are really saying that they do not want any more advice from the person. A smilie, or two, can help to lighten the tone of the e-mail, and show that it should not be taken too seriously:

```
Well, thanks for your advice, but it didn't help because I didn't get
the job (:-<). So I don't think I'll be asking you again ;-).

-- Fred (:->)
```

The Internet has also brought Netiquette, which is a whole new language. Users of chat programs often use acronyms for commonly used words. When I first started using the Internet I was lost for a while, as users kept saying things like LOL and ROFL. I didn't know that they actually thought that I was saying something funny. I thought that they were offering me a lolly (in the UK, a kind of Popsicle), and asking about my roof. Soon I realized that BTW was By The Way, and not some new fast food meal. Well, here goes:

```
IMHO, BTW, IRL I LOL when I C Some1 FTF . GTG. EOD. HAND

-- WB.
```

One of the great advantages of e-mail, is that people who contact me get my name right. A good old Scottish name like Buchanan (pro. Bu-cannon), can come out in many different ways over the telephone, my Top 5 are:

1. *Butch-an-an-a (I don't know where the last -a comes from, but I think it just comes out)*
2. *Butch-an-in.*
3. *Buk-hana-an.*
4. *Butch-anon.*
5. *Buchan-on.*

I must admit I love e-mail, because it is really the art of the written word. Many people though, abuse the great gift of e-mail, which is a real shame. The four top abusers are:

- ***Mr or Mrs Anonymous.** These are people who either generate anonymous e-mail accounts and then use these to send abusive e-mails to someone or are people who send e-mails from their own account but use another person's e-mail address as the sender. It is relatively simple to set up the sending e-mail address of a user to anything that a user wants (and pretend to be anyone that they want). Governments of the worlds are quickly realizing that this type of activity is a serious crime, and laws will soon be put in place to try and reduce the number of people who do this. I have seen several abuses of the e-mail system with this method. Luckily all of them were traced, as the users could be traced from their IP address, and the time that they sent the e-mail. So remember, systems keep trace of when users log in, and their IP address. Once an administrator has this information it is relatively easy to catch an abuser. There are many clues in the header of the e-mail address which can trace the original sender of an e-mail message, as the header contains the original server, as well as times and dates of transfers. Several e-mail systems now detect that a user is using a different e-mail address. A level of security is also provided by some ISPs who will only allow an e-mail message to be sent from a specific telephone number.*
- ***Spammers R Us.** Spammers are everywhere; they either do it unintentionally or intentionally. The worst type is the one which intentionally sends out e-mails to many users at a time, without first checking to see if they really want the information.*

- ***Mr and Mrs IdiotUser.*** *These two people tend to be the type that reply back to an e-mail message, and rather than sending to the person who sent the e-mail, they send it to everyone on the circulation list. This can be very embarrassing, and, for the recipients, annoying.*
- ***Mr LackofManners.*** *This tends to the type of user who writes e-mail as he speaks. I've seen lots of examples of this, and a few tips that I have are: always refer to someone using either their first name, or their official title; always reply with a courteous response (even although you may be fuming and wish to punch them in the face). Words such as: 'Thank you for your e-mail', 'I respect your option', 'I agree with you to a certain extent', are much better than: 'I've just wasted two hours of my precious life reading, and replying to your abusive e-mail', 'You're the next highest life form to a slug', 'I wanted to send you a more fulsome response, but your firewall would probably not allow through some of the words that I would like to say'.*

E-mail is still a relatively new technology, and unlike many technological advances, it is actually changing the way we work. Thus it will take a while to truly evolve, and like TCP/IP it is one of the true liberators in the world, as it allows the people of the world to intercommunicate, and share ideas. In the past governments have build virtual walls around their country in order to control the information that flows into and out of a country. TCP and IP are, of course, open to anyone reading the messages contained in their communications, but with encryption, not even a space alien could read the message (unless, of course, it was destined for them!).

Smilie	*Description*	*Smilie*	*Description*
`:-)`	smile	`:->`	sarcastic
`;-)`	wink	`:-)))`	laughing or double chin
`:.-)`	laughing tears	`;-)=)`	grin
`:-D`	laughing	`:-}`	wry smile
`:-P`	tongue	`:-(`	sad, angry
`:-<`	sad	`:-I`	indifferent/sad
`:.-(`	weeping	`:-II`	angry
`:-@`	angry	`}-)`	evil
`:-X`	mute	`:-()`	talking
`:-O`	surprised/shocked	`=:-)`	shocked
`O:-)`	halo	`:-3`	has eaten a lemon
`:-/`	skeptical	`:-Z`	sleeping
`:-x`	kissing	`:-*`	sorry, I didn't want to say that
`?-(`	sorry, I don't know what went wrong	`:*)`	drunk (red nose)
`%-)`	stared too long at monitor	`#-)`	dead
`X-)`	unconscious	`:-Q`	smoking
`(:-)`	bald	`.-)`	one-eyed
`-:-)`	punk	`<:-)`	stupid question (donkey's hat)
`<|-)`	chinese	`@:-)`	arab
`8:-)`	little girl	`:-)-8`	big girl
`[:-]`	robot	`::-)`	wearing glasses
`8-)`	wearing glasses/wide-eyed grin	`B-)`	horn-rimmed glasses
`B:-)`	sunglasses on head	`.^)`	side view
`:<)`	moustache	`_O-)`	aquanaut
`{:-)`	wig	`:-E`	vampire
`:-[`	vampire	`(-:`	left-handed
`:o)`	boxer's nose	`:)`	happy
`[:]`	robot	`:]`	gleep, friendly
`=)`	variations on a theme	`:}`	(what should we call these?)
`:>`	(what?)	`:@`	(what?)
`:D`	laughter	`:I`	hmmm...
`:(`	sad		

Acronym	*Description*	*Acronym*	*Description*
2U2	to you, too	AAMOF	as a matter of fact
AFAIK	as far as I know	AFK	away from keyboard
ASAP	as soon as possible	BBL	be back later
BOT	back on topic	BRB	be right back
BTW	by the way	BYORL	bring your own rocket launcher
C4N	ciao for now	CFD	call for discussion
CFV	call for vote	CU	see you
CUL	see you later	CYA	see ya
DIY	do it yourself	EOD	end of discussion
EOT	end of transmission	F2F	face to face
FAI	frequently argued issue	FAQ	frequently asked questions
FOAF	friend of a friend	FWIW	for what it's worth
FYI	for your information	GAL	get a life
GFC	going for coffee	GRMBL	grumble
GTG	got to go	HAND	have a nice day
HTH	hope this helps	IAC	in any case
IC	I see	IDGI	I don't get it
IMHO	in my humble opinion	IMNSHO	in my not so humble opinion
IMO	in my opinion	IMPE	in my previous/personal experience
IMVHO	in my very humble opinion	IOW	in other words
IRL	in real life	KISS	keep it simple stupid
LOL	laughing out loud	NC	no comment
ONNA	oh no, not again!	OOTC	obligatory on-topic content
OTOH	on the other hand	REHI	hello again (re-Hi!)
ROFL	rolling on the floor laughing	RTDox	read the documentation
SHTSI	somebody had to say it	SO	significant other
THX	thanks	TIA	thanks in advance
TLA	three letter acronym	TOS	terms of service
TTFN	ta-ta for now	TTYL	talk to you later
WIIWD	what it is we do	WWDWIIWD	when we do what it is we do
YGWYPF	you get what you pay for		

IMO YGWYPF and THX 4 buying this book. Well, C4N. GTG to the next chapter ;->.

WWW

20.1 Introduction

The World-Wide Web (WWW) and the Internet have more jargon words and associated acronyms than anything else in modern life. Words, such as

gopher, ftp, telnet, TCP/IP stack, intranets, Web servers, clients, browsers, hypertext, URLs, Internet access providers, dial-up connections, UseNet servers, firewalls

have all become common in the business vocabulary.

The WWW was initially conceived in 1989 by CERN, the European particle physics research laboratory in Geneva, Switzerland. Its main objective was:

to use the hypermedia concept to support the interlinking of various types of information through the design and development of a series of concepts, communications protocols, and systems

One of its main characteristics is that stored information tends to be distributed over a geographically wide area. The result of the project has been the worldwide acceptance of the protocols and specifications used. A major part of its success was due to the full support of the National Center for Supercomputing Applications (NCSA), which developed a family of user interface programs known collectively as Mosaic.

The WWW, or Web, is basically an infrastructure of information. This information is stored on the WWW on Web servers and it uses the Internet to transmit data around the world. These servers run special programs that allow information to be transmitted to remote computers which are running a Web browser, as illustrated in Figure 20.1. The Internet is a common connection in which computers can communicate using a common addressing mechanism (IP) with a TCP/IP connection.

The information is stored on Web servers and is accessed by means of pages. These pages can contain text and other multimedia applications such as graphic images, digitized sound files and video animation. There are several standard media files (with typical file extensions):

- **GIF/JPEG files for compressed images** (GIF or JPG).
- **QuickTime movies for video** (QT or MOV).
- **Audio** (AU, SND or WAV).
- **MPEG files for compressed video** (MPG).
- **MS video** (AVI).
- **Postscript files** (PS or EPS).
- **Compressed files** (ZIP, Z or GZ).
- **Java/JavaScript** (JAV or JS).

Each page contains text known as hypertext, which has specially reserved keywords to represent the format and the display functions. A standard language known as HTML (Hypertext Markup Language) has been developed for this purpose. Hypertext pages, when interpreted by a browser program, display an easy-to-use interface containing formatted text, icons, pictorial hot spots, underscored words, and so on. Each page can also contain links to other related pages.

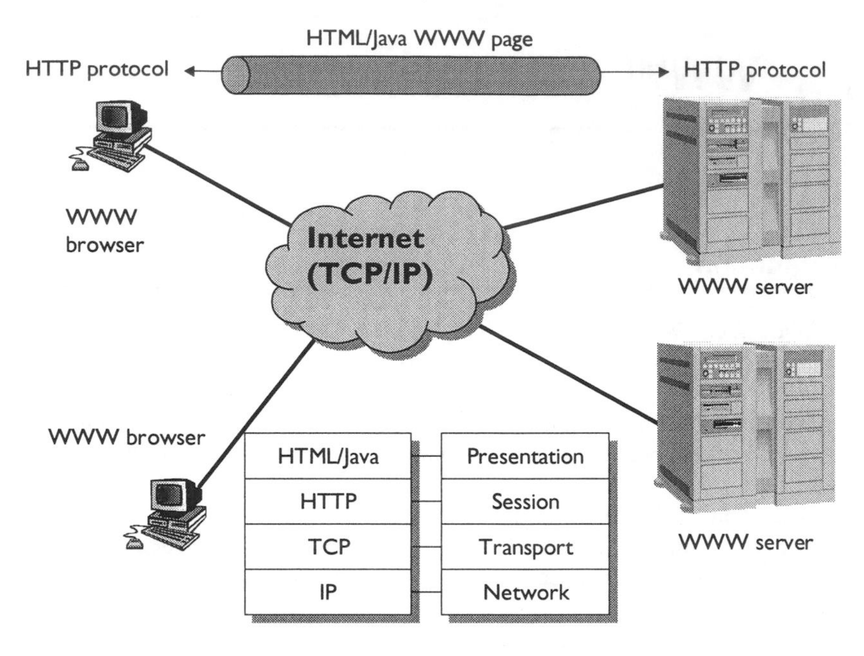

Figure 20.1 Web servers and browsers

The topology and power of the Web now allows for distributed information, where information does not have to be stored locally. To find information on the Web the user can use powerful search engines to search for related links. Figure 20.2 shows an example of Web connections. The user initially accesses a page on a German Web server, this then contains a link to a Japanese server. This server contains links to UK and USA servers. This type of arrangement leads to the topology that resembles a spider's web, where information is linked from one place to another.

20.2 Advantages and disadvantages of the WWW

The WWW and the Internet tend to produce a polarization of views. Thus, before analysing the WWW for its technical specification, a few words must be said on some of the subjective advantages and disadvantages of the WWW and the Internet. It should be noted that some of these disadvantages could be seen as advantages to some people, and vice-versa. For example, freedom of information will be seen as an advantage to a freedom-of-speech group but often a disadvantage to security organizations. Table 20.1 outlines some of the advantages and disadvantages.

Table 20.1 Advantages and disadvantages of the Internet and the WWW

	Advantages	*Disadvantages*
Global information flow	Less control of information by the media, governments and large organizations.	Lack of control on criminal material, such as certain types of pornography and terrorist activity.

Table 20.1 (cont.)

Global transmission	Communication between people and organizations in different countries which should create the Global Village.	Data can easily get lost or state secrets can be easily transmitted over the world.
Internet connections	Many different types of connections are possible, such as dial-up facilities (perhaps over a modem or with ISDN) or through frame relays. The user only has to pay for the service and the local connection.	Data once on the Internet is relatively easy to tap into and possibly easy to change.
Global information	Creation of an ever-increasing global information database.	Data is relatively easy to tap into and possibly easy to change.
Multimedia integration	Tailor-made applications with good presentation tools.	Lack of editorial control leads to inferior material, which is hacked together.
Increasing WWW usage	Helps to improve its chances of acceptance into the home.	Increased traffic swamps the global information network and slows down commercial traffic.
WWW links	Easy to set up and leads users from one place to the next in a logical manner.	WWW links often fossilize where the link information is out-of-date or doesn't even exist.
Education	Increased usage of remote teaching with full multimedia education.	Increase in surface learning and lack of deep research. It may lead to an increase in time-wasting (too much surfing and too little learning).

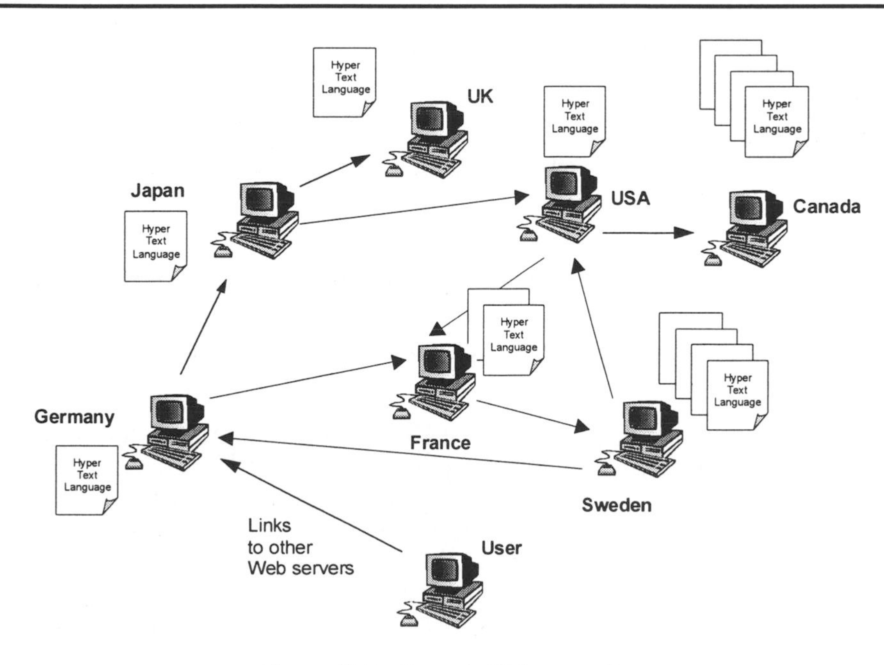

Figure 20.2 Example Web connections

20.3 Client/server architecture

The WWW is structured with clients and servers, where a client accesses services from the server. These servers can either be local or available through a global network connection. A local connection normally requires the connection over a local area network but a global connection normally requires connection to an Internet service provider. These providers are often known as Internet access providers (ISPs), sometimes as Internet connectivity providers (ICP). They provide the mechanism to access the Internet and have the required hardware and software to connect from the user to the Internet. This access is typically provided through one of the following:

- Connection to a client computer through a dial-up modem connection (typically at 28.8 kbps or 56 kbps).
- Connection to a client computer through a dial-up ISDN connection (typically at 64 kbps or 128 kbps).
- Connection of a client computer to a server computer which connects to the Internet through a frame relay router (typically 56 kbps or 256 kbps).
- Connection of a client computer to a local area network which connects to the Internet through a T1, 1.544 Mbps router.

These connections are illustrated in Figure 20.3. A router automatically routes all traffic to and from the Internet whereas the dial-up facility of a modem or ISDN link requires a connection to be made over a circuit-switched line (that is, through the public telephone network). Home users and small businesses typically use modem connections (although ISDN connections are becoming more common). Large corporations which require global Internet services tend to use frame routers. Note that an IAP may be a commercial organization (such as CompuServe or America On-line) or a support organization (such as giving direct connection to government departments or educational institutions). A commercial IAP organization is likely to provide added services, such as electronic mail, search engines, and so on.

An Internet Presence Provider (IPP) allows organizations to maintain a presence on the Internet without actually having to invest in the Internet hardware. The IPPs typically maintain WWW pages for a given charge (they may also provide sales and support information).

20.4 Web browsers

Web browsers interpret special hypertext pages which consist of the hypertext markup language (HTML) and JavaScript. They then display it in the given format. There are currently four main Web browsers:

- **Netscape Navigator** – Navigator is one of the most widely used WWW browsers and is available in many different versions on many systems. It runs on PCs (running Microsoft Windows), UNIX workstations and Macintosh computers. Figure 20.4 shows Netscape Navigator Version 4. It has become the standard WWW browser and has many add-ons and enhancements, which have been added through continual development by Netscape. The basic package also has many compatible software plug-ins which are developed by third-party suppliers. These add extra functionality such as video players and sound support.
- **NSCA Mosaic** – Mosaic was originally the most popular Web browser when the Internet first started. It has now lost its dominance to Microsoft Internet Explorer and

Netscape Navigator. NSCA Mosaic was developed by the National Center for Supercomputing Applications (NCSA) at the University of Illinois.

- **Lynx** – Lynx is typically used on UNIX-based computers with a modem dial-up connection. It is fast to download pages but does not support many of the features supported by Netscape Navigator or Mosaic.
- **Microsoft Internet Explorer** – Explorer now comes as a standard part of Microsoft Windows and as this has become the most popular computer operating system then so will this browser.

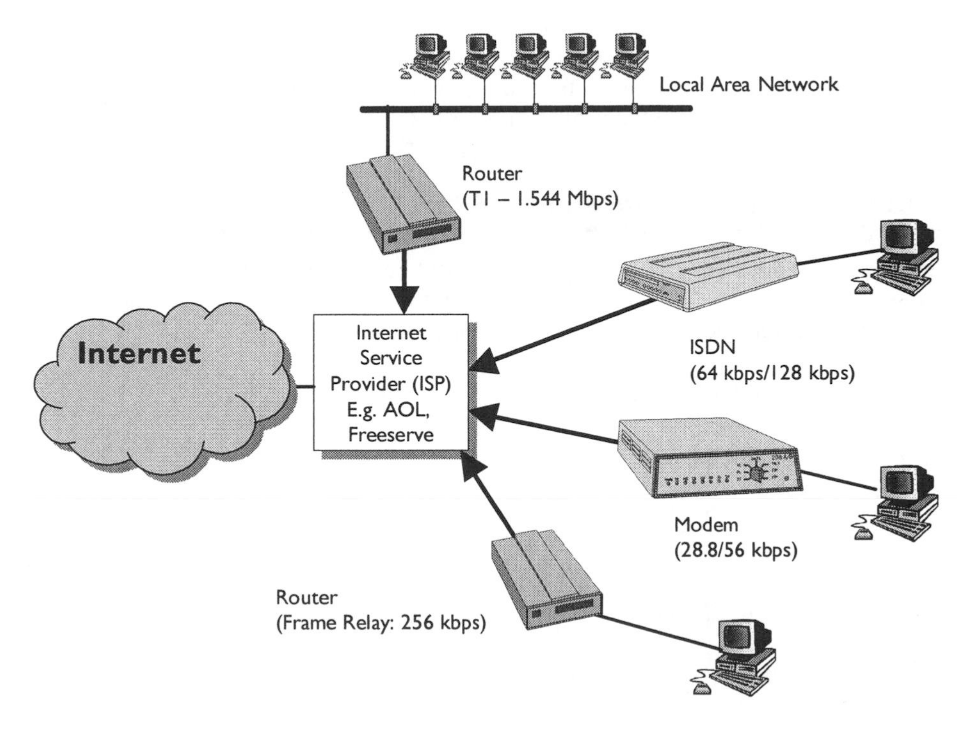

Figure 20.3 Example connections to the Internet

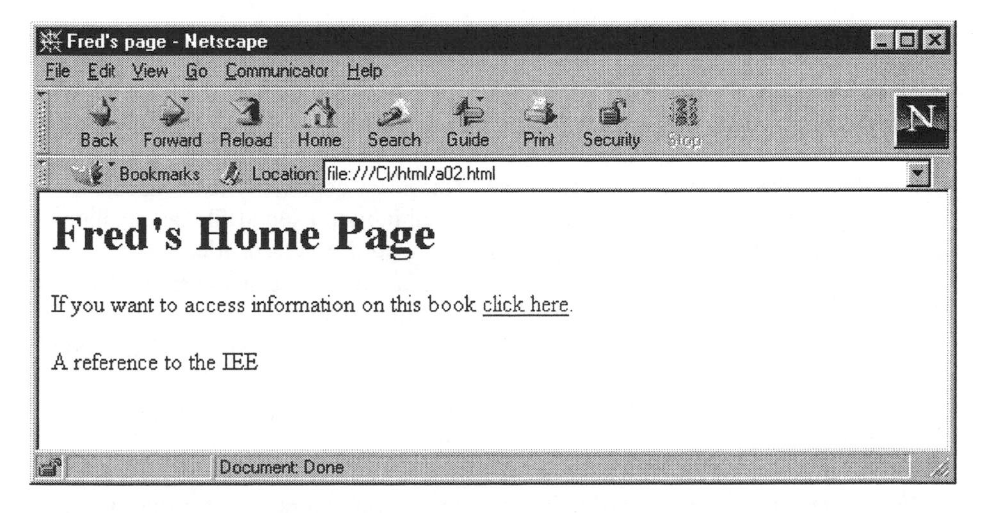

Figure 20.4 Netscape Navigator Version 4

20.5 Internet resources

The Internet expands by the day as the amount of servers and clients which connect to the global network increases and the amount of information contained in the network also increases. The three major services which the Internet provides are:

- The World Wide Web.
- Global electronic mail.
- Information sources.

The main information sources, apart from the WWW, are from FTP, Gopher, WAIS and UseNet servers. These different types of servers will be discussed in the next section.

20.6 Universal resource locators (URLs)

Universal resource locators (URLs) are used to locate a file on the WWW. They provide a pointer to any object on a server connected over the Internet. This link could give FTP access, hypertext references, and so on. URLs contain:

- The protocol of the file (the scheme).
- The server name (domain).
- The pathname of the file.
- The filename.

URL standard format is:

<scheme>:<scheme-specific-part>

and can be broken up into four parts (Figure 20.14). These are:

`aaaa://bbb.bbb.bbb/ccc/ccc/ccc?ddd`

where

`aaaa:` is the access method and specifies the mechanism to be used by the browser to communicate with the resource. The most popular mechanisms are:

- `http:`. HyperText Transfer Protocol. This is the most commonly used mechanism and is typically used to retrieve an HTML file, a graphic file, a sound file, an animation sequence file, a file to be executed by the server, or a word processor file.
- `https:`. HyperText Transfer Protocol. It is a variation on the standard access method and can be used to provide some level of transmission security.
- `file:`. Local File Access. This causes the browser to load the specified file from the local disk.
- `ftp:`. File Transport Protocol. This method allows files to be downloaded using an FTP connection.
- `mailto:`. E-Mail Form. This method allows access to a destination e-mail address. Normally the browser automatically generates an input form for entering the e-mail message.

- `news:`. USENET News. This method defines the access method for a news group.
- `nntp:`. Local Network News Transport Protocol.
- `wais:`. Wide Area Information Servers.
- `gopher:`. GOPHER protocol.
- `telnet:`. TELNET. The arguments following the access code are the login arguments to the telnet session as `user[ :password] @host`.
- `cid:`. Content identifiers for MIME body part.
- `mid:`. Message identifiers for electronic mail.
- `afs:`. AFS File Access.
- `prospero:`. Prospero Link.
- `x-exec:`. Executable Program.

`//bbb.bbb.bbb` is the Internet node and specifies the node on the Internet where the file is located. If a node is not given then the browser defaults to the computer which is running the browser. A colon may follow the node address and the port number (most browsers default to port 80, which is also the port that most servers use to reply to the browser).

`/ccc/ccc/ccc` is the file path (including subdirectories and the filename). Typically systems restrict the access to a system by allocating the root directory as a subdirectory of the main file system.

`?ddd` is the argument which depends upon the access method, and the file accessed. For example, with an HTML document a '#' identifies the fragment name internal to an HTML document which is identified by the A element with the NAME attribute.

An example URL is:

```
http://www.toytown.anycor.co/fred/index.html
```

where `http` is the file protocol (Hypertext Translation Protocol), `www.toytown.anycor.co` is the server name, `/fred` is the path of the file and the file is named `index.html`.

20.6.1 Files

A file URL scheme allows files to be assessed. It takes the form:

`file://`*<host>*/*<path>*

where *<host>* is the fully qualified domain name of the system to be accessed, *<path>* is the full path name, and takes the form of a directory path, such as *<directory>/<directory>/.../<name>*.

For a file:

```
C:\DOCS\NOTES\NETWORKS\NET_CHAP13.DOC
```

would be accessed as from `dummy.com` with:

```
file://dummy.com/C|DOCS/NOTES/NETWORKS/NET_CHAP13.DOC
```

Note, that if the host is defined as `localhost` or is an empty string then the host is assumed to be the local host. The general format is:

```
fileurl  = "file://" [ host | "localhost" ] "/" fpath
```

20.6.2 Electronic mail address

The `mailto` scheme defines a link to an Internet e-mail address. An example is:

```
mailto: fred.bloggs@toytown.ac.uk
```

When this URL is selected then an e-mail message will be sent to the e-mail address `fred.bloggs@toytown.ac.uk`. Normally, some form of text editor is called and the user can enter the required e-mail message. Upon successful completion of the text message it is sent to the addressee.

20.6.3 File Transfer Protocol (FTP)

The `ftp` URL scheme defines that the files and directories specified are accessed using the FTP protocol. In its simplest form it is defined as:

```
ftp://<hostname>/<directory-name>/<filename>
```

The FTP protocol normally requests a user to log into the system. For example, many public domain FTP servers use the login of:

```
anonymous
```

and the password can be anything (but it is normally either the user's full name or their Internet e-mail address). Another typical operation is changing directory from a starting directory or the destination file directory. To accommodate this, a more general form is:

```
ftp://<user>:<password>@<hostname>:<port>/<cd1>/<cd2>/
.../<cdn>/<filename>
```

where the user is defined by *<user>* and the password by *<password>*. The host name, *<hostname>*, is defined after the @ symbol and change directory commands are defined by the *cd* commands. The node name may take the form `//user [ :password] @host`. Without a user name, the user `anonymous` is used.

For example the reference to the standard related to HTML Version 2 can be downloaded using the URL:

```
ftp://ds.internic.net/rfc/rfc1866.txt
```

and draft Internet documents from:

```
ftp://ftp.isi.edu/internet-drafts/
```

The general format is:

```
ftpurl     = "ftp://" login [ "/" fpath [ ";type=" ftptype ]]
fpath      = fsegment *[ "/" fsegment ]
fsegment   = *[ uchar | "?" | ":" | "@" | "&" | "=" ]
ftptype    = "A" | "I" | "D" | "a" | "i" | "d"
```

20.6.4 Hypertext Transfer Protocol (HTTP)

HTTP is the protocol which is used to retrieve information connected with hypermedia links.

The client and server initially perform a negotiation procedure before the HTTP transfer takes place. This negotiation involves the client sending a list of formats it can support and the server replying with data in the required format.

Users generally move from a link on one server to another server. Each time the user moves from one server to another, the client sends an HTTP request to the server. Thus the client does not permanently connect to the server, and the server views each transfer as independent from all previous accesses. This is known as a stateless protocol.

An HTTP URL takes the form:

```
http://<host>:<port>/<path>?<searchpart>
```

Note that, if the *<port>* is omitted, port 80 is automatically used (HTTP service), *<path>* is an HTTP selector and *<searchpart>* is a query string.

The general format is:

```
httpurl      = "http://" hostport [ "/" hpath [ "?" search ]]
hpath        = hsegment *[ "/" hsegment ]
hsegment     = *[ uchar | ";" | ":" | "@" | "&" | "=" ]
search       = *[ uchar | ";" | ":" | "@" | "&" | "=" ]
```

20.6.5 Gopher Protocol

Gopher is widely used over the Internet and is basically a distribution system for the retrieval and delivery of documents. Users retrieve documents through a series of hierarchical menus, or through keyword searches. Unlike HTML documents they are not based on the hypertext concept.

20.6.6 Wide area information servers (WAIS)

WAIS is a public domain, fully text-based, information retrieval system over the Internet which performs text-based searches. The communications protocol used is based on the ANSI standard Z39.50, which is designed for networking library catalogues.

WAIS services include index generation and search engines. An indexer generates multiple indexes for organizations or individuals that offer services over the Internet. A WAIS search engine searches for particular words or text string indexes located across multiple Internet-attached information servers of various types.

20.6.7 News

UseNet or NewsGroup servers are part of the increasing use of general discussion news groups which share text-based news items. The news URL scheme defines a link to either a news group or individual articles with a group of UseNet news.

A news URL takes one of two forms:

```
news:<newsgroup-name>
news:<message-id>
```

where *<newsgroup-name>* is a period-delimited hierarchical name, such as 'news.inter', and *<message-id>* takes the full form of the message-ID, such as:

```
<message-ID>@<full_domain_name>
```

The general form is:

```
newsurl         = "news:" grouppart
grouppart       = "*" | group | article
group           = alpha *[ alpha | digit | "-" | "." | "+" | "_" ]
article         = 1*[ uchar | ";" | "/" | "?" | ":" | "&" | "=" ] "@" host
```

20.7 Web browser design

The Web browser is a carefully engineered software package which allows the user to efficiently find information on the WWW. Most are similar in their approach, but differ in their presentation. Figure 20.5 shows the tool bar for Microsoft Internet Explorer. This has been designed to allow the user to smoothly move through the WWW.

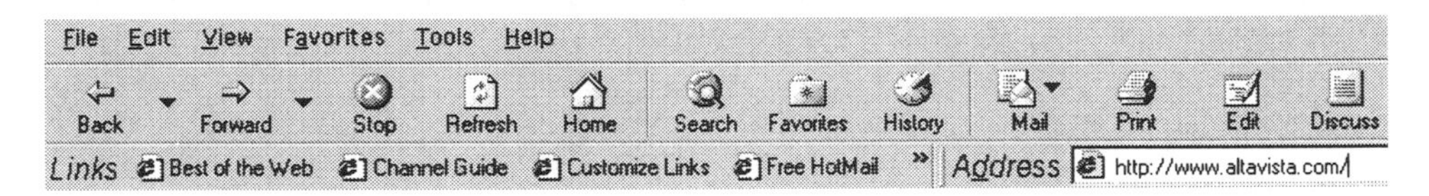

Figure 20.5 Microsoft Explorer tool bar

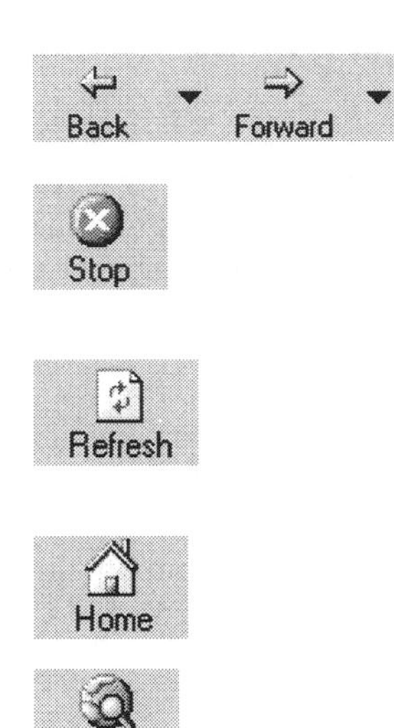

The Back and Forward options allow the user to traverse backwards and forwards through links. This allows the user to trace back to a previous link and possibly follow it.

The Stop option is used by the user to interrupt the current transfer. It is typically used when the user does not want to load the complete page. This often occurs when the browser is loading a graphics image.

The Web browser tries to reduce data transfer by holding recently accessed pages in a memory cache. This cache is typically held on a local disk. The Refresh forces the browser to re-load the page from the remote location.

Often a user wishes to restart a search and can use the Home option to return to it. The home page of the user is set up by one of the options.

The Search option is used to connect to a page which has access to the search programs. Microsoft Explorer typically connects to `http://home.microsoft.com/access/allinone.asp` which displays most of the available search engines. An example screen from Microsoft search facility is given in Figure 20.7. It can be seen that this links to the most commonly used search engines, such as Yahoo, Lycos, Magallan and eXcite.

Often a user has a list of favorite Web pages (see Figure 20.6). This can be automatically called from the Favorites option. A new favorite can be added with the Add To Favorites ... option. These favorites can either be selected from the Favorites menu option (such as Internet Start) or from within folders (such as Channels and Links). The favorites are organized using the Organize Favorites... option.

Other typically accessed sites can be recalled from Links, such as:

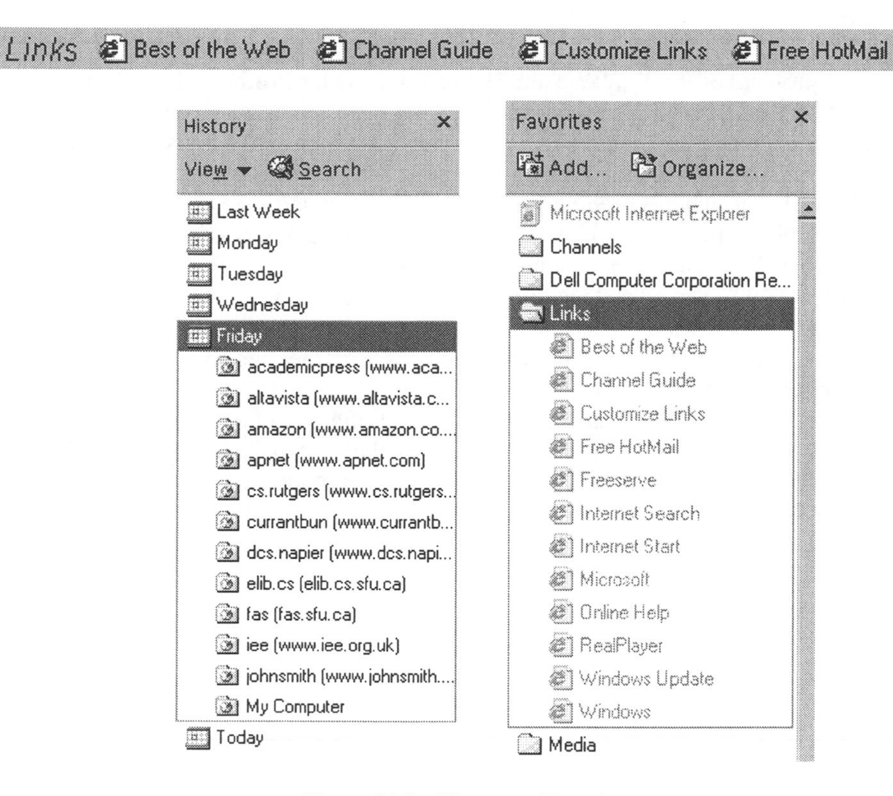

Figure 20.6 History and favorites menu

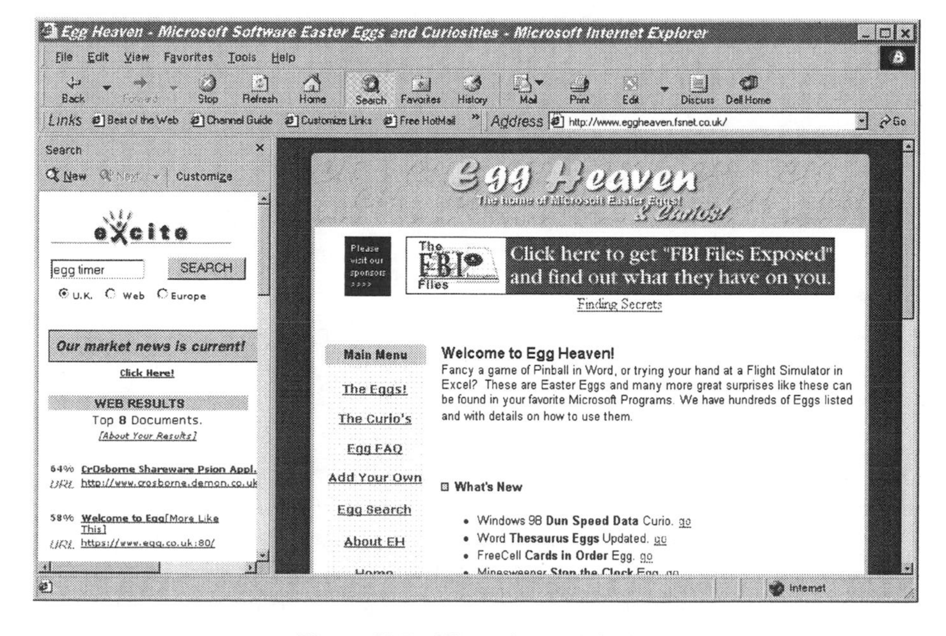

Figure 20.7 Microsoft search facility

20.8 HTTP

The foundation protocol of the WWW is the Hypertext Transfer Protocol (HTTP) which can be used in any client/server application involving hypertext. It is used on the WWW for transmitting information using hypertext jumps and can support the transfer of plaintext, hypertext, audio, images, or any Internet-compatible information. The most recently defined standard is HTTP 1.1, which has been defined by the IETF standard.

HTTP is a stateless protocol where each transaction is independent of any previous transactions. Thus when the transaction is finished the TCP/IP connection is disconnected, as illustrated in Figure 20.8. The advantage of being stateless is that it allows the rapid access of WWW pages over several widely distributed servers. It uses the TCP protocol to establish a connection between a client and a server for each transaction then terminates the connection once the transaction completes.

HTTP also supports many different formats of data. Initially a client issues a request to a server which may include a prioritized list of formats that it can handle. This allows new formats to be easily added and also prevents the transmission of unnecessary information.

A client's WWW browser (the user agent) initially establishes a direct connection with the destination server which contains the required WWW page. To make this connection the client initiates a TCP connection between the client and the server. After this is established the client then issues an HTTP request, such as the specific command (the method), the URL, and possibly extra information such as request parameters or client information. When the server receives the request, it attempts to perform the requested action. It then returns an HTTP response, which includes status information, a success/error code, and extra information. After the client receives this, the TCP connection is closed.

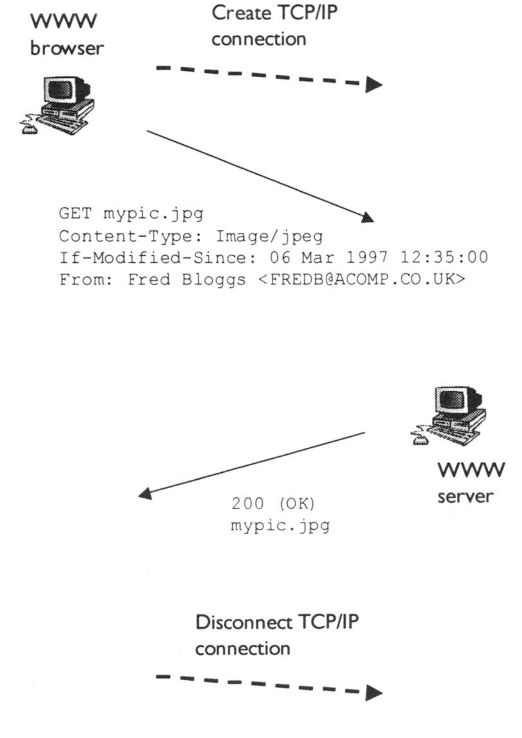

Figure 20.8 Example HTTP transaction

20.8.1 Caches, tunnels and user agents

In a computer system, a cache is an area of memory that stores information likely to be accessed in a fast access memory area. For example, a cache controller takes a guess on which information the process is likely to access next. When the processor wishes to access the disk then, if it has guessed right, the cache controller will load from the electronic memory rather than loading it from the disk. A WWW cache stores cacheable responses so that there is a reduction in network traffic and an improvement in access times. Figure 20.9 shows an example use of a cache. Initially (1) the client sends out a request for a page, along with the date that it was last accessed. If the page has not changed then the server sends back a message saying that it has not been changed. The client will then use the page that is stored in the local cache.

Often Internet Service Providers use a network cache which stores pages that users have recently accessed. If a request is made for a page which has already been accessed, and is stored in the cache, the network cache can be used to provide the page to the requester. In the future, network caches could be used to considerably speed-up WWW page download. In this, popular pages would be regularly downloaded to the cache, and sent to the clients when required. An important factor is keeping the cache regularly updated. For this HTTP requests must still be sent with the date and time that the page was last updated. If there is a change, the new updated page should be sent (and obviously also stored in the cache).

Some WWW browsers can also be setup so that they do not re-access a previously loaded page, if it is re-requested within a given time. This can be annoying for the user, especially if the page is updating itself within a short interval (such as with a WWW camera). A Reload button is typically used to force the WWW browser to re-request the page.

Tunnels are intermediary devices which act as a blind relay between two connections. When the tunnel becomes active, it is not seen to be part of the HTTP communications. When the connection is closed by both sides, the tunnel ceases to exit.

A user agent, in HTTP, is a client which initiates requests to a server. Typically this is a WWW browser or a WWW spider (an automated WWW trailing program).

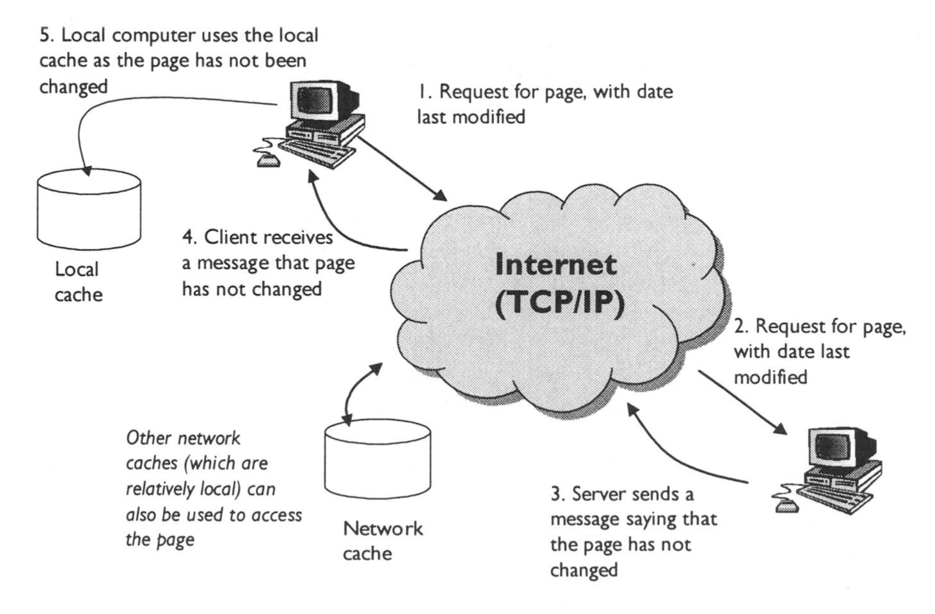

Figure 20.9 Using caches

20.8.2 HTTP messages

HTTP messages are either requests from clients to servers or responses from servers to clients (Figure 20.10). The message is either a simple-request, a simple-response, full-request or a full-response. HTTP Version 0.9 defines the simple request/ response messages whereas HTTP Version 1.1 defines full requests/responses.

HTTP is a stateless protocol where each transaction is independent of any previous transactions. The advantage of being stateless is that it allows the rapid access of WWW pages over several widely distributed servers. It uses the TCP protocol to establish a connection between a client and a server for each transaction then terminates the connection once the transaction completes.
HTTP also supports many different formats of data. Initially a client issues a request to a server which may include a prioritized list of formats that it can handle. This allows new formats to be easily added and also prevents the transmission of unnecessary information.

A client's WWW browser (the user agent) initially establishes a direct connection with the destination server which contains the required WWW page.
To make this connection the client initiates a TCP connection between the client and the server. After this is established the client then issues an HTTP request, such as the specific command (the method), the URL, and possibly extra information such as request parameters or client information.
When the server receives the request, it attempts to perform the requested action.
It then returns an HTTP response, which includes status information, a success/error code, and extra information itself.
After the client receives this, the TCP connection is closed.

Figure 20.10 HTTP operation

Simple requests/responses
The simple request is a `GET` command with the requested URI such as:

```
GET  /info/dept/courses.html
```

The simple response is a block containing the information identified in the URI (called the entity-body).

Full requests/responses
Very few security measures or enhanced services are built into the simple requests/responses. HTTP Version 1.0/1.1 improves on the simple requests/responses by adding many extra requests and responses, as well as adding extra information about the data supported. Each message header consists of a number of fields which begin on a new line and consist of the field name followed by a colon and the field value. This follows the format of RFC822 (as shown in Section 19.5.4) and allows for MIME encoding. It is thus similar to MIME-encoded e-mail. A full request starts with a request line command (such as `GET`, `MOVE` or `DELETE`) and is then followed by one or more of the following:

- General-headers which contain general fields that do not apply to the entity being transferred (such as MIME version, date, and so on).
- Request-headers which contain information on the request and the client (e.g. the client's name, its authorization, and so on).
- Entity-headers which contain information about the resource identified by the request and entity-body information (such as the type of encoding, the language, the title, the time when it was last modified, the type of resource it is, when it expires, and so on).

- Entity-body which contains the body of the message (such as HTML text, an image, a sound file, and so on).

A full response starts with a response status code (such as `OK`, `Moved Temporarily`, `Accepted`, `Created`, `Bad Request`, and so on) and is then followed by one or more of the following:

- General-headers, as with requests, contain general fields which do not apply to the entity being transferred (MIME version, date, and so on).
- Response-headers which contain information on the response and the server (e.g. the server's name, its location and the time the client should retry the server).
- Entity-headers, as with request, which contain information about the resource identified by the request and entity-body information (such as the type of encoding, the language, the title, the time when it was last modified, the type of resource it is, when it expires, and so on).
- Entity-body, as with requests, which contains the body of the message (such as HTML text, an image, a sound file, and so on).

The following example shows an example request. The first line is always the request method; in this case it is `GET`. Next there are various headers. The general-header field is `Content-Type`, the request-header fields are `If-Modified-Since` and `From`. There are no entity parts to the message as the request is to get an image (if the command had been to `PUT` then there would have been an attachment with the request). Notice that a single blank line delimits the end of the message as this indicates the end of a request/response. Note that the headers are case sensitive, thus `Content-Type` with the correct types of letters (and `GET` is always in uppercase letters).

Example HTTP request

```
GET mypic.jpg
Content-Type: Image/jpeg
If-Modified-Since: 06 Mar 1997 12:35:00
From: Fred Bloggs <FREDB@ACOMP.CO.UK>
```

Request messages

The most basic request message is to `GET` a URI. HTTP/1.1 (Figure 20.11) adds many more requests including:

```
COPY    DELETE    GET    HEAD     POST      LINK
MOVE    OPTIONS   PUT    TRACE    UNLINK    WRAPPED
```

As before, the `GET` method requests a WWW page. The `HEAD` method tells the server that the client wants to read only the header of the WWW page. If the `If-Modified-Since` field is included then the server checks the specified date with the date of the URI and verifies whether it has not changed since then.

A `PUT` method requests storage of a WWW page and `POST` appends to a named resource (such as electronic mail). `LINK` connects two existing resources and `UNLINK` breaks the link. A `DELETE` method removes a WWW page. The request-header fields are mainly used to define the acceptable type of entity that can be received by the client; they include (Figure 20.12):

```
Accept                   Accept-Charset              Accept-Encoding
Accept-Language          Authorization               From
Host                     If-Modified-Since           Referer
Proxy-Authorization      Range
Unless                   User-Agent
```

The `Accept` field is used to list all the media types and ranges that the client can accept. An `Accept-Charset` field defines a list of character sets acceptable to the server and `Accept-Encoding` is a list of acceptable content encodings (such as the compression or encryption technique). The `Accept-Language` field defines a set of preferred natural languages.

The `Authorization` field has a value which authenticates the client to the server. A `From` field defines the e-mail address of the user who is using the client (e.g. `From: fred.blogg@anytown.uk`) and the `Host` field specifies the name of the host of the resource being requested.

A useful field is the `If-Modified-Since` field, used with the `GET` method. It defines a date and time parameter and specifies that the resource should not be sent if it has not been modified since the specified time. This is useful when a client has a local copy of the resource in a local cache and, rather than transmitting the unchanged resource, it can use its own local copy.

The `Proxy-Authorization` field is used by the client to identify itself to a proxy when the proxy requires authorization. A `Range` field is used with the `GET` message to get only a part of the resource.

The `Referer` field defines the URI of the resource from which the `Request-URI` was obtained and enables the server to generate lists of back-links. An `Unless` field is used to make a comparison based on any entity-header field value rather than a date/time value (as with `GET` and `If-Modified-Since`).

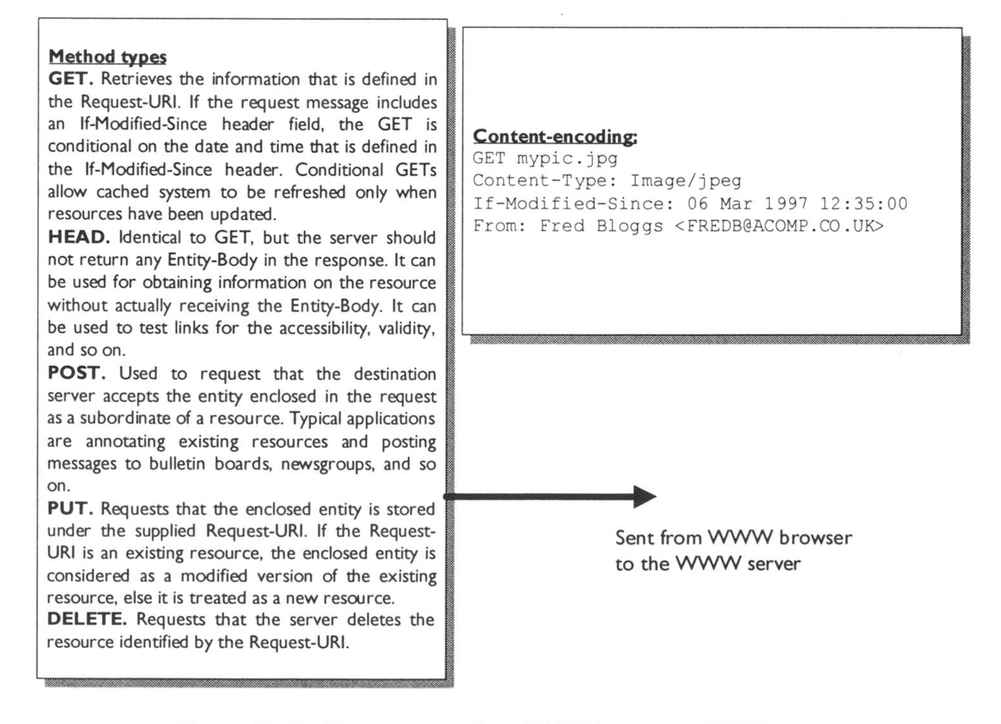

Figure 20.11 Messages sent from WWW browser to WWW server

Content-language:
The Language field defines the language of the resource. Its format is *Language-tag*, where *Language* is the language (such as en for English), and *tag* is the dialect. For example:

Content-type: audio/basic
Content-Language: en-scottish

Content-encoding
The Content-Encoding field describes the type of encoding used in the entity. It indicates the additional content coding that has been applied to the resource. In most cases the encoding is 'gzip' or 'x-compress' (or 'x-gzip' and 'compress'). The Content-Type header field defines the underlying content. An example of this field is:

Content-Encoding: x-gzip
Content-Type = text/plain

where the content type is plain text, which has been zipped with gzip (a Lempel-Ziv based compression program). The x-compress program uses Lempel-Ziv-Welsh (LZW) compression.

From:
The From field contains an Internet e-mail address of the owner of the requesting user agent, such as:

From: fred_b@myserver.com

It is typically used for logging purposes.

Last modified:
The Last-Modified indicates the date and time the resource was last modified. An example of the field is:

Last-Modified: Wed, 23 Dec 1998 03:20:25 GMT

Content-type:
The Content-Type field defines the media type of the Entity-Body sent to the recipient.
An example of the field is:

Content-type: text/plain; charset = ISO-8859-1

Content-Types are of the format *type/subject*. Typical types are 'application', 'image', 'text', 'audio' and 'video', and example content types are:

text/plain text/enriched image/gif
image/jpeg audio/basic video/mpeg
application/postscript application/msword

Figure 20.12 HTTP header types

The `User-Agent` field contains information about the user agent originating this request.

Response messages

In HTTP/0.9 the response from the server was either the entity or no response. HTTP/1.1 includes many other responses (Figure 20.13). These include:

```
Accepted                       Bad Gateway
Bad Request                    Conflict
Continue                       Created
Forbidden                      Gateway Timeout
Gone                           Internal Server Error
Length Required                Method Not Allowed
Moved Permanently              Moved Temporarily
Multiple Choices               No Content
Non-Authoritative Info         None Acceptable
Not Found                      Not Implemented
Not Modified                   OK
Partial Content                Payment Required
Proxy Authorization Required   Request Timeout
Reset Content                  See Other
Service Unavailable            Switching Protocols
Unauthorized                   Unless True
Use Proxy
```

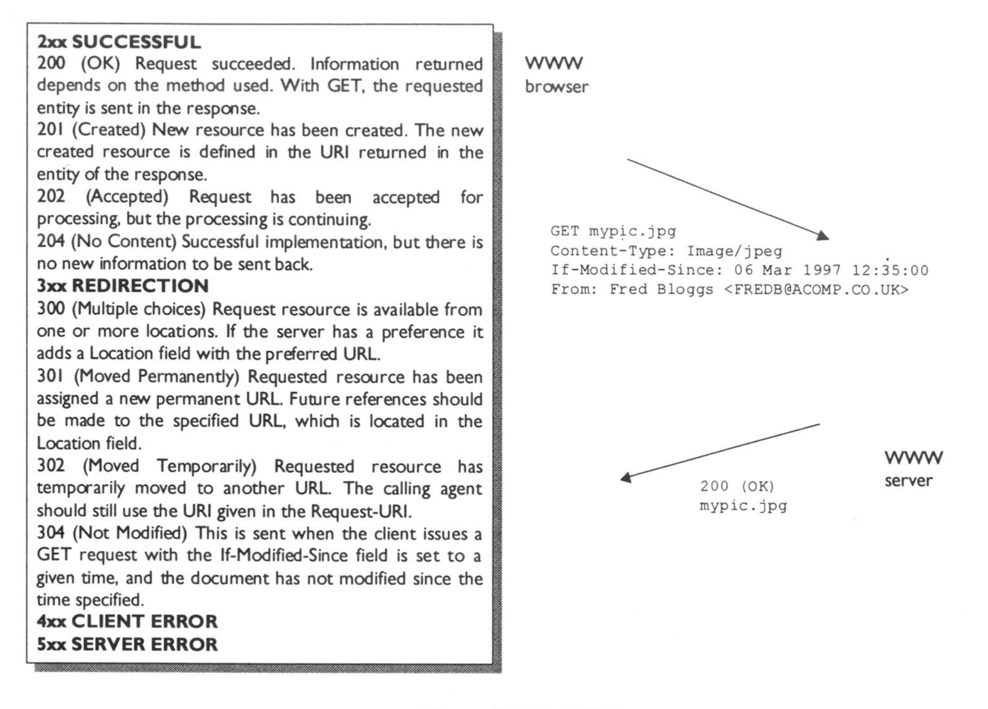

2xx SUCCESSFUL
200 (OK) Request succeeded. Information returned depends on the method used. With GET, the requested entity is sent in the response.
201 (Created) New resource has been created. The new created resource is defined in the URI returned in the entity of the response.
202 (Accepted) Request has been accepted for processing, but the processing is continuing.
204 (No Content) Successful implementation, but there is no new information to be sent back.
3xx REDIRECTION
300 (Multiple choices) Request resource is available from one or more locations. If the server has a preference it adds a Location field with the preferred URL.
301 (Moved Permanently) Requested resource has been assigned a new permanent URL. Future references should be made to the specified URL, which is located in the Location field.
302 (Moved Temporarily) Requested resource has temporarily moved to another URL. The calling agent should still use the URI given in the Request-URI.
304 (Not Modified) This is sent when the client issues a GET request with the If-Modified-Since field is set to a given time, and the document has not modified since the time specified.
4xx CLIENT ERROR
5xx SERVER ERROR
WWW browser
GET mypic.jpg
Content-Type: Image/jpeg
If-Modified-Since: 06 Mar 1997 12:35:00
From: Fred Bloggs <FREDB@ACOMP.CO.UK>
WWW server
200 (OK)
mypic.jpg

Figure 20.13 HTTP responses

These responses can be put into five main groupings:

- Client error – `Bad Request`, `Conflict`, `Forbidden`, `Gone`, `Payment required`, `Not Found`, `Method Not Allowed`, `None Acceptable`, `Proxy Authentication Required`, `Request Timeout`, `Length Required`, `Unauthorized`, `Unless True`.
- Informational – `Continue`, `Switching Protocol`.
- Redirection – `Moved Permanently`, `Moved Temporarily`, `Multiple Choices`, `See Other`, `Not Modified`, `User Proxy`.
- Server error – `Bad Gateway`, `Internal Server Error`, `Not Implemented`, `Service Unavailable`, `Gateway Timeout`.
- Successful – `Accepted`, `Created`, `OK`, `Non-Authoritative Info`. The `OK` field is used when the request succeeds and includes the appropriate response information.

The response header fields are:

```
Location            Proxy-Authenticate          Public
Retry-After         Server                      WWW-Authenticate
```

The `Location` field defines the location of the resource identified by the `Request-URI`. A `Proxy-Authenticate` field contains the status code of the `Proxy Authorization Required` response.

The `Public` field defines non-standard methods supported by this server. A `Retry-After` field contains values which define the amount of time a service will be unavailable (and is thus sent with the `Service Unavailable` response).

The `WWW-Authenticate` field contains the status code for the `Unauthorized` response.

General-header fields

General-header fields are used either within requests or within responses; they include:

```
Cache-Control   Connection     Date          Forwarded
Keep-Alive      MIME-Version   Pragma        Upgrade
```

The `Cache-Control` field gives information on the caching mechanism and stops the cache controller from modifying the request/response. A `Connection` field specifies the header field names that apply to the current TCP connection.

The `Date` field specifies the date and time at which the message originated; this is obviously useful when examining the received message as it gives an indication of the amount of time the message took to arrive at its destination. Gateways and proxies use the `Forwarded` field to indicate intermediate steps between the client and the server. When a gateway or proxy reads the message, it can attach a `Forwarded` field with its own URI (this can help in tracing the route of a message).

The `Keep-Alive` field specifies that the requester wants a persistent connection. It may indicate the maximum amount of time that the sender will wait for the next request before closing the connection. It can also be used to specify the maximum number of additional requests on the current persistent connection.

The `MIME-Version` field indicates the MIME version (such as `MIME-Version: 1.0`). A `Pragma` field contains extra information for specific applications.

In a request the `Upgrade` field specifies the additional protocols that the client supports and wishes to use, whereas in a response it indicates the protocol to be used.

Entity-header fields

Depending on the type of request or response, an entity-header can be included:

```
Allow             Content-Encoding                Content-Language
Content-Length    Content-MD5                     Content-Range
Content-Type      Content-Version                 Derived-From
Expires           Last-Modified                   Link
Title             Transfer-encoding
URI-Header extension-header
```

The `Allow` field defines the supported methods supported by the resource identified in the `Request-URI`. A `Content-Encoding` field indicates content encodings, such as ZIP compression, that have been applied to the resource (`Content-Encoding: zip`).

The `Content-Language` field identifies natural language(s) of the intended audience for the enclosed entity (e.g. `Content-language: German`) and the `Content-Length` field defines the number of bytes in the entity.

The `Content-Range` field designates a portion of the identified resource that is included in this response, while `Content-Type` indicates the media type of the entity body (such as `Content-Type=text/html`, `Content-Type=text/plain`, `Content-Type=image/gif` or `Content-type=image/jpeg`). The version of the entity is defined in the `Content-Version` field.

The `Expires` field defines the date and time when the entity is considered stale. The `Last-Modified` field is the date and time when the resource was last modified.

The `Link` field defines other links and the `Title` field defines the title for the entity. A `Transfer-Encoding` field indicates the transformation type that is applied so the entity can be transmitted.

20.9 Exercises

The following questions are multiple choice. Please select from a–d.

20.9.1 Where was the WWW first conceived:
(a) UCL (b) UMIST
(c) MIT (d) CERN

20.9.2 An MPEG file is what type of file:
(a) Image (b) Motion video
(c) Sound file (d) Generally compressed file

20.9.3 A GIF file is what type of file:
(a) Image (b) Motion video
(c) Sound file (d) Generally compressed file

20.9.4 A ZIP file is what type of file:
(a) Image (b) Motion video
(c) Sound file (d) Generally compressed file

20.9.5 An AU file is what type of file:
(a) Image (b) Motion video
(c) Sound file (d) Generally compressed file

20.9.6 Where is the information on the Web stored:
(a) Web clients and servers (b) Web clients
(c) Web servers (d) Web gateways

20.9.7 Which field is used to define the media type (such as JPEG, GIF, MPEG, and so on) of the entity:
(a) Content-Encoding (b) Content-Range
(c) Content-Type (d) Content-Language

20.9.8 Which field is used to define the additional compression (zip) that has been applied to the entity:
(a) Content-Encoding (b) Content-Range
(c) Content-Type (d) Content-Language

20.9.9 Which field does the GET method use to determine if a resource has changed since it was last accessed:
(a) Accept (b) If-Modified-Since
(c) Accept-Encoding (d) Authorization

20.9.10 If possible, search for the following subjects on the WWW:
(a) Ethernet (b) EaStMAN (c) Edinburgh
(d) Taxation (e) Intel

20.9.11 Discuss the limitation of simple requests and responses with HTTP.

20.9.12 Discuss request messages and the fields that are set.

20.9.13 Discuss response messages and the fields that are set.

20.9.14 If possible, access the following WWW servers:

Site	*Accessed (YES/NO)*	*Comments*
`http://www.microsoft.com`		
`http://www.intel.com`		
`http://www.ieee.com`		
`http://www.winzip.com`		
`http://www.netscape.com`		
`http://www.realaudio.com`		
`http://www.cyrix.com`		
`http://www.compaq.com`		
`http://www.psion.com`		
`http://www.amd.com`		
`http://www.cnn.com`		
`http://www.w3.org`		
`http://www.macromedia.com`		
`http://www.epson.co.uk`		
`http://www.euronec.com`		
`http://www.casio.com`		
`http://www.hayes.com`		
`http://www.lotus.com`		
`http://www.adobe.com`		
`http://www.corel.com`		
`http://www.symantec.co.uk`		
`http://www.fractal.com`		
`http://www.quarterdeck.com`		
`http://www.guinness.ie`		

20.9.15 Discuss the methods that users use to find information on the Web.

20.10 Research

The WWW and the Internet are powerful tools, but in Education they must be used with caution. Investigate the current uses of the WWW and the Internet within your own organization, and possibly within other organizations. Typical questions that can be asked are:

Sample Question	*Possible options (delete, as necessary)*	*Comments*
Does your organization (University/College/Business/etc) have its own WWW site? If not, visit another site, until you find one.	[YES]/[NO]	Name of site (e.g. `www.mysite.edu`):
Is there education material on the WWW site?	[YES]/[NO]	Sample locations (e.g. `www.mysite.edu/~fred`):
Are the Lecturers/Teachers/Trainers available to be contacted by e-mail?	[YES]/[NO]	Sample e-mail addresses (e.g. `fred@mysite.edu`):

What is the quality of the education material?	Rating (best is 10): [1] [2] [3] [4] [5] [6] [7] [8] [9] [10]	Sample page (e.g. www.mysite.edu/edu1.html):
What is the main content source of the education material?	[HTML] [HTML/Java] [PDF] [Documents] [Other]	Example page (e.g. www.mysite.edu/java.html):
Does the organization have a formal policy on WWW-based material, or is it left to individuals to generate the education material?	[YES][NO][N/A]	
Does the material integrate with traditional teaching methods (such as Lecturers/Tutorials/ Experiments)	Rating (best is 10): [1] [2] [3] [4] [5] [6] [7] [8] [9] [10]	Example of inte.g.ration (e.g. www.mysite.edu/course1.html):
Does the education material actually enhance the education learning?	Rating (best is 10): [1] [2] [3] [4] [5] [6] [7] [8] [9] [10] [Don't Know]	Good example (e.g. www.mysite.edu/tele.html):
Can the WWW-based material be accessed by people who are external to the organization?	[YES][NO][N/A]	
What percentage of Lecturer/ Teachers/Trainers have WWW-based material?	[0–10%] [10–20%] [30–40%] [40–50%] [60–70%] [80–90%] [90–100%]	Names (e.g. Fred Bloggs):
Find a site which, in your opinion, has excellent education material. What characteristics does it have and is it clear how it integrates into existing courses/programmes?		Name of site (e.g. www.mysite.edu): Reason it is excellent:
Find a site which, in your opinion, has poor education material. What characteristics does it have?		Name of site (e.g. www.mysite.edu): Reason it is poor:

Are there multi-media files (such as sound and video)? Do they enhance the understanding of the subject?		Example file (e.g. `www.mysite.edu/demo.avi`):
What are the main usages of the local Internet connection?		
What would be the main recommendations on changing/enhancing the educational material on the WWW site?		

20.11 Surveys

1. Section 21.2 outlines some of the advantages and disadvantages of the WWW and the Internet. The usage of WWW/Internet by companies and individuals varies from one to the next. Conduct a survey within a class, University/College, workplace, etc, on the best applications of the WWW/Internet. Each person is allowed 12 votes. Ten points should be allocated to their favorite application, five for their second favorite, three for their third, two for their fourth and one for their fifth. People should also give –3 to their three least favorite applications. Total the votes and fill in the table given next, and thus determine the Top 10 favorite applications of the WWW/Internet.

Application	*Vote*	*Application*	*Vote*	*Application*	*Vote*
Electronic Mail		Downloading Hardware Drivers		Video Conferencing	
Distributed information		Multimedia Education		Remote Access (such as TELNET)	
Application Software Downloads		Client/server Processing		News Groups	
Product Information		Remote Control/ Transmission of Data		Chat Programs	
Daily News Events		Information Archives		On-line Libraries	
Electronic Commerce		Search Facilities (Educational)		Search Facilities (Product Information)	
Search Facilities (People/ White Pages)		General WWW Surfing		Sampling Material (such as Music, Videos, etc)	
Source Code Download		Direct Access to Experts (such as through e-mail)		Special Interest Groups (Chess/etc)	
On-line Help		Software Registration		Distributed Databases	
Bulletin Boards		Internet Telephone		On-line Games	

2. Conduct a survey of the main disadvantages of the WWW/Internet. Every person is allowed 12 votes. Ten points should be allocated to, in their opinion, the main disadvantage, five for their second, three for their third, two for their fourth and one for their fifth. Total the votes and determine the Top 10 disadvantages of the WWW/Internet.

Application	*Vote*	*Application*	*Vote*	*Application*	*Vote*
Hacking		Time wastage		Poor Presentation of Material in Many WWW Sites	
Lack of Good Educational Material		Access to Pornography and Related Material		Too Much Remote Access	
Spying on Electronic Mail		Spying on WWW/ Internet usage		Fossilization of Links	
Poor On-line Support		Search Facilities are Swamped by Non-related Material		Security of Electronic Commerce	
Security of All Transmissions		Lack of Speed		Cost of Internet Service Provider	
On-line Internet Costs (such as Telephone Bills/ Network Maintenance/etc)		Lack of Interaction		Lack of Parental Control on WWW accesses	
Incompatibility of WWW Browsers		Incompatibility of WWW pages (such as Java/HTML incompatibilities)		Difficult to Locate Expert Help	
Lack of Real-Time Control		Poor Quality Video Conferencing		Poor Quality Audio	
Connection Varies in its Responsiveness		Lack of Editorial Control on Material		Cost (Physical and Time) of Maintaining a WWW Site	
Lack of Government Control		Lack of Organizational Control		Increased Alienation of People	
Too Much Surface Learning, Not Enough Deep Learning		Reduction in Educational Standards (through Remote Teaching/ On-line Help/etc)		Difficult to Learn from a Computer Screen	
Access to Violent Material		Other:		Other:	

3. Construct your own survey of the main advantages of the WWW/Internet.

20.12 HTTP reference

20.12.1 Method definitions

The main methods are:

- **GET.** Retrieves the information that is defined in the Request-URI. If the request message includes an If-Modified-Since header field, the GET is conditional on the date and time that is defined in the If-Modified-Since header. Conditional GETs allow cached system to be refreshed only when resources have been updated.
- **HEAD.** Identical to GET, but the server should not return any Entity-Body in the response. It can be used for obtaining information on the resource without actually receiving the Entity-Body. It can be used to test links for the accessibility, validity, and so on.
- **POST.** Used to request that the destination server accept the entity enclosed in the request as a subordinate of a resource. Typical applications are annotating existing resources and posting message to bulletin boards, newsgroups, and so on.
- **PUT.** Requests that the enclosed entity is stored under the supplied Request-URI. If the Request-URI is an existing resource, the enclosed entity is considered as a modified version of the existing resource, else it is treated as a new resource.
- **DELETE.** Requests that the server deletes the resource identified by the Request-URI.
- **LINK.** Establishes one or more links between the existing resources and resources identified by the Request-URI.
- **UNLINK.** Opposite of LINK.

20.12.2 HTTP message

HTTP messages are text-based and are either requests from the client to the server, or are responses from the server to the client. They can either be Simple-Requests/Responses (HTTP/0.9), such as:

GET <SPACE> Request-URI <CRLF>
or Full-Requests/Responses (HTTP/1.0 or HTTP/1.1), such as:

GET <SPACE> Request-URI <SPACE> HTTP-Version <CRLF>
HEAD <SPACE> Request-URI <SPACE> HTTP-Version <CRLF>
POST <SPACE> Request-URI <SPACE> HTTP-Version <CRLF>

Simple-Request format is discouraged as it prevents a server from identifying the media type of the returned entity. Both types can include optional header fields and an entity body. Entity bodies are separated from the headers by a null line.

HTTP header fields consist of a name followed by a colon (':'), then a space followed by the field values. They include:

- **General-Headers.** These are applicable for both request and response messages, but they do not apply to the entity being transferred.
- **Request-Header.** These allow clients to pass extra information about the request, and about itself, to the server. They act as request modifiers.
- **Entity-Header.** These define optional metainformation about the Entity-Body.
- **Response-Header.** These allow the server to pass additional information about the response which cannot be placed in the Status-Line.

The Date field represents the date and time of the transmitted message (in the same format as RFC822). An example is:

Date: Wed, 30 Dec 1998 15:22:15 GMT

The Pragma field allows a client to refresh a cached copy which is known to be corrupted or stale and is used to include implementation-specific directives that may apply to any recipient along the request/response chain. An example is:

pragma-directive = no-cache

The Authorization field authenticates a user agent to a server. The server sends a challenge to the user agent, based on a user-ID and a password for each realm (a sting). The user agent must return back a valid user-ID and password for the Request-URI (this is the basic authentication scheme).

When the server receives an unauthorized request for a URI, the server responds with a challenge, such as:

WWW-Authenticate: Basic realm="FredServer"

where "FredServer" is the realm to identify the protection space of the Request-URI. The client must then send back its user-ID and password, separated by a single colon character, and using base64 encoded (see Section 19.6.4). An example may be:

Authorization: Basic A=IYaacZZP===GSpKQarcc31pacbu

The From field is typically used for logging purposes and contains an Internet e-mail address of the owner of the requesting user agent, such as:

From: fred_b@myserver.com

The If-Modified-Since field is used with the GET to define that a GET is conditional on the date that the resource was last modified. If it was not modified since the specified date then the resource is not sent in the Entity-Body and a 304 response (Not Modified) is sent. An example of this field is:

If-Modified-Since: Wed, 30 Dec 1998 15:22:15 GMT

The Referer field allows the client to specify the address (URI) from which the Request-URI was obtained. This allows the server to trace referenced links. An example of this field is:

Referer: http://www.w3.org/hypertext/DataSources/Overview.html

The Location field defines the resource location that was specified in the Request-URI. An example of this field is:

Location: http://www.myserver.com/test/newlink.html

The Server field contains information about the software which is used by the origin server to handle the request. An example of this field is:

Server: CERN/3.0 libwww/2.17

The Content-Encoding field describes the type of encoding used in the entity. It indicates the additional content coding has been applied to the resource. In most cases the encoding is 'gzip' or 'x-compress' (or 'x-gzip' and 'compress'). The Content-Type header field defines the underlying content. An example of this field is:

Content-Encoding: x-gzip
Content-Type=text/plain

where the content type is plain text, which has been zipped with gzip (a Lempel-Ziv based compression program). The x-compress program uses Lempel-Ziv-Welsh (LZW) compression.

The Content-Length field defines the number of bytes in the Entity-Body. An example is:

Content-Length: 4095

The Content-Type field defines the media type of the Entity-Body sent to the recipient. An example of the field is:

Content-type: text/plain; charset=ISO-8859-1

Content-Types are of the format *type/subject*. Typical types are 'application', 'image', 'text', 'audio' and 'video', and example content types are:

```
text/plain                  text/enriched           image/gif
image/jpeg                  audio/basic             video/mpeg
application/octet-stream    application/postscript
application/msword          application/rtf
```

The Expires field defines the date and time after which the entity should be considered as stale. An example of the field is:

Expires: Wed, 30 Dec 1998 15:22:15 GMT

The Last-Modified field indicates the date and time the resource was last modified. An example of the field is:

Last-Modified: Wed, 23 Dec 1998 03:20:25 GMT

The Language field defines the language of the resource. Its format is *Language-tag*, where *Language* is the language (such as en for English), and *tag* is the dialect. For example:

Content-type: audio/basic
Content-Language: en-scottish

20.12.3 Status codes

Informational 1xx
The Informational status codes indicate a provisional response. They are typically used for experimental purposes.

Successful 2xx
The successful status codes indicate that a client's request has been successful. These include:

200 (OK)	Request succeeded. Information returned depends on the method used. With GET, the requested entity is sent in the response.
201 (Created)	New resource has been created. The new created resource is defined in the URI returned in the entity of the response.
202 (Accepted)	Request has been accepted for processing, but the processing is continuing.
204 (No Content)	Successful implementation, but there is no new information to be sent back.

Redirection 3xx
The redirection status code indicates that further action is required to service the request. These include:

300 (Multiple Choices)	Request resource is available from one or more locations. If the server has a preference it adds a Location field with the preferred URL.
301 (Moved Permanently)	Requested resource has been assigned a new permanent URL. Future references should be made to the specified URL, which is located in the Location field.
302 (Moved Temporarily)	Requested resource has temporarily moved to another URL. The calling agent should still use the URI given in the Request-URI.
304 (Not Modified)	This is sent when the client issues a GET request with the If-Modified-Since field is set to a given time, and the document has not modified since the time specified.

Client Error 4xx
The client error status codes indicate that an error has occurred on the client. These include:

400 (Bad Request)	Request cannot be understood by the server due to bad syntax.
401 (Unauthorized)	Request requires user authentication. A response includes the WWW-Authenticate header field, which contains a challenge for the requested resource. The client must then repeat the request with a suitable Authorization header field.
403 (Forbidden)	Server will not service the request. It is typically used when the server does not want to reveal the reason why it is not servicing the request.
404 (Not Found)	Server did not find the required Request-URI.

Server Error 5xx
The server error status codes indicate that there is an error with the server.

500 (Internal Server Error)	Unexpected server condition in servicing the request.
501 (Not Implemented)	Server does not support the request.
502 (Bad Gateway)	Server, while acting as a gateway or proxy, received an invalid response from the upstream server it accessed in attempting to service the request.
503 (Service Unavailable)	Server is currently unavailable, possibly because of server maintenance or overloading.

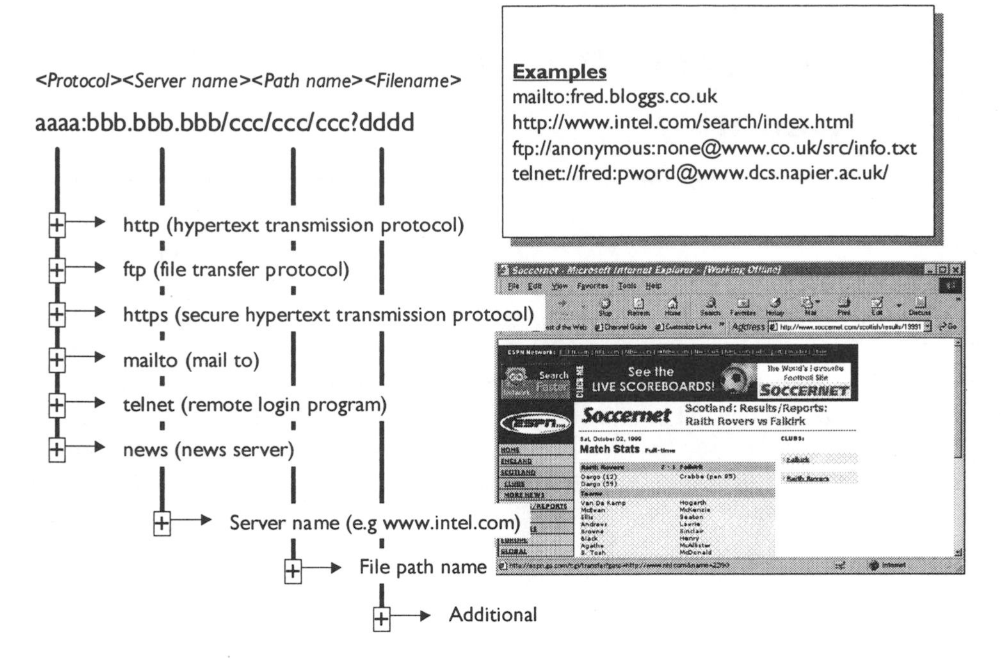

Figure 20.14 URL format

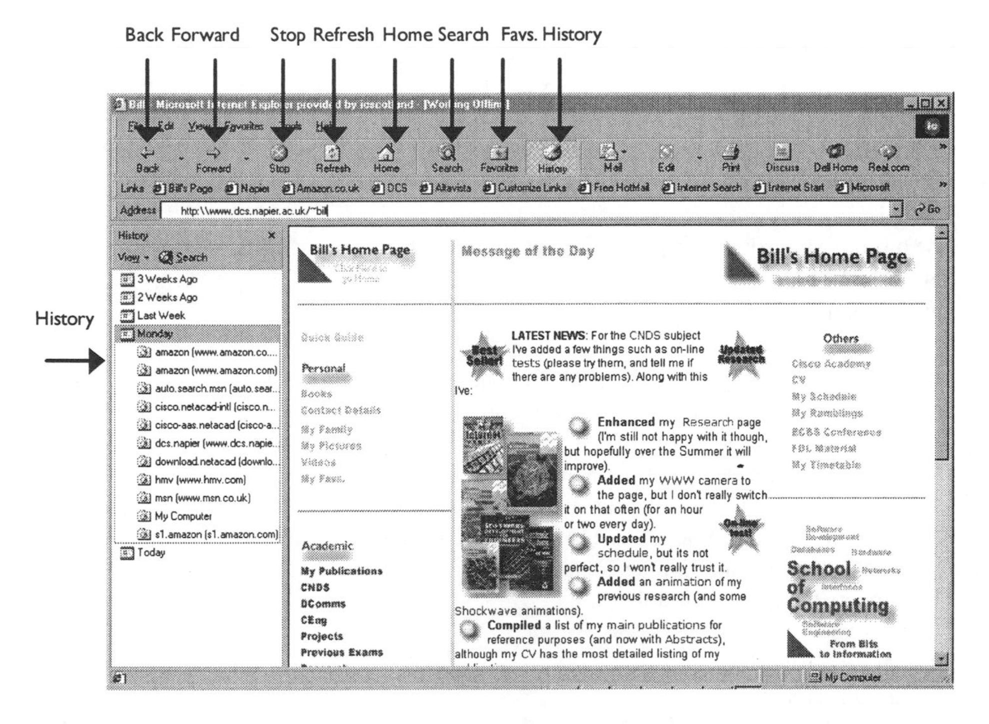

Figure 20.15 WWW Browser design

Firewalls, Proxy Servers and Security

Ring-fenced firewall

Many organizations with several interconnected sites use a single firewall for all incoming and outgoing Internet traffic, as the network administrator can properly configure the firewall for the required security level, and also monitor any incoming and outgoing network traffic.

Note: Gateways into and out of a network, these days, are typically routers. At one time they were basically computers that connected to the internal and external network that filtered the traffic.

21.1 Introduction

An organization may experience two disadvantages in having a connection to the WWW and the Internet:

- The possible usage of the Internet for non-useful applications (by employees).
- The possible connection of non-friendly users from the global connection into the organization's local network.

For these reasons many organizations have shied away from connection to the global network and have set up intranets. These are in-house, tailor-made internets for use within the organization and provide limited access (if any) to outside services and also limit the external traffic into the intranet (if any). An intranet might have access to the Internet but there will be no access from the Internet to the organization's Intranet.

Organizations which have a requirement for sharing and distributing electronic information normally have three choices:

- Use a propriety groupware package, such as Lotus Notes.
- Set up an intranet.
- Set up a connection to the Internet.

Groupware packages normally replicate data locally on a computer whereas intranets centralize their information on central servers which are then accessed by a single browser package. The stored data is normally open and can be viewed by any compatible WWW browser. Intranet browsers have the great advantage over groupware packages in that they are available for a variety of clients, such as PCs, UNIX workstations, Macs, and so on. A client browser also provides a single GUI interface which offers easy integration with other applications, such as electronic mail, images, audio, video, animation, and so on.

The main elements of an intranet are:

- Intranet server hardware.
- Intranet server software.
- TCP/IP stack software on the clients and server.
- WWW browsers.
- A firewall.

Typically the intranet server consists of a PC running the Linux (PC-based UNIX-like) operating system. The TCP/IP stack is software installed on each computer and allows communications between a client and a server using TCP/IP.

A firewall is the routing computer which isolates the intranet from the outside world. Another method is to use an intermediate system which isolates the intranet from the external Internet. These intermediate systems include:

- A proxy. This connects to a number of clients; it acts on behalf of clients and sends requests from the clients to a server. It thus acts as a client when it communicates with a server, but as a server when communicating with a client. A proxy is typically used for security purposes where the client and server are separated by a firewall. The proxy connects to the client side of the firewall and the server to the other side of the firewall. Thus the server must authenticate itself to the firewall before a connection can be made with the proxy. Only after this has been authenticated will the proxy pass requests through the firewall. A proxy can also be used to convert between different versions of the protocols that use TCP/IP, such as HTTP.
- A gateway.
- A tunnel.

Virus signatures and footprints

All viruses have a signature, or create a footprint when they are executed. Virus decoders must determine the operation of the virus in order for an anti-virus program not to make a mistake in thinking that a valid piece of program is virus code. Thus there is very little guesswork with a virus-seeking program, and most viruses must be carefully analyzed before an anti-virus program can be updated to eradicate it.

Each intermediate system is connected by TCP and acts as a relay for the request to be sent out and returned to the client. Figure 21.1 shows the set-up of the proxies and gateways.

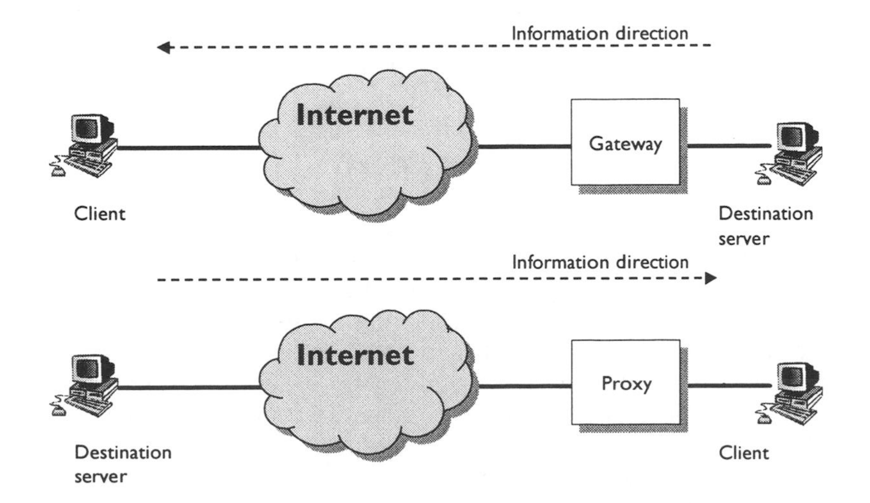

Figure 21.1 Usage of proxies and gateways

21.2 Firewalls

A firewall (or security gateway) protects a network against intrusion from outside sources. They tend to differ in their approach, but can be characterized either as firewalls which block traffic or firewalls which permit traffic. They can be split into three main types:

- **Network-level firewalls** (packet filters). This type of firewall examines the parameters of the TCP/IP packet to determine if it should be dropped or not. This can be done by examining the destination address, the source address, the destination port, the source port, and so on. The firewall must thus contain a list of barred IP addresses or allowable IP addresses. Typically a system manager will determine IP addresses of sites which are barred and add them to the table. Certain port numbers will also be barred, typically TELNET and FTP ports are barred and SMTP is allowed, as this allows mail to be routed into and out of the network, but no remote connections.
- **Application-level firewalls.** This type of firewall uses an intermediate system (a proxy server) to isolate the local computers from the external network. The local computer communicates with the proxy server, which in turn communicates with the external network, the external computer then communicates with the proxy which in turn communicates with the local computers. The external network never actually communicates directly with the local computers. The proxy server can then be set-up to be limited to certain types of data transfer, such as allowing HTTP (for WWW access), SMTP (for electronic mail), outgoing FTP, but blocking incoming FTP.
- **Circuit-level firewalls.** A circuit-level firewall is similar to an application-level firewall but it does not bother about the transferred protocol.

Viruses at the speed of light

Before the advent of networks and the Internet, the most common mechanism for spreading a virus was through floppy disks and CD-ROM disks. As the spread of the viruses was by manual insertion from one machine to another, the anti-virus programs could easily keep up-to-date with the latest viruses, and modify their databases. Viruses would spread around the world over a period of months, if not years. Networks and the Internet have changed all this, where a virus can now be transmitted over a LAN in a fraction of a second, and around the world in less than a second. Thus a virus can be created and transmitted around the world before an anti-virus programmer can even identify its make-up.

21.2.1 Network-level firewalls

The network-level firewall (or packet filter) is the simplest form of firewall and is also known as a screen router. It basically keeps a record of allowable source and destination IP addresses, and deletes all packets which do not have them. This technique is known as address filtering. The packet filter keeps a separate source and destination table for both directions, that is, into and out of the intranet. This type of method is useful for companies which have geographically spread sites, as the packet filter allows incoming traffic from other friendly sites, but blocks other non-friendly traffic. This is illustrated by Figure 21.2.

Unfortunately, this method suffers from the fact that IP addresses can be easily forged. For example, a hacker might determine the list of good source addresses and then add one of them to any packets which are addressed to the intranet. This type of attack is known as address spoofing and is the most common method of attacking a network.

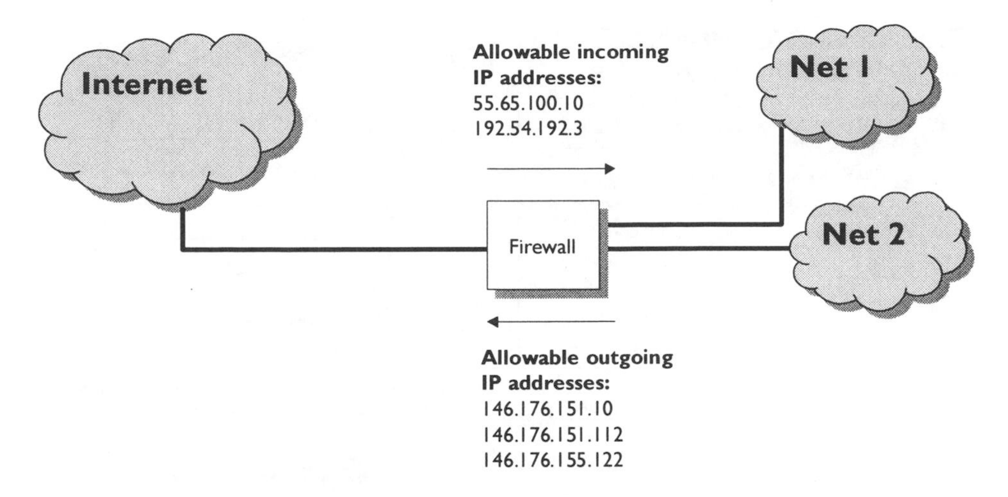

Figure 21.2 Packet filter firewalls

21.2.2 Application-level firewall

The application-level firewall uses a proxy server to act as an intermediate system between the external network and the local computer. Normally the proxy only supports a given number of protocols, such as HTTP (for WWW access) or FTP. It is thus possible to block certain types of protocols, typically outgoing FTP (Figure 21.3).

The proxy server thus isolates the local computer from the external network. The local computer communicates with the proxy server, which in turn communicates with the external network, the external computer then communicates with the proxy, which in turn, communicates with the local computer. The external network never actually communicates directly with the local computer. The left-hand window of Figure 21.4 shows a WWW browser is set up to communicate with a proxy server to get its access. In the advanced options (right-hand side of Figure 21.4) different proxy servers can be specified. In this case, for HTTP (WWW access), FTP, Gopher, Secure and Socks (Windows Sockets). It can also be seen that a proxy server can be bypassed by specifying a number of IP addresses (or DNS).

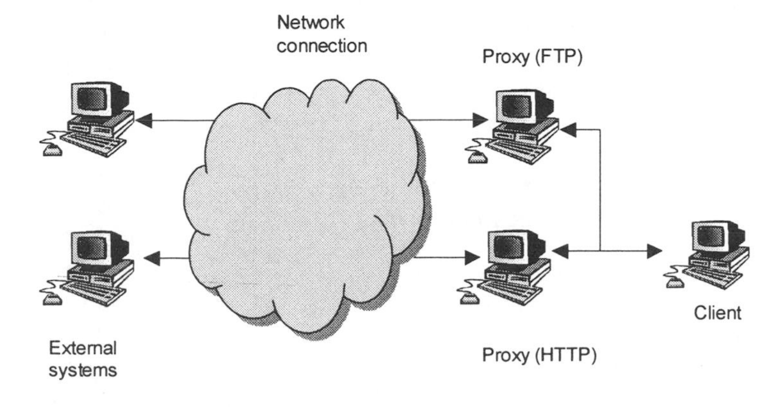

Figure 21.3 Application-level firewall

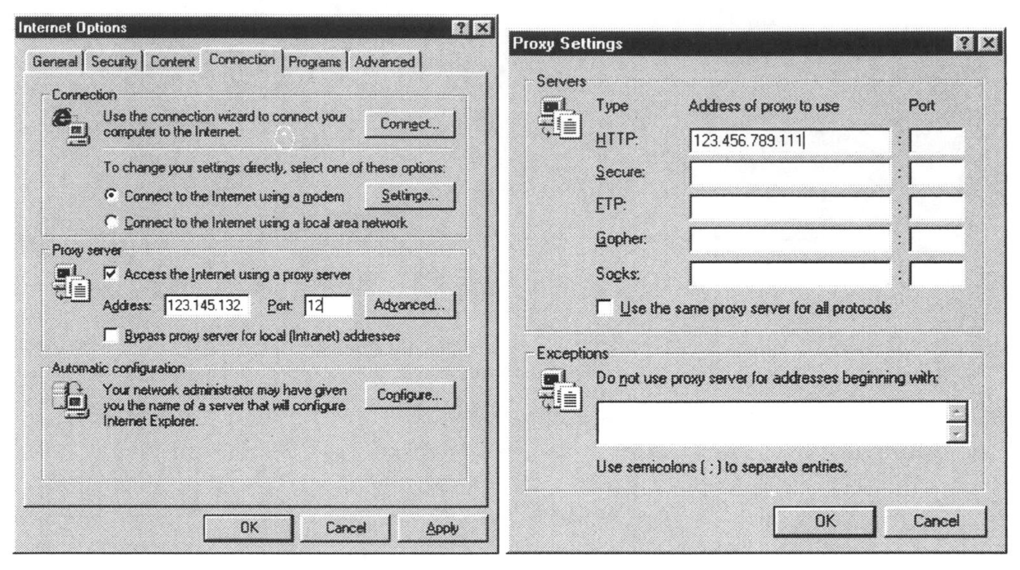

Figure 21.4 Internet options showing proxy server selection and Proxy settings

21.3 Application-level gateways

Application-level gateways provide an extra layer of security when connecting an intranet to the Internet. They have three main components:

- A gateway node.
- Two firewalls which connect on either side of the gateway and only transmit packets which are destined for or to the gateway.

Figure 21.5 shows the operation of an application level gateway. In this case, Firewall A discards anything that is not addressed to the gateway node, and discards anything that is not sent by the gateway node. Firewall B similarly discards anything from the local network that is not addressed to the gateway node, and discards anything that is not sent by the gateway node.

Thus, to transfer files from the local network into the global network, the user must do the following:

- Log onto the gateway node.
- Transfer the file onto the gateway.
- Transfer the file from the gateway onto the global network.

To copy a file from the network, an external user must:

- Log onto the gateway node.

I Love You

The Love Bug worm poked its head above the surface on the fourth day of May 2000, a day that will go down in history as the day that the world said 'I Love You'. Unfortunately it was not a message which would bring world peace, nor was it a sign of affection. Never before had such a global threat grown so quickly and attack so many systems. The first sign of the message appeared in Europe and then spread as the sun rose around the world. Soon, the National Infrastructure Protection Center (NIPC) issued a warning that the LoveLetter or LoveBug virus had now spread around the global e-mail system. On hearing of warning, IT managers around the world started shutting down their IT systems, while virus decoders were already working hard on methods that could be used to defeat the new menace.

- Transfer from the global network onto the gateway.
- Transfer the file from the gateway onto the local network.

A common strategy in organizations is to allow only electronic mail to pass from the Internet to the local network. This specifically disallows file transfer and remote login. Unfortunately, electronic mail can be used to transfer files. To overcome this problem the firewall can be designed specifically to disallow very large electronic mail messages, so it will limit the ability to transfer files. This tends not to be a good method as large files can be split up into small parts, then sent individually.

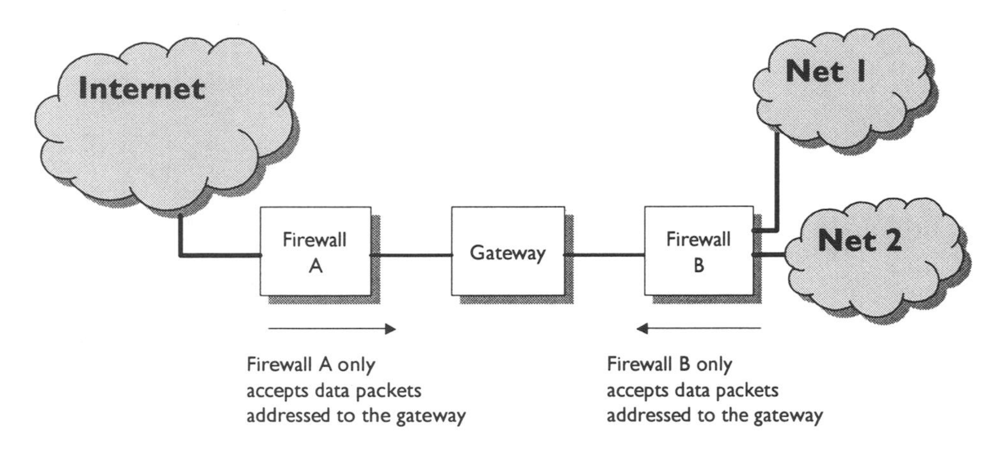

Figure 21.5 Application level gateway

21.4 Ring-fenced firewalls

Many large organizations have several sites which are spread over a geographically large area. This causes a major problem for security, as the different sites would have to be administered separately. Any small breaches on a single site could cause the whole organizational network to become threatened. One solution is to centralize the gateway into and out of the organization network. An example is given in Figure 21.6 where the three sites each have a firewall which protects security breaches between each of the sites. These then connect to a common router, which then connects to a strong firewall. At the gateway to each of the sites and at the gateway to the entire organization network, there is an audit-monitoring computer, which will log all the incoming and outgoing traffic over time. This could monitor incoming and outgoing IP addresses, domain names, transport protocols, and so on. Any security breaches can be easily detected by examining these logs. The audit monitor could also be used against staff if it shows that employees have been acting incorrectly, such as copying files from the organizational network to an external network, or accessing inappropriate WWW sites.

21.5 Encrypted tunnels

Packet filters and application level gateways suffer from insecurity, which can allow non-friendly users into the local network. Packet filters can be tricked with fake IP addresses and application level gateways can be hacked into by determining the password of certain users of the gateway then transferring the files from the network to the firewall, on to the gateway, on to the next firewall and out. The best form of protection for this type of attack is to allow only a limited number of people to transfer files onto the gateway.

The best method of protection is to encrypt the data leaving the network then to decrypt it on the remote site. Only friendly sites will have the required encryption key to receive and send data. This has the extra advantage that the information cannot be easily tapped-into. Only the routers which connect to the Internet require to encrypt and decrypt, as illustrated in Figure 21.7.

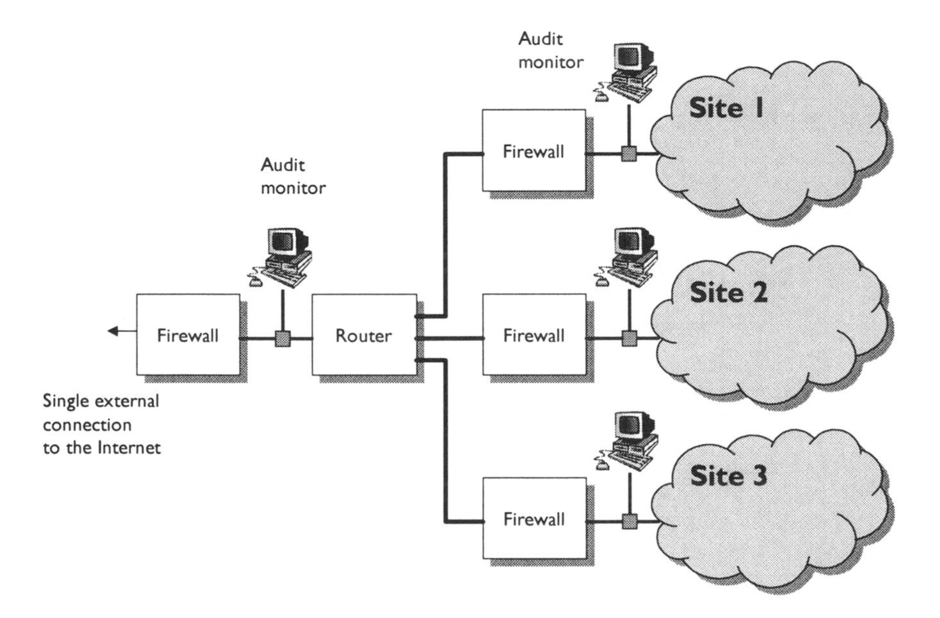

Figure 21.6 Ring-fenced firewall

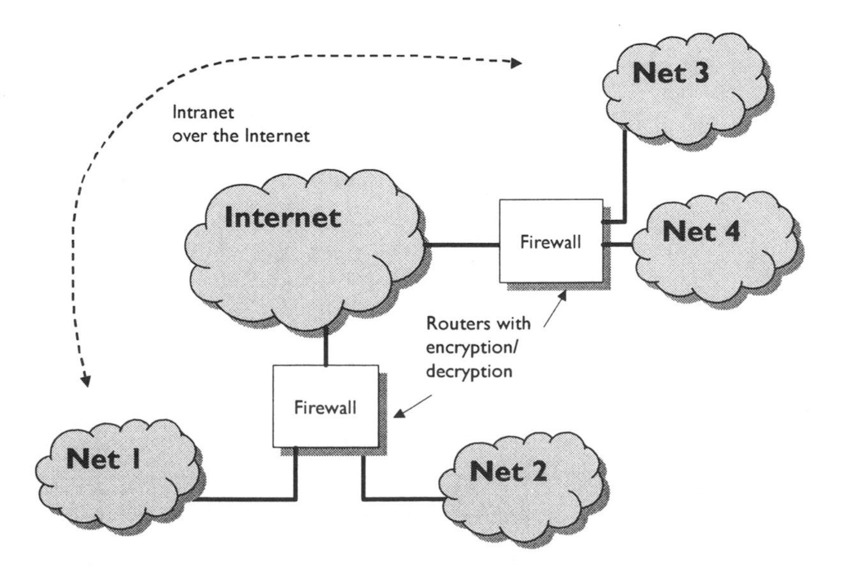

Figure 21.7 Encryption tunnels

Typically, remote users connect to a corporation intranet by connecting over a modem which is connected to the corporation intranet, and using a standard Internet connection protocol, such as Point-to-Point Protocol (PPP). This can be expensive in both phone calls or in providing enough modems for all connected users. These costs can be drastically reduced if the user connects to an ISP, as they provide local rate charges. For this a new protocol, called Point-to-Point Tunneling Protocol (PPTP) has been developed to allow remote users connections to intranets from a remote connection (such as from a modem or ISDN). It operates as follows:

- Users connect to an ISP, using a protocol such as Point-to-Point Protocol (PPP) and request that the information is sent to an intranet. The ISP has special software and hardware to handle PPTP.
- The data sent to the ISP, using PPTP, is encrypted before it is sent into the Internet.
- The ISP sends the encrypted data (wrapped in an IP packet) to the intranet.
- Data is passed through the firewall, which has the software and hardware to process PPTP packets.
- Next, the user logs in using Password Authentication Protocol (PAP) or Challenge Handshake Authentication (CHAP).
- Finally, the intranet server reads the IP packet and decrypts the data.

21.6 Filtering routers

Filtering routers run software which allows many different parameters of the incoming and outgoing packets to be examined, as illustrated in Figure 21.8, such as:

- **Source IP address.** The router will have a table of acceptable source IP addresses. This will limit the access to the external network as only authorized users will be granted IP addresses. This unfortunately is prone to IP spoofing, where a local user can steal an authorized IP address. Typically, it is done by determining the IP address of a computer and waiting until there is no-one using that computer, then using the unused IP address. Several users have been accused of accessing unauthorized material because other users have used their IP address. A login system which monitors IP addresses and the files that they are accessing over the Internet cannot be used as evidence against the user, as it is easy to steal IP addresses.
- **Destination IP address.** The router will have a table of acceptable outgoing destination IP addresses, addresses which are not in the table are blocked. Typically, this will be used to limit the range of destination addresses to the connected organizational intranet, or to block certain addresses (such as pornography sites).
- **Protocol.** The router holds a table of acceptable protocols, such as TCP and/or UDP.
- **Source port.** The router will have a table of acceptable TCP ports. For example, electronic mail (SMTP) on port 25 could be acceptable, but remote login on port 543 will be blocked.
- **Destination port.** The router will have a table of acceptable TCP ports. For example, ftp on port 21 could be acceptable, but telnet connections on port 23 will be blocked.
- **Rules.** Other rules can be added to the system which define a mixture of the above. For example, a range of IP addresses can be allowed to transfer on a certain port, but another range can be blocked for this transfer.

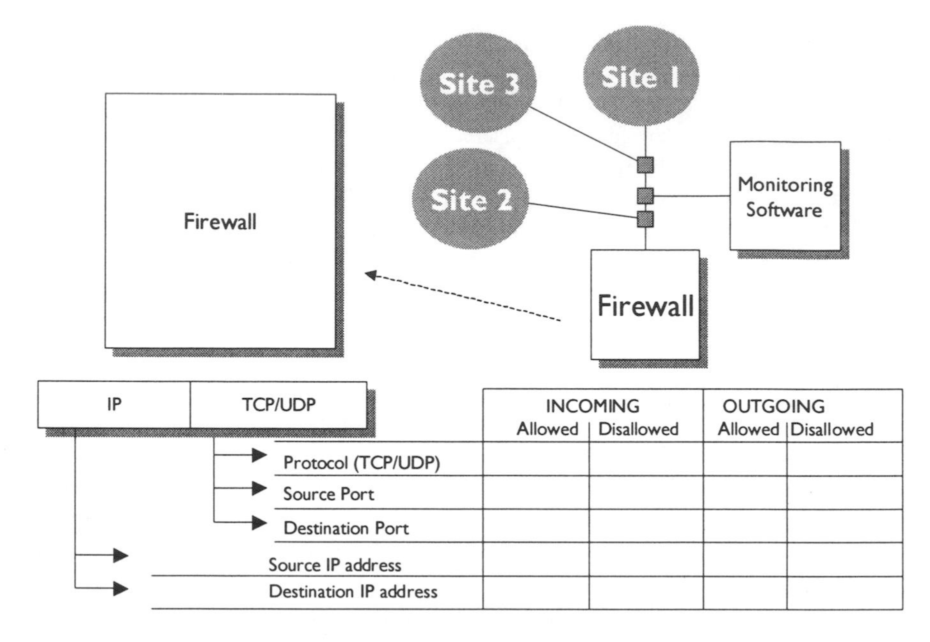

Figure 21.8 Filtering router

Filter routers are either tightly bound when they are installed and then relaxed, or are relaxed and then bound. The type depends on the type of organization. For example, a financial institution will have a very strict router which will allow very little traffic, apart from the authorized traffic. The router can be opened-up when the systems have been proved to be secure (they can also be closed quickly when problems occur).

An open organization, such as an education institution will typically have an open system, where users are allowed to access any location on any port, and external users are allowed any access to the internal network. This can then be closed slowly when internal or external users breach the security or access unauthorized information. For example, if a student is accessing a pornographic site consistently then the IP address for that site could be blocked (this method is basically closing the door after the horse has bolted).

To most users the filtering router is an excellent method of limited traffic access, but to the determined hacker it can be easily breached, as the hacker can fake both IP addresses and also port addresses. It is extremely easy for a hacker to write their own TCP/IP driver software to address whichever IP address, and port numbers that they want.

21.7 Security

Security involves protecting the system hardware and software from both internal attack and from external attack (hackers). An internal attack normally involves uneducated users causing damage, such as deleting important files, crashing systems. Another attack can come from internal fraud, where employees may intentionally attack a system for their own gain, or through some dislike for something within the organization. There are many cases of users who have grudges against other users, causing damage to systems, by misconfiguring systems. This effect can be minimized if the system manager properly protects the system. Typical actions are to limit the files that certain users can access and also the actions they can perform on the system.

Most system managers have seen the following:

- Users sending a file of the wrong format to the system printer (such as sending a binary file). Another typical one is where there is a problem on a networked printer (such as lack of paper), but the user keeps re-sending the same print job.
- Users deleting the contents of sub-directories, or moving files from one place to another (typically, these days, with the dragging of a mouse cursor). Regular backups can reduce this problem.
- Users deleting important system files (in a PC, these are normally AUTOEXEC.BAT and CONFIG.SYS). This can be overcome by the system administrator protecting important system files, such as making them read-only or hidden.
- Users telling other people their user passwords or not changing a password from the initial default one. This can be overcome by the system administrator forcing the user to change their password at given time periods.

Security takes many forms, such as:

- **Data protection.** This is typically where sensitive or commercially important information is kept. It might include information databases, design files or source code files. One method of reducing this risk is to encrypt important files with a password, and another is to encrypt data with a secret electronic key (files are encrypted with a commonly known public key, and decrypted with a secret key, which is only known by user who has the rights to access the files).
- **Software protection.** This involves protecting all the software packages from damage or from being misconfigured. A misconfigured software package can cause as much damage as a physical attack on a system, because it can take a long time to find the problem.
- **Physical system protection.** This involves protecting systems from intruders who might physically attack the systems. Normally, important systems are locked in rooms and then within locked rack-mounted cabinets.
- **Transmission protection.** This involves a hacker tampering with a transmission connection. It might involve tapping into a network connection or total disconnection. Tapping can be avoided by many methods, including using optical fibers which are almost impossible to tap into (as it would typically involve sawing through a cable with hundreds of fiber cables, which would each have to be connected back as they were connected initially). Underground cables can avoid total disconnection, or its damage can be reduced by having redundant paths (such as different connections to the Internet).
- **Using an audit log file.** Many secure operating systems, such as Windows NT/2000, have an audit file, which is a text file that the system maintains and updates daily. This is a text file that can record all of the actions of a specific user, and is regularly updated. It can include the dates and times that a user logs into the system, the files that were accessed, the programs that were run, the networked resources that were used, and so on. By examining this file the system administrator can detect malicious attacks on the system, whether it is by internal or external users.

21.7.1 Hacking methods

The best form of protection is to disallow hackers into the network in the first place. Organizational networks are hacked for a number of reasons and in a number of ways. The most common methods are:

- **IP spoofing attacks.** This is where the hacker steals an authorized IP address, as illustrated in Figure 21.9. Typically, it is done by determining the IP address of a computer and waiting until there is no-one using that computer, then using the unused IP address. Several users have been accused of accessing unauthorized material because other users have used their IP address. A login system which monitors IP addresses and the files that they are accessing over the Internet cannot be used as evidence against the user, as it is easy to steal IP addresses.
- **Packet-sniffing.** This is where the hacker listens to TCP/IP packets which come out of the network and steals the information in them. Typical information includes user logins, e-mail messages, credit card number, and so on. This method is typically used to steal an IP address, before an IP spoofing attack. Figure 21.10 shows an example where a hacker listens to a conversation between a server and a client. Most TELNET and FTP programs actually transmit the user name and password as text values; these can be easily viewed by a hacker, as illustrated in Figure 21.11.
- **Passwords attacks.** This is a common weak-point in any system, and hackers will generally either find a user with an easy password (especially users which have the same password as their login name) or will use a special program which cycles through a range of passwords. This type of attack is normally easy to detect. The worst nightmare of this type of attack is when a hacker determines the system administrator password (or a user who has system privileges). This allows the hacker to change system set-ups, delete files, and even change user passwords.
- **Sequence number prediction attacks.** Initially, in a TCP/IP connection, the two computers exchange a start-up packet which contains sequence numbers (Section 15.4.2). These sequence numbers are based on the computer's system clock and then run in a predictable manner, which can be determined by the hacker.
- **Session hi-jacking attacks.** In this method, the hacker taps into a connection between two computers, typically between a client and a server. The hacker then simulates the connection by using its IP address.

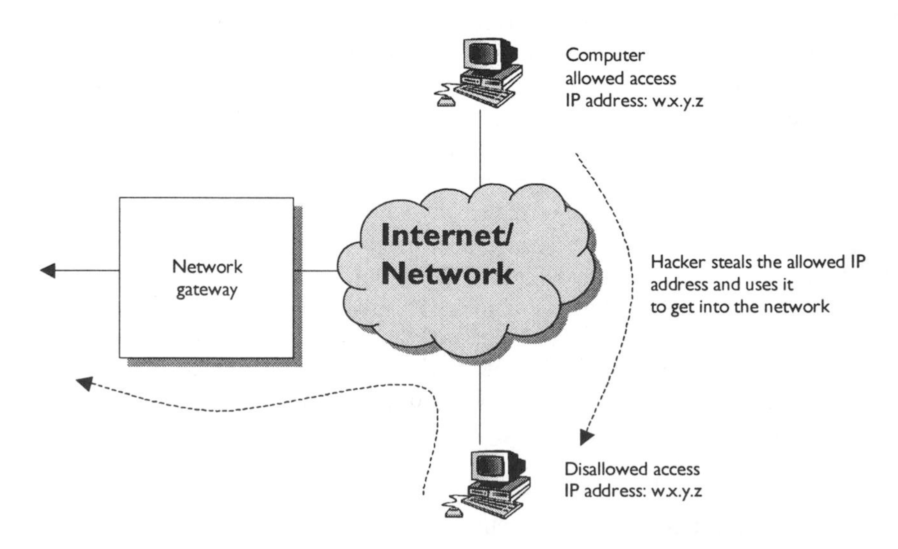

Figure 21.9 IP spoofing

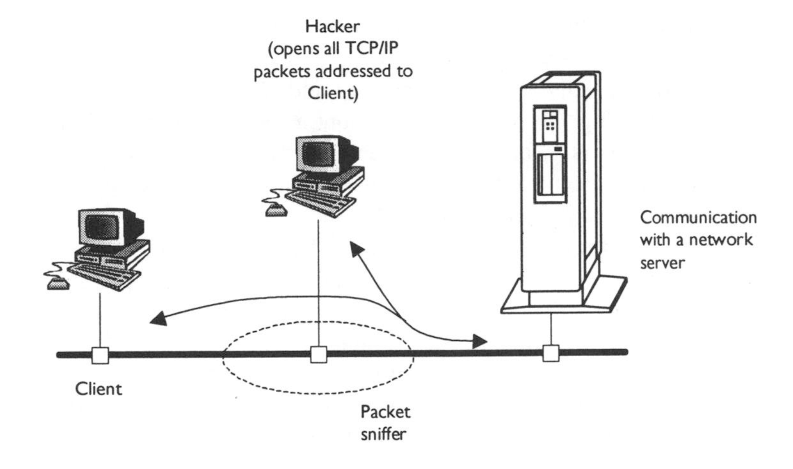

Figure 21.10 Packet sniffing

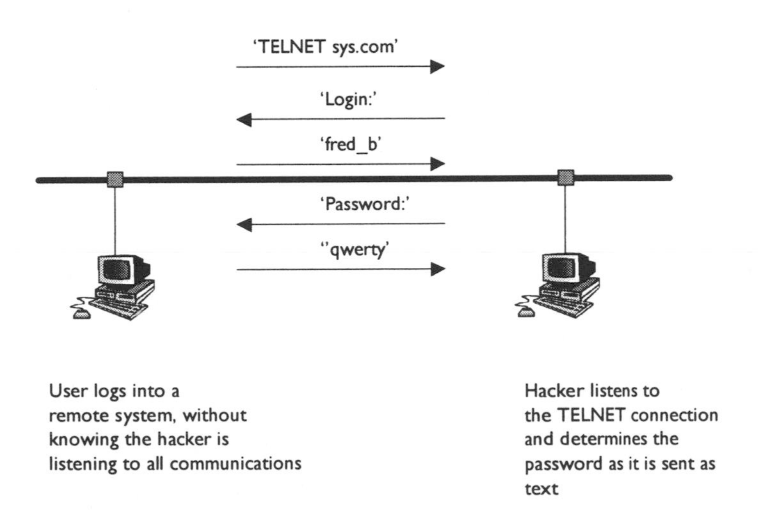

Figure 21.11 Packet sniffing on a TELNET connection

- **Shared library attacks.** Many systems have an area of shared library files. These are called by applications when they are required (for input/output, networking, graphics, and so on). A hacker may replace standard libraries for ones that have been tampered with, which allows the hacker to access system files and to change file privileges. Figure 21.12 illustrates how a hacker might tamper with dynamic libraries (which are called as a program runs), or with static libraries (which are used when compiling a program). This would allow the hacker to possibly do damage to the local computer, send all communications to a remote computer, or even view everything that is viewed on the user screen. The hacker could also introduce viruses and cause unpredictable damage to the computer (such as remotely rebooting it, or crashing it at given times).
- **Technological vulnerability attack.** This normally involves attacking some part of the system (typically the operating system) which allows a hacker to access the system. A typical one is for the user to gain access to a system and then run a program which re-boots the system or slows it down by running a processor intensive program. This can

be overcome in operating systems such as Microsoft Windows and UNIX by granting re-boot rights only to the system administrator.

- **Trust-access attacks.** This allows a hacker to add their system to the list of systems which are allowed to log into the system without a user password. In UNIX this file is the *.rhosts* (trusted hosts) which is contained in the user's home directory. A major problem is when the trusted hosts file is contained in the root directory, as this allows a user to log in as the system administrator.
- **Social engineering attacks.** This type of attack is aimed at users who have little understanding of their computer system. A typical attack is where the hacker sends an e-mail message to a user, asking for their password. Many unknowing users are tricked by this attack. A few examples are illustrated in Figure 21.13. From the initial user login, the hacker can then access the system and further invade the system. In one research study it was found that when telephoned by an unknown person and asked what their password was, 90% of users immediately gave it, without asking any questions.

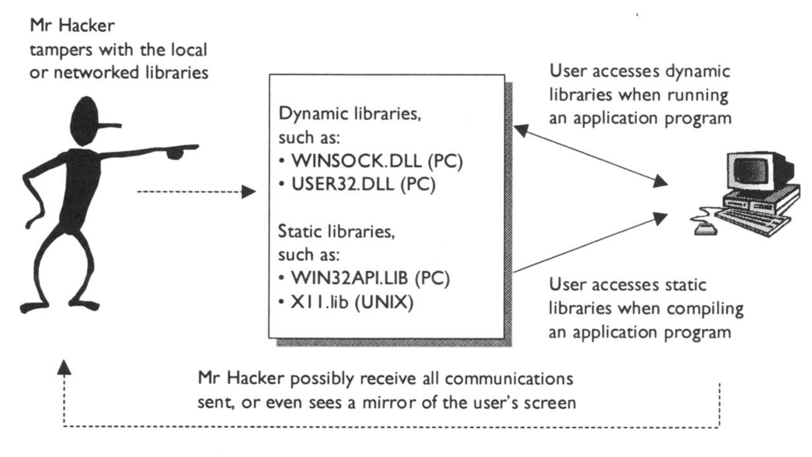

Figure 21.12 Shared library attack

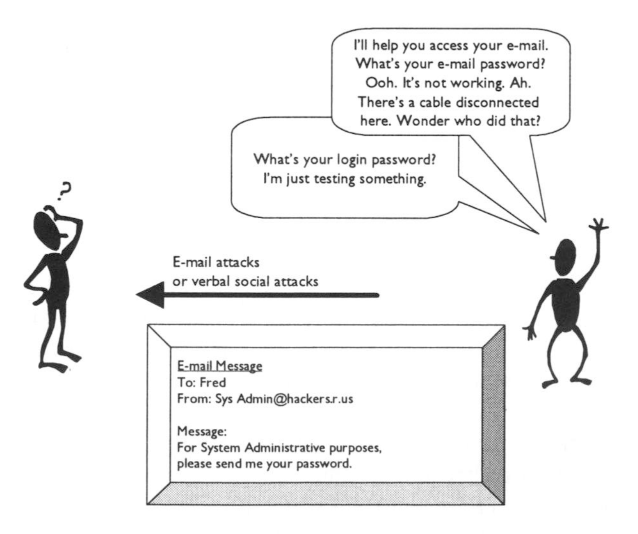

Figure 21.13 Social Engineering attack

21.7.2 Security policies

A well-protected system depends mainly on the system manager. It is up to the manager to define security policies which define how users can operate the system. A good set of policies would be:

- **Restrictions on users who can use a given account.** The system administrator needs to define the users who can login on a certain account.
- **Password requirements and prohibitions.** This defines the parameters of the password, such as minimum password size, time between password changes, and so on.
- **Internet access/restrictions.** This limits whether or not a user is allowed access to the Internet.
- **User account deletion.** The system administrator automatically deletes user accounts which are either not in use or users have been moved to another system.
- **Application program rules.** This defines the programs which a user is allowed to run (typically games can be barred for some users).
- **Monitoring consent.** Users should be informed about how the system monitors their activities. It is important, for example, to tell users that their Internet accesses are being monitored. This gives the user no excuse when they are found to be accessing restricted sites.

21.7.3 Passwords

Passwords are normally an important part of any secure network. They can be easily hacked with the use of a program which continually tries different passwords within a given range (normally called directory-based attacks). These can be easily overcome by only allowing a user three bad logins before the system locks the user out for a defined time. Novell NetWare and Windows NT/2000 both use this method, but UNIX does not. The system manager, though, can determine if an attack has occurred with the BADLOG file. This file stores a list of all the bad logins for a user and the location of the user.

Passwords are a basic system for providing security on a network, and they are only as secure as the user makes them. Good rules for passwords are:

- Use slightly unusual names, such as *vinegarwine*, *dancertop* or *helpcuddle*. Do not use names of a wife, husband, child or pet. Many users, especially ones who know the user, can easily guess the user's password.
- Use numbers after the name, such as *vinedrink55* and *applefox32*. This makes the password difficult to crack as users are normally only allowed a few chances to login correctly before they are logged out (and a bad login event written to a bad login file).
- Have several passwords which are changed at regular intervals. This is especially important for system managers. Every so often, these passwords should be changed to new ones.
- Make the password at least six characters long. This stops 'hackers' from watching the movement of the user's fingers when they login, or from running a program which tries every permutation of characters. Every character added multiplies the number of combinations by a great factor (for example, if just the characters from 'a' to 'z' and '0' to '9' are taken then every character added increases the number of combinations by a factor of 36).
- Change some letters for numbers, or special characters. Typically, 'o' becomes a 0 (zero), 'i' becomes 1 (one), 's' becomes 5 (five), spaces become '$', 'b' becomes '6', and so on. So a password of 'silly password' might become '5illy$pa55w0rd' (the user

makes a rule for 's' and 'o'). The user must obviously remember the rule that has been used for changing the letters to other characters. This method overcomes the technique of hackers and hacker programs, where combinations of words from a dictionary are hashed to try and make the hashed password.

The two main protocols used are:

- **Password Authentication Protocol** (PAP). This provides for a list of encrypted passwords.
- **Challenge Handshake Authentication Protocol** (CHAP). This is a challenge-response system which requires a list of unencrypted passwords. When a user logs into the system a random key is generated and sent to the user for encrypting the password. The user then uses this key to encrypt the password, and the encrypted password is sent back to the system. If it matches its copy of the encrypted password then it lets the user login. The CHAP system then continues to challenge the user for encrypted data. If the user gets these wrong then the system disconnects the login.

21.7.4 Hardware security

Passwords are a simple method of securing a system. A better method is to use a hardware-restricted system which either bar users from a specific area or even restricts users from login into a system. Typical methods are:

- **Smart cards**. With this method a user can only gain access to the system after they have inserted their personal smart card into the computer and then entered their PIN code.
- **Biometrics**. This is a better method than a smart card where a physical feature of the user is scanned. The scanned parameter requires to be unchanging, such as fingerprints or retina images.

21.7.5 Hacker problems

Once a hacker has entered into a system, there are many methods which can be used to further penetrate into the system, such as:

- **Modifying search paths.** All systems set up a search path in which the system looks into to find the required executable. For example, in a UNIX system, a typical search path is `/bin`, `/usr/bin`, and so on. A hacker can change the search paths for a user and then replace standard programs with ones that have been modified. For example, the hacker could replace the e-mail program for one that sends e-mails directly to the hacker or any directory listings could be sent to the hacker's screen.
- **Modifying shared libraries.** As discussed previously.
- **Running processor intensive tasks** which slows the system down; this task will be run in the background and will generally not be seen by the user. The hacker can then further attack the system by adding the processor intensive task to the system start-up file (such as the `rc` file on a UNIX system).
- **Running network intensive tasks** which will slow the network down, and typically slow down all the connected computers. As with the processor intensive task, the networking intensive task can be added to the system start-up file.
- **Infecting the system with a virus or worm.**

Most PCs now have virus scanners which test the memory and files for viruses. This makes viruses easy to detect. A more sinister virus is spread over the Internet, such as the Internet worm which was released in November 1988.

21.7.6 Viruses

Before the advent of LANs and the Internet, the most common mechanism for spreading a virus was through floppy disks and CD-ROM disks. Anti-virus programs can easily keep up-to-date with the latest viruses, and modify their databases. This is a relatively slow method of spreading a virus and will take many months, if not years, to spread a virus over a large geographical area. Figure 21.14 illustrates the spread of viruses.

LANs and the Internet have changed all this. A virus can now be transmitted over a LAN in a fraction of a second, and around the world in less than a second. Thus a virus can be created and transmitted around the world before an anti-virus program can even detect that it is available.

A worm is a program which runs on a computer and creates two threads. A thread in a program is a unit of code that can get a time slice from the operating system to run concurrently with other code units. Each process consists of one or more execution threads that identify the code path flow as it is run on the operating system. This enhances the running of an application by improving throughput and responsiveness. With a worm, the first thread searches for a network connection and when it finds a connection it copies itself, over the network, to that computer. Next, the worm makes a copy of itself, and runs it on the system. Thus, a single copy will become two, then four, eight, and so on. This continues until the system, and the other connected systems, will be shutdown. The only way to stop the worm is to shutdown all the effected computers at the same time and then restart them. Figure 21.15 illustrates a worm virus.

Macro viruses

One of the most common viruses is the macro virus, which attacks macro or scripting facilities which are available in word processors (such as Microsoft Word), spread-sheets (such as Microsoft Excel), and remote transfer programs. A typical virus is the WM/CAP virus which modifies macros within Microsoft Word. When a macro is executed it can cause considerable damage, such as: deleting files, corrupting files, and so on.

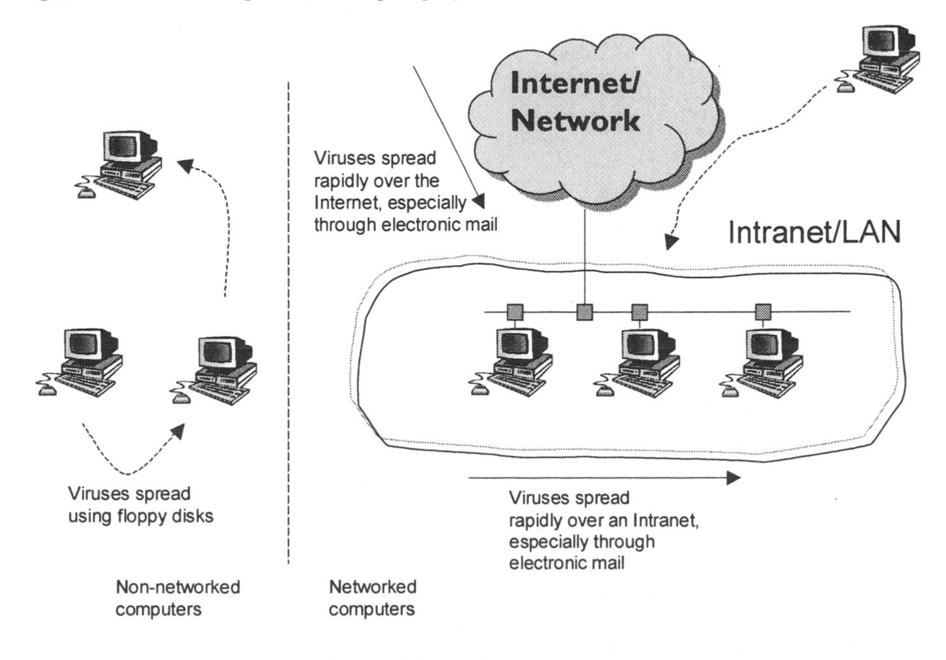

Figure 21.14 Spread of viruses

The greatest increase in macro viruses is the number of viruses which use Microsoft Visual Basic for Applications (VBA) as this integrates with Microsoft Office (although recent releases have guarded against macro viruses). A macro virus in Microsoft Word is spread by:

- The file is either transmitted over email, over a LAN or from floppy disk.
- The infected file is opened and the `normal.dot` main template file is modified so that it contains the modified macros. Any files that are opened or created will now have the modified macros. The WM/CAP macro virus does not do much damage and simply overwrites existing macros.
- VBA is made to be event-driven, so operations, such as File Open or File Close, can have an attached macro. This makes it easy for virus programmers to write new macros.

Figure 21.16 shows an example of a macro created by Visual Basic programming. The developed macro (`Macro1()`) simply loads a file called AUTHOR.doc, selects all the text, converts the text to bold, and then saves the file as AUTHOR.rtf. It can be seen that this macro is associated with `normal.dot`.

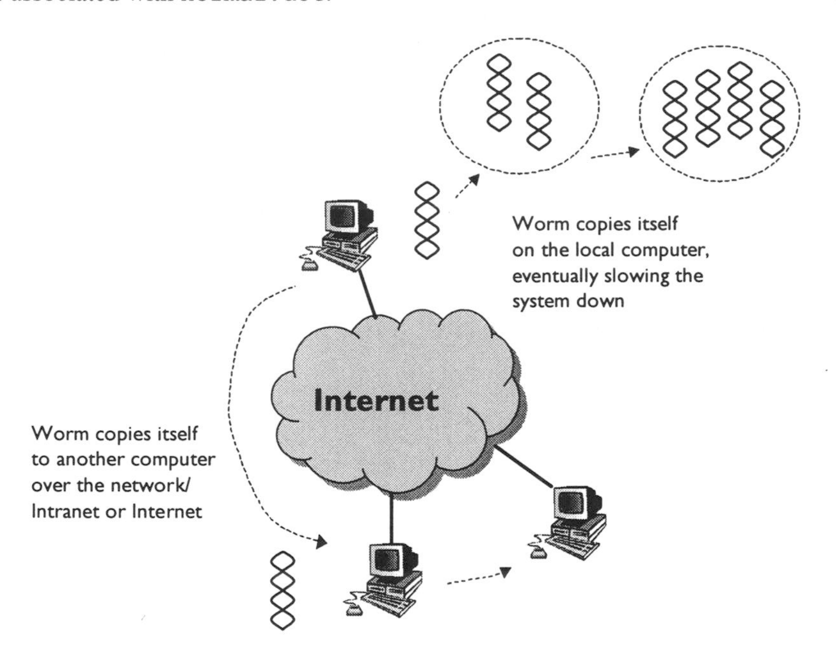

Figure 21.15 Worm viruses

21.8 Exercises

The following questions are multiple choice. Please select from a–d.

21.8.1 Which of the following best describes an Intranet:

(a) A company specific network using company designed tools
(b) A local internet which is isolated from the Internet
(c) A totally incompatible system to the Internet
(d) A faster version of the Internet

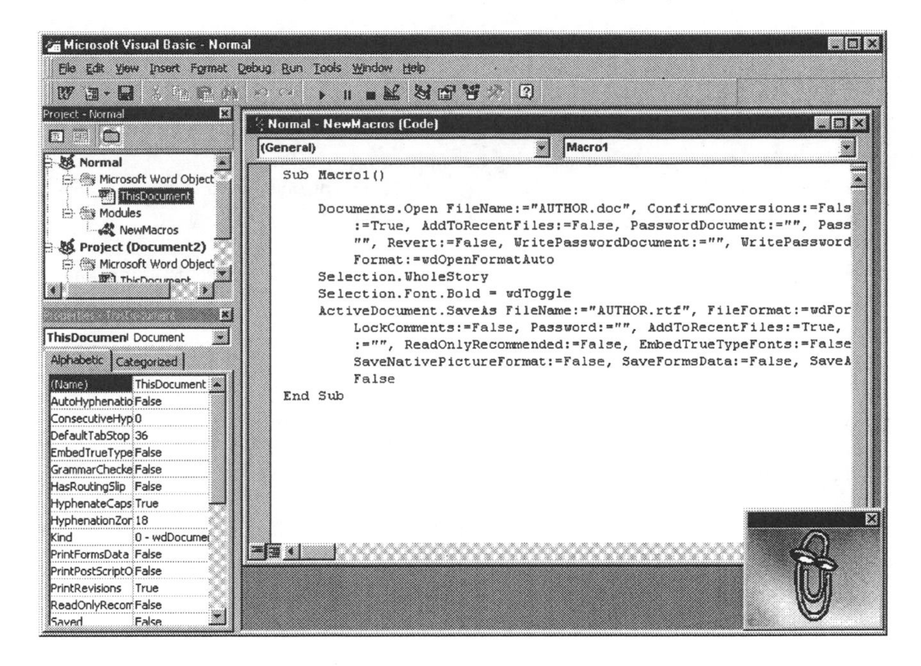

Figure 21.16 Sample macro using VB programming

21.8.2 The main function of a firewall is:
- (a) To disallow unwanted users into the network and allow wanted traffic
- (b) To allow users access to the Internet
- (c) To allow faster transfer of data between the Intranet and the Internet
- (d) Convert one type of network to connect to another type

21.8.3 Which of the following best describes a proxy:
- (a) It connects to a number of clients and acts on behalf of other clients and sends requests from the clients to a server
- (b) A server that acts as if it is the destination server
- (c) It passes messages to the client or server without modifying them
- (d) It stores responses

21.8.4 Which of the following best describes a gateway:
- (a) It connects to a number of clients and acts on behalf of other clients and sends requests from the clients to a server
- (b) A server that acts as if it is the destination server
- (c) It passes messages to the client or server without modifying them
- (d) It stores responses

21.8.5 Which of the following best describes a tunnel:
- (a) It connects to a number of clients and acts on behalf of other clients and sends requests from the clients to a server
- (b) A server that acts as if it is the destination server
- (c) It passes messages to the client or server without modifying them
- (d) It stores responses

21.8.6 Which TCP/IP application would be blocked if port 23 was blocked:
- (a) TELNET
- (b) FTP
- (c) WWW (HTTP)
- (d) Electronic Mail (SMTP)

21.8.7 Which TCP/IP application would be blocked if port 80 was blocked:
(a) TELNET (b) FTP
(c) WWW (HTTP) (d) Electronic Mail (SMTP)

21.8.8 Which is the best method of securing transmitted/received data:
(a) Firewalls (b) Encryption
(c) Proxy servers (d) Leased lines

21.8.9 Discuss the main techniques that a firewall would use to filter incoming and outgoing TCP/IP packets.

21.8.10 Which, with respect to a dictionary-based password hacking program, of the following is the least secure password:
(a) foxmilk (b) ele_phant_forest
(c) aaappplle (d) g00d%pa55w0rd
(e) ant_fox (f) ugrangwad

21.8.11 Assuming an alphabet of 26 letters ('a' to 'z') and a password of 4 letters, how many password permutations are possible:
(a) 26 (b) 104 (26×4)
(c) 456976 ($26 \times 26 \times 26 \times 26$) (d) 358800 ($26 \times 25 \times 24 \times 23$)

21.8.12 Modify the following passwords so that they are difficult to crack with a dictionary-based cracking program (Hints: change letters to numbers, spaces to other characters, mix upper and lower case letters, and so on):
(a) apple tree (b) hard disk drive
(c) simple password (d) keyboard connector

21.9 Research

1. Search on the Internet for information on the Data Protection Act (either for individuals or organizations).
2. Search on the Internet for information on security breaches, and the methods that were used to hack into the system.
3. Security policies are important in protecting the system from abuse. For a known system, investigate the following:

Subject	*Typical questions*
Restrictions on users who can use a given account	Are their individual logins or global logins? Have groups been defined?
Password requirements and prohibitions	Is there a minimum limit on the number of characters in a password? Does a password have a given lifetime?

	Is the user allowed to change their password so that it's the same as the previous one?
Internet access/restrictions	Is external Internet access allowed? Are there any restrictions on the types of access?
Application program rules	What applications are available over the network? Are there any restrictions on their usage?
Monitoring	Are network accesses monitored? Are Internet accesses monitored? Are WWW accesses monitored?

21.10 Survey

Many users use easy-to-remember names for their password. These are typically names of spouses, car types, and so on. Conduct a survey within a class, University/College, work-place, etc, to see if their current, or previous, passwords fits one of the following given below (ignore extra numbers and characters). Total the votes and fill-in the table given next, and thus determine the Top 10 favorite password classifications. NOTE: Don't tell anyone your current password, and withhold your current password's classification, if you want.

Password classification	*Vote*	*Password classification*	*Vote*
Name of a pet (e.g. Fido, Tiddles)		Car type or manufacturer (e.g. Ford, Mustang)	
Nonsense word (e.g. rippleworden)		Computer type or manufacturer (e.g. IBM)	
Default password which was initially allocated (e.g. QWERTY)		Parent's/Spouse's name (e.g. Bert)	
Name of a friend/ relation (e.g. Martin)		Pop group/singer (e.g. Beatles)	
Movie star name (e.g. M_Monroe)		Film title (e.g. Casablanca)	
Computer game (e.g. Doom)		Cartoon Character (e.g. Bugs_Bunny)	

Computer game character (e.g. Sonic)		Historical event (e.g. world_war)	
City/town (e.g. Edinburgh)		Sport team name (e.g. Celtic)	
Government Administration Number (LL 123 456 A)		Country name (e.g. England)	
Children's names (e.g. Lisa)		Historical character (e.g. Shakespeare)	
TV program (e.g. Simpsons)		Transport (e.g. Train)	
Children's book (e.g. WillyWonka)		Other books (e.g. Cuckoo'sNest)	
Comedian (e.g. Crosby)		Composer (e.g. Beethoven)	
Old/current school-related name (e.g. GraemeHigh)		Old/current college/university/company-related name (e.g. Cambridge)	
Church-related name (e.g. priest)		Musical Instrument (e.g. guitar)	
Animal name (e.g. tiger)		Food name (e.g. spaghetti)	
Sports person (e.g. Pele)		Computer peripherals (e.g. diskdrive)	
Someone's date of birth (27/11/90)		Bank details (e.g., 114466)	
Other:		Other:	
Other:		Other:	
Other:		Other:	

21.11 Note from the Author

So what's the big problem with the Internet? Well it's a totally open system which connects computers around the world. At one time it was relatively easy to secure networks from external attack as the networks did not have any public connection. These days, with more people working from a mobile base or from home, there is a greater need for external connections into a network. Many companies also have several sites which connect to the Internet to make an Intranet. At one time the connection between the sites would have been achieved with a leased line. These days it is less expensive to use the Internet to create the connection. Thus organizations must try and make their systems secure for many reasons. One is that data which they hold on their databases must be secure and there are restrictions on who can view the data and who can edit it. They must also guard against malicious attacks where their systems can be damaged in some one.

External attacks can be reduced with strong firewalls, and data encryption, but internal attacks can be difficult to protect against as users already have access to the organization network. There have been several cases of employees doing damage to computer systems so that they could either cause the organization to lose business or to keep themselves in a job.

Legal laws on Internet access will take some time to catch-up with the growth in technology, so in many cases it is the moral responsibility of site managers and division leaders to try and reduce the amount of objectionable material. If users want to download objection material or set-up their own WWW server with offensive material, they should do this at home and not at work or within an educational establishment. Often commercial organizations are strongly protected sites, but in open-systems, such as Schools and Universities, it is often difficult to protect against these problems.

These days, viruses tend to be more annoying than dangerous. In fact, they tend to be more embarrassing than annoying, because a macro virus sent to a business colleague can often lead to a great deal of embarrassment (and a great deal of loss of business confidence).

A particularly-worrying source of viruses is the Internet, either through the running of programs over Internet, the spread of worms or the spreading of macro viruses in electronic mail attachments. The best way to avoid getting these viruses is to purchase an up-to-date virus scanner which checks all Internet programs and documents when they are loaded.

So, with viruses and worms, it is better to be safe than sorry. Let the flu be the only virus you catch, and the only worms you have to deal with are the ones in your garden.

22 Networking Operating Systems

22.1 Introduction

A networking operating system is one which allows hosts to intercommunicate using operating system support. Thus networking is built into the operating system, and not just an add-on. Many early versions of operating systems from Microsoft, including DOS and Microsoft Windows Version 3, had networking as an add-on to the operating system, thus proved unreliable and difficult to setup. Recent versions of Microsoft Windows have successfully integrated networking, and also support mixed, or hybrid networks. The most successful networking operating systems are:

- **Microsoft Windows.** The de-facto standard PC operating system which supports many applications. It supports a client/server architecture and also peer-to-peer architecture. Windows NT/2000 provides a robust networking technology.
- **Novell NetWare.** A PC-based system which provides an excellent file server support and a print server. It has been enhanced to provide corporate networks using NDS.
- **UNIX.** A robust and well-tested networking operating system which supports most of the industry-standard protocols. UNIX tends to run on high-powered workstations.

STOP PRESS:
Cntrl-Alt-Del defeats Trojan Horses

Trojan horse viruses pretend to be valid programs and can either present a common user interface or pretend to be useful programs. One Trojan horse virus, which is available over the WWW, is said to contain over 100 active viruses. Someone running this program will quickly be infected with these viruses.

The Happy99 virus typically attaches itself to e-mails and when the user runs the file it shows a lovely display of on-screen fireworks. Unfortunately it also replaces the existing TCP/IP stack with its own version, and copies itself to the most used e-mail addresses in a users address book (which is obviously embarrassing, as they tend to be friends or business colleagues).

One method of determining a user's login ID and password is to create a program which displays the user login screen. When the user enters their password it can be sent to the hacker. Windows NT/2000 overcomes this by having a login screen which can only be displayed when the Ctrl-Alt-Del keystrokes are used. It is extremely difficult to overrule this, as many programs use these keystrokes to reboot the computer.

22.2 Microsoft Windows

Windows NT has provided an excellent network operating system. It communicates directly with many different types of networks, protocols and computer architectures. Windows NT and Windows 95/98 have the great advantage over other operating systems that they have integrated network support. Operating systems now use networks to make peer-to-peer connections and also connections to servers for access to file systems and print servers. The three most used operating systems are MS-DOS, Microsoft Windows and UNIX. Microsoft Windows comes in many flavors; the main versions in current use are:

- Microsoft Windows 3.xx – 16-bit PC-based operating system with limited multi-tasking. It runs from MS-DOS and thus still uses MS-DOS functionality and file system structure.
- Microsoft Windows 95/98 – robust 32-bit multi-tasking operating system (although there are some 16-bit parts in it) which can run MS-DOS applications, Microsoft Windows 3.xx applications and 32-bit applications.
- Microsoft Windows NT/2000 – robust 32-bit multi-tasking operating systems with integrated networking. Networks are built with NT/2000 servers and clients. As with Microsoft Windows 95/98 they can run MS-DOS, Microsoft Windows 3.x applications and 32-bit applications. In this chapter Windows NT/2000 will be simply referred to as Microsoft Windows.

22.2.1 Novell NetWare networking

Novell NetWare is one of the most popular systems for PC LANs and provides file and print server facilities. The protocol used is SPX/IPX. This is also used by Windows NT to communicate with other Windows NT nodes and with NetWare networks. The Internet Packet Exchange (IPX) protocol is a network layer protocol for transportation of data between computers on a Novell network. IPX is very fast and has a small connectionless datagram protocol. Sequenced Packet Interchange (SPX) provides a communications protocol which supervises the transmission of the packet and ensures its successful delivery.

NetWare uses the Open Data-Link Interface (ODI) standard to simplify network driver development and to provide support for multiple protocols on a single network adapter. It allows Novell NetWare drivers to be written without concern for the protocol that

Best Practice for User Accounts

1. **GUEST ACCOUNT.** A guest account should always have a password, and should only be used in low-security domains.
2. **RENAME ADMIN.** If the network connects to the Internet, the administrator account should be renamed to deter hackers.
3. **LIMIT ADMINISTRATOR.** Only log on as an administrator when required. This stops the administrator from accidentally making changes which are incorrect, as the administrator has the right to do anything (every user, no matter how good they are, has deleted something that they didn't intend to).
4. **PASSWORDS FOR ALL.** User accounts should always have a password. On medium-security and high-security domains, the password should expire after a given time, and will require a completely new password (not just the same one as given previously). Some systems remember the best few passwords, and bar the user from using any of them.
5. **CHANGE WHEN FIRST.** New users should change their password after they first log onto a domain. This forces users to protect their own account.
6. **RANDOM NEW PASSWORDS.** In medium-security and high-security networks, initial passwords should be random assigned.
7. **BAD LOCK-OUTS.** User accounts should be locked-out after a given number of bad logins. In low-security domains this should be a simple time out for a number of minutes, but on medium-security and high-security domains this should set to forever (that is, until the system administrator has reset the account, possibly after investigating the cause).
8. **PASSWORD SIZE.** On medium-security and high-security domains, passwords should be at least a given number of characters, and should typically not include words from a standard dictionary, and also include a number. Typically passwords are at least six characters long.
9. **GROUPS.** The user must be assigned to a well-defined group, as members of their group tend to have a high-privilege to the user's resources as any other user.
10. **DELETE OLD ACCOUNTS.** User accounts should have a defined time limit before they become inactive. Users who leave an organization should be deleted as quickly as possible.

will be used on top of them (similar to NDIS in Microsoft Windows). The link support layer (LSL or LSL.COM) provides a foundation for the MAC layer to communicate with multiple protocols (similar to NDIS in Windows NT). The IPX.COM (or IPXODI.COM) program normally communicates with the LSL and the applications. The MAC driver is a device driver or NIC driver. It provides low-level access to the network adapter by supporting data transmission and some basic adapter management functions. These drivers also pass data from the physical layer to the transport protocols at the network and transport layers.

22.2.2 Microsoft Windows networking

Networks must use a protocol to transmit data. Typical protocols are:

- IPX/SPX – used with Novell NetWare, it accesses file and printer services.
- TCP/IP – used for Internet access and client/server applications.
- SNA DLC – used mainly by IBM mainframes and minicomputers.
- AppleTalk – used by Macintosh computers.
- NetBEUI – used in some small LANs (stands for NetBIOS Extended User Interface).

Novell NetWare is installed in many organizations to create local area networks of PCs. It uses IPX/SPX for transmitting data and allows access to file servers and network printing services. TCP/IP is the standard protocol used when accessing the Internet and also for client/server applications (such as remote file transfer and remote login).

A major advantage of Microsoft Windows is that networking is built into the operating system. Figure 22.1 shows how it is organized in relation to the OSI model. Microsoft Windows has the great advantage of being protocol-independent and will work with most standard protocols, such as TCP/IP, IPX/SPX, NetBEUI, DLC and AppleTalk. The default protocol is NetBEUI.

Best Practices for in Administration

1. **REGULAR BACKUPS.** No network is secure from loss of data, either through hardware/software failure, accidental deletion, and external hackers. The only sure way to recover the data on a network is to backup the system, and restore it, if required.
2. **PROPERLY DEFINE AUDIT POLICY.** This should relate to security policies, resource usage, and so on.
3. **PROPERLY DEFINE USERS AND GROUPS.** Domains which split into proper groups are often easier to administer and control than domains which have a few loosely defined groups.
4. **SECURE THE SERVER.** The server is likely to be the most important computer within the domain, as any downtime can affect the whole domain. The server should thus be secure against attack or accidental damage.
5. **MAKE SERVERS ROBUST.** Servers should be protected against failure, especially through mains spikes, and power outages. This typically requires UPS and RAID technology.
6. **DEFINE DOMAINS.** Each domain has at least one server. The larger the domain becomes the more difficult it is to administer it, and the slower it becomes. If possible, define the limits of the domain for effective sharing of information.
7. **DEFINE HOW RESOURCES ARE SHARED.** It is important that resources are shared properly, and certain users should be restricted from certain resources.
8. **SETUP BACKUP RESOURCES.** Key resources should have a backup which will guard against failure. Typically this will involve a backup server, which contains a mirror of the data on the main server.
9. **MAKE NETWORK ROBUST.** The network should be designed so that failures are confined to small areas. Typically routers and bridges are used to segment the network, and contain faults.
10. **LIMIT EXTERNAL CONNECTIONS.** On secure domains the number of external connections should be limited. Many organizations do not allow modems to be used to connect to a computer, as this could be used by an external user to gain access to the network, without first going through the organizational firewall.

There are two main boundaries in Microsoft Windows and NDIS and TDI. The Network Device Interface Standard (NDIS) boundary layer interfaces to several network interface adapters (such as Ethernet, Token Ring, RS-232, modems, and so on) with different protocols. It allows for an unlimited number of network interface cards (NICs) and protocols to be connected to be used with the operating system. In Microsoft Windows, a single software module, `NDIS.SYS` (the NDIS wrapper), interfaces with the manufacturer-supplied NDIS NIC device driver. The wrapper provides a uniform interface between the protocol drivers (such as TCP/IP or IPX/SPX) and the NDIS device driver.

IPX/SPX and AppleTalk

Novell NetWare networks use SPX/IPX and are supported through Microsoft Windows using the NWLink protocol stack. The AppleTalk protocol allows Windows NT/2000 to share a network with Macintosh clients. It can also act as an AppleShare server.

NetBEUI

NetBEUI (NetBIOS Extended User Interface) has been used with network operating systems, such as Microsoft LAN manager and OS/2 LAN server. In Microsoft Windows, the NetBEUI frame (NBF) protocol stack gives backward compatibility with existing NetBEUI implementations and also provides for enhanced implementations. NetBEUI is the standard technique that NT clients and servers use to intercommunicate.

NBF is similar to TCP/IP and SPX/IPX. It is used to establish a session between a client and a server, and also to provide the reliable transport of the data across the connection-oriented session. Thus NetBEUI tries to provide reliable data transfer through error checking and acknowledgement of each successfully received data packet. In the standard form of NetBEUI each packet must be acknowledged after its delivery. This is wasteful in time. Windows NT uses NBF which improves NetBEUI as it allows several packets to be sent before requiring an acknowledgement (called an adaptive sliding window protocol).

Each NetBEUI is assigned a 1-byte session number and thus allows a maximum of 254 simultaneously active sessions (as two of the connection numbers are reserved). NBF enhances this by allowing 254 connections to computers with 254 sessions for each connection (thus there is a maximum of 254×254 sessions).

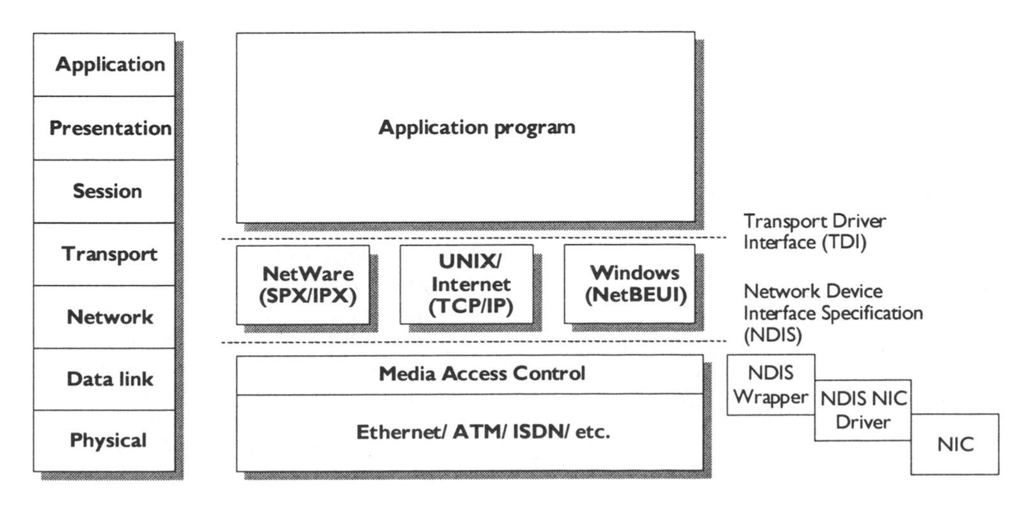

Figure 22.1 Microsoft Windows network interfaces

22.2.3 Windows sockets

A Windows socket (WinSock) is a standard method that allows nodes over a network to communicate with each other using a standard interface. It supports internetworking protocols such as TCP/IP, IPX/SPX, AppleTalk and NetBEUI. WinSock communicates through the TDI interface and uses the file `WINSOCK.DLL` or `WINSOCK32.DLL`. These DLLs (dynamic link libraries) contain a number of networking functions which are called in order to communicate with the transport and network layers (such as TCP/IP or SPX/IPX). As it communicates with these layers WinSock is independent of the networking interface (such as Ethernet or FDDI).

22.2.4 Robust networking

Microsoft Windows servers provide fault tolerance in a number of ways. These are outlined in the following sections.

Disk mirroring

Network servers normally support disk mirroring which protects against hard disk failure. It uses two partitions on different disk drives which are connected to the same controller. Data written to the first (primary) partition is mirrored automatically to the secondary partition. If the primary disk fails then the system uses the partition on the secondary disk. Mirroring also allows unallocated space on the primary drive to be allocated to the secondary drive. On a disk mirroring system the primary and secondary partitions have the same drive letter (such as C: or D:) and users are unaware that disks are being mirrored.

Disk duplexing

Disk duplexing means that mirrored pairs are controlled by different controllers. This provides for fault tolerance on both disk and controller. Unfortunately, it does not support multiple controllers connected to a single disk drive.

Best Practices for in High-Security

1. **BAN EXTERNAL CONNECTIONS.** In a highly secure network, all external traffic should go through a strong firewall. There should be no other external connections on the network. If possible, telephone lines should be monitored to stop data being transferred over without going through firewall.
2. **BAN FLOPPY DISKS AND DATA STORAGE DEVICES.** Employees should not be able to enter or leave the organization with any data on disk. Some organizations remove floppy disk drives from their computers to try and limit the possibility of transferring data.
3. **NO USER CAN INSTALL SOFTWARE.** Viruses can be easily spread if users are allowed to install their own software.
4. **SECURE ACCESS TO RESOURCES.** Typically users must use swipe cards, or some biometric technique to gain access to a restricted domain.
5. **LIMIT INTERNET ACCESS.** Only key personnel should be given rights to access the external Internet. If possible the computers which access the Internet should be well protected against malicious programs.
6. **FIREWALLS USED BETWEEN DOMAINS.** Internal hackers can be as big a problem as external hackers. Thus firewalls should be used between domains to limit access.
7. **BASE AUTHENTICATION ON MAC ADDRESSES.** Network addresses do not offer good authentication of a user, as they can be easily spoofed. An improved method is to check the MAC address of the computer (as no two computers have the same MAC address).
8. **EVERY FILE AND OBJECT SHOULD HAVE UNIQUELY DEFINED PRIVILEGES.** Every file and resource should have uniquely setup for user privileges which can limit access.
9. **EMPLOY SECURITY MANAGER.** The security manager will be responsible for the design of the initial security model, and any changes to it.
10. **LOG EVERY EVENT.** All the important security related events should be monitored within each domain. If possible they should be recorded over a long period of time. Software should be used to try and determine incorrect usage.

Striping with parity

Network servers normally support disk striping with parity. This technique is based on RAID 5 (Redundant Array of Inexpensive Disks), where a number of partitions on different disks are combined to make one large logical drive. Data is written in stripes across all of the disk drives and additional parity bits. For example, if a system has four disk drives then data is written to the first three disks and the parity is written to the fourth drive. Typically the stripe is 64 KB, thus 64 KB will be written to Drive 1, the same to Drive 2 and Drive 3, then the parity of the other three to the fourth. The following example illustrates the concept of RAID where a system writes the data `110`, `000`, `111`, `100` to the first three drives, which gives parity bits of `1`, `1`, `0` and `0`.

If one of the disk drives fails then the addition of the parity bit allows the bits on the failed disk to be recovered. For example, if disk 3 fails then the bits from the other disk are simply XOR-ed together to generate the bits from the failed drive. If the data on the other disk drives is `111` then the recovered data gives `0`, `001` gives `0`, and so on.

Disk 1	*Disk 2*	*Disk 3*	*Disk 4 (Odd parity)*
1	1	0	1
0	0	0	1
1	1	1	0
1	0	0	0

The 64 KB stripes of data are also interleaved across the disks. The parity block is written to the first disk drive, then in the next block to the second, and so on. A system with four disk drives would store the following data:

Disk 1	*Disk 2*	*Disk 3*	*Disk 4*
`Parity block 1`	`Data block A`	`Data block B`	`Data block C`
`Data block D`	`Parity block 2`	`Data block E`	`Data block F`
`Data block G`	`Data block H`	`Parity block 3`	`Data block I`

Each of the data blocks will be 64 KB, which is also equal to the parity block. The interlacing of the data ensures that the parity stripes are not all on the same disk. Thus there is no single point of failure for the set.

Striping of data improves reading performance when each of the disk drives has a separate controller, because the data is simultaneously read by each of the controllers and simultaneously passed to the systems. It thus provides fast reading of data but only moderate writing performance (because the system must calculate the parity block).

The main advantages of RAID 5 can be summarized as:

- It recovers data when a single disk drive or controller fails (RAID level 0 does not use a parity block thus it cannot regenerate lost data).
- It allows a number of small partitions to be built into a large partition.
- Several disks can be mounted as a single drive.
- Performance can be improved with multiple disk controllers.

The main disadvantages of RAID 5 are:

- It requires increased memory because of the parity block.
- Performance is reduced when one of the disks fails, because of the need to regenerate the failed data.
- It increases the amount of disk space as it has an overhead due to the parity block (although the overhead is normally less than disk mirroring, which has a 50% overhead).

- It requires at least three disk drives.

UPS services

Microsoft Windows servers provide services to uninterruptable power supplies (UPSs). UPS systems provide power, from batteries, to a computer system when there is a glitch in the supply, power sags or power failure. The operating system detects signals from a UPS unit and performs an orderly shutdown of applications, services and file systems as the stored energy in the UPS is depleted.

22.2.5 Security model

Microsoft Windows treats all its resources as objects that can only be accessed by authorized users and services. Examples of objects are directories, printers, processes, ports, devices and files. On an NTFS partition the access to an object is controlled by the security descriptor (SD) structure which contains an access control list (ACL) and security identifier (SI). The SD contains the user (and group) accounts that have access and permissions to the object. The system always checks the ACL of an object to determine whether the user is allowed to access it.

The main parts of the SI are:

OWNER	Indicates the user account for the object.
GROUP	Indicates the group the object belongs to.
User ACL	The user-controller ACL.
System ACL	System manager controlled ACL.

The ACL file access rights are:

Full control	(All)	Change	(RWXD)
Read	(RX)	Add	(WX)
List	(RX)	Change Permissions	(P)

Where R identifies read access, W identifies write access, X identifies execute, D identifies delete, and P identifies change permissions. There is also another attribute named O, which is take ownership.

Full control gives all access to all file permissions, and takes ownership of the NTFS volume. The Change rights allows: creation of folders and adding files to them, changing data in files, appending data to files, changing file attributes, deleting folders and files, and performing all tasks permitted by the Read permission. The Read permission allows the display of folder and file names, display of the data and attributes of a file, run program files, and access to other subfolders. For example a directory could have the following permissions:

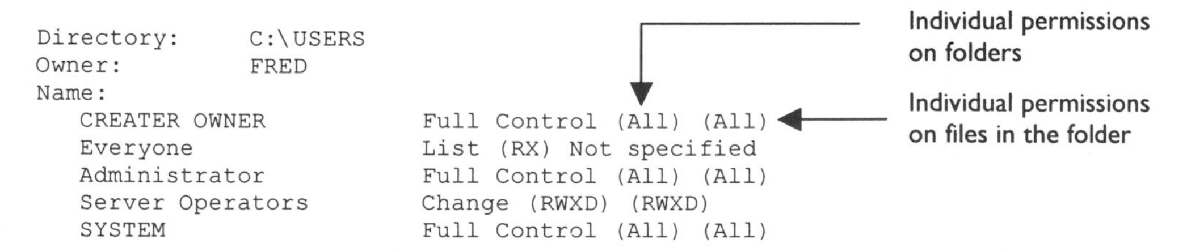

Directory: C:\USERS
Owner: FRED
Name:
CREATER OWNER Full Control (All) (All)
Everyone List (RX) Not specified
Administrator Full Control (All) (All)
Server Operators Change (RWXD) (RWXD)
SYSTEM Full Control (All) (All)
Individual permissions on folders
Individual permissions on files in the folder

It can be seen in this example, that the owner has full control over the directory, but everyone else (apart from the Administrator, Server Operators and SYSTEM) have only List rights (that is, they can only view or run programs, or get access to subfolders).

22.2.6 Workgroups and domains

Microsoft Windows assigns users to workgroups, which are collections of users who are grouped together with a common purpose. This purpose might be to share resources such as file systems or printers, and each workgroup has its own unique name. With workgroups each Microsoft Windows workstation interacts with a common group of computers on a peer-to-peer level. Each workstation then manages its own resources and user accounts. Workgroups are useful for small groups where a small number of users require to access resources on other computers.

A domain in Microsoft Windows is a logical collection of computers sharing a common user accounts database and security policy. Thus each domain must have at least one Microsoft Windows server. Each computer in the domain is assigned a unique name.

Microsoft Windows is designed to operate with either workgroups or domains. Figure 22.2 illustrates the difference between domains and workgroups. In this case the name of the domain is `my_d` (which must be provided when logging into the domain), along with the user name and the associated password. This domain contains a number of computers, such as `freds_pc` and `bills_pc`. The top level of the domain is \\, and files can be referred to with *computer_name*\ *directory**filename*.

Domains have the advantages that:

- Each domain forms a single administrative unit with shared security and user account information. This domain has one database containing user and group information and security policy settings.
- They segment the resources of the network so that users, by default, can view all networks for a particular domain.
- User accounts are automatically validated by the domain controller. This stops invalid users from gaining access to network resources. The domain can also be setup so that the server does not validate users, but this is not recommended as it can lead to security problems.

A Microsoft Windows server provides many client services (as illustrated in Figure 22.3), including:

- **User profiles.** The server can store profiles for the user, so that they can be easily changed and stored in a central source. This service can allow a user to get a consistent range of settings, no matter which computer they use (this is known as a roaming user profile). For example, an office may have a number of computers which can be logged into by any of the users within the domain. No matter which computer a user logs into, they will see the same settings, for their desktop, e-mail settings, and so on. User profiles typically related to display settings, regional settings, network connections, printer connections, mouse settings and sounds. A roaming user profile is made up of two parts: roaming personal user profile, which is the part of the profile that the user can change, and the roaming mandatory user profile, which is the part of the profile that cannot be changed by the user. The mandatory user profile is typically used to create a standard desktop configuration.
- **Hardware profiles.** This allows different hardware profiles to be setup.
- **Internet Information Server.** This provides HTTP and FTP access.
- **Directory Services.** Allows for a directory database which allows for network login, and a centralized point of administration and access to resources within the domain.
- **Management Tools.** This includes user and group account management.

- **Additional Network Services.** These include DNS (for domain name resolution), WINS (similar to DCHP and resolves domain names to IP addresses) and DHCP (which dynamically assigns IP addresses).

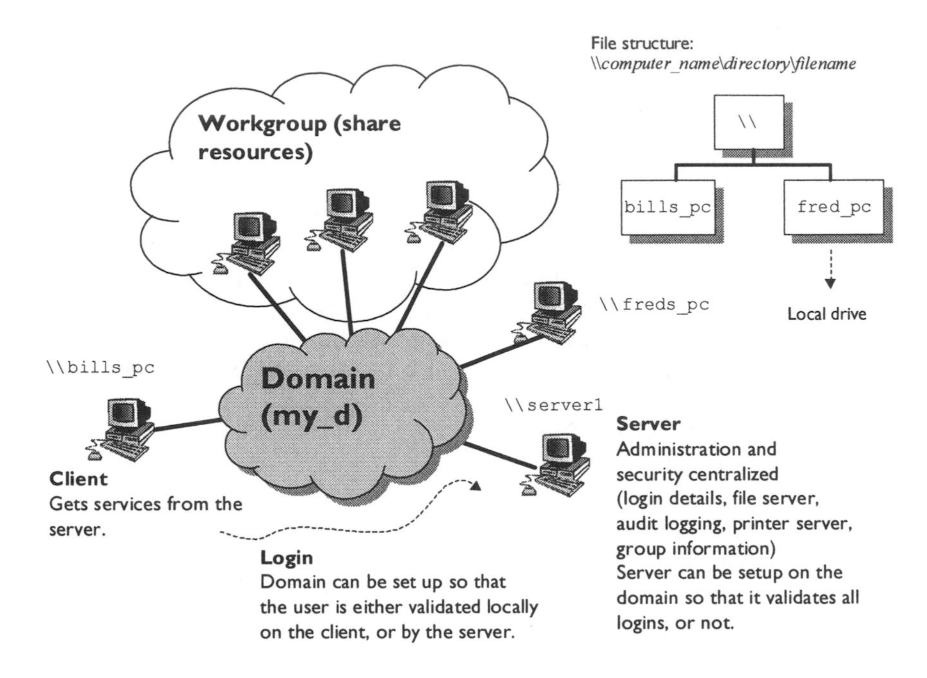

Figure 22.2 Workgroups and domains

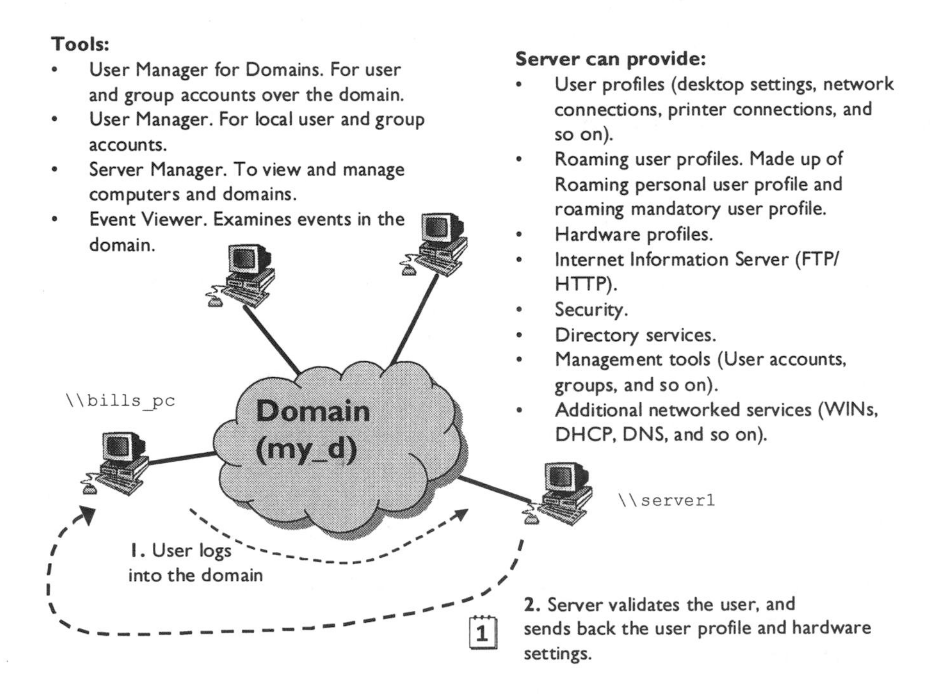

Figure 22.3 Services from the server

22.2.7 Audit policy

Audit policies allow certain events to be monitored, as illustrated in Figure 22.4. Typically this will be related to user login/logout (times that the user logs in and logs out, either successfully, or not), file and object access (files that are being accessed and objectives, such as printers, backup storage devices, and so), use of user rights, user and group management, security policy changes, restart/shutdown (when the computer is started and shutdown) and process tracking (the program which the user runs). These events can be monitored for good events (successful operations) and/or bad events (unsuccessful operations). The audit policy can either be setup on a local computer (for a local audit policy) or over the domain (the domain audit policy). The audit log can then be examined to determine how users have been using the resource within the domain. The system administrator might use the audit log to:

- **Unauthorized logins.** This may point to an external hacker trying to log into a valid account, and the administrator should audit users logging in and out of the domain.
- **Monitor out-of-hours logins.** Security problems typically occur out of normal hours. A system administrator can determine if someone is logging into the domain after normal hours, and possibly trace their operations.
- **Monitor the access to resources.** The administrator can monitor the access to networked resources, such as printers. If a user is using the resource too much, the administrator can limit the access rights to it for that user.
- **Monitoring how file and object permission are changed.** This may point to a breach of security where a user tries to change the permissions on objects and/or files. The administrator can monitor both successful and unsuccessful changes.
- **Monitor the processes that a user is running.** This might be used if it is thought that a user is using a certain package too often, especially if it is not related to their job function. For example, an administrator could monitor the access to a WWW browser, and the times in which it was used. If it was used too much, for non-work-related work, the administrator could either limit the access to it, or ban its usage for that user.
- **Times that the user uses the computer.** This allows the system administrator the chance to monitor how users use their computer and the normal times they use it. Any accesses outside these times can identify a security breach.

Microsoft Windows Security

The US Government defines certain security levels: D, C1, C2, B1, B2, B3 and A1, which are published in the *Trusted Computer Security Evaluation Criteria* books (each which have different colored cover to define their function), these include:

Orange Book. Describes system security.
Red Book. Interpretation of the Orange book in a network context.
Blue Book. Application of Orange book to systems not covered in the original book.

Microsoft Windows NT/2000 uses the C2 security level. It has the following features:

Object control. Users own certain objects and they have control over how they are accessed.
User names and passwords.
No object reuse. Once a user or a group has been deleted, the user and group numerical IDs are not used again. New users or groups are granted a new ID number.
Security auditing system. This allows the system administrator to trace security aspects, such as user login, bad logins, program access, file access, and so on.
Defined keystroke for system access. In Windows NT/2000, the CNTRL-ALT-DEL keystroke is used by a user to log into the system.

Failures in the audit typically identify security breaches, whereas successful accesses can be used to determine the amount of usage of a resource, and can be used for resource planning. (See Questions 22.6.1 to 22.6.7 for policy settings.) An example of audit policies for high-security, medium-security and minimum-security domains are:

High security	Medium security	Minimum security
Successful and unsuccessful user logins	Successful use of key resources	Successful user of resource, for planning purposes
Successful and unsuccessful user of all resources	Successful and unsuccessful administrative and security policy changes	
Successful and unsuccessful administrative and security policy changes	Successful use of sensitive and confidential data, such as accounting information	Successful use of sensitive and confidential data, such as accounting information

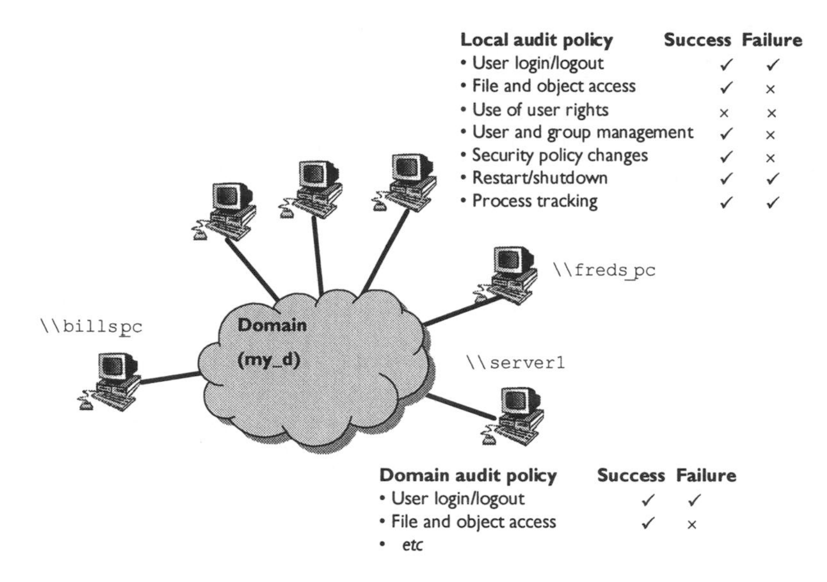

Figure 22.4 Local and domain audit policies

22.2.8 File systems

Microsoft Windows supports three different types of file system:

- **FAT** (file allocation table) – as used by MS-DOS, OS/2 and Windows NT. A single volume can be up to 2 GB (now increased to over 4 GB). It has no built-in security but can be accessed through Windows 95/98, MS-DOS and Windows NT/2000.
- **HPFS** (high performance file system) – a UNIX-style file system which is used by OS/2 and Windows NT. A single volume can be up to 8 GB. MS-DOS applications cannot access files.
- **NTFS** (NT file system) – as used by Windows NT. A single volume can be up to 64 TB (based on current hardware, but, theoretically, 16 exabytes). It has built-in security and also supports file compression/decompression. MS-DOS applications, themselves, cannot access the file system but they can when run with Windows NT/2000, nor can Windows 95/98.

The FAT file system is widely used and supported by a variety of operating systems, such as MS-DOS, Windows NT and OS/2. If a system is to use MS-DOS it must be installed with a FAT file system.

Four Levels of Security

The Orange Book produced by the US Department of Defense (DOD) defines levels of security for systems. There are four main divisions, which split into seven main security ratings. Division D is the lowest security level and Division A is the highest. The ratings are:

Division D. This rating provides no protection on files or for users. For example, a DOS-based computer has no real security on files and users, thus it has a Division D rating.

Division C. This rating splits into two groups: C1 rating and C2 rating. C1 contains a trust computing base (TCB) which separates users and data. It suffers from the fact that all the data on the system has the same security level. Thus, users cannot make distinctions between highly secure data and not-so secure data. A C1 system has user names and passwords, as well as some form of control of users and objects. C2 has a higher level of security and provides for some form of accountability and audit. This allows events to be logged and traced, for example, it might contain a list of user logins, network address logins, resource accesses, bad logins, and so on.

Division B. This rating splits into three groups: B1, B2 and B3. Division B rated systems have all the security of a C2 rating, but have more security because they have a different level of security for all system accesses. For example, each computer can have a different security level, each printer can also have different security levels, and so on. Each object (such as a computer, printer, and so on) has a label associated with it. It is with this label that the security is set by. Non-labeled resources cannot be connected to the system. In a B2 rated system, users are notified of any changes of an object that they are using. The TCB also includes separate operator and administrator functions. In a B3 rated system the TCB excludes information which is not related to security. The system should also be designed to be simple to trace, but also well tested to prevent external hackers. It should also have a full-time administrator, audit trails and system recovery methods.

Division A. This is the highest level of security. It is similar to B3, but has formal methods for the systems security policy. The system should also have a security manager, who should document the installation of the system, and any changes to the security model.

FAT

The standard MS-DOS FAT file and directory-naming structure allows an 8-character file name and a 3-character file extension with a dot separator (.) between them (the 8.3 file name). It is not case sensitive and the file name and extension cannot contain spaces and other reserved characters, such as:

```
" / \ : ; | = , ^ * ? .
```

With Windows NT/2000 and Windows 95/98 the FAT file system supports long file names which can be up to 255 characters. The name can also contain multiple spaces and dot separators. File names are not case sensitive, but the case of file names is preserved (a file named `FredDocument.XYz` will be displayed as `FredDocument.XYz` but can be accessed with any of the characters in upper or lower case.

Each file in the FAT table has four attributes (or properties): read-only, archive, system and hidden. The FAT uses a linked list where the file's directory entry contains its beginning FAT entry number. This FAT entry in turn contains the location of the next cluster if the file is larger than one cluster, or a marker that designates this is included in the last cluster. A file which occupies 12 clusters will have 11 FAT entries and 10 FAT links.

HPFS (high-performance file system)

HPFS is supported by OS/2 and is typically used to migrate from OS/2 to Microsoft Windows. It allows long file names of up to 254 characters with multiple extensions. As with Microsoft Windows FAT system the file names are not case sensitive but preserve the case. HPFS uses B-tree format to store the file system directory structure. The B-tree format stores directory entries in an alphabetic tree, and binary searches are used to search for the target file in the directory list.

NTFS (NT file system)

NTFS is the preferred file system for Windows NT/2000 as it makes more efficient usage of the disk and it offers increased security. It allows for file systems up to 16 EB (16 exabytes, or 1 billion gigabytes, or 2^{64} bytes). As with HPFS it uses B-tree format for storing the file system's directory structure. Its main objectives are:

- To increase **reliability**. NTFS automatically logs all directory and file updates which can be used to redo or undo failed operations resulting from system failures such as power losses, hardware faults, and so on.
- To provide sector sparing (or **hot fixing**). When NTFS finds errors in a bad sector, it causes the data in that sector to be moved to a different section and the bad sector to be marked as bad. No other data is then written to that sector. Thus, the disk fixes itself as it is working and there is no need for disk repair programs (FAT only marks bad areas when formatting the disk).
- Increases file system size (up to 16 EB).
- To enhance security permissions.
- To support POSIX requirements, such as case-sensitive naming, addition of a time stamp to show the time the file was last accessed and hard links from one file (or directory) to another.

22.3 UNIX

UNIX is an extremely popular operating system and dominates in the high-powered, multitasking workstation market. It is relatively simple to use and to administer, and also has a high degree of security. UNIX computers use TCP/IP communications to mount disk resources from one machine onto another. UNIX's main characteristics are:

- **Multi-user.**
- **Memory management with paging** (organizing programs so that the program is loaded into pages of memory) **and swapping** (which involves swapping the contents of memory to disk storage).
- **Pre-emptive multitasking.**
- **Multiprocessing.**
- **Multithreaded applications.**

The two main families of UNIX are UNIX System V and BSD (Berkeley Software Distribution) Version 4.4. System V is the operating system most often used and has descended from a system developed by the Bell Laboratories; it was recently sold to SCO (Santa Cruz Operation).

An initiative by several software vendors has resulted in a common standard for the user interface and the operation of UNIX. The user interface standard is defined by the common desktop environment (**CDE**). This allows software vendors to write calls to a standard CDE API (application program interface). The common UNIX standard has been defined as Spec 1170 APIs. Compliance with the CDE and Spec 1170 API are certified by X/Open, which is a UNIX standard organization.

Another important UNIX-like operating system is Linux, which was developed by Linus Torvalds at the University of Helsinki in Finland. It was first made public in 1991 and most of it is available free-of-charge. The most widely available version was developed by the Free Software Foundation's GNU project. It runs on most Intel-based, SPARC and

Alpha-based computers. Modern UNIX-based systems tend to be based around four main components:

- **UNIX operating system.**
- **TCP/IP communications.**
- **Network file system** (NFS). NFS allows disk drives to be linked together to make a global file system.
- **X-Windows interface.** X-Windows presents a machine-independent user interface for client/server applications.

22.3.1 File attributes

UNIX provides system security by assigning each file and directory with a set of attributes. These give the privileges on the file usage and the `ls -l` command displays their settings, such as:

- File attributes.
- Owner of the files. Person (user ID) who owns the file.
- Group information. The group name defines the name of the group to which the owner belongs.
- Size of file. The size of the file in bytes.
- Date and time created or last modified. This gives the date and time the file was last modified. If it was modified in a different year then the year and date are given, but no time information is given.
- Filename.

Figure 22.5 defines the format of the extended file listing. The file attributes contain the letters `r`, `w`, `x` which denote read, write and executable. If the attribute exists then the associated letter is placed at a fixed position, else a `-` appears. The definition of these attributes are as follows:

- Read (`r`). File can be copied, viewed, and so on, but it cannot be modified.
- Write (`w`). File can be copied, viewed and changed, if required.
- Executable (`x`). File can be executed.

The file attributes split into four main sections. The first position identifies if it is a directory or a file. A `d` character identifies a directory, else it is a file. Positions 2–4 are the owner attributes, positions 5–7 are the group's attributes and positions 8–10 are the rest of the world's attributes. The attributes are:

```
Owner     Group     Public
r w x     r w x     r w x
```

The owner is the person who created the file and the group is a collection of several users, such as research, development, production, admin groups, and so on. The public is for everyone else on the system.

The `r` attribute stands for read privilege and if it is set then the file can be read (that is, listed or copied) by that type of user. The `w` attribute stands for write privilege and if it is set then the file can be written to (that is, listed, copied or modified) by that type of user. The `x` attribute stands for execute privilege and if it is set then the file can be executed (that is, it can be run) by that type of user.

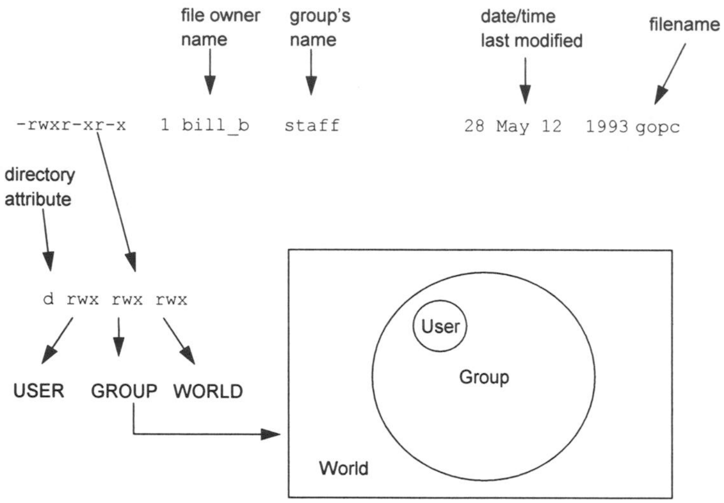

Figure 22.5 Extended file listing

For example `-rw-r--r--` is a file that the owner can read or write but the group and the public can only read this file. Another example is `-r-x--x--x`; with these attributes the owner can only read the file. No one else can read the file. No one can change the contents of the file. Everyone can execute the file. The `ls -al` listing gives the file attributes. Table 22.1 lists some examples.

Table 22.1 Example file attributes

Attributes	*Description*
`-r-x--x---`	This file can be executed by the owner and his group (e.g. staff, students, admin, research, system, and so on). It can be viewed by the owner but no-one else. No other privileges exist.
`drwxr-xr-x`	This directory cannot be written to by the members of group and others. All other privileges exist.
`-rwxrwxrwx`	This file can be read and written to by everyone and it can also be executed by everyone (beware of this).

Changing attributes of a file

The `chmod` command can be used by the owner of the file to change any of the attributes. Its general format is:

`chmod` *settings filename*

where *settings* define how the attributes are to be changed and the part of the attribute to change. The permission can be set using the octal system. If an attribute exists a 1 is set, if not it is set to a 0. For example, `rw-r--r--` translates to 110 100 100, which is 644 in octal. For example:

to set to	`rwx--x---`	use 710	to set to	`r-x------`	use 500
to set to	`rwxrwxrwx`	use 777	to set to	`--x------`	use 100
to set to	`rw-rw-rw-`	use 666			

The other method used is symbolic notation. The characters which define which part to modify are u (user), g (group), o (others), or a (all). The characters for the file attributes are a sign (+, - or =) followed by the characters r, w, x. A '+' specifies that the attribute is to be added, a '-' specifies that the attribute is to be taken away, and the '=' defines the actual attributes. They are defined as:

u	user permission	g	group permission
o	others (public) permission	a	all of user, group and other permissions
=	assign a permission	+	add a permission
-	take away permission	r	read attribute
w	write attribute	x	execute attribute

In sample session 22.1 [2] the owner of the file changes the execute attribute for the user. Sample session 22.2 makes the file `file.txt` into `rw-rw-r--`, and the file `Run_prog` into `--x--x---`. Some examples of setting and resetting attributes are:

`chmod u+x prog1.c`	owner has executable rights added
`chmod a=rwx prog2.c`	sets read, write, execute for all
`chmod g-r cprogs`	resets read option for group

Sample session 22.1

```
[ 1:/user/bill_b/shells ] % ls -l
-rw-r--r--    1 bill_b   staff            988 Nov  7 10:20 Cshrc
-rw-r--r--    1 bill_b   staff             43 Nov  7 10:20 Login
-rwxr-xr-x    1 bill_b   staff             28 May 12  2000 gopc
[ 2:/user/bill_b/shells ] % chmod u+x Login
[ 3:/user/bill_b/shells ] % ls -l
-rw-r--r--    1 bill_b   10           988 Nov  7 10:20 Cshrc
-rwxr--r--    1 bill_b   10            43 Nov  7 10:20 Login
-rwxr-xr-x    1 bill_b   10            28 May 12  2000 gopc
```

Sample session 22.2

```
[ 4:/user/bill_b ] %  chmod 664  file.txt
[ 5:/user/bill_b ] %  chmod 110  Run_prog
[ 6:/user/bill_b ] % ls -al
 ---x--x--- 4 bill staff  1102 Jun  4 12:05 Run_prog
 drw-r--r-- 2 bill staff    52 Jun  4 14:20 cprogs
 -rw-rw-r-- 4 bill staff   102 Jun  1 11:13 file.txt
```

22.3.2 TCP/IP protocols

UNIX uses the normal range of TCP/IP protocols, grouped into transport, routing, network addresses, user services, gateway and other protocols.

Routing

Routing protocols manage the addressing of the packets and provide a route from the source to the destination. Packets may also be split up into smaller fragments and reassembled at the destination. The main routing protocols are:

- **ICMP** (Internet Control Message Protocol) which supports status messages for the IP protocol. These may be errors or network changes that can affect routing.
- **IP** (Internet Protocol) which defines the actual format of the IP packet.
- **RIP** (Routing Information Protocol) which is a route determining protocol.

Popular UNIX systems are:

- **AIX** (on IBM workstations and mainframes).
- **HP-UX** (on HP workstations).
- **Linux** (on PC-based systems).
- **OSF/1** (on DEC workstations).
- **Solaris** (on Sun workstations).

Transport

The transport protocols are used by the transport layer to transport a packet around a network. The protocols used are:

- **TCP** (Transport Control Protocol) which is a connection-based protocol where the source and the destination make a connection and maintain the connection for the length of the communications.
- **UDP** (User Datagram Protocol) which is a connectionless service where there is no connection setup between the source and the destination.

Network and User addresses

The network address protocols resolve IP addresses with their symbolic names, and vice versa. These are:

- **ARP** (Address Resolution Protocol) determines the IP address of nodes on a network.
- **DNS** (Domain Name System) which determines IP addresses from symbolic names (such as `anytown.ac.uk` might be resolved to 112.123.33.22).

User services are applications to which users have direct access.

- **BOOTP** (Boot Protocol) which is typically used to start up a diskless networked node. Thus rather than reading boot information from its local disk it reads the data from a server. Typically it is used by X-Windows terminals.
- **FTP** (File Transfer Protocol) which is used to transfer files from one node to another.
- **TELNET** which is used to remotely log into another node.

Gateway and other protocols

The gateway protocols provide help for the routing process. These protocols include:

- **EGP** (Exterior Gateway Protocol) which transfers routing information for an external network.
- **GGP** (Gateway-to-Gateway Protocol) which transfers routing information between Internet gateways.
- **IGP** (Interior Gateway Protocol) which transfers routing information for internal networks.

Other important services provide support for networked files systems, electronic mail and time synchronization as well as helping maintain a global network database. The main services are:

- **NFS** (Network File System) which allows disk drives on remote nodes to be mounted on a local node and thus create a global file system. See Section 7.3.
- **NIS** (Network Information Systems) which maintains a network-wide database for user accounts and thus allows users to log into any computer on the network. Any changes to a user's account are made over the whole network. See Section 7.4.1.
- **NTP** (Network Time Protocol) which is used to synchronize clocks of nodes on the network.

- **RPC** (Remote Procedure Call) which enables programs running on different nodes on a network to communicate with each other using standard function calls.
- **SMTP** (Simple Mail Transfer Protocol) which is a standard protocol for transferring electronic mail messages.
- **SNMP** (Simple Network Management Protocol) which maintains a log of status messages about the network.

22.3.3 XDR format

As previously mentioned **XDR** is a standard technique which is used to describe and encode data. This standard form allows for transferring data between different computer architectures. It fits into the presentation layer and uses a language, which is similar to C, to describe data formats. See Section 22.8 for format.

22.4 Novell NetWare

Novell NetWare is one of the most popular network operating systems for PC LANs and provides file and print server facilities. Its default network protocol is normally **SPX/IPX**. This can also be used with Windows NT to communicate with other Windows NT nodes and with NetWare networks. The Internet Packet Exchange (**IPX**) protocol is a network layer protocol for transportation of data between computers on a NetWare network. IPX is very fast and has a small connectionless datagram protocol. The Sequenced Packet Interchange (**SPX**) provides a communications protocol which supervises the transmission of the packet and ensures its successful delivery.

NetWare is typically used in organizations and works well on a local network. Network traffic which travels out on the Internet or that communicates with UNIX networks must be in TCP/IP form. IP tunneling encapsulates the IPX packet within the IP packet. This can then be transmitted into the Internet network. When the IP packet is received by the destination NetWare gateway, the IP encapsulation is stripped off. IP tunneling thus relies on a gateway into each IPX-based network that also runs IP. The NetWare gateway is often called an IP tunnel peer.

22.4.1 NetWare architecture

NetWare provides many services, such as file sharing, printer sharing, security, user administration and network management. The interface between the network interface card (NIC) and the SPX/IPX stack is ODI (Open Data-link Interface). NetWare clients run software which connects them to the server; the supported client operating systems are DOS, Windows, Windows NT, UNIX, OS/2 and Macintosh.

With NetWare Version 3, DOS and Windows 3.*x* clients use a NetWare shell called NETx.COM. This shell is executed when the user wants to log into the network and stay resident. It acts as a command redirector and processes requests which are either generated by application programs or from the keyboard. It then decides whether they should be handled by the NetWare network operating system or passed to the client's local DOS operating system. NETx builds its own tables to keep track of the location of network-attached resources rather than using DOS tables. Figure 22.6 illustrates the relationship between the NetWare shell and DOS, in a DOS-based client. Note that Windows 3.*x* uses the DOS operating system, but Windows NT/2000 and 95 have their own operating systems and only emulate DOS. Thus, Windows NT/2000 and 95 do not need to use the NETx program.

The ODI allows NICs to support multiple transport protocols, such as TCP/IP and IPX/SPX, simultaneously. Also, in an Ethernet interface card, the ODI allows simultaneous support of multiple Ethernet frame types such as Ethernet 802.3, Ethernet 802.2, Ethernet II, and Ethernet SNAP.

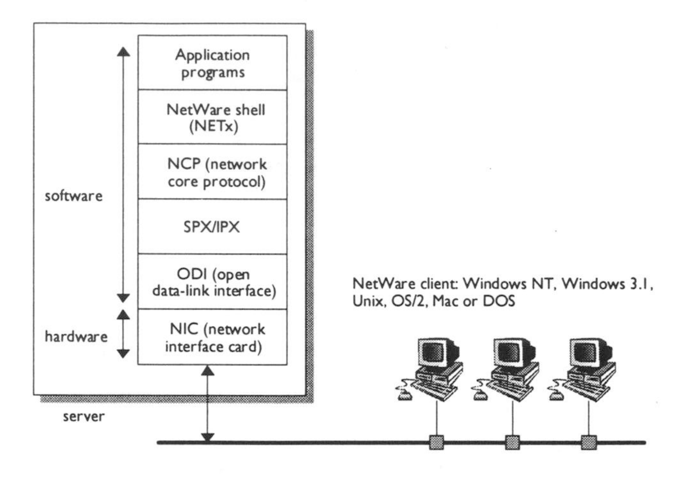

Figure 22.6 NetWare architecture

To install NetWare, the server must have a native operating system, such as DOS or Windows NT, and it must be installed on its own disk partition. NetWare then adds a partition in which the NetWare partition is added. This partition is the only area of the disk the NetWare kernel can access.

NetWare loadable modules (NLMs)

NetWare allows enhancements from third-party suppliers using NLMs. The two main categories are:

- **Operating systems enhancements** – these allow extra operating system functions, such as a virus checker and also client hardware specific modules, such as network interface drivers.
- **Application programs** – these programs actually run on the NetWare server rather than on the client machine.

Bindery services

NetWare must keep track of users and their details. Typically, NetWare must keep track of:

- User names and passwords.
- Groups and group rights.
- File and directory rights.
- Print queues and printers.
- User restrictions (such as allowable login times, the number of times a user can simultaneously login to the network).
- User/group administration and charging (such as charging for user login).
- Connection to networked peripherals.

This information is kept in the bindery files. Whenever a user logs into the network their login details are verified against the information in the bindery files.

The bindery is organized with objects, properties and values. Objects are entities that are controlled or managed, such as users, groups, printers (servers and queues), disk drives, and so on. Each object has a set of properties, such as file rights, login restrictions, restrictions to printers, and so on. Each property has a value associated with it. Here are some examples:

Object	*Property*	*Value*
User	Login restriction	Wednesday 9 am till 5pm
User	Simultaneous login	2
Group	Access to printer	No

Objects, properties and values are stored in three separate files which are linked by pointers on every NetWare server:

1. `NET$OBJ.SYS` (contains object information).
2. `NET$PROP.SYS` (contains property information).
3. `NET$VAL.SYS` (contains value information).

If multiple NetWare servers exist on a network then bindery information must be exchanged manually between the servers so that the information is the same on each server. In a multiserver NetWare 3.*x* environment, the servers send SAP (service advertising protocol) information between themselves to advertise available services. Then the bindery services on a particular server update their bindery files with the latest information regarding available services on other reachable servers. This synchronization is difficult when just a few servers exist but is extremely difficult when there are many servers. Luckily, NetWare 4.1 has addressed this problem with NetWare directory services; this will be discussed later.

22.4.2 NetWare protocols

NetWare uses IPX (Internet Packet Exchange) for the network layer and either SPX (Sequenced Packet Exchange) or **NCP** (NetWare Core Protocols) for the transport layer. The routing information protocol (**RIP**) is also used to transmit information between NetWare gateways. These protocols are illustrated in Figure 22.7.

IPX

IPX performs a network function that is similar to IP. The higher information is passed to the IPX layer which then encapsulates it into IPX envelopes. It is characterized by:

- A connectionless connection – each packet is sent into the network and must find its own way through the network to the final destination (connections are established with SPX).
- It is unreliable – as with IP it only basic error checking and no acknowledgement (acknowledgements are achieved with SPX).

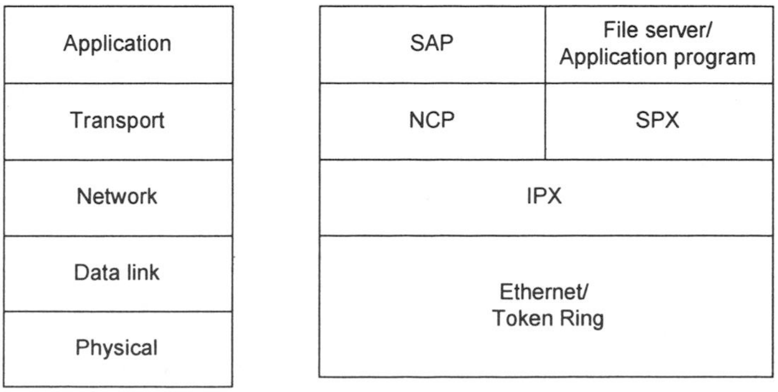

OSI	NetWare	
Application	SAP	File server/ Application program
Transport	NCP	SPX
Network	IPX	
Data link	Ethernet/ Token Ring	
Physical		

Figure 22.7 NetWare reference model

IPX uses a 12-byte station address (whereas IP uses a 4-byte address). The IPX fields are:

- **Checksum** (2 bytes) – this field is rarely used in IPX, as error checking is achieved in the SPX layer. The lower-level data link layer also provides an error detection scheme (both Ethernet and Token Ring support a frame check sequence).
- **Length** (2 bytes) – this gives the total length of the packet in bytes (i.e. header + DATA). The maximum number of bytes in the DATA field is 546, thus the maximum length will be 576 bytes (2 + 2 + 1 + 1 + 12 + 12 + 546).
- **Transport control** (1 byte) – this field is incremented every time the frame is processed by a router. When it reaches a value of 16 it is deleted. This stops packets from traversing the network for an infinite time. It is also typically known as the time-to-live field or hop counter.
- **Packet type** (1 byte) – this field identifies the upper layer protocol so that the DATA field can be properly processed.
- **Addressing** (12 bytes) – this field identifies the address of the source and destination station. It is made up of three fields: a network address (4 bytes), a host address (6 bytes) and a socket address (2 bytes). The 48-bit host address is the 802 MAC LAN address. NetWare supports a hierarchical addressing structure where the network and host addresses identify the host station and the socket address identifies a process or application and thus supports multiple connections (up to 50 per node).

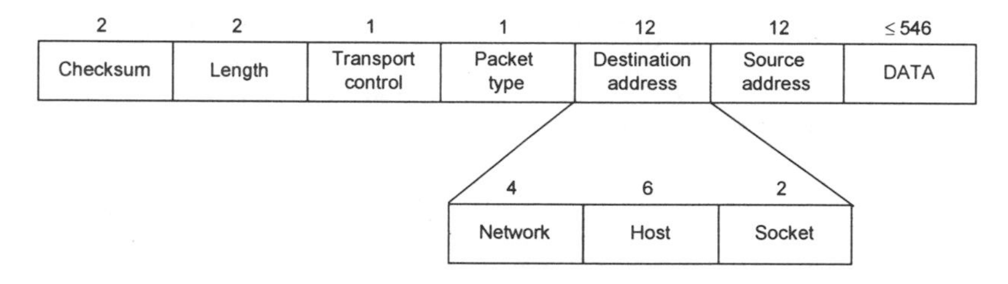

Figure 22.8 IPX packet format

SPX

On a NetWare network the level above IPX is either NCP or SPX. The SPX protocol sets up a virtual circuit between the source and the destination (just like TCP). Then all SPX packets follow the same path and will thus always arrive in the correct order. This type of connection is described as connection-oriented.

SPX also allows for error checking and an acknowledgement to ensure that packets are received correctly. Each SPX packet has flow control and also sequence numbers. Figure 22.9 illustrates the SPX packet.

The fields in the SPX header are:

- **Connection control** (1 byte) – this is a set of flags which assist the flow of data. These flags include an acknowledgement flag and an end-of-message flag.
- **Datastream type** (1 byte) – this byte contains information which can be used to determine the protocol or information contained within the SPX data field.
- **Destination connection ID** (2 bytes) – the destination connection ID allows the routing of the packet through the virtual circuit.
- **Source connection ID** (2 bytes) – the source connection ID identifies the source station when it is transmitted through the virtual circuit.

* **Sequence number** (2 bytes) – this field contains the sequence number of the packet sent. When the receiver receives the packet, the destination error checks the packet and sends back an acknowledgement with the previously received packet number in it.
* **Acknowledgement number** (2 bytes) – this acknowledgement number is incremented by the destination when it receives a packet. It is in this field that the destination station puts the last correctly received packet sequence number.
* **Allocation number** (2 bytes) – this field informs the source station of the number of buffers the destination station can allocate to SPX connections.
* **Data** (up to 534 bytes).

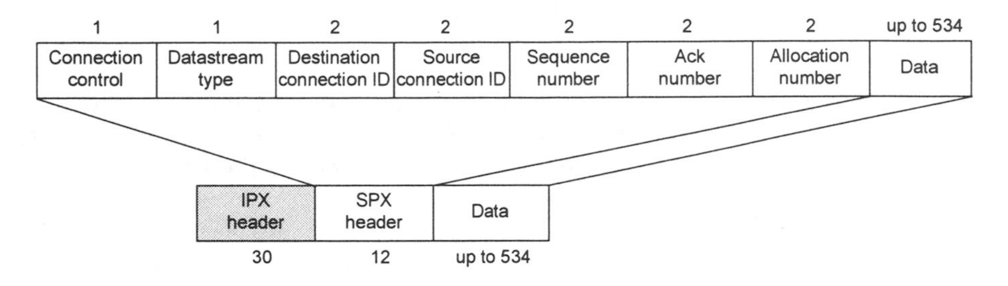

Figure 22.9 SPX packet format

RIP

The NetWare Routing Information Protocol (RIP) is used to keep routers updated on the best routes through the network. RIP information is delivered to routers via IPX packets. Figure 22.10 illustrates the information fields in an RIP packet. The RIP packet is contained in the field which would normally be occupied by the SPX packet.

Routers are used within networks to pass packets from one network to another in an optimal way (and error-free with a minimal time delay). A router reads IPX packets and examines the destination address of the node. If the node is on another network then it routes the packet in the required direction. This routing tends not to be fixed as the best route will depend on network traffic at given times. Thus the router needs to keep the routing tables up to date; RIP allows routers to exchange their current routing tables with other routers.

The RIP packet allows routers to request or report on multiple reachable networks within a single RIP packet. These routes are listed one after another (Figure 22.10 shows two routing entries). Thus each RIP packet has only one operation field, but has multiple entries of the network number, the number of router hops, and the number of tick fields, up to the length limit of the IPX packet.

The fields are:

* **Operation** (2 bytes) – this field indicates that the RIP packet is either a request or a response.
* **Network number** (4 bytes) – this field defines the assigned network address number to which the routing information applies.
* **Number of router hops** (2 bytes) – this field indicates the number of routers that a packet must go through in order to reach the required destination. Each router adds a single hop.
* **Number of ticks** (2 bytes) – this field indicates the amount of time (in 1/18 second) that it takes a packet to reach the given destination. Note that a route which has the fewest hops may not necessarily be the fastest.

RIP packets add to the general network traffic as each router broadcasts its entire routing table every 60 seconds. This shortcoming has been addressed by NetWare 4/5.

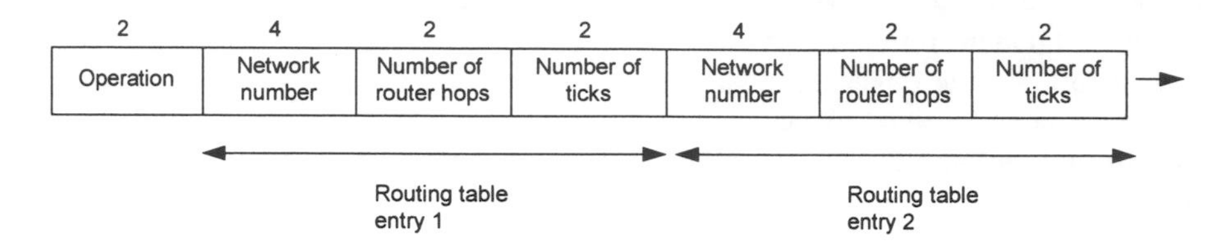

Figure 22.10 RIP packet format

SAP

Every 60 seconds each server transmits a **SAP** (Service Advertising Protocol) packet which gives its address and tells other servers which services it offers. These packets are read by special agent processes running on the routers which then construct a database that defines which servers are operational and where they are located.

When the client node is first booted it transmits a request in the network asking for the location of the nearest server. The agent on the router then reads this request and matches it up to the best server. This choice is then sent back to the client. The client then establishes an **NCP** (NetWare Core Protocol) connection with the server, from which the client and server negotiate the maximum packet size. After this, the client can access the networked file system and other NetWare services.

Figure 22.11 illustrates the contents of a SAP packet. It can be seen that each SAP packet contains a single operation field and data on up to seven servers. The fields are:

- **Operation type** (2 bytes) – defines whether the SAP packet is server information request or a broadcast of server information.
- **Server type** (2 bytes) – defines the type of service offered by a server. These services are identified by a binary pattern, such as:

File server	0000 1000	Job server	0000 1001
Gateway	0000 1010	Print server	0000 0111
Archive server	0000 1001	SNA gateway	0010 0001
Remote bridge server	0010 0100	TCP/IP gateway	0010 0111
NetWare access server	1001 1000		

- **Server name** (48 bytes) – which identifies the actual name of the server or host offering the service defined in the service type field.
- **Network address** (4 bytes) – which defines the address of the network to which the server is attached.
- **Node address** (6 bytes) – which defines the actual MAC address of the server.
- **Socket address** (6 bytes) – which defines the socket address on the server assigned to this particular type of service.
- **Hops to server** (2 bytes) – which indicates the number of hops to reach the particular service.

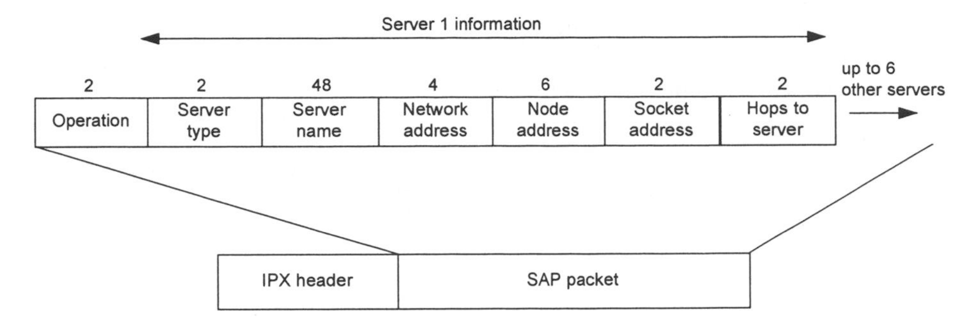

Figure 22.11 SAP packet format

NCP

The clients and servers communicate using the NetWare Core Protocols (NCPs). They have the following operation:

- The NETx shell reads the application program request and decides whether it should direct it to the server.
- If it does redirect, then it sends a message within an NCP packet, which is then encapsulated within an IPX packet and transmitted to the server.

Figure 22.12 illustrates the packet layout and encapsulation of an NCP packet. The fields are:

- **Request type** (2 bytes) – which gives the category of NCP communications. Among the possible types are:

Busy message	1001 1001 1001 1001
Create a service	0001 0001 0001 0001
Service request from workstation	0010 0010 0010 0010
Service response from server	0011 0011 0011 0011
Terminate a service connection	0101 0101 0101 0101

For example the create-a-service request is initiated at login time and a terminate-a-connection request is sent at logout.

- **Sequence number** (1 byte) – which contains a request sequence number. The client reads the sequence number so that it knows the request to which the server is responding.
- **Connection number** (1 byte) – a unique number which is assigned when the user logs into the server.
- **Task number** (1 byte) – which identifies the application program on the client which issued the service request.
- **Function code** (1 byte) – which defines the NCP message or commands. Example codes are:

Close a file	0100 0010	Create a file	0100 1101
Delete a file	0100 0100	Get a directory entry	0001 1111
Get file size	0100 0000	Open a file	0100 1100
Rename a file	0100 0101	Extended functions	0001 0110

Extended functions can be defined after the 0001 0110 field.

- **NCP message** (up to 539 bytes) – the NCP message field contains additional information which is passed between the clients and servers. If the function code contains 0001 0110 then this field will contain subfunction codes.

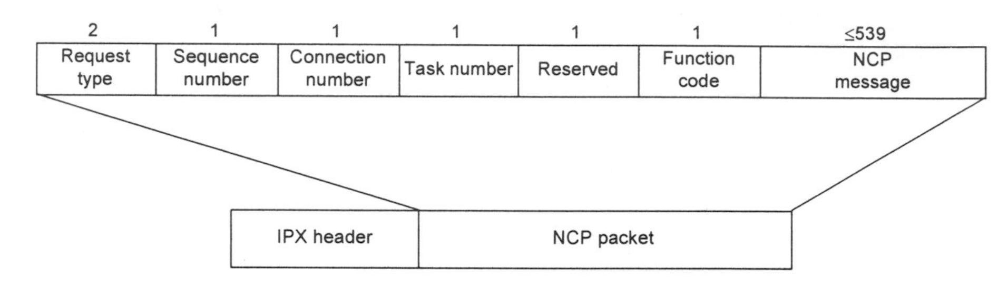

Figure 22.12 NCP packet format

22.4.3 Novell NetWare set-up

NetWare 3.x and 4/5 use the Open Data-Link Interface (**ODI**) to interface NetWare to the NIC. Figure 22.13 shows how the NetWare fits into the OSI model. ODI is similar to NDIS in Windows NT and was developed jointly between Apple and Novell. It provides a standard vendor-independent method to interface the software and the hardware.

A typical login procedure for a NetWare 3.x network is:

```
LSL.COM
NE2000
IPXODI
NETx /PS=EECE_1
F:
LOGIN
```

The program `LSL` (link support layer) provides a foundation for the MAC layer to communicate with multiple protocols. An interface adapter driver (in this case `NE2000`) provides a MAC layer driver and is used to communicate with the interface card. This driver is known as a multilink interface driver (**MLID**). After this driver is installed, the program `IPXODI` is then installed. This program normally communicates with `LSL` and applications.

The `NETx` program communicates with the server and sets up a connection with the server `EECE_1`. This then sets up a local disk partition of `F:` (onto which the user's network directory will be mounted). Next the user logs into the network with the command `LOGIN`.

ODI

ODI allows users to load several protocol stacks (such as TCP/IP and SPX/IPX) simultaneously for operation with a single NIC. It also allows support to link protocol drivers to adapter drivers. Figure 22.14 shows the architecture of the ODI interface. The **LSL** layer supports multiple protocols and it reads from a file `NET.CFG`, which contains information on the network adapter and the protocol driver, such as the interface adapter, frame type and protocol.

A sample `NET.CFG` file is:

```
Link Driver NE2000
      Int #1 11
      Port #1 320
      Frame Ethernet_II
      Frame Ethernet_802.3
      Protocol IPX 0 Ethernet_802.3
        Protocol Ethdev 0 Ethernet_II
```

This configuration file defines the interface adapter as using interrupt line 11, having a base address of 320h, operating IPX, Ethernet 802.3 frame type and following the Ethernet II protocol. Network interface card drivers (such as NE2000, from the previous set-up) are referred to as a multilink interface driver (MLID).

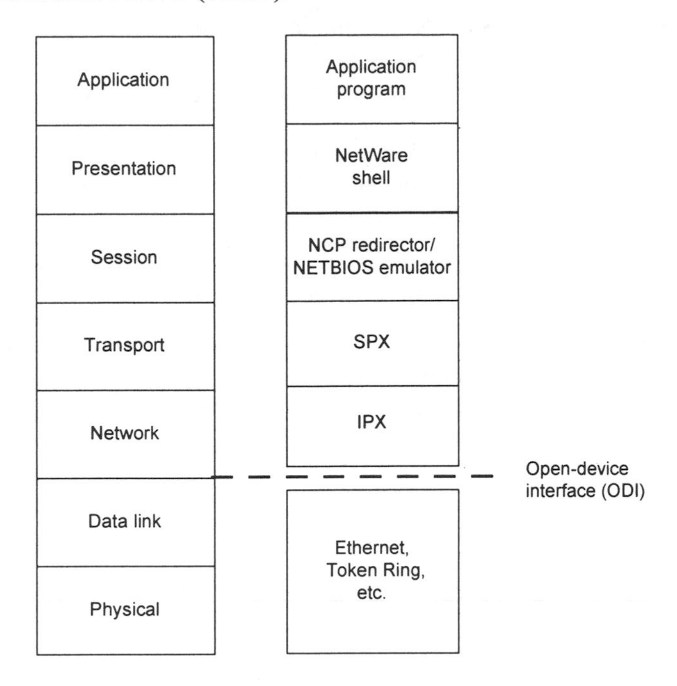

Figure 22.13 OSI model and NetWare 3.x

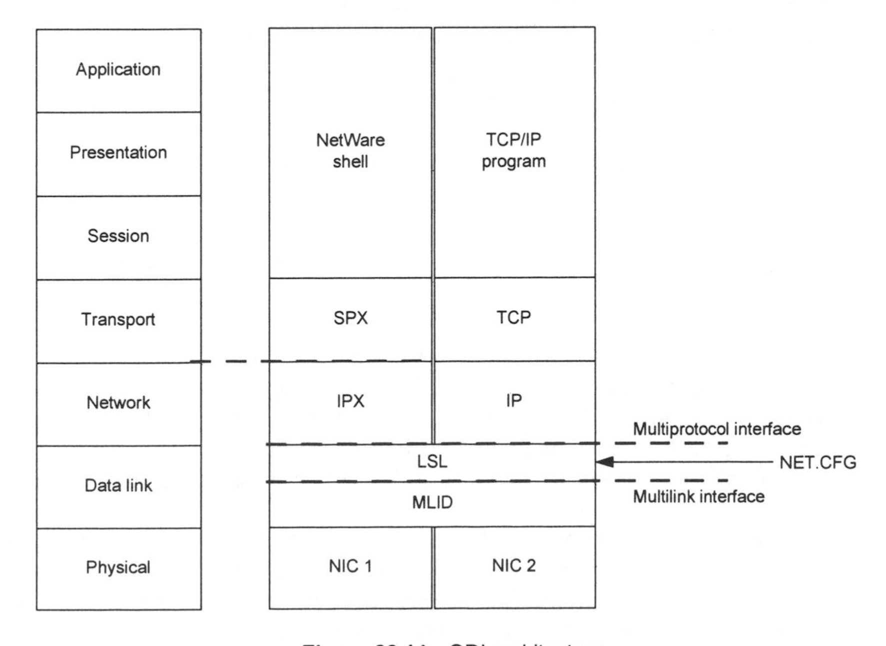

Figure 22.14 ODI architecture

22.5 NDS

The main disadvantages of NetWare 3.x are:

- It uses SPX/IPX which is incompatible with TCP/IP traffic.
- It is difficult to synchronize servers with user information.
- The file structure is local to individual servers.
- Server architecture is flat and cannot be organized into a hierarchical structure.

These were addressed with NetWare 4/5, in which the bindery was replaced by Novell Directory Services (NDS). NDS is a combination of features from OSI X.500 and Banyan StreeTalk. Its main characteristics are:

- Hierarchical server structure.
- Network-wide users and groups.
- Global objects. NDS integrates users, groups, printers, servers, volumes and other physical resources into a hierarchical tree structure.
- System-wide login with a single password. This allows users to access resources which are connected to remote servers.
- NDS processes logins between NetWare 3.1 and NetWare 4/5 servers, if the login names and passwords are the same.
- Supports distributed file system.
- Synchronization services. NDS allows for directory synchronization, which allows directories to be mirrored on different partitions or different servers. This provides increased reliability in that if a server develops a fault then the files on that server can be replicated by another server.
- Standardized organizational structure for applications, printers, servers and services. This provides a common structure across different organizations.
- It integrates most of the administrative tasks in Windows-based NWADMIN.EXE program.
- It is a truly distributed system where the directory information can be distributed around the tree.
- Unlimited number of licenses per server. NetWare 3.1 limits the number of licenses to 250 per server.
- Support for NFS server for UNIX resources.
- Multiple login scripts, as opposed to system and user login scripts in NetWare 3.1.
- Windows NT support.

NDS is basically a common, distributed Directory database of logical and physical resources made to look like a single information system. Many other applications have used Directory databases, such as electronic mail and network management. NDS servers within a network access the Directory database for the connected resources and details of how they are accessed. Thus application programs do not need to know the physical location and on which server it is connected, only its logical name.

The main reason to upgrade to NDS is that it better reflects the organizational structure of networked equipment within the organization. NetWare 3.1 is a server-based approach where resources are grouped around servers. This leads to increased maintenance around these servers, thus updates to one server may have to be updated on other servers. NDS allows for a central administration with a structure that reflects organizational structures.

22.5.1 NetWare directory services (NDS)

One of the major changes between NetWare 3.x and NetWare 4/5 is NDS. A major drawback of the NetWare 3.x bindery files is that they were independently maintained on each server. NDS addresses this by setting up a single logical database, which contains information on all network-attached resources. It is logically a single database, but may be physically located on different servers over the network. As the database is global to the network, a user can log into all authorized network-attached resources, rather than requiring to log into each separate server. Thus, administration is focused on the single database.

As with NetWare 3.x bindery services, NDS organizes network resources by objects, properties, and values. NDS differs from the bindery services in that it defines two types of object:

- **Leaf objects** – which are network resources such as disk volumes, printers, printer queues, and so on.
- **Container objects** – which are cascadable organization units that contain leaf objects. A typical organizational unit might be a company, department or group.

NDS organizes networked resources in a hierarchical or tree structure (as most organizations are structured in this way). The top of the tree is the root object, to which there is only a single root for an entire global NDS database. Servers then use container objects to connect to branches coming off the root object. This structure is similar to the organization of a directory file structure and can be used to represent the hierarchical structure of an organization. Figure 22.15 illustrates a sample NDS database with root, container and leaf objects. In this case, the organization splits into four main containers: Electrical, Mechanical, Production and Administration. Each of these containers has associated leaf objects, such as disk volumes, printer queues, and so on. This is a similar approach to Workgroups in Microsoft Windows.

To improve fault tolerance, NDS allows branches of the tree (or partitions) to be stored on multiple file servers. These mirrors are then synchronized to keep them up to date. Another advantage of replicating partitions is that local copies of files can be stored so that network traffic is reduced.

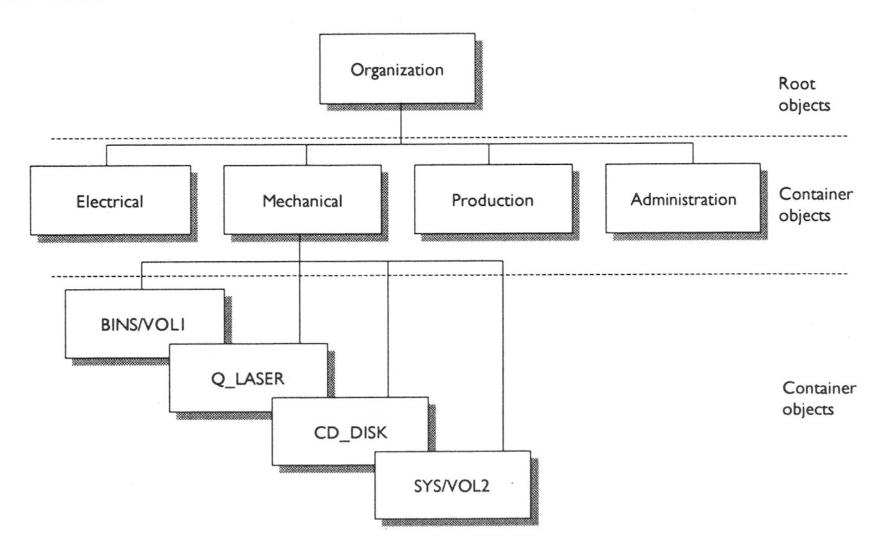

Figure 22.15 NDS structure

The container objects are:

[ROOT]. This is the top level of the inverted tree and contains all the objects within the organizational structure.

Organization. This object class defines the organizational name (such as FRED_AND_CO). It is normally the next level after [ROOT] (or below the C=Country object).

User. This object defines an individual user. The first user created in a NetWare 4 system is the ADMIN user, which is typically the only user with rights to add and delete objects on the whole of the NDS structure.

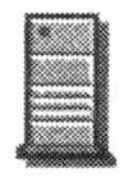

NCP (NetWare Control Protocol) **Server**. This appears for all NetWare 4 servers.

Volume. This identifies the mounted volume for file services. A network file system data links to the Directory tree through Volume objects.

The most commonly used objects are:

Bindery. These allow compatibility with existing Bindery-based NetWare 3, NetWare 3 clients and NetWare 4 servers which do not completely implement NDS. They display any object that isn't a user, group, queue, profile or print server, which was created using the bindery services.

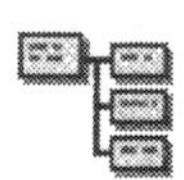

Organizational unit. This object represents the OU part of the NDS tree. These divide the NDS tree into subdivisions, which can represent different geographical sites, different divisions or workgroups. Different divisions might be PRODUCTION, ACCOUNT, RESEARCH, and so on. Each Organizational Unit has its own login script.

Organization role. This object represents a defined role within an organization object. It is thus easy to identify users who have an administrative role within the organization.

Group. This object represents a grouping of users. All users within a group inherit the same access rights.

Directory map. This object points to a file system directory on a mounted volume. It is typically used to create a global file system which has physically separate parts.

Alias. This identifies an object with another name. For example, a print queue which is called NET_PRINT1 might have an alias name of HP _LASER_JET_6.

Printer. This can either be connected to the printer port of a PC, or connected to a NetWare server.

Print queue. This object represents the queue of print jobs.

Profile. This object defines a special scripting file. This can be a global login script, a location login script or a special login script.

Print server. This object allows print jobs to be queued, waiting to be serviced by the associated printer.

22.5.2 NDS tree

Figure 22.16 shows the top levels of the NDS tree. These are:

- **[ROOT].** This is the top level of the tree. The top of the NDS tree is the [ROOT] object.
- **C=Country.** This object can be used, or not, to represent different countries, typically where an organization is distributed over two or more countries. If it is used then it must be placed below the [ROOT] object. NDS normally does not use the Country object and uses the Organization Unit to define the geographically located sites, such as SALES_UK.[ROOT], SALES_USA.[ROOT], and so on.
- **L=Locality.** This object defines locations within other objects, and identifies network portions. The Country and Locality objects are included in the X.500 specification, but they are not normally used, because many NetWare 4 utilities do not recognize it. When used, it must be placed below the [Root] object, Country object, Organization object, or Organizational Unit object.
- **LP=Licensed Product**. This object is automatically created when a license certificate is installed. When used, it must be placed below the [Root] object, Country object, Organization object, or Organizational Unit object.
- **O=Organization**. This object represents the name of the organization, a company division or a department. Each NDS Directory tree has at least one Organization object, and it must be placed below the [Root] object (unless the tree uses the Country or Locality object).
- **OU=Organization Unit**. This object normally represents the name of the organizational unit within the organization, such as Production, Accounts, and so on. At this level, User objects can be added and a system level login script is created. It is normally placed below the Organizational object.

The structure of the NDS should reflect the organization of the company, for its organizational structure, its locations and the organization of its networks. Normally there is only one Organization object as this makes it easier to merge the NDS tree with other organizations. With every Organization object, there are normally several Organization Units.

Apart from the container objects (C, O, OU, and so on) there are leaf objects. These are assigned a CN (for Common Name). They include:

CN=AFP Server	CN=Bindery	CN=Bindery Queue
CN=Computer	CN=Directory Map	CN=Group
CN=Organizational Role	CN=Print Queue	CN=Print Server
CN=Printer	CN=Profile	CN=Server
CN=User	CN=Volume	

If possible, the NDS tree depth should have between four and eight levels. This makes management easier and allows resources to be easily accessed.

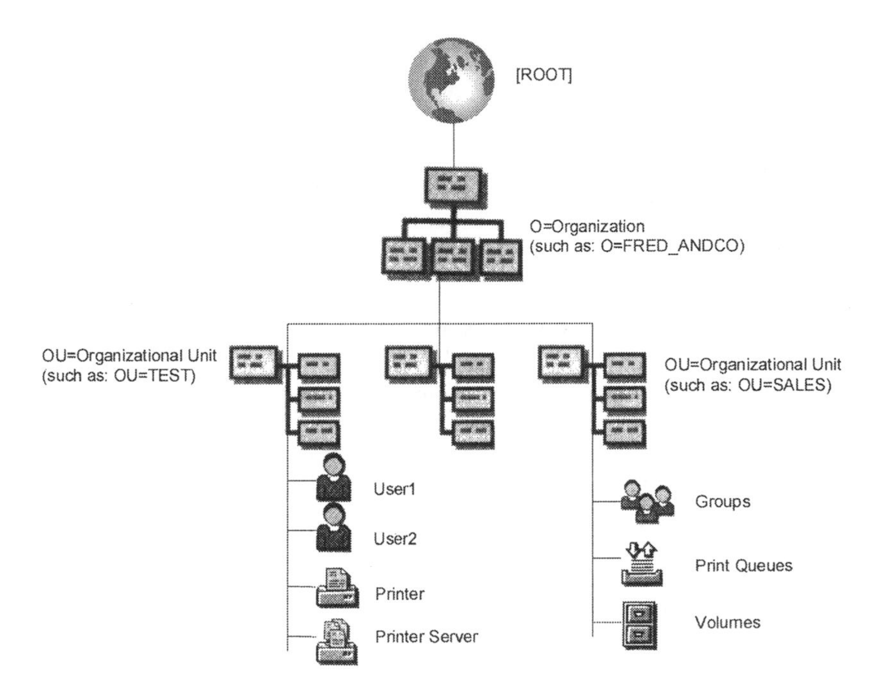

Figure 22.16 NDS structure

22.5.3 Typical naming syntax

The NDS tree can use many different naming formats, but a standardized naming structure has been developed. These are:

	Syntax	*Example*
[ROOT]	*company*_TREE	`FRED_TREE`
Organization	*company_name*	`O=FRED`
Organization Units	*location* (or *department*)	`OU=SALES`
Servers	*location-department*-SRV#	`SALES-SRV1`
Printer Servers	*location-department*-PS#	`SALES-LZ5-PS3`
Printers	*printer*-P#	`HPLJ5-P2`
Print Queues	*type*-P#	`HPLJ5-P2`
Volumes	*server_volume*	`SALES-SRV1_DATA`

22.5.4 Object names

The location at which an object is placed is called its context. Two objects which are placed in the same container have the same context. For example, if the user FRED_B works for the Fred & Co. (O=FRED_AND_CO), within the Test Department (OU=TEST), which is within the Engineering Unit (OU=ENGINEERING) then his context will be:

```
OU=TEST.OU=ENGINEERING.O=FRED_AND_CO
```

An object is either identified by its distinguishing name (such as `LP_LASER5`) or by its complete name (CN). In the name, periods separate the objects (these periods are similar to back slashes or forward slashes, which is common in many operating systems). For a complete name, which is referred to from the [ROOT] object, a leading period is used, whereas a relative name does not have a leading period. For example, a complete name for a User object `FRED_B` could be:

```
.CN=FRED_B.OU=TEST.OU=ENGINEERING.O=FRED_AND_CO
```

This defines a User, which has an Organization of `FRED_AND_CO`, which has an Organization Unit called `ENGINEERING`, there is then a subdivision below this called `TEST`. It is also possible to define a relative distinguishing name (RDN) which defines the relative path with respect to the current context.

Periods can be added to the start or the end of the context. They have the following definitions:

- Leading period. NDS ignores the current context of the object and resolves the name at the [ROOT] object.
- Trailing period. NDS selects a new context when resolving an object's complete name at the [ROOT] object.

For example, the partial name for the User object `FRED_B` relative to other objects in `OU=TEST` would be:

```
.CN=FRED_B.
```

The partial name of the User object `FRED_B` that has a complete name of:

```
.CN=FRED_B.OU=TEST.OU=ENGINEERING.O=FRED_AND_CO
```

relative to a server object with a complete name of:

```
.CN=OU=SALES-SRV1.OU=SALES.O=FRED_AND_CO
```

is:

```
CN=FRED_B.OU=TEST.OU=ENGINEERING.
▲
'------------------------------------------------------ Relative name.
```

The `HPLJ5-P2` printer object which has the complete name of:

```
.CN=HPLJ5-P2.OU=TEST.OU=ENGINEERING.O=FRED_AND_CO
```

would be referred to, within the `OU=TEST.OU=ENGINEERING.O=FRED_AND_CO` container, as:

```
CN=HPLJ5-P2
```

Notice that a relative name has a trailing period to identify that it is a partial name. It is also possible not to include the object types (such as CN for common name, OU for Organizational Unit and O for Organization). This is called a **typeless** name, and NDS makes a guess as to the object types. For example:

```
FRED_B.TEST.ENGINEERING.FRED_AND_CO
```

is the same as one of the previous examples. When guessing NDS uses the following rules:

- The object which is furthest to the left is assumed to be a common name (leaf object).
- The object which is furthest to the right is assumed to be the organization (container object).
- All other objects are assumed to be Organizational Units (container objects).

22.5.5 CX

The `CX` (Change conteXt) command is used to display or modify the context, or to view containers and leaf objects in the Directory tree. In a Command Prompt window, the following can be used:

Command	*Description*
`CX`	display current context
`CX /?`	display help manual
`CX /CONT`	display all containers in the current context
`CX /T`	display all containers at and below the current context
`CX .`	move up one level
`CX ..`	move up two levels
`CX /CONT`	display containers in the [ROOT]
`CX` *content*	display context for *content*
`CX /R`	change current context to [ROOT]
`CX /A`	display all containers and objects in the current context
`CX /R /A /T`	display all containers and objects, from the [ROOT] down

For example to set the current context to the `TEST.ENGINEERING.FRED_AND_CO` container:

```
CX TEST.ENGINEERING.FRED_AND_CO
```

Then to change the context to `ENGINEERING`:

```
CX ENGINEERING.FRED_AND_CO
```

or

```
CX .
```

22.5.6 Startup files and scripts

Much of the initialization of a client is done with startup files and scripts. The main startup files are:

- `CONFIG.SYS` and `AUTOEXEC.BAT`. These are standard startup files for the PC and normally set up the environment of the computer. The `AUTOEXEC.BAT` file should include the `STARTNET.BAT` file.
- `STARTNET.BAT`. Provides a network connection.
- `NET.CFG`. Customizes the NetWare setup, such as setting ODI and VLM settings.

The login scripts are:

- Container Login Scripts. These set up the Organization and Organizational Unit properties, and generally replace System login scripts.
- Profile Login Scripts. These set up the environment of User groups.
- User Login Scripts. These customize the User environment. If no User login script exists then a Default Login Script is executed.

The NET.CFG file is similar to NetWare 3 but has extra lines to define the NetWare 4 options. An example file is:

```
Link Driver NE2000
         Int #1 11
         Port #1 320
         Frame Ethernet_II
         Frame Ethernet_802.3
         Protocol IPX 0 Ethernet_802.3

NetWare DOS Requester
      NAME CONTEXT="OU=electrical.OU=engineering.O=napier"
      PREFERRED SERVER = EEE-SRV1
      FIRST NETWORK DRIVE = G
      NETWARE PROTOCOL = NDS, BIND
```

This defines the name context for the user (with `NAME CONTEXT`) and that the preferred server is `EEE-SRV1`. The first network drive will be G: and the NetWare protocol is NDS and Bindery.

Drive disks can be mounted by adding lines to the Login Script (such as NETSTART.BAT, which is started from the AUTOEXEC.BAT file). For example, to mount the F:, G: and M: drives then the following could be added:

```
MAP ROOT F:= .EEE-SRV1.ENGINEERING.NAPIER\SYS:APPS
MAP ROOT G:= .CRAIGLOCKHART_1.MAJOR.NAPIER.AC.UK\SYS:APPS
MAP ROOT M:= .CRAIGLOCKHART_3.MAJOR.NAPIER.AC.UK\SYS:MAIL
```

22.5.7 Volume mapping

Volumes can be mounted as drives using the syntax:

`MAP` *drive_letter*`:=CN=`*servername_volumename.context*`:`

For example, to map the DATA volume of the TEST server to drive letter F: then the following is used:

```
MAP F:=CN=TEST_DATA.OU=TEST.:
```

22.5.8 Country object

The country object is commonly not used as it fixes the geographical location of objects. It has the advantage, though, that it fits into a common Internet naming structure (such as, `www.eece.napier.ac.uk`) or X.500 names. Most networks though have the Organizational Unit following the [ROOT] level. For example an educational organization in the UK will have a country object of UK and the organization object of AC (as defined in the Internet name). The Organization Unit would then be the name of the academic organization (in this case, Napier). Next, the facilities and departments are defined, as follows:

```
[ Root]
   c=uk
      o=ac
         ou=napier
            ou=Arts
            ou=Business
            ou=Engineering
               ou=electrical
               ou=mechanical
               ou=computing
```

Thus, the context name for a printer (`LJET5`) in the Electrical department would be:

```
CN=LJET5.OU=ELECTRICAL.OU=ENGINEERING.O=NAPIER.C=UK
```

This is obviously similar to the Internet name for the device, which would be:

```
ljet5.electrical.engineering.napier.ac.uk
```

22.5.9 User class

NDS has an object-oriented database, where each object (such as Users, Printers, and so on) has associated properties. The User object has the following properties:

- **Login Name.** Normally the first character of the first name followed by the last name, such as BBuchanan (for Bill Buchanan). [Required]; **Given Name.** User's first name. [Required]
- **Last Name.** User's last name. [Required]; **Full Name.** User's full name. [Required]
- **Generation Qualification.** [Optional]; **Middle Initial.** [Required]
- **Other Name.** [Optional]; **Title.** Job title. [Required]
- **Description.** [Optional]; **Location.** City or location. [Required]
- **Department.** [Optional]; **Telephone Number.** Full telephone number [Required]; **Fax Number.** Full Fax number [Required]; **Language.** Spoken language. [Optional]
- **Network address.** [System adds this]; **Default server**. Server that the user initially logs into. [Optional]
- **Home Directory**. Volume:subdirectory\user, such as DATA:HOME\BILL_B [Required]; **Required Password.** Force password, or not. [Required]
- **Account Balance.** [Optional]; **Login Script.** [Optional]; **Print Job Configuration.** [Optional]
- **Post Office Box.** [Optional]; **Street.** [Optional]; **City.** [Optional]
- **State or Province.** [Optional]; **Zip Code.** [Optional]; **See also.** [Optional]

22.5.10 Bindery services

In NetWare 4/5, and further versions, the bindery services have been replaced by NetWare Directory Services (NDS). Many networks take some time to be fully upgraded from bindery to NDS, thus NetWare 4/5 supports bindery. This allows a NetWare 3 server to be upgraded to NetWare 4/5, but still run a bindery. It also allows NetWare 4/5 to integrate with NetWare 3 servers (typically, which run print services).

NetWare 4/5 servers support the following bindery-based resources:

- **Bindery objects class.**
- **Groups.**
- **Bindery programs.**
- **Print servers.**

- **Profiles.**
- **Users.**
- **Queues.**
- **Bindery-based NetWare client software.**

On a bindery service, NDS supports a flat structure for leaf objects in an Organization or Organizational Unit object. All objects within the specified container can then be accessed by NDS objects and by bindery-based servers and client workstations. On a bindery enabled server, a client gets its login script from that server. The login and script are not automatically transmitted to other servers (as they would be with NDS).

Figure 22.17 shows an example of a bindery-enabled server which is an Organizational Unit object (OU).

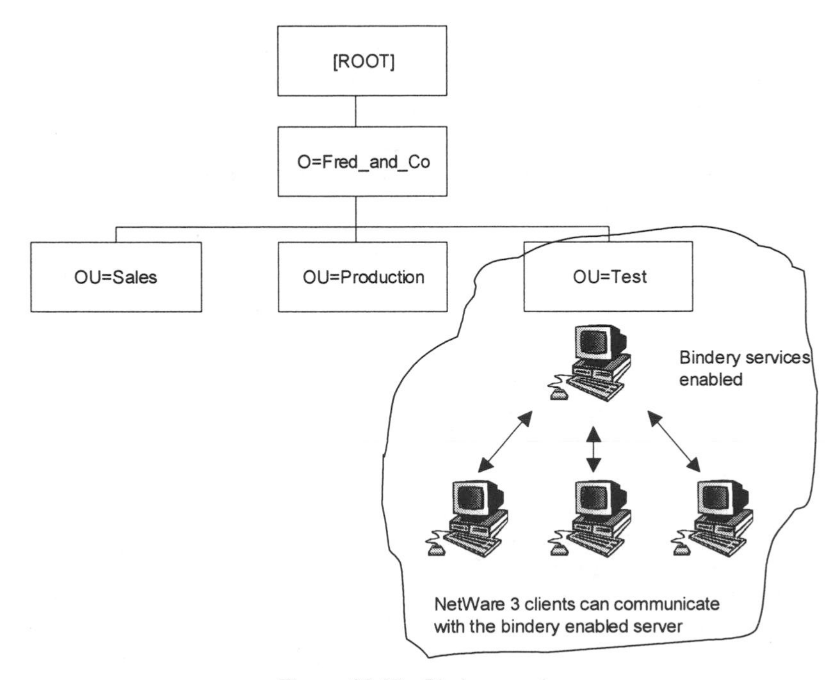

Figure 22.17 Bindery services

22.5.11 Time synchronization

Time synchronization between servers is important as it allows NDS events and modifications to be accurately time stamped. There are two main options:

- **Single reference configuration.** This provides a single source of time reference and is typically used with networks with less than thirty servers.
- **Time provider group configuration.** This provides a single reference primary server and, at least, two other primary time providers. It is typically used when there are more than thirty servers connected to the network. The other primary time providers allow for a system failure of the reference primary server.

The time synchronization is provided with:

- **SAP.** With SAP the SAP type is 0000 0010 0110 1011 (026Bh). The main disadvantage with this method is that SAP generally adds to network traffic and, as SAP is self configuring, an incorrect set up can cause the incorrect time to be transmitted.

- **Configured List Communication.** This method allows each server to keep a list of servers which it can communicate with. This will generally lead to less network traffic as it does not use an SAP broadcast. It also stops incorrectly setup servers from transmitting incorrect time information (as the server will only communicate with preferred time servers).

NetWare uses the TIMESYNC.NLM (NetWare Loadable Module) module to synchronize their local time. The server then calculates Universal Coordinated Time (UTC) which provides a world standard time. UTC is a machine-independent time standard. It assumes that there are 86 400 seconds each day ($24 \times 60 \times 60$) and once every year or two an extra second is added (a 'leap' second). This is normally added on 31 December or 30 June.

Most computer systems define time with GMT (Greenwich Mean Time), which is UT (Universal Time). UTC is based on an atomic clock, whereas GMT is based on astronomical observations. Unfortunately, because of the earth's rotation GMT is not uniform, and is not as accurate as UTC. UTC is calculated by:

$$\text{UTC} = \text{LOCAL TIME} + \textit{timezone\_offset} + \textit{current\_daylight\_adjustment}$$

With TIMESYNC.NLM the system time is not actually changed, the local clock is either speeded up (if the time is behind) or slowed down (if it is ahead). This makes for gradual changes in the system time. This is especially important for server synchronization where directories and files are kept up to date between servers. An incorrectly set time on one server could cause an older file to replace a newer file. Every NDS object and its associated properties have a timestamp associated with them.

NDS timestamp

Particular problems caused with time on computer systems are the Year 2000 bug (where dates are referenced to just the last two digits of the year) and where there is a roll-over in the counter value which stores the system time. The Year 2000 bug was easily eradicated by making sure that all references to time take into account the full year format.

The PC contains a 32-bit counter which is updated every second and is referenced to 1 January 1970 (the starting date for the PC). This provides for 4,294,967,296 seconds (715,827,882 minutes, 11,930,465 hours, 497,103 days and 1361 years). The format of the NDS time-

Standard NT accounts

Administrator. Used for administration of a domain.
Guest. Designed for limited-time or occasional user. On medium-security and high-security domains, this account should be disabled. Guests should be given unique accounts.
System. Used to run many of the server processes and for assigning file access permission.

Standard NT domain groups

Domain Admins. Used to assign the administrators group within the domain.
Domain Users. Used to assign the users accounts in the domain.
Domain Guests. Used to assign the guest accounts in the domain.

Standard NT local groups

Administrators. Contains the Administrators account and the Domain Admins domain group.
Account Operators, **Backup Operators**, **Print Operators** and **Server Operators.** Less privileged than the Administrators but more than user accounts. Each perform a specific task for an administrative function.
Replicators. Used by the Directory Replicator Service, which allows for automatic copying of files between systems within a domain.
Users. A group which holds ordinary users.
Guests. A group which holds guest accounts for the local domain.

stamp uses this format and adds other fields to define the place the event occurred and an indication of events that occur within a single second. It uses 64 bits and its format is:

- **Seconds** (32 bits). This stores the number of seconds since 1/1/1970. This allows for 4 billion seconds, which is approximately 1371 years, before a roll-over occurs.
- **Replica Number** (16 bits). This is a unique number which defines where the event occurred and the timestamp issued.
- **Event ID** (16 bits). Defines each event that occurs within a second a different Event ID. This is required as many events can occur within a single second. This value is reset on every second, and thus allows up to 65,536 events each second.

NDS always uses the most recently time stamped object or properties for any updates. When an object is deleted its property is marked as 'not present'. It will only be deleted once the replica synchronization process propagates the change to all other replicas.

Time server types
NetWare 4/5 servers are set up as time servers when they are installed. They can either be:

- **Primary time servers.** A primary time server provides time information to others, but must contact at least the primary (or reference) server for their own time.
- **Secondary time servers.** These are time consumers, which receive their time from other servers (such as from a primary, reference or single reference time server).
- **Reference time servers.** These servers do not need to contact any other servers, and provide a time source for other primary time servers. This is a good option where there is a large network, as the primary time servers can provide local time information (this is called a time provider group).
- **Single reference time servers.** These servers do not need to contact other time servers to get their own time and are used as a single source of time. This is normally used in a small network, where there is a single reference time server with one or more secondary time servers. The single reference time server and reference time server normally get their local time information from another source, such as Internet time, radio or satellite time. This is the default condition for installation.

22.5.12 Virtual loadable modules (VLMs)

The NETx redirector shell has been replaced with DOS client software known as the requester. Its main advantage is that it allows NetWare clients to easily add or update their functionality by using VLMs. This is controlled through the DOS-based VLM management program (VLM.EXE). It differs from NETx in that the requester uses DOS tables of network-attached resources rather than creating and maintaining its own. The main difference between NETx and the requester is that it is the DOS system which controls whether the NetWare DOS request is called to handle network requests.

Various VLM modules can be added onto the client, such as:

- **Bindery-based services.**
- **File management.**
- **IPX and NCP protocol stacks.**
- **NDS services.**
- **NetWare support for multiple protocol stacks** (e.g. TCP/IP, SPX/IPX).
- **NETx shell emulation.**
- **Printer redirector to network print queues.**
- **TCP/IP and NCP protocol stacks.**

22.5.13 Fault tolerance

NetWare 4/5 allows disk mirroring of partitions when a disk drive fails. Thus if one of the disk drives fails, it is possible to switch to the mirror drive. Another major fault occurs when a server becomes inoperative. NetWare 4/5 uses a novel technique, known as SFT III, which allows server duplexing. In this technique, the contents of the disk, memory and CPU are synchronized between primary and duplexed servers. When the primary server fails then the duplexed server takes over transparently. These servers are synchronized using the mirror server link (MSL), a dedicated link between the two servers, as illustrated in Figure 22.18. The MSL is a dedicated link because it prevents general network traffic from swamping the data.

It may seem expensive to have a backup server doing nothing apart from receiving data, but if it is costed with the loss of business or data when the primary server goes down then it is extremely cheap.

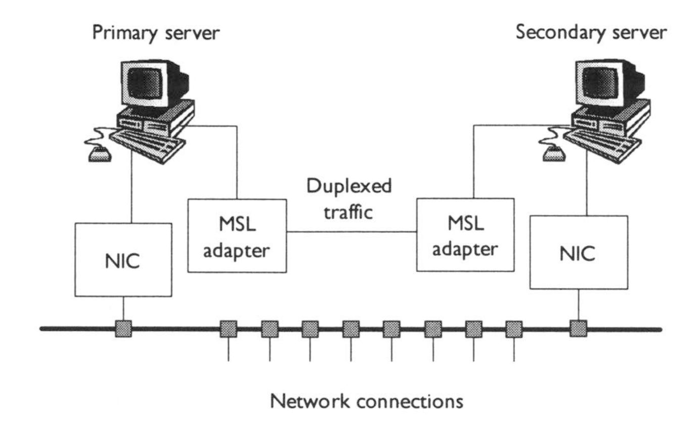

Figure 22.18 Mirror server link

22.5.14 Communications protocols

NetWare 4/5 has improved existing protocols and added the support for other standard network protocols, especially TCP/IP. These are:

- **TCP/IP.** TCP/IP is supported with NetWare/IP which is included with NetWare 4/5; NetWare/IP servers can support IP, IPX or IP and IPX traffic.
- **Large IPX packets.** Most networks have become less prone to error. Thus larger data packets can be transmitted with a low risk of errors occurring. **LIP** allows NetWare clients to increase the size of their data field by negotiating with routers as to the size of the IPX frame (normally its has a maximum of 576 bytes). Unfortunately, an error in the packet causes the complete packet to be retransmitted (thus causing inefficiencies). In addition, the router must support the use of LIP. The software-based Novell router has a multiprotocol router which supports LIP. Unfortunately, other vendors may not support the LIP protocol.
- **NetWare Link-State Routing Protocol (NLSP).** NLSP overcomes the problems of RIP (File servers transmit their routing table every 60 seconds. This can have a great effect on the network loading, especially for interconnected networks. RIP also only supports 16 hops before an RIP packet is discarded, thus limiting the physical size of the internetwork linking NetWare LAN). With the routing table, NLSP only broadcasts when a

change occurs, with a minimum update of once every 2 hours. This can significantly reduce the router-to-router traffic. As with LIP, Novell routers support NLSP, but other vendors may not necessarily support it. NLSP supports an increased hop size. A great advantage with NLSP is that it can coexist with RIP and is thus backward compatible. This allows a gradual migration of network segments to NLSP.

22.5.15 NetWare 4.1 SMP

One of the great improvements in computer processing and power will be achievable through the use of parallel processing. This processing can either be realized using multiple local processors or network processors, called symmetrical multiprocessing (SMP). To maintain compatibility with a previous release, NetWare 4.1 **SMP** loads the SMP kernel which works co-operatively with the operating system kernel. The main processor runs the main operating system while the SMP kernel runs the second, third and fourth processors.

22.5.16 Other enhancements

Other enhancements have been added, such as:

- File compression, which is controllable on a file-by-file basis.
- Increased supervisor security.
- Increased support for printers (up to 255 can be connected).

22.6 Exercises

22.6.1 For an audit policy, which would be audited for which programs users are actually using (select one):

(a) Users logging on and off (b) User of folder and file resources
(c) Use of user rights (d) User and group management
(e) Security policy changes (f) Restarting/shutting down a system
(g) Process tracking

22.6.2 For an audit policy, which would be audited for changes made for tampering with a server (select one):

(a) Users logging on and off (b) User of folder and file resources
(c) Use of user rights (d) User and group management
(e) Security policy changes (f) Restarting/shutting down a system
(g) Process tracking

22.6.3 For an audit policy, which would be audited for changes made to user rights or audit policy (select one):

(a) Users logging on and off (b) User of folder and file resources
(c) Use of user rights (d) User and group management
(e) Security policy changes (f) Restarting/shutting down a system
(g) Process tracking

22.6.4 For an audit policy, which would be audited for unauthorized access to resources (select one):

(a) Users logging on and off (b) User of folder and file resources
(c) Use of user rights (d) User and group management
(e) Security policy changes (f) Restarting/shutting down a system
(g) Process tracking

22.6.5 For an audit policy, which would be audited for system tasks performed by the user:

(a)	Users logging on and off	(b)	User of folder and file resources
(c)	Use of user rights	(d)	User and group management
(e)	Security policy changes	(f)	Restarting/shutting down a system
(g)	Process tracking		

23.6.6 For an audit policy, which would be audited for unauthorized login attempts (select one):

(a)	Users logging on and off	(b)	User of folder and file resources
(c)	Use of user rights	(d)	User and group management
(e)	Security policy changes	(f)	Restarting/shutting down a system
(g)	Process tracking		

22.6.7 For an audit policy, which would be audited for changes made to user and group accounts (select one):

(a)	Users logging on and off	(b)	User of folder and file resources
(c)	Use of user rights	(d)	User and group management
(e)	Security policy changes	(f)	Restarting/shutting down a system
(g)	Process tracking		

22.6.8 Which networking protocol does Novell NetWare use:

(a)	SPX/IPX	(b)	NetBEUI
(c)	AppleTalk	(d)	TCP/IP

22.6.9 Which networking protocol does UNIX use:

(a)	SPX/IPX	(b)	NetBEUI
(c)	AppleTalk	(d)	TCP/IP

22.6.10 Contrast the file attributes used with NTFS with UNIX-type attributes. Which is simpler and which allows better control over security?

22.6.11 Outline the main protocols used on networked UNIX systems.

22.6.12 Explain how NFS is set up on a UNIX system and describe the protocols it uses.

22.6.13 Explain how NIS is set up on a UNIX system and the files it uses.

23.6.14 For a known NetWare network determine the following:

(a) Its version number.
(b) Its architecture.
(c) The connected peripherals (such as printers, tape backups, and so on).
(d) The number of user logins.
(e) File servers.
(f) The connections it makes with other NetWare servers.
(g) The connections it makes with the Internet.
(h) The location of bridges, routers or gateways.

22.6.15 Discuss the format of an IPX packet, and the SPX packet.

22.6.16 Discuss the RIP, SAP and NCP packets.

22.6.17 Outline the main advantages that NetWare 4/5 has over NetWare 3.1.

22.6.18 Define the difference between typeless names and typeful names. Also define the difference between leading and trailing periods.

22.6.19 Draw how the NDS tree connects to the user MIKE_A for the following context name:

```
.CN=MIKE_A.OU=TEST_ON_LINE.OU=TEST.O=TEST_R_US
```

Also, what is the typeless name for the above context name?

22.6.20 If a server's time is behind the UTC time, what happens to the system time on the server after receiving a synchronization time update?

22.6.21 Explain how NetWare 4/5 uses NDS.

22.6.22 Explain how NetWare 4 allows server fault tolerance.

22.7 Note from the Author

The Microsoft Windows operating system has carved a massive market, especially as it integrates networking, an operating system and a graphical user interface into a single package. It has proved popular, but many versions have been unreliable when compared with UNIX. Microsoft Windows, though, works well as a stand-alone operating system, which mounts network drives as local drives. It also supports a whole host of different peripherals from many manufacturers. It took a long time for Microsoft Windows to properly support network. Early versions of Microsoft Windows were a nightmare, and even Microsoft Windows Version 3 was basically just a graphical user interface which sat on top of the horrendous DOS. The first real Microsoft networking system was Microsoft NT Version 3, which was quickly followed by Windows 95. These two operating systems were built to support networking. Their great advantage was that they supported many different network protocols, such as TCP/IP, SPX/IPX, AppleTalk, IBM DLC and even the old Microsoft Windows network protocol, NetBEUI. This allowed the Microsoft Windows operating system to co-exist with other, existing, network operating systems. This was a great strategy as it allowed organizations to gradually migrate their existing networking operating systems towards Microsoft Windows. It has been a strategy which has been extremely successful, especially when Microsoft Windows NT Version 4 was released, which had the robustness of NT, added to the slick user interface of Windows 95. Further versions have enhanced networking, and Windows 2000 is likely to become the standard networking operating system for most organizations.

The great strength of UNIX is its networking protocols, many of which have become industry standard. Protocols such as TCP (for reliable connections), IP (for addressing), UDP (for unreliable connections), NFS (for connecting file systems over a network), ARP (to determine a MAC address from a known IP address) and DNS (for naming systems) have all grown up within the UNIX operating system. It is an operating system which has always supported networking, and it shows it. The big problem with UNIX, though, is that it is relatively difficult to set up, but once it is set up it will generally run reliably without any problems. UNIX systems also tend to be set up to operate with a global file system, thus when one of the disk drives becomes unavailable it can have a great effect on the rest of the system. UNIX is the last great defender against a global domination by Microsoft Windows. Its success depends on many things, including its ease-of-use, its robustness, its support, its support for standard protocols, and its non-Microsoftness (I made that one up!). Apple has found that the PC is a difficult beast to fight against. There are just too many developers making hardware and software, and there are too many great packages to dismiss it as a top purchase for any organization. But along with Microsoft Windows, it has led a privileged existence.

The PC is an amazing device, and has allowed computers to move from technical specialists to, well, anyone. However, they are also one of the most annoying of pieces of technology of all time, in terms of their software, their operating system, and their hardware.

If we bought a car and it failed at least a few times every day, we would take it back and demand another one. When that failed, we would demand our money back. Or, sorry, I could go on forever here, imagine a toaster that failed half way through making a piece of toast, and we had to turn the power off, and restart it. We just wouldn't allow it.

So why does the PC lead such a privileged life. Well it's because it's so useful and multi-talented, although it doesn't really excel at much. Contrast a simple games computer against the PC and you find many lessons in how to make a computer easy-to-use, and to configure. One of the main reasons for many of its problems is the compatibility with previous systems both in terms of hardware compatibility and software compatibility (and dodgy software, of course). The big change on the PC was the introduction of proper 32-bit software, Windows 95/NT/2000.

In the future systems will be configured by the operating system, and not by the user. How many people understand what an IRQ is, what I/O addresses are, and so on. Maybe if the PC faced some proper competition it would become easy to use and become totally reliable. Then when they were switched on they would configure themselves automatically, and you could connect any device you wanted and it would understand how to configure (we're nearly there, but it's still not perfect). Then we would have a tool which could be used to improve creativity and you didn't need a degree in computer engineering to use one (in your dreams!). But, anyway, it's keeping a lot of technical people in a job, so, don't tell anyone our little secret. The Apple Macintosh was a classic example of a well-designed computer that was designed as a single unit. When initially released it started up with messages like I'm glad to be out of that bag and Hello, I am Macintosh. Never trust a computer you cannot lift.

One of the classic comments of all time was by Ken Olson at DEC, who stated that there is no reason anyone would want a computer in their home. This seems farcical now, but at the time, in the 1970s, there were no CD-ROMs, no microwave ovens, no automated cash dispensers, and no Internet. Few people predicted them, so, predicting the PC was also difficult. But the two best comments were:

Computers in the future may weigh no more than 1.5 tons. Popular Mechanics.
I think there is a world market for maybe five computers, Thomas Watson, chairman of IBM, 1943.

Novell NetWare has, over the years, proved to be a reliable networking operating system. It is still extensively used and works well. Many large organizations use Novell NetWare as their core corporate networking operating system. With NDS, organizations can control resources and users around the network without having to set up each server. This allows the network to reflect the setup of the organization, rather than organization resources around servers. Microsoft Windows supports SPX/IPX, as this allows organizations to use Microsoft Windows to communicate with a Novell NetWare server, and it uses a different protocol to communicate with a Microsoft Windows server. This type of approach allows organizations to slowly migrate their systems away from Novell NetWare towards an integrated Microsoft Windows networks.

22.8 XDR format

The basic definition of the blocks is this:

- Items are defined in multiples of four bytes (32 bits) of data.
- These bytes are numbered from 0 to $n-1$.
- Bytes are read (or written) to a byte stream so that byte m always precedes byte $m+1$.

If the number of bytes (n) in the data is not divisible by 4 then the bytes are followed by enough (0 to 3) residual zero bytes (r) to make the count a multiple of 4.

The basic data types are defined in this section.

22.8.1 Unsigned Integer and Signed Integer

A signed integer has 32 bits and thus has a range from –2 147 483 648 to +2 147 483 647. It uses a 2's complement notation with the first byte the most significant and byte 3 the least significant. Integers are declared as follows:

```
int identifier;
```

and can be represented as:

```
(MSB)                      (LSB)
+-------+-------+-------+-------+
|byte 0 |byte 1 |byte 2 |byte 3 |
+-------+-------+-------+-------+
<------------32 bits------------>
```

An unsigned integer has 32 bits and thus has a range from 0 to +4 294 967 295. The most and least significant bytes are 0 and 3, respectively. Unsigned integers are declared as follows:

```
unsigned int identifier;
```

22.8.2 Enumeration

Enumerations have the same representation as signed integers and are useful in defining subsets of the integers. They are declared as follows:

```
enum { name-identifier = constant, ... } identifier;
```

For example, three menu options (FILE, EDIT and VIEW) could be described by an enumerated type:

```
enum { FILE = 1, EDIT = 2, VIEW = 3 } menu_options
```

22.8.3 Boolean

Booleans are declared as follows:

```
bool val;
```

which is equivalent to:

```
enum { FALSE = 0, TRUE = 1 } val;
```

22.8.4 Hyper Integer and Unsigned Hyper Integer

A hyper integer is a 64-bit value and allows greater ranges for integer values. The signed integer format uses 2's completed. In a hyper integer the most significant byte is 0 and the least significant is 7. They are declared as:

```
hyper identifier; unsigned hyper identifier;
```

and can be represented by:

```
  (MSB)                                                      (LSB)
+-------+-------+-------+-------+-------+-------+-------+-------+
|byte 0 |byte 1 |byte 2 |byte 3 |byte 4 |byte 5 |byte 6 |byte 7 |
+-------+-------+-------+-------+-------+-------+-------+-------+
<----------------------------64 bits---------------------------->
```

22.8.5 Floating-point

A float data type has 32 bits and uses the standard IEEE standard for normalized single-precision floating-point numbers. It has three fields:

- **S** (sign). A 1-bit value which represents a positive number as a 0 and a negative number as a 1.
- **E** (exponent). An 8-bit value which represents the exponent of the number in base 2, minus 127.
- **F** (fractional part). A 23-bit value which represents the base-2 fractional part of the number's mantissa.

The floating-point value is thus represented by:

$$\text{Value} = -1^{S} \times 2^{(E-127)} \times 1.F$$

It is declared as follows:

```
float identifier;
```

22.8.6 Double-precision Floating-point

A double data type has 64 bits and uses the standard IEEE standard for normalized double-precision floating-point numbers. It has three fields:

- **S** (sign). A 1-bit value which represents a positive number as a 0 and a negative number as a 1.
- **E** (exponent). An 11-bit value which represents the exponent of the number in base 2, minus 1023.
- **F** (fractional part). A 52-bit value which represents the base-2 fractional part of the number's mantissa.

The floating-point value is thus represented by:

$$\text{Value} = -1^{S} \times 2^{(E-1023)} \times 1.F$$

It is declared as follows:

```
double identifier;
```

and can be represented by:

```
+------+------+------+------+------+------+------+------+
|byte 0|byte 1|byte 2|byte 3|byte 4|byte 5|byte 6|byte 7|
S|    E   |                    F                        |
+------+------+------+------+------+------+------+------+
1|<--11-->|<-----------------52 bits------------------->|
<-----------------------64 bits------------------------>
```

22.8.7 Fixed-length and Variable-length Opaque Data

Opaque data is uninterpreted data and consists of a number of bytes (either fixed or variable). It is declared as following

```
opaque identifier[ n] ;
```

where the constant n is the (static) number of bytes necessary to contain the opaque data. If n is not divisible by 4 then a number of residual bytes are added. This can be represented as follows:

```
    0        1     ...
+--------+--------+...+--------+--------+...+--------+
| byte 0 | byte 1 |...|byte n-1|    0   |...|    0   |
+--------+--------+...+--------+--------+...+--------+
|<-----------n bytes---------->|<------r bytes------>|
|<-----------n+r (where (n+r) mod 4 = 0)------------>|
```

Variable-length opaque data is defined as a sequence of n (numbered 0 through n−1) arbitrary bytes. The first four bytes define the number (as an unsigned integer) of encoded bytes in the sequence. If this value is not divisible by 4 then a number of residual bytes are added. It is declared as following:

```
opaque identifier<m>;
```

where the constant m denotes an upper bound of the number of bytes that the sequence may contain. It can be represented by:

```
   0     1     2     3     4     5   ...
+-----+-----+-----+-----+-----+-----+...+-----+-----+...+-----+
|        length n       |byte0|byte1|...| n-1 |  0  |...|  0  |
+-----+-----+-----+-----+-----+-----+...+-----+-----+...+-----+
|<-------4 bytes------->|<------n bytes------>|<---r bytes--->|
                        |<----n+r (where (n+r) mod 4 = 0)---->|
```

22.8.8 String

A string contains a number of ASCII bytes (numbered 0 through n−1). The first value is an unsigned 4-byte integer which is the number of bytes in the string. If this value is not divisible by 4 then a number of residual bytes are added. It is declared as following:

```
string object<m>;
```

It can be represented by:

```
   0     1     2     3     4     5   ...
+-----+-----+-----+-----+-----+-----+...+-----+-----+...+-----+
|        length n       |byte0|byte1|...| n-1 |  0  |...|  0  |
+-----+-----+-----+-----+-----+-----+...+-----+-----+...+-----+
|<-------4 bytes------->|<------n bytes------>|<---r bytes--->|
                        |<----n+r (where (n+r) mod 4 = 0)---->|
```

22.8.9 Fixed-length Array

Fixed-length arrays of homogeneous elements are declared as follows:

```
type-name identifier[ n] ;
```

where the elements are numbered from 0 to n−1 and each element contains 4 bytes. It can be represented by:

```
+---+---+---+---+---+---+---+---+...+---+---+---+---+
|   element 0   |   element 1   |...|  element n-1  |
+---+---+---+---+---+---+---+---+...+---+---+---+---+
|<--------------------n elements------------------->|
```

22.8.10 Variable-length Array

A variable-length array is represented by:

```
type-name identifier<m>;
```

where m specifies the maximum acceptable element count of an array. The first 4 bytes of the array contains the number of elements in the array. It can be represented as:

```
  0  1  2  3
+--+--+--+--+--+--+--+--+--+--+--+--+...+--+--+--+--+
|     n     | element 0 | element 1 |...|element n-1|
+--+--+--+--+--+--+--+--+--+--+--+--+...+--+--+--+--+
|<-4 bytes->|<--------------n elements------------->|
```

22.8.11 Structure

Structures are declared as follows:

```
struct {
   first-declaration;
   second-declaration;
   ...
} identifier;
```

Each component has four bytes and can be represented as:

```
+-----------------+-----------------+...
| 1st declaration | 2nd declaration |...
+-----------------+-----------------+...
```

22.8.12 Others

XDR declares several other data types, which include:

- **Void.** This is a 0-byte quantity which can be used for describing operations that take no data as input or no data as output.
- **Constant.** This allows the definition of a constant value. Its syntax is:

```
const name-identifier = n;
```

- **Typedef.** This is used to declared a different data type. Its syntax is:

typedef declaration;

Encryption

ENCRYPTION PUZZLES

```
H|%Qju%45%vjkbv?

4+%Lwlv)%Bjj%Bjj%Ajjiv
7+%Udwdkjla)%Gidfn%Vdggdqm
6+%Cjw%Vmdwl)%Vmdrk%Hpiilkv
1+%Cw``%Cdiilkb)%Qjh%U`qq|
0+%Mds`%L%qjia%|jp:)%Sdk%Hjwwlvjk
3+%Vmlhh`w)%Vmdrk%Hpiilkv
2+%Rjka`wrdii)%Jdvlv
=+%Fjwq`□%qm`%Nlii`w)%K`li%\jpkb
<+%Rlvm%\jp%R`w`%M`w`)%Ulkn%Cij|a
45+@s`w|gja|%Mpwqv)%W@H

((%mqqu?**rrr+afv+kdul`w+df+pn*{glii
```

23.1 Introduction

The increase in electronic mail has also increased the need for secure data transmission. An electronic mail message can be easily intercepted as it transverses the world's communication networks. Thus there is a great need to encrypt the data contained in it. Traditional mail messages tend to be secure as they are normally taken by a courier or postal service and transported in a secure environment from source to destination. Over the coming years more individuals and companies will be using electronic mail systems and these must be totally secure.

Data encryption involves the science of cryptographics (note that the word *crytopgraphy* is derived from the Greek words which means hidden, or secret, writing). The basic object of cryptography is to provide a mechanism for two people to communicate without any other person being able to read the message.

Encryption is mainly applied to text transmission as binary data can be easily scrambled so it becomes almost impossible to unscramble. This is because text-based information contains certain key pointers:

- Most lines of text have the words 'the', 'and', 'of' and 'to'.
- Every sentence has a full stop.
- Words are separated by a space (the space character is the most probable character in a text document).
- The characters 'e', 'a' and 'i' are more probable than 'q', 'z' and 'x'.

Thus to decode a message an algorithm is applied and the decrypted text is then tested to determine whether it contains standard English (or the required language).

23.2 Encryption and the OSI model

It is possible to encrypt data at any level of the OSI model, but typically it is encrypted when it is passed from the application program. This must occur at the presentation layer of the model, as illustrated in Figure 23.1. Thus, an external party will be able to determine the data at the session, transport, network and data link layer, but not the originally transmitted application data. Thus encryption is useful in hiding data from external parties, but cannot be used (with standard protocols) to hide:

- The session between the two parties. This will give information on the type of session used (such as FTP, TELNET or HTTP).
- The transport layer information. This will give information on the data packets, such as, with TCP, port and socket numbers, and acknowledgements.

- The network address of the source and the destination (all IP packets can be examined and the source and destination address can be viewed).
- The source and destination MAC address. The actually physical addresses of both the source and the destination can be easily examined.

Most encryption techniques use a standard method to encrypt the message. This method is normally well known and the software which can be used to encrypt or decrypt the data is widely available. The thing that makes the encryption process different is an electronic key, which is added into the encryption process. This encryption key could be private so that both the sender and receiver could use the same key to encrypt and decrypt the data. Unfortunately this would mean that each conversation with a user would require a different key. Another disadvantage is that a user would have to pass the private key through a secret channel. There is no guarantee that this channel is actually secure, and there is no way of knowing that an external party has a secret key. Typically public keys are changed at regular intervals, but if the external party knows how these change, they can also change their own keys. These problems are overcome with public-key encryption.

Most encryption is now public-key encryption (as illustrated in Figure 23.1). This involves each user having two encryption keys. One is a public-key which is given to anyone that requires to send the user some encrypted data. This key is used to encrypt any data that is sent to the user. The other key is a private-key which is used to decrypt the received encrypted data. No one knows the private-key (apart from the user who is receiving data encrypted with their public-key).

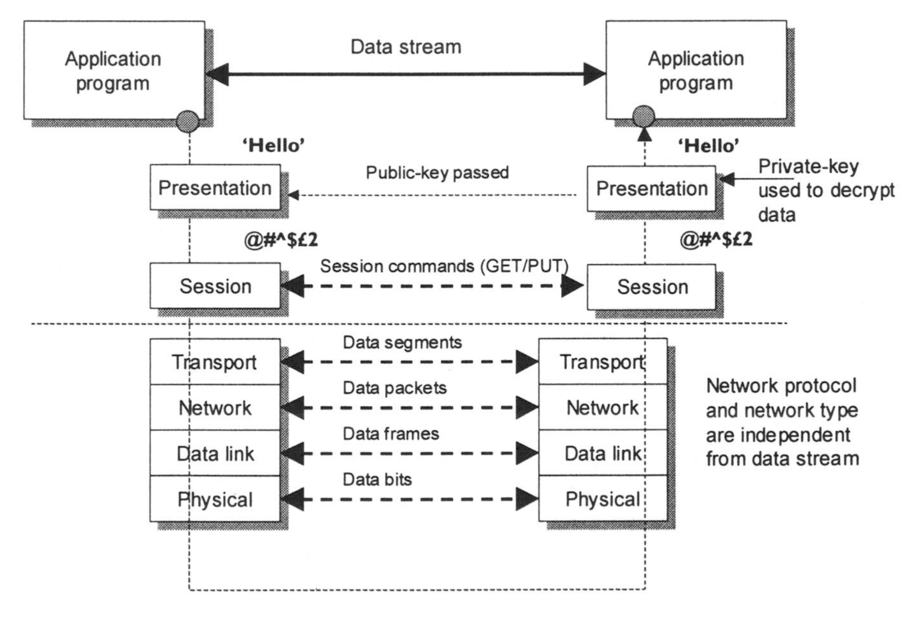

Figure 23.1 Encryption and the OSI model

23.3 Cryptography

The main object of cryptography is to provide a mechanism for two (or more) people to communicate without anyone else being able to read the message. Along with this it can provide other services, such as:

- Giving a reassuring integrity check – this makes sure the message has not been tampered with by non-legitimate sources.
- Providing authentication – this verifies the sender identity.

Initially plaintext is encrypted into ciphertext, it is then decrypted back into plaintext, as illustrated in Figure 23.2. Cryptographic systems tend to use both an algorithm and a secret value, called the key. The requirement for the key is that it is difficult to keep devising new algorithms and also to tell the receiving party that the data is being encrypted with the new algorithm. Thus, using keys, there are no problems with everyone having the encryption/decryption system, because without the key it is very difficult to decrypt the message.

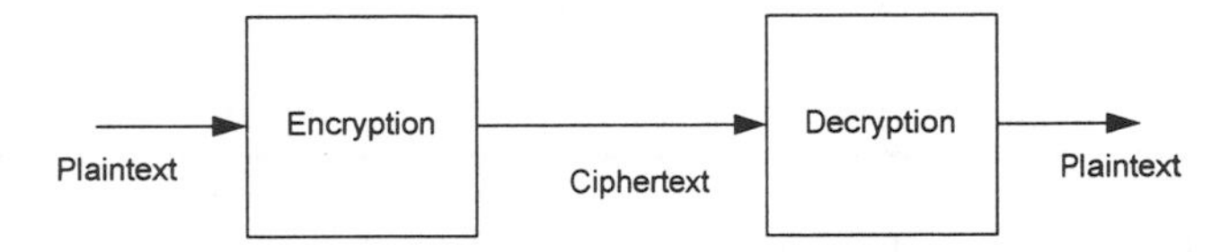

Figure 23.2 Encryption/decryption process

```
Shw'67'soni`t'sh'lihp
fehrs'ibsphult=

6)'NW'fccubttbt
5)'Pnichpt'IS(5777
4)'wni`'dhjjfic
3)'Bjfnk'tbsrw
2)'Sufdbuhrsb'dhjjfic
1)'Uhrsbu'dhian`rufsnhi
0)'IND'nitsfkkfsnhi
?)'RIN_'ibsphulni`
>)'Dknbis(tbuqbu'fudonsbdsrub
67)Bsobuibs'ibsphulni`
```

23.3.1 Public key versus private key

The encryption process can either use a public key or a secret key. With a secret key the key is only known to the two communicating parties. This key can be fixed or can be passed from the two parties over a secure communications link (perhaps over the postal network or a leased line). The two most popular private key techniques are DES (Data Encryption Standard) and IDEA (International Data Encryption Algorithm).

In public-key encryption, each user has both a public and a private key. The two users can communicate because they know each other's public keys. Normally in a public-key system, each user uses a public enciphering transformation which is widely known and a private deciphering transform which is known only to that user. The private transformation is described by a private key, and the public transformation by a public key derived from the private key by a one-way transformation. The RSA (after its inventors Rivest, Shamir and Adleman) technique is one of the most popular public-key techniques and is based on the difficulty of factoring large numbers.

23.3.2 Computational difficulty

Every code is crackable and the measure of the security of a code is the amount of time it takes persons not addressed in the code to break that code. Normally to break the code a computer tries all the possible keys until it finds a match. Thus a 1-bit code would only have 2 keys, a 2-bit code would have 4 keys, and so on. Table 23.1 shows the number of keys as a function of the number of bits in the key. For example it can be seen that a 64-bit code has 18 400 000 000 000 000 000 different keys. If one key is tested every 10 μs then it would take 1.84×10^{14} seconds (5.11×10^{10} hours or 2.13×10^{8} days or 5 834 602 years). So, for example, if it takes 1 million years for a person to crack the code then it can be considered safe. Unfortunately the performance of computer systems increases by the year. For example if a computer takes 1 million years to crack a code, then assuming an increase in computing power of a factor of 2 per year, then it would only take 500 000 years the next year. Table 23.2 shows that after almost 20 years it would take only 1 year to decrypt the same message.

Table 23.1 Number of keys related to the number of bits in the key

Code size	*Number of keys*	*Code size*	*Number of keys*	*Code size*	*Number of keys*
1	2	12	4 096	52	4.5×10^{15}
2	4	16	65 536	56	7.21×10^{16}
3	8	20	1 048 576	60	1.15×10^{18}
4	16	24	16 777 216	64	1.84×10^{19}
5	32	28	2.68×10^{8}	68	2.95×10^{20}
6	64	32	4.29×10^{9}	72	4.72×10^{21}
7	128	36	6.87×10^{10}	76	7.56×10^{22}
8	256	40	1.1×10^{12}	80	1.21×10^{24}
9	512	44	1.76×10^{13}	84	1.93×10^{25}
10	1 024	48	2.81×10^{14}	88	3.09×10^{26}

The increasing power of computers is one factor in reducing the processing time; another is the increasing usage of parallel processing. Data decryption is well suited to parallel processing as each processor or computer can be assigned a number of keys to check the encrypted message. Each of them can then work independently of the other (this differs from many applications in parallel processing which suffer from interprocess(or) communication). Table 23.3 gives typical times, assuming a doubling of processing power each year, for processor arrays of 1, 2, 4 ... 4096 elements. It can be seen that with an array of 4096 processing elements it takes only seven years before the code is decrypted within two years. Thus an organization which is serious about deciphering messages will have the resources to invest in large arrays of processors or networked computers. It is likely that many governments have computer systems with thousands or tens of thousands of processors operating in parallel. A prime use of these systems will be in decrypting messages.

Table 23.2 Time to decrypt a message assuming an increase in computing power

Year	*Time to decrypt (years)*	*Year*	*Time to decrypt (years)*
0	1 million	10	977
1	500 000	11	489
2	250 000	12	245
3	125 000	13	123
4	62 500	14	62
5	31 250	15	31
6	15 625	16	16
7	7 813	17	8
8	3 907	18	4
9	1 954	19	2

23.4 Government pressure

Many institutions and individuals read data which is not intended for them; they include:

- Government departments. Traditionally governments around the world have reserved the right to tap into any communications which they think may be against the national interest.
- Spies who tap into communications for industrial or governmental information.
- Individuals who like to read other people's messages.

- Individuals who 'hack' into systems and read secure information.
- Criminals who intercept information in order to use it for crime, such as intercepting PIN numbers on bankcards.

Table 23.3 Time to decrypt a message with increasing power and parallel processing

Processors	*Year 0*	*Year 1*	*Year 2*	*Year 3*	*Year 4*	*Year 5*	*Year 6*	*Year 7*
1	1 000 000	500 000	250 000	125 000	62 500	31 250	15 625	7 813
2	500 000	250 000	125 000	62 500	31 250	15 625	7 813	3 907
4	250 000	125 000	62 500	31 250	15 625	7 813	3 907	1 954
8	125 000	62 500	31 250	15 625	7 813	3 907	1 954	977
16	62 500	31 250	15 625	7 813	3 907	1 954	977	489
32	31 250	15 625	7 813	3 907	1 954	977	489	245
64	15 625	7 813	3 907	1 954	977	489	245	123
128	7 813	3 907	1 954	977	489	245	123	62
256	3 906	1 953	977	489	245	123	62	31
512	1 953	977	489	245	123	62	31	16
1 024	977	489	245	123	62	31	16	8
2 048	488	244	122	61	31	16	8	4
4 096	244	122	61	31	16	8	4	2

Governments around the world tend to be against the use of encryption as it reduces their chances to tap into information and determine messages. It is also the case that governments do not want other countries to use encryption because it also reduces their chances of reading their secret communications (especially military maneuvers). In order to reduce this threat they must do either of the following:

- Prevent the use of encryption.
- Break the encryption code.
- Learn everyone's cryptographic keys.

Many implementations of data encryption are in hardware, but increasingly it is implemented in software (especially public-key methods). This makes it easier for governments to control their access. For example the US government has proposed to beat encryption by trying to learn everyone's cryptographic key with the Clipper chip. The US government keeps a record of all the serial numbers and encryption keys for each Clipper chip manufactured.

23.5 Legal issues

'zko wzznaooao kjhu 1 iacwxupa kb nwi xay-wqoa
sa ywjjkp eiwceja wju wllheywpekjo jaazejc ikna.',
ieynkokbp, 1980, kj pda zarahkliajp kb zko

Patent laws and how they are implemented vary around the world. Like many good ideas, patents cover most of the cryptographic techniques. The main commercial techniques are:

- DES (Data Encryption Standard) which is patented but royalty-free.
- IDEA (International Data Encryption Algorithm) which is also patented and royalty-free for the non-commercial user.

Access to a global network normally requires the use of a public key. The most popular public-key algorithm is one developed at MIT and is named RSA (after its inventors Rivest, Shamir and Adleman). All public-key algorithms are patented, and most of the important patents have been acquired by Public Key Partners (PKP). As the US government funded much of the work, there are no license fees for US government use. RSA is only patented in the US, but Public Key Partners (PKP) claim that the international Hellman-Merkle patent also covers RSA. The patent on RSA runs out in the year 2000. Public keys are generated by licensing software from a company called RSA Data Security Inc. (RSADSI).

The other widely used technique is Digital Signature Standard (DSS). It is freely licensable but in many respects it is technically inferior to RSA. The free licensing means that it is not necessary to reach agreement with RSADSI or PKP. Since it was announced, PKP have claimed the Hellman-Merkle patent covers all public-key cryptography. It has also strengthened its position by acquiring rights to a patent by Schnorr which is closely related to DSS.

23.6 Cracking the code

A cryptosystem converts plaintext into ciphertext using a key. There are several methods that a hacker can use to crack a code, including:

- **Known plaintext attack.** Where the hacker knows part of the ciphertext and the corresponding plaintext. The known ciphertext and plaintext can then be used to decrypt the rest of the ciphertext.
- **Chosen-ciphertext.** Where the hacker sends a message to the target, this is then encrypted with the target's private-key and the hacker then analyses the encrypted message. For example, a hacker may send an e-mail to the encryption file server and the hacker spies on the delivered message.
- **Exhaustive search.** Where the hacker uses brute force to decrypt the ciphertext and tries every possible key.
- **Active attack.** Where the hacker inserts or modifies messages.
- **Man-in-the-middle.** Where the hacker is hidden between two parties and impersonates each of them to the other.
- **The replay system.** Where the hacker takes a legitimate message and sends it into the network at some future time.
- **Cut and paste.** Where the hacker mixes parts of two different encrypted messages and, sometimes, is able to create a new message. This message is likely to make no sense, but may trick the receiver into doing something that helps the hacker.
- **Time resetting.** Some encryption schemes use the time of the computer to create the key. Resetting this time or determining the time that the message was created can give some useful information to the hacker.
- **Time attack.** This involves determining the amount of time that a user takes to decrypt the message; from this the key can be found.

```
'rkj mxqj ... yi yj weet
veh?',
udwyduuh qj jxu qtlqdsut
secfkjydw ioijuci tylyiyed
ev yrc, 1968, seccudjydw
ed jxu cyshesxyf.
```

```
'q pidm bzidmtml bpm
tmvobp ivl jzmilbp wn
bpqa
kwcvbz› ivl bitsml eqbp
bpm jmab xmwxtm, ivl q
kiv iaaczm ›wc bpib libi
xzwkmaaqvo qa i nil
bpib ewv'b tiab wcb bpm
›miz',
mlqbwz, xzmvbqkm pitt,
1957
```

23.7 Random number generators

One way to crack a code is to exploit a weakness in the generation of the encryption key. The hacker can then guess which keys are more likely to occur. This is known as a statistical attack.

Many programming languages use a random number generator which is based on the current system time (such as `rand()`). This method is no good in data encryption as the hacker can simply determine the time that the message was encrypted and the algorithm used.

An improved source of randomness is the time between two keystrokes (as used in PGP – pretty good privacy). However this system has been criticized as a hacker can spy on a user over a network and determine the time between keystrokes. Other sources of true randomness have also been investigated, including noise from an electronic device and noise from an audio source.

23.8 Letter probabilities

The English language has a great deal of redundancy in it, thus common occurrences in text can be coded with short bit sequences. The probability of each letter also varies. For example the letter '*e*' occurs many more times than the letter '*z*'. Program 23.7 in Section 24.15 gives a simple C program which determines the probability of letters within a text file. This program can be used to determine typical letter probabilities. Sample run 23.1 shows a sample run using some sample text. It can be seen that the highest probability is with the letter '*e*', which occurs, on average, 94.3 times every 1000 letters. Table 23.4 lists the letters in order of their probability. Notice that the letters which are worth the least in the popular board game Scrabble (such as, '*e*', '*t*', '*a*', and so on) are the most probable and the letters with the highest scores (such as '*x*', '*z*' and '*q*') are the least probable.

Scrabble™ letter values (placing in Table 23.4):

A	1 (3)	E	1 (1)
I	1 (4)	L	1 (9)
N	1 (6)	O	1 (5)
R	1 (8)	S	1 (7)
T	1 (2)	U	1 (14)
D	2 (11)	G	2 (15)
B	3 (19)	C	3 (12)
M	3 (13)	P	3 (17)
F	4 (16)	H	4 (10)
V	4 (21)	W	4 (18)
Y	4 (20)	K	5 (22)
J	8 (26)	X	8 (23)
Q	10 (24)	Z	10 (25)

Sample run 23.1

```
Char. Occur. Prob.          Char. Occur. Prob.
  a    1963   0.0672          b     284    0.0097
  c     914   0.0313          d     920    0.0315
  e    2752   0.0943          f     471    0.0161
  g     473   0.0162          h     934    0.0320
  i    1680   0.0576          j      13    0.0004
  k      96   0.0033          l     968    0.0332
  m     724   0.0248          n    1541    0.0528
  o    1599   0.0548          p     443    0.0152
  q      49   0.0017          r    1410    0.0483
  s    1521   0.0521          t    2079    0.0712
  u     552   0.0189          v     264    0.0090
  w     383   0.0131          x      57    0.0020
  y     278   0.0095          z      44    0.0015
  .     292   0.0100          SP   4474    0.1533
  ,     189   0.0065
```

23.8.1 Frequency analysis

Frequency analysis involves measuring the occurrences of the letters in the ciphertext. This can give many clues as the English language contains certain key features for decypering, such as:

Table 23.4 Letters and their occurrence in a sample text file

Character	*Occurrences*	*Probability*	*Character*	*Occurrences*	*Probability*
SPACE	4474	0.1533	g	473	0.0162
e	2752	0.0943	f	471	0.0161
t	2079	0.0712	p	443	0.0152
a	1963	0.0672	w	383	0.0131
i	1680	0.0576	.	292	0.0100
o	1599	0.0548	b	284	0.0097
n	1541	0.0528	y	278	0.0095
s	1521	0.0521	v	264	0.0090
r	1410	0.0483	,	189	0.0065
l	968	0.0332	k	96	0.0033
h	934	0.0320	x	57	0.0020
d	920	0.0315	q	49	0.0017
c	914	0.0313	z	44	0.0015
m	724	0.0248	j	13	0.0004
u	552	0.0189			

- Determine the probabilities of ciphertext letters. The least probable should be 'j', 'k', 'x' and 'z'. These have an accumulated occurrence of less than 1 per cent. One of the letters, an 'e', should have an occurrence of more than 10 per cent. Next the ciphertext letter probabilities should be measured against standard English language letter probabilities. If the two do not tie-up, it is likely that the text was written in another language.
- If the single letters do not yield the code, then two letter occurrences of the same letter should be examined. The most common ones are: ss, ee, tt, ff, ll, mm and oo. If the ciphertext contains repeated letters, it may relate to one of these sequences.
- If there are spaces between the words, the two letter words can be examined. The most popular two letter words are: an, as, at, am, be, by, do, of, to, in, it, is, so, we, he, or, on, if, me, up, go, no and us (see Section 24.17).
- If possible, the list of letter probabilities should be related to the type of message that is being sent. For example, military communications tend to omit pronouns and articles (excluding words like he, a and I).
- Try and identify whole phrases, such as 'Hello who are you'. This can be used as a crowbar to get the rest of the code.
- If the ciphertext corresponds to correct letter probabilities, but the deciphered text is still unreadable, it may be that the code is a transpositional cipher, where the letters have had their positions changed. For example, every two cipher characters have been swapped around.

23.9 Basic encryption principles

Encryption codes have been used for many centuries. They have tended to be used in military situations where secret messages have to be sent between troops without the risk of them being read by the enemy.

23.9.1 Alphabet shifting (Caesar code)

A simple encryption code is to replace the letters with a shifted equivalent alphabet. For example moving the letters two places to the right gives:

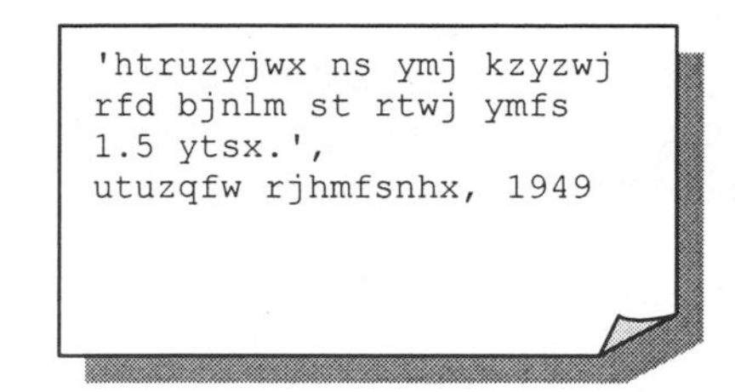

'htruzyjwx ns ymj kzyzwj
rfd bjnlm st rtwj ymfs
1.5 ytsx.',
utuzqfw rjhmfsnhx, 1949


```
abcdefghijklmnopqrstuvwxyz
YZABCDEFGHIJKLMNOPQRSTUVWX
```

Thus a message:

```
the boy stood on the burning deck
```

would become:

```
RFC ZMW QRMMB ML RFC ZSPLGLE BCAI
```

This code has the problem of being reasonably easy to decode, as there are only 26 different code combinations. The first documented use of this type of code was by Julius Caesar who used a 3-letter shift.

23.9.2 Vigenère code

A Caesar-type code shifts the alphabet by a number of places (as given in Table 23.5). An improved code was developed by Vigenère, but as a shifted alphabet is not very secure. In this code, a different row is used for each encryption. The way that the user moves between the rows must be agreed before encryption. This can be achieved with a code word, which defines the sequence of the rows. For example the codeword GREEN could be used which defined that the rows used were: Row 6 (G), Row 17 (R), Row 4 (E), Row 4 (E), Row 13 (N), Row 6 (G), Row 17 (R), and so on.

Table 23.5 Character-shifted alphabets

Plain	a	b	c	d	e	f	g	h	i	j	k	l	m	n	o	p	q	r	s	t	u	v	w	x	y	z
1	B	C	D	E	F	G	H	I	J	K	L	M	N	O	P	Q	R	S	T	U	V	W	X	Y	Z	A
2	C	D	E	F	G	H	I	J	K	L	M	N	O	P	Q	R	S	T	U	V	W	X	Y	Z	A	B
3	D	E	F	G	H	I	J	K	L	M	N	O	P	Q	R	S	T	U	V	W	X	Y	Z	A	B	C
4	E	F	G	H	I	J	K	L	M	N	O	P	Q	R	S	T	U	V	W	X	Y	Z	A	B	C	D
5	F	G	H	I	J	K	L	M	N	O	P	Q	R	S	T	U	V	W	X	Y	Z	A	B	C	D	E
6	G	H	I	J	K	L	M	N	O	P	Q	R	S	T	U	V	W	X	Y	Z	A	B	C	D	E	F
7	H	I	J	K	L	M	N	O	P	Q	R	S	T	U	V	W	X	Y	Z	A	B	C	D	E	F	G
8	I	J	K	L	M	N	O	P	Q	R	S	T	U	V	W	X	Y	Z	A	B	C	D	E	F	G	H
9	J	K	L	M	N	O	P	Q	R	S	T	U	V	W	X	Y	Z	A	B	C	D	E	F	G	H	I
10	K	L	M	N	O	P	Q	R	S	T	U	V	W	X	Y	Z	A	B	C	D	E	F	G	H	I	J
11	L	M	N	O	P	Q	R	S	T	U	V	W	X	Y	Z	A	B	C	D	E	F	G	H	I	J	K
12	M	N	O	P	Q	R	S	T	U	V	W	X	Y	Z	A	B	C	D	E	F	G	H	I	J	K	L
13	N	O	P	Q	R	S	T	U	V	W	X	Y	Z	A	B	C	D	E	F	G	H	I	J	K	L	M
14	O	P	Q	R	S	T	U	V	W	X	Y	Z	A	B	C	D	E	F	G	H	I	J	K	L	M	N
15	P	Q	R	S	T	U	V	W	X	Y	Z	A	B	C	D	E	F	G	H	I	J	K	L	M	N	O
16	Q	R	S	T	U	V	W	X	Y	Z	A	B	C	D	E	F	G	H	I	J	K	L	M	N	O	P
17	R	S	T	U	V	W	X	Y	Z	A	B	C	D	E	F	G	H	I	J	K	L	M	N	O	P	Q
18	S	T	U	V	W	X	Y	Z	A	B	C	D	E	F	G	H	I	J	K	L	M	N	O	P	Q	R
19	T	U	V	W	X	Y	Z	A	B	C	D	E	F	G	H	I	J	K	L	M	N	O	P	Q	R	S
20	U	V	W	X	Y	Z	A	B	C	D	E	F	G	H	I	J	K	L	M	N	O	P	Q	R	S	T
21	V	W	X	Y	Z	A	B	C	D	E	F	G	H	I	J	K	L	M	N	O	P	Q	R	S	T	U
22	W	X	Y	Z	A	B	C	D	E	F	G	H	I	J	K	L	M	N	O	P	Q	R	S	T	U	V
23	X	Y	Z	A	B	C	D	E	F	G	H	I	J	K	L	M	N	O	P	Q	R	S	T	U	V	W
24	Y	Z	A	B	C	D	E	F	G	H	I	J	K	L	M	N	O	P	Q	R	S	T	U	V	W	X
25	Z	A	B	C	D	E	F	G	H	I	J	K	L	M	N	O	P	Q	R	S	T	U	V	W	X	Y

Thus the message:

Keyword	`GREENGREENGREEN`
Plaintext	`hellohowareyou`
Ciphertext	`NVPPBNFAEEKPSY`

The great advantage of this type of code is that the same plaintext character will be encrypted with different values, depending on the position of the keyword. For example, if the keyword is GREEN, 'e' can be encrypted as 'K' (for G), 'V' (for R), 'I' (for E) and 'R' (for N). The greater the size of the code word, the more the rows that will be included in the encryption process. It is not possible to decipher the code by a frequency analysis, as letters will change their coding depending on the current position of the keyword. It is also safe from analysis of common two- and three-letter occurrences. For example 'ee' could be encrypted with 'KV' (for GR), 'VI' (for RE), 'II' (for EE), 'IR' (for EN) and 'RK' (for NG). A longer keyword would generate more combinations.

The Vigenère code is *polyalphabetic*, as it uses a number of cipher alphabets.

23.9.3 Homophonic substitution code

A homophonic substitution code overcomes the problems of frequency analysis of code, as it assigns a number of codes to a character which relates to the probability of the characters. For example the character 'e' might have 10 codes assigned to it, but 'z' would only have one. An example code is given in Table 23.6.

Each of the codes is assigned at random to each of the letters, with the number of codes assigned related to the probability of their occurrence. Thus, using the code table in Table 23.6, the code mapping would be:

Plaintext	h	e	l	l	o	e	v	e	r	y	o	n	e
Ciphertext:	19	25	42	81	16	26	22	28	04	55	30	00	32

In this case there are four occurrences of the letter 'e', and each one has a different code. As the number of codes depends on the number of occurrences of the letter, each code will roughly have the same probability, thus it is not possible to determine the code mapping from the probabilities of codes. Unfortunately the code isn't perfect as the English language still contains certain relationships which can be traced. For example the letter 'q' normally is represented by a single code, and three codes represent a 'u'. Thus, if the ciphertext contains a code followed by one of three codes, then it is likely that the plaintext is a 'q' and a 'u'.

Table 23.6 Example homophonic substitution

a	b	c	d	e	f	g	h	i	j	k	l	m	n	o	p	q	r	s	t	u	v	w	x	y	z
07	11	17	10	25	08	44	19	02	18	41	42	40	00	16	01	15	04	06	05	13	22	45	12	55	47
31	64	33	27	26	09	83	20	03			81	52	43	30	62		24	34	23	14		46		93	
50		49	51	28			21	29			86		80	61			39	56	35	36					
63			76	32			54	53			95		88	65			58	57	37						
66				48			70	68					89	91			71	59	38						
77				67			87	73						94			00	90	60						
84				69										96					74						
				72															78						
				75															92						
				79																					
				82																					
				85																					

A homophonic cipher is a monoalphabetic code, as it only uses one translation for the code mappings (even though several codes can be used for a single plaintext letter). This alphabet remains constant, whereas a polyalphabet can change its mapping depending on a variable keyword.

23.9.4 Code mappings

```
'n ymnsp ymjwj nx f
btwqi rfwpjy ktw
rfdgj knaj
htruzyjwx',
ymtrfx bfyxts,
hmfnwrfs tk ngr,
1943
```

Code mappings can have no underlying mathematical relationship and simply use a codebook to represent the characters. This is known as a *monoalphabetic* code, as only one cipher alphabet is used. An example could be:

Input: `abcdefghijklmnopqrstuvwxyz`
Encrypted: `MGQOAFZBCDIEHXJKLNTQRWSUVY`

Program 23.1 shows a C program which uses this code mapping to encrypt entered text and Sample run 23.3 shows a sample run.

The number of different character maps can be determined as follows:

- Take the letter 'A' then this can be mapped to 26 different letters.
- If 'A' is mapped to a certain letter then 'B' can only map to 25 letters.
- If 'B' is mapped to a certain letter then 'C' can be mapped to 24 letters.
- Continue until the alphabet is exhausted.

Thus, in general, the number of combinations will be:

$$26 \times 25 \times 24 \times 23 \ldots 4 \times 3 \times 2 \times 1$$

Thus the code has 26! different character mappings (approximately 4.03×10^{26}). It suffers from the fact that the probabilities of the mapped characters will be similar to those in normal text. Thus if there is a large amount of text then the character having the highest probability will be either an 'e' or a 't'. The character with the lowest probability will tend to be a 'z' or a 'q' (which is also likely be followed by the character map for a 'u').

```
'pdana eo jk nawokj
wjukja skqhz swjp w
ykilqpan ej pdaen
dkia.', gaj khokj, lna-
oezajp,
ydweniwj wjz bkqjzan kb
zecepwh amqeliajp yknl.,
1977.
```

Program 23.1

```
#include <stdio.h>
#include <ctype.h>

int   main(void)
{
int   key,ch,i=0,inch;
char  text[ BUFSIZ] ;
char input[ 26] ="abcdefghijklmnopqrstuvwxyz";
char output[ 26] ="mgqoafzbcdiehxjklntqrwsuvy";

      printf("Enter text >>");
      gets(text);

      ch=text[ 0] ;
      do
      {
         if (ch!=' ')    inch=output[ (tolower(ch)-'a')] ;
         else inch='#';

         putchar(inch);
```

```
        i++;
        ch=text[ i] ;
    } while (ch!=NULL);
    return(0);
}
```

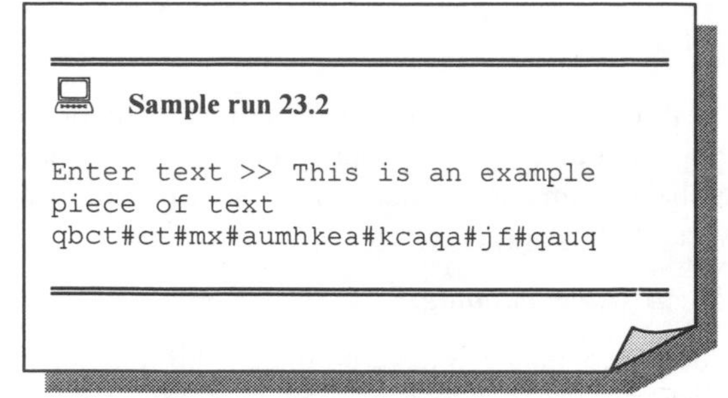

Sample run 23.2
Enter text >> This is an example
piece of text
qbct#ct#mx#aumhkea#kcaqa#jf#qauq

A code mapping encryption scheme is easy to implement but unfortunately, once it has been 'cracked', it is easy to decrypt the encrypted data. Normally this type of code is implemented with an extra parameter which changes its mapping, such as changing the code mapping over time depending on the time of day and/or date. Only parties which are allowed to decrypt the message know the mappings of the code to time and/or date. For example, each day of the week could have a different code mapping.

23.9.5 Applying a key

To make it easy to decrypt, a key is normally applied to the text. This makes it easy to decrypt the message if the key is known, but difficult to decrypt the message if the key is not known. An example of a key operation is to take each of the characters in a text message and then exclusive-OR (XOR) the character with a key value. For example the ASCII character 'A' has the bit pattern:

100 0001

and if the key had a value of 5 then 'A' exclusive-OR'ed with 5 would give:

'A'	100 0001
Key (5)	000 0101
Ex-OR	100 0100

The bit pattern 100 0100 would be encrypted as character 'D'. Program 23.3 is a C program which can be used to display the alphabet of encrypted characters for a given key. In this program the ^ operator represents exclusive-OR. Sample run 23.4 shows a sample run with a key of 5. The exclusive-OR operator has the advantage that when applied twice it results in the original value (it thus changes a value, but does not lose any information when it operates on it).

Program 23.2

```
#include <stdio.h>

int   main(void)
{
int   key,ch;

      printf("Enter key value >>");
      scanf("%d",&key);

      for (ch='A';ch<='Z';ch++)
         putchar(ch^key);

      return(0);
}
```

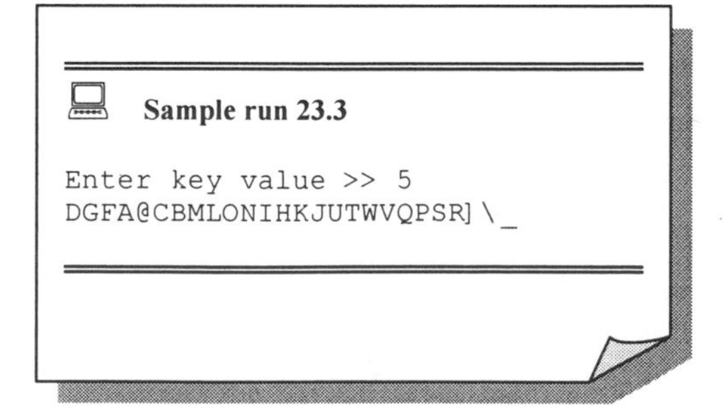

Sample run 23.3
Enter key value >> 5
DGFA@CBMLONIHKJUTWVQPSR] _

Program 23.3 is an encryption program which reads some text from the keyboard, then encrypts it with a given key and saves the encrypted text to a file. Program 23.4 can then be used to read the encrypted file for a given key; only the correct key will give the correct results.

Program 23.3

```
/* Encryt.c */
#include <stdio.h>

int   main(void)
{
FILE *f;
char  fname[ BUFSIZ] ,str[ BUFSIZ] ;
int   key,ch,i=0;

      printf("Enter output file name >>");
      gets(fname);

      if ((f=fopen(fname,"w"))==NULL)
      {
         puts("Cannot open input file");
         return(1);
      }
      printf("Enter text to be save to file>>");
      gets(str);

      printf("Enter key value >>");
      scanf("%d",&key);

      ch=str[ 0] ;

      do
      {
         ch=ch^key; /* Exclusive-OR character with itself */
         putc(ch,f);
         i++;
         ch=str[ i] ;
      } while (ch!=NULL); /* test if end of string */
      fclose(f);
      return(0);
}
```

Sample run 23.4

```
Enter output filename >> out.dat
Enter text to be saved to file>> The boy
stood on the burning deck
Enter key value >> 3
```

File listing 23.1 gives a file listing for the saved encrypted text. One obvious problem with this coding is that the SPACE character is visible in the coding. As the SPACE character is 010 0000, the key can be determined by simply XORing 010 0000 with the '#' character, thus:

SPACE	010 0000
'#'	010 0011
Key	000 0011

Thus the key is 000 0011 (decimal 3).

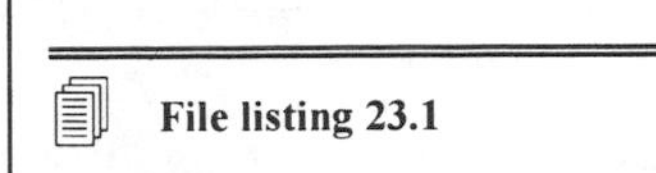

File listing 23.1

```
Wkf#alz#pwllg#lm#wkf#avqmjmd#gf`h
```

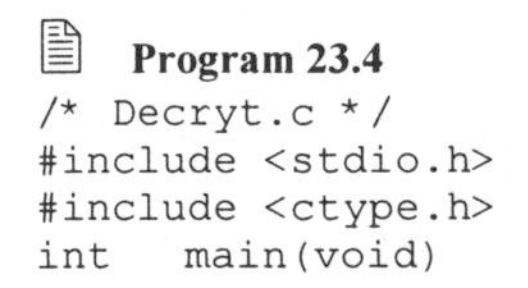

Program 23.4

```
/* Decryt.c */
#include <stdio.h>
#include <ctype.h>
int   main(void)
```

```
{
FILE  *f;
char  fname[BUFSIZ];
int   key,ch;

      printf("Enter encrypted filename >>");
      gets(fname);
      if ((f=fopen(fname,"r"))==NULL)
      {
         puts("Cannot open input file");
         return(1);
      }

      printf("Enter key value >>");
      scanf("%d",&key);

      do
      {
         ch=getc(f);
         ch=ch^key;
         if (isascii(ch)) putchar(ch); /* only print ASCII char */
      } while (!feof(f));
      fclose(f);
      return(0);
}
```

Program 23.5 uses the exclusive-OR operator and reads from an input file and outputs to an output file. The format of the run (assuming that the source code file is called `key.c`) is:

`key` *infile.dat outfile.enc*
where *infile.dat* is the name of the input file (text or binary) and *outfile.enc* is the name of the output file.

The great advantage of this program is that the same program is used for encryption and for decryption. Thus:

`key` *outfile.enc newfile.dat*

converts the encrypted file back into the original file.

Program 23.5

```
#include <stdio.h>

int main(int argc, char *argv[])
{
FILE *in,*out;
char fname[BUFSIZ],key,ch,fout[BUFSIZ],fext[BUFSIZ],*str;


      printf("Enter key >>");
      scanf("%c",&key);

      if ((in=fopen(argv[1],"rb"))==NULL)
      {
         printf("Cannot open");
         return(1);
      }

      out=fopen(argv[2],"wb");
```

```
        do
        {
            fread(&ch,1,1,in); /* read a byte from the file */
            ch=((ch & 0xff) ^ (key & 0xff)) & 0xff;
            if (!feof(in)) fwrite(&ch,1,1,out); /* write a byte */

        } while (!feof(in));

        fclose(in); fclose(out);
}
```

23.9.6 Applying a bit shift

A typical method used to encrypt text is to shift the bits within each character. For example ASCII characters only use the lower 7 bits of an 8-bit character. Thus, shifting the bit positions one place to the left will encrypt the data to a different character. For a left shift a 0 or a 1 can be shifted into the least significant bit; for a right shift the least significant bit can be shifted into the position of the most significant bit. When shifting more than one position a rotate left or rotate right can be used. Note that most of the characters produced by shifting may not be printable, thus a text editor (or viewer) cannot be used to view them. For example, in C the characters would be processed with:

```
ch=ch << 1;
```

which shifts the bits of `ch` one place to the left, and decrypted by:

```
ch=ch >> 1;
```

which shifts the bits of `ch` one place to the right.

Program 23.6 gives an example of a program that reads in a text file (or any file), and reads it one byte at a time. For each byte the program rotates the bits 2 places to the left (with `rot_left`) and saves the byte.

Program 23.6

```
#include <stdio.h>

unsigned char rot_left(unsigned char ch);
unsigned char rot_right(unsigned char ch);

int main(int argc, char *argv[])
{
unsigned char ch;
int i;
FILE *in,*out;
char fname[BUFSIZ],fout[BUFSIZ],fext[BUFSIZ],*str;

        if ((in=fopen(argv[1],"rb"))==NULL)
        {
            printf("Cannot open");
            return(1);
        }

        out=fopen(argv[2],"wb");

        do
        {
            fread(&ch,1,1,in);      /* read a byte from the file */
            ch=rot_left(ch);        /* perform two left rotates */
```

```
        ch=rot_left(ch);
        if (!feof(in)) fwrite(&ch,1,1,out); /* write a byte */
    } while (!feof(in));

    fclose(in); fclose(out);
}

// rotate bits to the left
unsigned char rot_left(unsigned char ch)
{
unsigned char bit8;

    bit8=(ch & 0x80) & 0x80;
    ch=ch << 1;
    ch = ch | ((bit8>>7) & 0x01);
    return(ch);
}
/* rotate bits to the right */
unsigned char rot_right(unsigned char ch)
{
unsigned char bit1;

    bit1=(ch & 1) & 0x01;
    ch=ch >> 1;
    ch = ch | ((bit1<<7) & 0x80);
    return(ch);
}
```

For example the text:

```
Hello. This is some sample text.

Fred.
```

becomes:

```
!•±±½¸□Q¡¥Í□¥Í□Í½µ•□Í…µÁ±•□Ñ•áÑ¸4(4(□É•'¸
```

This can then be decrypted by changing the left rotates (`rot_left`) to right rotates (`rot_right`).

23.10 Message hash

A message hash is a simple technique which basically mixes up the bits within the message, using exclusive-OR operations, bit-shifts or character substitutions.

- **Base-64 encoding.** This is used in electronic mail, and is typically used to change a binary file into a standard 7-bit ASCII form. It takes 6-bit characters and converts them to a printable character, as given in Table 19.5.
- **MD5.** This is used in several encryption and authentication methods. An example conversion is from:

```
Hello, how are you?
Are you feeling well?

Fred.
```

to:

```
518bb66a80cf187a20e1b07cd6cef585
```

23.11 Private-key

Encryption techniques can use either public keys or secret keys. Secret-key encryption techniques use a secret key which is only known by the two communicating parties, as illustrated in Figure 23.3. This key can be fixed or can be passed from the two parties over a secure communications link (for example over the postal network or a leased line). The two most popular private-key techniques are DES (Data Encryption Standard) and IDEA (International Data Encryption Algorithm) and a popular public-key technique is RSA (named after its inventors, Rivest, Shamir and Adleman). Public-key encryption uses two keys, one private and the other public.

```
gsqtyxiv qiwweki:
'ivvsv, rs oicfsevh,
tviww j1 xs
gsrxmryi.'
```

23.11.1 Survey of private-key cryptosystems

The main private-key cryptosystems include:

- **DES.** DES (Data Encryption Standard) is a block cipher scheme which operates on 64-bit block sizes. The private key has only 56 useful bits as eight of its bits are used for parity. This gives 2^{56} or 10^{17} possible keys. DES uses a complex series of permutations and substitutions, the result of these operations is XOR'ed with the input. This is then repeated 16 times using a different order of the key bits each time. DES is a very strong code and has never been broken, although several high-powered computers are now available which, using brute force, can crack the code. A possible solution is 3DES (or triple DES) which uses DES three times in a row. First to encrypt, next to decrypt and finally to encrypt. This system allows a key-length of more than 128 bits.

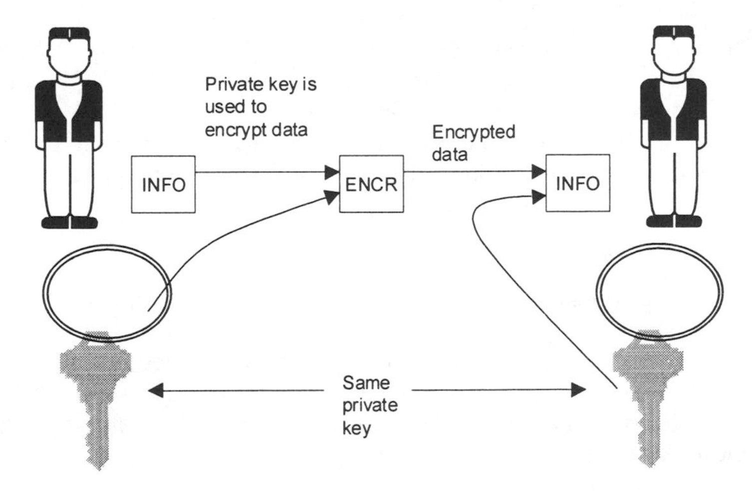

Figure 23.3 Private key encryption/decryption process

- **MOSS.** MOSS (MIME object security service) is an Internet RFC and is typically used for sound encryption. It uses symmetric encryption and the size of the key is not specified. The only public implementation is TIS/MOSS 7.1, which is basically an implementation of 56-bit DES code with a violation.

- **IDEA.** IDEA (International Data Encryption Algorithm) is similar to DES. It operates on 64-bit blocks of plaintext, uses a 128-bit key, and has over 17 rounds with a complicated mangler function. During decryption this function does not have to be reversed and can simply be applied in the same way as during encryption (this also occurs with DES). IDEA uses a different key expansion for encryption and decryption, but every other part of the process is identical. The same keys are used in DES decryption but in the reverse order. The key is devised in eight 16-bit blocks; the first six are used in the first round of encryption the last two are used in the second run. It is free for use in non-commercial version and appears to be a strong cipher.
- **RC4/RC5.** RC4 is a cipher designed by RSA Data Security, Inc and was a secret until information on it appeared on the Internet. The Netscape secure socket layer (SSL) uses RC4. It uses a pseudo random number generator where the output of the generator is XOR'ed with the plaintext. It is a fast algorithm and can use any key-length. Unfortunately the same key cannot be used twice. Recently a 40-bit key version was broken in eight days without special computer power. RC5 is a fast block cipher designed by Rivest for RSA Data Security. It has a parameterized algorithm with a variable block size (32, 64 or 128 bits), a variable key size (0 to 2048 bits) and a variable number of rounds (0 to 255). It has a heavy use of data dependent rotations and the mixture of different operations. This assures that RC5 is secure. Kaliski and Yin found that RC5 with a 64-bit block size and 12 or more rounds gives good security.
- **SAFER.** SAFER (Secure and Fast Encryption Routine) is a non-proprietary block-cipher developed by Massey in 1993. It operates on a 64-bit block size and has a 64-bit or 128-bit key size. SAFER has up to 10 rounds (although a minimum of 6 is recommended). Unlike most recent block ciphers, SAFER has a slightly different encryption and decryption procedure. The algorithm operates on single bytes at a time and it thus can be implemented on systems with limited processing power, such as on smart-cards applications. A typical implementation is SAFER K-64 which uses a 40-bit key and has been shown that it is immune from most attacks when the number of rounds is greater than six.
- **SKIPJACK.** Skipjack is a new block cipher which operates on 64-bit blocks. It uses an 80-bit key and has 32 rounds. The NSA has classified details of Skipjack and its algorithm is only available in hardware implementation called Clipper Chips. The name Clipper derives from an earlier implementation of the algorithm. Each transmission contains the session key encrypted in the header. The licensing of Clipper chips allows US government to decrypt all SKIPJACK messages.

23.11.2 Data Encryption Standard (DES)

In 1977, the National Bureau of Standards (now the National Institute of Standards and Technology) published the DES for commercial and unclassified US government applications. DES is based on an algorithm known as the Lucifer cipher designed by IBM. It maps a 64-bit input block to a 64-bit output block and uses a 56-bit key. The key itself is actually 64 bits long but as 1 bit in each of the 8 bytes is used for odd parity on each byte, the key only contains 56 meaningful bits.

DES overview

The main steps in the encryption process are as follows:

- Initially the 64-bit input is permutated to obtain a 64-bit result (this operation does little to the security of the code).

- Next, there are 16 iterations of the 64-bit result and the 56-bit key. Only 48 bits of the key are used at a time. The 64-bit output from each iteration is used as an input to the next iteration.
- After the 16th iteration, the 64-bit output goes through another permutation, which is the inverse of the initial permutation.

Figure 23.4 shows the basic operation of DES encryption.

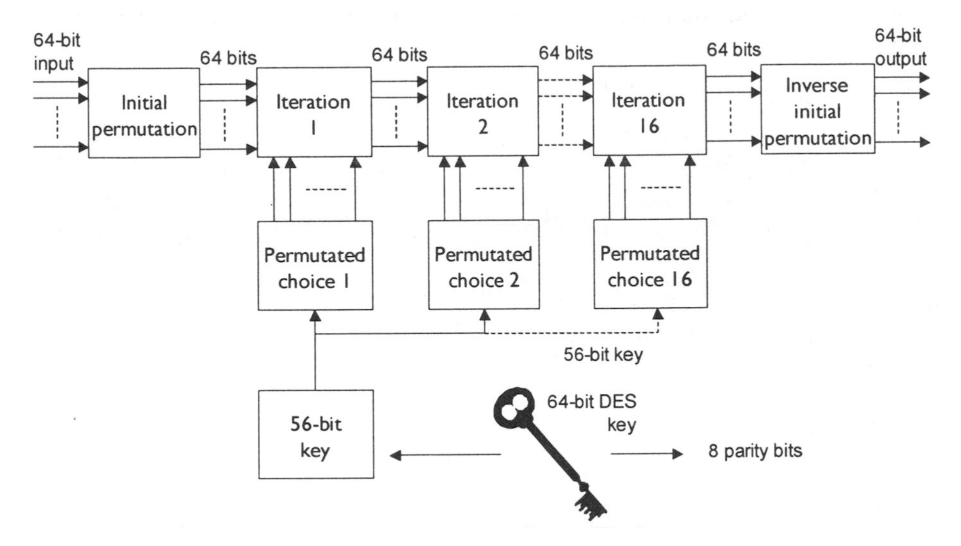

Figure 23.4 Overview of DES operation

Permutation of the data

Before the first iteration and after the last iteration, DES performs a permutation on the data. The permutation is as follows:

Initial permutation:

```
58 50 42 34 26 18 10 2  60 52 44 36 28 20 12 4  62 54 46 38 30 22 14 6
64 56 48 40 32 24 16 8  57 49 41 33 25 17 9  1  59 51 43 35 27 19 11 3
61 53 45 37 29 21 13 5  63 55 47 39 31 23 15 7
```

Final permutation:

```
40 8  48 16 56 24 64 32 39 7  47 15 55 23 63 31 38 6  46 14 54 22 62 30
37 5  45 13 53 21 61 29 36 4  44 12 52 20 60 28 35 3  43 11 51 19 59 27
34 2  42 10 50 18 58 26 33 1  41 9  49 17 57 25
```

These numbers specify the bit numbers of the input to the permutation and the order of the numbers corresponds to the output bit position. Thus, input permutation:

- Input bit 58 moves to output bit 1 (58 is in the 1st bit position).
- Input bit 50 moves to output bit 2 (50 is in the 2nd bit position).
- Input bit 42 moves to output bit 3 (42 is in the 3rd bit position).
- Continue until all bits are exhausted.

In addition, the final permutation could be:

- Input bit 58 moves to output bit 1 (1 is in the 58th bit position).
- Input bit 50 moves to output bit 2 (2 is in the 50th bit position).
- Input bit 42 moves to output bit 3 (3 is in the 42nd bit position).
- Continue until all bits are exhausted.

Thus, the input permutation is the reverse of the output permutation. Arranged as blocks of 8 bits, it gives:

```
58 50 42 34 26 18 10 2
60 52 44 36 28 20 12 4
62 54 46 38 30 22 14 6
 :                 :
61 53 45 37 29 21 13 5
63 55 47 39 31 23 15 7
```

It can be seen that the first byte of input gets spread into the 8th bit of each of the other bytes. The second byte of input gets spread into the 7th bit of each of the other bytes, and so on.

Generating the per-round keys

The DES key operates on 64-bit data in each of the 16 iterations. The key is made of a 56-bit key used in the iterations and 8 parity bits. A 64-bit key of:

$k_1 k_2 k_3 k_4 k_5 k_6 k_7 k_8 k_9 k_{10} k_{11} k_{12} k_{13 \ldots} k_{64}$

contains the parity k_8, k_{16}, $k_{32 \ldots} k_{64}$. The iterations are numbered I_1, I_2, ... I_{16}. The initial permutation of the 56 useful bits of the key is used to generate a 56-bit output. It divides into two 28-bit values, called C_0 and D_0. C_0 is specified as:

$k_{57} k_{49} k_{41} k_{33} k_{25} k_{17} k_9 k_1 k_{58} k_{50} k_{42} k_{34} k_{26} k_{18} k_{10} k_2 k_{59} k_{51} k_{43} k_{35} k_{27} k_{19} k_{11} k_3 k_{60} k_{52} k_{44} k_{36}$

And D_0 is:

$k_{63} k_{55} k_{47} k_{39} k_{31} k_{23} k_{15} k_7 k_{62} k_{54} k_{46} k_{38} k_{30} k_{22} k_{14} k_6 k_{61} k_{53} k_{45} k_{37} k_{29} k_{21} k_{13} k_5 k_{28} k_{20} k_{12} k_4$

Thus the 28-bit C_0 key will contain the 57th bit of the DES key as the first bit, the 49th as the second bit, and so on. Notice that none of the 28-bit values contains the parity bits.

Most of the rounds have a 2-bit rotate left shift, but rounds 1, 2, 9 and 16 have a single-bit rotate left (ROL). A left rotation moves all the bits in the key to the left and the bit which is moved out of the left-hand side is shifted into the right-hand end.

The key for each iteration (K_i) is generated from C_i (which makes the left half) and D_i (which makes the right half). The permutations of C_i that produces the left half of K_i is:

$c_{14} c_{17} c_{11} c_{24} c_1 c_5 c_3 c_{28} c_{15} c_6 c_{21} c_{10} c_{23} c_{19} c_{12} c_4 c_{26} c_8 c_{16} c_7 c_{27} c_{20} c_{13} c_2$

and the right half of K_i is:

$d_{41} d_{52} d_{31} d_{37} d_{47} d_{55} d_{30} d_{40} d_{51} d_{45} d_{33} d_{48} d_{44} d_{49} d_{39} d_{56} d_{34} d_{53} d_{46} d_{42} d_{50} d_{36} d_{29} d_{32}$

Thus the 56-bit key is made up of:

$c_{14}\, c_{17}\, c_{11}\, c_{24}\, c_{1}\, c_{5}\, c_{3}\, c_{28}\, c_{15}\, c_{6}\, c_{21}\, c_{10}\, c_{23}\, c_{19}\, c_{12}\, c_{4}\, c_{26}\, c_{8}\, c_{16}\, c_{7}\, c_{27}\, c_{20}\, c_{13}\, c_{2}\, d_{41}\, d_{52}\, d_{31}\, d_{37}\, d_{47}\, d_{55}\, d_{30}\, d_{40}$
$d_{51}\, d_{45}\, d_{33}\, d_{48}\, d_{44}\, d_{49}\, d_{39}\, d_{56}\, d_{34}\, d_{53}\, d_{46}\, d_{42}\, d_{50}\, d_{36}\, d_{29}\, d_{32}$

Figure 23.5 illustrates the process (note that only some of the bit positions have been shown).

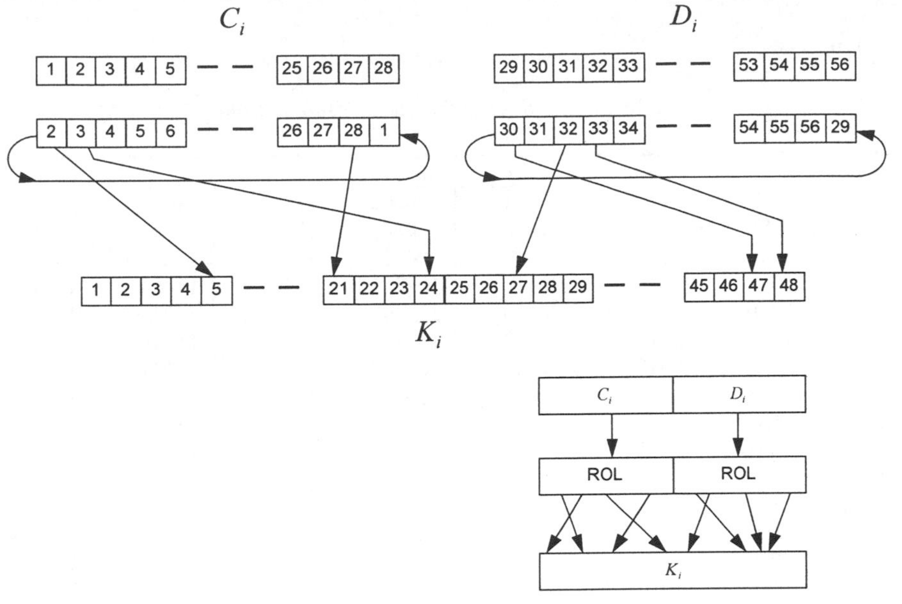

Figure 23.5 Generating the iteration key

Iteration operations
Each iteration takes the 64-bit output from the previous iteration and operates on it with a 56-bit per iteration key. Figure 23.6 shows the operation of each iteration. The 64-bit input is split into two parts, L_i and R_i. R_i is operated on with an expansion/permutation (E-table) to give 48 bits. The output from the E-table conversion is then exclusive-OR'ed with the permutated 48-bit key. Next a substitute/choice stage (S-box) is used to transform the 48-bit result to 32 bits. These are then XOR'ed with L_i to give the resulting R_{i+1} (which is R_i for the next iteration). The operation of expansion/XOR/substitution is often known as the mangler function. The R_i input is also used to produce L_{i+1}.

The mangler function takes the 32-bit R_i and the 48-bit K_i and produces a 32-bit output (which when XORed with L_i produces R_i+1). It initially expands R_i from 32 bits to 48 bits. This is done by splitting R_i into eight 4-bit chunks and then expanding each of the chunks into 6 bits by taking the adjacent bits and concatenating them onto the chunk. The leftmost and rightmost bits of R are considered adjacent. For example, if R_i is:

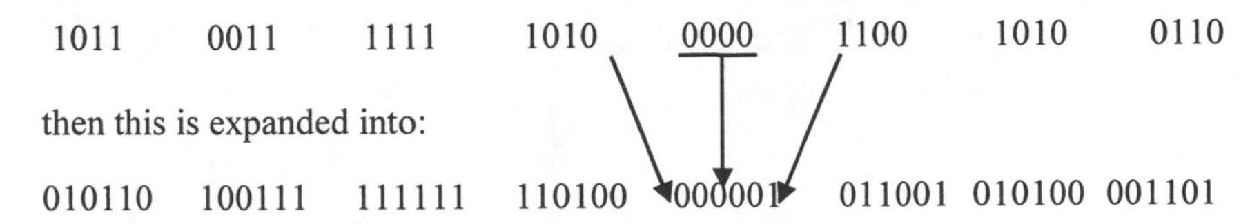

The output from the expansion is then XOR'ed with K_i and the output of this is fed into the S-box. Each 6-bit chunk of the 48-bit output from the XOR operation is then substituted with a 4-bit chunk using a lookup table. An S-box table for the first 6-bit chunk is given in Table 23.5. Thus, for example, the input bit sequence of:

000000 000001 *XXXXXX* ...

would be converted to:

1110 0000 *xxxx* ...

Table 23.7 S-box conversion for first 6-bit chunk

Input	*Output*	*Input*	*Output*	*Input*	*Output*	*Input*	*Output*
000000	1110	010000	0011	100000	0100	110000	1111
000001	0000	010001	1010	100001	1111	110001	0101
000010	0100	010010	1010	100010	0001	110010	1100
000011	1111	010011	0110	100011	1100	110011	1011
000100	1101	010100	0100	100100	1110	110100	1001
000101	0111	010101	1100	100101	1000	110101	0011
000110	0001	010110	1100	100110	1000	110110	0111
000111	0100	010111	1011	100111	0010	110111	1110
001000	0010	011000	0101	101000	1101	111000	0011
001001	1110	011001	1001	101001	0100	111001	1010
001010	1111	011010	1001	101010	0110	111010	1010
001011	0010	011011	0101	101011	1001	111011	0000
001100	1011	011100	0000	101100	0010	111100	0101
001101	1101	011101	0011	101101	0001	111101	0110
001110	1000	011110	0111	101110	1011	111110	0000
001111	0001	011111	1000	101111	0111	111111	1101

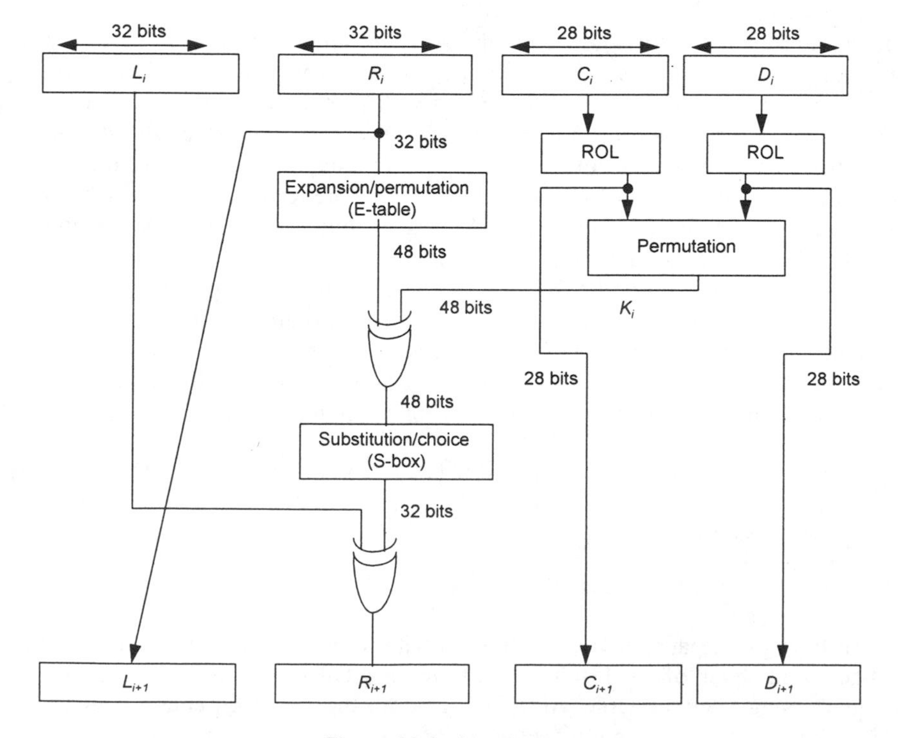

Figure 23.6 Iteration step

The design of the DES scheme was constructed behind closed doors so there are few pointers to the reasons for the construction of the encryption. One of the major weaknesses of the scheme is the usage of a 56-bit key, which means there are only 2^{56} or 7.2×10^{16} keys. Thus, as the cost of hardware reduces and the power of computers increases, the time taken to exhaustively search for a key becomes smaller each year.

In the past, there was concern about potential weaknesses in the design of the eight S-boxes. This appears to have been misplaced as no one has found any weaknesses yet. Indeed several researchers have found that swapping the S-boxes significantly reduces the security of the code.

A new variant, called Triple DES, has been proposed by Tuchman and has been standardized in financial applications. The technique uses two keys and three executions of the DES algorithm. A key, K_1, is used in the first execution, then K_2 is used and finally K_1 is used again. These two keys give an effective key length of 112 bits, that is 2×64 key bits minus 16 parity bits. The Triple DES process is illustrated in Figure 23.7.

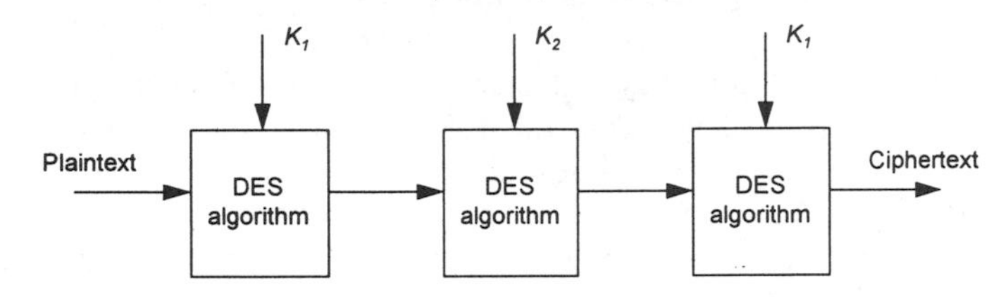

Figure 23.7 Triple DES process

23.11.3 IDEA

IDEA (International Data Encryption Algorithm) is a private-key encryption process which is similar to DES. It was developed by Xuejia Lai and James Massey of ETH Zuria and is intended for implementation in software. IDEA operates on 64-bit blocks of plaintext; using a 128-bit key, it converts them into 64-bit blocks of ciphertext. Figure 23.8 illustrates the basic encryption process.

IDEA operates over 17 rounds with a complicated mangler function. During decryption this function does not have to be reversed and can simply be applied in the same way as during encryption (this also occurs with DES). IDEA uses a different key expansion for encryption and decryption, but every other part of the process is identical. The same keys are used in DES decryption but in the reverse order.

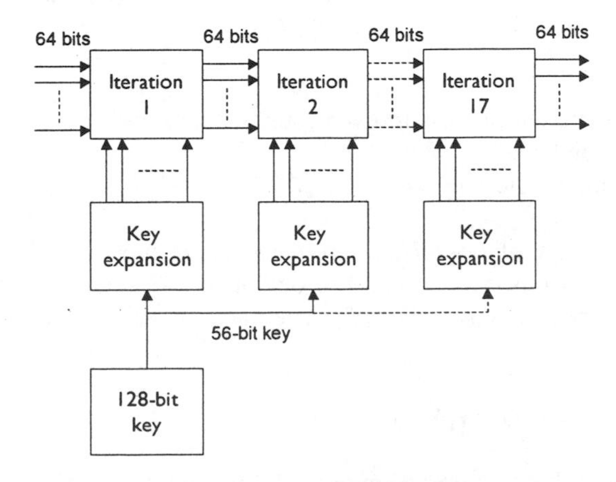

Figure 23.8 IDEA encryption

Operation

Each primitive operation in IDEA maps two 16-bit quantities into a 16-bit quantity. IDEA uses three operations, all easy to compute in software, to create a mapping. The three basic operations are:

- Exclusive-OR (⊕).
- Slightly modified add (+), and ignore any bit carries.
- Slightly modified multiply (⊗) and ignore any bit carries. Multiplying involves first calculating the 32-bit result, then taking the remainder when divided by $2^{16}+1$ (mod $2^{16}+1$).

Key expansions

The 128-bit key is expanded into fifty-two 16-bit keys, K_1, K_2, ... K_{52}. The key is generated differently for encryption than for decryption. Once the 52 keys are generated, the encryption and decryption processes are the same.

The 52 encryption keys are generated as follows:

- Keys 1–8: write out the 128-bit key and, starting from the left, chop off 16 bits at a time. This generates eight 16-bit keys. Thus the 128-bit key of *AAAAAAAAAAAAAAAA*... *HHHHHHHHHHHHHHHH* will generate eight keys of *AAAAAAAAAAAAAAAA*, *BBBBBBBBBBBBBBBB*, and so on.
- Keys 9–16: the next eight keys are generated at bit 25, and wrapped around to the beginning when at the end.
- Keys 17–24: the next eight keys are generated at bit 50, and wrapped around to the beginning when at the end.
- The rest of the keys are generated by offsetting by 25 bits and wrapped around to the beginning until the end.

The 64-bit (or 32-bit) per round, keys used are made up of 4 (or 2) of the encryption keys:

Key 1:	$K_1K_2K_3K_4$	Key 2:	K_5K_6	Key 3:	$K_7K_8K_9K_{10}$
Key 4:	$K_{11}K_{12}$	Key 5:	$K_{13}K_{14}K_{15}K_{16}$	Key 6:	$K_{17}K_{18}$
Key 7:	$K_{19}K_{20}K_{21}K_{22}$	Key 8:	$K_{23}K_{24}$	Key 9:	$K_{25}K_{26}K_{27}K_{28}$
Key 10:	$K_{29}K_{30}$	Key 11:	$K_{31}K_{32}K_{33}K_{34}$	Key 12:	$K_{35}K_{36}$
Key 13:	$K_{37}K_{38}K_{39}K_{40}$	Key 14:	$K_{41}K_{42}$	Key 15:	$K_{43}K_{44}K_{45}K_{46}$
Key 16:	$K_{47}K_{48}$	Key 17:	$K_{49}K_{50}K_{51}K_{52}$		

Iteration

Odd rounds have a different process to even rounds. Each odd round uses a 64-bit key and even rounds use a 32-bit key.

Odd rounds are simple; the process is:

If the input is a 64-bit key of $I_1 I_2 I_3 I_4$, where the I_1 is the most significant 16 bits, I_2 is the next most significant 16 bits, and so on. The output of the iteration is also a 64-bit key of $O_1 O_2 O_3 O_4$ and the applied key is $K_a K_b K_c K_d$. The iteration for the odd iteration is then:

$$O_1 = I_1 \otimes K_a \qquad O_2 = I_3 + K_c$$
$$O_3 = I_2 + K_b \qquad O_4 = I_4 \otimes K_d$$

An important feature is that this operation is totally reversible: multiplying O_1 by the inverse of K_a gives I_1, and multiplying O_4 by the inverse of K_d gives I_4. Adding O_2 to the negative of K_c gives I_3, and adding O_3 to the negative of K_b gives I_2.

Even rounds are less simple, the process is as follows. Suppose the input is a 64-bit key of $I_1I_2I_3I_4$, where I_1 is the most significant 16 bits, I_2 is the next most significant 16 bits, and so on. The output of the iteration is also a 64-bit key of $O_1O_2O_3O_4$ and the applied key is 32 bits of K_aK_b. The iteration for the even round performs a mangler function of:

$$A = I_1 \otimes I_2 \qquad B = I_3 \otimes I_4$$
$$C = ((K_a \otimes A) + B)) \otimes K_b) \qquad D = (K_a \otimes A) + C)$$

$$O_1 = I_1 \oplus C \qquad O_2 = I_2 \oplus C$$
$$O_3 = I_3 \oplus D \qquad O_4 = I_4 \oplus D$$

The most amazing thing about this iteration is that the inverse of the function is simply the function itself. Thus, the same keys are used for encryption and decryption (this differs from the odd round, where the key must be either the negative or the inverse of the encryption key).

IDEA security

There are no known methods that can be used to crack IDEA, apart from exhaustive search. Thus, as it has a 128-bit code it is extremely difficult to break, even with modern high-performance computers.

23.12 Public-key

Public-key algorithms use a secret element and a public element to their code. One of the main algorithms is RSA. Compared with DES it is relatively slow but it has the advantage that users can choose their own key whenever they need one. The most commonly used public-key cryptosystems are covered in the next sections.

Private-key systems are not feasible for large-scale networks, such as the Internet or electronic commerce, as this would involve organizations creating hundreds or thousands of different private keys. Each conversation with an organization or even an individual within a company would require a separate key. Thus, public-key methods are much better suited to the Internet and, therefore, Intranets.

Figure 23.9 shows that a public-key system has two keys, a private key and a public key. The private key is secret to the user and is used to decrypt messages that have been encrypted with the user's public key. The public key is made available to anyone who wants to send an encrypted message to the person. Someone sending a message to the user will use the user's public key to encrypt the message and it can only be decrypted using the user's private key (as the private and public keys are linked in a certain way). Once the message has been encrypted not even the sender can decrypt it.

Typical public-key methods are:

- **RSA.** RSA stands for Rivest, Shamir and Adelman, and is the most commonly used public-key cryptosystem. It is patented only in the USA and is secure for key-length of over 728 bits. The algorithm relies on the fact that it is difficult to factorize large numbers. Unfortunately, it is particularly vulnerable to chosen plaintext attacks and a new timing attack (spying on keystroke time) was announced on the 7 December 1995. This attack would be able to break many existing implementations of RSA.

- **Elliptic curve.** Elliptic curve is a new kind of public-key cryptosystem. It suffers from speed problems, but this has been overcome with modern high-speed computers.
- **DSS.** DSS (digital signature standard) is related to the DSA (digital signature algorithm). This standard has been selected by the NIST and the NSA, and is part of the Capstone project. It uses 512-bit or 1024-bit key size. The design presents some lack in key-exchange capability and is slow for signature-verification.
- **Diffie-Hellman.** Diffie-Hellman is commonly used for key-exchange. The security of this cipher relies on both the key-length and the discrete algorithm problem. This problem is similar to the factorizing of large numbers. Unfortunately, the code can be cracked and the prime number generator must be carefully chosen.
- **LUC.** Peter Smith developed LUC which is a public-key cipher that uses Lucas functions instead of exponentiation. Four other algorithms have also been developed, these are: LUCDIF (a key-negotiation method); LUCELG PK (equivalent to EL Gamel encryption); LUCELG DS (equivalent to EL Gamel data signature system) and LUCDSA (equivalent to the DSS).

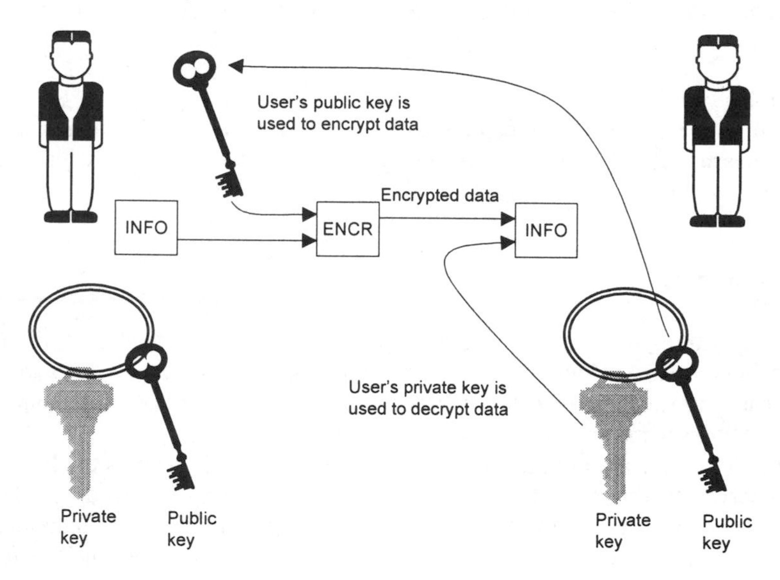

Figure 23.9 Overview of public-key systems

23.12.1 RSA

RSA is a public-key encryption/decryption algorithm and is much slower than IDEA and DES. The key length is variable and the block size is also variable. A typical key length is 512 bits. RSA uses a public-key and a private key, and uses the fact that large prime numbers are extremely difficult to factorize. The following steps are taken to generate the public and private keys:

1. Select two large prime numbers, a and b (each will be roughly 256 bits long). The factors a and b remain secret and n is the result of multiplying them together. Each of the prime numbers is of the order of 10^{100}.
2. Next, the public-key is chosen. To do this a number e is chosen so that e and $(a-1)\times(b-1)$ are relatively prime. Two numbers are relatively prime if they have no common factor greater than 1. The public-key is then $<e, n>$ and results in a key which is 512 bits long.

3. Next the private key for decryption, d, is computed so that:

   $$d = e^{-1} \bmod [(a-1) \times (b-1)]$$

   This then gives a private key of $<d, n>$. The values a and b can then be discarded (but should never be disclosed to anyone).

The encryption process to ciphertext, c, is then defined by:

$$c = m^e \bmod n$$

The message, m, is then decrypted with:

$$m = c^d \bmod n$$

It should be noted that the message block m must be less than n. When n is 512 bits then a message which is longer than 512 bits must be broken up into blocks of 512 bits.

Encryption/decryption keys

When two parties, P_1 and P_2, are communicating they encrypt data using a pair of public/private key pairs. Party P_1 encrypts its message using P_2's public-key. Then party P_2 uses its private key to decrypt this data. When party P_2 encrypts a message it sends to P_1 using P_1's public-key and P_1 decrypts this using its private key. Notice that party P_1 cannot decrypt the message that it has sent to P_2 as only P_2 has the required private key.

A great advantage of RSA is that the key has a variable number of bits. It is likely that, in the coming few years, powerful computer systems will determine all the factors to 512-bit values. Luckily the RSA key has a variable size and can easily be changed. Many users are choosing keys with 1024 bits.

Simple RSA example

Initially the PARTY1 picks two prime numbers. For example:

$$a = 11 \text{ and } b = 3$$

Next, the n value is calculated. Thus:

$$n = a \times b = 11 \times 3 = 33$$

Next *PHI* is calculated by:

$$PHI = (a-1)(b-1) = 20$$

The public exponent e is then generated so that the greatest common divisor of e and *PHI* is 1 (e is relatively prime with PHI). Thus, the smallest value for e is:

$$e = 3$$

The n (33) and the e (3) values are the public keys. The private key (d) is the inverse of e modulo *PHI*.

$$d = e^{-1} \bmod [(a-1)\times(b-1)]$$

This can be calculated by using extended Euclidian algorithm, to give the private key, *d* of 7.

Thus n=33, e=3 and d=7.

The PARTY2 can be given the public keys of *e* and *n*, so that PARTY2 can encrypt the message with them. PARTY1, using *d* and *n* can then decrypt the encrypted message.

For example, if the message value to decrypt is 4, then:

$$c = m^e \bmod n = 4^3 \bmod 33 = 31$$

Therefore, the encrypted message (c) is 31.

The encrypted message (c) is then decrypted by PARTY1 with:

$$m = c^d \bmod n = 31^7 \bmod 33 = 4$$

which is equal to the message value.

Simple RSA program

An example program which has a limited range of prime numbers is given next.

```
#include <stdio.h>
#include <math.h>

#define  TRUE  1
#define  FALSE 0

void   get_prime( long *val);
long   getE( long PHI);
long   get_common_denom( long e, long PHI);
long   getD( long e,  long PHI);
long   decrypt(long c,long n, long d);

int    main(void)
{
long   a,b,n,e,PHI,d,m,c;

       get_prime(&a);
       get_prime(&b);
       n=a*b;
       PHI=(a-1)*(b-1);
       e=getE(PHI);

       d= getD(e,PHI);
       printf("Enter input value >> "); scanf("%ld",&m);

       printf("a=%ld b=%ld n=%ld PHI=%ld\n",a,b,n,PHI);

       c=(long)pow(m,e) % n; /* note, this may overflow with large numbers */
                             /* when e is relatively large */

       printf("e=%ld d=%ld c=%ld\n",e,d,c);

       m=decrypt(c,n,d); /* this function required as c to           */
                         /*the power of d causes an overflow          */
```

```
      printf("Message is %ld ",m);
      return(0);
}

long  decrypt(long c,long n, long d)
{
long  i,g,f;

if (d%2==0) g=1; else g=c;

      for (i=1;i<=d/2;i++)
      {
         f=c*c % n;
         g=f*g % n;
      }
 return(g);
}

long getD( long e,  long PHI)
{
long u[3]={1,0,PHI};
long v[3]={0,1,e};
long q,temp1,temp2,temp3;

      while (v[2]!=0)
      {
         q=floor(u[2]/v[2]);
         temp1=u[0]-q*v[0];
         temp2=u[1]-q*v[1];
         temp3=u[2]-q*v[2];
         u[0]=v[0];
         u[1]=v[1];
         u[2]=v[2];
         v[0]=temp1;
         v[1]=temp2;
         v[2]=temp3;
      }
      if (u[1]<0) return(u[1]+PHI);
      else return(u[1]);
}

long  getE( long PHI)
{
 long great=0, e=2;

      while (great!=1)
      {
         e=e+1;
         great = get_common_denom(e,PHI);
      }
      return(e);
}

long get_common_denom(long e, long PHI)
{
long great,temp,a;

      if (e >PHI)
      {
         while (e % PHI != 0)
         {
            temp= e % PHI;
            e =PHI;
            PHI = temp;
```

```
			}
			great = PHI;
		} else
		{
			while (PHI % e != 0)
			{
				a = PHI % e;
				PHI = e;
				e = a;
			}
			great = e;
		}
		return(great);
}

void get_prime( long *val)
{
#define NO_PRIMES 11
long primes[ NO_PRIMES] ={ 3,5,7,11,13,17,19,23,29,31,37} ;
long prime,i;
		do
		{
			prime=FALSE;
			printf("Enter a prime number >> ");
			scanf("%ld",val);
			for (i=0;i<NO_PRIMES;i++)
				if (*val==primes[ i] ) prime=TRUE;
		} while (prime==FALSE);
}
```

A sample run of the program is given next.

```
Enter a prime number >> 11
Enter a prime number >> 3
Enter input value >> 4
a=11 b=3 n=33 PHI=20
e=3 d=7 c=31
Message is 4
```

23.12.2 PGP

PGP (Pretty Good Privacy) uses the RSA algorithm with a 128-bit key. It was developed by Phil Zimmermann and gives encryption, authentication, digital signatures and compression. Its source code is freely available over the Internet and its usage is also free of charge, but it has encountered two main problems:

- The source code is freely available on the Internet causing the US government to claim that it violates laws which relate to the export of munitions. Current versions have since been produced outside of the US to overcome this problem.
- It uses algorithms which have patents, such as RSA, IDEA and MD5.

Figure 23.10 shows the basic encryption process. The steps taken are:

A. Sender hashes the information using the MD5 algorithm.
B. Hashed message is then encrypted using RSA with the sender's private key (this is used to authenticate the sender as the sender's public key will be used to decrypt this part of the message).
C. Encrypted message is then concatenated with the original message.

D. Message is compressed using LZ compression.
E. A 128-bit IDEA key (K_M) is generated by some random input, such as the content of the message and the typing speed.
F. K_M is then used with the IDEA encryption. K_M is also encrypted with the receiver's public key.
G. Output from IDEA encryption and the encrypted K_M key are concatenated together.
H. Output is encoded as ASCII characters using Base-64 (Section 19.6.4 on Electronic Mail).

To decrypt the message the receiver goes through the following steps:

A. Receiver reverses the Base-64 conversion.
B. The receiver decrypts the K_M key using its own private RSA key.
C. The K_M key is then used with the IDEA algorithm to decode the message.
D. The message is then decompressed using an UNZIP program.
E. The two fragments produced after decompression will be the plaintext message and an MD5/RSA encrypted message. The plaintext message is the original message, whereas the MD5/RSA encrypted message can be used to authenticate the sender. This is done by applying the sender's public key to the decompressed encrypted part of the message. This should produce the original plaintext message.

PGP allows for three main RSA key sizes. There are:

- 384 bits. This is intended for the casual user and can be cracked by serious crackers.
- 512 bits. This is intended for the commercial user and can only be cracked by organizations with a large budget and extensive computing facilities.
- 1024 bits. This is intended for military uses and, at present, cannot be cracked by anyone. This is the recommended key size for most users of reasonably powerful computers (386/486/Pentium/*etc*). In the future, a 2048-bit code may be used.

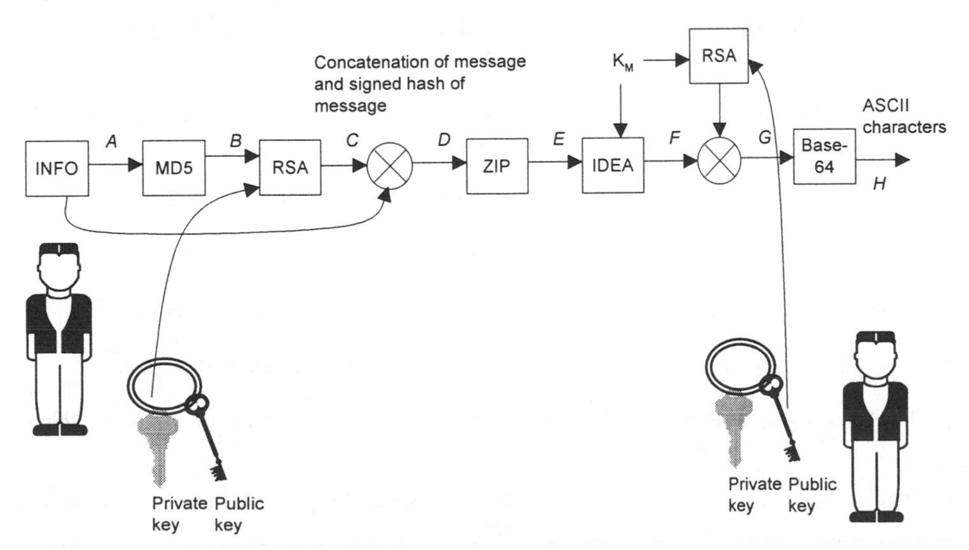

Figure 23.10 PGP encryption

23.12.3 Example PGP encryption

The Pretty Good Privacy (PGP) program developed by Philip Zimmermann is widely available over the WWW. It runs as a stand-alone application, and uses various options to use the package. Table 23.8 outlines some of the options.

To produce output in ASCII for e-mail or to publish over the Internet, the –a option is used with other options. Table 23.9 shows the key management functions.

Table 23.8 PGP options

Option	*Description*
pgp -e textfile her_userid [other userids]	Encrypts a plaintext file with the recipient's public key. In this case, it produces a file named textfile.pgp.
pgp -s textfile [-u your_userid]	Sign a plaintext file with a secret key. In this case, it produces a file named textfile.pgp.
pgp -es textfile her_userid [other userids] [-u your_userid]	Signs a plaintext file with the sender's secret key, and then encrypt it with recipient's public key. In this case, it produces a file named textfile.pgp.
pgp -c textfile	Encrypt with conventional encryption only.
pgp ciphertextfile [-o plaintextfile]	Decrypt or check a signature for a ciphertext (.pgp) file.

Table 23.9 PGP key management options

Option	*Description*
pgp –kg	Generate a unique public and private key.
pgp –ka keyfile [keyring]	Adds key file's contents to the user's public or secret key ring.
pgp –kr userid [keyring]	Removes a key or a user ID from the user's public or secret key ring.
pgp –ke your_userid [keyring]	Edit user ID or pass phrase.
pgp –kx userid keyfile [keyring]	Extract a key from the public or secret key ring.
pgp –kv[v] [userid] [keyring]	View the contents of the public-key ring.
pgp –kc [userid] [keyring]	Check signatures on the public-key ring.
pgp –ks her_userid [–u your_userid] [keyring]	Sign someone else's public-key on your public-key ring.
pgp –krs userid [keyring]	Remove selected signatures from a userid on a keyring.

RSA Key Generation

Both the public and the private keys are generated with:

```
pgp -kg
```

Initially, the user is asked about the key sizes. The larger the key the more secure it is. A 1024 bit key is very secure.

```
C:\pgp> pgp -kg
Pretty Good Privacy(tm) 2.6.3i - Public-key encryption for the masses.
(c) 1990-96 Philip Zimmermann, Phil's Pretty Good Software. 1996-01-18
International version - not for use in the USA. Does not use RSAREF.
Current time: 1998/12/29 23:13 GMT
Pick your RSA key size:
    1)    512 bits- Low commercial grade, fast but less secure
    2)    768 bits- High commercial grade, medium speed, good security
    3)  1024 bits- "Military" grade, slow, highest security
Choose 1, 2, or 3, or enter desired number of bits: 3
Generating an RSA key with a 1024-bit modulus.
```

Next, the program asks for a user ID, which is normally the user's name and his/her password. This ID helps other users to find the required public key.

```
You need a user ID for your public key.  The desired form for this
user ID is your name, followed by your E-mail address enclosed in
<angle brackets>, if you have an E-mail address.
For example:  John Q. Smith <12345.6789@compuserve.com>
Enter a user ID for your public key:
Fred Bloggs <fred_b@myserver.com>
```

Next PGP asks for a pass phrase, which is used to protect the private key if another person gets hold of it. No person can use the secret key file, unless they know the pass phrase. Thus, the pass phase is like a password but is typically much longer. The phase is also required when the user is encrypting a message with his/her private key.

```
You need a pass phrase to protect your RSA secret key.
Your pass phrase can be any sentence or phrase and may have many
words, spaces, punctuation, or any other printable characters.
Enter pass phrase: fred bloggs
Enter same pass phrase again: fred bloggs
Note that key generation is a lengthy process.
```

The public and private keys are randomly derived by measuring the intervals between keystrokes. For this, the software asks for the user to type a number of keys.

```
We need to generate 384 random bits.  This is done by measuring the
time intervals between your keystrokes.  Please enter some random text
on your keyboard until you hear the beep:

We need to generate 384 random bits.  This is done by measuring the
time intervals between your keystrokes.  Please enter some random text
on your keyboard until you hear the beep:
<keyboard typing>

   0 * -Enough, thank you.
...................................****
....................................****
Pass phrase is good.  Just a moment....
Key signature certificate added.
Key generation completed.
```

This has successfully generated the public and private keys. The public-key is placed on the public key ring (PUBRING.PGP) and the private key is place on the user's secret key ring (SECRING.PGP).

```
C:\pgp> dir *.pgp
SECRING  PGP          518  12-29-98 11:20p secring.pgp
PUBRING  PGP          340  12-29-98 11:20p pubring.pgp
```

The –kx option can be used to extract the new public key from the public-key ring and place it in a separate public key file, which can be sent to people who want to send an encrypted message to the user.

```
C:\pgp> pgp -kx fred_b
Extracting from key ring: 'pubring.pgp', userid "fred_b".
Key for user ID: Fred Bloggs <fred_b@myserver.com>

1024-bit key, key ID CD5AE745, created 1998/12/29
Extract the above key into which file? mykey
Key extracted to file 'mykey.pgp'.
```

The public-key file (`mykey.pgp`) can be sent to other users, and can be added to their public key rings. Care must be taken never to send anyone a private key, but even if it is sent then it is still protected by the pass phase.

Often a user wants to publish their public key on their WWW page or transmit it by e-mail. Thus, it requires to be converted into an ASCII format. For this the –kxa option can be used, such as:

```
C:\pgp> pgp -kxa fred_b
Extracting from key ring: 'pubring.pgp', userid "fred_b".
Key for user ID: Fred Bloggs <fred_b@myserver.com>
1024-bit key, key ID CD5AE745, created 1998/12/29

Extract the above key into which file? mykey
Transport armor file: mykey.asc
Key extracted to file 'mykey.asc'.

Extract the above key into which file? mykey
Transport armor file: mykey.asc
Key extracted to file 'mykey.asc'.
```

The file `mykey.asc` now contains an ASCII form of the key, such as:

```
Type Bits/KeyID    Date       User ID
pub  1024/CD5AE745 1998/12/29 Fred Bloggs <fred_b@myserver.com>
-----BEGIN PGP PUBLIC KEY BLOCK-----
Version: 2.6.3i
mQCNAzaJY84AAAEEAK0nvnuYcwGEaNdeqcDGXD6IrMFwX3iKtdGkZgyPyiENLb+C
bGX7P2zSG0z1d8c4f5OKYR/RgxzN4ILsAKthGaweGD0FJRgeIvn6FHJxEzmdBWIh
ME/8h2HZfegSXta8hFAMc8o9ASamolk5KBL0YWfsQlDNbR+dMJpPqQ7NWudFAAUT
tCFGcmVkIEJsb2dncyA8ZnJlZF9iQG15c2VydmVyLmNvbT6JAJUDBRA2iWPOmk+p
Ds1a50UBAfkoA/4gO5DllYko4DfjPnq4ItDtN55SgoE3upPWL52R5RQZF1BoJEF6
eLT/kejD5b7gli/yP1S456bh/k8ifi9RwSPUFN/zFUsVVYrSjZKD3kzClV1/QgTy
YmlDHHHgou6rYFXk7mGEtWc4g4D1rzds+ppc/UjN8uNp5KQUg1FsVatvPA==
=X5Xx
-----END PGP PUBLIC KEY BLOCK-----
```

Now, someone's public-key can be added to Fred's public-key ring. In this case, Fred Bloggs wants to send a message to Bert Smith. Bert's public key, in an ASCII form, is:

```
Type Bits/KeyID    Date       User ID
pub  1024/770CA60D 1998/12/30 Bert Smith <Bert_s.otherserver.com>
-----BEGIN PGP PUBLIC KEY BLOCK-----
Version: 2.6.3i
mQCNAzaKE5AAAAEEAN+5td9acGlPcTKp5J42UpwbDqz6mHOaxcO1lp6CoPE3+AXT
jfREEQ+TC0ZxMP6cCcwtEMnjVqu2M7F6li3v/AVqQIRZZkFsEOZ+8hlseHB0FR8Y
f8FDpmgld6wNpp8ocOyVul/sBQl549u0C/KnVQ6LtXo7UlsBtnbua9J3DKYNAAUR
tCNCZXJ0IFNtaXRoIDxCZXJ0X3Mub3RoZXJzZXJ2ZXIuY29tPokAlQMFEDaKE5B2
7mvSdwymDQEB2xkEANLMEDncVrFjR7labUIWHqquEFK+sqnOHPbHyIBni18x03UM
jeQJM1WA9/uIPqzeABJdD6anX4oK3yiByQjI5CT5+OdmU0y4e2+k1ab5mxxUWs7S
Tib3K5LLvPGxsOInOdunjFKaBLkrfU/L+zid3iW9FV6Zy8P07yDL2SmobRbh
=6rTj
-----END PGP PUBLIC KEY BLOCK-----
```

Fred can add Bert's key onto his public-key ring with the –ka option:

```
C:\pgp> pgp -ka bert.pgp

Looking for new keys...
pub  1024/770CA60D 1998/12/30  Bert Smith <Bert_s.otherserver.com>

Checking signatures...
pub  1024/770CA60D 1998/12/30 Bert Smith <Bert_s.otherserver.com>
sig!      770CA60D 1998/12/30  Bert Smith <Bert_s.otherserver.com>

Keyfile contains:
   1 new key(s)

One or more of the new keys are not fully certified.
Do you want to certify any of these keys yourself (y/N)?
```

Bert's key has been added to Fred's public-key ring. This ring can be listed with the –kv, as given next:

```
C:\pgp> pgp -kv

Key ring: 'pubring.pgp'
Type Bits/KeyID    Date       User ID
pub  1024/770CA60D 1998/12/30 Bert Smith <Bert_s.otherserver.com>
pub  1024/CD5AE745 1998/12/29 Fred Bloggs <fred_b@myserver.com>
2 matching keys found.
```

Next, a message can be sent to Bert, using his public-key.

```
C:\pgp>edit message.txt
Bert,

This is a secret message. Please
delete it after you have read it!

Fred.

C:\pgp>pgp -e message.txt

Recipients' public key(s) will be used to encrypt.
```

```
A user ID is required to select the recipient's public key.
Enter the recipient's user ID: bert smith
Key for user ID: Bert Smith <Bert_s.otherserver.com>
1024-bit key, key ID 770CA60D, created 1998/12/30

WARNING:  Because this public key is not certified with a trusted
signature, it is not known with high confidence that this public key
actually belongs to: "Bert Smith <Bert_s.otherserver.com>".

Are you sure you want to use this public key (y/N)? y
.
Ciphertext file: message.pgp
```

If the message needs to be transmitted by electronic mail or via a WWW page, it can be converted into text format with the –ea option, as given next:

```
C:\pgp>pgp -ea message.txt

Recipients' public key(s) will be used to encrypt.
A user ID is required to select the recipient's public key.
Enter the recipient's user ID: bert smith

Key for user ID: Bert Smith <Bert_s.otherserver.com>
1024-bit key, key ID 770CA60D, created 1998/12/30

WARNING:  Because this public key is not certified with a trusted
signature, it is not known with high confidence that this public key
actually belongs to: "Bert Smith <Bert_s.otherserver.com>".
But you previously approved using this public key anyway.
.
Transport armor file: message.asc
```

The `message.asc` file is now in a form which can be transmitted in ASCII characters. In this case, it is:

```
-----BEGIN PGP MESSAGE-----
Version: 2.6.3i

hIwDdu5r0ncMpg0BBAC7jOUx74vLb701lOCO0/5Fkc6pDJinqpA7isJH+JYbFkDj
wSv6vF/jAEonEPL8RVtqWncNDwjjwwV9OVPEZeaZ0qgZTWdbdSUilfqxZsaBo8Uz
dmmbzxd7CDTpnSYEyFWosPyzdxJqlsICig79Loh7l1BdJXEhKnMy+1VMieNYtKYA
AABrB8LTMj2lkk9t6JfS2yOc1t9EfpVMLX+rxtPZ+Tq1aCOwfid4E77FyiKN260N
APzF8J6elXhBgNM3zesA8fR8KdEnrI2BYC2XsBzTxOiKnpqoLMwWl0A7TTyhv24L
1PhwFi/YQ2SPhemdpqY=
=ooNT
-----END PGP MESSAGE-----
```

Bert can now simply decrypt the received message.

```
C:\pgp\bert> pgp message.pgp

File is encrypted.  Secret key is required to read it.
Key for user ID: Bert Smith <Bert_s.otherserver.com>
1024-bit key, key ID 770CA60D, created 1998/12/30

You need a pass phrase to unlock your RSA secret key.
Enter pass phrase: Bert Smith
Pass phrase is good.  Just a moment......
Plaintext filename: message
```

Or Bert can convert the ASCII form into a binary format with the –da option, and decrypt the message as before.

```
C:\pgp\bert> pgp -da message.asc
Stripped transport armor from 'message.asc', producing 'message.pgp'.
```

23.13 Authentication

It is obviously important to encrypt a transmitted message, but how can it be proved that the user who originally encrypted the message sent the message. This is achieved with message authentication. The two users who are communicating are sometimes known as the principals. It should be assumed that an intruder (hacker) could intercept and listen to messages at any part of the communication, whether it is the initial communication between the two parties and their encryption keys or when the encrypted messages are sent. The intruder could thus playback any communications between the parties and pretend to be the other.

23.13.1 Shared secret-key authentication

With this approach a secret key, K_{12} (between Fred and Bert) is used by both users. This would be transmitted through a secure channel, such as a telephone call, personal contact, mail message, and so on. The conversation will then be:

- The initiator (Fred) sends a challenge to the responder (Bert) which is a random number.
- The responder transmits it back using a special algorithm and the secret key. If the initiator receives back the correctly encrypted value then it knows that the responder is allowed to communicate with the user.

The random number should be large enough so that it is not possible for an intruder to listen to the communication and repeat it. There is little chance of the same 128-bit random number occurring within days, months or even years.

This method has validated Bert to Fred, but not Fred to Bert. Thus, Bert needs to know that the person receiving his communications is Fred. Thus, Bert initiates the same procedure as before, sending a random number to Fred, who then encrypts it and sends it back. After Bert has successfully received this, encrypted communications can begin.

23.13.2 Diffie-Hellman key exchange

In the previous section, a private key was passed over a secure line. The Diffie-Hellman method allows for keys to be passed electronically. For this, Fred and Bert pick two large prime numbers:

a=Prime Number 1
b=Prime Number 2

where:

$(a-1)/2$ is also prime. The values of a and b are public keys. Next, Fred picks a private key (c) and Bert picks a private key (d). Fred sends the values of:

$(a, b, b^c \bmod a)$

Bert then responds by sending:

(b^d mod a)

For example:

a=43 (first prime number), b=7 (second prime number), c=9 (Fred's private key), d=8 (Bert's private key). Note, that the value of a (43) would not be used as $(a-1)/2$ is not prime (21).

Thus, the values sent by Fred will be:

(43, 7, 42)

The last value is 42 as 7^9 is 40,353,607, and 40,353,607 mod 43, is 42. Bert will respond back with:

(6)

as 7^8 is 5,764,801, and 5,764,801 mod 43 is 6.

Next Fred and Bert will calculate:

b^{cd} mod a

and both use this as their secret key. Figure 23.11 shows an example of the interchange. It is difficult for an intruder to determine the values of c and d, when the values of a, b, c and d are large.

Unfortunately, this method suffers from the man-in-the-middle attack, where the intruder incepts the communications between Fred and Bert. Figure 23.12 shows an interceptor (Bob) who has chosen a private key of e. Thus, Fred thinks he is talking to Bert, and vice-versa, but Bob is acting as the man-in-the-middle. Bob then uses two different keys when talking with Fred and Bert.

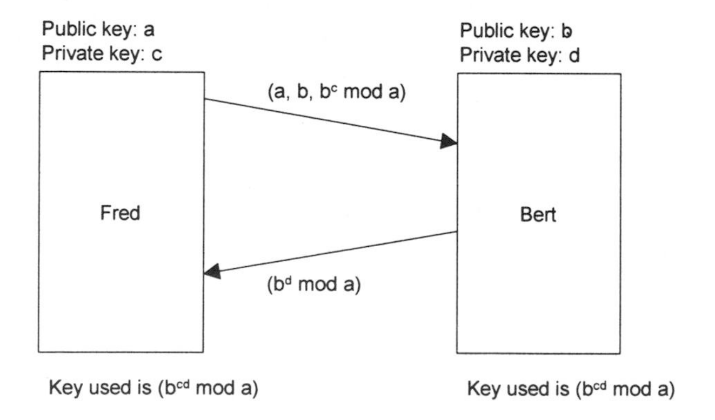

Figure 23.11 Diffie-Hellman key exchange

23.13.3 Key distribution center

The Diffie-Hellman method suffers from the man-in-the-middle attack, it also requires a separate key for each communication channel. A KDC (key distribution center) overcomes these problems with a single key and a secure channel for authentication. In a KDC, the authentication and session keys are managed through the KDC. One method is the wide-mouth protocol, which does the following:

- Fred selects a session key (K_{SESS}).
- Fred sends an encrypted message which contains the session key. The message is encrypted with K_{KDC1}, which is the key that Fred uses to pass messages to and from the KDC.
- The KDC decrypts this encrypted message using the K_{KDC1} key. It also extracts the session key (K_{SESS}). This session key is added to the encrypted message and then encrypted with K_{KDC2}, which is the key that Bert uses to pass messages to and from the KDC.

This method is relatively secure as there is a separate key used between the transmissions of Fred and the KDC, and Bert and the KDC. These keys are secret to Fred and the KDC, and between Bert and the KDC. The drawback with the method is that if the intruder determines secret key used for Fred to communicate with the KDC then it is possible to trick the KDC that it is communicating with Fred. As the key is unchanging, the theft of a key may take some time to discover and the possible damage widespread. The intruder can simply choose a new session key each time there is a new session.

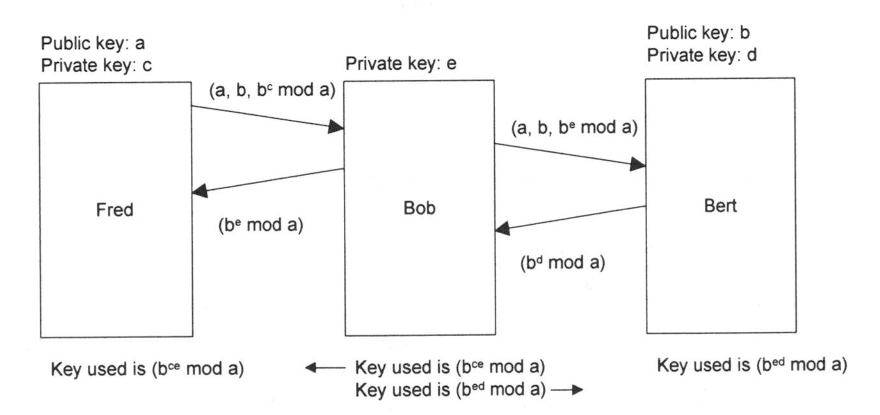

Figure 23.12 Man-in-the-middle attack

23.13.4 Digital signatures

Digital signatures provide a way of validating an electronic document, in the same way as a hand-written signature does on a document. It must provide:

- Authentication of the sender. This is important as the recipient can verify the sender of the message.
- Authentication of the contents of the message. This is important, as the recipient knows that a third party has not modified the original contents of the message. Normally, this is also time-stamped.
- Authentication that the contents have not been changed by the recipient. This is important in legal cases where the recipient can prove that the message was as the original.

Secret-key signatures

The secret-key signature involves a user selecting a secret key which is passed to a central authority, who keeps the key private. When Fred wants to communicate with Bert, he passes the plaintext to the central authority and encrypts it with a secret key and the time-stamp. The central authority then passes an encrypted message to Bert using the required secret key. A time-stamp is added to the message that is sent to Bert. This provides for a legal verification of the time the message was sent, and also stops intruders from replaying a transmitted message. The main problem with this method is that the central authority (typically, banks, government departmental or legal professionals) must be trustworthy and reliable. They can also read all of the transmitted messages.

Message digests

Public- and private-key signatures provide for both authentication and secrecy, but in many cases, all that is required is that a text message is sent with the required authentication. A method of producing authentication is a message digest, which generates a unique message digest for every message. The most common form of message digest is MD5 (RFC1321, R.Rivest). It is designed to be relatively fast to compute and does not require any large substitution tables. In summary, its operation is:

- It takes as input a message of arbitrary length.
- Produces a 128-bit 'fingerprint' (or message digest) of the input.
- It is not possible to produce two messages which have the same message digest, or to produce any message from a prespecified target message digest.

MD5 algorithm

Initially, the message with *b* bits is arranged as follows:

$m_0\ m_1\ m_2\ m_3\ m_4\ m_5\ m_6\ \ldots\ m_{b-1}$

Next five steps are performed:

- Adding padding bits. The message is padded so that its length is 64 bits less than being a multiple of 512 bits. For example, if the message is 900 bits long, then an extra 60 bits will be added so that it is 64 bits short of 1024 bits. The padded bits are a single '1' bit followed by '0' bits. At least one bit must be added and, at the most, 512 bits are added.

- Append Length. A 64-bit representation of *b* (the length of the message before the padding bits were added) is appended to the result of the previous step. The resulting message will thus be a multiple of 512 bits, or:

  $m_0\ m_1\ m_2\ m_3\ m_4\ m_5\ m_6\ \ldots\ m_{n-1}$

  where *n* is a multiple of 512.

- MD Buffer initialized. A four-word buffer (A, B, C, D) is used to compute the message digest. These are initialized to the following hexadecimal values (low-order bytes first):

  A: 01 23 45 67h (0000 0001 0010 ... 0111) B: 89 ab cd efh
  C: fe dc ba 98h (1111 1110 1101 ... 1000) D: 76 54 32 10h

- Message processed in 16-word blocks. Next four auxiliary functions are defined which operate on three 32-bit words and produce a single 32-bit word. These are:

$$F(x,y,z) = X.Y + \overline{X}.Z$$
$$G(x,y,z) = X.Z + Y.\overline{Z}$$
$$H(x,y,z) = X \oplus Y \oplus Z$$
$$I(x,y,z) = Y \oplus \left(X + \overline{Z}\right)$$

This step also involves a 64-element table T[1 ... 64] which is made of a function, where T[i] is equal to the integer part of 4294967296 times abs(sin(i)), where i is in radians.

The algorithm is as follows:

```
/* Process each 16-word block. */
For i = 0 to N/16-1 do
    /* Copy block i into X. */
    For j = 0 to 15 do
        Set X[j] to M[i*16+j].
    end /* of loop on j */

    /* Save A as AA, B as BB, C as CC, and D as DD. */
    AA = A   BB = B
    CC = C   DD = D

    /* Round 1. */
    /* Let [abcd k s i] denote the operation  a = b + ((a + F(b,c,d) + X[k] + T[i]) <<< s). */
    /* Do the following 16 operations. */
    [ABCD  0  7  1] [DABC  1 12  2] [CDAB  2 17  3] [BCDA  3 22  4]
    [ABCD  4  7  5] [DABC  5 12  6] [CDAB  6 17  7] [BCDA  7 22  8]
    [ABCD  8  7  9] [DABC  9 12 10] [CDAB 10 17 11] [BCDA 11 22 12]
    [ABCD 12  7 13] [DABC 13 12 14] [CDAB 14 17 15] [BCDA 15 22 16]

    /* Round 2. */
    /* Let [abcd k s i] denote the operation   a = b + ((a + G(b,c,d) + X[k] + T[i]) <<< s). */

    /* Do the following 16 operations. */
    [ABCD  1  5 17] [DABC  6  9 18] [CDAB 11 14 19] [BCDA  0 20 20]
    [ABCD  5  5 21] [DABC 10  9 22] [CDAB 15 14 23] [BCDA  4 20 24]
    [ABCD  9  5 25] [DABC 14  9 26] [CDAB  3 14 27] [BCDA  8 20 28]
    [ABCD 13  5 29] [DABC  2  9 30] [CDAB  7 14 31] [BCDA 12 20 32]

    /* Round 3. */
    /* Let [abcd k s t] denote the operation    a = b + ((a + H(b,c,d) + X[k] + T[i]) <<< s). */

    /* Do the following 16 operations. */
    [ABCD  5  4 33] [DABC  8 11 34] [CDAB 11 16 35] [BCDA 14 23 36]
    [ABCD  1  4 37] [DABC  4 11 38] [CDAB  7 16 39] [BCDA 10 23 40]
    [ABCD 13  4 41] [DABC  0 11 42] [CDAB  3 16 43] [BCDA  6 23 44]
    [ABCD  9  4 45] [DABC 12 11 46] [CDAB 15 16 47] [BCDA  2 23 48]

    /* Round 4. */
    /* Let [abcd k s t] denote the operation   a = b + ((a + I(b,c,d) + X[k] + T[i]) <<< s). */
    /* Do the following 16 operations. */
    [ABCD  0  6 49] [DABC  7 10 50] [CDAB 14 15 51] [BCDA  5 21 52]
    [ABCD 12  6 53] [DABC  3 10 54] [CDAB 10 15 55] [BCDA  1 21 56]
    [ABCD  8  6 57] [DABC 15 10 58] [CDAB  6 15 59] [BCDA 13 21 60]
    [ABCD  4  6 61] [DABC 11 10 62] [CDAB  2 15 63] [BCDA  9 21 64]

    /* Then perform the following additions */

    A = A + AA   B = B + BB   C = C + CC   D = D + DD
end
/* of loop on i */
```

Note that the <<< symbol represents the rotate left operation, where the bits are rotated to the left.

- Output. The message digest is produced from A, B, C and D, where A is the low-order byte and D the high-order byte.

Standard test results give the following message digests:

Message	*Message digest*
“”	`d41d8cd98f00b204e9800998ecf8427e`
“a”	`0cc175b9c0f1b6a831c399e269772661`
“abc”	`900150983cd24fb0d6963f7d28e17f72`
“abcdefghijklmnopqrstuvwxyz”	`f96b697d7cb7938d525a2f31aaf161d0`
“ABCDEFGHIJKLMNOPQRSTUVWX YZabcdefghijklmnopqrstuvwxyz012345 6789”	`c3fcd3d76192e4007dfb496cca67e13b`
“12345678901234567890123456789012345678901234567890123456789012345678901234567890”	`57edf4a22be3c955ac49da2e2107b67a`

23.13.5 PGP authentication

As well as encryption, PGP provides for a digital signature. This public-key method is an excellent way of authentication as it does not require the exchange of keys over a secure channel. In PGP, the sender's own private key is used to encrypt the message, thus signing it. This digital signature is then checked by the recipient by using the sender's public key to decrypt it. As previously mentioned, the advantages are:

- Authenticates the sender.
- Authentication of the contents of the message.
- Authentication that the message has not been modified by the recipient or any third party.
- Sender cannot undo a signature, once applied (there are no erasers with digital signatures).
- Message integrity.
- Allows signatures to be stored separately from messages, without actually revealing the contents of the message.

Thus, encryption is achieved by encrypting the message with the recipient's public key, and authentication by signing the message with the sender's private key. To make a digital signature, PGP encrypts using the secret key. It does not encrypt the whole message, only a message digest, which is a 128-bit extract of the message (a bit like a checksum). The MD5 algorithm is used for this, and provides a fingerprint that is extremely difficult to forge.

The MD5 algorithm is a standard algorithm and can be easily replicated, thus the senders private key is used to provide authentication, as the sender encrypts the message digest with his secret key. The steps are:

- Sender uses the message to determine the message digest.
- Sender's secret key encrypts the message digest and an electronic timestamp, forming a digital signature, or signature certificate.

- Sender sends the digital signature along with the message.
- Recipient reads the message and the digital signature. The message is decrypted by the receiver using the recipient's private key.
- Recipient recovers the original message digest from the digital signature by decrypting it with the sender's public key.
- Recipient calculates a new message digest from the message, and then checks this against the recovered message digest from the received information. If they are the same then it authenticates the message and the sender.

A hacker, if they tried to modify the message in any way, would have to do the following:

- Recreate another identical message digest (which is not really possible)
- Produce an altered message which produces an identical message digest (which again is not really possible).
- Create a new message with a different message digest (which is not really possible without knowing the sender's private key).

With the PGP program, the –sa option can be used to generate the digital signature. For example:

```
C:\pgp\bert> pgp -sa message.txt
A secret key is required to make a signature.
You specified no user ID to select your secret key,
so the default user ID and key will be the most recently
added key on your secret keyring.

You need a pass phrase to unlock your RSA secret key. bert smith
Key for user ID: Bert Smith <Bert_s.otherserver.com>
1024-bit key, key ID 770CA60D, created 1998/12/30

Enter pass phrase: Pass phrase is good.  Just a moment....
Output file 'message.asc' already exists.  Overwrite (y/N)? y

Transport armor file: message.asc
```

The encrypted digital signature, in this case, is:


```
-----BEGIN PGP MESSAGE-----
Version: 2.6.3i

owHrZJjKzMpg1jVtTdm77EvlPMt4GRn/KDP/Yr7JWPN8/Zani1pkBBmWznvTsdl4
g1yN9+WoRz+WqFb9L3vC1jElv3xZ1v45jLdMCuTqdz5ZyKe5WFV06/yfTbu7ZZ78
nP93yhIb4XW8qSu2sdrHfbhjr9TNY+Om/6FzveeigymfyzZvcOT/l2PKky3iqCGi
WC2+4jnjbLUXKVNfWc7NfPbr4ZqMJO7c1OLixPRUvZKKEgYgcEotKtHh5eLlCsnI
LFYAokSF4tTkotQSBZhChYCc1MTiVF6ulNSc1JJUhcwShcS0ktQihcr8UoWMxLJU
haLUxBSgsCLIGLei1BQ9Xi4A
=iGfx
-----END PGP MESSAGE-----
```


The message and the digital signature can be produced using the –sea option. In this case, it produces the file:

```
-----BEGIN PGP MESSAGE-----
Version: 2.6.3i

hIwDdu5r0ncMpg0BBACF3bP4h0vt9ajaD3Vgf4aSUds03jfB9xXzZY9YjzjHyFBX
dO8IzMyDB6KdeX2cJk1pdPWhHi0cRQ2ddxoEBdS38XCJtjuTf0DYkwid+0dClt69
ntkwy0Lc4Y6QoDk9BHnVtDTkUu8J12KJrkoRx4DikumVbGB+CCAfCTOcr1U2vqYA
AAEMsURzRPLqwXDToFkzXA11EAfQ5ECJPFbsejBJhkbZAZ0aswVMYgX52wEnWxcI
MRmz0IdRLDtXtZ9SJvFzWMpPzVygOmOMDKhiDuEOI89D/HOomMlBaRH41Zx6xqf4
8LuhtJSwNdgHE07jiGAmvKkxRobUeOmZoEqs6BrU8hveJwGE4n0OVwWIzXbqH2BL
GTD8nAMFgqbh1LGfc3SV6bIst7z13HdFMSg1ZonbQj39i/ZTv8qzHY5rqN7uBPJb
eHU02wjCo3Dyc1atohPApcNEYmgkzaSQYkKeL9Zo3JRlk9xGbjZdtSk6+fxYU2WF
BQrW/AQheT51M68uDLe7OJ2+ny9m4nNEnwwDGqNaWg==
=f4f8
-----END PGP MESSAGE-----
```

The file can then be validated and decrypted as follows:

```
C:\pgp> pgp message
File has signature.  Public key is required to check signature.
.
Good signature from user "Bert Smith <Bert_s.otherserver.com>".
Signature made 1998/12/30 21:08 GMT using 1024-bit key, key ID 770CA60D

WARNING:  Because this public key is not certified with a trusted
signature, it is not known with high confidence that this public key
actually belongs to: "Bert Smith <Bert_s.otherserver.com>".
But you previously approved using this public key anyway.
Plaintext filename: message
```

23.14 Exercises

23.14.1 How many keys are used in the public-key system:

(a) 1 (b) 3
(c) 2 (d) 4

23.14.2 A typical public-key system is:

(a) IDE (b) IDA
(c) PGP (d) IDEA

23.14.3 In the PGP program, which option is used to create a public and a private key:

(a) –key (b) –ka
(c) –kg (d) –k

23.14.4 How many keys are used in the private-key system:

(a) 1 (b) 3
(c) 2 (d) 4

23.14.5 How many keys are used in the public-key system:

(a) 1 (b) 3
(c) 2 (d) 4

Morse code:
(Note this is not encryption and is basically an alternative alphabet)

A	•—	W	•——
B	—••	X	—••—
C	—•—•	Y	—•——
D	—••	Z	——••
E	•	1	•————
F	••—•	2	••———
G	——•	3	•••——
H	••••	4	•••—
I	••	5	•••••
J	•———	6	—••••
K	—•—	7	——•••
L	•—••	8	———••
M	——	9	————•
N	—•	10	——————
O	———	.	•—•—•—
P	•——•	,	——••——
Q	——•—	?	••——••
R	•—•	:	———•••
S	•••	;	—•—•—•
T	—	-	—•••—
U	••—	/	—••—•
V	•••—		

23.14.6 A typical private-key system is:

(a) IDE (b) IDA
(c) PGP (d) IDEA

23.14.7 How many possible keys are there with a 16-bit key:

(a) 16 (b) 65,536
(c) 256 (d) 4,294,967,296

23.14.8 How many possible keys are there with a 32-bit key:

(a) 32 (b) 1,048,576
(c) 1024 (d) 4,294,967,296

23.14.9 How many actual encryption key bits does IDEA have:

(a) 56 (b) 64
(c) 128 (d) 256

23.14.10 If it takes 10 ns (10×10^{-9} s) to test a key, determine the amount of time it would take, on average, to decrypt a message with a 32-bit key:

(a) 21.48 seconds (b) 43 seconds
(c) 21.48 minutes (d) 43 minutes

23.14.11 Which key does the recipient use to decrypt the main message:

(a) Recipient's public key (b) Recipient's private key
(a) Sender's public key (b) Sender's private key

23.14.12 Which key does the recipient use to authenticate the sender:

(a) Recipient's public key (b) Recipient's private key
(a) Sender's public key (b) Sender's private key

23.14.13 How many bits does the message digest have in PGP:

(a) Depends on the message (b) 16
(c) 128 (d) 256

23.14.14 What bitwise operator is used in encryption, as it always preserves the contents of the information:

(a) Exclusive-OR'ed (b) AND
(c) NOR (d) OR

23.14.15 What happens when a value is Exclusive-OR'ed by the same value, twice:

(a) Value becomes all 0's (b) Value becomes all 1's
(c) Same value results (d) Double the value

23.14.16 Using a shifted alphabet, which is the encrypted message for 'help':

(a) gdho (b) ifnq
(c) ebim (d) qatp

23.14.17 If it takes 100 days to crack an encrypted message, and assuming that computing speed increases by 100% each year, determine how long it will take to crack the message after 2 years:

(a) 25 days (b) 44.44... days
(c) 50 days (d) 100 days

23.14.18 If it takes 100 days to crack an encrypted message, and assuming that computing speed increases by 50% each year, determine how long it will take to crack the message after 2 years:

(a) 25 days (b) 44.44... days
(c) 50 days (d) 100 days

23.14.19 Search for the PGP program on the Internet, and, if possible, download it. Next, do the following:

(a) Generate a public and a private key ring.
(b) Generate an ASCII form for your public key.
(c) Create the following message:

```
Help me! I'm on fire.
Martin.
```

(d) Create another user and a new public key, or determine another person's public key.
(e) Encrypt the message and give it to the other user.
(f) Decrypt the message.

23.14.20 Explain why public-key methods tend to be more secure than private-key methods. The discussion should include:

- Ease of changing the key.
- Ease of distribution.
- Additional extra bits to the key to make it more secure.
- Crackability.
- Compression techniques (especially in the PGP method).
- etc.

23.14.21 The RSA encryption process involves factorizing a prime number within a given range. Write an algorithm which determines whether a number is prime. If possible, implement it using a software language such as C, Pascal or BASIC. An outline of a C program is given next. With this program, identify how this program could be run faster, and benchmark the time it takes for the following ranges:

0 to 1,000,000 and 0 to 1,000,000,000

Note, for integers in these ranges the `int` data type must be changed to a `long`.

```
#include <stdio.h>
#define  TRUE  1
#define  FALSE 0

int main(void)
{
unsigned int i, j, prime;
   for (i=3;i<1000;i++)
   {
      prime=TRUE;
      for (j=2;j<=i/2;j++) {
         if ((i%j)==0)
         {
            prime=FALSE;
            break;
         }
      }
      if (prime) printf("%u ",i);
   }
   return(0);
}
```

A sample run is given next.

```
3 5 7 11 13 17 19 23 29 31 37 41 43 47 53 59 61 67 71 73 79 83 89 97 101
103 107 109 113 127 131 137 139 149 151 157 163 167 173 179 181 191 193
197 199 211 223 227 229 233 239 241 251 257 263 269 271 277 281 283 293
307 311 313 317 331 337 347 349 353 359 367 373 379 383 389 397 401 409
419 421 431 433 439 443 449 457 461 463 467 479 487 491 499 503 509 521
523 541 547 557 563 569 571 577 587 593 599 601 607 613 617 619 631 641
643 647 653 659 661 673 677 683 691 701 709 719 727 733 739 743 751 757
761 769 773 787 797 809 811 821 823 827 829 839 853 857 859 863 877 881
883 887 907 911 919 929 937 941 947 953 967 971 977 983 991 997
```

23.14.22 Determine the number of iterations that a prime number factorizing program would require for the following ranges:

1 to 10   1 to 10^2 1 to 10^3
1 to 10^6 1 to 10^8

23.14.23 Modify the program in Section 24.12 so that it checks that the entered value is a prime number (rather than using a look-up table).

23.14.24 The program in Section 24.12 causes a mathematical overflow when the prime number (e) is chosen to be large (such as 49). Investigate other methods of implementing this program so that there are no mathematical overflows. Also, check the range of integer values that can be implemented.

23.14.25 How might a hacker spy on a sender who is using PGP, and determine the encryption code (K_M).

23.14.26 Send me (*William Buchanan*) an e-mail message that has been encrypted with my public key, which is available on my WWW page, a copy of which is:

```
Type Bits/KeyID     Date       User ID
pub  1024/AC7612DD 1998/12/30 William Buchanan <w.buchanan@napier.ac.uk>
-----BEGIN PGP PUBLIC KEY BLOCK-----
Version: 2.6.3i

mQCNAzaKKSoAAAEEALQzETdRSHMO8QhyaEJe2bPP7suheyj3q2Wa3Xq8g34V6DS0
+APHRKLWmikt4SFqq8Y0q67Zq+NHhDGCjMEI3OlVXZiRjOiqKabkqheZTFf5eCJI
Ugq5hPcStb6bBjnXX0CTO9PW13XaSA0SJNALzRtOD2Ag+4i5tz7Wg2CsdhLdAAUT
tCpXaWxsaWFtIEJ1Y2hhbmFuIDx3LmJ1Y2hhbmFuQG5hcGllci5hYy51az6JAJUD
BRA2iikqPtaDYKx2Et0BAYM8A/wKbqPeNpRApfr+RaG0WxYVEGDUQIItzjFiR3+v
bjxJrUfK7vQ0STKVTPLaAQY7bBUaoaF9RtiT/pBbvLvMJMmUsmC3JJOZueFVsMf6
wY1BTYufTJ7OroFUxcNhiXRyvbVneR4xPsVlyoeqkPSnipyVpfjE48L/P1qs8PYu
PZzjLg==
=5bQj
-----END PGP PUBLIC KEY BLOCK-----
```

Note, if you want an encrypted message in reply, remember to send your public key.

23.14.27 Show that it will take 5849 years to search all the keys for a 64-bit encryption key. Assume it takes 10 ns (10×10^{-9} s) to test a key. How might this time be drastically reduced?

23.14.28 Show, by an example, that if $C = A \oplus B$ then $B = A \oplus C$. Use an 8-bit example, a 16-bit example and a 32-bit example.

23.14.29 Create your own encryption algorithm and implement it. For example:

- Read file, one byte at a time.
- Exclusive-OR the byte with a value of AAh (0xAA, in C).
- Rotate bits by 2 positions to the right.
- Exclusive-OR'ed the byte with a value of 73h.
- Rotate bits by 3 positions to the left.
- Save byte.
- Read next byte, and so on.

Note that when decrypting the file, the reverse should be implemented, that is:

- Read file, one byte at a time.
- Rotate bits by 3 positions to the right
- Exclusive-OR'ed the byte with 73h.
- Rotate bits by 2 positions to the left.
- Exclusive-OR'ed the byte with AAh.
- Save byte.
- Read next byte, and so on.

An outline is:

```
fread(&ch,1,1,in); /* read a byte from the file */
ch=((ch & 0xff) ^ 0xAA) & 0xff; /* X-OR with AAh */
ch=rot_right(ch); /* perform two right rotates */
ch=rot_right(ch);
/*********** Add code here ************/
if (!feof(in)) fwrite(&ch,1,1,out); /* write a byte */
```

23.14.30 If it currently takes 1 million years to decrypt a message then complete Table 23.10 assuming a 40% increase in computing power each year.

Table 23.10 Time to decrypt a message assuming an increase in computing power

Year	*Time to decrypt (years)*	*Year*	*Time to decrypt (years)*
0	1 million	10	
1	714,286	11	
2		12	
3		13	
4		14	
5		15	
6		16	
7		17	
8		18	
9		19	

23.14.31 The following messages were encrypted using the code mapping:

Input: `abcdefghijklmnopqrstuvwxyz`
Encrypted: `mgqoafzbcdiehxjklntqrwsuvy`

(i) `qnv#mxo#oaqjoa#qbct#hattmza`
(ii) `zjjogva#mxo#fmnasaee#jxa#mxo#mee`
(iii) `oaqjoa#qbct#mx#vjr#bmwa#fcxctbao#qbct#lratqcjx`

Decrypt them and determine the message. (Note that a '#' character has been used as a SPACE character.)

23.14.32 The following messages were encrypted using a shifted alphabet. Decrypt them by determining the number of shifts. (Note that a '#' character has been used as a SPACE character.)

(i) XLMW#MW#ER#IBEQTPI#XIBX
(ii) ROVZ#S#KW#NBYGXSXQ#SX#DRO#COK
(iii) ZVOKCO#MYWO#AESMU#WI#RYECO#SC#YX#PSBO
(iv) IJ#D#YJ#IJO#RVIO#OJ#BJ#OJ#OCZ#WVGG

23.14.33 The following messages were encrypted using a numeric key and the XOR operation. Decrypt them by identifying the SPACE character.

(i)]a`z)`z)hg)lqhdyel)}lq} Hint: ')' is a SPACE
(ii) Onv!v`ri!xnts!i`oer/ Hint: '!' is a SPACE
(iii) Xhddir+Odd'+|cnyn+jyn+rd~ Hint: '+' is a SPACE
(iv) Cftclagf"Fcvc"Amooq# Hint: '"' is a SPACE

23.14.34 If you have access to a software development package, write a program in which the user enters a line of text. Encrypt it by shifting the bits in each character one position to the left. Save them to a file. Also write a decryption program.

23.14.35 The following text is a character-mapped encryption. The common 2-letter words in the text are:

to it is to in as an

and the common 3-letter words are:

for and the

and the only 1-letter word is *a*. Table 23.4 gives a table of letter probabilities from a sample piece of text. If required, this table can be compared with the probabilities in the encrypted text.

```
tzf hbcq boybqtbmf ja ocmctbe tfqzqjejmv jyfl bqbejmrf cn tzbt ocmctbe ncmqben
blf efnn baafqtfo gv qjcnf. bqv rqwbqtfo ocntjltcjq boofo tj b ncmqbe cn
ofnqlcgfo bn qjcnf. tzcn qjreo gf mfqflbtfo gv futflqbe firckhfqt kljorqcqm
bclgjlqf ntbtcq, aljh jtzfl ncmqben qjrkecqm cqtj tzf ncmqbe'n kbtz (qljnn-tbed),
aljh wctzcq fefqtlcqbe qjhkjqfqtn, aljh lfqjlocqm bqo kebvgbqd hfocb, bqo nj jq.
b qjhkblbtjl jrtkrtn b zcmz efyfe ca tzf ncmqbe yjetbmf cn mlfbtfl tzbq tzf
tzlfnzjeo yjetbmf, fenf ct jrtkrtn b ejw. ca tzf qjcnf yjetbmf cn efnn tzbq tzf
tzlfnzjeo yjetbmf tzfq tzf qjcnf wcee qjt baafqt tzf lfqjyflfo ncmqbe. fyfq ca
tzf qjcnf cn mlfbtfl tzbq tzcn tzlfnzjeo tzflf blf tfqzqcirfn wzcqz qbq lforqf
ctn faafqt. ajl fubhkef, futlb gctn qbq gf boofo tj tzf obtb fctzfl tj oftfqt
flljln jl tj qjllfqt tzf gctn cq flljl.
      eblmf bhjrqtn ja ntjlbmf blf lfirclfo ajl ocmctbe obtb. ajl fubhkef,
nfyfqtv hcqrtfn ja zcac irbectv hrncq lfirclfn jyfl ncu zrqolfo hfmgvtfn ja obtb
ntjlbmf. tzf obtb jqqf ntjlfo tfqon tj gf lfecbgef bqo wcee qjt ofmlbof jyfl tchf
(futlb obtb gctn qbq benj gf boofo tj qjllfqt jl oftfqt bqv flljln). tvkcqbeev,
tzf obtb cn ntjlfo fctzfl bn hbmqftcq acfeon jq b hbmqftcq ocnd jl bn kctn jq bq
jktcqbe ocnd. tzf bqqrlbqv ja ocmctbe nvntfhn ofkfqon jq tzf qrhgfl ja gctn rnfo
ajl fbqz nbhkef, wzflfbn bq bqbejmrf nvntfh'n bqqrlbqv ofkfqon jq qjhkjqfqt
tjeflbqqf. bqbejmrf nvntfhn benj kljorqf b ocaaflcqm lfnkjqnf ajl ocaaflfqt
nvntfhn wzflfbn b ocmctbe nvntfh zbn b ofkfqobgef lfnkjqnf.
      ct cn yflv ocaacqret (ca qjt chkjnncgef) tj lfqjyfl tzf jlcmcqbe bqbejmrf
ncmqbe batfl ct cn baafqtfo gv qjcnf (fnkfqcbeev ca tzf qjcnf cn lbqojh). hjnt
```

```
hftzjon ja lforqcqm qjcnf cqyjeyf njhf ajlh ja acetflcqm jl nhjjtzcqm ja tzf
ncmqbe.  b mlfbt boybqtbmf ja ocmctbe tfqzqjejmv cn tzbt jqqf tzf bqbejmrf obtb
zbn gffq qjqyfltfo tj ocmctbe tzfq ct cn lfebtcyfev fbnv tj ntjlf ct wctz jtzfl
krlfev ocmctbe obtb. jqqf ntjlfo cq ocmctbe ct cn lfebtcyfev fbnv tj kljqfnn tzf
obtb gfajlf ct cn qjqyfltfo gbqd cqtj bqbejmrf.
      bq boybqtbmf ja bqbejmrf tfqzqjejmv cn tzbt ct cn lfebtcyfev fbnv tj ntjlf.
ajl fubhkef, ycofj bqo brocj ncmqben blf ntjlfo bn hbmqftcq acfeon jq tbkf bqo b
kcqtrlf cn ntjlfo jq kzjtjmlbkzcq kbkfl. tzfnf hfocb tfqo tj boo qjcnf tj tzf
ncmqbe wzfq tzfv blf ntjlfo bqo wzfq lfqjyflfo (nrqz bn tbkf zcnn).
rqajltrqbtfev, ct cn benj qjt kjnncgef tj oftfqt ca bq bqbejmrf ncmqbe zbn bq
flljl cq ct.
```

23.14.36 The following is a piece of character-mapped encrypted text. The common 2-letter words in the text are:

to it is to in as an

and the common 3-letter words are:

for and the

```
ixq rnecq ja geie bjhhrtqbeiqjtn etg bjhkriqw tqisjwzn qn qyqw qtbwqenqtc. qi qn
jtq ja ixq aqs iqbxtjmjcqbem ewqen sxqbx fwqtcn fqtqaqin ij hjni ja ixq bjrtiwqqn
etg ixq kqjkmqn ja ixq sjwmg. sqixjri qi hetv qtgrniwqqn bjrmg tji quqni. qi qn
ixq jfdqbiqyq ja ixqn fjjz ij gqnbrnn geie bjhhrtqbeiqjtn qt e wqegefmq ajwh ixei
nirgqtin etg kwjaqnnqjtemn emm jyqw ixq sjwmg bet rtgqwnietg.
      qt ixq keni, hjni qmqbiwjtqb bjhhrtqbeiqjt nvniqhn iwetnhqiiqg etemjcrq
nqctemn. jt et etemjcrq iqmqkxjtq nvniqh ixq yjmiecq mqyqm awjh ixq kxjtq yewqqn
sqix ixq yjqbq nqctem. rtsetiqg nqctemn awjh quiqwtem njrwbqn qenqmv bjwwrki
ixqnq nqctemn. qt e gqcqiem bjhhrtqbeiqjt nvniqh e nqwqqn ja gqcqiem bjgqn
wqkwqnqtin ixq etemjcrq nqctem. ixqnq ewq ixqt iwetnhqiiqg en jtqn etg oqwjn.
gqcqiem qtajwheiqjt qn mqnn mqzqmv ij fq eaaqbiqg fv tjqnq etg xen ixrn fqbjhq
ixq hjni kwqgjhqteti ajwh ja bjhhrtqbeiqjtn.
      gqcqiem bjhhrtqbeiqjt emnj jaaqwn e cwqeiqw trhfqw ja nqwyqbqn, cwqeiqw
iweaaqb etg emmjsn ajw xqcx nkqqg bjhhrtqbeiqjtn fqisqqt gqcqiem qlrqkhqti. ixq
rnecq ja gqcqiem bjhhrtqbeiqjtn qtbmrgqn befmq iqmqyqnqjt, bjhkriqw tqisjwzn,
aebnqhqmq, hjfqmq gqcqiem wegqj, gqcqiem ah wegqj etg nj jt.
```

23.14.37 Implement a program which converts a text file using Base-64 encoding.

23.14.38 Which of the following is not an advantage of public-key authentication:

(a) Provides message integrity
(b) Signatures can be stored, independently of the message
(c) No party can change the contents of an encrypted message, even the sender
(d) It makes the encrypted message, even more secure

23.14.39 Using the PGP program, pass a message with your digital and check that the recipient has validated your signature.

23.14.40 Explain how PGP adds a digital signature to a message.

23.14.41 Outline the advantages of authentication. Describe why a digital signature is a more secure method of authentication than a normal hand-written signature.

23.14.42 What advantages does PGP authentication have over private-key methods.

23.14.43 Show that in Diffie-Hellam that if a=37 (first prime number), b=11 (second prime number), c=5 (Fred's private key), d=8 (Bert's private key), the values sent be:

(37, 11, 27)

and

(10)

23.14.44 Write a program (or use a spreadsheet or a calculator) to search for the solution to:

$11^x \bmod 7 = 2$

Prove that the value of x is 5.

23.14.45 Determine values of a so that a is prime and the following is also prime:

(a–1)/2

23.15 Letter probability program

Program 23.7

```
#include <stdio.h>
#include <string.h>
#include <ctype.h>

#define  NUM_LETTERS 29

int   get_occurances(char c, char txt[ ]);

int   main(void)
{
char  ch, fname[ BUFSIZ] ;
int   occ[ NUM_LETTERS] ={ 0,0,0,0,0,0,0,0,0,0,0,0,0,0,0,
                                        0,0,0,0,0,0,0,0,0,0,0,0,0,0} ;
unsigned int   total,i;
FILE           *in;

      printf("Enter text file>>");
      gets(fname);
      if ((in=fopen(fname,"r"))==NULL)
      {
         printf("Can't find file %s\n",fname);
         return(1);
      }
      do
      {
         ch=tolower(getc(in));
         if (isalpha(ch))
         {
            (occ[ ch-'a'])++;
            total++;
         }
         else if (ch=='.') {  occ[ NUM_LETTERS-3] ++;  total++; }
         else if (ch==' ') {  occ[ NUM_LETTERS-2] ++;  total++; }
         else if (ch==',') {  occ[ NUM_LETTERS-1] ++;  total++; }
      } while (!feof(in));

      fclose(in);
      puts("Char. Occur. Prob.");
      for (i=0;i<NUM_LETTERS;i++)
```

```
        {
           printf("  %c  %5d %5.4f\n",'a'+i,occ[ i] ,(float)occ[ i] /(float)total);
        }
        return(0);
}

int     get_occurances(char c, char txt[ ] )
{
int     occ=0,i;

        for (i=0;i<strlen(txt);i++) if (c==txt[ i] ) occ++;
        return(occ);
}
```

23.16 Note from the Author

Sxeolf-nhb hqfubswlrq phwkrgv duh wkh nhb wr wkh vhfxulwb ri wkh Lqwhuqhw, dqg dqb lqirupdwlrq wkdw lv wudqvplwwhg ryhu wkh Lqwhuqhw. Lw vr hdvb wr fkdqjh iru d xvhu wr fkdqjh sxeolf dqg sulydwh nhbv vr wkdw wkh wudqvplwwhg gdwd lv pruh vhfxuh, dqg wkhq wr sxeolflch wkh qhz sxeolf-nhb. Pdqb jryhuqphqwv duh reylrxvob djdlqvw wklv pryh dv lw doorzv iru wkh wrwdo vhfxulwb ri pdqb wbshv ri gdwd. Lq idfw, vhyhudo jryhuqphqwv kdyh edqqhg wkh xvdjh ri zlghob dydlodeoh sxeolf-nhb vrxufh frgh. Vr zkr lv uljkw? L wklqn L nqrz.

Jryhuqphqwv ri wkh zruog zrxog olnh xv doo wr xvh sulydwh nhbv, dqg wkhb zrxog olnh wr nqrz hyhubrqh'v nhb, vr wkdw wkhb fdq lqwhuurjdwh wkh frqwhqwv ri phvvdjhv. Wkhuh duh pdqb jryhuqphqwv durxqg wkh zruog zkr wds skrqh phvvdjhv. Pdqb uhfrq wkdw vhyhudo jryhuqphqwv durxqg wkh zruog dfwxdoob wds prvw ri wkh whohskrqh phvvdjhv wkdw wudyho rq wkh sxeolf skrqh vhuylfh. Reylrxvob wkhuh lv qr zdb wkdw kxpdq rshudwhv frxog olvwhq wr doo wkhvh fdoov, dv wkhb rffxu lq uhdo-wlph, wkxv wkhb xvh vriwzduh zklfk ghwhfwv fhuwdlq nhb zrugv, vxfk dv: whuurulvw, erpe, LUD, dqg vr rq. Wkhvh rffxuuhqfhv fdq wkhq eh orjjhg dqg fkhfnhg eb d kxpdq.

Dv zh pryh wrzdugv d wrwdoob lqwhjudwhg gljlwdo qhwzrun, wkdw lv edvhg rq LS dgguhvvlqj, lq wkh ixwxuh, pdqb irupv ri gljlwdo lqirupdwlrq zloo eh hqfubswhg. Wklv frxog lqfoxgh whohskrqh frqyhuvdwlrqv, ylghr frqihuhqflqj, ZZZ dffhvvhv, uhprwh orjlq, iloh wudqvihu, hohfwurqlf frpphufh, uxqqlqj olfhqvhg dssolfdwlrqv, dqg vr rq. Hohfwurqlf frpphufh ghshqgv rq hqfubswlrq, dv ihz xvhuv zrxog wuxvw wkh dffhvv li lw zhuh qrw vhfxuh, dqg prvw edqnv zrxog qrw zlvk wr jhw lqwr d wudqvdfwlrq zlwk d xvhu li lw zhuh qrw vhfxuh.

Rqh plqru iodz lv wkdw hqfubswhg gdwd fdq vwloo eh wudfhg iru wkh wudqvplwwhu dqg wkh uhfhlyhu, dv wkh LS dgguhvv ri wkh wudqvplvvlrq fdq vwloo eh hadplqhg. Wkh surwrfro ehlqj xvhg fdq dovr eh ghwhuplqhg. Wkxv dq hawhuqdo sduwb frxog ghwhuplqh wkh ghvwlqdwlrq ri d frppxqlfdwlrq, dqg lwv wbsh, exw zrxog kdyh gliilfxowb lq uhdglqj wkh vhqg lqirupdwlrq. Plolwdub rshudwlrqv duh olnhob wr kdyh hqg-wr-hqg hqfubswlrq zkhuh wkh gdwd olqn odbhuv duh dfwxdoob hqfubswhg.

D sduwlfxodu dwwdfn rq hqfubswlrq lv iru wkh hawhuqdo sduwb wr ordg d surjudp rq wkh xvhuv frpsxwhu, zklfk zloo prqlwru doo wkh surfhvvhv zklfk duh uxqqlqj. Rqh ri wkh surfhvvhv pdb eh wkh rqh zklfk uhdgv wkh sulydwh-nhb iurp wkh glvn gulyh. Wklv nhb frxog wkhq eh vhqw wr wkh hawhuqdo sduwb. Iurp wkhq rq wkh hawhuqdo sduwb frxog wds lqwr dq hqfubswhg frppxqlfdwlrq wkdw zdv vhqw xvlqj wkh xvhuv sxeolf-nhb. Wkhvh phwkrgv duh srvvleoh dv wkh hawhuqdo sduwb mxvw kdv wr uhsodfh wkh WFS/LS vwdfn (vxfk dv ZLQVRFN.GOO) zlwk wkhlu rzq vwdfn zklfk prqlwruv dqb frppxqlfdwlrqv ryhu wkh Lqwhuqhw.

Dqbrqh zkr eholhyhv wkdw d zulwwhq vljqdwxuh lv d jrrg phwkrg ri vhfxulwb lv zurqj. L kdyh vhhq pdqb rffxuuhqfhv ri shrsoh irujlqj dqrwkhu shuvrq'v vljqdwxuh. Dovr, prghuq hohfwurqlf vfdqqlqj dqg qhdu-shuihfw txdolwb sulqwlqj doorzv iru hdvb irujhulhv. Iru hadpsoh, lw lv hawuhphob hdvb wr vfdq-lq d zkroh grfxphqw, frqyhuw lw wr whaw (xvlqj rswlfdo fkdudfwhu uhfrjqlwlrq) dqg wkhq wr uhsulqw d fkdqjhg yhuvlrq. Wkxv, gljlwdo dxwkhqwlfdwlrq lv wkh rqob uhdo zdb wr dxwkhqwlfdwh wkh vhqghu dqg wkh frqwhqwv ri d phvvdjh. Exw, xqiruwxqdwhob, wkh ohjdo vbvwhp whqgv wr wdnh d orqj wlph wr fdwfk xs zlwk whfkqrorjb, exw lw zloo kdsshq vrphgdb.

23.17 Occurrences of English letters, digrams, trigrams and words

Letters (%)		*Digrams (%)*		*Trigrams (%)*		*Words (%)*	
E	13.05	TH	3.16	THE	4.72	THE	6.42
T	9.02	IN	1.54	ING	1.42	OF	4.02
O	8.21	ER	1.33	AND	1.13	AND	3.15
A	7.81	RE	1.30	ION	1.00	TO	2.36
N	7.28	AN	1.08	ENT	0.98	A	2.09
I	6.77	HE	1.08	FOR	0.76	IN	1.77
R	6.64	AR	1.02	TIO	0.75	THAT	1.25
S	6.46	EN	1.02	ERE	0.69	IS	1.03
H	5.85	TI	1.02	HER	0.68	I	0.94
D	4.11	TE	0.98	ATE	0.66	IT	0.93
L	3.60	AT	0.88	VER	0.63	FOR	0.77
C	2.93	ON	0.84	TER	0.62	AS	0.76
F	2.88	HA	0.84	THA	0.62	WITH	0.76
U	2.77	OU	0.72	ATI	0.59	WAS	0.72
M	2.62	IT	0.71	HAT	0.55	HIS	0.71
P	2.15	ES	0.69	ERS	0.54	HE	0.71
Y	1.51	ST	0.68	HIS	0.52	BE	0.63
W	1.49	OR	0.68	RES	0.50	NOT	0.61
G	1.39	NT	0.67	ILL	0.47	BY	0.57
B	1.28	HI	0.66	ARE	0.46	BUT	0.56
V	1.00	EA	0.64	CON	0.45	HAVE	0.55
K	0.42	VE	0.64	NCE	0.43	YOU	0.55
X	0.30	CO	0.59	ALL	0.44	WHICH	0.53
J	0.23	DE	0.55	EVE	0.44	ARE	0.50
Q	0.14	RA	0.55	ITH	0.44	ON	0.47
Z	0.09	RO	0.55	TED	0.44	OR	0.45

23.18 Internet Security

As more information is stored on the Internet, and the amount of secure information, such as credit transfers and database transfers, increases, the need for a secure transmission mechanism also increases. The Internet has outgrown its founding protocol, HTTP. There thus has to be increased security in:

- Data encryption of WWW pages. This provides for secret information to be encrypted with a secret key.
- Message integrity. This provides a method of validating that the transmitted message is valid and has not been changed, either in transmission or in storage.
- Server authentication. This provides a method in which a server is authenticated to a client, to stop hackers pretending that they are the accessed server.
- Client authentication. This provides a method in which the client is authenticated to the server, to stop hackers from accessing a restricted server.

The main methods used are Secure Socket Layer (SSL) which was developed by Netscape, and Secure-HTTP (S-HTTP) which was developed by Enterprise Integration Technologies. Both are now being considered as International standards.

Two main problems are:

- Protection of transmitted information. For example, a person could access a book club over the Internet and then send information on the book and also credit-card information. The Book Club is likely to be a reputable company but criminals who can simply monitor the connection between the user and the Book Club could infiltrate the credit-card information.
- Protection of the client's computer. The Internet has little inherent security for the programs which can be downloaded from it. Thus, with no security, programs could be run or files can be download which could damage the local computer.

23.18.1 Secure Socket Layer (SSL)

Many WWW sites now state that they implement SSL. SSL was developed by Netscape and has now been submitted to the W3 Consortium for its acceptance as an International standard.

SSL allows the information stored and the server to be authenticated. Its advantages are:

- Open and non-propriety protocol.
- Data encryption, server authentication and message integrity.
- Firewall compatibility.
- Tunneling connections.
- Supports S-MIME (Secure-MIME).

Figure 23.13 shows how SSL fits into the OSI model. It can be seen that it interfaces directly to TCP/IP. This thus has the advantage that it makes programs and high-level protocols secure, such as ftp, telnet, SMTP and HTTP. Most browsers now support SSL, and many servers also support it. Figure 23.14 shows that the current browser supports SSL Version 2 and also Version 3.

Digital certificates

SSL supports independent certificate publishing authorities for server certificate authentication. These certificates have been validated by a reputable body and verify that the site is secure and genuine. These certificates can include Network server authentication, Network client authentication, Secure e-mail and Software Publishing, and can be viewed when connecting to the server. Typical certifying authorities are:

- ATT Certificate Services and ATT Directory Services.
- GTE Cyber Trust Root.

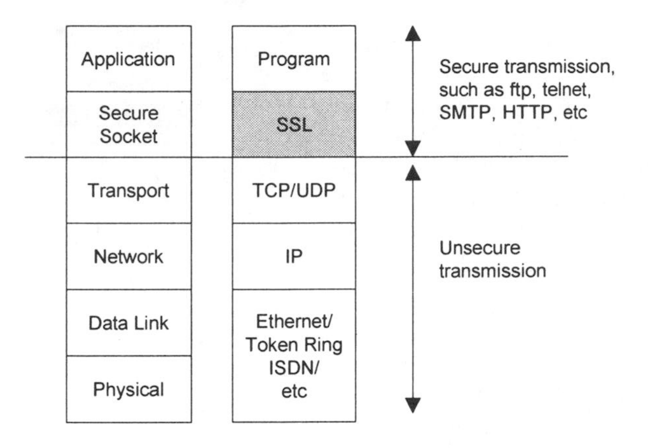

Figure 23.13 SSL and the OSI model

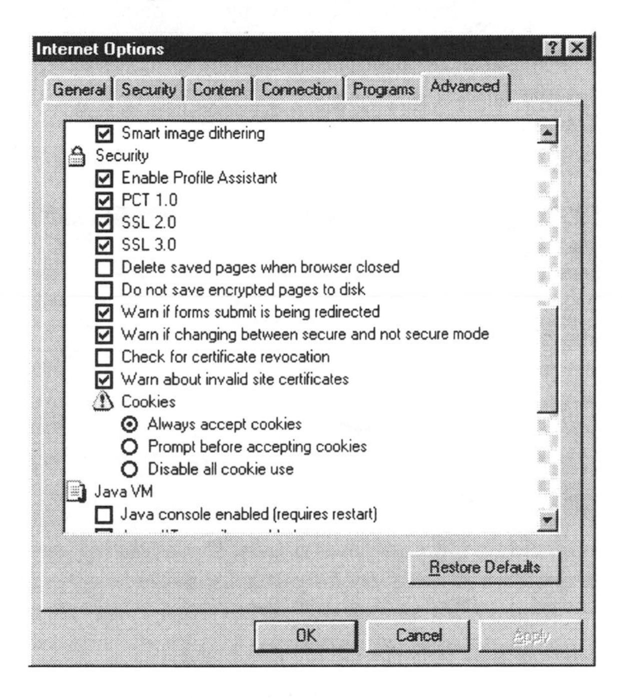

Figure 23.14 Security options

- internetMCI.
- Keywitness Canada.
- Microsoft Authenticode (TM) Root, Microsoft Root Authority, Microsoft Root SGC Authority and Microsoft Timestamp Root.
- Thawte Personal Basic CA, Thawte Personal Free-mail CA, Thawte Personal Premium CA and Thawte Personal Server CA.
- VeriSign Class 1 Primary CA, VeriSign Class 2 Primary CA, VeriSign Class 2 Primary CA, VeriSign Class 2 Primary CA, VeriSign Commercial Software Publishers CA, VeriSign Individual Software Publishers CA, VeriSign Time Stamping CA, VeriSign/RSA Commercial CA and VeriSign/RSA Secure Server CA.

The WWW browser checks the certificate to see if it is valid. This includes checking the date that the certificate was issued, the site fingerprint (as show in Figure 23.15). The current date should be later than the issue date of the certificate. There is also an expiry date. If any of the information is not current and valid, the browser displays a warning.

For example, for VeriSign/RSA Commercial:

Subject:	US, "RSA Data Security, Inc.", Commercial Certification Authority
Issuer:	US, "RSA Data Security, Inc.", Commercial Certification Authority
Effective date:	04/11/94 18:58:34
Expiration date:	03/11/99 18:58:34
Fingerprint:	5A:0B:DD:42:9E:B2:B4:62:97:32:7F:7F:0A:AA:9A:39

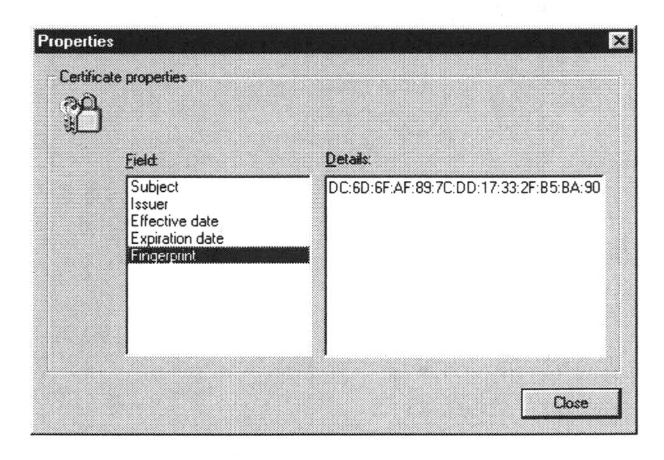

Figure 23.15 Fingerprint

23.18.2 S-HTTP

HTTP supports the communication of multimedia information, such as audio, video, graphics and text. Unfortunately it is not a secure method of transmission and an external person can tap it into. To overcome this, S-HTTP has been developed to provide improved security. Its features are:

- It is an extension to HTTP and uses HTTP-style headers.
- It incorporates encryption.
- It supports digital signatures for authentication.
- It supports certificates and key signing.

The Internet Engineering Task Force (IETF) is now considering S-HTTP for a standard method of transmitting secure HTTP over the Internet.

Message creation
An HTTP message contains both the message header and the message body. The header defines the type coding used in the body, its format, and so on. S-HTTP uses an encrypted message body and a message header, which includes the method that can be used to decrypt the message body. The process is as follows:

- The S-HTTP server (such as a WWW server) obtains the plain-text message of the information to send to the client (either locally or via a proxy). The message is either an HTTP message or some other data object (such as, a database entry or a graphical image).
- The server then encrypts the message body using the client's cryptographic preferences and the key provided by the client. The preference and key is passed in the initial handshake connection. This preference must also match the server's supported encryption techniques. The client normally sends a list of supported techniques and the server picks the most preferred one.
- The server sends the encrypted message to the client, in the same way as an HTTP transaction would occur, from which the client recovers the original message.

Message recovery
The encrypted file S-HTTP message is transmitted to the client, who then reads the message header to determine encryption technique. The client then decrypts it using the required private key, as shown in Figure 23.16. After this, the client displays the encapsulated HTTP or other data within the client's browser.

Normally in an HTTP transaction, the server terminates the connection after the transmission of the HTTP data. In S-HTTP this does not occur, the server does not terminate the connection until the browser tells it so. This is because there is no need to perform a handshake, and thus the encryption key remains valid.

Browser and servers
SSL supports server authentication and also secure connections. This is achieved with:

- Secure transmission. Secure sites have URLs which begins with `https://`, where the `s` stands for secure. Other secure protocols are:

  - HTTPS (for HTTP).
  - NNTPS (NNTP, news server).

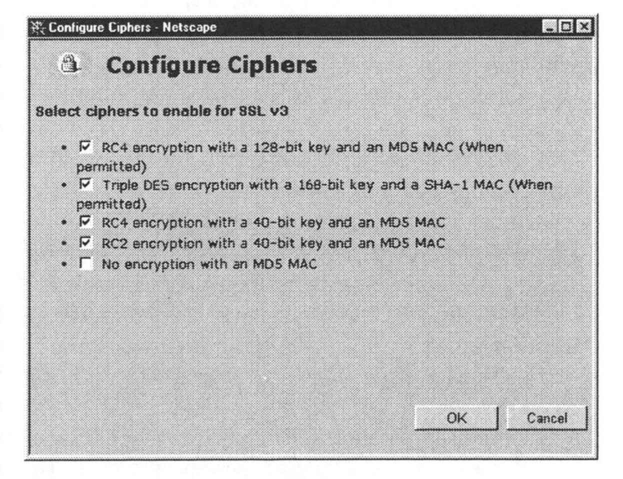

Netscape Navigator and Microsoft Internet Explorer both show the connection to secure site. Navigator shows it with key at the bottom of the window (a broken key shows a non-secure site), and Netscape Communicator and Internet Explorer displays a padlock. The figure on the right-hand side shows the configuration for SSL Version 3.

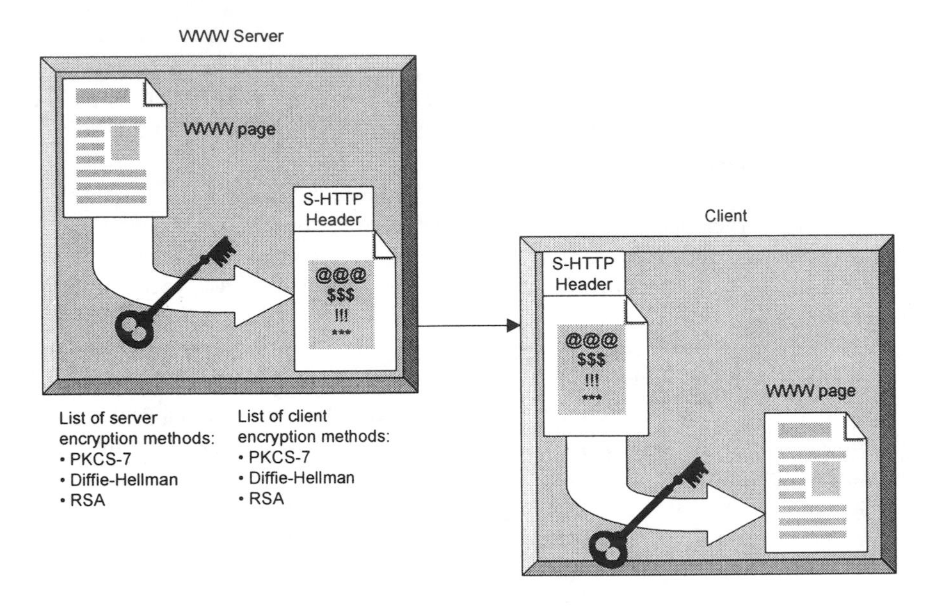

Figure 23.16 Encryption of WWW page

23.18.3 Content advisor

A major problem with the Internet is the access to an almost unlimited amount of information, some of which may be unsuitable for certain users. An important consideration is the protection of children from unsuitable material, especially those of a violent or sexual content. Content advisors, which are built into browsers, allow a method of controlling the access to certain WWW pages. The ratings of WWW pages include four content types: language, nudity, sex and violence. The level of content can also be set and Table 23.11 gives different levels within each of the types. For example language is split into five levels: Inoffensive slang (least likely to offend), mild expletives, moderate expletives, obscene gestures and explicit or crude language (most likely to offend). These ratings are industry-standards and have been defined independently by the Platform for Internet Content Selection (PICS) committee. It should be noted that not all Internet content is rated, although hopefully, in the future, all the material from certified sites will have a rating. This is the only true way to protect users from objectionable content. Every parent whose child has access to the Internet, and site managers (especially in schools) should have knowledge of these ratings and set the system up so that it protects innocent minds. Children should not be left to their own devices on what they can, and what they cannot, be able to view. National laws cannot properly legislate against the content of material which is located in a foreign country.

Methods, which can be used to reduce the problem, are:

- Restrict access of WWW browsers to certified sites and set the content ratings for these sites.
- Log all DNS access. Many sites have a local DNS which tries to resolve IP addresses to domain names. Many of these programs allow a log to be kept to log DNS enquires and the IP address of the computer which accessed the site. The system manager can occasionally view the file to determine if certain users are abusing their privileges. A real-time trace can then be applied to catch these users (often a warning is the best medicine, followed by formal procedures if the accesses continue).

- Issue clear statements to all users, normally with large typed notices, which clearly state the sites and the type of content which should not be accessed. This is important from a legal point of view, and could be used as evidence that the organization has a clear statement on Internet access. Organizations who fail to do this are in danger of being held responsible for objectionable accesses.

The best policing policy is one of trust and of educating users. Unfortunately, a minority of users are tempted by the ease of access to objectionable information.

Table 23.11 Content advisor level settings

Level	*Language*	*Nudity*	*Sex*	*Violence*
0	Inoffensive slang	None	No sexual activity portrayed/Romance.	No aggressive violence; No natural or accidental violence.
1	Mild expletives. Or mild terms for body functions.	Revealing attire	Passionate kissing	Creatures injured or killed; damage to realistic objects.
2	Expletives; non-sexual anatomical references.	Partial nudity	Clothed sexual touching	Humans or creatures injured or killed. Rewards injuring non-threatening creatures.
3	Strong, vulgar language; obscene gestures. Use of epithets.	Frontal nudity	Non-explicit sexual touching	Humans injured or killed.
4	Extreme hate, speech or crude language. Explicit sexual references.	Provocative nudity	Explicit sexual actual	Wanton and gratuitous violence

23.18.4 Security zones

Security zones support different levels of security for different areas of the Web to protect your computer. Each zone has a suitable security level, this is shown on the right-hand side of the Internet Explorer status bar. The browser checks the WWW sites zone when a page is opened or downloaded. The four zones are:

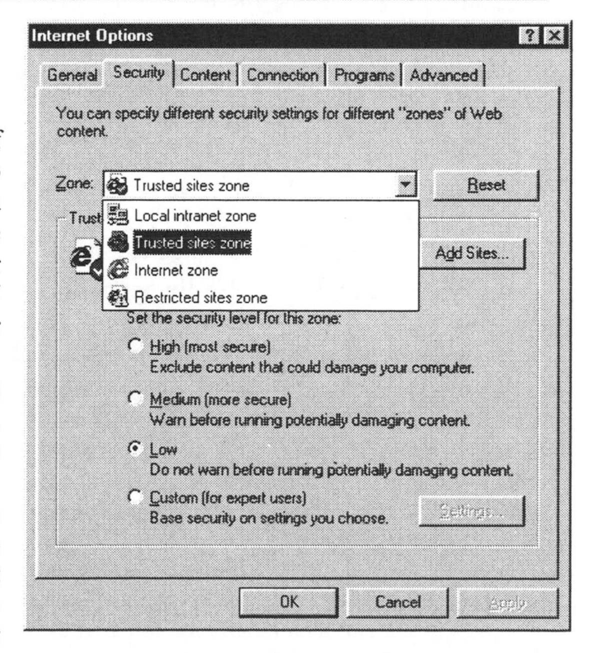

- **Local Intranet.** This is any address that doesn't require a proxy server. The default security level for this zone is Medium.
- **Trusted Sites.** This defines the sites that are most trusted, in which files can be run or downloaded without worrying about damaging the computer. The default security level for this zone is Low.

- **Restricted Sites.** This defines the sites that cannot be trusted. The default security level for this zone is High.
- **Internet.** This zone contains anything that is not on your computer or an intranet, or assigned to any other zone. The default security level for this zone is Medium.

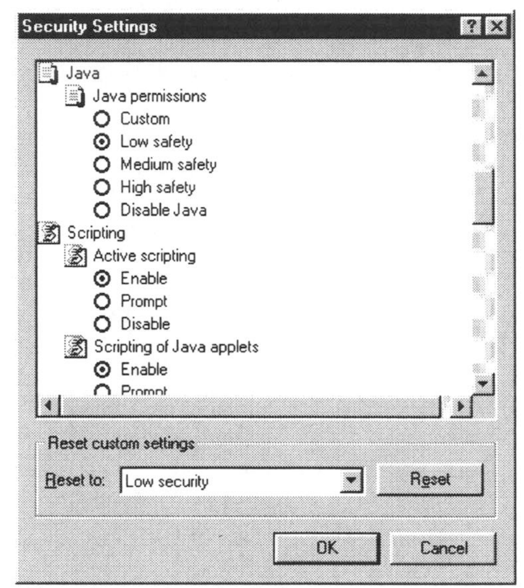

Sites with a low security rate will generally be safe to run or download programs, whereas, they should be avoided in restricted sites. The security ratings are:

- High (most secure). Exclude content that could damage the local computer.
- Medium (more secure). Warn before running potentially damaging content.
- Low. Do not warn before running potentially damaging content.
- Custom. Allows users to set site security.

Security settings include:

- Active X controls and plugins.
- Java. Sets Java permissions. See Figure on the right-hand side.
- Downloads. Enable file downloads (enable or disable) and font downloads (enable, disable or prompt).
- User authentication. Logon details, such as Automatic logon, Anonymous login, Prompt for user name and password and Automatic Logon with current user name and password.
- Miscellaneous. Such as drag-and-drop and installation of desktop files.

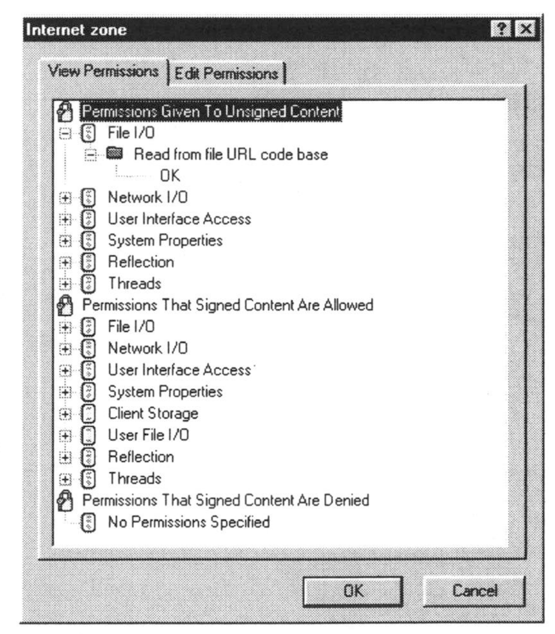

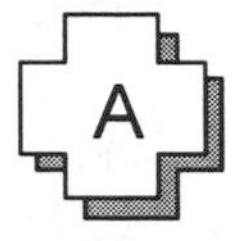

Glossary

100Base-FX IEEE-defined standard for 100 Mbps Ethernet using multimode fiber-optic cable.

100Base-TX (802.3u) IEEE-defined standard for 100 Mbps Ethernet using two pairs of Cat-5 twisted-pair cable.

100VG-AnyLAN HP-derived network architecture based on the IEEE 802.12 standard that uses 100 Mbps transmission rates. It uses a centrally controlled access method referred to as the Demand Priority Protocol (DPP), where the end node requests permission to transmit and the hub determines which node may do so, depending on the priority of the traffic.

10BASE-T IEEE-defined standard for 10 Mbps Ethernet using twisted-pair cables.

802.10 IEEE-defined standard for LAN security. It is sometimes used by network switches as a VLAN protocol and uses a technique where frames on any LAN carry a virtual LAN identification. For large networks this can be modified to provided security over the Internet.

802.12 Demand Priority Protocol
IEEE-defined standard of transmitting 100 Mbps over voice grade (telephone) twisted-pair cabling. *See* 100VG-AnyLAN.

802.1d IEEE-defined bridging standard for Spanning Tree protocol that is used to determine factors on how bridges (or switches) forward packets and avoid networking loops. Networks which use redundant loops (for alternative routes) need to implement the IEEE 802.1d standard to stop packets from looping forever.

802.2 A set of IEEE-defined specifications for Logical Link Control (LLC) layer. It provides some network functions and interfaces the IEEE 802.5, or IEEE 802.3, standards to the transport layer.

802.3 IEEE-defined standard for CSMA/CD networks. IEEE 802.3 is the most popular implementation of Ethernet.

802.3u IEEE-defined standard for 100 Mbps Fast Ethernet. It also covers a technique called autosensing which allows 100 Mbps devices to connecting to 10 Mbps devices.

802.4 IEEE-defined token bus specifications.

802.5 IEEE-defined standard for token ring networks.

AAL ATM adaptation layer. A service-dependent sublayer of the data link layer, which accepts data from different applications and presents it to the ATM layer as a 48-byte ATM payload segment. AALs have two sublayers: CS and SAR. There are four types of AAL, recommended by the ITU-T, these are: AAL1, AAL2, AAL3/4, and AAL5.

AAL1 ATM adaptation layer 1. Connection-oriented, delay-sensitive services requiring constant bit rates, such as uncompressed video and other isochronous traffic.

AAL2 ATM adaptation layer 2. Connection-oriented services that support a variable bit rate, such as some isochronous video and voice traffic.

AAL3/4 ATM adaptation layer 3/4. Connectionless and connection-oriented links, but is primarily used for the transmission of SMDS packets over ATM networks.

AAL5 ATM adaptation layer 5. Connection-oriented, VBR services, and is used predominantly for the transfer of classical IP over ATM and LANE traffic.

ABM Asynchronous Balanced Mode. An HDLC communication mode supporting peer-oriented, point-to-point communications between two nodes, where either station can initiate transmission.

ABR Available bit rate. A QoS class defined by the ATM Forum for ATM networks. ABR is used for connections that do not require a timing relationships between source and destination. It also provides no guarantees in terms of cell loss or delay, providing only best-effort service.

Access method The method that network devices use to access the network medium.

Acknowledgment Notification sent from one network device to another to acknowledge an event.

Active hub Multiported device that amplifies LAN transmission signals.

Active monitor A device which is responsible for managing a Token Ring. A node becomes the active monitor if it has the highest MAC address on the ring, and is responsible for such management tasks, such as ensuring that tokens are not lost, or that frames do not circulate indefinitely.

Adapter Device which usually connects a node onto a network, normally called a network interface adapter (NIC).

Adaptive cut-through switching
A forwarding technique on a switch which determines when the error count on frames received has exceeded the pre-configured limits. When this count is exceeded, it modifies its own operating state so that it no longer performs cut-through switching and goes into a store-and-forward mode. The cut-through method is extremely fast but suffers from the inability to check the CRC field. Thus if incorrect frames are transmitted they could have severe effects on the network segment. This is overcome with an adaptive cut-through switch by checking the CRC as the frame moves through the switch. When errors become too great the switch implements a store-and-forward method.

Adaptive delta modulation PCM
Similar to delta modulation PCM, but uses a number of bits to code the slope of the signal.

Adaptive Huffman coding Uses a variable Huffman coding technique which responds to local changes in probabilities.

Address aging The time that a dynamic address stays in the address routing table of a bridge or switch.

ARP Address Resolution Protocol. TCP/IP process which dynamically binds an IP address to a hardware address (such as an Ethernet MAC address). It can only operate across a single network segment.

Address resolution Resolves the data link layer address from the network layer address.

Address tables These are used by routers, switches and hubs to store either physical (such as MAC addresses) or higher-level addresses (such as IP addresses). The tables map node addresses to network addresses or physical domains. These address tables are dynamic and change due to nodes moving around the network.

Address A unique label for the location of data or the identity of a communications device. This address can either be numeric or alphanumeric.

Administrative distance Rating of the trustworthiness of a routing information source (typically between 0 and 255). The higher the value, the lower the trustworthiness rating.

Advertising Method used by routers where routing or service updates are sent at specified intervals so that other routers on the network can maintain lists of usable routes.

Agent A program which allows users to configure or fault-find nodes on a network, and also a program that processes queries and returns replies on behalf of an application.

Aging The removal of an address from the address table of a router or switch that is no longer referenced to forward a packet.

A-law The ITU-T companding standard used in the conversion between analog and digital signals in PCM systems. Used mainly in European telephone networks.

Alignment error In Ethernet, an error that occurs when the total number of bits of a received frame is not divisible by eight.

AM Amplitude modulation. Modulation technique which represents the data as the amplitude of a carrier signal.

ANSI American National Standards Institute. ANSI is a non-profit making organization which is made up of expert committees that publish standards for national industries.

ASCII American Standard Code for Information Interchange. An ANSI-defined character alphabet which has since been adopted as a standard international alphabet for the interchange of characters.

AM Amplitude modulation. Information is contained in the amplitude of a carrier.

ASK Amplitude-Shift Keying. Uses two, or more, amplitudes to represent binary digits. Typically used to transmit binary over speech-limited channels.

AppleTalk Series of communications protocols designed by Apple Computer.

Application layer The highest layer of the OSI model.

ARP Address Resolution Protocol. Internet protocol used to map an IP address to a MAC address.

ARPA Advanced Research Projects Agency. Research and development organization that is part of DoD. ARPA evolved into DARPA, but have since changed back to ARPA.

ARPANET Advanced Research Projects Agency Network, which was developed in the 1970s (funded by ARPA, then DARPA).

ASN.1 Abstract Syntax Notation One. OSI language for describing data types independent of particular computer structures and representation techniques.

Asynchronous transmission Transmission where individual characters are sent one-by-one. Normally each character is delimited by a start and a stop bit. With asynchronous communications the transmitter and receiver only have to be roughly synchronized.

Asynchronous Communication which does not depend on a clock.

ATM Asynchronous Transfer Mode. Networking technology which involves sending 53-byte fast packets (ATM cell), as specified by the ANSI T1S1 subcommittee. The first 5 bytes are the header and the remaining bytes are the information field which can hold 48 bytes of data. Optionally the data can contain a 4-byte ATM adaptation layer and 44 bytes of actual data. The ATM adaptation layer field allows for fragmentation and reassembly of cells into larger packets at the source and destination respectively. The control field also contains bits which specify whether this is a flow control cell or an ordinary data cell, a bit to indicate whether this packet can be deleted in a congested network, and so on.

ATM Forum Promotes standards-based implementation agreements for ATM technology.

ATM layer Service-independent sublayer of the data link layer in an ATM network. The ATM layer receives the 48-byte payload segments from the AAL and attaches a 5-byte header to each, producing standard 53-byte ATM cells.

Attenuation Loss of communication signal energy.

AUI Attachment unit interface. In Ethernet, it is the interface between an MAU and a NIC (network interface card).

Automatic broadcast control Technique which minimizes broadcast and multicast traffic flooding through a switch. A switch acts as a proxy server and screens previously resolved ARP. This eliminates broadcasts associated with them.

Autonegotiation Technique used by an IEEE 802.3u node which determines whether a device that it is receiving or transmitting data in one of a number of Ethernet modes (100Base-TX, 100Base-TX Full Duplex, 10Base-T, 10Base-T Full Duplex or 100Base-T4). When the mode is learned, the device then adjusts to the required transmission speed.

Autonomous system A collection of networks which have a common administration and share a common routing strategy. Each autonomous system is assigned a unique 16-bit number by the IANA.

Autosensing Used by a 100Base-TX device to determine if the incoming data is transmitted at 10 Mbps or 100 Mbps.

Back pressure Technique which slows the incoming data rate into the buffer of a 802.3 port preventing it from receiving too much data. Switches which implement back pressure will transmit a jam signal to stop data input.

Backbone cabling Cabling interconnects wiring closets, wiring closets, and between buildings.

Backbone The primary path for networked traffic.

Backoff The retransmission delay enforced when a collision occurs.

BACP Bandwidth allocation control protocol. Protocol which monitors network traffic and allows or disallows access to users, depending on their needs. It is awaiting approval by the IETF.

Bandwidth In an analogue system it is defined as the range of frequencies contained in a signal. As an approximation it is the difference between the highest and lowest frequency in the signal. In a digital transmission system it is normally quoted as bits per second.

Baseband Data transmission using unmodulated signals.

BRI Basic rate interface. Connection between ISDN and the user. It has three separate channels, one D-channel (which carries control information) and two B channels (which carry data).

Baud rate The number of signaling elements sent per second with RS-232, or modem, communications. In RS-232 the baud rate is equal to the bit-rate. With modems, two or more bits can be encoded as a single signaling element, such as 2 bits being represented by four different phase shifts (or one signaling element). The signaling element could change its amplitude, frequency or phase-shift to increase the bit-rate. Thus the bit-rate is a better measure of information transfer.

BER Bit error rate. The ratio of received bits that contain errors.

BGP Border Gateway Protocol. Interdomain routing protocol that replaces EGP.

Big-endian Method of storing or transmitting data in which the most significant bit or byte is presented first.

Bit stuffing The insertion of extra bits to prevent the appearance of a defined sequence. In HDLC the bit sequence 01111110 delimits the start and end of a frame. Bit stuffing stops this bit sequence from occurring anywhere in the frame by the receiver inserting a 0 whenever there are five consecutive 1's transmitted. At the receiver if five consecutive 1's are followed by a 0 then the 0 is deleted.

BNC A commonly used connector for coaxial cable.

BOOTP A standard TCP/IP protocol which allows nodes to be dynamically allocated an IP address from an Ethernet MAC address.

Border gateway Router that communicates with routers in other autonomous systems.

Bridge A device which physically links two or more networks using the same communications protocols, such as Ethernet/Ethernet or token ring/token ring. It allows for the filtering of data between network segments.

Broadband Data transmission using multiplexed data using an analogue signal or high-frequency electromagnetic waves.

Broadcast address Special address reserved for sending a message to all stations. Generally, a broadcast address is a MAC destination address of all ones.

Broadcast domain Network where broadcasts can be reported to all nodes on the network bounded by routers. Broadcast packets cannot traverse a router.

Broadcast storm Flood of broadcast packets generated by a broadcast transmission where high numbers of receivers are targeted for a long period of time.

Broadcast Data packet that will be sent to all nodes on a network. Broadcasts are identified by a broadcast address.

BSD Berkeley Standard Distribution. Term used to describe any of a variety of UNIX-type operating systems.

Buffer A temporary-storage space in memory.

Bus A network topology where all nodes share a common transmission medium.

Byte A group of eight bits.

Capacity The maximum data rate in Mbps.

Cat-1 cable Used for telephone communications and is not suitable for transmitting data.

Cat-2 cable Used for transmitting data at speeds up to 4 Mbps.

Cat-3 cable An EIA/TIA-568 wiring standard for unshielded or shielded twisted pair cables. Up to 10 Mbps.

Cat-4 cable Used in Token Ring networks and can transmit data at speeds up to 16 Mbps.

Cat-5 cable An EIA/TIA-568 wiring standard for unshielded or shielded twisted-pair cables for the transmission of over 100 Mbps.

CBR Constant bit rate. QOS class defined by the ATM Forum for ATM networks and is used for connections that depend on precise clocking to ensure undistorted delivery.

CCITT Consultative Committee for International Telegraph and Telephone. International organization responsible for the development of communications standards. Now named ITU-T.

CDP Cisco Discovery Protocol. Used in Cisco routers, bridges and switches to pass information on the connected networks.

Cell relay Networking technology based on the use of small, fixed-size packets, or cells.

Cell The basic unit for ATM switching and multiplexing. Cells contain identifiers that specify the data stream to which they belong. Each cell consists of a 5-byte header and 48 bytes of payload.

CEPT Conférence Européenne des Postes et des Télécommunications. Association

CHAP Challenge-handshake authentication protocol. Identification method used by PPP to determine the originator of a connection.

Cheapernet IEEE 802.3 10Base2 standard.

Checksum An error-detection scheme in which bits are grouped to form integer values which are then summated. Normally, the negative of this value is then added as a checksum. At the receiver, all the grouped values and the checksum are summated and, in the absence of errors, the result should be zero.

Circuit switching Switching system in which a dedicated physical circuit path must exist between sender and receiver for the call duration

Cisco IOS software Cisco Internetwork Operating System software. Provides an operating system for a Cisco router.

Client Node or program that connects to a server node or program.

CLP Cell loss priority. Field in the ATM cell header that determines the probability of a cell being dropped if the network becomes congested.

Coaxial cable A transmission medium consisting of one or more central wire conductors surrounded by an insulating layer and encased in either a wire mesh or extruded metal sheathing. It supports RF frequencies from 50 to about 500 MHz. It comes in either a 10-mm diameter (thick coax) or a 5-mm diameter (thin coax).

Collapsed backbone Nondistributed backbone in which all network segments are interconnected by way of an internetworking device.

Collision domain The network area within which frames that have collided are propagated. Repeaters and hubs propagate collisions, but switches, bridges and routers do not.

Collision Occurs when one or more devices try to transmit over an Ethernet network simultaneously.

Connectionless Describes data transfer without the existence of a virtual circuit.

Connection-oriented Describes data transfer that requires the establishment of a virtual circuit. See also connectionless.

Contention Access method in which network devices compete to get access the physical medium.

Convergence The speed and ability of a group of internetworking devices running a specific routing protocol to agree on the topology of an internetwork after a change in that topology.

CDDI Copper distributed data interface. FDDI over copper.

Cost An arbitrary value used by routers to compare different routes. Typically it is measured by hop counts, typical time delays or bandwidth.

Count to infinity Occurs in routing algorithms that are slow to converge, where routers continuously increment the hop count to particular networks. It is typically overcome by setting an arbitrary hop-count limit.

CRC Cyclic Redundancy Check. An error-detection scheme.

Cross-talk Interference noise caused by conductors radiating electromagnetic radiation to couple into other conductors.

CSMA/CD Carrier sense multiple access collision detect. Media-access method in which nodes contend to get access to the common bus. If the bus is free of traffic (Carrier Sense) any of the nodes can transmit (Multiple Access). If two nodes gain access at the same time then a collision occurs (Collision Detection). A collision then occurs, and the nodes causing the collision then wait for a random period of time before they retransmit. CSMA/CD access is used by Ethernet and IEEE 802.3.

Cut sheet Rough diagram indicating where cable runs are located and the numbers of rooms they lead to.

Cut-through switching Technique where a switching device directs a packet to the destination port(s) as soon as it receives the destination and source address scanned from the packet header.

DARPA Defense Advanced Research Projects Agency. US government agency that funded research for and experimentation with the Internet.

Data link layer Second layer of the OSI model which is responsible for link, error and flow control. It normally covers the framing of data packets, error control and physical addressing. Typical data link layers include Ethernet and FDDI.

Data stream All data transmitted through a communications line in a single read or write operation.

Datagram Logical grouping of information sent as a network layer unit over a transmission medium without prior establishment of a virtual circuit. IP datagrams are the primary information units in the Internet.

DCE Data communications equipment. These are devices and connections of a communications network that comprise the network end of the user-to-network interface, such as modems and cables.

Decorative raceway Wall-mounted channel with removable cover used to support horizontal cabling.

Delta modulation PCM Uses a single-bit code to represent the analogue signal. A 1 is transmission when the current sample increases its level, else a 0 is transmitted. Delta modulation PCM requires a higher sampling rate that the Nyquist rate, but the actual bit rate is normally lower.

Destination MAC address A 6-byte data unique of the destination MAC address. It is normally quoted as a 12-digit hexadecimal number (such as A5:B2:10:64:01:44).

Destination network address A unique Internet Protocol (IP) or Internet Packet Exchange (IPX) address of the destination node.

Differential encoding Source coding method which is used to code the difference between two samples. Typically used in real-time signals where there is limited change between one sample and the next, such as in audio and speech.

Distance vector routing algorithm Routing algorithms which use the number of hops in a route to find a shortest-path spanning tree. With distance vector routing algorithms, each router to send its entire routing table in each update, but only to its neighbors. They be prone to routing loops, but are relatively simple as compared with link state routing algorithms.

DNS Domain Naming System. Used on the Internet to translated domain names into IP addresses.

Dot address Notation for IP addresses in the form <w.x.y.z> where each number represents, in decimal, 1 byte of the 4-byte IP address.

DQDB Distributed Queue Dual Bus. Data link layer communication protocol, specified in the IEEE 802.6 standard, designed for use in MANs.

DTE Data terminal equipment. Device at the user end of a user-network interface that is a data source, destination, or both.

Dual homing Topology where devices connect to the network by two independent access points (points of attachment). One gives the primary connection, and the other is the standby connection that is activated in the event of a failure of the primary connection.

Dynamic address resolution Use of an address resolution protocol to determine and store address information on demand.

DHCP Dynamic host control protocol. It manages a pool of IP addresses for computers without a known IP address. This allows a finite number of IP addresses to be reused quickly and efficiently by many clients.

Dynamic routing Routing that adjusts automatically to network topology or traffic changes.

E1 Wide-area digital transmission scheme that is used in Europe to carry data at a rate of 2.048 Mbps.

Early token release Used in Token Ring networks that allows stations to release the token onto the ring immediately after transmitting, instead of waiting for the first frame to return.

EGP Exterior Gateway Protocol. Internet protocol for exchanging routing information between autonomous systems (RFC904). Replaced by BGP.

EIA Electronic Industries Association. Specifies electrical transmission standards.

EIA/TIA-232 Physical layer interface standard that supports unbalanced circuits at signal speeds of up to 64 kbps.

EIA/TIA-449 Physical layer interface for rates up to 2 Mbps.

EIA/TIA-568 Characteristics and applications for UTP cabling.

EIA/TIA-606 Standard for the telecommunications infrastructure of commercial buildings, such as terminations, media, pathways, spaces and grounding.

Encapsulation Wrapping of data in a particular protocol header.

End system An end-user device on a network.

Entity An individual, manageable network device.

Entropy coding Coding scheme which does not take into account the characteristics of the data and treats all the bits in the same way. It produces lossless coding. Typical methods used are statistical encoding and suppressing repetitive sequences.

Equalization Used to compensate for communications channel distortions.

Ethernet address 48-bit number that identifies a node on an Ethernet network. Ethernet addresses are assigned by the Xerox Corporation.

Ethernet A local area network which uses coaxial, twisted-pair or fiber-optic cable as a communication medium. It transmits at a rate of 10 Mbps and was developed by DEC, Intel and Xerox Corporation. The IEEE 802.3 network standard is based upon Ethernet.

ETSI European Telecommunication Standards Institute. Created by the European PTTs and the European Community (EC) for telecommunications standards in Europe.

Even parity An error-detection scheme where defined bit-groupings have an even number of 1's.

EBCDIC Extended Binary Coded Decimal Interchange Code. An 8-bit code alphabet developed by IBM allowing 256 different bit patterns for character definitions.

Exterior gateway protocol Any internetwork protocol that exchanges routing information between autonomous systems.

Fast Ethernet See IEEE 802.3u standard.

Fat pipe Term used to indicate a high level of bandwidth the defined port.

FDDI Fiber Distributed Data Interface. A standard network technology that uses a dual counter-rotating token-passing fiber ring. It operates at 100 Mbps and provides for reliable backbone connections.

File server Computer that allows the sharing of files over a network.

FTP File transfer protocol. A protocol for transmitting files between host computers using the TCP/IP protocol.

Firewall Device which filters incoming and outgoing traffic.

Flow control Procedure to regulate the flow of data between two nodes.

Forward adaptive bit allocation This technique is used in audio compression and makes bit allocation decisions adaptively, depending on signal content.

Fragment free cut-through switching A modified cut-through switching technique where a switch or switch module waits until it has received a large enough packet to determine if it is error free.

FCS Frame check sequence. Standard error detection scheme.

Frame Normally associated with a packet which has layer 2 information added to it. Packets are thus contained within frames. Frames and packets have variable lengths as opposed to cells which have fixed lengths.

FSK Frequency-shift Keying. Uses two, or more, frequencies to represent binary digits. Typically used to transmit binary data over speech-limited channels.

Full duplex Simultaneous, two-way communications.

Gateway A device that connects networks using different communications protocols, such as between Ethernet and FDDI. It provides protocol translation, in contrast to a bridge which connects two networks that are of the same protocol.

GIF Standard image compression technique which is copyrighted by CompuServe Incorporated. It uses LZW compression and supports a palette of 256 24-bit colors (16.7M colors). GIF support local and global color tables and animated images.

Half-duplex (HDX) Two-way communications, one at a time.

Handshake Messages or signals exchanged between two or more network devices to ensure transmission synchronization.

Handshaking A reliable method for two devices to pass data.

HCC Horizontal cross-connect. Wiring closet where the horizontal cabling connects to a patch panel which is connected by backbone cabling to the main distribution facility.

HDLC ISO standard for the data link layer.

Hello packet Message transmitted from a root bridge to all other bridges in the network to constantly verify the Spanning Tree setup.

Heterogeneous network Network consisting of dissimilar devices that run dissimilar protocols.

Hierarchical routing Routing based on a hierarchical addressing system. IP has a hierarchical structure as they use network numbers, subnet numbers, and host numbers.

Holddown A router state where they will not advertise information on a specific route, nor accept advertisements about the route for a specific length of time (the holddown period). This time is used to flush bad information about a route from all routers in the network, or when a fault occurs on a route.

Hop count Used by the RIP routing protocol to measure the distance between a source and a destination.

Hop The number of gateways and routers in a transmission path.

Host number Part of an IP address which identifies the node on a subnetwork.

Host A computer that communicates over a network. A host can both initiate communications and respond to communications that are addressed to it.

Hub A hub is a concentration point for data and repeats data from one node to all other connected nodes. Hubs can be active (where they repeat signals sent through them) or passive (where they do not repeat, but merely split, signals sent through them).

Huffman coding Uses a variable length code for each of the elements within the data. It normally analyses the probability of the element in the data and codes the most probable with fewer bits than the least probable.

Hybrid network Internetwork made up of more than one type of network technology.

HTML Hypertext markup language. Standard language that allows the integration of text and images over a distributed network.

IAB Internet Architecture Board. A group that discusses important matters relating to the Internet.

IANA Internet Assigned Numbers Authority. Organization which delegates authority for IP address-space allocation and domain-name assignment to the NIC and other organizations.

ICMP Internet Control Message Protocol. Used to report errors and provides other information relevant to IP packet processing.

IETF Internet Engineering Task Force. Consists of a number of working groups which are responsible for developing Internet standards.

IGP Interior Gateway Protocol. Used to exchange routing information within an autonomous system.

IGRP Interior Gateway Routing Protocol. Developed by Cisco for large and heterogeneous networks.

ISDN Integrated systems digital network. Communication technology that contains two data channels (2B) and a control channel (H). It supports two 64 kbps data channels and sets up a circuit-switched connection.

ITU-TSS International Telegraph Union Telecommunications Standards Sector. Organization which has replaced the CCITT.

Internet address An address that conforms to the DARPA-defined Internet protocol. A unique, four byte number identifies a host or gateway on the Internet. This consists of a network number followed by a host number. The host number can be further divided into a subnet number.

IETF Internet Engineering Task Force. A committee that reviews and supports Internet protocol proposals.

Internet Connection of nodes on a global network which use a DARPA-defined Internet address.

internet Two or more connected networks that may, or may not, use the same communication protocol.

Intranet A company specific network which has additional security against external users.

Inverse ARP Inverse Address Resolution Protocol. This is a method of building dynamic routes in a network, and allows an access server to discover the network address of a device associated with a virtual circuit.

IP (Internet Protocol) Part of the TCP/IP which provides for node addressing.

IP address An address which is used to identify a node on the Internet.

IP multicast Addressing technique that allows IP traffic to be propagated from one source to a group of destinations.

IPX Internet Packet Exchange. Novell NetWare communications protocol which is similar to the IP protocol. The packets include network addresses and can be routed from one network to another.

IPX address Station address on a Novell NetWare network. It consists of two fields: a network number field and a node number field. The node number is the station address of the device and the network number is assigned to the network when the network is started up. It is written in the form: NNNNNNNN:XXXXXX-XXXXXX, where N's represent the network number and X's represent the station address. An example of an IPX address is: DC105333:542C10-FF1432.

ISO International Standards Organization.

Isochronous transmission Asynchronous transmission over a synchronous data link. Isochronous signals require a constant bit rate for reliable transport.

ITU-T The Consultative Committee for International Telephone and Telegraph (now known at the ITU-TSS) is an advisory committee established by the United Nations. It attempts to establish standards for inter-country data transmission on a worldwide basis.

Jabber Occurs when the transmission of network signals exceeds the maximum allowable transmission time (20 ms to 150 ms). The medium becomes overrun with data packets. This is caused by a faulty node or wiring connection.

Jitter Movement of the edges of pulse over time, that may introduce error and loss of synchronization.

JPEG Image compression technique defined by the Joint Photographic Expert Group (JPEG), a subcommittee of the ISO/IEC. It uses a DCT, quantization, run-length and Huffman coding.

Keepalive interval Time period between each keepalive message.

Latency Defines the amount of time between a device receiving data and it being forwarded on. Hubs have the lowest latency (less than 10 μs), switches the next lowest (between 40 μs and 60 μs), then bridges (200 μs to 300 μs) and routers have the highest latency (around 1000 μs).

Learning bridge Bridge which learns the connected nodes to it. It uses this information to forward or drop frames.

Leased line A permanent telephone line connection reserved exclusively by the leased customer. There is no need for any connection and disconnection procedures.

Lempel-Ziv coding Coding method which takes into account repetition in phases, words or parts of words. It uses pointers to refer to previously defined sequences.

LZW coding Lempel-Ziv Welsh coding. Coding method which takes into account repetition in phases, words or parts of words. It builds up a dictionary of previously sent (or stored) sequences.

Line driver A device which converts an electrical signal to a form that is transmittable over a transmission line. Typically, it provides the required power, current and timing characteristics.

Link layer Layer 2 of the OSI model.

Link segment A point-to-point link terminated on either side by a repeater. Nodes cannot be attached to a link segment.

Link state routing algorithm Routing algorithm where each router broadcasts or multicasts information regarding the cost of reaching each of its neighbors to all nodes in the internetwork. These algorithms create a consistent view of the network but are much more complete that distance vector routing algorithms).

Little-endian Storage method in which the least byte is stored first.

LLC Logical Link Control. Higher of the two data link layer sublayers defined by the IEEE, which provides error control, flow control, framing, and MAC-sublayer addressing (IEEE 802.2).

Lossless compression Where information, once uncompressed, is identical to the original uncompressed data.

Lossy compression Where information, once uncompressed, cannot be fully recovered.

LSA Link-state advertisement. Used by link-state protocols to advertise information about neighbors and path costs.

MAC address A 6-byte data unique data-link layer address. It is normally quoted as a 12-digit hexadecimal number (such as A5:B2:10:64:01:44).

Masking effect Where noise is only heard by a person when there are no other sounds to mask it.

MDI Medium Dependent Interface. The IEEE standard for the twisted-pair interface to 10Base-T (or 100Base-TX).

MAC Media Access Control. Media-specific access-control for Token Ring and Ethernet.

MIC Media Interface Controller. Media-specific access-control for Token Ring and Ethernet.

MAU Medium Attachment Unit. Method of converting digital data into a form which can be transmitted over a band-limited channel. Methods use either ASK, FSK, PSK or a mixture of ASK, FSK and PSK.

Microsegmentation Division of a network into smaller segments. This helps to increase aggregate bandwidth to network devices.

Modem Modulator-Demodulator. A device which converts binary digits into a form which can be transmitted over a speech-limited transmission channel.

MTU Maximum Transmission Unit. The largest packet that the IP protocol will send through the selected interface or segment.

Multicast Packets which are sent to all nodes on a subnet of a group within a network. This differs from a broadcast which forwards packet to all users on the network.

Multimode fiber Fiber-optic cable that has the ability to carry more than one frequency (mode) of light at a time.

NDIS Network driver interface specification. Software specification for network adapter drivers. It supports multiple protocols and multiple adapters, and is used in many operating systems, such as Windows 95/88/NT.

Network layer Third layer of the OSI model, which is responsible for ensuring that data passed to it from the transport layer is routed and delivered through the network. It provides end-to-end addressing and routing. It provides support for a number of protocols, including IP, IPX, CLNP, X.25, or DDP.

NT1 Network termination. Network termination for ISDN.

NFS Network File System. Standard defined by Sun Microsystems for accessing remote file systems over a network.

NIS Network Information Service. Standard defined by Sun Microsystems for the administration of network-wide databases.

NLM NetWare Loadable Module. Program that can be loaded into the NetWare NOS.

Node Any point in a network which provides communications services or where devices interconnect.

N-series connectors Connector used with thick coaxial cable.

Octet Same as a byte, a group of eight bits (typically used in communications terminology).

Odd parity An error-detection scheme where a defined bit-grouping has an odd number of 1's.

ODLI Open Data-Link Interface. Software specification for network adapter drivers used in NetWare and Apple networks. It supports multiple protocols and multiple adapters.

Optical repeater A device that receives, restores, and re-times signals from one optical-fiber segment to another.

Packet switching — Network switching in which data is processed in units of whole packets rather than attempting to process data by dividing packets into fixed-length cells.

Packet — A sequence of binary digits that is transmitted as a unit in a computer network. A packet usually contains control information and data. They normally are contained with data link frames.

PAP — Password authentication protocol. Protocol which checks a user's password.

Patch panel — An assembly of pin locations and ports which are typically mounted on a rack or wall bracket in the wiring closet.

PLL — Phase-Locked Loop. Tunes into a small range of frequencies in a signal and follows any variations in them.

PSK — Phase-Shift Keying. Uses two, or more, phase-shifts to represent binary digits. Typically used to transmit binary data over speech-limited channels.

Physical layer — Lowest layer of the OSI model which is responsible for the electrical, mechanical, and handshaking procedures over the interface that connects a device to a transmission medium

Ping — Standard protocol used to determine if TCP/IP nodes are alive. Initially a node sends an ICMP (Internet Control Message Protocol) echo request packet to the remote node with the specified IP address and waits for echo response packets to return.

POP — Point of presence. Physical access point to a long distance carrier interchange.

PPP — Point-to-point protocol. Standard protocol to transfer data over the Internet asynchronously or synchronously.

Port — Physical connection on a bridge or hub that connects to a network, node or other device.

POST — Power-on self test. Hardware diagnostics that runs on a hardware device when that device is powered up.

Protocol — Specification for coding of messages exchanged between two communications processes.

Quantization — Involves converting an analogue level into a discrete quantized level. The number of bits used in the quantization process determines the number of quantization levels.

Quartet signaling — Signaling technique used in 100VG-AnyLAN networks that allows data transmission at 100 Mbps over frame pairs of UTP cabling.

Repeater — A device that receives, restores, and re-times signals from one segment of a network and passes them on to another. Both segments must have the same type of transmission medium and share the same set of protocols. A repeater cannot translate protocols.

RARP — Reverse address resolution protocol. The opposite of ARP which maps an IP address to a MAC address.

RJ-45 — Connector used with US telephones and with twisted-pair cables. It is also used in ISDN networks, hubs and switches.

RMON — An SNMP MIB that specifies the types of information listed in a number of special MIB groups that are commonly used for traffic management. Some of the popular groups used are Statistics, History, Alarms, Hosts, Hosts Top N, Matrix, Filters, Events, and Packet Capture.

Routing node A node that transmits packets between similar networks. A node that transmits packets between dissimilar networks is called a gateway.

RS-232C EIA-defined standard for serial communications.

RS-422, 423 EIA-defined standard which uses higher transmission rates and cable lengths than RS-232.

RS-449 EIA-defined standard for the interface between a DTE and DCE for 9- and 37-way D-type connectors.

RS-485 EIA-defined standard which is similar to RS-422 but uses a balanced connection.

RLE Run-length encoding. Coding technique which represents long runs of a certain bit sequence with a special character.

SAP Service Access Point. Field defined by the IEEE 802.2 specification that is part of the address specification.

SAP Service Advertisement Protocol. Used by the IPX protocol to provide a means of informing network clients, via routers and servers of available network resources and services.

Segment A segment is any length of LAN cable terminated at both ends. In a bus network, segments are electrically continuous pieces of the bus, connected by repeaters. It can also be bounded by bridges and routers.

SLIP Serial line internet protocol. A standard used for the point-to-point serial connections running TCP/IP.

Simplex One-way communication.

SNMP Simple Network Management Protocol. Standard protocol for managing network devices, such as hubs, bridges, and switches.

Source encoding Coding method which takes into account the characteristics of the information. Typically used in motion video and still image compression.

Statistical encoding Where the coding analyses the statistical pattern of the data. Commonly occurring data is coded with a few bits and uncommon data by a large number of bits.

Suppressing repetitive sequences Compression technique where long sequences of the same data is compressed with a short code.

Switch A very fast, low-latency, multiport bridge that is used to segment local area networks.

Synchronous Data which is synchronized by a clock.

T1 Digital WAN carrier facility for 1.544 Mbps transmission.

TCP Part of the TCP/IP protocol which provides an error-free connection between two cooperating programs.

TCP/IP Internet An Internet is made up of networks of nodes that can communicate with each other using TCP/IP protocols.

Telnet Standard program which allows remote users to log into a station using the TCP/IP protocol.

TIFF Graphics format that supports many different types of images in a number of modes. It is supported by most packages and, in one mode, provides for enhanced high-resolution images with 48-bit color.

Time to live A field in the IP header which defines the number of routers that a packet is allowed to traverse before being discarded.

Token A token transmits data around a token ring network.

Topology The physical and logical geometry governing placement of nodes on a network.

Transceiver A device that transmits and receives signals.

Transform encoding Source-encoding scheme where the data is transformed by a mathematical transform in order to reduce the transmitted (or stored) data. A typical technique is the discrete cosine transform (DCT) and the fast Fourier transform (FFT).

Transport layer Fourth layer of the OSI model. It allows end-to-end control of transmitted data and the optimized use of network resources.

UART Universal asynchronous receiver transmitter. Device which converts parallel data into a serial form, which can be transmitted over a serial line, and vice-versa.

V.24 ITU-T-defined specification, similar to RS-232C.

V.25bis ITU-T specification describing procedures for call set-up and disconnection over the DTE-DCE interface in a PSDN.

V.32/V.32bis ITU-T standard serial communication for bi-directional data transmissions at speeds of 4.8 or 9.6 Kbps, or 14.4 Kbps for V.32bis.

V.34 Improved v.32 specification with higher transmission rates (28.8 Kbps) and enhanced data compression.

V.35 ITU-T standard describing a synchronous, physical layer protocol used for communications between a network access device and a packet network.

V.42 ITU-T standard protocol for error correction.

VLC-LZW code Variable-length-code LZW code. Uses a variation of LZW coding where variable-length codes are used to replace patterns detected in the original data.

Vertical cabling Backbone cabling.

Virtual circuit Logical circuit which connects two networked devices together.

Workgroup Collection of nodes on a LAN which exchange data with each other.

X.121 ITU-T standard for an addressing scheme used in X.25 networks.

X.21 ITU-T-defined specification for the interconnection of DTEs and DCEs for synchronous communications.

X.21bis ITU-T standard for the physical layer protocol for communication between DCE and DTE in an X.25 network.

X.25 ITU-T-defined for packet-switched network connections.

X.28 ITU-T recommendation for terminal-to-PAD interface in X.25 networks.

X.29 ITU-T recommendation for control information in the terminal-to-PAD interface used in X.25 networks.

X.3 ITU-T recommendation for PAD parameters used in X.25 networks.

X.400 ITU-T recommendation for electronic mail transfer.

X.500 ITU-T recommendation for distributed maintenance of files and directories.

X3T9.5 ANSI Task Group definition of FDDI.

X-ON/ X-OFF The Transmitter On/ Transmitter Off characters are used to control the flow of information between two nodes.

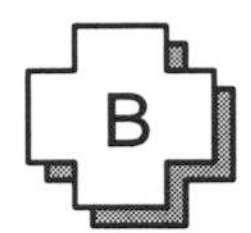

Quick Reference

B.1 Abbreviations

AA	auto-answer
AAL	ATM adaptation layer
AAN	autonomously attached network
ABM	asynchronous balanced mode
AbMAN	Aberdeen MAN
ABNF	augmented BNF
AC	access control
ACAP	application configuration access protocol
ACK	acknowledge
ACL	access control list
ADC	analogue-to-digital converter
ADPCM	adaptive delta pulse code modulation
ADPCM	adaptive differential pulse code modulation
AEP	AppleTalk Echo Protocol
AES	audio engineering society
AFI	authority and format identifier
AGENTX	agent extensibility protocol
AGP	accelerated graphics port
AM	amplitude modulation
AMI	alternative mark inversion
ANSI	American National Standards Institute
APCM	adaptive pulse code modulation
API	application program interface
ARM	asynchronous response mode
ARP	address resolution protocol
ARPA	Advanced Research Projects Agency
AS	Autonomous system
ASCII	American standard code for information exchange
ASK	amplitude-shift keying
AT	attention
ATM	asynchronous transfer mode
AUI	attachment unit interface
BCC	blind carbon copy
BCD	binary coded decimal
BGP	border gateway protocol
BIOS	basic input/output system
B-ISDN	broadband ISDN
BMP	bitmapped
BNC	British Naval Connector
BOM	beginning of message
BOOTP	bootstrap protocol
BPDU	bridge protocol data units
bps	bits per second

BVCP	Banyan Vines control protocol
CAD	computer-aided design
CAN	concentrated area network
CASE	common applications service elements
CATNIP	common architecture for the Internet
CC	carbon copy
CCITT	International Telegraph and Telephone Consultative
CD	carrier detect
CD	compact disk
CDE	common desktop environment
CDFS	CD file system
CD-R	CD-recordable
CD-ROM	compact disk – read-only memory
CF	control field
CGI	common gateway interface
CGM	computer graphics metafile
CHAP	challenge handshake authentication protocol
CHAP	Challenge Handshake Authentication Protocol
CHARGEN	character generator protocol
CIF	common interface format
CMC	common mail call
CMOS	complementary MOS
CN	common name
COM	continuation of message
CON-MD5	content-MD5 header field
CPCS	convergence protocol communications sublayer
CPI	common part indicator
CPSR	computer professionals for social responsibility
CPU	central processing unit
CRC	cyclic redundancy check
CRLF	carriage return, line feed
CRT	cathode ray tube
CSDN	circuit-switched data network
CSMA	carrier sense multiple access
CSMA/CA	CSMA with collision avoidance
CSMA/CD	CSMA with collision detection
CS-MUX	circuit-switched multiplexer
CSPDN	circuit-switched public data network
CTS	clear to send
DA	destination address
DAA	digest access authentication
DAC	digital-to-analogue converter
DAC	dual attachment concentrator
DARPA	Defense Advanced Research Projects Agency
DAS	dual attachment station
DASS	distributed authentication security
DAT	digital audio tape
DAYTIME	daytime protocol
dB	decibel
DBF	NetBEUI frame
DC	direct current
DCC	digital compact cassette
DCD	data carrier detect
DCE	data circuit-terminating equipment

DC-MIB	dial control MIB
DCT	discrete cosine transform
DD	double density
DDE	dynamic data exchange
DENI	Department of Education for Northern Ireland
DES	data encryption standard
DHCP	dynamic host configuration program
DIB	device-independent bitmaps
DIB	directory information base
DISC	disconnect
DISCARD	discard protocol
DLC	data link control
DLL	dynamic link library
DM	disconnect mode
DMA	direct memory access
DNS	domain name server
DNS-SEC	domain name system security extensions
DOS	disk operating system
DPCM	differential PCM
DPSK	differential phase-shift keying
DQDB	distributed queue dual bus
DR	dynamic range
DRAM	dynamic RAM
DSN	delivery status notifications
DSP	domain specific part
DSS	digital signature standard
DTE	data terminal equipment
DTR	data terminal ready
EASE	embedded advanced sampling environment
EaStMAN	Edinburgh/Stirling MAN
EBCDIC	extended binary coded decimal interchange code
EBU	European Broadcast Union
ECHO	echo protocol
ECP	extended communications port
EEPROM	electrically erasable PROM
EF	empty flag
EFF	electronic frontier foundation
EFM	eight-to-fourteen modulation
EGP	exterior gateway protocol
EIA	Electrical Industries Association
EISA	extended international standard interface
EMF	enhanced metafile
ENQ	inquiry
EOM	end of message
EOT	end of transmission
EPP	enhanced parallel port
EPROM	erasable PROM
EPS	encapsulated postscript
ESP	IP encapsulating security payload
ETB	end of transmitted block
ETHER-MIB	ethernet MIB
ETX	end of text
FAT	file allocation table
FATMAN	Fife and Tayside MAN

FAX	facsimile
FC	frame control
FCS	frame check sequence
FDDI	fiber distributed data interface
FDDI-MIB	FDDI management information base
FDM	frequency division multiplexing
FDX	full duplex
FEC	forward error correction
FF	full flag
FFIF	file format for internet fax
FIFO	first in, first out
FINGER	finger protocol
FM	frequency modulation
FRMR	frame reject
FS	frame status
FSK	frequency-shift keying
FTP	file transfer protocol
FYI	for your information
GFI	group format identifier
GGP	gateway-gateway protocol
GIF	graphics interface format
GQOS	guaranteed quality of service
GSSAP	generic security service application
GUI	graphical user interface
HAL	hardware abstraction layer
HD	high density
HDB3	high-density bipolar code no. 3
HDLC	high-level data link control
HDTV	high-definition television
HDX	half duplex
HEFCE	Higher Education Funding Councils of England
HEFCW	Higher Education Funding Councils of Wales
HF	high frequency
HMUX	hybrid multiplexer
HPFS	high performance file system
HTML	Hypertext Mark-up Language
HTTP	Hypertext Transfer Protocol
Hz	Hertz
I/O	input/output
IA5	international alphabet no. 5
IAB	Internet Advisory Board
IAP	internet access provider
IARP	inverse ARP
IBM	International Business Machines
ICMP	internet control message protocol
ICP	internet connectivity provider
IDEA	international data encryption algorithm
IDENT	identification Protocol
IDI	initial domain identifier
IDP	initial domain part
IDPR	inter-domain policy routing
IEEE	Institute of Electrical and Electronic Engineers
IEFF	Internet Engineering Task Force
IFS	installable file system

IGMP	Internet group management protocol
IGMP	Internet group multicast protocol
IGP	interior gateway protocol
ILD	injector laser diode
IMAC	isochronous MAC
IMAP	Internet message access protocol
IOS	input/output supervisor
IP	Internet protocol
IP-ARC	IP over ARCNET networks
IP-ARPA	IP over ARPANET
IP-ATM	IP over ATM
IP-CMPRS	IP with compressed headers
IP-DC	IP over DC Networks
IP-E	IP over ethernet networks
IP-EE	IP over experimental ethernet networks
IP-FDDI	IP over FDDI networks
IP-FR	IP over frame relay
IP-HC	IP over hyperchannel
IP-HIPPI	IP over HIPPI
IP-IEEE	IP over IEEE 802
IP-IPX	IP over IPX networks
IP-MTU	path MTU discovery
IP-NETBIOS	IP over NETBIOS
IPNG	IP next generation
IPP	internet presence provider
IP-SLIP	IP over serial lines
IP-SMDS	IP datagrams over SMDS
IP-TR-MC	IP Multicast over token-ring LANs
IPV6-FDDI	IPv6 over FDDI
IPv6-Jumbo	IPv6 Jumbograms
IPV6-PPP	IPv6 over PPP
IP-WB	IP over wideband network
IPX	Internet packet exchange
IP-X.25	IP over ISDN
IPX-IP	IPX over IP
IRQ	interrupt request
ISA	international standard interface
ISDN	integrated services digital network
IS-IS	immediate system to intermediate system
ISO	International Standards Organization
ISP	internet service provider
ITOT	ISO transport service on top of TCP
ITU	International Telecommunications Union
JANET	joint academic network
JFIF	JPEG file interchange format
JISC	Joint Information Systems Committee
JPEG	Joint Photographic Expert Group
KDC	key distribution centre
KERBEROS	Kerberos network authentication service
LAN	local area network
LAPB	link access procedure balanced
LAPD	link access procedure
LCN	logical channel number
LDAP-URL	LDAP URL Format

LD-CELP	low-delay code excited linear prediction
LED	light emitting diode
LGN	logical group number
LIP	large IPX packets
LLC	logical link control
LRC	longitudinal redundancy check
LSL	link support level
LSP	link state protocol
LSRR	loose source and record route
LZ	Lempel-Ziv
LZW	LZ-Welsh
MAC	media access control
MAIL-MIB	mail monitoring MIB
MAN	metropolitan area network
MAP	messaging API
MAU	multi-station access unit
MD	message digest
MDCT	modified discrete cosine transform
MDI	media dependent interface
MHS	message handling service
MIB-II	management information base-II
MIC	media interface connector
MIME	multi-purpose internet mail extension
MLID	multi-link interface driver
MODEM	modulation/demodulator
MOS	metal oxide semiconductor
MPEG	Motion Picture Experts Group
MPI	multi-precision integer
MSL	maximum segment lifetime
MTP	multicast transport protocol
NAK	negative acknowledge
NCP	NetWare control protocols
NCSA	National Center for Supercomputer Applications
NDIS	network device interface standard
NDS	Novell Directory Services
NETBEUI	NetBIOS extended user interface
NETFAX	network file format for the exchange of images
NHRP	next hop resolution protocol
NIC	network interface card
NICNAME	whois protocol
NIS	network information system
NLSP	netware link-state routing protocol
NNTP	network news transfer protocol
NRZI	non-return to zero with inversion
NSAP	network service access point
NSCA	National Center for Supercomputing Applications
NSM-MIB	network services monitoring MIB
NSS	named service server
NTE	network terminal equipment
NTFS	NT file system
NTP	network time protocol
NTSC	National Television Standards Committee
ODI	open data-link interface
OH	off-hook

ONE-PASS	one-time password system
OSI	open systems interconnection
OSI-UDP	OSI TS on UDP
OSPF	open shortest path first
OUI	originator's unique identifier
PA	point of attachment
PAL	phase alternation line
PAP	password authentication protocol
PC	personal computer
PCM	pulse code modulation
PCT	personal communications technology
PDN	public data network
PGP	pretty good privacy
PHY	physical layer protocol
PING	packet Internet gopher
PISO	parallel-in-serial-out
PKP	public-key partners
PLL	phase-locked loop
PLS	physical signaling
PMA	physical medium attachment
PMD	physical medium dependent
POP3	post office protocol, Version 3
POP-URL	POP URL Scheme
PPP	point-to-point protocol
PPP-AAL	PPP over AAL
PPP-CCP	PPP compression control protocol
PPP-CHAP	PPP challenge handshake authentication
PPP-EAP	PPP extensible authentication protocol
PPP-HDLC	PPP in HDLC framing
PPP-IPCP	PPP control protocol
PPP-ISDN	PPP over ISDN
PPP-LINK	PPP link quality monitoring
PPP-MP	PPP multilink protocol
PPP-NBFCP	PPP NetBIOS frames control protocol
PPP-SNACP	PPP SNA control protocol
PPP-SONET	PPP over SONET/SDH
PPP-X25	PPP in X.25
PPSDN	public packet-switched data network
PS	postscript
PSDN	packet-switched data network
PSE	packet switched exchange
PSK	phase-shift keying
PSTN	public-switched telephone network
QAM	quadrature amplitude modulation
QCIF	quarter common interface format
QIC	quarter inch cartridge
QoS	quality of service
QT	quicktime
QUOTE	quote of the day protocol
RADIUS	remote authentication dial-in service
RAID	redundant array of inexpensive disks
RAM	random-access memory
RD	receive data
REJ	reject

RFC	request for comment
RGB	red, green and blue
RI	ring in
RIF	routing information field
RIP	routing information protocol
RIP2-MD5	RIP-2 MD5 Authentication
RIP2-MIB	RIP Version 2 MIB Extension
RIPNG-IPV6	RIPng for IPv6
RIP-TRIG	Trigger RIP
RLE	run-length encoding
RMON	remote monitoring
RMON-MIB	remote network monitoring MIB
RNR	receiver not ready
RO	ring out
ROM	read-only memory
RPC	remote procedure call
RPSL	routing policy specification language
RR	receiver ready
RSA	Rivest, Shamir and Adleman
RSVP	resource reservation protocol
RTF	rich text format
RTMP	routing table maintenance protocol
RTP	real-time transport protocol
RTSP	real-time streaming protocol
S/PDIF	Sony/Philips digital interface format
SA	source address
SABME	set asynchronous balanced mode extended
SAC	single attachment concentrator
SAP	service advertising protocol
SAPI	service access point identifier
SAR	segment and reassemble
SARPDU	segmentation and reassembly protocol data unit
SAS	single attachment station
SASL	simple authentication and security layer
SASL-ANON	anonymous SASL mechanism
SB-ADCMP	sub-band ADPCM
SCMS	serial copy management system
SCSI	small computer systems interface
SCSP	server cache synchronization protocol
SD	sending data
SD	start delimiter
SDH	synchronous digital hierarchy
SDIF	Sony digital interface
SDLC	synchronous data link control
SDNSDU	secure domain name system dynamic update
SDP	session description protocol
SECAM	séquential couleur à mémoire
SEL	selector/extension local address
SHEFC	Scottish Higher Education Funding Council
SIPO	serial-in parallel-out
SIPP	simple Internet protocol plus
SLM-APP	system-level managed objects for applications
SLP	service location protocol
SMDS	switched multi-bit data stream

SMI	structure of management information
SMP	symmetrical multiprocessing
SMT	station management
SMTP	simple mail transfer protocol
SNA	serial number arithmetic
SNA	systems network architecture (IBM)
SND	send
SNMP	simple network management protocol
SNMP-AT	SNMP over AppleTalk
SNMP-IPX	SNMP over IPX
SNMP-OSI	SNMP over OSI
SNR	signal-to-noise ratio
SONET	synchronous optical network
SPKM	simple public-key GSS-API mechanism
SPX	sequenced packet exchange
SQTV	studio-quality television
SRAM	static RAM
SSL	secure socket layer
SSM	single sequence message
SSRR	strict source and record route
STA	spanning-tree architecture
STM	synchronous transfer mode
STP	shielded twisted-pair
SVGA	super VGA
TCB	transmission control block
TCC	transmission control code
TCP	transmission control protocol
TDAC	time-division aliasing cancellation
TDM	time-division multiplexing
TEI	terminal equipment identifier
TELNET	telnet protocol
TFTP	trivial file transfer protocol
TIFF	tag image file format
TIFF	tagged input file format
TIME	time server protocol
TIP	transaction internet protocol
TMUX	transport multiplexing protocol
TOS	type of service
TP-TCP	ISO transport service on top of the TCP
TR	transmit data
TSR	terminate and stay resident
TTL	time-to-live
TUBA	TCP and UDP with bigger addresses
UDP	user datagram protocol
UI	unnumbered information
UNI	universal network interface
UNI	user network interface
UPS	uninterruptable power supplies
URI	universal resource identifier
URL	uniform resource locator
USB	universal serial bus
USERS	active users protocol
UTF-8	UTF-8 transformation format of ISO 10646
UTP	unshielded twisted pair

UV	ultra violet
VCI	virtual circuit identifier
VCO	voltage controller oscillator
VCR	video cassette recorder
VDD	virtual device driver
VGA	variable graphics adapter
VIM	vendor-independent messaging
VLC-LZW	variable-length-code LZW
VLM	virtual loadable modules
VMM	virtual machine manager
VRC	vertical redundancy check
VRRP	virtual router redundancy protocol
WAIS	wide area information servers
WAN	wide area network
WIMPs	Windows, icons, menus and pointers
WINS	Windows Internet name service
WINSOCK	windows sockets
WORM	write-once read many
WWW	World Wide Web
XDR	external data representation
XOR	exclusive-OR
ZIP	Zone Information Protocol

B.2 Miscellaneous

NetBIOS name types

Microsoft networks identify computers by their NetBIOS name. Each is 16 characters long, and the 16th character represents the purpose of the name. An example list of a WINS database is:

Name		*Type*	*Status*
FRED	<00>	UNIQUE	Registered
BERT	<00>	UNIQUE	Registered
STAFF	<1C>	GROUP	Registered
STAFF	<1E>	GROUP	Registered

The values for the 16th byte are:

00	Workstation	03	Message service
06	RAS server service	1B	Domain master browser
1C	Domain group name	1D	Master browser's name
1E	Normal group name (workgroup)	1F	NetDDE service
20	Server service	21	RAS client
BE	Network Monitor Agent	BF	Network Monitor Utility

On Microsoft Windows, the names in the WINS database can be shown with the `nbstat` command.

Windows NT TCP/IP setup

Microsoft Windows uses the files LMHOSTS, HOSTS and NETWORKS to map TCP/IP names and network addresses. These are stored in the *<winNT root>*`\SYSTEM32\DRIVERS\ETC`. `LMHOSTS` maps IP addresses to a computer name. An example format is:

```
#IP-address          host-name
146.176.1.3          bills_pc
146.176.144.10       fred_pc      #DOM:STAFF
```

where comments have a preceding '#' symbol. To preserve compatibility with previous versions of Microsoft LAN Manager, special commands have been included after the comment symbol. These include:

```
#PRE
#DOM:domain
#INCLUDE fname
#BEGIN_ALTERNATE
#END_ALTERNATE
```

where

`#PRE` specifies that the name is preloaded into the memory of the computer and no further references to the `LMHOSTS` file will be made.

`#DOM:`*domain* specifies the name of the domain that the node belongs to.

`#BEGIN_ALTERNATE` and `#END_ALTERNATE` are used to group multiple `#include`'s

`#include` *fname* specifies other `LMHOST` files to include.

The `HOSTS` file format is IP address followed by the fully qualified name (FQDN) and then any aliases. Comments have a preceding '#' symbol. For example:

```
#IP Address        FQDN            Aliases
146.176.1.3        superjanet      janet
146.176.144.10     hp
146.176.145.21     mimas
146.176.144.11     mwave
146.176.144.13     vax
146.176.146.23     oberon
146.176.145.23     oberon
```

Microsoft Windows TCP/IP commands (quick reference)

Command	*Description*	*Examples*
`arp`	Modifies Address Resolution Protocol tables. `-s` *IP-address* [*MAC-address*] ; manually modify `-a` [*IP-address*] ; display ARP entry `-d` *IP-address* ; delete entry	`arp -s 146.176.151.10 FF-AA-10-3F-A1-3F`
`finger`	Queries users on a remote computer. *@hostname* ; name of remote computer `-l` ; extend list	`finger -l fred@miranda` `finger @moon`
`ftp`	Remote file transfer. After connection the following commands can be used:	`ftp intel.com`

	`ascii binary bye cd`
	`dir get hash help`
	`lcd ls mget mput`
	`open prompt pwd quit`
	`remote help user`	

`hostname`	Displays the TCP/IP hostname of the local node.	`hostname`
`ipconfig`	Displays the TCP/IP settings on the local computer. `/all` ; show all settings	`ipconfig /all`
`lpq`	Sends a query to a TCP/IP host or printer. `-S` *print_server* `-P` *printer*	`lpq -p lp_laser` `lpq -s mirands -p dot_matrix`
`lpr`	Prints to a TCP/IP-based printer. `-S` *print_server* `-P` *printer*	`lpr -p lp_laser file.ps`
`nbstat`	Displays mapping of NetBIOS names to IP addresses. `-a` *NetBIOS-name* ; display name table for ; computer `-A` *IP-address* ; display name table for ; computer `-n` ; display NetBIOS table of ; local computer	`nbstat -A freds`
`netstat`	Displays status of TCP/IP connections. `-p` *protocol* ; display for given protocol `-r` ; show routing tables `-s` ; display statistics `-R` ; reload `HMHOSTS` `-S` ; display NetBIOS sessions by ; NetBIOS names `-s` ; display NetBIOS sessions by ; IP addresses	*See Section 5.9.5*
`nslookup`	Queries DNS servers. After connection the following commands can be used: `help` `finger` [*username*] `port=`*port* `querytype=`*type* ; *type* can be A (address), ; CNAME (canonical name which is an alias for ; another host), MX (mail exchanger which ; handles mail for a given host), NS (name server ; for the domain), PTR (pointer record which ; maps an IP address to a hostname), SOA (start ; of authority record) or ANY.	*See Section 5.9.3*

<table>
<tr><td><code>ping</code></td><td>Test TCP/IP connectivity.

<code>-a</code> ; resolve IP addresses to hostnames
<code>-n</code> count ; set number of echo packets
<code>-l</code> size ; specify packet size
<code>-t</code> ; continuously ping
<code>-i</code> ttl ; set time-to-live field
<code>-w</code> timeout ; specify timeout in ms</td><td></td></tr>
<tr><td><code>rcp</code></td><td>Remote copy.

[hostname[. username]]
<code>-a</code> ; ASCII copy
<code>-b</code> ; binary copy
<code>-h</code> ; also hidden files
<code>-r</code> ; recursively copy</td><td><code>rcp -r *.txt</code>
<code>miranda.bill/home</code></td></tr>
<tr><td><code>rexec</code></td><td>Execute remote command.</td><td><code>rexec miranda -l bill "ls -</code>
<code>l"</code></td></tr>
<tr><td><code>route</code></td><td>Manipulates TCP/IP routing table.
<code>-f</code> ; delete all routes
<code>-p</code> ; make a permanent route
<code>add</code> ; add a route
<code>change</code> ; modify an existing route
<code>delete</code> ; delete a route
<code>gateway</code> ; specifies gateway
<code>mask</code> netmask ; define subnet mask
<code>print</code> ; print current table</td><td><code>route gateway</code>
<code>146.151.176.12</code></td></tr>
<tr><td><code>rsh</code></td><td>Executes remote shell.

<code>-l</code> username ; user name
<code>command</code> ; command to execute</td><td><code>rsh -l bill "ls -l"</code></td></tr>
<tr><td><code>telnet</code></td><td>Remote login.</td><td><code>telnet www.intel.com</code></td></tr>
<tr><td><code>tftp</code></td><td>Trivial FTP (uses UDP).</td><td></td></tr>
<tr><td><code>tracert</code></td><td>Trace route.

<code>-d</code> ; do not resolve IP addresses
<code>-h</code> max_hops ; maximum number of hops
<code>-w</code> timeout ; specify timeout</td><td></td></tr>
</table>

Windows NT system administration commands (quick reference)

<table>
<tr><th>Command</th><th>Description</th><th>Examples</th></tr>
<tr><td><code>at</code></td><td>Runs commands at a specified time. Options include:

<code>\\computer-name</code>
<code>time</code>
<code>/every:date</code> ; such as day of the week such
; as M/T/W/Th/F/S/Su or day
; of the month</td><td><code>at 14:00 \\freds</code>
<code>"cmd ping miranda ></code>
<code>log"</code>

<code>at 00:00 /every:M/W/F</code>
<code>"cmd lpr log.txt"</code></td></tr>
<tr><td><code>attrib</code></td><td>Displays or changes file attributes. Attributes include:</td><td><code>attrib +h test.txt</code></td></tr>
</table>

	`+r`, `-r`, (read) `+a`, `-a`, (archive) `+s`, `-s`, (system) `+h`, `-h`, (hidden) `/s` (include sub-directories)	
`backup`	Backup program.	
`cacls`	Command-line Access Control Lists (ACLs). `/g` *username*:*right* ; grant user the following ; rights: r (read), c (change), ; f (full control). `/p` *username* ; replace rights, these are as ; above, but n (none) is added `/r` *username* ; delete all rights `/t` ; recursive change	`calcs list.txt /g fred:cf` `calcs *.* /r bill /t`
`chkdsk`	Checks disk. Options include: `/f` ; automatically fix errors	`chkdsk c: /f`
`cmd`	Run command-line shell.	
`convert`	Converts drive partition from FAT to NTFS.	`convert d: /fs:ntfs`

`convlog`	Converts files from Microsoft Information Server, FTP server and Gopher servers, and produces log files in NSCA or EMWAC format.
	
	`-t` [*emwave* | *ncsa*] ; specify EMWAC or
	; NCSA
	`-s` [*f* | *w* | *g*] ; specify FTP (f), WWW (w) or
	; Gopher (g)
	`-o` *outdir* ; specify output directory
		`convlog -sg -ncsa -o c:\temp`
		
		`*.log`

`diskperf`	Toggles the disk performance counter.	
`ipxroute`	IPX routing. `servers` ; list NetWare servers	`ipxsroute servers`

`jetpack`	Compacts WINS databases.	`net stop wins`
		`jetpack win.mdb tmp.mdb`
		`net start wins`

`netmon`	Network monitoring tool.	
`ntbackup`	Backup file system.	
`rasadmin`	Remote Access Server (RAS) administration.	
`rasautou`	Remote Access Server (RAS) debugging.	
`rasdial`	Remote Access Server (RAS) dial-up. `/phone` *tel*; telephone number	`rasdial miranda` `/phone:1112222`

`rasphone`	Edit RAS phonebook.
`rdisk`	Create emergency repair disk.
`regedit`	Edit registry.
`restore`	Restores files after a backup.
`start`	Starts applications from the command line.
`winnt`	16-bit Window NT installation program. `/r` *dir* ; specify install directory `/s` *dir* ; installation source files
`winnt32`	32-bit Window NT installation program.

Microsoft Windows control services commands (quick reference)

Command	*Description*	*Example*
`net accounts`	Controls account settings `/domain` *dom* ; specify default domain	
`net computer`	Adds or deletes computers from current domain. `\\`*computer-name* `/add`; add computer `/del`; delete computer	`net computer \\freds /add` `net computer \\bills /del`
`net config server`	Configure server.	
`net config workstation`	Configure workstation.	
`net continue`	Unpauses a command that was paused with `net pause`.	
`net file`	Closes an opened file. When used on its own without arguments it gives the ID of all opened files. The `/close` option is used with the ID number to close a given file. `/close` ; close file	

`net group`	Creates, edits or deletes groups.
	
	`/add`
	; add new group or users to the named group
	; specified group
	`/delete`
	; delete group or users to the named group
	; specified group
		`net group "Staff" /add`
		`net group "Staff" /add fred`

`net help`	Help messages for net.	
`net helpmsg`	Detailed help for a given error message.	

Command	Description	Example
`net localgro`	Create or deletes local groups or local users.	
`net name`	Administers list of names for the Messenger service.	
`net pause`	Pauses a service.	`net pause lpdsvc` *; pause print service*
`net print`	Administers print queues. `\\`*computer-name* `/delete` ; delete job	
`net send`	Sends a text message to users or computers.	`net send bill "Hello"`
`net session`	Displays information of a current session.	
`net share`	Administers networks sharing.	
`net start`	Starts a service.	`net start snmp` *; start SNMP*
`net statisti`	Displays service statistics.	
`net stop`	Stops a service.	`net stop lpdsvc` *; stop print server*
`net time`	Sets or queries time on a remote computer.	
`net use`	Administers networked resources.	
`net user`	Administers user accounts. `password` ; prompts for password `/active:[y/n]` ; active status `/add` ; add user `/delete` ; delete user `/expires:[`*date*`|NEVER]` ; expire time `/fullname: "`*name*`"` ; full name `/homedir:` *homedirpath* ; home directory `/passwdchg: [y |n ]` ; password change `/times: [`*times* `| ALL]` ; login times	`net user bill_c /add` `net user bill_c /active:y`
`net view`	Displays networked resources. `\\`*computer-name* `/domain [`*domain*`]` ; list of domain or ; computers within the ; specified domain	`net view \\freds` `net view /domain`

FTP commands

ABOR	Abort previous command	**ACCT**	Specify account
ALLO	Allocate storage	**APPE**	Append to a file
CDUP	Go to directory above	**CWD**	Change working directory
DELE	Delete a file	**HELP**	Show help information
LIST	List directory (ls -l)	**MKD**	Make directory
MDTM	Show last modification time	**MODE**	Specify data transfer mode
NLST	Give name of list of files	**NOOP**	No operation (to prevent disconnecti

PASS	Specify password	**PASV**	Prepare for server-to-server transfer
PORT	Specify port	**PWD**	Display current working directory
QUIT	Quit session	**REST**	Restart incomplete session
RETR	Retrieve a file	**RMD**	Remove a directory
RNFR	Specify rename-from filename	**RNTO**	Spread rename-to filename
SITE	Non-standard commands	**SIZE**	Return size of file
STOR	Store a file	**STOU**	Store a file with a unique name
STRU	Specify data transfer structure	**SYST**	Show operation system type
TYPE	Specify data transfer type	**USER**	Specify user name
XCUP	Change of parent of current working directory	**XCWD**	Change working directory
XMKD	Make a directory	**XPWD**	Print a directory
XRMD	Remove a directory		

The `ftpd` daemon is run from the Internet daemon (`inetd` and `inetd.conf`). It is automatically run when there is a request from the ftp port which is specified in the `/etc/services` file (typically port 21). Users who use ftp must have an account in the `/etc/password` file and any users who are barred from ftp access should be listed in the `/etc/ftpusers` file.

Anonymous FTP logins are where the user logs in as `anonymous` (typically using their e-mail address as the password). The user then uses the `~ftp` directory for all transfers. Typically the structure is:

`~ftp/bin`	Contains a copy of the `/bin/ls` file.
`~ftp/etc`	Contains the `passwd`, `group` and `logingroup` files for the users which own files within ~ftp.
`~ftp/etc/passwd`	Defines users who own files within `~ftp`.
`~ftp/etc/group`	Defines groups who own files within `~ftp`.
`~ftp/pub`	General area where files can be uploaded to or downloaded from.

Note that anonymous FTP is inherently dangerous to system security. To protect against damage the `-l` option in the `ftpd` daemon can be used to log all accesses to the system log. The `-t` option in the `ftpd` daemon defines the timeout time for a session.

In order to permit anonymous FTP, a line similar to the following must be added to the `/etc/password` file:

```
ftp:*:400:10:anonymousFTP:/user/ftp:/bin/false
```

where, in this case, 400 is the unique user ID for the ftp login, 10 is the group ID number for the guest group and `/user/ftp` specifies the home directory for the anonymous ftp account.

Telnet

The Internet daemon (`inetd`) starts the `telnetd` daemon when there is a request from the telnet port which is specified in the `/etc/services` files (typically port 23). To start the `telnetd` the `/etc/inetd.conf` file must contain the line:

```
telnet stream tcp nowait root /etc/telnetd telnetd -b/etc/issue
```

/etc/inetd.conf

The /etc/inetd.conf is called from the Internet daemon (inet). Its general format is:

<service_name> <sock_type> <proto> <flags> <user> <server> <args>

where

service_name	Name of the specified service and is defined in the /etc/services file.
sock_type	Defines whether it is connection oriented (stream) or connectionless (datagram).
proto	Defines the protocol (TCP or UDP) and is defined in the /etc/protocols.
flags	Defines the action which should be taken upon detection of a connection. nowait defines that there should be no wait in starting up the service daemon, whereas wait informs it that it should wait for the service to start.
user	Defines the owner of the service.
server	Defines the name of the service.
arg	Defines addition arguments.

For example:

```
ftp     stream  tcp   nowait   root       /etc/ftpd     ftpd
telnet  stream  tcp   nowait   root       /etc/telnetd  telnetd
finger  stream  tcp   nowait   guest      /etc/fingerd  fingerd
login   stream  tcp   nowait   root       /etc/login    logind
```

/etc/inittab

The /etc/inittab file is called when the system is initially started. Its general format is:

label : *run_level* : *action* : *process*

where

label	Specified name.
run_level	Defines run level, such as: 0 (power down), 1 (administratively down), s (single user), 2 (multi-user), 3 (remote file sharing), 4 (unused), 5 (firmware) and 6 (system restart).
action	Control method that init uses over the execution of the process, such as boot (start up the specified process), bootwait (start up the process, but wait until the process has started before parsing the configuration file), initdefault (set the run level that init enters at system start up), off (if process is not running, ignore this line), once (start process once, and then leave it to run to its end), respawn (restart the process, if it has stopped) and wait (wait until the process completes before continuing).
process	Program that should be used.

An example is given next:

```
mess: 2:bootwait:/bin/cat /etc/message_of_the_day > /dev/console
r0: 1:wait:/etc/rc1
r1: 2: wait:/etc/rc2
go1:2:off:/etc/getty tty1
```

The Definition of an Upgrade: Take old bugs out, put new ones in.

B.3 Windows NT architecture

Windows NT/2000 uses two modes:

- Kernel mode. This is a privileged mode of operation and allows all code direct access to the hardware and memory, including memory allocated to user mode processes. Kernel mode processes also have a higher priority over user mode processes.
- User mode. This is a lower privileged mode than kernel mode. It has no direct access to the hardware or to memory. It interfaces to the operating system through well-defined API (Application Program Interface) calls.

Figure B.1 shows an outline of the architecture of NT/2000. It can be seen that only the kernel mode has access to the hardware. This kernel includes executive services which include managers (for I/O, interprocess communications, and so on) and device drivers (which control the hardware). Its parts include:

- Microkernel. Controls basic operating system services, such as interrupt handling and scheduling.
- HAL (Hardware Abstraction Layer). This is a library of hardware-specific programs which give a standard interface between the hardware and software. This can either be Microsoft-written or manufacturer-provided. They have the advantage of allowing for transportability of programs across different hardware platforms.
- Win32 Window Manager. Supports Win32, MS-DOS and Windows 3.*x* applications.

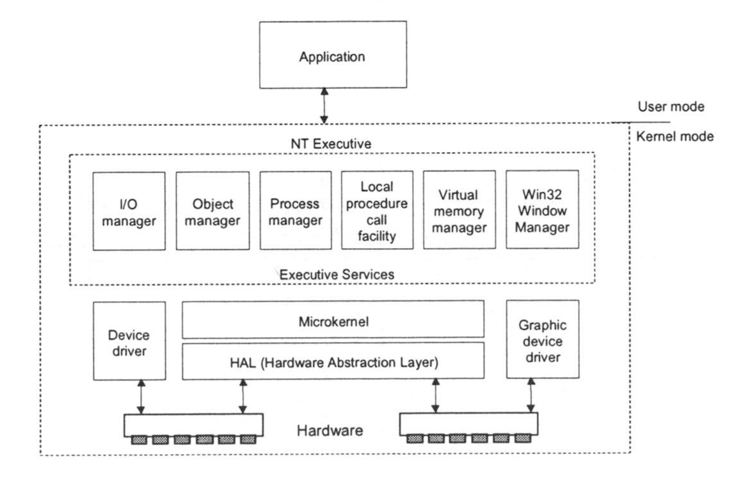

Figure B.1 NT architecture

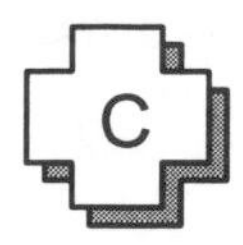

ASCII and WWW References

C.1 ASCII and Extended ASCII

ANSI defined a standard alphabet known as ASCII. This has since been adopted by the CCITT as a standard, known as IA5 (International Alphabet No. 5). The following tables define this alphabet in binary, as a decimal value, as a hexadecimal value and as a character. Unfortunately, standard ASCII character has 7 bits and the basic set ranges from 0 to 127. This code is rather limited as it does not contain symbols such as Greek letters, lines, and so on. For this purpose the extended ASCII code has been defined, which uses character numbers 128 to 255.

Binary	*Decimal*	*Hex*	*Character*	*Binary*	*Decimal*	*Hex*	*Character*
00000000	0	00	NUL	00010000	16	10	DLE
00000001	1	01	SOH	00010001	17	11	DC1
00000010	2	02	STX	00010010	18	12	DC2
00000011	3	03	ETX	00010011	19	13	DC3
00000100	4	04	EOT	00010100	20	14	DC4
00000101	5	05	ENQ	00010101	21	15	NAK
00000110	6	06	ACK	00010110	22	16	SYN
00000111	7	07	BEL	00010111	23	17	ETB
00001000	8	08	BS	00011000	24	18	CAN
00001001	9	09	HT	00011001	25	19	EM
00001010	10	0A	LF	00011010	26	1A	SUB
00001011	11	0B	VT	00011011	27	1B	ESC
00001100	12	0C	FF	00011100	28	1C	FS
00001101	13	0D	CR	00011101	29	1D	GS
00001110	14	0E	SO	00011110	30	1E	RS
00001111	15	0F	SI	00011111	31	1F	US
00100000	32	20	SPACE	00110000	48	30	0
00100001	33	21	!	00110001	49	31	1
00100010	34	22	"	00110010	50	32	2
00100011	35	23	#	00110011	51	33	3
00100100	36	24	$	00110100	52	34	4
00100101	37	25	%	00110101	53	35	5
00100110	38	26	&	00110110	54	36	6
00100111	39	27	/	00110111	55	37	7
00101000	40	28	(	00111000	56	38	8
00101001	41	29	)	00111001	57	39	9
00101010	42	2A	*	00111010	58	3A	:
00101011	43	2B	+	00111011	59	3B	;
00101100	44	2C	,	00111100	60	3C	<
00101101	45	2D	-	00111101	61	3D	=
00101110	46	2E	.	00111110	62	3E	>
00101111	47	2F	/	00111111	63	3F	?
01000000	64	40	@	01010000	80	50	P
01000001	65	41	A	01010001	81	51	Q
01000010	66	42	B	01010010	82	52	R
01000011	67	43	C	01010011	83	53	S
01000100	68	44	D	01010100	84	54	T
01000101	69	45	E	01010101	85	55	U
01000110	70	46	F	01010110	86	56	V

01000111	71	47	G	01010111	87	57	W
01001000	72	48	H	01011000	88	58	X
01001001	73	49	I	01011001	89	59	Y
01001010	74	4A	J	01011010	90	5A	Z
01001011	75	4B	K	01011011	91	5B	[
01001100	76	4C	L	01011100	92	5C	\
01001101	77	4D	M	01011101	93	5D	]
01001110	78	4E	N	01011110	94	5E	`
01001111	79	4F	O	01011111	95	5F	_
01100000	96	60		01110000	112	70	p
01100001	97	61	a	01110001	113	71	q
01100010	98	62	b	01110010	114	72	r
01100011	99	63	c	01110011	115	73	s
01100100	100	64	d	01110100	116	74	t
01100101	101	65	e	01110101	117	75	u
01100110	102	66	f	01110110	118	76	v
01100111	103	67	g	01110111	119	77	w
01101000	104	68	h	01111000	120	78	x
01101001	105	69	i	01111001	121	79	y
01101010	106	6A	j	01111010	122	7A	z
01101011	107	6B	k	01111011	123	7B	{
01101100	108	6C	l	01111100	124	7C	:
01101101	109	6D	m	01111101	125	7D	}
01101110	110	6E	n	01111110	126	7E	~
01101111	111	6F	o	01111111	127	7F	DEL
10000000	128	80	Ç	10010000	144	90	É
10000001	129	81	ü	10010001	145	91	æ
10000010	130	82	é	10010010	146	92	Æ
10000011	131	83	â	10010011	147	93	ô
10000100	132	84	ä	10010100	148	94	ö
10000101	133	85	à	10010101	149	95	ò
10000110	134	86	å	10010110	150	96	û
10000111	135	87	ç	10010111	151	97	ù
10001000	136	88	ê	10011000	152	98	ÿ
10001001	137	89	ë	10011001	153	99	Ö
10001010	138	8A	è	10011010	154	9A	Ü
10001011	139	8B	ï	10011011	155	9B	¢
10001100	140	8C	î	10011100	156	9C	£
10001101	141	8D	ì	10011101	157	9D	¥
10001110	142	8E	Ä	10011110	158	9E	₧
10001111	143	8F	Å	10011111	159	9F	ƒ
10100000	160	A0	á	10110000	176	B0	░
10100001	161	A1	í	10110001	177	B1	▒
10100010	162	A2	ó	10110010	178	B2	▓
10100011	163	A3	ú	10110011	179	B3	│
10100100	164	A4	ñ	10110100	180	B4	┤
10100101	165	A5	Ñ	10110101	181	B5	╡
10100110	166	A6	ª	10110110	182	B6	╢
10100111	167	A7	º	10110111	183	B7	╖
10101000	168	A8	¿	10111000	184	B8	╕
10101001	169	A9	⌐	10111001	185	B9	╣
10101010	170	AA	¬	10111010	186	BA	║

10101011	171	AB	½	10111011	187	BB	╗
10101100	172	AC	¼	10111100	188	BC	╝
10101101	173	AD	¡	10111101	189	BD	╜
10101110	174	AE	«	10111110	190	BE	╛
10101111	175	AF	»	10111111	191	BF	┐
11000000	192	C0	└	11010000	208	D0	╨
11000001	193	C1	┴	11010001	209	D1	╤
11000010	194	C2	┬	11010010	210	D2	╥
11000011	195	C3	├	11010011	211	D3	╙
11000100	196	C4	─	11010100	212	D4	╘
11000101	197	C5	┼	11010101	213	D5	╒
11000110	198	C6	╞	11010110	214	D6	╓
11000111	199	C7	╟	11010111	215	D7	╫
11001000	200	C8	╚	11011000	216	D8	╪
11001001	201	C9	╔	11011001	217	D9	┘
11001010	202	CA	╩	11011010	218	DA	┌
11001011	203	CB	╦	11011011	219	DB	█
11001100	204	CC	╠	11011100	220	DC	▄
11001101	205	CD	═	11011101	221	DD	▌
11001110	206	CE	╬	11011110	222	DE	▐
11001111	207	CF	╧	11011111	223	DF	▀
11100000	224	E0	α	11110000	240	F0	Ξ
11100001	225	E1	ß	11110001	241	F1	±
11100010	226	E2	Γ	11110010	242	F2	≥
11100011	227	E3	π	11110011	243	F3	≤
11100100	228	E4	Σ	11110100	244	F4	⌠
11100101	229	E5	σ	11110101	245	F5	⌡
11100110	230	E6	µ	11110110	246	F6	÷
11100111	231	E7	τ	11110111	247	F7	≈
11101000	232	E8	Φ	11111000	248	F8	°
11101001	233	E9	Θ	11111001	249	F9	·
11101010	234	EA	Ω	11111010	250	FA	·
11101011	235	EB	δ	11111011	251	FB	√
11101100	236	EC	φ	11111100	252	FC	n
11101101	237	ED	ϕ	11111101	253	FD	2
11101110	238	EE	E	11111110	254	FE	■
11101111	239	EF	Λ	11111111	255	FF	

C.2 Additional WWW material

Additional WWW material is available at:

`http://www.dcs.napier.ac.uk/~bill/dist/winsock.html`	Winsock development
`http://www.dcs.napier.ac.uk/~bill/dist/agent.html`	Agent technologies
`http://www.dcs.napier.ac.uk/~bill/dist/corba.html`	DCOM/CORBA

and several other additional chapters.

The following diagrams show the EaStMAN and SuperJANET MANs in the UK (see WWW reference for more details).

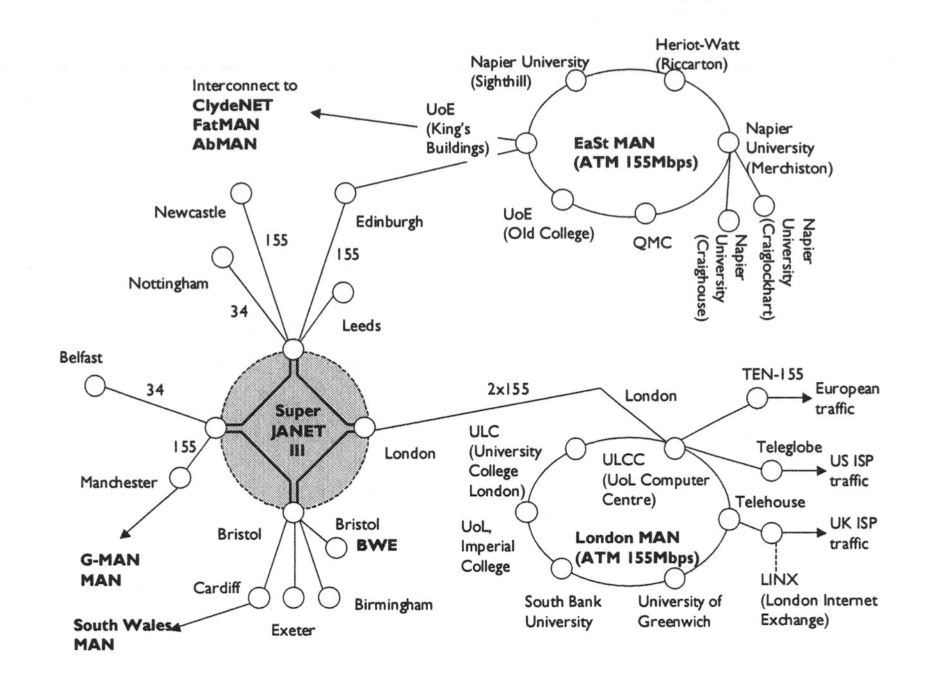
Interconnect to
ClydeNET
FatMAN
AbMAN
UoE
(King's
Buildings)
Napier University
(Sighthill)
Heriot-Watt
(Riccarton)
EaSt MAN
(ATM 155Mbps)
Napier
University
(Merchiston)
UoE
(Old College)
QMC
Napier
University
(Craighouse)
Napier
University
(Craiglockhart)
Newcastle
155
Edinburgh
155
Nottingham
34
Leeds
Belfast
34
Super
JANET
III
2x155
London
155
Manchester
London
G-MAN
MAN
Bristol
Bristol
BWE
Cardiff
South Wales
MAN
Exeter
Birmingham
TEN-155
European
traffic
ULC
(University
College
London)
ULCC
(UoL Computer
Centre)
Teleglobe
US ISP
traffic
Telehouse
UoL,
Imperial
College
London MAN
(ATM 155Mbps)
UK ISP
traffic
South Bank
University
University of
Greenwich
LINX
(London Internet
Exchange)

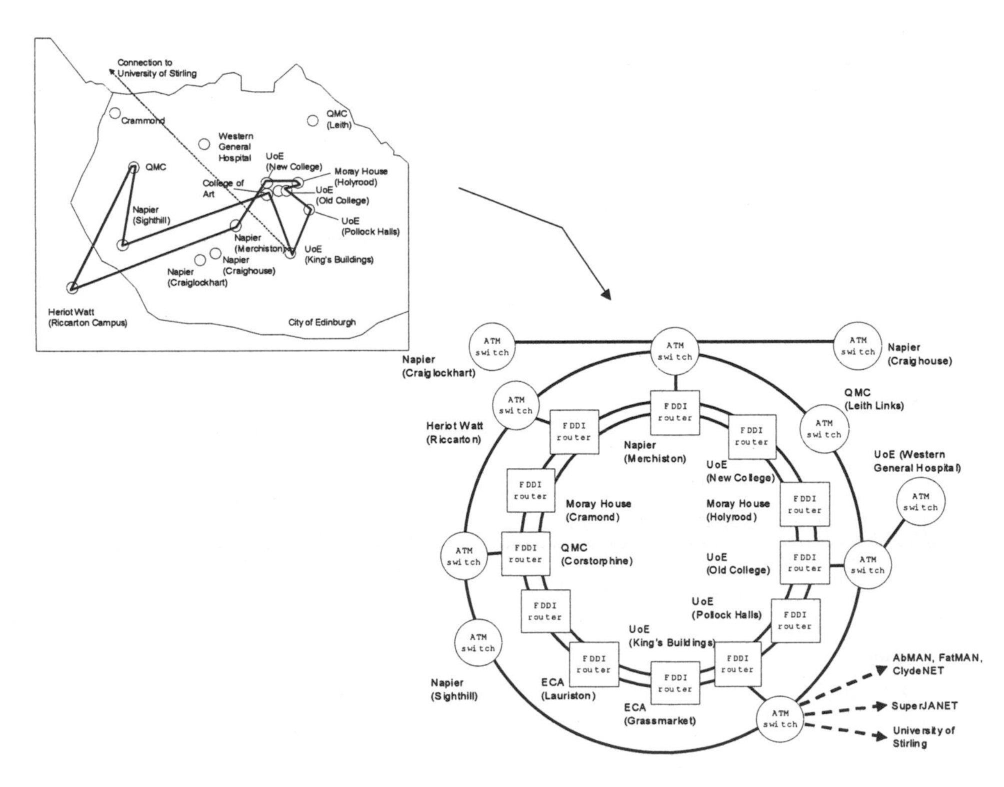
Connection to
University of Stirling
Crammond
QMC
(Leith)
Western
General
Hospital
QMC
UoE
(New College)
Moray House
(Holyrood)
College of
Art
UoE
(Old College)
Napier
(Sighthill)
UoE
(Pollock Halls)
Napier
(Merchiston)
UoE
(King's Buildings)
Napier
(Craighouse)
Napier
(Craiglockhart)
Heriot Watt
(Riccarton Campus)
City of Edinburgh
ATM
switch
FDDI
router
Napier
(Craiglockhart)
Napier
(Craighouse)
QMC
(Leith Links)
Heriot Watt
(Riccarton)
Napier
(Merchiston)
UoE
(New College)
UoE (Western
General Hospital)
Moray House
(Cramond)
Moray House
(Holyrood)
QMC
(Corstorphine)
UoE
(Old College)
UoE
(Pollock Halls)
UoE
(King's Buildings)
AbMAN, FatMAN,
ClydeNET
Napier
(Sighthill)
ECA
(Lauriston)
ECA
(Grassmarket)
SuperJANET
University of
Stirling

Index